Springer Finance

Editorial Board
Marco Avellaneda
Giovanni Barone-Adesi
Francesca Biagini
Bruno Bouchard
Mark Broadie
Emanuel Derman
Paolo Guasoni
Mathieu Rosenbaum

Springer Finance

Springer Finance is a programme of books addressing students, academics and practitioners working on increasingly technical approaches to the analysis of financial markets. It aims to cover a variety of topics, not only mathematical and computational finance but foreign exchange, term structure, risk management, portfolio theory, equity derivatives, energy finance and commodities, financial economics.

More information about this series at http://www.springer.com/series/3674

Ernst Eberlein • Jan Kallsen

Mathematical Finance

Ernst Eberlein
Department of Mathematical Stochastics
University of Freiburg
Freiburg im Breisgau
Germany

Jan Kallsen
Department of Mathematics
Kiel University
Kiel
Germany

ISSN 1616-0533 ISSN 2195-0687 (electronic)
Springer Finance
ISBN 978-3-030-26108-5 ISBN 978-3-030-26106-1 (eBook)
https://doi.org/10.1007/978-3-030-26106-1

Mathematics Subject Classification (2010): 91G20, 91G80, 60G51, 60G44, 60H05, 60J75, 60H10, 91G10, 91G30, 93E20

JEL Classification: G13, G11, D52, C61

© Springer Nature Switzerland AG 2019
This work is subject to copyright. All rights are reserved by the Publisher, whether the whole or part of the material is concerned, specifically the rights of translation, reprinting, reuse of illustrations, recitation, broadcasting, reproduction on microfilms or in any other physical way, and transmission or information storage and retrieval, electronic adaptation, computer software, or by similar or dissimilar methodology now known or hereafter developed.
The use of general descriptive names, registered names, trademarks, service marks, etc. in this publication does not imply, even in the absence of a specific statement, that such names are exempt from the relevant protective laws and regulations and therefore free for general use.
The publisher, the authors, and the editors are safe to assume that the advice and information in this book are believed to be true and accurate at the date of publication. Neither the publisher nor the authors or the editors give a warranty, expressed or implied, with respect to the material contained herein or for any errors or omissions that may have been made. The publisher remains neutral with regard to jurisdictional claims in published maps and institutional affiliations.

This Springer imprint is published by the registered company Springer Nature Switzerland AG.
The registered company address is: Gewerbestrasse 11, 6330 Cham, Switzerland

To Carl-Philipp, Friederike, Sophie and to Birgit, Dörte, Matthies

Preface

Mathematical finance provides a quantitative description of financial markets, more specifically markets for exchange-traded assets, using mostly dynamic stochastic models. It is used to tackle three basic issues.

- *Valuation of assets*
  What can reasonably be said about the price of a financial asset? As opposed to economic theory, mathematical finance focuses mainly on *relative* valuation of securities in comparison to other assets. This is particularly useful and in fact indispensable for derivative securities, which are by definition strongly linked to corresponding underlying quantities in the market.
- *Optimal or at least reasonable portfolio selection*
  How shall an investor choose her portfolio of liquid securities? Here, the focus is on *hedging*, i.e. on minimising the risk which arises, for example, from selling derivative contracts to customers.
- *Quantification of risk*
  The random nature of asset prices naturally involves the risk of losses. How can it be quantified reasonably?

Mathematical finance has grown into a field which is by far too broad to be covered in a single book. Markets, products and risks are diverse and so are the mathematical models and methods which they require.

The starting point and focal point of this present monograph is continuous-time stochastic processes allowing for jumps. Most textbooks on mathematical finance are limited to diffusion-type setups, which cannot easily account for abrupt price movements. Such changes, however, play an important role in real markets, which is why models with jumps have become an established tool in the statistics and mathematics of finance. Just as importantly, purely discontinuous processes lead to a much wider variety of, at the same time flexible and tractable, models. For example, their marginal laws are often known explicitly, which is typically not the case for diffusions.

Compared to the abundant literature on continuous models, such as [29, 78, 149, 187, 204, 223, 223, 279], and many more, there still seems to be a scarcity of textbooks allowing for processes with jumps. Notable exceptions are the monographs [60] and [38, 160, 276]. Other useful texts such as [143, 263], address more specific questions rather than general principles of financial mathematics.

Our goal is twofold:

- to give an account of general semimartingale theory, stochastic control and specific classes of processes to the extent needed for the applications in the second part
- to introduce basic concepts such as arbitrage theory, hedging, valuation principles, portfolio choice and term structure modelling

In a single monograph, we cannot give a comprehensive overview of stochastic models with and without jumps in mathematical finance. Rather, we provide an introduction to the basic building blocks and principles, helping the reader to understand the advanced research literature and to come up with concrete models and solutions in more specific situations.

The book is divided into two parts. Part I introduces the stochastic analysis of general semimartingales along with the basics of stochastic control theory. We do not cover the whole theory with complete proofs, which can be found in a number of excellent mathematical monographs. Rather, we focus on concepts and results that are needed to apply the theory to questions in mathematical finance. Proofs are mostly replaced by informal illustrations along with references to the literature. Nevertheless, we made an effort to provide mathematically rigorous definitions and theorems.

Part II turns to both advanced models and basic principles of mathematical finance. It differs in style from Part I in the sense that results are stated as engineering-style *rules* rather than precise mathematical theorems with all the technical assumptions. For example, we do not distinguish between local and true martingales, and questions of existence and uniqueness are swept under the rug. This is done deliberately in order to make basic concepts accessible to the mathematically less inclined reader who wants to apply advanced stochastic models in practice and also to the non-specialist who wants to get an overview of the general ideas before delving more thoroughly into the subject.

The theory of Parts I and II simplifies occasionally if one focuses on stochastic processes without jumps. Major changes are summarised in Sect. A.7 for the convenience of the reader. Mathematical finance in the broad sense has produced some insights, which may seem counterintuitive and hence surprising to the novice in the field. We collect links to such results in Sect. A.8. Otherwise, the appendix mostly contains mathematical tools that are needed in the main part on the text.

This book could not have appeared in the present form without the help of many people. An incomplete list includes Aleš Černý, Sören Christensen, Friedrich Hubalek, Simon Kolb, Paul Krühner, Matthias Lenga, Johannes Muhle-Karbe, Arnd Pauwels, and Richard Vierthauer with whom we had long discussions, which had an effect on the contents of the book. Funda Akyüz and Britta Ramthun-Kasan assisted with preparing the manuscript. Partial financial support through *DFG Sachbeihilfe 1682/4-1* is gratefully acknowledged. We also benefited from the environment provided by the Freiburg Institute for Advanced Studies (FRIAS). We thank Catriona Byrne and Marina Reizakis from Springer-Verlag for their interest, encouragement and patience.

Errors can hardly be avoided in a text of this size. Since they will be discovered only gradually, we refer to www.math.uni-kiel.de/finmath/~book for an updated list of corrections. On this page, you can also find the *Scilab* code that we have used to generate the figures and numerical examples. Of course, any comments and in particular hints to errors are welcome.

Freiburg im Breisgau, Germany
Kiel, Germany
June 2019

Ernst Eberlein
Jan Kallsen

Contents

Part II Mathematical Finance

Part I
Stochastic Calculus

Overview

A large part of mathematics for finance is written in the language of *stochastic processes*, i.e. of random functions of time. The calculus of these processes is introduced in this first part of the monograph. Some of its concepts naturally generalise notions from ordinary calculus, others are intrinsically linked to their probabilistic nature.

Even though this book deals with continuous-time models, we devote the first chapter to stochastic calculus in discrete time. The results themselves will not be needed in the sequel but they help to understand the intuition behind the corresponding concepts in continuous time.

The remaining chapters generalise important notions from ordinary calculus to the random case. *Lévy processes* can be viewed as the stochastic counterparts of linear functions. They are of interest in their own right but they also appear as building blocks of more general classes of processes. In Chap. 3 we cover the theory of *stochastic integration*, which is indispensable for mathematical finance. By contrast, it seems less obvious whether and how differentiation can be transferred to the random case. In Chaps. 4 and 5 we discuss *semimartingale characteristics* and *infinitesimal generators* as two natural candidates for a stochastic "differentiation". If Lévy processes and semimartingale characteristics generalise linear functions and derivatives, respectively, *affine Markov processes* correspond to solutions of linear ordinary differential equations. They play an important role in finance because of their flexibility and analytical tractability. Finally, we introduce the basic concepts of *stochastic optimal control* in Chap. 7 because many questions in Mathematical Finance are explicitly or implicitly related to optimisation.

Informal differential notation and arguments are used occasionally in the physics literature and to some extent in finance as well. We mimic such reasoning here in a few so-called *physicist's corners*. While these remarks may be insightful to some readers, they could confuse others with a more formal mathematical background. In the latter case they can be skipped altogether because they are primarily meant to illustrate separately stated rigorous mathematical statements.

Chapter 1
Discrete Stochastic Calculus

The theory of stochastic processes deals with random functions of time such as asset prices, interest rates, and trading strategies. As is also the case for Mathematical Finance, it can be developed in both discrete and continuous time. Actual calculations are often easier and more transparent in continuous-time models, which is why we focus on the latter in this book. However, there is a price to be paid. A completely sound treatment of the continuous case requires considerably more complex mathematical arguments, which are beyond the scope of this monograph. On the other hand, the phenomena and formulae in discrete and continuous time resemble each other quite closely. Therefore we use the simpler discrete case as a means to motivate the technically more demanding results in the subsequent chapters.

1.1 Processes, Stopping Times, Martingales

The natural starting point in probability theory is provided by a *probability space* $(\Omega, \mathscr{F}, P)$. The more or less abstract *sample space* Ω stands for the possible outcomes of a random experiment. For example, it could contain all conceivable sample paths of a stock price process. The *probability measure* P states probabilities of sets of outcomes. For measure-theoretic reasons it is typically impossible to assign probabilities to *all* subsets of Ω in a consistent manner. As a way out one specifies a *σ-field* $\mathscr{F}$, i.e. a collection of subsets of Ω which is closed under countable set operations $\cap, \cup, \setminus, {}^C$. If the probability $P(F)$ is defined only for *events* $F \in \mathscr{F}$, one can avoid the paradoxes involved in considering arbitrary sets.

Random variables X are functions of the outcome $\omega \in \Omega$. Typically its values $X(\omega)$ are numbers but they may also be vectors or even functions, in which case X is a *random vector* resp. *process*. We denote by $E(X)$ and $\mathrm{Var}(X)$ the expected value

© Springer Nature Switzerland AG 2019
E. Eberlein, J. Kallsen, *Mathematical Finance*, Springer Finance,
https://doi.org/10.1007/978-3-030-26106-1_1

and variance, respectively, of a real-valued random variable. Accordingly, $E(X)$ and $\mathrm{Cov}(X)$ denote the expectation vector and covariance matrix, respectively, of a random vector X.

For static random experiments one needs to consider only two states of information. *Before* the experiment nothing precise is known about the outcome, only probabilities and expected values can be assigned. *After* the experiment the outcome is completely determined. In dynamic random experiments such as stock markets the situation is more involved. In the process of observation, some random events (e.g. yesterday's stock returns) have already happened and can be considered as deterministic whereas others (e.g. tomorrow's stock returns) still belong to the unknown future. As time passes, more and more information is accumulated.

This increasing knowledge is expressed mathematically in terms of a **filtration** $\mathbf{F} = (\mathscr{F}_t)_{t\geq 0}$, i.e. an increasing sequence of sub-σ-fields of $\mathscr{F}$. The collection of events $\mathscr{F}_t$ stands for the observable information up to time t. Specifically, the statement $F \in \mathscr{F}_t$ means that the random event F (e.g. $F = \{$stock return positive at time $t-1\}$) is no longer random at time t. We know for sure whether it is true or not. For example, if our observable information is given by the evolution of the stock price, then $\mathscr{F}_t$ contains all events that can be expressed in terms of the stock price up to time t. The quadruple $(\Omega, \mathscr{F}, \mathbf{F}, P)$ is called a **filtered probability space**. We consider it to be fixed during most of the following. Often one assumes $\mathscr{F}_0 = \{\varnothing, \Omega\}$, i.e. $\mathscr{F}_0$ is the *trivial σ-field* corresponding to no prior information.

As time passes, not only the observable information but also probabilities and expectations of future events change. For example, our conception of the terminal stock price evolves gradually from vague ideas to perfect knowledge. This is modelled mathematically in terms of conditional expectations. The *conditional expectation* $E(X|\mathscr{F}_t)$ of a random variable X is its expected value given the information up to time t. As such, it is not a number but itself a random variable which may depend on the randomness up to time t, e.g. on the stock price up to t in the above example. Mathematically speaking, $Y = E(X|\mathscr{F}_t)$ is *$\mathscr{F}_t$-measurable*, which means that $\{Y \in B\} := \{\omega \in \Omega : Y(\omega) \in B\} \in \mathscr{F}_t$ for any reasonable (mathematically phrased: *Borel*) set B. Accordingly, the *conditional probability* $P(F|\mathscr{F}_t)$ denotes the probability of an event $F \in \mathscr{F}$ given the information up to time t. As is true for conditional expectation, it is not a number but an $\mathscr{F}_t$-measurable random variable.

Formally, the **conditional expectation** $E(X|\mathscr{F}_t)$ is defined as the unique $\mathscr{F}_t$-measurable random variable Y such that $E(XZ) = E(YZ)$ for any bounded, $\mathscr{F}_t$-measurable random variable Z. It can also be interpreted as the best prediction of X given $\mathscr{F}_t$. Indeed, if $E(X^2) < \infty$, then $E(X|\mathscr{F}_t)$ minimises the mean squared difference $E((X-Z)^2)$ among all $\mathscr{F}_t$-measurable random variables Z. Strictly speaking, $E(X|\mathscr{F}_t)$ is unique only up to a set of probability 0, i.e. any two *versions* $Y, \widetilde{Y}$ satisfy $P(Y \neq \widetilde{Y}) = 0$. In this book we do not make such fine distinctions. Equations, inequalities etc. are always meant to hold only almost surely, i.e. up to a set of probability 0.

A few rules on conditional expectations are used over and over again. For instance, we have

$$E(X|\mathscr{F}_t) = E(X) \tag{1.1}$$

if $\mathscr{F}_t = \{\varnothing, \Omega\}$ is the trivial σ-field representing no information on random events. More generally, (1.1) holds if X and $\mathscr{F}_t$ are *stochastically independent*, i.e. if

$$P(\{X \in B\} \cap F) = P(X \in B)P(F)$$

for any Borel set B and any $F \in \mathscr{F}_t$. On the other hand we have $E(X|\mathscr{F}_t) = X$ and more generally

$$E(XY|\mathscr{F}_t) = XE(Y|\mathscr{F}_t)$$

if X is $\mathscr{F}_t$-measurable, i.e. known at time t. The **law of iterated expectations** tells us that

$$E\big(E(X|\mathscr{F}_t)\big|\mathscr{F}_s\big) = E(X|\mathscr{F}_s)$$

for $s \leq t$. Almost as a corollary we have

$$E\big(E(X|\mathscr{F}_t)\big) = E(X).$$

Finally, the conditional expectation shares many properties of the expectation, e.g. it is linear and monotone in X and it satisfies monotone and dominated convergence, Fatou's lemma, Jensen's inequality, etc.

Recall that the probability of a set can be expressed as the expectation of an indicator function via $P(F) = E(1_F)$. This suggests to use the relation

$$P(F|\mathscr{F}_t) := E(1_F|\mathscr{F}_t) \tag{1.2}$$

to define conditional probabilities in terms of conditional expectation. Of course, we would like $P(F|\mathscr{F}_t)$ to be a probability measure when it is considered as a function of F. This property, however, is not as evident as it seems because of the null sets involved in the definition of conditional expectation. We do not worry about technical details here and assume instead that we are given a **regular conditional probability**, i.e. a version of $P(F|\mathscr{F}_t)(\omega)$ which, for any fixed ω, is a probability measure when viewed as a function of F. Such a regular version exists in all instances where it is used in this book.

In line with (1.2) we denote by

$$P^{X|\mathscr{F}_t}(B) := P(X \in B|\mathscr{F}_t) := E(1_B(X)|\mathscr{F}_t)$$

the **conditional law** of X given $\mathscr{F}_t$. A useful rule states that

$$E(f(X,Y)|\mathscr{F}_t) = \int f(x,Y)P^{X|\mathscr{F}_t}(dx) \tag{1.3}$$

for real-valued measurable functions f and $\mathscr{F}_t$-measurable random variables Y. If X is stochastically independent of $\mathscr{F}_t$, we have $P^{X|\mathscr{F}_t} = P^X$, i.e. the conditional law of X coincides with the law of X. In this case, (1.3) turns into

$$E(f(X,Y)|\mathscr{F}_t) = \int f(x,Y)P^X(dx) \tag{1.4}$$

for $\mathscr{F}_t$-measurable random variables Y.

A **stochastic process** $X = (X(t))_{t\geq 0}$ is a collection of random variables $X(t)$, indexed by time t. In this chapter the time set is assumed to be $\mathbb{N} = \{0, 1, 2, \dots\}$, afterwards we consider continuous time $\mathbb{R}_+ = [0, \infty)$. As noted earlier, a stochastic process $X = (X(t))_{t\geq 0}$ can be interpreted as a random function of time. Indeed, $X(\omega, t)$ is a function of t (or sequence in the current discrete case) for fixed ω. Sometimes, it is also convenient to interpret a process X as a real-valued function on the product space $\Omega \times \mathbb{N}$ or $\Omega \times \mathbb{R}_+$, respectively. In the discrete time case we use the notation

$$\Delta X(t) := X(t) - X(t-1).$$

Moreover, we denote by $X_- = (X_-(t))_{t\geq 0}$ the process

$$X_-(t) := \begin{cases} X(t-1) & \text{for } t \geq 1, \\ X(0) & \text{for } t = 0. \end{cases}$$

We will only consider processes which are consistent with the information structure, i.e. $X(t)$ is supposed to be observable at time t. Mathematically speaking, we assume $X(t)$ to be $\mathscr{F}_t$-measurable for any t. Such processes X are called **adapted** to the filtration $\mathbf{F}$.

There is in fact a minimal filtration $\mathbf{F}$ such that X is $\mathbf{F}$-adapted. Formally, this filtration is given by

$$\mathscr{F}_t = \sigma(X(s) : s \leq t), \tag{1.5}$$

i.e. $\mathscr{F}_t$ is the smallest σ-field such that all $X(s)$, $s \leq t$, are $\mathscr{F}_t$-measurable. Intuitively, this means that the only available information on random events is coming from observing the process X. One calls $\mathbf{F}$ the **filtration generated by** X.

For some processes one actually needs a stronger notion of measurability than adaptedness, namely *predictability*. A stochastic process X is called **predictable** if $X(t)$ is known already one period in advance, i.e. $X(t)$ is $\mathscr{F}_{t-1}$-measurable. The use of this notion will become clearer in Sect. 1.2.

Example 1.1 (Random Walks and Geometric Random Walks) We call an adapted process X with $X(0) = 0$ a **random walk** (relative to $\mathbf{F}$) if the increments $\Delta X(t), t \geq 1$, are identically distributed and independent of $\mathscr{F}_{t-1}$. We obtain such a process if $\Delta X(t), t \geq 1$ are independent and identically distributed (i.i.d.) random variables and if the filtration $\mathbf{F}$ is generated by X.

Similarly, we call a positive adapted process X with $X(0) = 1$ a **geometric random walk** (relative to $\mathbf{F}$) if the *relative increments*

$$\frac{\Delta X(t)}{X(t-1)} = \frac{X(t)}{X(t-1)} - 1 \tag{1.6}$$

are identically distributed and independent of $\mathscr{F}_{t-1}$ for $t \geq 1$. A process X is a geometric random walk if and only if $\log X$ is a random walk or, equivalently,

$$X(t) = \exp(Y(t))$$

for some random walk Y. Indeed, the random variables in (1.6) are identically distributed and independent of $\mathscr{F}_{t-1}$ if and only if this holds for

$$\Delta(\log X(t)) = \log X(t) - \log X(t-1) = \log\left(\frac{\Delta X(t)}{X(t-1)} + 1\right), \quad t \geq 1.$$

Random walks and geometric random walks represent processes of constant growth in an additive or multiplicative sense, respectively. Simple asset price models are often of geometric random walk type.

A **stopping time** τ is a random variable whose values are times, i.e. are in $\mathbb{N} \cup \{\infty\}$ in the discrete case. Additionally one requires that τ is consistent with the information structure $\mathbf{F}$. More precisely, one assumes that $\{\tau = t\} \in \mathscr{F}_t$ (or equivalently $\{\tau \leq t\} \in \mathscr{F}_t$) for any t. Intuitively, this means that the decision to say "stop!" right now can only be based on our current information. As an example consider the first time τ when an observed stock price hits the level 100. Even though this time is random and not known in advance, we obviously know τ in the instant it occurs. The situation is different if we define τ to be the instant *one period before the stock hits 100*. Since we cannot look into the future, we only know τ one period after it has happened. Consequently, this random variable is not a stopping time. Stopping times occur naturally in finance, e.g. in the context of American options, but they also play an important technical role in stochastic calculus.

As indicated above, the time when some adapted process first hits a given set is a stopping time:

Proposition 1.2 *Let X be some adapted process and B a Borel set. Then*

$$\tau := \inf\{t \geq 0 : X(t) \in B\}$$

is a stopping time.

Proof By adaptedness, we have $\{X(s) \in B\} \in \mathscr{F}_s \subset \mathscr{F}_t$, $s \leq t$ and hence

$$\{\tau \leq t\} = \bigcup_{s=0}^{t} \{X(s) \in B\} \in \mathscr{F}_t. \qquad \square$$

Occasionally, it turns out to be important to "freeze" a process at a stopping time. For any adapted process X at any stopping time τ, the **process stopped at** τ is defined as

$$X^\tau(t) := X(\tau \wedge t),$$

where we use the notation $a \wedge b := \min(a, b)$ as usual. The stopped process X^τ remains constant on the level $X(\tau)$ after time τ. It is easy to see that it is adapted as well.

The σ-field $\mathscr{F}_t$ represents the information up to time t. Sometimes we also need the concept of information up to τ, where τ now denotes a stopping time rather than a fixed number. The corresponding σ-field $\mathscr{F}_\tau$ can be interpreted as for fixed t: an event F belongs to $\mathscr{F}_\tau$ if we must wait at most until τ in order to decide whether F occurs or not. Formally, this σ-field is defined as

$$\mathscr{F}_\tau := \left\{F \in \mathscr{F} : F \cap \{\tau \leq t\} \in \mathscr{F}_t \text{ for any } t \geq 0\right\}.$$

Although it may not seem evident at first glance that this definition truly implements the above intuition, one can at least check that some intuitive properties hold:

Proposition 1.3 *Let σ, τ, τ_n be stopping times.*

1. *$\mathscr{F}_\tau$ is a σ-field.*
2. *If $\tau = t$ is a constant stopping time, $\mathscr{F}_\tau = \mathscr{F}_t$.*
3. *$\sigma \leq \tau$ implies that $\mathscr{F}_\sigma \subset \mathscr{F}_\tau$.*
4. *τ is $\mathscr{F}_\tau$-measurable.*
5. *Infima and suprema of finitely or countably many stopping times are again stopping times.*
6. *$\tau_n \downarrow \tau$ implies $\mathscr{F}_\tau = \bigcap_{n \in \mathbb{N}} \mathscr{F}_{\tau_n}$.*

Proof

1. It is straightforward to verify the axioms.
2. Since $\{\tau \leq s\} = \Omega$ for $s \leq t$ and $\varnothing$ for $s > t$, this follows immediately from the definition of $\mathscr{F}_\tau$.
3. For $F \in \mathscr{F}_\sigma$ we have $F \cap \{\tau \leq t\} = (F \cap \{\sigma \leq t\}) \cap \{\tau \leq t\} \in \mathscr{F}_t$.
4. We need to show that $\{\tau \leq s\} \in \mathscr{F}_\tau$ for any $s \geq 0$. Since $\{\tau \leq s\} \cap \{\tau \leq t\} = \{\tau \leq s \wedge t\} \in \mathscr{F}_{s \wedge t} \subset \mathscr{F}_t$ for any $t \geq 0$, this follows from the definition of $\mathscr{F}_\tau$.
5. For stopping times τ_i, $i \in I$ and $t \geq 0$ we have $\{\sup_{i \in I} \tau_i \leq t\} = \bigcap_{i \in I} \{\tau_i \leq t\} \in \mathscr{F}_t$ and $\{\inf_{i \in I} \tau_i \leq t\} = \bigcup_{i \in I} \{\tau_i \leq t\} \in \mathscr{F}_t$.

6. The inclusion $\subset$ follows from 3. If $F \in \mathscr{F}_{\tau_n}$ for all n, we have

$$F \cap \{\tau \leq t\} = \bigcup_{n \in \mathbb{N}} (F \cap \{\tau_n \leq t\}) \in \mathscr{F}_t,$$

which yields the claim. □

If X is a process and τ a stopping time, we denote by $X(\tau)$ the random variable $X(\omega, \tau(\omega))$. Since this does not make sense if $\tau(\omega) = \infty$, we consider $X(\tau)1_{\{\tau<\infty\}}$ if this may happen.

Proposition 1.4 *If X is an adapted process and τ a stopping time, $X(\tau)1_{\{\tau<\infty\}}$ is $\mathscr{F}_\tau$-measurable.*

Proof For Borel sets $B \subset \mathbb{R} \setminus \{0\}$ we have

$$\{X(\tau)1_{\{\tau<\infty\}} \in B\} \cap \{\tau \leq t\} = \bigcup_{s \leq t} \big(\{X(s) \in B\} \cap \{\tau = s\}\big) \in \mathscr{F}_t$$

as desired. □

The concept of martingales is central to stochastic calculus and finance. A **martingale** (resp. **submartingale, supermartingale**) is an adapted process X that is **integrable** in the sense that $E(|X(t)|) < \infty$ for any t and satisfies

$$E(X(t)|\mathscr{F}_s) = X(s) \quad (\text{resp.} \geq X(s), \leq X(s)) \tag{1.7}$$

for $s \leq t$. If X is a martingale, the best prediction for future values is the present level. For example, if the price process of an asset follows a martingale, it is neither going up nor down on average. In that sense it corresponds to a *fair game*. By contrast, submartingales (resp. supermartingales) may increase (resp. decrease) on average. They correspond to favourable (resp. unfavourable) games.

If ξ denotes an integrable random variable, then it naturally induces a martingale X, namely

$$X(t) = E(\xi|\mathscr{F}_t).$$

X is called the **martingale generated by** ξ. If the time horizon is finite, i.e. we consider the time set $\{0, 1, \ldots, T-1, T\}$ rather than $\mathbb{N}$, any martingale is generated by some random variable, namely by $X(T)$. This ceases to be true for infinite time horizons. For instance, random walks are not generated by a single random variable unless they are constant.

Example 1.5 (Density Process) A probability measure Q on $(\Omega, \mathscr{F})$ is called *equivalent* to P (written $Q \sim P$) if the events of probability 0 are the same under P and Q. By the Radon–Nikodym theorem, Q has a *P-density* and vice versa, i.e.

there are some unique random variables $\frac{dQ}{dP}, \frac{dP}{dQ}$ such that

$$Q(F) = E_P\left(1_F \frac{dQ}{dP}\right), \quad P(F) = E_Q\left(1_F \frac{dP}{dQ}\right)$$

for any set $F \in \mathscr{F}$, where E_P, E_Q denote expectation under P and Q, respectively. P, Q are in fact equivalent if and only if such mutual densities exist, in which case we have $\frac{dP}{dQ} = 1/\frac{dQ}{dP}$.

The martingale Z generated by $\frac{dQ}{dP}$ is called the **density process** of Q, i.e. we have

$$Z(t) = E_P\left(\frac{dQ}{dP}\middle|\mathscr{F}_t\right).$$

One easily verifies that $Z(t)$ coincides with the density of the restricted measures $Q|_{\mathscr{F}_t}$ relative to $P|_{\mathscr{F}_t}$, i.e. $Z(t)$ is $\mathscr{F}_t$-measurable and

$$Q(F) = E_P(1_F Z(t))$$

holds for any event $F \in \mathscr{F}_t$. Note further that Z and the density process Y of P relative to Q are reciprocal to each other because

$$Z(t) = \frac{dQ|_{\mathscr{F}_t}}{dP|_{\mathscr{F}_t}} = 1\Big/\frac{dP|_{\mathscr{F}_t}}{dQ|_{\mathscr{F}_t}} = 1/Y(t).$$

The density process Z can be used to compute conditional expectations relative to Q. Indeed, the **generalised Bayes' rule**

$$E_Q(\xi|\mathscr{F}_t) = \frac{E_P(\xi\frac{dQ}{dP}|\mathscr{F}_t)}{Z(t)}$$

(e.g. [226, Lemma 8.6.2], [154, III.3.9]) holds for sufficiently integrable random variables ξ because

$$\begin{aligned}
E_Q(\xi\zeta) &= E_P\left(\xi\zeta\frac{dQ}{dP}\right) \\
&= E_P\left(E_P\left(\xi\zeta\frac{dQ}{dP}\middle|\mathscr{F}_t\right)\right) \\
&= E_P\left(E_P\left(\frac{\xi\zeta\frac{dQ}{dP}}{Z(t)}\middle|\mathscr{F}_t\right)Z(t)\right) \\
&= E_Q\left(E_P\left(\frac{\xi\zeta\frac{dQ}{dP}}{Z(t)}\middle|\mathscr{F}_t\right)\right) \\
&= E_Q\left(\frac{E_P(\xi\frac{dQ}{dP}|\mathscr{F}_t)}{Z(t)}\zeta\right)
\end{aligned}$$

for any bounded $\mathscr{F}_t$-measurable ζ. Similarly, one shows

$$E_Q(\xi|\mathscr{F}_s) = \frac{E_P(\xi Z(t)|\mathscr{F}_s)}{Z(s)} \tag{1.8}$$

for $s \leq t$ and $\mathscr{F}_t$-measurable random variables ξ.

For later use, we note that a supermartingale with constant expectation is actually a martingale.

Proposition 1.6 *If X is a supermatingale and $T \geq 0$ with $E(X(T)) = E(X(0))$, then (1.7) holds with equality for any $s \leq t \leq T$.*

Proof The supermartingale property means that $E((X(t) - X(s))1_F) \leq 0$ for any $s \leq t$ and any $F \in \mathscr{F}_s$. Since

$$\begin{aligned}
0 &= E(X(T)) - E(X(0)) \\
&= E(X(T) - X(t)) + E\big((X(t) - X(s))1_F\big) + E\big((X(t) - X(s))1_{F^C}\big) \\
&\quad + E(X(s) - X(0)) \\
&\leq 0
\end{aligned}$$

for any $s \leq t \leq T$ and any event $F \in \mathscr{F}_s$, the four nonpositive summands must actually be 0. This yields $E((X(t) - X(s))1_F) = 0$ and hence the assertion. □

The following technical result is used in Sect. 1.5.

Lemma 1.7 *Let X be a supermartingale, Y a martingale, $t \leq T$, and $F \in \mathscr{F}_t$ with $X(t) = Y(t)$ on F and $X(T) \geq Y(T)$. Then $X(s) = Y(s)$ on F for $t \leq s \leq T$. The statement remains to hold if we only require $X - X^t$, $Y - Y^t$ instead of X, Y to be a supermartingale resp. martingale.*

Proof From

$$X(s) - Y(s) \geq E(X(T) - Y(T)|\mathscr{F}_s) \geq 0$$

and

$$E((X(s) - Y(s))1_F) \leq E((X(t) - Y(t))1_F) = 0$$

it follows that $(X(s) - Y(s))1_F = 0$. □

The concept of a martingale is "global" in the sense that (1.7) must be satisfied for any $s \leq t$. If we restrict attention to the case $s = t - 1$, we obtain the slightly more general "local" counterpart. A **local martingale** (resp. **submartingale**,

supermartingale) is an adapted process X which satisfies $E(|X(0)|) < \infty$, $E(|X(t)||\mathscr{F}_{t-1}) < \infty$ and

$$E(X(t)|\mathscr{F}_{t-1}) = X(t-1) \quad (\text{resp.} \geq X(t-1), \leq X(t-1)) \tag{1.9}$$

for $t = 1, 2, \ldots$ In discrete time the difference between martingales and local martingales is minor:

Proposition 1.8 *Any* integrable *local martingale (in the sense that $E(|X(t)|) < \infty$ for any t) is a martingale. An analogous statement holds for sub- and supermartingales.*

Proof This follows by induction from the law of iterated expectations. □

Integrability in Proposition 1.8 holds, for instance, if X is a nonnegative local supermartingale.

The above classes of processes are *stable under stopping* in the sense of the following proposition, which has a natural economic interpretation: you cannot turn a fair game into, say, a favourable one by stopping play at some reasonable time.

Proposition 1.9 *Let τ denote a stopping time. If X is a martingale (resp. sub-/supermartingale), so is X^τ. A corresponding statement holds for local martingales and local sub-/supermartingales.*

Proof We start by verifying the integrability conditions. For martingales (resp. sub-supermartingales) $E(|X(t)|) < \infty$ implies $E(|X^\tau(t)|) \leq E(\sum_{s=0}^t |X(s)|) < \infty$. For local martingales (resp. local sub-/supermartingales) $E(|X(t)||\mathscr{F}_{t-1}) < \infty$ yields

$$\begin{aligned} E(|X^\tau(t)||\mathscr{F}_{t-1}) &\leq \sum_{s=0}^{t} E(|X(s)||\mathscr{F}_{t-1}) \\ &= \sum_{s=0}^{t-1} |X(s)| + E(|X(t)||\mathscr{F}_{t-1}) < \infty. \end{aligned}$$

In order to verify (1.9), observe that $\{\tau \geq t\} = \{\tau \leq t-1\}^C \in \mathscr{F}_{t-1}$ implies

$$\begin{aligned} E(X^\tau(t)1_{\{\tau\geq t\}}|\mathscr{F}_{t-1}) &= E(X(t)1_{\{\tau\geq t\}}|\mathscr{F}_{t-1}) \\ &= E(X(t)|\mathscr{F}_{t-1})1_{\{\tau\geq t\}} \\ &= X(t-1)1_{\{\tau\geq t\}} \\ &= X^\tau(t-1)1_{\{\tau\geq t\}} \end{aligned}$$

(resp. $\geq$, $\leq$ in the sub-/supermartingale case). For $s < t$ we have $\{\tau = s\} \in \mathscr{F}_s \subset \mathscr{F}_{t-1}$ and hence

$$\begin{aligned} E(X^\tau(t)1_{\{\tau=s\}}|\mathscr{F}_{t-1}) &= E(X(s)1_{\{\tau=s\}}|\mathscr{F}_{t-1}) \\ &= X(s)1_{\{\tau=s\}} \\ &= X^\tau(t-1)1_{\{\tau=s\}}. \end{aligned}$$

Together we obtain

$$\begin{aligned} E(X^\tau(t)|\mathscr{F}_{t-1}) &= \sum_{s=0}^{t-1} E(X^\tau(t)1_{\{\tau=s\}}|\mathscr{F}_{t-1}) + E(X(t)1_{\{\tau\geq t\}}|\mathscr{F}_{t-1}) \\ &= \sum_{s=0}^{t-1} X^\tau(t-1)1_{\{\tau=s\}} + X^\tau(t-1)1_{\{\tau\geq t\}} = X^\tau(t-1) \end{aligned}$$

(resp. $\geq$, $\leq$). □

An alternative more complicated definition of local martingales uses stopping times, which turns out to be useful in continuous time.

Proposition 1.10 *An adapted process X is a local martingale if and only if there exists a sequence $(\tau_n)_{n\in\mathbb{N}}$ of stopping times, increasing to ∞ almost surely, such that the stopped processes X^{τ_n} are martingales for any n. A corresponding statement holds for sub-/supermartingales.*

Proof In order to show the *if* statement suppose that X^{τ_n} is a martingale for any n. Note that $\{\tau_n \geq t\} = \{\tau_n \leq t-1\}^C \in \mathscr{F}_{t-1}$ and integrability of $X^{\tau_n}(t)$ imply

$$\begin{aligned} E(|X(t)||\mathscr{F}_{t-1})1_{\{\tau_n\geq t\}} &= E(|X(t)|1_{\{\tau_n\geq t\}}|\mathscr{F}_{t-1}) \\ &= E(|X^{\tau_n}(t)|1_{\{\tau_n\geq t\}}|\mathscr{F}_{t-1}) < \infty \end{aligned}$$

and

$$\begin{aligned} E(X(t)|\mathscr{F}_{t-1})1_{\{\tau_n\geq t\}} &= E(X(t)1_{\{\tau_n\geq t\}}|\mathscr{F}_{t-1}) \\ &= E(X^{\tau_n}(t)1_{\{\tau_n\geq t\}}|\mathscr{F}_{t-1}) \\ &= E(X^{\tau_n}(t)|\mathscr{F}_{t-1})1_{\{\tau_n\geq t\}} \\ &= X^{\tau_n}(t-1)1_{\{\tau_n\geq t\}} \\ &= X(t-1)1_{\{\tau_n\geq t\}}. \end{aligned}$$

Therefore $E(|X(t)||\mathscr{F}_{t-1}) < \infty$ and (1.9) hold on $\cup_{n\geq 1}\{\tau_n \geq t\}$ and hence almost surely.

For the *only if* statement we define a sequence of stopping times $\tau_n := \inf\{t \geq 0 : E(|X(t+1)||\mathscr{F}_t) \geq n\}$. By Proposition 1.9, X^{τ_n} is a local martingale for any n. Since

$$
\begin{aligned}
E(|X^{\tau_n}(t+1)|) &= E\big(|X(0)|1_{\{\tau_n=0\}}\big) + \sum_{s=0}^{t-1} E\big(|X(s+1)|1_{\{\tau_n=s+1\}}\big) \\
&\quad + E\big(|X(t+1)|1_{\{\tau_n>t\}}\big) \\
&\leq E(|X(0)|) + \sum_{s=0}^{t} E\big(|X(s+1)|1_{\{\tau_n\leq s\}^C}\big) \\
&= E(|X(0)|) + \sum_{s=0}^{t} E\big(E(|X(s+1)||\mathscr{F}_s)1_{\{\tau_n\leq s\}^C}\big) \\
&\leq E(|X(0)|) + (t+1)n < \infty,
\end{aligned}
$$

X^{τ_n} is a martingale by Proposition 1.8. The assertion for sub-/supermartingales follows along the same lines. □

If X is a martingale, the martingale property (1.7) holds also for bounded and sometimes even general stopping times.

Theorem 1.11 (Doob's Stopping Theorem) *If X is a martingale, we have*

$$E(X(\tau)|\mathscr{F}_\sigma) = X(\sigma)$$

for any two bounded stopping times $\sigma \leq \tau$. If the martingale is generated by some random variable $X(\infty)$, we need not require σ, τ to be bounded.

For a supermartingale X, we have accordingly

$$E(X(\tau)|\mathscr{F}_\sigma) \leq X(\sigma)$$

for any two bounded stopping times $\sigma \leq \tau$.

Proof

Step 1: We show $E(X(r)|\mathscr{F}_\sigma) = X(\sigma)$ (resp. "$\leq$" in the supermartingale case). Indeed, $E(X(r)|\mathscr{F}_\sigma) = X(\sigma)$ means $E((X(r)-X(\sigma))1_F) = 0$ for any $F \in \mathscr{F}_\sigma$. For such F we have $F \cap \{\sigma = t\} = (F \cap \{\sigma \leq t\}) \setminus (F \cap \{\sigma \leq t-1\}) \in \mathscr{F}_t$ and hence $E((X(r) - X(\sigma))1_F) = \sum_{t=0}^{r} E((X(r) - X(t))1_{F\cap\{\sigma=t\}}) = 0$ because $X(t) = E(X(r)|\mathscr{F}_t)$.

Step 2: In view of Proposition 1.10, we can apply step 1 to the stopped process X^τ and obtain $E(X(\tau)|\mathscr{F}_\sigma) = E(X^\tau(r)|\mathscr{F}_\sigma) = X^\tau(\sigma) = X(\sigma)$ (resp. "$\leq$" in the supermartingale case). The statement for unbounded σ, τ follows for $r = \infty$. □

Sometimes one needs statements on uniform integrability of martingales. Here, inequalities such as the following one prove to be useful.

Theorem 1.12 (Doob's Inequality) *We have*

$$E\Big(\sup_{s\leq t} X(s)^2\Big) \leq 4E\big(X(t)^2\big), \quad t = 0, 1, 2, \ldots$$

for any martingale X.

Proof The proof can be found, for example, in [275, Corollary VII.3.2]. □

One easily verifies that an integrable random walk X is a martingale if and only if the increments $\Delta X(t)$ have expectation 0. An analogous result holds for integrable geometric random walks whose relative increments $\Delta X(t)/X(t-1)$ have vanishing mean. For the martingale property to hold, one does not actually need the increments resp. relative increments of X to be identically distributed.

Martingales are expected to stay on the current level on average. More general processes may show an increasing, decreasing or possibly variable trend. This fact is expressed formally by a variant of *Doob's decomposition*. The idea is to decompose the increment $\Delta X(t)$ of an arbitrary process into a predictable trend component $\Delta A^X(t)$ and a random deviation $\Delta M^X(t)$ from this short time prediction.

Theorem 1.13 (Canonical Decomposition) *Any integrable adapted process X (i.e. with $E(|X(t)|) < \infty$ for any t) can be uniquely decomposed as*

$$X = X(0) + M^X + A^X \tag{1.10}$$

with some martingale M^X and some predictable process A^X satisfying $M^X(0) = A^X(0) = 0$. We call A^X the ***compensator*** *of X.*

Proof Define $A^X(t) = \sum_{s=1}^t \Delta A^X(s)$ by $\Delta A^X(s) := E(\Delta X(s)|\mathscr{F}_{s-1})$ and set $M^X := X - X(0) - A^X$. Predictability of A^X is obvious. The integrability of X implies that of A^X and thus of M^X. The latter is a martingale because

$$\begin{aligned} E(M^X(t)|\mathscr{F}_{t-1}) &= M^X(t-1) + E(\Delta X(t) - \Delta A^X(t)|\mathscr{F}_{t-1}) \\ &= M^X(t-1) + E(\Delta X(t)|\mathscr{F}_{t-1}) - E(E(\Delta X(t)|\mathscr{F}_{t-1})|\mathscr{F}_{t-1}) \\ &= M^X(t-1). \end{aligned}$$

Conversely, for any decomposition as in Theorem 1.13 we have

$$E(\Delta X(t)|\mathscr{F}_{t-1}) = E(\Delta M^X(t)|\mathscr{F}_{t-1}) + E(\Delta A^X(t)|\mathscr{F}_{t-1}) = \Delta A^X(t),$$

which means that it coincides with the decomposition in the first part of the proof. □

For instance, the compensator of an integrable random walk X equals

$$A^X(t) = \sum_{s=1}^{t} E(\Delta X(s)|\mathscr{F}_{s-1}) = tE(\Delta X(1)).$$

If X is a submartingale (resp. supermartingale), then A^X is increasing (resp. decreasing). This is the case commonly referred to as **Doob's decomposition**.

Note that uniqueness of the decomposition still holds if we require M^X to be only a *local* martingale. In this relaxed sense, it suffices to assume $E(|X(t)||\mathscr{F}_{t-1}) < \infty$ for any t in order to obtain (1.10) and to define the **compensator** A^X.

1.2 Stochastic Integration

Gains from trade in dynamic portfolios can be expressed in terms of *stochastic integrals*, which are nothing else than sums in discrete time. Consider an adapted process X and a predictable—or at least adapted—process φ. The **stochastic integral** of φ relative to X is the adapted process $\varphi \bullet X$ defined as

$$\varphi \bullet X(t) := \sum_{s=1}^{t} \varphi(s)\Delta X(s). \tag{1.11}$$

If both $\varphi = (\varphi_1, \dots, \varphi_d)$ and $X = (X_1, \dots, X_d)$ are vector-valued processes, we define $\varphi \bullet X$ to be the real-valued process given by

$$\varphi \bullet X(t) := \sum_{s=1}^{t}\sum_{i=1}^{d} \varphi_i(s)\Delta X_i(s). \tag{1.12}$$

In order to motivate this definition, let us interpret $X(t)$ as the price of a stock at time t. We invest in this stock using the trading strategy φ, i.e. $\varphi(t)$ denotes the number of shares we own at time t. Due to the price move from $X(t-1)$ to $X(t)$ our wealth changes by $\varphi(t)(X(t) - X(t-1)) = \varphi(t)\Delta X(t)$ in the period between $t-1$ and t. Consequently, the integral $\varphi \bullet X(t)$ stands for the cumulative gains from trade up to time t. If we invest in a portfolio of several stocks, both the trading strategy φ and the price process X are vector-valued. $\varphi_i(t)$ now stands for the number of shares of stock i and $X_i(t)$ is its price. In order to compute the total gains of the portfolio, we must sum up the gains $\varphi_i(t)\Delta X_i(t)$ in each single stock, which leads to (1.12).

For the above reasoning to make sense, one must be careful about the order in which things happen at time t. If $\varphi(t)(X(t) - X(t-1))$ is meant to stand for the gains at time t, we obviously have to buy the portfolio $\varphi(t)$ *before* prices change from $X(t-1)$ to $X(t)$. Put differently, we must choose $\varphi(t)$ at the end of period $t-1$, right after the stock price has attained the value $X(t-1)$. This choice can only be based on information up to time $t-1$ and in particular not on $X(t)$, which is as yet unknown. This motivates why one typically requires trading strategies to be predictable rather than adapted. The purely mathematical definition of $\varphi \bullet X$, however, makes sense regardless of any measurability assumption.

The **covariation process** $[X, Y]$ of adapted processes X, Y is defined as

$$[X, Y](t) := \sum_{s=1}^{t} \Delta X(s)\Delta Y(s). \tag{1.13}$$

Its compensator

$$\langle X, Y\rangle(t) = \sum_{s=1}^{t} E(\Delta X(s)\Delta Y(s)\,|\mathscr{F}_{s-1})$$

is called the **predictable covariation process** if it exists. In the special case $X = Y$ one refers to the **quadratic variation** resp. **predictable quadratic variation** of X. If X, Y are martingales, their predictable covariation can be viewed as a dynamic analogue of the covariance of two random variables.

We are now ready to state a few properties of stochastic integration:

Proposition 1.14 *For adapted processes X, Y, Z and predictable processes φ, ψ we have:*

1. *$\varphi \bullet X$ is linear in φ and X.*
2. *$[X, Y]$ and $\langle X, Y\rangle$ are symmetric and linear in X and Y.*
3. *$\psi \bullet (\varphi \bullet X) = (\psi\varphi) \bullet X$.*
4. *$[\varphi \bullet X, Y] = \varphi \bullet [X, Y]$.*
5. *$\langle \varphi \bullet X, Y\rangle = \varphi \bullet \langle X, Y\rangle$ whenever the predictable covariations are defined.*
6. ***(Integration by parts)***

$$\begin{aligned} XY &= X(0)Y(0) + X_- \bullet Y + Y \bullet X \\ &= X(0)Y(0) + X_- \bullet Y + Y_- \bullet X + [X, Y]. \end{aligned} \tag{1.14}$$

7. *If X is a local martingale, then so is $\varphi \bullet X$.*
8. *If $\varphi \geq 0$ and X is a local sub-/supermartingale, $\varphi \bullet X$ is a local sub-/supermartingale as well.*
9. *$A^{\varphi \bullet X} = \varphi \bullet A^X$ if the compensator A^X exists in the relaxed sense following Theorem 1.13.*
10. *If X, Y are martingales with $E(|X(t)Y(t)|) < \infty$ for any t, the process $XY - \langle X, Y\rangle$ is a martingale, i.e. $\langle X, Y\rangle$ is the compensator of XY.*
11. *$[X, [Y, Z]](t) = [[X, Y], Z](t) = \sum_{s=1}^{t} \Delta X(s)\Delta Y(s)\Delta Z(s)$.*

Proof

1. This is obvious from the definition.
2. This is obvious from the definition as well.
3. This follows from

$$\Delta(\psi \bullet (\varphi \bullet X))(t) = \psi(t)\Delta(\varphi \bullet X)(t) = \psi(t)\varphi(t)\Delta X(t) = \Delta((\psi\varphi) \bullet X)(t).$$

4. This follows from

$$\Delta[\varphi \bullet X, Y](t) = \varphi(t)\Delta X(t)\Delta Y(t) = \Delta(\varphi \bullet [X, Y])(t).$$

5. Predictability of φ yields

$$\begin{aligned}\Delta\langle\varphi \bullet X, Y\rangle(t) &= E(\Delta(\varphi \bullet X)(t)\Delta Y(t) \,|\mathscr{F}_{t-1}) \\ &= E(\varphi(t)\Delta X(t)\Delta Y(t) \,|\mathscr{F}_{t-1}) \\ &= \varphi(t)E(\Delta X(t)\Delta Y(t) \,|\mathscr{F}_{t-1}) \\ &= \Delta(\varphi \bullet \langle X, Y\rangle)(t).\end{aligned}$$

6. The first equation is

$$\begin{aligned}X(t)Y(t) &= X(0)Y(0) + \sum_{s=1}^{t} \big(X(s)Y(s) - X(s-1)Y(s-1)\big) \\ &= X(0)Y(0) + \sum_{s=1}^{t} \big(X(s-1)(Y(s) - Y(s-1))\big) \\ &\quad + \sum_{s=1}^{t} \big((X(s) - X(s-1))Y(s)\big) \\ &= X(0)Y(0) + \sum_{s=1}^{t} \big(X(s-1)\Delta Y(s) + Y(s)\Delta X(s)\big) \\ &= X(0)Y(0) + X_- \bullet Y(t) + Y \bullet X(t).\end{aligned}$$

The second follows from

$$Y \bullet X(t) = Y_- \bullet X(t) + (\Delta Y) \bullet X(t) = Y_- \bullet X(t) + [X, Y](t).$$

7. Predictability of φ and (1.9) yield

$$\begin{aligned} E(\varphi \bullet X(t)|\mathscr{F}_{t-1}) &= E(\varphi \bullet X(t-1) + \varphi(t)(X(t) - X(t-1))|\mathscr{F}_{t-1}) \\ &= \varphi \bullet X(t-1) + \varphi(t)(E(X(t)|\mathscr{F}_{t-1}) - X(t-1)) \\ &= \varphi \bullet X(t-1). \end{aligned}$$

8. This follows along the same lines as 7.
9. This follows from statement 7 because $\varphi \bullet X = \varphi \bullet M^X + \varphi \bullet A^X$ is the canonical decomposition of $\varphi \bullet X$.
10. This follows from statements 6 and 7.
11. This follows from the definition. □

If they make sense, the above rules hold for vector-valued processes as well, e.g.

$$\psi \bullet (\varphi \bullet X) = (\psi\varphi) \bullet X$$

if both φ, X are $\mathbb{R}^d$-valued.

Itō's formula is probably the most important rule in continuous-time stochastic calculus. This motivates why we state its simple discrete-time counterpart here.

Proposition 1.15 (Itō's Formula) *If X is an $\mathbb{R}^d$-valued adapted process and $f : \mathbb{R}^d \to \mathbb{R}$ a differentiable function, then*

$$\begin{aligned} f(X(t)) &= f(X(0)) + \sum_{s=1}^{t} (f(X(s)) - f(X(s-1))) \\ &= f(X(0)) + Df(X_-) \bullet X(t) \\ &\quad + \sum_{s=1}^{t} \Big(f(X(s)) - f(X(s-1)) - Df(X(s-1))^\top \Delta X(s) \Big), \end{aligned} \tag{1.15}$$

where $Df(x)$ denotes the derivative or gradient of f in x.

Proof The first statement is obvious. The second follows from the definition of the stochastic integral. □

If the increments $\Delta X(s)$ are small and f is sufficiently smooth, we may use the second-order Taylor expansion

$$\begin{aligned} f(X(s)) &= f(X(s-1) + \Delta X(s)) \\ &\approx f(X(s-1)) + f'(X(s-1))\Delta X(s) + \frac{1}{2} f''(X(s-1))(\Delta X(s))^2 \end{aligned}$$

in the univariate case, which leads to

$$f(X(t)) \approx f(X(0)) + f'(X_-) \bullet X(t) + \sum_{s=1}^{t} \frac{1}{2} f''(X(s-1))(\Delta X(s))^2$$
$$= f(X(0)) + f'(X_-) \bullet X(t) + \frac{1}{2} f''(X_-) \bullet [X, X](t).$$

If X is vector-valued, we obtain accordingly

$$f(X(t)) \approx f(X(0)) + Df(X_-) \bullet X(t) + \frac{1}{2} \sum_{i,j=1}^{d} D_{ij} f(X_-) \bullet [X_i, X_j](t). \tag{1.16}$$

Processes of multiplicative structure are called *stochastic exponentials.*

Definition 1.16 Let X be an adapted process. The unique adapted process Z satisfying

$$Z = 1 + Z_- \bullet X$$

is called the **stochastic exponential** of X and it is written $\mathscr{E}(X)$.

The stochastic exponential can be motivated from a financial point of view. Suppose that 1€ earns the possibly random interest ΔX_t in period t, i.e. 1€ at time $t-1$ turns into $(1 + \Delta X_t)$€ at time t. Then 1€ at time 0 runs up to $\mathscr{E}(X)(t)$ € at time t. It is easy to compute $\mathscr{E}(X)$ explicitly:

Proposition 1.17 *We have*

$$\mathscr{E}(X)(t) = \prod_{s=1}^{t} (1 + \Delta X(s)),$$

where the empty product for $t = 0$ *is set to* 1.

Proof For $Z(t) = \prod_{s=1}^{t}(1 + \Delta X(s))$ we have

$$\Delta Z(t) = Z(t) - Z(t-1) = Z(t-1)\Delta X(t)$$

and hence

$$Z(t) = Z(0) + \sum_{s=1}^{t} \Delta Z(s) = 1 + \sum_{s=1}^{t} Z(s-)\Delta X(s) = 1 + Z_- \bullet X(t). \qquad \square$$

We note in passing that $Z = c\mathscr{E}(X)$ is the unique solution to $Z = c + Z_- \bullet X$ for $c \in \mathbb{R}$.

The previous proposition implies that the stochastic exponential of a random walk $\widetilde{X}$ with increments $\Delta\widetilde{X}(t) > -1$ is a geometric random walk. More specifically, one can write any geometric random walk Z alternatively in exponential or stochastic exponential form, namely

$$Z = e^X = \mathscr{E}(\widetilde{X})$$

with random walks X, $\widetilde{X}$, respectively. X and $\widetilde{X}$ are related to each other via

$$\Delta\widetilde{X}(t) = e^{\Delta X(t)} - 1 \quad \text{resp.} \quad \Delta X(t) = \log(1 + \Delta\widetilde{X}(t)).$$

If the increments $\Delta X(s)$ are small enough, we can use the approximation

$$\log(1 + \Delta X(s)) \approx \Delta X(s) - \frac{1}{2}(\Delta X(s))^2$$

and obtain

$$\begin{aligned}
\mathscr{E}(X)(t) &= \prod_{s=1}^{t} (1 + \Delta X(s)) \\
&= \exp\left(\sum_{s=1}^{t} \log(1 + \Delta X(s))\right) \\
&\approx \exp\left(\sum_{s=1}^{t} \left(\Delta X(s) - \frac{1}{2}(\Delta X(s))^2\right)\right) \\
&= \exp\left(X(t) - X(0) - \frac{1}{2}[X, X](t)\right).
\end{aligned}$$

The product of stochastic exponentials is again a stochastic exponential. Observe the similarity of the following result to the rule $e^x e^y = e^{x+y}$ for the exponential function.

Proposition 1.18 (Yor's Formula)

$$\mathscr{E}(X)\mathscr{E}(Y) = \mathscr{E}(X + Y + [X, Y])$$

holds for any two adapted processes X, Y.

Proof Let $Z := \mathscr{E}(X)\mathscr{E}(Y)$. Integration by parts and the other statements of Proposition 1.14 yield

$$\begin{aligned} Z &= Z(0) + \mathscr{E}(X)_- \bullet \mathscr{E}(Y) + \mathscr{E}(Y)_- \bullet \mathscr{E}(X) + [\mathscr{E}(X), \mathscr{E}(Y)] \\ &= 1 + (\mathscr{E}(X)_-\mathscr{E}(Y)_-) \bullet Y + (\mathscr{E}(Y)_-\mathscr{E}(X)_-) \bullet X + (\mathscr{E}(X)_-\mathscr{E}(Y)_-) \bullet [X, Y] \\ &= 1 + Z_- \bullet (X + Y + [X, Y]), \end{aligned}$$

which implies that $Z = \mathscr{E}(X + Y + [X, Y])$. □

If an adapted process Z does not attain the value 0, it can be written as

$$Z = Z(0)\mathscr{E}(X)$$

with some unique process X satisfying $X(0) = 0$. This process X is naturally called the **stochastic logarithm** $\mathscr{L}(Z)$ of Z. We have

$$\mathscr{L}(Z) = \frac{1}{Z_-} \bullet Z.$$

Indeed, $X = \frac{1}{Z_-} \bullet Z$ satisfies

$$\frac{Z_-}{Z(0)} \bullet X = \frac{Z_-}{Z(0)} \bullet \left(\frac{1}{Z_-} \bullet Z\right) = \frac{1}{Z(0)} \bullet Z = \frac{Z}{Z(0)} - 1$$

and hence

$$\frac{Z}{Z(0)} = \mathscr{E}(X)$$

as claimed. Observe that the same notation $\mathscr{L}(Z)$ is used for the stochastic logarithm of a process and for the law of a random variable Z. It should be evident from the context which one we are referring to.

Changes of the underlying probability measure play an important role in Mathematical Finance. Since the notion of a martingale involves expectation, it is not invariant under such measure changes.

Proposition 1.19 *Let $Q \sim P$ be a probability measure with density process Z. An adapted process X is a Q-martingale (resp. Q-local martingale) if and only if XZ is a P-martingale (resp. P-local martingale).*

Proof X is a Q-local martingale if and only if

$$E_Q(X(t)|\mathscr{F}_{t-1}) = X(t-1) \tag{1.17}$$

for any t. By Bayes' rule (1.8) the left-hand side equals $E(X(t)Z(t)|\mathscr{F}_{t-1})/Z(t-1)$. Hence (1.17) is equivalent to $E(X(t)Z(t)|\mathscr{F}_{t-1}) = X(t-1)Z(t-1)$, which is the local martingale property of XZ relative to P. The integrability property for martingales (cf. Proposition 1.8) is shown similarly. □

A martingale X may possibly show a trend under the new probability measure Q. This trend can be expressed in terms of a predictable covariation.

Proposition 1.20 *Let $Q \sim P$ be a probability measure with density process Z. Moreover, suppose that X is a P-martingale. If X is Q-integrable, its Q-compensator equals $\langle \mathscr{L}(Z), X\rangle$, where the angle bracket is computed relative to P.*

Proof Denote the Q-compensator of X by A. In view of the comment after Theorem 1.13, we have that $X - X(0) - A$ is a Q-local martingale. By Proposition 1.19 this means that $Z(X - X(0) - A)$ is a P-local martingale. Integration by parts yields

$$Z(X-X(0)-A) = Z_- \bullet X + (X-X(0))_- \bullet Z + [Z, X-X(0)] - Z_- \bullet A - A \bullet Z.$$

The integrals relative to X and Z are local martingales. Hence

$$[Z, X] - Z_- \bullet A = [Z, X - X(0)] - Z_- \bullet A$$

is a P-local martingale as well, which implies $Z_- \bullet A = \langle Z, X\rangle$. Consequently, $A = \frac{1}{Z_-} \bullet \langle Z, X\rangle = \langle \mathscr{L}(Z), X\rangle$ as desired. □

The continuous-time analogue of the following representation theorem plays an important role in Mathematical Finance.

Proposition 1.21 (Martingale Representation) *Suppose that X is a random walk such that the increments $\Delta X(t)$ have only two values $a, -b$, attained with probabilities p and $1 - p$, respectively. If X is a martingale and if the filtration is generated by X, any martingale M can be written as a stochastic integral $M = M(0) + \varphi \bullet X$ with some predictable process φ.*

Proof The martingale property of X implies $ap - b(1 - p) = 0$. Since $\Delta M(t)$ is $\sigma(\Delta X(1), \dots, \Delta X(t))$-measurable, there is some function $f_t : \{-b, a\}^t \to \mathbb{R}$ such that $\Delta M(t) = f_t(\Delta X(1), \dots, \Delta X(t))$. The martingale property of M and (1.4) yield

$$\begin{aligned} 0 &= E(\Delta M(t)|\mathscr{F}_{t-1}) \\ &= pf_t(\Delta X(1), \dots, \Delta X(t-1), a) + (1-p)f_t(\Delta X(1), \dots, \Delta X(t-1), -b) \end{aligned}$$

and hence

$$\frac{1}{a} f_t(\Delta X(1), \dots, \Delta X(t-1), a) = -\frac{1}{b} f_t(\Delta X(1), \dots, \Delta X(t-1), -b) =: \varphi(t).$$

This implies

$$\varphi(t)\Delta X(t) = f_t(\Delta X(1), \dots, \Delta X(t-1), a) = \Delta M(t)$$

for $\Delta X(t) = a$ and likewise for $\Delta X(t) = -b$. □

The following statement and its continuous-time counterpart play a key role in Mathematical Finance.

Theorem 1.22 (Optional Decomposition) *Let S denote an $\mathbb{R}^d$-valued process and $\mathcal{Q}$ the set of all $Q \sim P$ such that S is a Q-local martingale (in the sense that $S_1, \dots, S_d$ are Q-local martingales). Moreover, let X be a process which is a Q-local supermartingale relative to all $Q \in \mathcal{Q}$. If $\mathcal{Q}$ is not empty, there exists a predictable $\mathbb{R}^d$-valued process φ such that $C := X(0) + \varphi \bullet S - X$ is increasing.*

Proof This is stated in [114, Theorem 2] if the time horizon is finite, i.e. if S and X are constant after some deterministic time $T < \infty$. The general case follows from [114, Theorem 1].

We illustrate the statement in the case of a finite time horizon $T \in \mathbb{N}$ and a finite sample space Ω whose elements have strictly positive probability.

Step 1: The proof will be based on an application of the separating hyperplane theorem A.14. To this end, let U be the finite-dimensional space of adapted processes $x = (x(t))_{t=0,\dots,T}$ with $x(0) = 0$. Moreover, we consider $V := U$ as the dual space of U via $y(x) := E(\sum_{t=1}^T x(t)y(t))$. Set

$$K := \left\{(\varphi(t)^\top \Delta S(t))_{t=0,\dots T} : \varphi \text{ predictable and } \mathbb{R}^d\text{-valued}\right\}$$

and

$$M := \left\{\left(\lambda\Delta X(t)+(1-\lambda)x(t)\right)_{t=0,\dots,T} : \lambda \in [0,1], (x(t))_{t=0,\dots,T} \text{ nonnegative}\right.$$
$$\left.\text{adapted process with } x(0) = 0 \text{ and } E\left(\sum_{t=1}^T x(t)\right) = 1\right\}.$$

One easily verifies that K is a subspace and M a compact convex subset of U.

Step 2: In steps 2–4 we show by contradiction that $K \cap M \neq \varnothing$. Otherwise the separating hyperplane theorem A.14 yields the existence of $y \in V$ with

$$E\left(\sum_{t=1}^T x(t)y(t)\right) = 0, \quad x \in K, \tag{1.18}$$

$$E\left(\sum_{t=1}^T x(t)y(t)\right) > 0, \quad x \in M. \tag{1.19}$$

Since $x \in M$ for

$$x(s) := \begin{cases} 1_F/E(F) & \text{for } s = t, \\ 0 & \text{otherwise} \end{cases}$$

with fixed $t \in \{1, \dots, T\}$ and $F \in \mathscr{F}_t$, we conclude from (1.19) that y is strictly positive. Define the martingale

$$N(t) := \sum_{s=1}^{t} \left(\frac{y(s)}{E(y(s)|\mathscr{F}_{s-1})} - 1 \right)$$

and $Z = \mathscr{E}(N)$. Since $\Delta N > -1$, we have that Z is the density process of some probability measure $Q \sim P$.

Step 3: We show that $Q \in \mathscr{Q}$. By Proposition 1.19 this follows if $(\varphi \bullet S)Z$ is a martingale for any predictable $\mathbb{R}^d$-valued φ. Defining the predictable, positive process $a(t) := E(y(t)|\mathscr{F}_{t-1})$, $t = 1, \dots, T$, it suffices to show that $V := \frac{a}{Z_-} \bullet ((\varphi \bullet S)Z)$ is a martingale, cf. Proposition 1.14(7). Fix $t \in \{0, \dots, T\}$ and $F \in \mathscr{F}_t$. Since

$$(\varphi \bullet S)Z = (\varphi \bullet S)_- \bullet Z + Z_- \bullet (\varphi \bullet S) + [\varphi \bullet S, Z]$$

and $Z = \mathscr{E}(N)$, we have

$$(\varphi \bullet S)Z = Z_- \bullet \big((\varphi \bullet S)_- \bullet N + \varphi \bullet S + [\varphi \bullet S, N]\big). \tag{1.20}$$

Hence V equals

$$\begin{aligned} a \bullet \big(\varphi \bullet S + [\varphi \bullet S, N]\big)(t) &= \sum_{s=1}^{t} a(s) \left(\varphi(s)^\top \Delta S(s) + \varphi(s)^\top \Delta S(s) \left(\frac{y(s)}{a(s)} - 1 \right) \right) \\ &= \sum_{s=1}^{t} \varphi(s)^\top \Delta S(s) y(s) =: \widetilde{V}(t) \end{aligned}$$

up to a martingale. Observe that

$$\begin{aligned} E\Big((\widetilde{V}(T) - \widetilde{V}(t))1_F\Big) &= E\Big(\sum_{s=t+1}^{T} \varphi(s)^\top \Delta S(s) y(s) 1_F \Big) \\ &= E\Big(\sum_{s=1}^{T} \widetilde{\varphi}(s)^\top \Delta S(s) y(s) \Big) \end{aligned} \tag{1.21}$$

for the predictable process

$$\widetilde{\varphi}(s) := \begin{cases} \varphi(s)1_F & \text{for } s > t, \\ 0 & \text{for } s \leq t. \end{cases}$$

By (1.18) we have that (1.21) equals 0, which in turn means that $\widetilde{V}$ and hence also V are martingales as desired.

Step 4: By assumption X is a Q-supermartingale. Since

$$X(s)Z(s) \geq E_Q(X(t)|\mathscr{F}_s)Z(s) = E(X(t)Z(t)|\mathscr{F}_s)$$

for $s \leq t$ by (1.8), we have that XZ is a supermartingale relative to P. By Proposition 1.14(8) this in turn implies that $\widetilde{S} := \frac{a}{Z_-} \bullet (XZ)$ is a supermartingale. Essentially the same calculations as in (1.20–1.21) yield

$$0 \geq E(\widetilde{S}(T) - \widetilde{S}(0)) = E\left(\sum_{s=1}^{T} \Delta X(s)y(s)\right)$$

in contradiction to (1.19). It follows that $K \cap M = \varnothing$ cannot hold, cf. step 2.

Step 5: Since $K \cap M \neq \varnothing$, there exists some predictable φ, some $\lambda \in [0, 1]$ and some nonnegative adapted x with $x(0) = 0$ and $P(\sum_{t=1}^{T} x(t) > 0) > 0$ such that

$$\lambda \Delta X(t) + (1 - \lambda)x(t) = \varphi(t)^\top \Delta S(t), \quad t = 1, \dots, T. \tag{1.22}$$

Assume by contradiction that $\lambda = 0$. This implies $\sum_{t=1}^{T} x(t) = \varphi \bullet S(T)$. For any $Q \in \mathscr{Q}$ we have $E_Q(\varphi \bullet S(T)) = 0$. On the other hand, $\sum_{t=1}^{T} x(t) \geq 0$ and $Q(\sum_{t=1}^{T} x(t) > 0) > 0$ imply that $E_Q(\varphi \bullet S(T)) > 0$ and hence a contradiction to (1.19). Consequently $\lambda > 0$.

Step 6: From (1.22) and $\lambda > 0$ we obtain

$$\Delta X(t) = \frac{\varphi(t)^\top}{\lambda} \Delta S(t) - \frac{1 - \lambda}{\lambda} x(t), \quad t = 1, \dots T,$$

whence $X = X(0) + \frac{\varphi}{\lambda} \bullet S - C$ for the increasing process $C(t) = \sum_{s=1}^{t} \frac{1-\lambda}{\lambda} x(s)$. This yields the assertion. □

A prime example for a process X as in Theorem 1.22 is given by

$$X(t) = \operatorname*{ess\,sup}_{Q \in \mathscr{Q}} E_Q(H|\mathscr{F}_t), \quad t \geq 0, \tag{1.23}$$

where H denotes a bounded random variable.

Proposition 1.23 *X in (1.23) is a Q-supermartingale for all $Q \in \mathscr{Q}$.*

Proof

Step 1: Fix $t \geq 0$. We show that $(E_Q(H|\mathscr{F}_t))_{Q\in\mathscr{Q}}$ has the lattice property in the sense of Lemma A.3. To this end, take $Q_1, Q_2 \in \mathscr{Q}$ with density processes Y_1, Y_2 and consider $F := \{E_{Q_1}(H|\mathscr{F}_t) \leq E_{Q_2}(H|\mathscr{F}_t)\} \in \mathscr{F}_t$. Using Proposition 1.19 it is easy to verify that

$$\widetilde{Y}(s) := \begin{cases} Y_1(s) & \text{for } s \leq t, \\ Y_1(s) & \text{on } F^C \text{ for } s > t, \\ \frac{Y_1(t)}{Y_2(t)} Y_2(s) & \text{on } F \text{ for } s > t \end{cases}$$

is the density process of some probability measure $\widetilde{Q} \in \mathscr{Q}$. Since

$$\begin{aligned} E_{\widetilde{Q}}(H|\mathscr{F}_t) &= E_{Q_1}(H|\mathscr{F}_t) 1_{F^C} + E_{Q_2}(H|\mathscr{F}_t) 1_F \\ &= E_{Q_1}(H|\mathscr{F}_t) \vee E_{Q_2}(H|\mathscr{F}_t), \end{aligned}$$

the lattice property holds.

Step 2: Fix $Q \in \mathscr{Q}$ and $t \geq 0$. By Lemma A.3 there is a sequence of probability measures $Q_n \in \mathscr{Q}$ such that $E_{Q_n}(H|\mathscr{F}_t) \uparrow \operatorname{ess\,sup}_{\widetilde{Q}\in\mathscr{Q}} E(H|\mathscr{F}_t)$ as $n \to \infty$. Denote by Y_n and Y the density processes of Q_n and Q, respectively. Using Proposition 1.19 and the generalised Bayes' rule (1.8) it is easy to verify that the process

$$\widetilde{Y}_n(s) := \begin{cases} Y(s) & \text{for } s \leq t, \\ \frac{Y(t)}{Y_n(t)} Y_n(s) & \text{for } s > t \end{cases}$$

is the density process of some probability measure $\widetilde{Q}_n \in \mathscr{Q}$ satisfying $\widetilde{Q}_n|_{\mathscr{F}_t} = Q|_{\mathscr{F}_t}$ and $E_{\widetilde{Q}_n}(H|\mathscr{F}_t) = E_{Q_n}(H|\mathscr{F}_t)$. Together with the monotone convergence theorem for conditional expectations we conclude that

$$\begin{aligned} E_Q(X(t)|\mathscr{F}_s) &= E_Q\left(\operatorname*{ess\,sup}_{\widetilde{Q}\in\mathscr{Q}} E_{\widetilde{Q}}(H|\mathscr{F}_t) \middle| \mathscr{F}_s \right) \\ &= E_Q\left(\sup_{n\in\mathbb{N}} E_{Q_n}(H|\mathscr{F}_t) \middle| \mathscr{F}_s \right) \\ &= \sup_{n\in\mathbb{N}} E_Q\big(E_{Q_n}(H|\mathscr{F}_t)\big|\mathscr{F}_s\big) \\ &= \sup_{n\in\mathbb{N}} E_{\widetilde{Q}_n}\left(E_{\widetilde{Q}_n}(H|\mathscr{F}_t)\middle|\mathscr{F}_s\right) \end{aligned}$$

$$= \sup_{n\in\mathbb{N}} E_{\widetilde{Q}_n}(H|\mathscr{F}_s)$$

$$\leq \operatorname*{ess\,sup}_{\widetilde{Q}\in\mathscr{Q}} E_{\widetilde{Q}}(H|\mathscr{F}_s)$$

$$= X(s)$$

for $s \leq t$. □

1.3 Jump Characteristics

The distribution of a concrete stochastic process *as a whole* is often unknown in the first place. However, we usually have a local conception of its dynamics, i.e. at time $t - 1$ we are aware of the distribution of the next value $X(t)$ given the history up to $t-1$. This information can be used in a second step to derive unconditional expected values $E(X(t))$, probabilities etc. We consider two concepts of local descriptions of the process, jump characteristics in the present section and generators of Markov processes in the next.

If X is an adapted process with values in $E \subset \mathbb{R}^d$, we call the mapping

$$K^X(t, B) := P(\Delta X(t) \in B|\mathscr{F}_{t-1}) := E\big(1_B(\Delta X(t))\big|\mathscr{F}_{t-1}\big) \tag{1.24}$$

for $t = 1, 2, \ldots$ and $B \in \mathscr{B}(E)$ the **jump characteristic** of X. It is nothing else than the conditional law of the increments of the process. The name is inspired by *semimartingale characteristics*. This more involved continuous-time counterpart is discussed in Chap. 4.

In applications the jump characteristic depends only on the present value of the process. Specifically, we say an adapted process X is **of Markov type** if $K^X(t, B)$ depends on ω, t only through $X(t-1)(\omega)$, i.e. more precisely if it is of the form

$$K^X(t, B) = \kappa(X(t-1), B) \tag{1.25}$$

with some function κ that is a probability measure in its second argument. This Markovian case is studied more thoroughly in the next section.

Random walks have particularly simple jump characteristics.

Proposition 1.24 (Random Walk) *An adapted process X with $X(0) = 0$ is a random walk if and only if its jump characteristic is of the form*

$$K^X(t, B) = \nu(B)$$

for some probability measure ν which does not depend on (ω, t). In this case ν is the law of $\Delta X(1)$. Note that X is of Markov type.

Proof If X is a random walk, we have $P^{\Delta X(t)|\mathscr{F}_{t-1}} = P^{\Delta X(t)} = P^{\Delta X(1)}$ since $\Delta X(t)$ is independent of $\mathscr{F}_{t-1}$ and has the same law for all t.

Conversely,

$$\begin{aligned} P(\{\Delta X(t) \in B\} \cap F) &= E(1_B(\Delta X(t))1_F) \\ &= \int E\big(1_B(\Delta X(t))\big|\mathscr{F}_{t-1}\big)1_F dP \\ &= \int \nu(B)1_F dP \\ &= \nu(B)P(F) \end{aligned}$$

for $F \in \mathscr{F}_{t-1}$. For $F = \Omega$ we obtain $P(\Delta X(t) \in B) = \nu(B) = P(\Delta X(1) \in B)$. Hence $\Delta X(t)$ is independent of $\mathscr{F}_{t-1}$ and has the same law for all t. □

Geometric random walks turn out to be of Markov type as well.

Example 1.25 The jump characteristic of a geometric random walk X is given by

$$K^X(t, B) = \varrho\big(\{x \in \mathbb{R}^d : X(t-1)(x-1) \in B\}\big),$$

where ϱ denotes the distribution of $X(1)/X(0) = X(1)$. Indeed, we have

$$\begin{aligned} E\big(1_B(\Delta X(t))\big|\mathscr{F}_{t-1}\big) &= E\left(1_B\left(X(t-1)\left(\frac{X(t)}{X(t-1)} - 1\right)\right)\middle|\mathscr{F}_{t-1}\right) \\ &= \int 1_B(X(t-1)(x-1))\varrho(dx) \end{aligned}$$

by (1.4) and the fact that $X(t)/X(t-1)$ has law ϱ and is independent of $\mathscr{F}_{t-1}$.

Since adapted processes are invariant under stochastic integration, stopping, application of continuous mappings and measure changes, it makes sense to discuss the effect of these operations on the jump characteristic. The following propositions are provided as a motivation for similar rules in Chap. 4.

Proposition 1.26 (Stochastic Integration) *If X is an adapted process with jump characteristic K^X and φ is a predictable process, the jump characteristic $K^{\varphi \bullet X}$ of $\varphi \bullet X$ is given by*

$$K^{\varphi \bullet X}(t, B) = \int 1_B(\varphi(t)x)K^X(t, dx).$$

Proof By (1.3) we have

$$E\big(1_B(\Delta(\varphi \bullet X(t)))\big|\mathscr{F}_{t-1}\big) = E\big(1_B(\varphi(t)\Delta X(t))\big|\mathscr{F}_{t-1}\big)$$
$$= \int 1_B(\varphi(t)x)P^{\Delta X(t)|\mathscr{F}_{t-1}}(dx),$$

which yields the claim because $P^{\Delta X(t)|\mathscr{F}_{t-1}} = K^X(t,\cdot)$. □

Proposition 1.27 (Stopping) *If X is an adapted process with jump characteristic K^X and τ is a stopping time, the jump characteristic K^{X^τ} of X^τ is given by*

$$K^{X^\tau}(t,B) = K^X(t,B)1_{\{\tau\geq t\}} + \varepsilon_0(B)1_{\{\tau<t\}}.$$

Here, ε_0 denotes the Dirac measure in 0.

Proof Since $X^\tau = X(0) + \varphi \bullet X$ with $\varphi(t) := 1_{\{\tau\geq t\}}$, the assertion follows from the previous proposition. □

Proposition 1.28 (Functions) *If X is an adapted process with jump characteristic K^X and f is a real- or vector-valued function, then the jump characteristic $K^{f(X)}$ of the process $f(X)$ is given by*

$$K^{f(X)}(t,B) = \int 1_B\big(f(X(t-1)+x) - f(X(t-1))\big)K^X(t,dx).$$

Proof By (1.3) we have

$$E\big(1_B(f(X(t)) - f(X(t-1)))\big|\mathscr{F}_{t-1}\big)$$
$$= \int 1_B\big(f(X(t-1)+x) - f(X(t-1))\big)P^{\Delta X(t)|\mathscr{F}_{t-1}}(dx)$$
$$= \int 1_B\big(f(X(t-1)+x) - f(X(t-1))\big)K^X(t,dx),$$

which yields the claim. □

Proposition 1.29 (Change of Measure) *Suppose that $Q \sim P$ denotes a probability measure with density process $Z = Z(0)\mathscr{E}(N)$. Let X be an adapted process and denote by $K^{(X,N)}$ the characteristic of the bivariate process (X,N). Then the jump characteristic $K^{X,Q}$ of X relative to Q is given by*

$$K^{X,Q}(t,B) = \int 1_B(x)(1+y)K^{(X,N)}(t,d(x,y)).$$

Proof By (1.8) we have

$$\begin{aligned}E_Q\big(1_B(\Delta X(t))\big|\mathscr{F}_{t-1}\big) &= E\big(1_B(\Delta X(t))\tfrac{Z(t)}{Z(t-1)}\big|\mathscr{F}_{t-1}\big)\\ &= E\big(1_B(\Delta X(t))(1+\Delta N(t))\big|\mathscr{F}_{t-1}\big).\end{aligned}$$

Note that

$$E\big(f(\Delta X(t),\Delta N(t))\big|\mathscr{F}_{t-1}\big) = \int f(x,y)K^{(X,N)}(t,d(x,y))$$

holds by definition for indicator functions $f(x,y) = 1_C(x,y)$ and hence by standard arguments for arbitrary functions f, cf. the proof of Proposition 1.38. Considering $f(x,y) = 1_C(x)(1+y)$ yields the claim. □

Propositions 1.26–1.28 are easy to memorise: if x stands for the jump size of X at t, then $\varphi \bullet X$ jumps by $\varphi(t)x$ (needed for Proposition 1.26), X^τ jumps by x if $\tau \geq t$ and by 0 if $\tau < t$ (needed for Proposition 1.27), and $f(X)$ by $f(X(t-1)+x) - f(X(t-1))$ (for Proposition 1.28). Proposition 1.29 may seem less obvious. It can be understood by viewing the transition from P to Q as a composition of one-period measure changes with conditional density $1 + \Delta N(t)$.

In order to illustrate some of these rules, we consider the following

Example 1.30 We have already observed that geometric random walks can be written as ordinary or alternatively as stochastic exponentials of random walks, i.e.

$$Z = \exp(X) = \mathscr{E}(\widetilde{X}).$$

This can also be seen by computing their characteristics. Proposition 1.28 for $f(x) = e^x$ yields

$$\begin{aligned}K^Z(t,B) &= \int 1_B\big(e^{X(t-1)+x} - e^{X(t-1)}\big)K^X(t,dx)\\ &= \int 1_B\big(Z(t-1)(e^x-1)\big)K^X(t,dx),\end{aligned}$$

where K^X, K^Z denote the jump characteristics of X and Z, respectively. Since $\widetilde{X} = \frac{1}{Z_-} \bullet Z$, we have

$$\begin{aligned}K^{\widetilde{X}}(t,B) &= \int 1_B(x/Z(t-1))K^Z(t,dx)\\ &= \int 1_B(e^x-1)K^X(t,dx) \qquad (1.26)\end{aligned}$$

for the jump characteristic of $\widetilde{X}$ by Proposition 1.26. From Proposition 1.24 we observe that $\widetilde{X}$ is a random walk if and only if this holds for X.

Conversely, we could also have applied Proposition 1.26 to $Z = Z(0) + Z_- \bullet \widetilde{X}$ to obtain

$$K^Z(t, B) = \int 1_B\big(Z(t-1)x\big)K^{\widetilde{X}}(t, dx).$$

Proposition 1.28 for $f(x) = \log x$ now yields

$$\begin{aligned} K^X(t, B) &= \int 1_B\big(\log(Z(t-1)+z) - \log Z(t-1)\big)K^Z(t, dz) \\ &= \int 1_B(\log(1+x))K^{\widetilde{X}}(t, dx), \end{aligned}$$

which is equivalent to (1.26).

For the following we define the **identity process** I as

$$I(t) := t.$$

The characteristics can be used to compute the compensator of an adapted process.

Proposition 1.31 (Compensator) *If X is an adapted process, its compensator A^X and its jump characteristic K^X are related to each other via*

$$A^X = a^X \bullet I$$

with

$$a^X(t) := \int x K^X(t, dx)$$

provided that X is integrable or, more generally, if the integral a^X is finite.

Proof By definition of the compensator we have

$$\begin{aligned} \Delta A^X(t) = E(\Delta X(t)|\mathscr{F}_{t-1}) &= \int x P^{\Delta X(t)|\mathscr{F}_{t-1}}(dx) \\ &= \int x K^X(t, dx) = a^X(t) = \Delta(a^X \bullet I)(t). \end{aligned}$$ □

Since the predictable covariation is a compensator, it can also be expressed in terms of characteristics.

Proposition 1.32 (Predictable Covariation) *The predictable covariations of two adapted processes X, Y and of their martingale parts M^X, M^Y are given by*

$$\langle X, Y\rangle = \widetilde{c}^{X,Y} \bullet I,$$
$$\langle M^X, M^Y\rangle = \widehat{c}^{X,Y} \bullet I$$

with

$$\widetilde{c}^{X,Y}(t) := \int xyK^{(X,Y)}(t, d(x, y)),$$
$$\widehat{c}^{X,Y}(t) := \int xyK^{(X,Y)}(t, d(x, y)) - a^X(t)a^Y(t),$$

provided that the integrals $a^X, a^Y, \widetilde{c}^{X,Y}$ are finite.

Proof The first statement follows similarly as Proposition 1.31 by observing that

$$\Delta\langle X, Y\rangle(t) = E(\Delta X(t)\Delta Y(t)|\mathscr{F}_{t-1}).$$

The second in turn follows from the first and from Proposition 1.31 because

$$\begin{aligned}\Delta\langle M^X, M^Y\rangle(t) &= E(\Delta M^X(t)\Delta M^Y(t)|\mathscr{F}_{t-1})\\ &= E(\Delta X(t)\Delta Y(t)|\mathscr{F}_{t-1}) - E(\Delta X(t)|\mathscr{F}_{t-1})\Delta A^Y(t)\\ &\quad - \Delta A^X(t)E(\Delta Y(t)|\mathscr{F}_{t-1}) + \Delta A^X(t)\Delta A^Y(t)\\ &= \Delta\langle X, Y\rangle(t) - \Delta A^X(t)\Delta A^Y(t)\\ &= \int xyK^{(X,Y)}(t, d(x, y)) - a^X(t)a^Y(t).\end{aligned}$$

□

Let us rephrase the integration by parts rule in terms of characteristics.

Proposition 1.33 *For adapted processes X, Y we have*

$$a^{XY}(t) = X(t-1)a^Y(t) + Y(t-1)a^X(t) + \widetilde{c}^{X,Y}(t),$$

provided that X, Y and XY are integrable or, more generally, if the integrals $a^X, a^Y, \widetilde{c}^{X,Y}$ are finite.

Proof Computing the compensators of

$$XY = X(0)Y(0) + X_- \bullet Y + Y_- \bullet X + [X, Y]$$

yields

$$\begin{aligned} a^{XY} \bullet I &= A^{XY} \\ &= X_- \bullet A^Y + Y_- \bullet A^X + A^{[X,Y]} \\ &= (X_- a^Y) \bullet I + (Y_- a^X) \bullet I + \widetilde{c}^{X,Y} \bullet I \end{aligned}$$

by Propositions 1.14(3, 9) and 1.32. Considering increments yields the claim. □

1.4 Markov Processes

Fix a state space $E \subset \mathbb{R}^d$ containing the possible values of the stochastic processes under consideration. We call an adapted E-valued process X a **(homogeneous) Markov process** if

$$P(X(s+t) \in B|\mathscr{F}_s) = P(t, X(s), B), \quad s, t \geq 0, \ B \in \mathscr{B}(E) \tag{1.27}$$

holds for some family $(P(t, x, \cdot))_{t \geq 0, x \in E}$ of probability measures on E. Intuitively, this means that the future evolution $X(s+t)$ given the past up to s depends on this past only through the present value $X(s)$. In this sense the process has *no memory*. Moreover, its dynamics do not depend explicitly on time. If the process attains the value x at time t, it evolves in the future as if it had started afresh at this point. Note that both sides of (1.27) are random variables, the right-hand side through the starting value $X(s)$.

The function $(t, x, B) \mapsto P(t, x, B)$ is called the *transition function* of X. The value $P(t, x, B)$ is the probability of ending up in B if one started t periods ago in x. The transition function "almost" satisfies $P(0, x, B) = \varepsilon_0(B)$ and the **Chapman–Kolmogorov equation**

$$P(s+t, x, B) = \int P(t, y, B) P(s, x, dy) \tag{1.28}$$

for $s, t \geq 0, x \in E, B \in \mathscr{B}(E)$. Indeed, for arbitrary $r \geq 0$ we have

$$\begin{aligned} P(s+t, X(r), B) &= P(X(r+s+t) \in B|\mathscr{F}_r) \\ &= E\big(P(X(r+s+t) \in B|\mathscr{F}_{r+s})\big|\mathscr{F}_r\big) \\ &= E\big(P(t, X(r+s), B)\big|\mathscr{F}_r\big) \\ &= \int P(t, y, B) P(s, X(r), dy), \end{aligned} \tag{1.29}$$

where the last line follows because $P(s, X(r), \cdot)$ is the law of $X(r+s)$ given the information up to time r. The restriction *almost* above refers to the fact that (1.29) yields (1.28) only up to some set of points x that is visited with probability 0 by X. To be more precise, we call $(P(t, x, \cdot))_{t\geq 0, x\in E}$ a **(homogeneous) transition function** only if (1.28) holds for any x without exception. Moreover, one generally requires the family in the above definition of a Markov process to be a transition function in this sense. Of course, it is unique only up to some set of points x that are never visited by the process.

In discrete time it suffices to consider one-step transitions in order to verify the Markov property:

Proposition 1.34 *Suppose that $P(X(t+1) \in B|\mathscr{F}_t) = Q(X(t), B)$, $t \geq 0$, $B \in \mathscr{B}(E)$ for some family $(Q(x, \cdot))_{x\in E}$ of probability measures on E. Then X is a Markov process relative to the transition function defined recursively by $P(0, x, B) = \varepsilon_x(B)$ and $P(t+1, x, B) = \int Q(y, B)P(t, x, dy)$.*

Proof Equation (1.27) follows by induction because

$$\begin{aligned}
P(X(s+t+1) \in B|\mathscr{F}_s) &= E\big(P(X(s+t+1) \in B|\mathscr{F}_{s+t})\big|\mathscr{F}_s\big)\\
&= E\big(Q(X(s+t), B)\big|\mathscr{F}_s\big)\\
&= \int Q(y, B)P(t, X(s), dy)\\
&= P(t+1, X(s), B).
\end{aligned}$$

Similarly, (1.28) is obtained by induction from

$$\begin{aligned}
P(s+t+1, x, B) &= \int Q(y, B)P(s+t, x, dy)\\
&= \iint Q(y, B)P(t, z, dy)P(s, x, dz)\\
&= \int P(t+1, z, B)P(s, x, dz).
\end{aligned}$$ □

Proposition 1.35 *An adapted process X is a Markov process if and only if it is of Markov type in the sense of Sect. 1.3.*

Proof Suppose that X is a Markov process. By (1.3) we have

$$\begin{aligned}
K^X(t, B) &= E\big(1_B(X(t) - X(t-1))\big|\mathscr{F}_{t-1}\big)\\
&= \int 1_B(x - X(t-1))P^{X(t)|\mathscr{F}_{t-1}}(dx)\\
&= \int 1_B(x - X(t-1))P(1, X(t-1), dx),
\end{aligned}$$

which is a function of $X(t-1)$ and B as claimed.

Conversely, assume that $K^X(t, B) = \kappa(X(t-1), B)$ with some function κ that is a probability measure in its second argument. Again by (1.3) we have

$$\begin{aligned} P(X(s+1) \in B|\mathscr{F}_s) &= E\big(1_B(X(s) + \Delta X(s+1))\big|\mathscr{F}_s\big) \\ &= \int 1_B(X(s)+x)P^{\Delta X(s+1)|\mathscr{F}_s}(dx) \\ &= \int 1_B(X(s)+x)\kappa(X(s), dx) \qquad (1.30) \\ &=: Q(X(s), B) \qquad (1.31) \end{aligned}$$

for a family $(Q(x, \cdot))_{x\in E}$ of probability measures on E. The Markov property follows from Proposition 1.34. □

Corollary 1.36 (Random Walks and Geometric Random Walks) *Random walks and geometric random walks are Markov processes.*

Proof In view of Proposition 1.35, this follows immediately from Proposition 1.24 and Example 1.25. □

Together with the law of $X(0)$, the transition function determines the distribution of the whole process uniquely, as the following formula shows.

Proposition 1.37 *For any $0 \le t_1 \le \cdots \le t_n$ and any bounded or nonnegative measurable function $f : E^n \to \mathbb{R}$, we have*

$$\begin{aligned} E(f(X(t_1), \ldots, X(t_n))) = \int \cdots \int f(x_1, \ldots, x_n)P(t_n - t_{n-1}, x_{n-1}, dx_n) \\ \cdots P(t_2 - t_1, x_1, dx_2)P(t_1, x_0, dx_1)P^{X(0)}(dx_0). \end{aligned}$$

Proof We proceed by induction on n. For $n = 1$ we have

$$\begin{aligned} E(f(X(t_1))) &= E\big(E(f(X(t_1))|\mathscr{F}_0)\big) \\ &= E\left(\int f(x_1)P^{X(t_1)|\mathscr{F}_0}(dx_1)\right) \\ &= E\left(\int f(x_1)P(t_1, X(0), dx_1)\right) \\ &= \iint f(x_1)P(t_1, x_0, dx_1)P^{X(0)}(dx_0), \end{aligned}$$

where the third equation follows from the Markov property (1.27). Suppose now that the assertion holds for $n-1$. Again by (1.27), we obtain

$$\begin{aligned} E(f(X(t_1),\dots,X(t_n))) &= E\big(E(f(X(t_1),\dots,X(t_n))|\mathscr{F}_{t_{n-1}})\big) \\ &= E\Big(\int f(X(t_1),\dots,X(t_{n-1}),x_n)P^{X(t_n)|\mathscr{F}_{t_{n-1}}}(dx_n)\Big) \\ &= E\Big(\int f(X(t_1),\dots,X(t_{n-1}),x_n)P(t_n-t_{n-1},X(t_{n-1}),dx_n)\Big) \\ &= E(g(X(t_1),\dots,X(t_{n-1}))) \end{aligned} \tag{1.32}$$

with

$$g(x_1,\dots,x_{n-1}) := \int f(x_1,\dots,x_{n-1},x_n)P(t_n-t_{n-1},x_{n-1},dx_n).$$

By assumption we have that (1.32) equals

$$\int\cdots\int g(x_1,\dots,x_{n-1})P(t_{n-1}-t_{n-2},x_{n-2},dx_{n-1}) \\ \cdots P(t_2-t_1,x_1,dx_2)P(t_1,x_0,dx_1)P^{X(0)}(dx_0),$$

which yields the claim. □

The integrator

$$P(t_n-t_{n-1},x_{n-1},dx_n)\cdots P(t_2-t_1,x_1,dx_2)P(t_1,x_0,dx_1)P^{X(0)}(dx_0)$$

stands for the joint law of $X(0),X(t_1),\dots,X(t_n)$. Specifically, the probability of observing $X(0),X(t_1),\dots,X(t_n)$ in $x_0,\dots,x_n$ is given by the probability of observing $X(0)$ in x_0, multiplied by the probability of moving from x_0 to x_1 in time t_1, multiplied by etc. up to the probability of moving from x_{n-1} to x_n in time t_n-t_{n-1}.

The transition function leads naturally to a family $(p_t)_{t\geq 0}$ of operators on bounded measurable functions $f: E\to\mathbb{R}$, called the **transition semigroup**. These operators are defined via

$$p_t f(x) := \int f(y)P(t,x,dy)$$

for $t\geq 0$, bounded measurable $f: E\to\mathbb{R}$, and $x\in E$. Put differently, $p_t f(x)$ is the expected value of $f(X(s+t))$ given that $X(s)=x$. In view of

$$p_t 1_B(x) = P(t,x,B),$$

the transition function can be recovered from $(p_t)_{t\geq 0}$. The family $(p_t)_{t\geq 0}$ is called *semigroup* because

$$p_{s+t} = p_s \circ p_t, \tag{1.33}$$

which means that $p_{s+t}f = p_s(p_t f)$ for any bounded measurable function f. Equation (1.33) follows easily from the Chapman–Kolmogorov equation.

The Chapman–Kolmogorov equation or Proposition 1.34 shows that in discrete time all p_t and hence the distribution of the whole process can be recovered from p_1. Alternatively, the transition probabilities can be derived from the **generator** G of the Markov process, which is another operator mapping bounded measurable functions $f : E \to \mathbb{R}$ on the like. It is defined as $Gf := p_1 - p_0$, i.e.

$$Gf(x) = p_1 f(x) - f(x). \tag{1.34}$$

In terms of the probability measures $Q(x,\cdot)$ in Proposition 1.34 or the function κ in (1.25), we can write it as

$$\begin{aligned} Gf(x) &= \int f(y)Q(x,dy) - f(x) \\ &= \int (f(x+y) - f(x))\kappa(x,dy), \end{aligned}$$

cf. (1.30, 1.31). For the random walk in Proposition 1.24 we have

$$Gf(x) = \int \big(f(x+y) - f(x)\big)\nu(dy).$$

The generator of the geometric random walk in Example 1.25 satisfies

$$Gf(x) = \int \big(f(xy) - f(x)\big)\varrho(dy).$$

Observe that the generator has the property that

$$\begin{aligned} M(t) &:= f(X(t)) - f(X(0)) - Gf(X_-) \bullet I \\ &= f(X(t)) - f(X(0)) - \sum_{s=1}^{t} Gf(X(s-1)) \end{aligned}$$

is a martingale for bounded measurable functions f because

$$\begin{aligned} E(M(t)|\mathscr{F}_{t-1}) - M(t-1) &= E\big(f(X(t)) - f(X(t-1)) - Gf(X(t-1))\big|\mathscr{F}_{t-1}\big) \\ &= p_1 f(X(t-1)) - f(X(t-1)) - Gf(X(t-1)) \\ &= 0. \end{aligned}$$

Moments $E(f(X(t)))$ of $X(t)$ can be computed by successive application of the generator G:

Proposition 1.38 (Backward Equation) *For bounded measurable $f : E \to \mathbb{R}$ the function $u(t,x) := p_t f(x)$ satisfies*

$$u(0,x) = f(x), \quad u(t,x) = u(t-1,x) + Gu(t-1,x),$$

where we use the notation

$$Gu(t-1,x) := (Gu(t-1,\cdot))(x). \tag{1.35}$$

Proof $u(0,x) = f(x)$ follows from $p_0 f = f$. By definition we have

$$\begin{aligned} u(t-1,x) + Gu(t-1,x) &= \int u(t-1,y)P(1,x,dy) \\ &= \iint f(z)P(t-1,y,dz)P(1,x,dy). \end{aligned}$$

The Chapman–Kolmogorov equation (1.28) yields

$$\iint f(z)P(t-1,y,dz)P(1,x,dy) = \int f(z)P(t,x,dz) \tag{1.36}$$

for $f = 1_B$. By standard arguments from measure theory, (1.36) actually holds for arbitrary bounded f. Indeed, both sides are linear in f and arbitrary measurable f can be approximated by linear combinations of indicator functions.

Since the right-hand-side of (1.36) is $p_t f(x) = u(t,x)$, we are done. □

In Mathematical Finance the previous proposition can, for instance, be applied to compute call option prices $E((X(T) - K)^+)$ for a stock following a geometric random walk.

Conversely, the law of $X(t)$ can be obtained from the law of $X(0)$ by successive application of the adjoint A of the generator G. To make this precise, denote by $B(E)$ the set of bounded measurable functions $E \to \mathbb{R}$. Any probability measure μ on E can be viewed as a linear mapping $B(E) \to \mathbb{R}$ via $\mu f := \int f d\mu$. We denote by $\mathscr{M}(E)$, $B(E)'$ the set of probability measures on E and the set of linear mappings $B(E) \to \mathbb{R}$, respectively. Moreover, we define the **adjoint operator** $A : \mathscr{M}(E) \to B(E)'$ of G by $(A\mu)f := \mu(Gf)$.

Proposition 1.39 (Forward Equation) *The laws $\mu_t := P^{X(t)}$, $t = 0, 1, 2, \ldots$ satisfy*

$$\mu_t - \mu_{t-1} = A\mu_{t-1}.$$

Proof We need to verify that $\mu_t f - \mu_{t-1} f = \mu_{t-1}(Gf)$ for any bounded measurable function $f : E \to \mathbb{R}$, i.e. $E(f(X(t))) - E(f(X(t-1))) = E(Gf(X(t-1)))$. Since $Gf(x) = p_1 f(x) - f(x)$ and $p_1 f(x) = \int f(y) P(1, x, dy)$, Proposition 1.37 yields

$$\begin{aligned} E(Gf(X(t-1))) &= E(p_1 f(X(t-1))) - E(f(X(t-1))) \\ &= \iiint f(x_2) P(1, x_1, dx_2) P(t-1, x_0, dx_1) P^{X(0)}(dx_0) - E(f(X(t-1))) \\ &= E(f(X(t))) - E(f(X(t-1))) \end{aligned}$$

and hence the claim. □

1.5 Optimal Control

In Mathematical Finance one often faces optimisation problems of various kinds, in particular when it comes to choosing trading strategies with in some sense maximal utility or minimal risk. Such problems can be tackled with different methods. We distinguish two approaches, which are discussed in the following sections and, for continuous time, in Chap. 7. As a motivation we first consider the simple situation of maximising a deterministic function of one or several variables.

Example 1.40

1. (Direct approach) Suppose that the goal is to maximise an *objective function*

$$(x, \alpha) \mapsto \sum_{t=1}^{T} f(t, x_{t-1}, \alpha_t) + g(x_T) \tag{1.37}$$

over all $x = (x_1, \ldots, x_T) \in (\mathbb{R}^d)^T$, $\alpha = (\alpha_1, \ldots, \alpha_T) \in A^T$ such that

$$\Delta x_t := x_t - x_{t-1} = \delta(x_{t-1}, \alpha_t), \quad t = 1, \ldots, T$$

for some given function $\delta : \mathbb{R}^d \times \mathbb{R}^m \to \mathbb{R}^d$. The number or vector α_t stands for a dynamic *control* which determines the *state* x_t of the system. The reward (1.37) in turn depends primarily on x but possibly also on the control α itself. The initial value $x_0 \in \mathbb{R}^d$, the state space of controls $A \subset \mathbb{R}^m$ and the functions $f : \{1, \ldots, T\} \times \mathbb{R}^d \times A \to \mathbb{R}$, $g : \mathbb{R}^d \to \mathbb{R}$ are assumed to be given. The approach in Sect. 1.5.1 below corresponds to finding the maximum directly, without relying on smoothness or convexity of the functions f, g, δ or on topological properties of A. Rather, the idea is to reduce the problem to a sequence of simpler optimisations in just one A-valued variable α_t.

2. (Lagrange multiplier approach) Since the problem above concerns constrained optimisation, Lagrange multiplier techniques may make sense. To this end, define the *Lagrange function*

$$L(x,\alpha,y) := \sum_{t=1}^{T} f(t,x_{t-1},\alpha_t) + g(x_T) - \sum_{t=1}^{T} y_t(\Delta x_t - \delta(x_{t-1},\alpha_t))$$

on $(\mathbb{R}^d)^T \times A^T \times (\mathbb{R}^d)^T$. The usual first-order conditions lead us to look for a candidate $x^\star \in (\mathbb{R}^d)^T$, $\alpha^\star \in A^T$, $y^\star \in (\mathbb{R}^d)^T$ satisfying

a) $\Delta x_t^\star = \delta(x_{t-1}^\star, \alpha_t^\star)$ for $t = 1,\dots,T$, where we set $x_0^\star := x_0$,
b) $y_T^\star = \nabla g(x_T^\star)$,
c) $\Delta y_t^\star = -\nabla_x H(t, x_{t-1}^\star, \alpha_t^\star)$ for $t = 1,\dots,T$, where we set $H(t,\xi,a) := f(t,\xi,a) + y_t^\star \delta(\xi,a)$ and $\nabla_x H$ denotes the gradient of H viewed as a function of its second argument,
d) $\alpha_t^\star$ maximises $a \mapsto H(t, x_{t-1}^\star, a)$ on A for $t = 1,\dots,T$.

Provided that some convexity conditions hold, a)–d) are in fact sufficient for optimality of $\alpha^\star$:

Proposition 1.41 *Suppose that the set A is convex, $\xi \mapsto g(\xi)$, $(\xi,a) \mapsto H(t,\xi,a)$, $t = 1,\dots,T$ are concave and $\xi \mapsto g(\xi)$, $\xi \mapsto H(t,\xi,a)$, $t = 1,\dots,T$, $a \in A$ are differentiable. If conditions a)–d) hold, $(x^\star,\alpha^\star)$ is optimal for the problem in Example 1.40(1).*

Proof We set $h(t,\xi) := \sup_{a\in A} H(t,\xi,a)$ for any competitor (x,α) satisfying the constraints. Condition d) yields $h(t,x_{t-1}^\star) = H(t,x_{t-1}^\star,\alpha_t^\star)$ for $t = 1,\dots,T$. We have

$$\begin{aligned}
&\sum_{t=1}^{T} f(t,x_{t-1},\alpha_t) + g(x_T) - \sum_{t=1}^{T} f(t,x_{t-1}^\star,\alpha_t^\star) - g(x_T^\star) \\
&= \sum_{t=1}^{T}\Big(H(t,x_{t-1},\alpha_t) - H(t,x_{t-1}^\star,\alpha_t^\star) - y_t^\star(\Delta x_t - \Delta x_t^\star)\Big) + g(x_T) - g(x_T^\star) \\
&\le \sum_{t=1}^{T}\Big((H(t,x_{t-1},\alpha_t) - h(t,x_{t-1})) + \big(h(t,x_{t-1}) - h(t,x_{t-1}^\star)\big) \\
&\quad - y_t^\star(\Delta x_t - \Delta x_t^\star)\Big) + \nabla g(x_T^\star)(x_T - x_T^\star) \\
&\le \sum_{t=1}^{T}\Big(\nabla_x h(t,x_{t-1}^\star)(x_{t-1} - x_{t-1}^\star) - y_t^\star(\Delta x_t - \Delta x_t^\star)\Big)\nabla g(x_T^\star)(x_T - x_T^\star)
\end{aligned} \tag{1.38}$$

$$= \sum_{t=1}^{T} \Big(-\Delta y_t^\star (x_{t-1} - x_{t-1}^\star) - y_t^\star (\Delta x_t - \Delta x_t^\star) \Big) + y_T^\star (x_T - x_T^\star) \tag{1.39}$$

$$= y_0^\star (x_0 - x_0^\star),$$

$$= 0$$

where the existence of $\nabla_x h(t, x_{t-1}^\star)$, inequality (1.38) as well as equation (1.39) follow from Proposition 1.42 below and the concavity of g. □

Under some more convexity (e.g. if δ is affine and $f(t, \cdot, \cdot)$ is concave for $t = 1, \dots, T$), the Lagrange multiplier solves some dual minimisation problem. This happens, for example, in the stochastic examples 1.71–1.76 in Sect. 1.5.4.

The following proposition is a version of the *envelope theorem* which concerns the derivative of the maximum of a parametrised function.

Proposition 1.42 *Let A be a convex set, $f : \mathbb{R}^d \times A \to \mathbb{R} \cup \{-\infty\}$ a concave function, and $\widetilde{f}(x) := \sup_{a \in A} f(x, a)$, $x \in \mathbb{R}^d$. Then $\widetilde{f}$ is concave. Suppose in addition that, for some fixed $x^\star \in \mathbb{R}^d$, the optimiser $a^\star := \arg\max_{a \in A} f(x^\star, a)$ exists and $x \mapsto f(x, a^\star)$ is differentiable in $x^\star$. Then $\widetilde{f}$ is differentiable in $x^\star$ with derivative*

$$D_i \widetilde{f}(x^\star) = D_i f(x^\star, a^\star), \quad i = 1, \dots, d. \tag{1.40}$$

Proof One easily verifies that $\widetilde{f}$ is concave. For $h \in \mathbb{R}^d$ we have

$$\widetilde{f}(x^\star + yh) \geq f(x^\star + yh, a^\star)$$

$$= f(x^\star, a^\star) + y \sum_{i=1}^{d} D_i f(x^\star, a^\star) h_i + o(y) \tag{1.41}$$

as $y \in \mathbb{R}$ tends to 0. In view of [144, Proposition I.1.1.4],

$$s(y) := \frac{\widetilde{f}(x^\star + yh) - \widetilde{f}(x^\star)}{y}$$

is decreasing in $y \in \mathbb{R} \setminus \{0\}$. Denoting its limits in 0 by $s(0-)$, $s(0+)$, we obtain $s(0+) \leq s(0-) \leq \sum_{i=1}^{d} D_i f(x^\star, a^\star) h_i \leq s(0+)$ from (1.41). Consequently, $\widetilde{f}$ is differentiable in $x^\star$ with derivative (1.40). □

In the remainder of this section we discuss optimisation in a dynamic stochastic setup.

1.5.1 Dynamic Programming

Since we consider discrete-time stochastic control in this introductory chapter, we work on a filtered probability space $(\Omega, \mathscr{F}, \mathbf{F}, P)$ with filtration $\mathbf{F} = (\mathscr{F}_t)_{t=0,1,\dots,T}$ for finite T. For simplicity, we assume $\mathscr{F}_0$ to be trivial, i.e. all $\mathscr{F}_0$-measurable random variables are deterministic. By (1.1) this implies $E(X|\mathscr{F}_0) = E(X)$ for any random variable X.

Our goal is to maximise some expected reward $E(u(\alpha))$ over controls $\alpha \in \mathscr{A}$. The set $\mathscr{A}$ of **admissible controls** is a subset of all $\mathbb{R}^m$-valued adapted processes and it is assumed to be *stable under bifurcation*, i.e. for any stopping time τ, any event $F \in \mathscr{F}_\tau$, and any $\alpha, \widetilde{\alpha} \in \mathscr{A}$ with $\alpha^\tau = \widetilde{\alpha}^\tau$, the process $(\alpha|\tau_F|\widetilde{\alpha})$ defined by

$$(\alpha|\tau_F|\widetilde{\alpha})(t) := 1_{F^C}\alpha(t) + 1_F\widetilde{\alpha}(t)$$

is again an admissible control. Intuitively, this means that the decision how to continue may depend on the observations so far. Moreover, we suppose that $\alpha(0)$ coincides for all controls $\alpha \in \mathscr{A}$. The reward is expressed by some **reward function** $u : \Omega \times (\mathbb{R}^m)^{\{0,1,\dots,T\}} \to \mathbb{R} \cup \{-\infty\}$. For fixed $\alpha \in \mathscr{A}$, we use the shorthand $u(\alpha)$ for the random variable $\omega \mapsto u(\omega, \alpha(\omega))$. The reward is meant to refer to the final time $T \in \mathbb{N}$, which is expressed mathematically by the assumption that $u(\alpha)$ is $\mathscr{F}_T$-measurable for any $\alpha \in \mathscr{A}$.

Example 1.43 Typically, the reward function is of the form

$$u(\alpha) = \sum_{t=1}^{T} f(t, X^{(\alpha)}(t-1), \alpha(t)) + g(X^{(\alpha)}(T)) \tag{1.42}$$

for some functions $f : \{1, \dots, T\} \times \mathbb{R}^d \times \mathbb{R}^m \to \mathbb{R} \cup \{-\infty\}$, $g : \mathbb{R}^d \to \mathbb{R} \cup \{-\infty\}$, and $\mathbb{R}^d$-valued adapted *controlled processes* $X^{(\alpha)}$. The controlled process is assumed to depend on the control only up to the present time, i.e. for $t = 0, \dots, T$ we have $X^{(\alpha)}(t) = X^{(\widetilde{\alpha})}(t)$ if $\alpha^t = \widetilde{\alpha}^t$.

We call $\alpha^\star \in \mathscr{A}$ an **optimal control** if it maximises $E(u(\alpha))$ over all $\alpha \in \mathscr{A}$, where we set $E(u(\alpha)) := -\infty$ if $E(u(\alpha)^-) := -\infty$. Moreover, the **value process** of the optimisation problem is the family $(\mathscr{V}(\cdot, \alpha))_{\alpha \in \mathscr{A}}$ of adapted processes defined via

$$\mathscr{V}(t, \alpha) := \operatorname{ess\,sup}\left\{E(u(\widetilde{\alpha})|\mathscr{F}_t) : \widetilde{\alpha} \in \mathscr{A} \text{ with } \widetilde{\alpha}^t = \alpha^t\right\} \tag{1.43}$$

for $t = 0, \dots, T$ and $\alpha \in \mathscr{A}$. To this end, recall that the *essential supremum* in (1.43) is the smallest $\mathscr{F}_t$-random variable that dominates the right-hand side outside some null set, cf. Sect. A.2. The right-hand side of (1.43) represents the optimisation problem if we follow the control α up to time t and behave optimally afterwards.

The value process is characterised by some martingale/supermartingale property:

Theorem 1.44

1. *If* $\mathscr{V}(0) := \sup_{\alpha\in\mathscr{A}} E(u(\alpha)) \neq \pm\infty$, *the following holds.*
   a) $\mathscr{V}(t,\widetilde{\alpha}) = \mathscr{V}(t,\alpha)$ *if* $\widetilde{\alpha}^t = \alpha^t$.
   b) $(\mathscr{V}(t,\alpha))_{t\in\{0,\dots,T\}}$ *is a supermartingale with terminal value* $\mathscr{V}(T,\alpha) = u(\alpha)$ *for any admissible control* α *with* $E(u(\alpha)) > -\infty$.
   c) *If* $\alpha^\star$ *is an optimal control, then* $(\mathscr{V}(t,\alpha^\star))_{t\in\{0,\dots,T\}}$ *is a martingale.*
2. *Suppose that* $(\widetilde{\mathscr{V}}(\cdot,\alpha))_{\alpha\in\mathscr{A}}$ *is a family of processes such that*
   a) $\widetilde{\mathscr{V}}(t,\widetilde{\alpha}) = \widetilde{\mathscr{V}}(t,\alpha)$ *if* $\widetilde{\alpha}^t = \alpha^t$. *The common value* $\widetilde{\mathscr{V}}(0,\alpha)$ *is denoted by* $\widetilde{\mathscr{V}}(0)$,
   b) $(\widetilde{\mathscr{V}}(t,\alpha))_{t\in\{0,\dots,T\}}$ *is a supermartingale with terminal value* $\widetilde{\mathscr{V}}(T,\alpha) = u(\alpha)$ *for any admissible control* α *with* $E(u(\alpha)) > -\infty$,
   c) $(\widetilde{\mathscr{V}}(t,\alpha^\star))_{t\in\{0,\dots,T\}}$ *is a submartingale—and hence a martingale—for some admissible control* $\alpha^\star$ *with* $E(u(\alpha^\star)) > -\infty$.

   Then $\alpha^\star$ *is optimal and* $\mathscr{V}(t,\alpha^\star) = \widetilde{\mathscr{V}}(t,\alpha^\star)$ *for* $t = 0,\dots,T$. *In particular,* $\mathscr{V}(0) = \widetilde{\mathscr{V}}(0)$.

Proof

1. The first statement, adaptedness, and the terminal value of $\mathscr{V}(\cdot,\alpha)$ are evident. In order to show the supermartingale property, let $t \in \{0,\dots,T\}$. Stability under bifurcation implies that the set of all $E(u(\widetilde{\alpha})|\mathscr{F}_t)$ with $\widetilde{\alpha} \in \mathscr{A}$ satisfying $\widetilde{\alpha}^t = \alpha^t$ has the lattice property. By Lemma A.3 there exists a sequence of admissible controls α_n with $\alpha_n^t = \alpha^t$ and $E(u(\alpha_n)|\mathscr{F}_t) \uparrow \mathscr{V}(t,\alpha)$. For $s \leq t$ we have

   $$E(E(u(\alpha_n)|\mathscr{F}_t)|\mathscr{F}_s) = E(u(\alpha_n)|\mathscr{F}_s) \leq \mathscr{V}(s,\alpha).$$

   The supermartingale property $E(\mathscr{V}(t,\alpha))|\mathscr{F}_s) \leq \mathscr{V}(s,\alpha)$ is now obtained by monotone convergence.
   Let $\alpha^\star$ be an optimal control. Since $\mathscr{V}(\cdot,\alpha^\star)$ is a supermartingale, the martingale property follows from

   $$\mathscr{V}(0,\alpha^\star) = \sup_{\alpha\in\mathscr{A}} E(u(\alpha)) = E(u(\alpha^\star)) = E(\mathscr{V}(T,\alpha^\star))$$

   and Proposition 1.6.
2. The supermartingale property implies that

   $$E(u(\alpha)) = E(\widetilde{\mathscr{V}}(T,\alpha)) \leq \widetilde{\mathscr{V}}(0,\alpha) = \widetilde{\mathscr{V}}(0) \tag{1.44}$$

   for any admissible control α. Since equality holds for $\alpha^\star$, we have that $\alpha^\star$ is optimal. By statement 1, $\mathscr{V}(\cdot,\alpha^\star)$ is a martingale with terminal value $u(\alpha^\star)$.

Since the same is true for $\widetilde{\mathscr{V}}(\cdot, \alpha^\star)$ by assumptions 2b,c), we have $\mathscr{V}(t, \alpha^\star) = E(u(\alpha^\star)|\mathscr{F}_t) = \widetilde{\mathscr{V}}(t, \alpha^\star)$ for $t = 0, \dots, T$. □

The previous theorem does not immediately lead to an optimal control but it often helps in order to verify that some candidate control is in fact optimal.

Remark 1.45

1. It may happen that the supremum in the definition of the optimal value is not a maximum, i.e. an optimal control does not exist. In this case Theorem 1.44 cannot be applied. Sometimes this problem can be circumvented by considering a certain closure of the set of admissible controls which does in fact contain the optimiser.
   If this is not feasible, a variation of Theorem 1.44(2) without assumption 2c) may be of interest. The supermartingale property 2b) of the candidate value process ensures that $\widetilde{\mathscr{V}}(0)$ is an upper bound of the optimal value $\mathscr{V}(0)$. If, for any $\varepsilon > 0$, one can find an admissible control $\alpha^{(\varepsilon)}$ with $\widetilde{\mathscr{V}}(0) \leq E(\widetilde{\mathscr{V}}(T, \alpha^{(\varepsilon)})) + \varepsilon$, then $\widetilde{\mathscr{V}}(0) = \mathscr{V}(0)$ and the $\alpha^{(\varepsilon)}$ yield a sequence of controls approaching this optimal value.
2. The conditions in statement 2 of Theorem 1.44 can also be used in order to obtain upper and lower bounds of the true value process $\mathscr{V}$. More specifically, if $(\widetilde{\mathscr{V}}(\cdot, \alpha))_{\alpha \in \mathscr{A}}$ denotes a family of processes such that 2a) and 2b) hold, then $\mathscr{V}(t, \alpha) \leq \widetilde{\mathscr{V}}(t, \alpha)$ for $t = 0, \dots, T$ and any admissible control α. We may even relax $\widetilde{\mathscr{V}}(T, \alpha) = u(\alpha)$ to $\widetilde{\mathscr{V}}(T, \alpha) \geq u(\alpha)$ in 2b).
   If, on the other hand, $(\widetilde{\mathscr{V}}(\cdot, \alpha))_{\alpha \in \mathscr{A}}$ denotes a family of processes such that 2a) and the submartingale property in 2c) hold for some control, then $\widetilde{\mathscr{V}}(0) \leq \mathscr{V}(0)$.
3. The above setup allows for a straightforward extension to the infinite time horizon $T = \infty$. However, one must be careful that *supermartingale* and *(sub-)martingale* in Theorem 1.44 refer to the time set $\{0, \dots, T\}$ as stated, i.e. including T. Specifically, $\mathscr{V}(\infty, \alpha)$ is defined and satisfies $E(\mathscr{V}(\infty, \alpha)|\mathscr{F}_t) \leq \mathscr{V}(t, \alpha)$ for $t \leq T$ etc.
4. In condition 2b) of the previous theorem it is enough to require that the process is a *local* supermartingale unless $T = \infty$, i.e. integrability need not be verified. Indeed, the inequality $E(u(\alpha)|\mathscr{F}_t) \leq \widetilde{\mathscr{V}}(t, \alpha)$ (1.44) is obtained recursively in time because we set $E(u(\alpha)) := -\infty$ if $E(u(\alpha)^-) := -\infty$.
5. Let us have a brief look at the intermediate optimisation problem (1.43) for later use. Fix $t \in \{0, \dots, T\}$. If $\widetilde{\mathscr{V}}(\cdot, \overline{\alpha}) - \widetilde{\mathscr{V}}(\cdot, \overline{\alpha})^t$ in Theorem 1.44(2) is a martingale for some admissible control $\overline{\alpha}$ with $E(u(\overline{\alpha})) > -\infty$, then $\mathscr{V}(s, \overline{\alpha}) = \widetilde{\mathscr{V}}(s, \overline{\alpha})$ and $\overline{\alpha}$ is the maximiser in the definition of $\mathscr{V}(s, \overline{\alpha})$ for $s = t, \dots, T$. This follows as in the proof of Theorem 1.44(2). In particular, the optimal control $\alpha^\star$ also maximises the conditional optimisation problem in the definition of $\mathscr{V}(t, \alpha^\star)$.

The value process can be obtained recursively starting from $t = T$:

Proposition 1.46 (Dynamic Programming Principle) *Suppose that $\mathscr{V}(0) \neq \pm\infty$. For $t = 1, \dots, T$ and any control $\alpha \in \mathscr{A}$ we have*

$$\mathscr{V}(t-1, \alpha) = \operatorname{ess\,sup}\left\{E(\mathscr{V}(t, \widetilde{\alpha})|\mathscr{F}_{t-1}) : \widetilde{\alpha} \in \mathscr{A} \text{ with } \widetilde{\alpha}^{t-1} = \alpha^{t-1}\right\}.$$

Proof The inequality "$\leq$" is obvious because $E(u(\widetilde{\alpha})|\mathscr{F}_t) \leq \mathscr{V}(t, \widetilde{\alpha})$ and hence $E(u(\widetilde{\alpha})|\mathscr{F}_{t-1}) = E(E(u(\widetilde{\alpha})|\mathscr{F}_t)|\mathscr{F}_{t-1}) \leq E(\mathscr{V}(t, \widetilde{\alpha})|\mathscr{F}_{t-1})$.

In order to verify the converse inequality "$\geq$" fix a control $\widetilde{\alpha} \in \mathscr{A}$ with $\widetilde{\alpha}^{t-1} = \alpha^{t-1}$. Let $(\alpha_n)_{n=1,2,\dots}$ be a sequence of controls with $\alpha_n^t = \widetilde{\alpha}^t$ and such that $E(u(\alpha_n)|\mathscr{F}_t) \uparrow \mathscr{V}(t, \widetilde{\alpha})$ as $n \to \infty$. For any n we have

$$E(E(u(\alpha_n)|\mathscr{F}_t)|\mathscr{F}_{t-1}) = E(u(\alpha_n)|\mathscr{F}_{t-1}) \leq \mathscr{V}(t-1, \alpha).$$

Monotone convergence yields $E(\mathscr{V}(t, \widetilde{\alpha})|\mathscr{F}_{t-1}) \leq \mathscr{V}(t-1, \alpha)$, which implies the desired inequality. □

Moreover, the value process has a certain minimality property, which has already been mentioned in Remark 1.45(2):

Proposition 1.47 *Suppose that $\mathscr{V}(0) \neq \pm\infty$. If a family of processes $(\widetilde{\mathscr{V}}(\cdot, \alpha))_{\alpha \in \mathscr{A}}$ satisfies 2a,b) in Theorem 1.44, we have $\mathscr{V}(\cdot, \alpha) \leq \widetilde{\mathscr{V}}(\cdot, \alpha)$ for $t = 0, \dots, T$ and any control $\alpha \in \mathscr{A}$.*

Proof For any control $\widetilde{\alpha}$ satisfying $\widetilde{\alpha}^t = \alpha^t$ we have

$$E(u(\widetilde{\alpha})|\mathscr{F}_t) = E(\widetilde{\mathscr{V}}(T, \widetilde{\alpha})|\mathscr{F}_t) \leq \widetilde{\mathscr{V}}(t, \widetilde{\alpha}) = \widetilde{\mathscr{V}}(t, \alpha),$$

which implies that

$$\mathscr{V}(t, \alpha) = \operatorname{ess\,sup}\left\{E(u(\widetilde{\alpha})|\mathscr{F}_t) : \widetilde{\alpha} \in \mathscr{A} \text{ with } \widetilde{\alpha}^t = \alpha^t\right\} \leq \widetilde{\mathscr{V}}(t, \alpha). \qquad \square$$

As an example we consider the *Merton problem* of maximising the expected logarithmic utility of terminal wealth.

Example 1.48 (Logarithmic Utility of Terminal Wealth) An investor trades in a market consisting of a constant bank account and a stock whose price at time t equals

$$S(t) = S(0)\mathscr{E}(X)(t) = S(0)\prod_{s=1}^{t}(1 + \Delta X(s))$$

with $\Delta X(t) > -1$. Given that $\varphi(t)$ denotes the number of shares in the investor's portfolio from time $t-1$ to t, the profits from the stock investment in this period are $\varphi(t)\Delta S(t)$. If $v_0 > 0$ denotes the investor's initial endowment, her wealth at any time t amounts to

$$V_\varphi(t) := v_0 + \sum_{s=1}^{t} \varphi(s)\Delta S(s) = v_0 + \varphi \bullet S(t). \tag{1.45}$$

We assume that the investor's goal is to maximise the expected logarithmic utility $E(\log V_\varphi(T))$ of wealth at time T. To this end, we assume that the stock price process S is exogenously given and the investor's set of admissible controls is

$$\mathscr{A} := \{\varphi \text{ predictable} : V_\varphi > 0 \text{ and } \varphi(0) = 0\}.$$

It turns out that the problem becomes more transparent if we consider the *relative portfolio*

$$\pi(t) := \varphi(t)\frac{S(t-1)}{V_\varphi(t-1)}, \quad t = 1, \dots, T, \tag{1.46}$$

i.e. the fraction of wealth invested in the stock at time $t-1$. Starting with v_0, the stock holdings $\varphi(t)$ and the wealth process $V_\varphi(t)$ are recovered from π via

$$V_\varphi(t) = v_0 \mathscr{E}(\pi \bullet X)(t) = v_0 \prod_{s=1}^{t} (1 + \pi(s)\Delta X(s)) \tag{1.47}$$

and

$$\varphi(t) = \pi(t)\frac{V_\varphi(t-1)}{S(t-1)} = \pi(t)\frac{v_0 \mathscr{E}(\pi \bullet X)(t-1)}{S(t-1)}.$$

Indeed, (1.47) follows from

$$\Delta V_\varphi(t) = \varphi(t)\Delta S(t) = \frac{V_\varphi(t-1)\pi(t)}{S(t-1)}\Delta S(t) = V_\varphi(t-1)\Delta(\pi \bullet X)(t).$$

If $T = 1$, a simple calculation shows that the investor should buy $\varphi^\star(1) = \pi^\star(1)v_0/S(0)$ shares at time 0, where the optimal fraction $\pi^\star(1)$ maximises the function $\gamma \mapsto E(\log(1 + \gamma\Delta X(1)))$. We guess that the same essentially holds for multi-period markets, i.e. we assume that the optimal relative portfolio is obtained as the maximiser $\pi^\star(\omega, t)$ of the mapping

$$\gamma \mapsto E(\log(1 + \gamma\Delta X(t))|\mathscr{F}_{t-1})(\omega). \tag{1.48}$$

For simplicity we suppose that this maximiser exists and that $\log(1 + \pi^\star(t)\Delta X(t))$ has finite expectation for $t = 1, \dots, T$. Both can be shown to hold if the maximal achievable utility $\sup_{\varphi\in\mathscr{A}} E(\log V_\varphi(T))$ is finite, e.g. based on duality results in [126, 197].

The corresponding candidate value process is

$$\begin{aligned}\mathscr{V}(t,\varphi) &:= E\left(\log\left(V_\varphi(t)\prod_{s=t+1}^{T}\left(1+\pi^\star(s)\Delta X(s)\right)\right)\middle|\mathscr{F}_t\right)\\ &= \log V_\varphi(t) + E\left(\sum_{s=t+1}^{T}\log\left(1+\pi^\star(s)\Delta X(s)\right)\middle|\mathscr{F}_t\right), \qquad (1.49)\end{aligned}$$

where empty products are set to one and empty sums to zero as usual. Observe that

$$\begin{aligned}E(\mathscr{V}(t,\varphi)|\mathscr{F}_{t-1}) &= \log V_\varphi(t-1)\\ &\quad + E\left(\log\left(1+\frac{\varphi(t)S(t-1)}{V_\varphi(t-1)}\Delta X(t)\right)\middle|\mathscr{F}_{t-1}\right)\\ &\quad + E\left(\sum_{s=t+1}^{T}\log\left(1+\pi^\star(s)\Delta X(s)\right)\middle|\mathscr{F}_{t-1}\right)\\ &\leq \log V_\varphi(t-1) + E\big(\log(1+\pi^\star(t)\Delta X(t))\big|\mathscr{F}_{t-1}\big)\\ &\quad + E\left(\sum_{s=t+1}^{T}\log\left(1+\pi^\star(s)\Delta X(s)\right)\middle|\mathscr{F}_{t-1}\right)\\ &= \mathscr{V}(t-1,\varphi)\end{aligned}$$

for any $t \geq 1$ and $\varphi \in \mathscr{A}$, with equality for the candidate optimiser $\varphi^\star$ satisfying $\varphi^\star(t) = \pi^\star(t)V_{\varphi^\star}(t-1)/S(t-1)$. By Theorem 1.44, we conclude that $\varphi^\star$ is indeed optimal.

Note that the optimiser or, more precisely, the optimal fraction of wealth invested in stock depends only on the local dynamics of the stock. This *myopic* property holds only for logarithmic utility.

The following variation of Example 1.48 considers utility of consumption rather than terminal wealth.

Example 1.49 (Logarithmic Utility of Consumption) In the market of the previous example the investor now spends $c(t)$ currency units at any time $t-1$. We assume that utility is derived from this consumption rather than terminal wealth, i.e. the goal is to maximise

$$E\left(\sum_{t=1}^{T}\log c(t) + \log V_{\varphi,c}(T)\right) \qquad (1.50)$$

subject to the affordability constraint that the investor's wealth should stay positive:

$$0 < V_{\varphi,c}(t) := v_0 + \varphi \bullet S(t) - \sum_{s=1}^{t} c(s). \tag{1.51}$$

The last term $V_{\varphi,c}(T)$ in (1.50) refers to consumption of the remaining wealth at the end. The investor's set of admissible controls is

$$\mathscr{A} := \{(\varphi, c) \text{ predictable} : V_{\varphi,c} > 0, (\varphi, c)(0) = (0, 0)\}.$$

We try to come up with a reasonable candidate $(\varphi^\star, c^\star)$ for the optimal control. As in the previous example, matters simplify in relative terms. We write

$$\kappa(t) = \frac{c(t)}{V_{\varphi,c}(t-1)}, \quad t = 1, \dots, T \tag{1.52}$$

for the fraction of wealth that is consumed and

$$\pi(t) = \varphi(t)\frac{S(t-1)}{V_{\varphi,c}(t-1) - c(t)} = \varphi(t)\frac{S(t-1)}{V_{\varphi,c}(t-1)(1-\kappa(t))} \tag{1.53}$$

for the relative portfolio. Since the wealth after consumption at time $t-1$ is now $V_{\varphi,c}(t-1) - c(t)$, the numerator in (1.53) had to be adjusted. Similarly to (1.47), the wealth is given by

$$V_{\varphi,c}(t) = v_0 \prod_{s=1}^{t} (1 - \kappa(s))(1 + \pi(s)\Delta X(s)) = v_0 \mathscr{E}(-\kappa \bullet I)(t) \mathscr{E}(\pi \bullet X)(t). \tag{1.54}$$

We guess that the same relative portfolio as in the previous example is optimal in this modified setup, which leads to the candidate

$$\varphi^\star(t) = \pi^\star(t)\frac{V_{\varphi^\star,c^\star}(t-1) - c^\star(t)}{S(t-1)}.$$

As before we assume that $\pi^\star$ maximising (1.48) exists and that $\log(1+\pi^\star(t)\Delta X(t))$ has finite expectation for $t = 1, \dots, T$. Moreover, it may seem natural that the investor tries to spread consumption of wealth evenly over time. This idea leads to $\kappa^\star(t) = 1/(T+2-t)$ and hence

$$c^\star(t) = \frac{V_{\varphi^\star,c^\star}(t-1)}{T+2-t}$$

because at time $t-1$ there are $T+2-t$ periods left for consumption. This candidate pair $(\varphi^\star, c^\star)$ corresponds to the candidate value process

$$\begin{aligned}
\mathscr{V}(t,(\varphi,c)) &:= \sum_{s=1}^{t} \log c(s) \\
&\quad + E\Bigg(\sum_{s=t+1}^{T} \log\Bigg(V_{\varphi,c}(t) \prod_{r=t+1}^{s-1} \Big((1-\kappa^\star(r))(1+\pi^\star(r)\Delta X(r))\Big)\kappa^\star(s)\Bigg) \\
&\quad + \log\Bigg(V_{\varphi,c}(t) \prod_{r=t+1}^{T} \Big((1-\kappa^\star(r))(1+\pi^\star(r)\Delta X(r))\Big)\Bigg)\Bigg|\mathscr{F}_t\Bigg) \\
&= \sum_{s=1}^{t} \log c(s) + (T+1-t)\log V_{\varphi,c}(t) \\
&\quad + \sum_{r=t+1}^{T} \Bigg((T+1-r) E\Big(\log\big(1+\pi^\star(r)\Delta X(r)\big)\Big|\mathscr{F}_t\Big) \\
&\quad + (T+1-r)\log\tfrac{T+1-r}{T+2-r} - \log(T+2-r)\Bigg),
\end{aligned}$$

which is obtained if, starting from $t+1$, we invest the candidate fraction $\pi^\star$ of wealth in the stock and consume at any time $s-1$ the candidate fraction $\kappa^\star(s) = 1/(T+2-s)$ of wealth. In order to verify optimality, observe that

$$\begin{aligned}
E\big(\mathscr{V}(t,(\varphi,c))\big|\mathscr{F}_{t-1}\big) &= \sum_{s=1}^{t-1} \log c(s) + (T+1-t)\log V_{\varphi,c}(t-1) \\
&\quad + (T+1-t)E\Big(\log\big(1+\pi(t)\Delta X(t)\big)\Big|\mathscr{F}_{t-1}\Big) \\
&\quad + (T+1-t)\log(1-\kappa(t)) + \log\kappa(t) + \log V_{\varphi,c}(t-1) \\
&\quad + \sum_{s=t+1}^{T} \Bigg((T+1-s) E\Big(\log\big(1+\pi^\star(s)\Delta X(s)\big)\Big|\mathscr{F}_{t-1}\Big) \\
&\quad + (T+1-s)\log\tfrac{T+1-s}{T+2-s} - \log(T+2-s)\Bigg) \\
&\leq \sum_{s=1}^{t-1} \log c(s) + (T+2-t)\log V_{\varphi,c}(t-1) \\
&\quad + (T+1-t)E\Big(\log\big(1+\pi^\star(t)\Delta X(t)\big)\Big|\mathscr{F}_{t-1}\Big)
\end{aligned}$$

$$
\begin{aligned}
&+(T+1-t)\log(1-\tfrac{1}{T+2-t})-\log(T+2-t)\\
&+\sum_{s=t+1}^{T}\Big((T+1-s)E\Big(\log\big(1+\pi^\star(s)\Delta X(s)\big)\Big|\mathscr{F}_{t-1}\Big)\\
&+(T+1-s)\log\tfrac{T+1-s}{T+2-s}-\log(T+2-s)\Big)\\
=\ &\mathscr{V}(t-1,(\varphi,c))
\end{aligned}
$$

for any admissible control (φ, c), where $\pi(t)$, $\kappa(t)$ are defined as in (1.53, 1.52). To wit, the inequality holds because $\pi^\star(t)$ maximises $\gamma \mapsto E(\log(1+\gamma\Delta X(t))|\mathscr{F}_{t-1})$ and $1/(T+2-t)$ maximises $\delta \mapsto (T+1-t)\log(1-\delta)+\log\delta$. Again, equality holds if $(\varphi, c) = (\varphi^\star, c^\star)$. By Theorem 1.44 we conclude that $(\varphi^\star, c^\star)$ is indeed optimal.

The optimal consumption rate changes slightly if the objective is to maximise $E(\sum_{t=1}^T e^{-\delta(t-1)}\log c(t) + e^{-\delta T}\log V_{\varphi,c}(T))$ with some *impatience rate* $\delta \geq 0$.

The two previous examples allow for a straightforward extension to $d > 1$ assets. We continue with a second example which is also motivated by Mathematical Finance.

Example 1.50 (Quadratic Hedging) In the context of option hedging, the question arises how to approximate a given random variable by the terminal value of a stochastic integral relative to a given process. The random variable represents the payoff of a contingent obligation and the stochastic integral stands for the profits of an investor as in Example 1.48, cf. Chap. 12 for details.

More specifically, let X denote a square-integrable random variable, v_0 a real number, and S a martingale with $E(S(t)^2) < \infty$ for $t = 0, \dots, T$. The aim is to minimise the so-called *expected squared hedging error*

$$
E\big((V_\varphi(T)-X)^2\big) \tag{1.55}
$$

where $V_\varphi(t) := v_0 + \varphi \bullet S(t)$ represents the wealth of the investor except for the obligation X and φ ranges over all admissible controls in the set

$$
\begin{aligned}
\mathscr{A} &:= \{\varphi \text{ predictable} : \varphi(0)=0 \text{ and } E((V_\varphi(T)-X)^2) < \infty\}\\
&= \{\varphi \text{ predictable} : \varphi(0)=0 \text{ and } E(V_\varphi(T)^2) < \infty\}\\
&= \{\varphi \text{ predictable} : \varphi(0)=0 \text{ and } E(\varphi^2 \bullet \langle S,S\rangle(T)) < \infty\}.
\end{aligned}
$$

The problem becomes simpler by introducing the martingale $V(t) := E(X|\mathscr{F}_t)$ generated by X and also the martingale $M_\varphi := V_\varphi - V$, which means that we have to maximise $E(u(\varphi))$ with $u(\varphi) := -M_\varphi(T)^2$. Integration by parts yields

$$\begin{aligned} M_\varphi(t)^2 &= M_\varphi(0)^2 + 2(M_\varphi)_- \bullet M_\varphi(t) + [M_\varphi, M_\varphi](t) \\ &= M_\varphi(0)^2 + N(t) + \langle M_\varphi, M_\varphi\rangle(t), \end{aligned}$$

where $N := 2(M_\varphi)_- \bullet M + [M_\varphi, M_\varphi] - \langle M_\varphi, M_\varphi\rangle$ is a martingale by Propositions 1.14(7) and 1.8. Consequently,

$$E(u(\varphi)|\mathscr{F}_t) = -M_\varphi(t)^2 - E\big(\langle M_\varphi, M_\varphi\rangle(T) - \langle M_\varphi, M_\varphi\rangle(t)\big|\mathscr{F}_t\big).$$

This expression is to be minimised in the definition (1.43) of the value process. Observe that $M_\varphi(t)$ only depends on φ^t. Moreover,

$$\begin{aligned} \langle M_\varphi, M_\varphi\rangle(T) - \langle M_\varphi, M_\varphi\rangle(t) &= \sum_{s=t+1}^{T} \Delta\langle M_\varphi, M_\varphi\rangle(s) \\ &= \sum_{s=t+1}^{T} \Big(\varphi(s)^2\Delta\langle S, S\rangle(s) - 2\varphi(s)\Delta\langle S, V\rangle(s) + \Delta\langle V, V\rangle(s)\Big) \end{aligned}$$

can be optimised separately for any s. It is easy to see that

$$\varphi(s) \mapsto \varphi(s)^2\Delta\langle S, S\rangle(s) - 2\varphi(s)\Delta\langle S, V\rangle(s) + \Delta\langle V, V\rangle(s) \tag{1.56}$$

is minimised by

$$\varphi^\star(s) := \frac{\Delta\langle S, V\rangle(s)}{\Delta\langle S, S\rangle(s)} \tag{1.57}$$

with minimal value

$$\Delta\langle V, V\rangle(s) - \varphi^\star(s)^2\Delta\langle S, S\rangle(s).$$

This leads to the ansatz

$$\mathscr{V}(t, \varphi) := -M_\varphi(t)^2 - \sum_{s=t+1}^{T} E\left(\Delta\langle V, V\rangle(s) - \varphi^\star(s)^2\Delta\langle S, S\rangle(s)\middle|\mathscr{F}_t\right) \tag{1.58}$$

for the value process, with $\varphi^\star$ defined in (1.57). Since $0 \leq \varphi^\star(s)^2\Delta\langle S, S\rangle(s) \leq \Delta\langle V, V\rangle(s)$, the control $\varphi^\star$ is admissible. Obviously, we have $\mathscr{V}(T, \varphi) = u(\varphi)$. For $t = 1, \dots, T$ we obtain

$$\begin{aligned} E(\mathscr{V}(t, \varphi)|\mathscr{F}_{t-1}) &= \mathscr{V}(t-1, \varphi) - E\big(M_\varphi(t)^2 - M_\varphi(t-1)^2|\mathscr{F}_{t-1}\big) \\ &\quad + \Delta\langle V, V\rangle(t) - \varphi^\star(t)^2\Delta\langle S, S\rangle(t) \\ &= \mathscr{V}(t-1, \varphi) - \big(\varphi(t)^2\Delta\langle S, S\rangle(t) - 2\varphi(t)\Delta\langle S, V\rangle(t) + \Delta\langle V, V\rangle(t)\big) \\ &\quad + \varphi^\star(t)^2\Delta\langle S, S\rangle(t) - 2\varphi^\star(t)\Delta\langle S, V\rangle(t) + \Delta\langle V, V\rangle(t). \end{aligned}$$

Since (1.56) is minimised by $\varphi^\star(s)$, we have $E(\mathscr{V}(t, \varphi)|\mathscr{F}_{t-1}) \leq \mathscr{V}(t-1, \varphi)$ with equality for $\varphi = \varphi^\star$. Theorem 1.44 yields that $\varphi^\star$ is indeed optimal. Moreover, the optimal value of the original control problem (1.55) amounts to

$$\begin{aligned} -\mathscr{V}(0, \varphi^\star) &= (V_{\varphi^\star}(0) - V(0))^2 + \sum_{t=1}^{T} E\Big(\Delta\langle V, V\rangle(t) - \varphi^\star(t)^2\Delta\langle S, S\rangle(t)\Big) \\ &= (v_0 - E(X))^2 + E\Big(\langle V, V\rangle(T) - (\varphi^\star)^2 \bullet \langle S, S\rangle(T)\Big). \end{aligned}$$

If this is to be minimised over the investor's initial capital v_0 as well, the optimal choice is obviously $v_0 = E(X)$, leading to the optimal value

$$E\Big(\langle V, V\rangle(T) - (\varphi^\star)^2 \bullet \langle S, S\rangle(T)\Big)$$

of the control problem.

1.5.2 *Optimal Stopping*

An important subclass of control problems concerns *optimal stopping*. Given some time horizon $T < \infty$ and some adapted process X with $E(\sup_{t\in\{0,\dots,T\}} |X(t)|) < \infty$, the goal is to maximise the *expected reward*

$$\tau \mapsto E(X(\tau)) \tag{1.59}$$

over all stopping times τ with values in $\{0, \dots, T\}$.

Remark 1.51

1. In the spirit of Sect. 1.5.1, a stopping time τ can be identified with the corresponding adapted process $\alpha(t) := 1_{\{t\leq\tau\}}$, and hence $X(\tau) = X(0) + \alpha \bullet X(T)$. Put differently, $\mathscr{A} := \{\alpha$ predictable: α $\{0, 1\}$-valued, decreasing, $\alpha(0) = 1\}$ and $u(\alpha) := X(0) + \alpha \bullet X(T)$ in Sect. 1.5.1 lead to the above optimal stopping problem.
2. Sometimes X may not be adapted in applications such as in Example 1.58 below. Then we can replace it with the adapted process $\widetilde{X}(t) := E(X(t)|\mathscr{F}_t)$. Indeed, we have

$$\begin{aligned} E(X(\tau)) &= \sum_{t=0}^{T} E\big(X(t)1_{\{\tau=t\}}\big) \\ &= \sum_{t=0}^{T} E\Big(E\big(X(t)\big|\mathscr{F}_t\big)1_{\{\tau=t\}}\Big) \\ &= \sum_{t=0}^{T} E\big(\widetilde{X}(t)1_{\{\tau=t\}}\big) \\ &= E(\widetilde{X}(\tau)) \end{aligned}$$

because $\{\tau = t\} \in \mathscr{F}_t$ for any stopping time τ.

The role of the value process in general control problems is now taken by the **Snell envelope** of X, which denotes the adapted process V given by

$$V(t) := \text{ess sup}\big\{E(X(\tau)|\mathscr{F}_t) : \tau \text{ stopping time with values in } \{t, t+1, \dots, T\}\big\}. \tag{1.60}$$

It represents the maximal expected reward if we start at time t and have not stopped yet. The following martingale criterion may be helpful to verify the optimality of a candidate stopping time. We will apply its continuous-time version in an example in Chap. 7.

Proposition 1.52

1. *Let τ be a stopping time with values in $\{0, \dots, T\}$. If V is an adapted process such that V^τ is a martingale, $V(\tau) = X(\tau)$, and $V(0) = M(0)$ for some martingale (or at least supermartingale) $M \geq X$, then τ is optimal for (1.59) and V coincides up to time τ with the Snell envelope of X.*
2. *More generally, let τ be a stopping time with values in $\{t, \dots, T\}$ and $F \in \mathscr{F}_t$. Suppose that M is a process with $M \geq X$ on F and such that $M - M^t$ is a martingale (or at least a supermartingale). If V is an adapted process such that*

$V^\tau - V^t$ is a martingale, $V(\tau) = X(\tau)$, and $V(t) = M(t)$ on F, then $V(s)$ coincides on F for $t \leq s \leq \tau$ with the Snell envelope of X and τ is optimal on F for (1.60), i.e. it maximises $E(X(\tau)|\mathscr{F}_t)$ on F.

Proof

1. This follows from the second statement.
2. Let $s \in \{t, \ldots, T\}$. We have to show that

$$V(s) = \operatorname{ess\,sup}\left\{E(X(\widetilde{\tau})|\mathscr{F}_s) : \widetilde{\tau} \text{ stopping time with values in } \{s, \ldots, T\}\right\}$$

holds on $\{s \leq \tau\} \cap F$.
"$\leq$": On $\{s \leq \tau\} \cap F$ we have

$$\begin{aligned} V(s) &= E(V^\tau(T)|\mathscr{F}_s) \\ &= E(X(\tau \vee s)|\mathscr{F}_s) \\ &\leq \operatorname{ess\,sup}\left\{E(X(\widetilde{\tau})|\mathscr{F}_s) : \widetilde{\tau} \text{ stopping time with values in } \{s, \ldots, T\}\right\}. \end{aligned} \tag{1.61}$$

"$\geq$": Note that $V^\tau(T) = V(\tau) = X(\tau) \leq M(\tau) = M^\tau(T)$ and Lemma 1.7 yield

$$V^\tau(s) = M^\tau(s) \geq E(M(\widetilde{\tau})|\mathscr{F}_s) \geq E(X(\widetilde{\tau})|\mathscr{F}_s) \tag{1.62}$$

on $\{s \leq \tau\} \cap F$ for any stopping time $\widetilde{\tau} \geq s$.
(1.61) and (1.62) yield that τ is optimal on F. Indeed, $E(X(\tau)|\mathscr{F}_t) = V(t) \geq E(X(\widetilde{\tau})|\mathscr{F}_t)$ holds for any stopping time $\widetilde{\tau} \geq t$. □

The following more common verification result corresponds to Theorem 1.44 in the framework of optimal stopping.

Theorem 1.53

1. *Let τ be a stopping time with values in $\{0, \ldots, T\}$. If $V \geq X$ is a supermartingale such that V^τ is a martingale and $V(\tau) = X(\tau)$, then τ is optimal for (1.59) and V coincides up to time τ with the Snell envelope of X.*
2. *More generally, let τ be a stopping time with values in $\{t, \ldots, T\}$. If $V \geq X$ is an adapted process such that $V - V^t$ is a supermartingale, $V^\tau - V^t$ is a martingale, and $V(\tau) = X(\tau)$, then $V(s)$ coincides for $t \leq s \leq \tau$ with the Snell envelope of X and τ is optimal for (1.60), i.e. it maximises $E(X(\tau)|\mathscr{F}_t)$.*
3. *If τ is optimal for (1.60) and V denotes the Snell envelope, they have the properties in statement 2.*

Proof

1. This follows from the second statement.
2. This immediately follows from choosing $M = V$ and $F = \Omega$ in statement 2 of Proposition 1.52 but we give a short direct proof as well. For any competing stopping time $\widetilde{\tau}$ with values in $\{t, t+1, \dots, T\}$, Proposition 1.9 yields

$$E(X(\widetilde{\tau})|\mathscr{F}_t) \leq E(V(\widetilde{\tau})|\mathscr{F}_t) = E(V^{\widetilde{\tau}}(T)|\mathscr{F}_t) \leq V^{\widetilde{\tau}}(t) = V(t),$$

with equality everywhere for $\widetilde{\tau} = \tau$.
3. $\tau \geq t$, $V \geq X$, and adaptedness of V are obvious. It remains to be shown that $V - V^t$ is a supermartingale, $V^\tau - V^t$ is a martingale, and $V(\tau) = X(\tau)$. Using the identification of Remark 1.51(1), we have $V(t) = \mathscr{V}(t, 1)$. Theorem 1.44(1) yields that V and hence also $V - V^t$ are supermartingales. In particular, $V(t) \geq E(V(\tau)|\mathscr{F}_t)$. But optimality of τ implies

$$V(t) = E(X(\tau)|\mathscr{F}_t) \leq E(V(\tau)|\mathscr{F}_t). \tag{1.63}$$

Hence, equality holds in (1.63), which yields $X(\tau) = V(\tau)$ and $E(V^\tau(T) - V^t(T)) = 0 = V^\tau(0) - V^t(0)$. The martingale property of $V^\tau - V^t$ now follows from Proposition 1.6. □

Remark 1.54 Statement 3 in Theorem 1.53 shows that the sufficient condition in Proposition 1.52(1) is necessary, i.e. for the Snell envelope V there exists a martingale M as in statement 1 of this proposition. Indeed, one may choose $M := V(0) + M^V$ if $V = V(0) + M^V + A^V$ denotes the Doob decomposition of V.

Hence, Proposition 1.52 can in principle be used to determine the Snell envelope numerically, namely by minimising $M(0)$ over all martingales dominating X. The resulting process coincides with the Snell envelope up to time τ. Cf. also Remark 7.20 in this context.

The following result helps to determine both the Snell envelope and an optimal stopping time. The first statement corresponds to the backward recursion of Proposition 1.46 in the context of optimal stopping, the third to Proposition 1.47.

Proposition 1.55 *Let V denote the Snell envelope.*

1. *V is obtained recursively by $V(T) = X(T)$ and*

$$V(t-1) = \max\left\{X(t-1), E(V(t)|\mathscr{F}_{t-1})\right\}, \quad t = T-1, \dots, 0. \tag{1.64}$$

2. *The stopping times*

$$\underline{\tau}_t := \inf\left\{s \in \{t, \dots, T\} : V(s) = X(s)\right\}$$

and

$$\overline{\tau}_t := T \wedge \inf\{s \in \{t, \dots, T-1\} : E(V(s+1)|\mathscr{F}_s) < X(s)\} \tag{1.65}$$

are optimal in (1.60), i.e. they maximise $E(X(\tau)|\mathscr{F}_t)$.

3. *V is the smallest supermartingale dominating X.*

Proof

1. and 2. Define V recursively as in statement 1 rather than as in (1.60) and set $\tau = \underline{\tau}_t$ or $\tau = \overline{\tau}_t$. We show that τ and V satisfy the conditions of Theorem 1.53(2). It is obvious that V is a supermartingale with $V \geq X$ and $V(\tau) = X(\tau)$. Fix $s \in \{t+1, t+2, \dots, T\}$. On the set $\{s \leq \tau\}$ we have $V(s-1) = E(V(s)|\mathscr{F}_{s-1})$ by definition of τ. Hence $V^\tau - V^t$ is a martingale.
3. In view of statement 1 or Theorem 1.53(3) it remains to be shown that $V \leq W$ for any supermartingale dominating X. To this end observe that

$$\begin{aligned} V(t) &= \operatorname{ess\,sup}\{E(X(\tau)|\mathscr{F}_t) : \tau \text{ stopping time in } \{t, t+1, \dots, T\}\} \\ &\leq \operatorname{ess\,sup}\{E(W(\tau)|\mathscr{F}_t) : \tau \text{ stopping time in } \{t, t+1, \dots, T\}\} \\ &\leq W(t) \end{aligned}$$

by Proposition 1.9. □

In some sense, $\underline{\tau}_t$ is the earliest and $\overline{\tau}_t$ the latest optimal stopping time for (1.60). Often the optimal stopping time is unique, in which case we have $\underline{\tau}_t = \overline{\tau}_t$. The solution to the original stopping problem (1.59) is obtained for $t = 0$. In particular, Proposition 1.55(2) yields that an optimiser of (1.59) exists.

Remark 1.56

1. In terms of the drift coefficient a^V of V as in Proposition 1.31, (1.64) and (1.65) can be reformulated as

$$\max\{a^V(t), X(t-1) - V(t-1)\} = 0, \quad t = 1, \dots, T \tag{1.66}$$

and

$$\begin{aligned} \overline{\tau}(t) &= T \wedge \inf\{s \in \{t, \dots T-1\} : a^V(t+1) < X(t) - V(t)\} \\ &= T \wedge \inf\{s \in \{t, \dots T-1\} : a^V(t+1) < 0\}, \end{aligned} \tag{1.67}$$

respectively. The second equality in (1.67) follows from (1.66).

2. If the reward process X is nonnegative, one may also consider the infinite time horizon $T = \infty$, in which case we set $X(\infty) := 0$. Propositions 1.52, 1.55(1,3) and Theorem 1.53 remain true in this case if we make sure that the martingale resp. supermartingale property always refers to the time set $\{0, \dots, T\}$ including

$T = \infty$, cf. Remark 1.45(3). However, it is not obvious how to obtain V from the "infinite recursion" (1.64) any more. Moreover, an optimal stopping time may fail to exist.

As an example we consider the price of a perpetual put option in the so-called Cox–Ross–Rubinstein model.

Example 1.57 (Perpetual Put) Suppose that $S(t)/S(0)$ is a geometric random walk such that $S(t)/S(t-1)$ has only two possible values $u > 1, d := 1/u < 1$ which are assumed with probabilities p resp. $1-p$. In particular, the process S moves only on the grid $S(0)u^{\mathbb{Z}}$. Consider the reward process $X(t) = e^{-rt}(K - S(t))^+$, $t = 0, 1, 2, \dots$ with some constants $r > 0$, $K > 0$. In the context of Mathematical Finance, $S(t)$ represents a stock price and $X(t)$ the discounted payoff of a perpetual put option on the stock that is exercised at time t. The goal is to maximise the expected discounted payoff $E(X(\tau))$ over all stopping times τ.

We make the natural ansatz that the Snell envelope is of the form

$$V(t) = e^{-rt}v(S(t)) \tag{1.68}$$

for some function v. If we guess that it is optimal to stop when $S(t)$ falls below some threshold $s^\star \leq K$, we must have $v(s) = (K-s)^+$ for $s \leq s^\star$. As long as $S(t) > s^\star$, the process $e^{-rt}v(S(t))$ should behave as a martingale, i.e.

$$\begin{aligned} 0 &= E(\Delta V(t)|\mathscr{F}_{t-1}) \\ &= e^{-rt}\big(pv(S(t-1)u) + (1-p)v(S(t-1)d) - e^r v(S(t-1))\big), \end{aligned} \tag{1.69}$$

which only holds if v grows in the right way for $s > s^\star$. A power ansatz for v turns out to be successful, i.e.

$$v(s) = \begin{cases} K - s & \text{for } s \leq s^\star, \\ cs^{-a} & \text{for } s > s^\star \end{cases}$$

with some constants $s^\star, a, c > 0$. In order for (1.69) to hold, a needs to be chosen such that $pu^{-a} + (1-p)u^a = e^r$, i.e.

$$u^{-a} = \frac{e^r - \sqrt{e^{2r} - 4p(1-p)}}{2p}.$$

Subsequently, we hope that some sort of contact condition at the boundary leads to the solution, which is often true for optimal stopping problems. More specifically, we choose the largest $c > 0$ such that the linear function $s \mapsto K - s$ and the decreasing convex function $s \mapsto cs^{-a}$ coincide at least in one element $s^\star$ in $S(0)u^{\mathbb{Z}}$, i.e. $s^\star = S(0)u^k$ for some integer k. Define the stopping time $\tau := \inf\{t \geq 0 : S(t) \leq s^\star\}$ and V as in (1.68) with $V(\infty) = 0$. We suppose that $p \leq 1/2$ so that τ is almost surely finite. It is now easy to verify that $V \geq X, V(\tau) = X(\tau)$, and V^τ is a martingale. By dominated convergence, the martingale property actually holds on $\{0, \ldots, \infty\} = \mathbb{N} \cup \{\infty\}$ rather than just $\mathbb{N}$. In view of Theorem 1.53, it remains to show that V is a supermartingale. To this end, fix $t \in \{1, \ldots, T\}$. On $\{S(t-1) > s^\star\}$ we have

$$E(V(t)|\mathscr{F}_{t-1}) = E(e^{-rt}cS(t)^{-a}|\mathscr{F}_{t-1}) = e^{-r(t-1)}cS(t-1)^{-a} = V(t-1).$$

Similarly, we have

$$E(V(t)|\mathscr{F}_{t-1}) \leq E(e^{-r(t)}cS(t)^{-a}|\mathscr{F}_{t-1}) = e^{-r(t-1)}cS(t-1)^{-a} = V(t-1)$$

on $\{S(t-1) = s^\star\}$. On $\{S(t-1) < s^\star\}$ we argue that

$$\begin{aligned}
E(V(t)|\mathscr{F}_{t-1}) &= E\big(e^{-rt}(K - S(t))\big|\mathscr{F}_{t-1}\big) \\
&= e^{-rt}K\left(1 - \frac{S(t-1)}{s^\star}\right) + \frac{S(t-1)}{s^\star}E\left(e^{-rt}\left(K - \frac{S(t)}{S(t-1)}s^\star\right)\middle|\mathscr{F}_{t-1}\right) \\
&\leq e^{-rt}K\left(1 - \frac{S(t-1)}{s^\star}\right) + \frac{S(t-1)}{s^\star}E\left(e^{-rt}c\left(\frac{S(t)}{S(t-1)}s^\star\right)^{-a}\middle|\mathscr{F}_{t-1}\right) \\
&= e^{-rt}K\left(1 - \frac{S(t-1)}{s^\star}\right) + \frac{S(t-1)}{s^\star}e^{-r(t-1)}c(s^\star)^{-a} \\
&= e^{-rt}K\left(1 - \frac{S(t-1)}{s^\star}\right) + \frac{S(t-1)}{s^\star}e^{-r(t-1)}(K - s^\star) \\
&\leq e^{-r(t-1)}(K - S(t-1)) = V(t-1),
\end{aligned}$$

which finishes the proof.

If u is small, we can compute approximations of $c, s^\star$. Indeed, at $s^\star$ the linear function $s \mapsto K - s$ is almost tangent to $s \mapsto cs^{-a}$, i.e. their derivatives coincide. The two conditions $cs^{\star-a} = K - s^\star$ and $-acs^{\star-a-1} \approx -1$ are solved by $s^\star \approx Ka/(1+a)$ and $c \approx K^{1+a}a^a/(1+a)^{1+a}$.

As another example we consider the famous so-called *secretary* or *marriage problem*.

Example 1.58 (Marriage Problem) Suppose that you are looking for a spouse, a flat, or an employee. We assume that time permits you to examine n candidates before a decision must be made. Moreover, you are supposed to be satisfied only with the best of all T aspirants. The problem is that you can inspect them only one by one and that you have to opt for or against any candidate before you can see the next one. The goal is to maximise the probability of choosing the best one.

In mathematical terms we assume the ranks $1, \ldots, T$ of the candidates to appear in totally random order. The rank of candidate t is denoted by $N(t)$. But when applicant t shows up, you only observe his or her *relative rank* $R(t)$ compared to all previous applicants. Since any permutation is assumed to be equally likely, it is not hard to show that the law of $R(t)$ is uniform on $1, \dots, t$, independently of the past, i.e. $R(t)$ is independent of $R(1), \dots, R(t-1)$. The decision to stop must be based on the observed relative ranks, i.e. on the filtration $\mathscr{F}_t := \sigma(R(1), \dots, R(t))$. The goal is to maximise $P(N(\tau) = 1)$ among all stopping times τ with values $1, 2, \dots, T$. In order to make this problem look as in (1.59), note that $P(N(\tau) = 1) = E(X(\tau))$ for $X(t) := 1_{\{N(t)=1\}}$. However, the reward process X is not adapted to our filtration. In line with Remark 1.51(2), we replace it with its conditional expectation $\widetilde{X}(t) = E(X(t)|\mathscr{F}_t)$. It is easy to see that

$$\widetilde{X}(t) = \begin{cases} 0 & \text{if } R(t) = 0, \\ t/T & \text{if } R(t) = 1. \end{cases}$$

Indeed, t cannot be globally optimal if not even the relative rank is 1. If, on the other hand, it has relative rank $R(t) = 1$, it is globally optimal if and only if the global optimiser is among the first t candidates. This happens with probability t/T, independently of the observed relative ranks $R(1), \dots, R(t)$.

We can now determine the Snell envelope recursively according to $V(T) = \widetilde{X}(T)$ and (1.64) with $\widetilde{X}$ instead of X. We obtain $V(T) = 1_{\{R(T)=1\}}$ and

$$\begin{aligned} V(T-1) &= \max\left\{\widetilde{X}(T-1), E(V(T)|\mathscr{F}_{T-1})\right\} \\ &= \max\left\{\frac{T-1}{T}1_{\{R(T-1)=1\}}, P(R(T)=1)\right\} \\ &= \max\left\{\frac{T-1}{T}1_{\{R(T-1)=1\}}, \frac{1}{T}\right\} \\ &= \frac{T-1}{T}1_{\{R(T-1)=1\}} + \frac{1}{T}1_{\{R(T-1)=0\}}. \end{aligned}$$

By induction, we show that—at least for n small enough—

$$V(T-n) = \frac{T-n}{T}1_{\{R(T-n)=1\}} + v_{T-n}1_{\{R(T-n)=0\}},$$

where v_t is given recursively by $v_T = 0$ and $v_{t-1} = \frac{1}{T} + \frac{t-1}{t}v_t$ for $t = T-1, T-2, \dots$ Indeed,

$$\begin{aligned} V(T-n) &= \max\left\{\widetilde{X}(T-n), E(V(T-n+1)|\mathscr{F}_{T-n})\right\} \\ &= \max\left\{\frac{T-n}{T}1_{\{R(T-n)=1\}}, \right. \\ &\quad \left. \frac{T-n+1}{T}P(R(T-n+1)=1) + v_{T-n+1}P(R(T-n+1)=0)\right\} \\ &= \max\left\{\frac{T-n}{T}1_{\{R(T-n)=1\}}, \frac{T-n+1}{T}\frac{1}{T-n+1} + v_{T-n+1}\frac{T-n}{T-n+1}\right\} \end{aligned}$$

$$= \max\left\{\frac{T-n}{T}1_{\{R(T-n)=1\}}, v_{T-n}\right\}$$

$$= \frac{T-n}{T}1_{\{R(T-n)=1\}} + v_{T-n}1_{\{R(T-n)=0\}}.$$

However, the last equality only holds as long as $v_{T-n} \leq \frac{T-n}{T}$. Otherwise the second term in the maximum is larger, regardless of the value of $R(T-n)$. Therefore we obtain

$$V(t) = \begin{cases} v_{t_0}, & t \leq t_0, \\ \frac{t}{T}1_{\{R(t)=1\}} + v_t 1_{\{R(t)=0\}}, & t > t_0, \end{cases}$$

where t_0 is the largest integer such that $\frac{t_0}{T} < v_{t_0}$. Proposition 1.55 together with the above derivation yields that it is optimal to stop at

$$\tau := \inf\{t > t_0 : R(t) = 1\} \wedge T$$

and $P(N(\tau) = 1) = E(\widetilde{X}(\tau)) = V(0) = v_{t_0}$.

The solution to the recursion for v_t is

$$v_t = \frac{t}{T}\sum_{s=t}^{T-1}\frac{1}{s}$$

as one can easily verify. Hence t_0 is the largest integer t with $\sum_{s=t}^{T-1}\frac{1}{s} > 1$. For large T we obtain the approximation

$$\sum_{s=t}^{T-1}\frac{1}{s} = \sum_{s=t}^{T-1}T^{-1}\frac{1}{(s/T)} \approx \int_{t/T}^{1}\frac{1}{s}ds = -\log\frac{t}{T}.$$

The threshold condition $\sum_{s=t_0}^{T-1}\frac{1}{s} \approx 1$ can then be rephrased as $-\log\frac{t_0}{T} \approx 1$ or

$$\frac{t_0}{T} \approx \frac{1}{e} \approx 0.368,$$

which leads to $v_{t_0} \approx -e^{-1}\log e^{-1}$ and hence

$$v_{t_0} \approx \frac{1}{e} \approx 0.368.$$

Put differently, it is approximately optimal to let 37% of all candidates pass and then pick the next one that is better than all which have been observed so far. This strategy leads to the globally optimal candidate with approximately 37% probability, which may seem surprisingly high if the total number T of candidates amounts to thousands or millions! With another 37% probability, however, one will be stuck with the last inspected candidate because the truly optimal choice happened to belong to the initial "training set" and hence no other candidate ever surpassed it.

The Snell envelope can also be obtained by minimisation over martingales rather than maximising over stopping times. This fact is sometimes used for numerical solution of optimal stopping problems.

Proposition 1.59 *The Snell envelope V of X satisfies*

$$V(0) = \min\left\{E\Big(\sup_{t=0,\dots,T}(X(t) - M(t))\Big) : M \text{ martingale with } M(0) = 0\right\}$$

and, more generally,

$$V(t) = \min\left\{E\Big(\sup_{s=t,\dots,T}(X(s) - M(s))\Big|\mathscr{F}_t\Big) : M \text{ martingale with } M(t) = 0\right\}$$

for any $t \in \{0, \dots, T\}$.

Proof The first statement follows from the second one.

"$\leq$": If τ is a stopping time with values in $\{t, \dots, T\}$, the optional sampling theorem yields

$$E(X(\tau)|\mathscr{F}_t) \leq E(X(\tau) - M(\tau)|\mathscr{F}_t) \leq E\Big(\sup_{s=t,\dots,T}(X(s) - M(s))\Big|\mathscr{F}_t\Big).$$

Since $V(t)$ is the supremum of all such expressions, we have $V(\tau) \leq \min\{\dots\}$.

"$\geq$": Since V is a supermartingale, the predictable part A^V in its Doob decomposition $V = V(0) + M^V + A^V$ is decreasing. The desired inequality follows from

$$\begin{aligned}
&E\Big(\sup_{s=t,\dots,T}\big(X(s) - M^V(s) + M^V(t)\big)\Big|\mathscr{F}_t\Big)\\
&\quad= E\Big(\sup_{s=t,\dots,T}\big(X(s) - V(s) + V(0) + A^V(s) + M^V(t)\big)\Big|\mathscr{F}_t\Big)\\
&\quad\leq E\Big(\sup_{s=t,\dots,T}\big(V(0) + A^V(s) + M^V(t)\big)\Big|\mathscr{F}_t\Big)\\
&\quad= E(V(t)|\mathscr{F}_t) = V(t).
\end{aligned}$$

□

1.5.3 The Markovian Situation

Above we have derived results that help to prove that a candidate solution is optimal. However, it is not yet obvious how to come up with a reasonable candidate. A standard approach in a Markovian setup is to assume that the value process is a function of the state variables and to derive a recursive equation for this function—the so-called *Bellman equation.*

We consider the setup of Sect. 1.5.1 with a reward function as in Example 1.43 such that $E(u(\alpha)) \neq -\infty$ for at least one admissible control α. Suppose that the jump characteristic $K^{(\alpha)}$ as in (1.24) of the controlled process $X^{(\alpha)}$ is of Markovian type in the sense that the conditional jump distribution $K^{(\alpha)}(t,\cdot)$ depends on ω, α only through $X^{(\alpha)}(t-1)$ and $\alpha(t)$, i.e.

$$K^{(\alpha)}(t,\cdot) = \kappa(\alpha(t), t, X^{(\alpha)}(t-1), \cdot)$$

for some deterministic function κ. Morally speaking, we assume that the set $\mathscr{A}$ of admissible controls contains—up to some measurability or integrability conditions—essentially all predictable processes with values in some set $A \subset \mathbb{R}^m$. Recall, however, that the initial value $\alpha(0)$ and hence also $x_0 := X^{(\alpha)}(0)$ is supposed to coincide for all controls $\alpha \in \mathscr{A}$. Under these conditions, it is natural to assume that the value process is of the form

$$\mathscr{V}(t,\alpha) = \sum_{s=1}^{t} f(s, X^{(\alpha)}(s-1), \alpha(s)) + v(t, X^{(\alpha)}(t)) \tag{1.70}$$

for some so-called **value function** $v : \{0, 1, \dots, T\} \times \mathbb{R}^d \to \mathbb{R}$ of the control problem. Indeed, what else should the optimal expected future reward depend on?

In order for the supermartingale/martingale conditions from Theorem 1.44 to hold, we need that

$$\begin{aligned}
&E(\mathscr{V}(t,\alpha) - \mathscr{V}(t-1,\alpha)|\mathscr{F}_{t-1}) \\
&\quad = f(t, X^{(\alpha)}(t-1), \alpha(t)) + E\big(v(t, X^{(\alpha)}(t)) - v(t-1, X^{(\alpha)}(t-1))\big|\mathscr{F}_{t-1}\big) \\
&\quad = f(t, X^{(\alpha)}(t-1), \alpha(t)) \\
&\qquad + \int \Big(v(t, X^{(\alpha)}(t-1)+y) - v(t, X^{(\alpha)}(t-1))\Big) \kappa(\alpha(t), t, X^{(\alpha)}(t-1), dy) \\
&\qquad + v(t, X^{(\alpha)}(t-1)) - v(t-1, X^{(\alpha)}(t-1)) \\
&\quad = (f^{(\alpha(t))} + G^{(\alpha(t))}v + \Delta_t v)(t, X^{(\alpha)}(t-1))
\end{aligned} \tag{1.71}$$

is nonpositive for any control α and zero for some control α, where we use the notation

$$(G^{(a)}u)(t,x) := \int u(t,x+y)\kappa(a,t,x,dy) - u(t,x), \tag{1.72}$$

$$\Delta_t u(t,x) := u(t,x) - u(t-1,x), \tag{1.73}$$

$$f^{(a)}(t,x) := f(t,x,a). \tag{1.74}$$

These considerations lead to the following result.

Theorem 1.60 *As noted above we assume that there is some deterministic function κ such that*

$$E\big(1_B(\Delta X^{(\alpha)}(t))\big|\mathscr{F}_{t-1}\big) = \kappa(\alpha(t),t,X^{(\alpha)}(t-1),B) \tag{1.75}$$

for any $\alpha \in \mathscr{A}$, $t \in \{1,\dots,T\}$, $B \in \mathscr{B}^d$. In addition, let $A \subset \mathbb{R}^m$ be such that any admissible control is A-valued. Suppose that

1. *the function $v : \{0,1,\dots,T\} \times \mathbb{R}^d \to \mathbb{R}$ satisfies*

$$v(T,x) = g(x) \tag{1.76}$$

and the **Bellman equation**

$$\sup_{a\in A}(f^{(a)} + G^{(a)}v + \Delta_t v)(t,x) = 0, \quad t = 1,2,\dots,T, \tag{1.77}$$

where we use the notation (1.70, 1.72–1.74) and assume that the integral in the definition of $G^{(a)}v$ is finite,

2. *for any $\alpha \in \mathscr{A}$ with $E(u(\alpha)^-) < \infty$ and any $t = 0,\dots,T$ there is some admissible control $\overline{\alpha}$ with $\overline{\alpha}^t = \alpha^t$ and*

$$\overline{\alpha}(s) = \arg\max_{a\in A}(f^{(a)} + G^{(a)}v + \Delta_t v)(s, X^{(\overline{\alpha})}(s-1)), \quad s = t+1, t+2, \dots, T. \tag{1.78}$$

Then (1.70) coincides with the value process for any $\alpha \in \mathscr{A}$ with $E(u(\alpha)^-) < \infty$ and $\overline{\alpha}$ is optimal for the problem (1.43). In particular, $\alpha^\star := \overline{\alpha}$ is an optimal control if $t = 0$ is chosen in 2.

Proof Define $(\mathscr{V}(\cdot,\alpha))_{\alpha\in\mathscr{A}}$ as in (1.70). By assumption on $X^{(\alpha)}$ we have $\mathscr{V}(t,\widetilde{\alpha}) = \mathscr{V}(t,\alpha)$ if $\widetilde{\alpha}^t = \alpha^t$. Equation (1.76) implies that $\mathscr{V}(T,\alpha) = u(\alpha)$. Equations (1.77, 1.78, 1.71) imply the martingale property of $\mathscr{V}(\cdot,\overline{\alpha}) - \mathscr{V}(\cdot,\overline{\alpha})^t$. Similarly, equations (1.77) and (1.71) yield that $\mathscr{V}(\cdot,\alpha)$ is a supermartingale for any admissible control. The assertion now follows from Theorem 1.44(2) and Remark 1.45(5). □

Remark 1.61 Theorem 1.60 can be extended to $T = \infty$ if one argues carefully. Suppose that $g = 0$ and consider the candidate value process $\mathscr{V}$ from (1.70). If the series (1.42) converges and $\lim_{t\to\infty} \mathscr{V}(t, \alpha) = u(\alpha)$ almost surely and if $E(\sup_{t\in\mathbb{R}_+} |\mathscr{V}(t, \alpha)|) < \infty$ holds for any $\alpha \in \mathscr{A}$, we can apply Remark 1.45(3) in order to obtain the statements of Theorem 1.60. Indeed, dominated convergence yields the (super-)martingale property of $\mathscr{V}(\cdot, \alpha)$ on $\{0, \dots, \infty\}$.

Note that the Bellman equation (1.77) allows us to compute the value function $v(t, \cdot)$ recursively starting from $t = T$, namely via $v(T, x) = g(x)$ and

$$v(t-1, x) = \sup_{a\in A} \left(f^{(a)}(t, x) + \int v(t, x+y)\kappa(a, t, x, dy) \right). \tag{1.79}$$

One may compare this to the general statement in Proposition 1.46. The minimality in Proposition 1.47 has a Markovian counterpart as well:

Remark 1.62 In the situation of Theorem 1.60, the value function is the smallest function v satisfying (1.76) and

$$(f^{(a)} + G^{(a)}v + \Delta_t v)(t, x) \leq 0, \quad t = 1, 2, \dots, T \tag{1.80}$$

for any $a \in A$. Indeed, this follows recursively starting from $t = T$ because (1.80) means that

$$v(t-1, x) \geq \sup_{a\in A} \left(f^{(a)}(t, x) + \int v(t, x+y)\kappa(a, t, x, dy) \right)$$

and equality holds for the value function.

If we are just interested in upper and lower bounds of the value process, we may consider sub- and supersolutions of the Bellman equation:

Remark 1.63 If a function $v : \{0, 1, \dots, T\} \times \mathbb{R}^d \to \mathbb{R}$ satisfies $v(T, x) \geq g(x)$ and

$$\sup_{a\in A}(f^{(a)} + G^{(a)}v + \Delta_t v)(t, x) \leq 0, \quad t = 1, 2, \dots, T,$$

we call it a **supersolution** to the Bellman equation. Using Remark 1.45(4) instead of Theorem 1.44 we conclude that the right-hand side of (1.70) is $\geq \mathscr{V}(t, \alpha)$, i.e. the supersolution provides an upper bound to the value process of the problem.

Conversely, we call $v : \{0, 1, \dots, T\} \times \mathbb{R}^d \to \mathbb{R}$ a **subsolution** to the Bellman equation if $v(T, x) \leq g(x)$ and

$$\sup_{a\in A}(f^{(a)} + G^{(a)}v + \Delta_t v)(t, x) \geq 0, \quad t = 1, 2, \dots, T,$$

and

$$\overline{\alpha}(t) = \underset{a \in A}{\arg\max}(f^{(a)} + G^{(a)}v + \Delta_t v)(t, X^{(\overline{\alpha})}(t-1)), \quad t = 1, 2, \dots, T$$

for some admissible control $\overline{\alpha}$ with $E(u(\overline{\alpha})^-) < \infty$. Again using Remark 1.45(4), we get $v(0, x_0) \leq \mathscr{V}(0)$, i.e. the supersolution yields a lower bound to the true optimal value of the problem.

As an illustration we consider variations of Examples 1.48 and 1.49 for power instead of logarithmic utility functions.

Example 1.64 (Power Utility of Terminal Wealth) We consider the same problem as in Example 1.48 but with power utility function $u(x) = x^{1-p}/(1-p)$ for some $p \in (0, \infty) \setminus \{1\}$, i.e. the investor's goal is to maximise $E(u(V_\varphi(T)))$ over all investment strategies φ. In order to derive explicit solutions we assume that the stock price $S(t) = S(0)\mathscr{E}(X)(t)$ is a multiple of a geometric random walk.

The set of controls in Example 1.48 is not of the "myopic" form assumed in the beginning of this section. Therefore we consider again relative portfolios $\pi(t)$, which represent the fraction of wealth invested in the stock, i.e. (1.46). By slight abuse of notation, we write $V_\pi(t)$ for the wealth generated from the investment strategy π, which evolves as

$$V_\pi(t) = v_0 \prod_{s=1}^{t} \big(1 + \pi(s)\Delta X(s)\big) = v_0 \mathscr{E}(\pi \bullet X)(t),$$

cf. (1.47). This is our controlled process. In order for wealth to stay positive, $\pi(t)$ needs to be A-valued for

$$A := \big\{a \in \mathbb{R} : 1 + ax > 0 \text{ for any } x \text{ in the support of } \Delta X(1)\big\},$$

i.e. we consider the set

$$\mathscr{A} = \big\{\pi \text{ predictable} : \pi(0) = 0 \text{ and } \pi(t) \in A, t = 1, \dots, T\big\}.$$

Since $E(1_B(\Delta V_\pi(t))|\mathscr{F}_{t-1}) = E(1_B(\Delta V_\pi(t))|\mathscr{F}_{t-1})$ by (1.4), we have that $\kappa(a, t, x, B) := E(1_B(xa\Delta X(1)))$ is the kernel satisfying (1.75) and (1.77) reads as

$$\sup_{a \in A} E\Big(v\big(t, x(1 + a\Delta X(1))\big)\Big) - v(t-1, x) = 0. \tag{1.81}$$

We guess that v factors in functions of t and x, respectively. Since the terminal condition (1.76) is $v(T,x) = u(x)$, this leads to the ansatz $v(t,x) = g(t)u(x) = g(t)x^{1-p}/(1-p)$ for some function g with $g(T) = 1$. Inserting into (1.81) yields

$$(1-p)\sup_{a\in A} E\left(\frac{(1+a\Delta X(1))^{1-p}}{1-p}\right) g(t) = g(t-1),$$

which is solved by $g(t) = \varrho^{T-t}$ for

$$\varrho := E\Big((1+a^\star \Delta X(1))^{1-p}\Big)$$

and

$$a^\star := \arg\max_{a\in A} E\left(\frac{(1+a\Delta X(1))^{1-p}}{1-p}\right).$$

The existence of $a^\star$ is warranted if $\Delta X(1)$ attains both positive and negative values and if $E(|\Delta X(1)|) < \infty$ for $p < 1$. Theorem 1.60 now yields that the constant deterministic strategy $\pi^\star(t) = a^\star$ is optimal and that the value function is indeed of product form $v(t,x) = \varrho^{T-t}u(x)$.

Summing up, it is optimal to invest a fixed fraction of wealth in the stock. The optimal fraction depends on both the law of $\Delta X(1)$ and the parameter p.

In the consumption case, one proceeds similarly.

Example 1.65 (Power Utility of Consumption) We consider now the counterpart of Example 1.49, again with power instead of logarithmic utility, i.e. the aim is to maximise

$$E\left(\sum_{t=1}^{\infty} e^{-\delta t}u(c_t)\right) \tag{1.82}$$

over all predictable processes (φ, c) such that the wealth $V_{\varphi,c}(t) = v_0 + \varphi \bullet S(t) - \sum_{s=1}^{t} c(s)$ is positive for $t \geq 0$. As in Example 1.64 the utility function is assumed to be of power type $u(x) = x^{1-p}/(1-p)$. The constant δ represents an impatience rate.

Also as above we need to consider relative portfolios and relative consumption rates, i.e. we consider $\pi(t)$ and $\kappa(t)$ as in (1.53, 1.52). Again by slight abuse of notation, we write $V_{\pi,\kappa}$ for the corresponding wealth. We obtain

$$V_{\pi,\kappa}(t) = v_0 \prod_{s=1}^{t}(1-\kappa(s))(1+\pi(s)\Delta X(s)) = v_0\mathscr{E}(-\kappa \bullet I)(t)\mathscr{E}(\pi \bullet X)(t)$$

as in (1.54) and (1.82) equals

$$E\left(\sum_{t=1}^{\infty} e^{-\delta t} u\big(V_{\pi,\kappa}(t-1)\kappa(t)\big)\right).$$

Similarly as in Example 1.64 the set of admissible controls is defined as

$$\mathscr{A} = \Bigg\{(\pi,\kappa) \text{ predictable} : (\pi,\kappa)(0) = 0 \text{ and } (\pi,\kappa)(t) \in A \text{ for } t \geq 1$$
$$\text{as well as } E\left(\sum_{t=1}^{\infty} e^{-\delta t}\big(V_{\pi,\kappa}(t-1)\kappa(t)\big)^{1-p}\right) < \infty\Bigg\}$$

with

$$A := \big\{(a,b) \in \mathbb{R} \times (0,1) : 1 + ax > 0 \text{ for any } x \text{ in the support of } \Delta X(1)\big\}.$$

The integrability condition warrants that the expected utility is finite for any admissible control.

The Bellman equation (1.77) now reads as

$$\sup_{(a,b)\in A} \Big(e^{-\delta t} u(xb) + E\big(v\big(t, x(1+a\Delta X(1))(1-b)\big)\big) - v(t-1,x)\Big) = 0$$

instead of (1.81). The separation ansatz

$$v(t,x) = g(t)u(x) = \frac{g(t)x^{1-p}}{1-p}$$

leads to

$$(1-p) \sup_{(a,b)\in A} \frac{e^{-\delta t} b^{1-p} + E\big((1+a\Delta X(1))^{1-p}\big)(1-b)^{1-p} g(t)}{1-p} = g(t-1).$$

Straightforward calculations yield that this is solved by

$$g(t) = \frac{e^{\delta t}}{(e^{\delta/p} - \varrho^{1/p})^p}$$

with

$$\varrho := (1-p) \max_{a\in A} E\left(\frac{(1+a\Delta X(1))^{1-p}}{1-p}\right)$$

and provided that $\varrho < e^{\delta}$, which we add as an assumption. The optimal controls are $\pi^\star(t) = a^\star$ and $\kappa^\star(t) = b^\star$ with

$$a^\star := \underset{a \in A}{\arg\max}\, E\left(\frac{(1 + a\Delta X(1))^{1-p}}{1-p}\right)$$

as in Example 1.64 and

$$b^\star := 1 - (\varrho e^{-\delta})^{1/p}.$$

The verification is based on Remark 1.61, where some regularity needs to be checked. If $p > 1$ one easily verifies that $\lim_{t\to\infty} v(t, V_{\pi,\kappa}(t)) \to 0$ almost surely and $E(\sup_{t\geq 0} |\mathscr{V}(t, (\pi,\kappa))|) < \infty$ for

$$\mathscr{V}(t, (\pi,\kappa)) = \sum_{s=1}^{t} e^{-\delta} u\big(V_{\pi,\kappa}(s-1)\kappa(s)\big) + v(t, V_{\pi,\kappa}(t))$$

and any admissible control (π, κ). For $p < 1$ this is not obvious. However, the supermartingale property of $(V(t, (\pi,\kappa)))_{t\geq 0}$ can actually be obtained by Fatou's lemma if we set $\mathscr{V}(\infty, (\pi,\kappa)) = \sum_{s=1}^{\infty} e^{-\delta} u(V_{\pi,\kappa}(s-1)\kappa(s))$. For our candidate control we have

$$\begin{aligned} E\big(V_{\pi^\star,\kappa^\star}(t)^{1-p} \big| \mathscr{F}_{t-1}\big) &= V_{\pi^\star,\kappa^\star}(t)\varrho(1-b^\star)^{1-p} \\ &= V_{\pi^\star,\kappa^\star}(t)e^{\delta}(1-b^\star), \end{aligned}$$

which implies

$$E\big(V_{\pi^\star,\kappa^\star}(t)^{1-p} \big| \mathscr{F}_{t-1}\big) = v_0^{1-p} e^{\delta t}(1-b^\star)^t$$

and hence

$$E\left(\sum_{t=1}^{\infty} e^{-\delta t} V_{\pi^\star,\kappa^\star}(t)^{1-p}\kappa(t)\right) = (v_0 b^\star)^{1-p} \sum_{t=1}^{\infty} (1-b^\star)^t < \infty$$

because $0 < b^\star < 1$. We conclude that $(\pi^\star, \kappa^\star)$ is in fact an admissible control.

Altogether, we observe that the fixed investment fraction from Example 1.64 is still optimal in the consumption case. The optimal consumption fraction is deterministic and constant as well.

Finally, we reformulate Theorem 1.60 for stopping problems as in Sect. 1.5.2.

Theorem 1.66 *Suppose that $X(t) = g(Y(t))$ for $g : \mathbb{R}^d \to \mathbb{R}$ and some $\mathbb{R}^d$-valued Markov process $(Y(t))_{t=1,2,\ldots}$ with generator G in the sense of (1.34). If the function $v : \{0, 1, \ldots, T\} \times \mathbb{R}^d \to \mathbb{R}$ satisfies*

$$v(T, y) = g(y)$$

and

$$\max\left\{g(y)-v(t-1,y),\ \Delta_t v(t,y)+Gv(t,y)\right\}=0,\quad t=1,2,\dots,T, \tag{1.83}$$

then $V(t):=v(t,Y(t))$, $t=0,\dots,T$ *is the Snell envelope of* X*. Here we use the notation* $Gv(t,y):=G(v(t,\cdot))(y)$ *and* $\Delta_t v(t,y)=v(t,y)-v(t-1,y)$ *similarly as in (1.35, 1.73). An optimal stopping time is obtained from Proposition 1.55(2).*

Proof Since $v(T,Y(T))=g(Y(T))$ and

$$\begin{aligned} v(t-1,Y(t-1)) &= v(t-1,Y(t-1)) \\ &\quad + \max\{g(Y(t-1))-v(t-1,Y(t-1)), \\ &\quad v(t,Y(t-1))-v(t-1,Y(t-1))+Gv(t,Y(t-1))\} \\ &= \max\{g(Y(t-1)),v(t,Y(t-1))+Gv(t,Y(t-1))\} \\ &= \max\{X(t-1),E(v(t,Y(t))|\mathscr{F}_{t-1})\}, \end{aligned}$$

the assertion follows from Proposition 1.55(1). □

As in Theorem 1.60, equation (1.83) allows us to compute $v(t,\cdot)$ recursively starting from $t=T$, namely via

$$v(t-1,y)=\max\left\{g(y),\int v(t,x)P(1,y,dx)\right\},\quad t=T,\dots,1, \tag{1.84}$$

where $P(\cdot)$ denotes the transition function of Y.

Remark 1.67 If g is nonnegative, Theorem 1.66 can be extended to the case $T=\infty$ as in Remark 1.56(2). To this end we assume that $v:\mathbb{N}\times\mathbb{R}_+\to\mathbb{R}_+$ satisfies (1.83) for $t=1,2,\dots$ and $E(\sup_{t\in\mathbb{N}}v(t,Y(t)))<\infty$. Moreover, $\tau:=\inf\{t\geq 0: V(t)=X(t)\}$ is assumed to be finite for $V(t)=v(t,Y(t))$. Then V is still the Snell envelope and τ an optimal stopping time. As an example one may consider the perpetual put, cf. Example 1.57.

Proof As in the proof of Theorem 1.66 we observe that V is a supermartingale and V^τ a martingale. From $E(V(\infty)|\mathscr{F}_t)=0\leq V(t)$ we conclude that V is actually a supermartingale on $\mathbb{N}\cup\{\infty\}$. Since $E(V^\tau(\infty)|\mathscr{F}_t)=V(t)$ follows from dominated convergence, V^τ is a martingale on $\mathbb{N}\cup\{\infty\}$. The assertion now follows from Theorem 1.53(1). □

The minimality of Proposition 1.55(3) resp. Remark 1.62 here read as follows:

Remark 1.68 The function $v : \{0, 1, \ldots, T\} \times \mathbb{R}^d \to \mathbb{R}$ in Theorem 1.66 is the smallest one that satisfies $v(t, \cdot) \geq g(\cdot)$, $t = 0, \ldots, T$ and $\Delta_t v + Gv \leq 0$. Indeed, this follows recursively for $t = T, \ldots, 0$ because these two properties imply

$$v(t-1, y) = \max\left\{g(y), \int v(t, x)P(1, y, dx)\right\},$$

with equality for the value function by (1.84).

1.5.4 The Stochastic Maximum Principle

We turn now to the counterpart of the first-order conditions in Example 1.40(2). As before, we consider the setup of Sect. 1.5.1 with a reward function as in Example 1.43. The set of admissible controls $\mathscr{A}$ is assumed to contain only predictable processes with values in some convex set $A \subset \mathbb{R}^m$. Moreover, we suppose that the function g is concave.

A sufficient condition for optimality reads as follows:

Theorem 1.69 *Suppose that Y is an $\mathbb{R}^d$-valued adapted process and $\alpha^\star$ an admissible control with the following properties, where as usual we suppress the argument $\omega \in \Omega$ in random functions.*

1. $E(|u(\alpha^\star)|) < \infty$.
2. *g is differentiable in $X^{(\alpha^\star)}(T)$ with derivative denoted as $\nabla g(X^{(\alpha^\star)}(T))$.*
3. *For any $\alpha \in \mathscr{A}$ we have*

$$\sum_{i=1}^{d}\left(Y_i(t-1)a^{X_i^{(\alpha)}}(t) + \widetilde{c}^{Y_i, X_i^{(\alpha)}}(t)\right) + f\big(t, X^{(\alpha)}(t-1), \alpha(t)\big) \tag{1.85}$$

$$= H(t, X^{(\alpha)}(t-1), \alpha(t)), \quad t = 1, \ldots, T$$

for some function $H : \Omega \times \{1, \ldots, T\} \times \mathbb{R}^d \times A \to \mathbb{R}$ such that $(x, a) \mapsto H(t, x, a)$ is concave and $x \mapsto H(t, x, \alpha^\star(t))$ is differentiable in $X^{(\alpha^\star)}(t-1)$. We denote the derivative by $\nabla_x H(t, X^{(\alpha^\star)}(t-1), \alpha^\star(t)) = (\partial_{x_i} H(t, X^{(\alpha^\star)}(t-1), \alpha^\star(t)))_{i=1,\ldots,d}$.

4. *The drift rate and the terminal value of Y are given by*

$$a^Y(t) = -\nabla_x H(t, X^{(\alpha^\star)}(t-1), \alpha^\star(t)), \quad t = 1, \ldots, T \tag{1.86}$$

and

$$Y(T) = \nabla g(X^{(\alpha^\star)}(T)). \tag{1.87}$$

5. *$\alpha^\star(t)$ maximises $a \mapsto H(t, X^{(\alpha^\star)}(t-1), a)$ over A for $t = 1, \dots, T$.*

Then $\alpha^\star$ is an optimal control, i.e. it maximises

$$E\left(\sum_{t=1}^{T} f(t, X^{(\alpha)}(t-1), \alpha(t)) + g(X^{(\alpha)}(T))\right)$$

over all $\alpha \in \mathscr{A}$.

Proof Let $\alpha \in \mathscr{A}$ be a competing control. By induction we show that

$$\begin{aligned}
&\sum_{s=1}^{T} E\left(f(s, X^{(\alpha)}(s-1), \alpha(s)) - f(s, X^{(\alpha^\star)}(s-1), \alpha^\star(s))\Big|\mathscr{F}_t\right) \\
&\qquad + E\big(g(X^{(\alpha)}(T)) - g(X^{(\alpha^\star)}(T))\big|\mathscr{F}_t\big) \\
&\qquad \leq \sum_{s=1}^{t} \left(f(s, X^{(\alpha)}(s-1), \alpha(s)) - f(s, X^{(\alpha^\star)}(s-1), \alpha^\star(s))\right) \\
&\qquad + Y(t)^\top \left(X^{(\alpha)}(t) - X^{(\alpha^\star)}(t)\right)
\end{aligned} \tag{1.88}$$

for $t = T, T-1, \dots, 0$. Since $\alpha(0) = \alpha^\star(0)$ implies $X^{(\alpha)}(0) = X^{(\alpha^\star)}(0)$, we obtain optimality of $\alpha^\star$ by inserting $t = 0$.

In order to verify (1.88) we introduce the notation

$$h(t, x) := \sup_{a \in A} H(t, x, a). \tag{1.89}$$

We start with $t = T$. Concavity of g implies

$$g(X^{(\alpha)}(T)) - g(X^{(\alpha^\star)}(T)) \leq Y(T)^\top \big(X^{(\alpha)}(T) - X^{(\alpha^\star)}(T)\big), \tag{1.90}$$

which in turn yields (1.88).

For the step from t to $t-1$ observe that

$$\begin{aligned}
&E\bigg(f(t, X^{(\alpha)}(t-1), \alpha(t)) - f(t, X^{(\alpha^\star)}(t-1), \alpha^\star(t)) \\
&\qquad + Y(t)^\top \big(X^{(\alpha)}(t) - X^{(\alpha^\star)}(t)\big)\Big|\mathscr{F}_{t-1}\bigg) \\
&\qquad - Y(t-1)^\top \big(X^{(\alpha)}(t-1) - X^{(\alpha^\star)}(t-1)\big) \\
&= f(t, X^{(\alpha)}(t-1), \alpha(t)) - f(t, X^{(\alpha^\star)}(t-1), \alpha^\star(t))
\end{aligned}$$

$$
\begin{aligned}
&+\sum_{i=1}^{d}\Big(X_i^{(\alpha)}(t-1)a^{Y_i}(t)+Y_i(t-1)a^{X_i^{(\alpha)}}(t)+\widetilde{c}^{Y_i,X_i^{(\alpha)}}(t)\\
&\quad-\Big(X_i^{(\alpha^\star)}(t-1)a^{Y_i}(t)+Y_i(t-1)a^{X_i^{(\alpha^\star)}}(t)+\widetilde{c}^{Y_i,X_i^{(\alpha^\star)}}(t)\Big)\Big)\\
&=H(t,X^{(\alpha)}(t-1),\alpha(t))-H(t,X^{(\alpha^\star)}(t-1),\alpha^\star(t))\\
&\quad+\big(X^{(\alpha)}(t-1)-X^{(\alpha^\star)}(t-1)\big)^\top a^Y(t)\\
&=H(t,X^{(\alpha)}(t-1),\alpha(t))-h(t,X^{(\alpha)}(t-1)) \qquad (1.91)\\
&\quad+\Big(h(t,X^{(\alpha)}(t-1))-h(t,X^{(\alpha^\star)}(t-1))\\
&\quad+\big(X^{(\alpha)}(t-1)-X^{(\alpha^\star)}(t-1)\big)^\top a^Y(t)\Big)\\
&\leq 0,
\end{aligned}
$$

where the first equality follows from integration by parts, cf. Proposition 1.33. The inequality holds because (1.91) is nonpositive by definition and

$$
\begin{aligned}
&h(t,X^{(\alpha)}(t-1))-h(t,X^{(\alpha^\star)}(t-1))+\big(X^{(\alpha)}(t-1)-X^{(\alpha^\star)}(t-1)\big)^\top a^Y(t)\\
&=h(t,X^{(\alpha)}(t-1))-h(t,X^{(\alpha^\star)}(t-1))\\
&\quad-\big(X^{(\alpha)}(t-1)-X^{(\alpha^\star)}(t-1)\big)^\top\nabla_x h(t,X^{(\alpha^\star)}(t-1))\\
&\leq 0
\end{aligned}
$$

holds by (1.86) and Proposition 1.42. □

Remark 1.70

1. One can do without differentiability of g in Theorem 1.69. It is only needed for the estimate in (1.90). Since g is concave, (1.87) can be replaced by the weaker requirement that $Y(T)$ belongs to the subdifferential of g in $X^{(\alpha^\star)}(T)$.
2. The proof shows that concavity of the mapping $(x,a)\mapsto H(t,x,a)$ is only needed in order to apply Proposition 1.42. Instead, we could assume that the mapping $x\mapsto h(t,x)$ is concave and moreover differentiable in $X^{(\alpha^\star)}(t-1)$ with derivative

$$\nabla_x h(t,X^{(\alpha^\star)}(t-1))=\nabla_x H(t,X^{(\alpha^\star)}(t-1),\alpha^\star(t)). \qquad (1.92)$$

These properties were obtained from Proposition 1.42 if $(x,a)\mapsto H(t,x,a)$ is concave. Since $\alpha^\star$ is the maximiser in (1.89), (1.92) can be expected to hold in more general circumstances. Indeed, consider a real-valued smooth function $f(x,y)$ of two variables $x,y\in\mathbb{R}$ and set $g(x):=\sup_y f(x,y)$. If the supremum is attained in $y(x)$, we have $g(x) = f(x, y(x))$. The first-order condition for $y(x)$ is $D_2 f(x, y(x)) = 0$. This in turn implies

$$g'(x) = D_1 f(x, y(x)) + D_2 f(x, y(x))y'(x) = D_1 f(x, y(x)),$$

which is the analogue of (1.92) in this toy setup.

The second condition in Theorem 1.69 primarily means that the random expression (1.85) depends on the control process α only via the present values $X^{(\alpha)}(t-1)$ and $\alpha(t)$. Moreover, this dependence is assumed to be concave unless we are in the more general situation of Remark 1.70(2). The function H is usually called the **Hamiltonian**. In the deterministic case it reduces to the function with the same name in Example 1.40(2). Similarly, the process Y can be viewed as a counterpart of the vector $y^\star$ in that example.

If additional convexity holds, the process Y solves a dual minimisation problem. We consider three versions of portfolio optimisation problems in Mathematical Finance where this is actually the case.

Example 1.71 (Expected Utility of Terminal Wealth) We consider the extension of Example 1.48 for a general increasing, strictly concave, differentiable utility function $u : \mathbb{R} \to [-\infty, \infty)$ (strictly speaking, differentiable in the interior of the half line where it is finite). The investor's goal is to maximise her expected utility of terminal wealth

$$E\big(u(V_\varphi(T))\big) \tag{1.93}$$

over all controls $\varphi \in \mathscr{A} := \{\varphi \text{ predictable } : \varphi(0) = 0 \text{ and } E(|u(V_\varphi(T))|) < \infty\}$. The process

$$V_\varphi(t) := v_0 + \varphi \bullet S(t), \quad t \in [0, T]$$

represents the investor's wealth as in Example 1.48. In this situation the process Y in Theorem 1.69 solves a dual convex minimisation problem.

Theorem 1.72

1. *If Y is a martingale and $\varphi^\star \in \mathscr{A}$ such that YS is a martingale as well and $Y(T) = u'(V_{\varphi^\star}(T))$, then Y and $\varphi^\star$ satisfy the conditions of Theorem 1.69. In particular, $\varphi^\star$ maximises (1.93).*
2. *Suppose that Y and the corresponding optimal control $\varphi^\star$ are as in Theorem 1.69. Then Y and YS are martingales satisfying $Y(T) = u'(V_{\varphi^\star}(T))$.*
3. *Denote the* convex dual *of u, here in the sense of (A.7), by*

$$\widetilde{u}(y) := \sup_{x\in\mathbb{R}}(u(x) - xy), \quad y \in \mathbb{R}. \tag{1.94}$$

Y in Theorem 1.69 minimises

$$Z \mapsto E(\widetilde{u}(Z(T))) + v_0 Z(0) \tag{1.95}$$

among all nonnegative martingales Z such that ZS is a martingale.

4. *The optimal values of the primal maximisation problem (1.93) and the dual minimisation problem (1.95) coincide in the sense that*

$$E\big(u(V_{\varphi^\star}(T))\big) = E(\widetilde{u}(Y(T))) + v_0 Y(0),$$

where Y and $\varphi^\star \in \mathscr{A}$ are processes as in Theorem 1.69 (provided that they exist).

Proof

1. Since $H(t, x, a) = (Y(t-1)a^S(t) + \widetilde{c}^{Y,S}(t))a$ does not depend on x, we have $-\nabla_x H(t, V_{\varphi^\star}(t-1), \varphi^\star(t)) = 0 = a^Y(t)$. If YS is a martingale, we have $0 = a^{YS}(t) = S(t-1)a^Y(t) + Y(t-1)a^S(t) + \widetilde{c}^{Y,S}(t)$ and hence $H(t, x, a) = 0$ for any a, cf. Proposition 1.33. Consequently, condition 5 in Theorem 1.69 holds as well.
2. As in statement 1, $H(t, x, a) = (Y(t-1)a^S(t) + \widetilde{c}^{Y,S}(t))a$ does not depend on x, which yields $a^Y = 0$. Moreover, it is linear in $a \in A := \mathbb{R}$. In order to have a maximum in $\varphi^\star(t)$ it must vanish, which in turn implies $Y(t-1)a^S(t)+\widetilde{c}^{Y,S}(t) = 0$ and hence $a^{YS}(t) = S(t-1)a^Y(t)+Y(t-1)a^S(t)+\widetilde{c}^{Y,S}(t) = 0$. Consequently, both Y and YS are local martingales. They are in fact martingales because they are nonnegative, cf. Proposition 1.8 and the following sentence.
3. If both Z and ZS are martingales, integration by parts implies

$$\begin{aligned} V_{\varphi^\star} Z &= (v_0 + \varphi^\star \bullet S)Z \\ &= v_0 Z(0) + V_{\varphi^\star -} \bullet Z + (Z_- \varphi^\star) \bullet S + \varphi^\star \bullet [S, Z] \\ &= v_0 Z(0) + V_{\varphi^\star -} \bullet Z + \varphi^\star \bullet (SZ - S_- \bullet Z) \end{aligned} \tag{1.96}$$

and hence $V_{\varphi^\star} Z$ is a local martingale by Proposition 1.14(7). Since

$$u(V_{\varphi^\star}(T)) - V_{\varphi^\star}(T)Z(T) \le \widetilde{u}(Z(T)),$$

we have $E((V_{\varphi^\star}(T)Z(T))^-) < \infty$ if $E(\widetilde{u}(Z(T))) < \infty$. In this case $V_{\varphi^\star} Z$ is a martingale and we obtain $E(V_{\varphi^\star}(T)Z(T)) = v_0 Z(0)$ and hence

$$\begin{aligned} & E(\widetilde{u}(Z(T))) + v_0 Z(0) \\ &\quad = E\Big(\sup_{x \in \mathbb{R}} (u(x) - xZ(T))\Big) + v_0 Z(0) \\ &\quad \ge E\big(u(V_{\varphi^\star}(T)) - V_{\varphi^\star}(T)Z(T)\big) + v_0 Z(0) \\ &\quad = E\big(u(V_{\varphi^\star}(T))\big). \end{aligned} \tag{1.97}$$

Concavity of u yields that $x \mapsto u(x) - xY(T)$ is maximised by $V_{\varphi^\star}(T)$ because the first-order condition $u'(V_{\varphi^\star}(T)) = Y(T)$ holds. Therefore we obtain equality in (1.97) for $Z = Y$.
4. This follows from the proof of statement 3. □

Remark 1.73

1. As a positive martingale, $Y/Y(0)$ is the density process of a probability measure $Q \sim P$. Since YS is a martingale, S is a Q-martingale by Proposition 1.19. Such measures Q are called *equivalent martingale measures* and they play a key role in Mathematical Finance, as will become clear in Part II.
2. Inspecting the proof shows that the *local* martingale property of Y and YS suffices for statement 1 in Theorem 1.72. The same is true for the corresponding statements in Theorems 1.75 and 1.77 below.

Example 1.74 (Expected Utility of Consumption) Parallel to Example 1.49 we can also consider expected utility of consumption. Given a constant bank account and a stock with price process $S \geq 0$, recall that the investor's wealth is given by

$$V_{\varphi,c}(t) := v_0 + \varphi \bullet S(t) - \sum_{s=1}^{t} c(s), \quad t \in [0, T],$$

where $\varphi(t)$ denotes her number of shares in the period from $t-1$ to t and $c(t)$ the consumption at time $t-1$. The investor's goal is to maximise her expected utility

$$E\left(\sum_{t=1}^{T} u(t-1, c(t)) + u(T, V_{\varphi,c}(T))\right) \tag{1.98}$$

over all admissible (φ, c), where $u(t, \cdot) : \mathbb{R} \to [-\infty, \infty), t = 0, \dots, T$ are increasing, strictly concave, differentiable *utility* functions as in the previous example. We denote the derivative of $u'(t, \cdot)$ and consider all elements of

$$\mathscr{A} := \Bigg\{ (\varphi, c) \text{ predictable} : (\varphi, c)(0) = (0, 0) \text{ and}$$
$$E\left(\sum_{t=1}^{T} |u(t-1, c(t))| + |u(T, V_{\varphi,c}(T))|\right) < \infty \Bigg\}$$

as admissible controls. The counterpart of Theorem 1.75 reads as follows.

Theorem 1.75

1. *If Y is a martingale and $(\varphi^\star, c^\star) \in \mathscr{A}$ such that YS is a martingale, $Y(t) = u'(t-1, c^\star(t))$, $t = 1, \dots, T$, and $Y(T) = u'(T, V_{\varphi^\star,c^\star}(T))$, then Y and $(\varphi^\star, c^\star)$ satisfy the conditions of Theorem 1.69. In particular, $(\varphi^\star, c^\star)$ maximises (1.98).*
2. *Suppose that Y and the corresponding optimal control $(\varphi^\star, c^\star)$ are as in Theorem 1.69. Then Y and YS are martingales satisfying $Y(t-1) = u'(t-1, c^\star(t))$, $t = 1, \dots, T$ and $Y(T) = u'(T, V_{\varphi^\star,c^\star}(T))$.*
3. *Denote the* convex dual *of $u(t, \cdot)$, $t = 0, \dots, T$, again in the sense of (A.7), by*

$$\widetilde{u}(t, y) := \sup_{x \in \mathbb{R}}(u(t, x) - xy), \quad y \in \mathbb{R}. \tag{1.99}$$

   Y in Theorem 1.69 minimises

$$Z \mapsto \sum_{t=0}^{T} E(\widetilde{u}(t, Z(t))) + v_0 Z(0) \tag{1.100}$$

   among all nonnegative martingales Z such that ZS is a martingale.
4. *The optimal values of the primal maximisation problem (1.98) and the dual minimisation problem (1.100) coincide in the sense that*

$$E\left(\sum_{t=1}^{T} u(t-1, c^\star(t)) + u(T, V_{\varphi^\star,c^\star}(T))\right) = \sum_{t=0}^{T} E(\widetilde{u}(t, Y(t))) + v_0 Y(0),$$

   where Y and $(\varphi^\star, c^\star) \in \mathscr{A}$ are processes as in Theorem 1.69 (provided that they exist).

Proof

1. Since

$$\begin{aligned} H(t, x, (a_1, a_2)) = {} & \big(Y(t-1)a^S(t) + \widetilde{c}^{Y,S}(t)\big)a_1 \\ & - \big(Y(t-1) - a^Y(t)\big)a_2 + u(t-1, a_2) \end{aligned}$$

   in the consumption case, the first statement is obtained similarly as in Theorem 1.72. Condition 5 in Theorem 1.69 follows from the fact that the first-order condition for optimality of

$$a_2 \mapsto H\big(t, V_{\varphi^\star,c^\star}(t-1), (a_1, a_2)\big) = -Y(t-1)a_2 + u(t-1, a_2)$$

   is $-Y(t-1) + u'(t-1, a_2) = 0$. This holds for $a_2 = c^\star(t)$.
2. The same reasoning as in the proof of Theorem 1.72 yields that Y and YS are martingales. Differentiating $H(t, V_{\varphi^\star,c^\star}(t-1), (a_1, a_2))$ with respect to a_2 yields the first-order condition $-Y(t-1) + u'(t-1, c^\star(t)) = 0$.

3. If Z and ZS are martingales, $(v_0 + \varphi^\star \bullet S)Z$ is a local martingale by (1.96). If the right-hand side of (1.100) is finite, a careful argument similar to the one in the proof of Theorem 1.72(3) shows that $E(((v_0 + \varphi^\star \bullet S(T))Z(T))^-) < \infty$, which implies that $(v_0 + \varphi^\star \bullet S)Z$ is actually a martingale. Using

$$V_{\varphi^\star,c^\star}(T) = v_0 + \varphi^\star \bullet S(T) - \sum_{t=1}^{T} c^\star(t)$$

we obtain

$$\begin{aligned}
\sum_{t=0}^{T} E(\widetilde{u}(t, Z(t))) + v_0 Z(0) &= \sum_{t=0}^{T} E\left(\sup_{x\in\mathbb{R}}(u(t,x) - xZ(t))\right) + v_0 Z(0) \\
&\geq \sum_{t=1}^{T} E\big(u(t-1, c^\star(t)) - c^\star(t)Z(t-1)\big) \\
&\quad + E\big(u(T, V_{\varphi^\star,c^\star}(T)) - V_{\varphi^\star,c^\star}(T)Z(T)\big) + v_0 Z(0) \\
&= \sum_{t=1}^{T} E\big(u(t-1, c^\star(t))\big) + E\big(u(T, V_{\varphi^\star,c^\star}(T))\big) \\
&\quad - E\left(Z(T)\left(V_{\varphi^\star,c^\star}(T) + \sum_{t=1}^{T} c^\star(t)\right)\right) + v_0 Z(0) \\
&= \sum_{t=1}^{T} E\big(u(t-1, c^\star(t))\big) + E\big(u(T, V_{\varphi^\star,c^\star}(T))\big) \\
&\quad - E\big(Z(T)(v_0 + \varphi^\star \bullet S(T)\big) + v_0 Z(0) \\
&= \sum_{t=1}^{T} E\big(u(t-1, c^\star(t))\big) + E\big(u(T, V_{\varphi^\star,c^\star}(T))\big).
\end{aligned}$$

For Y we have equality because of the first-order conditions listed in statement 2.

4. This follows from the proof of statement 3. □

As in the previous example, $Y/Y(0)$ is the density process of some equivalent martingale measure.

Example 1.76 (Expected Utility of Terminal Wealth Under Proportional Transaction Costs) We return to the above terminal wealth problem, but now in the presence of proportional transaction costs or, equivalently, differing bid and ask prices. As in Example 1.71 we consider two assets (*bond* and *stock*) with prices 1 and $S(t)$, respectively. We suppose that $(1+\varepsilon)S(t)$ must be paid for buying one share of stock at time t, while only $(1-\varepsilon)S(t)$ is received for selling one share. For simplicity we assume ε to be fixed.

The investor enters the market with initial endowment v_0, entirely invested in bonds. If $\varphi(t)$ denotes the investor's number of shares as before, the wealth invested in bonds evolves according to

$$X_0^{(\varphi)}(t) = v_0 - \sum_{s=1}^{t} \Big(\Delta\varphi(s)S(s-1) - \varepsilon|\Delta\varphi(s)|S(s-1) \Big),$$

while the wealth in stock amounts to

$$X_1^{(\varphi)}(t) = \varphi(t)S(t) = \sum_{s=1}^{t} \Delta\varphi(s)S(s-1) + \varphi \bullet S(t). \tag{1.101}$$

The second equality in (1.101) follows from the integration by parts rule (1.14). The investor's goal is to maximise expected utility of terminal wealth after liquidation of the stock position, i.e. to maximise

$$E\left(u\big(X_0^{(\varphi)}(T) + X_1^{(\varphi)}(T) - \varepsilon|X_1^{(\varphi)}(T)|\big) \right) \tag{1.102}$$

over all predictable processes φ starting at 0 and such that the expectation (1.102) exists and is finite. In contrast to Example 1.71, we consider the increments $\Delta\varphi(t)$ as controls and set $\mathscr{A} = \{\Delta\varphi \text{ predictable} : \Delta\varphi(0) = 0 \text{ and (1.102) exists and is finite}\}$.

In order to express the increments $\Delta X_1^{(\varphi)}(t)$ in terms of $\Delta\varphi(t)$ rather than both $\Delta\varphi(t)$ and $\varphi(t)$, observe that $\varphi(t) = X_1^{(\varphi)}(t)/S(t)$, which yields

$$\begin{aligned}
\Delta X_1^{(\varphi)}(t) &= \Delta\varphi(t)S(t-1) + \varphi(t)\Delta S(t) \\
&= \Delta\varphi(t)S(t-1) + (\varphi(t-1) + \Delta\varphi(t))\,\Delta S(t) \\
&= \Delta\varphi(t)S(t-1) + \left(\frac{X_1^{(\varphi)}(t-1)}{S(t-1)} + \Delta\varphi(t) \right) \Delta S(t).
\end{aligned}$$

Similarly as in Examples 1.71 and 1.74, the process Y in Theorem 1.69 can be related to a minimisation problem. More specifically, we obtain the following counterpart of Theorem 1.72.

Theorem 1.77

1. *Suppose that Y_0 is a martingale, $\widetilde{S}$ a $[(1-\varepsilon)S, (1+\varepsilon)S]$-valued process and $\Delta\varphi^\star \in \mathscr{A}$ such that $Y_0\widetilde{S}$ is a martingale,*

$$Y_0(T) = u'(v_0 + \varphi^\star \bullet \widetilde{S}(T)), \tag{1.103}$$

and

$$X_0^{(\varphi^\star)}(T) + X_1^{(\varphi^\star)}(T) - \varepsilon |X_1^{(\varphi^\star)}(T)| = v_0 + \varphi^\star \bullet \widetilde{S}(T). \tag{1.104}$$

Then $Y := (Y_0, \frac{Y_0 \widetilde{S}}{S})$ *and* $\Delta\varphi^\star$ *satisfy the conditions of Theorem 1.69 in the relaxed version of Remark 1.70(1). In particular,* $\varphi^\star$ *maximises (1.102).*

2. *Suppose that* $Y = (Y_0, Y_1)$ *and the corresponding optimal control* $\Delta\varphi^\star \in \mathscr{A}$ *are as in Theorem 1.69/Remark 1.70(1). Set* $\widetilde{S} := Y_1 S / Y_0$*. Then* Y_0 *and* $Y_0 \widetilde{S} = Y_1 S$ *are martingales,* $\widetilde{S}$ *is* $[(1-\varepsilon)S, (1+\varepsilon)S]$*-valued, and (1.103, 1.104) hold.*
3. *Let* $Y = (Y_0, Y_1)$ *be as in Theorem 1.69/Remark 1.70(1). If* $\widetilde{u}$ *denotes the convex dual of* u *as in (1.94),* (Y_0, Y_1) *minimises*

$$(Z_0, Z_1) \mapsto E(\widetilde{u}(Z_0(T))) + v_0 Z_0(0) \tag{1.105}$$

among all pairs (Z_0, Z_1) *such that* Z_0 *and* $Z_1 S$ *are nonnegative martingales and* Z_1/Z_0 *is* $[1-\varepsilon, 1+\varepsilon]$*-valued.*

4. *The optimal values of the primal maximisation problem (1.102) and the dual minimisation problem (1.105) coincide in the sense that*

$$E\Big(u\big(X_0^{(\varphi^\star)}(T) + X_1^{(\varphi^\star)}(T) - \varepsilon |X_1^{(\varphi^\star)}(T)|\big)\Big) = E(\widetilde{u}(Y_0(T))) + v_0 Y_0(0),$$

where $Y = (Y_0, Y_1)$ *and* $\Delta\varphi^\star \in \mathscr{A}$ *are processes as in Theorem 1.69 (provided that they exist).*

Proof

1. *Step 1:* Firstly, we verify that

$$X_0^{(\varphi^\star)}(T) + X_1^{(\varphi^\star)}(T) - \varepsilon |X_1^{(\varphi^\star)}(T)| \le v_0 + \varphi^\star \bullet \widetilde{S}(T) \tag{1.106}$$

for any adapted $[(1-\varepsilon)S, (1+\varepsilon)S]$-valued process $\widetilde{S}$. Indeed,

$$\begin{aligned}
&X_0^{(\varphi^\star)}(T) + X_1^{(\varphi^\star)}(T) - \varepsilon |X_1^{(\varphi^\star)}(T)| \\
&= v_0 - \sum_{t=1}^{T} \Delta\varphi^\star(t) S(t-1)(1 - \operatorname{sgn}(\Delta\varphi^\star(t))\varepsilon) \\
&\quad + \varphi^\star(T) S(T)(1 - \operatorname{sgn}(\varphi^\star(T))\varepsilon) \\
&\le v_0 - \sum_{t=1}^{T} \Delta\varphi^\star(t) \widetilde{S}(t-1) + \varphi^\star(T)\widetilde{S}(T) \\
&= v_0 - \widetilde{S}_- \bullet \varphi^\star(T) + \varphi^\star(T)\widetilde{S}(T) \\
&= v_0 + \varphi^\star \bullet \widetilde{S}(T),
\end{aligned}$$

where we used integration by parts for the last step. The above calculation also shows that equality holds in (1.106) if and only if

$$\Delta\varphi^\star(t)\begin{cases} > 0 & \text{only if } \widetilde{S}(t-1) = (1-\varepsilon)S(t-1), \\ < 0 & \text{only if } \widetilde{S}(t-1) = (1+\varepsilon)S(t-1) \end{cases} \tag{1.107}$$

for $t = 1, \dots, T$ and

$$\varphi^\star(T)\begin{cases} > 0 & \text{only if } \widetilde{S}(T) = (1-\varepsilon)S(T), \\ < 0 & \text{only if } \widetilde{S}(T) = (1+\varepsilon)S(T). \end{cases} \tag{1.108}$$

Step 2: Since $\varphi(t) = \Delta\varphi(t) + X_1^{(\varphi)}(t-1)/S(t-1)$, straightforward calculations yield that the Hamiltonian equals

$$\begin{aligned} H(t, (x_0, x_1), a) = {} & Y_0(t-1)S(t-1)(-a - |a|\varepsilon) \\ & + Y_1(t-1)a\big(S(t-1) + a^S(t)\big) \\ & + a\big(c^{Y_1,S}(t) + a^{Y_1}(t)S(t-1)\big) \\ & + \big(Y_1(t-1)a^S(t) + \widetilde{c}^{Y_1,S}(t)\big)\frac{x_1}{S(t-1)} \end{aligned} \tag{1.109}$$

in this setup.
Step 3: Note that $Y_1 = Y_0\widetilde{S}/S$ has values in $[(1-\varepsilon)Y_0, (1+\varepsilon)Y_0]$ and

$$Y_1(T) = \frac{\widetilde{S}(T)}{S(T)}Y_0(T) = \begin{cases} (1-\varepsilon)Y_0(T) & \text{if } X_1^{(\varphi^\star)}(T) > 0, \\ (1+\varepsilon)Y_0(T) & \text{if } X_1^{(\varphi^\star)}(T) < 0 \end{cases} \tag{1.110}$$

by (1.108).
Step 4: For the application of Theorem 1.69 we need to consider the function $g(x_0, x_1) := u(x_0 + x_1 - |x_1|\varepsilon)$. It is differentiable in x_0 with derivative $u'(x_0+x_1-|x_1|\varepsilon)$ and in $x_1 \neq 0$ with derivative $u'(x_0+x_1-|x_1|\varepsilon)(1-\operatorname{sgn}(x_1)\varepsilon)$. Let $\partial_1 g(x_0, 0)$ denote the subdifferential of the concave function $x_1 \mapsto g(x_0, x_1)$ in $x_1 = 0$. It is easy to see that $y_1 \in \partial_1 g(x_0, 0)$ holds if and only if $y_1 \in [u'(x_0 + x_1 - |x_1|\varepsilon)(1-\varepsilon), u'(x_0 + x_1 - |x_1|\varepsilon)(1+\varepsilon)]$. Together with (1.110) it follows that $Y_0(T)$ and $Y_1(T)$ satisfy the conditions in (1.87) resp. Remark 1.70(1).
Step 5: Since $a^{Y_0} = 0$ and $H(t, (x_0, x_1), a)$ does not depend on x_0, we obtain the first component of the vector-valued equation (1.86). From the martingale property of $Y_0\widetilde{S} = Y_1 S$ we conclude $0 = a^{Y_1 S}(t) = Y_1(t-1)a^S(t) + \widetilde{c}^{Y_1,S}(t) + S(t-1)a^{Y_1}(t)$, which implies $a^{Y_1}(t) = -(Y_1(t-1)a^S(t) + \widetilde{c}^{Y_1,S}(t))/S(t-1)$. This yields the second component of (1.86).
Step 6: Equation (1.107) can be rephrased as

$$\Delta\varphi^\star(t)\begin{cases} > 0 & \text{only if } Y_1(t-1) = (1-\varepsilon)Y_0(t-1), \\ < 0 & \text{only if } Y_1(t-1) = (1+\varepsilon)Y_0(t-1). \end{cases} \tag{1.111}$$

In view of (1.109), we obtain condition 4 in Theorem 1.69.

2. *Step 1:* The local martingale property of the process Y_0 follows from (1.86) because $H(z,(x_0,x_1),a)$ does not depend on x_0. The second component of (1.86) yields $a^{Y_1}(t) = -(Y_1(t-1)a^S(t) + \widetilde{c}^{Y_1,S}(t))/S(t-1)$ and hence $a^{Y_1S}(t) = Y_1(t-1)a^S(t) + \widetilde{c}^{Y_1,S}(t) + S(t-1)a^{Y_1}(t) = 0$. Therefore, $Y_0\widetilde{S} = Y_1S$ is a local martingale as well. Y_0, $Y_0\widetilde{S}$ are in fact martingales because they are nonnegative, cf. Proposition 1.8 and the following sentence.
   Step 2: Condition 4 in Theorem 1.69 together with (1.109) yields that $\widetilde{S}/S = Y_1/Y_0$ must have values in $[1-\varepsilon, 1+\varepsilon]$ and that (1.111) holds, which in turn implies (1.107).
   Step 3: From (1.87) and the arguments in step 4 above we conclude
$$Y_0(T) = X_0^{(\varphi^\star)}(T) + X_1^{(\varphi^\star)}(T) - \varepsilon|X_1^{(\varphi^\star)}(T)|$$
   and (1.110), which in turn implies (1.108). From step 1 of the first part of the proof we obtain (1.103, 1.104).
3. For (Z_0, Z_1) as in statement 3 of Theorem 1.69 set $S^Z := SZ_1/Z_0$. From step 1 of the first part of the proof and integration by parts we obtain
$$\begin{aligned} v_0 + \varphi^\star \bullet \widetilde{S}(T) &= v_0 - \sum_{t=1}^{T} \Delta\varphi^\star(t)\widetilde{S}(t-1) + \varphi^\star(T)\widetilde{S}(T) \\ &\leq v_0 - \sum_{t=1}^{T} \Delta\varphi^\star(t)S^Z(t-1) + \varphi^\star(T)S^Z(T) \\ &= v_0 + \varphi^\star \bullet S^Z(T). \end{aligned}$$
   Moreover, $E(\varphi^\star \bullet S^Z(T)Z_0(T)) = 0$ follows from the martingale property of $S^Z Z_0 = Z_1 S$ similarly as in the proof of statement 3 of Theorem 1.72. Together we obtain
$$\begin{aligned} &E(\widetilde{u}(Z_0(T))) + v_0 Z_0(0) \\ &= E\Big(\sup_{x\in\mathbb{R}}(u(x) - xZ_0(T))\Big) + v_0 Z_0(0) \\ &\geq E\big(u(v_0 + \varphi^\star \bullet \widetilde{S}(T)) - (v_0 + \varphi^\star \bullet \widetilde{S}(T))Z_0(T)\big) + v_0 Z_0(0) \qquad (1.112) \\ &\geq E\big(u(v_0 + \varphi^\star \bullet \widetilde{S}(T)) - (v_0 + \varphi^\star \bullet S^Z(T))Z_0(T)\big) + v_0 Z_0(0) \qquad (1.113) \\ &= E\Big(u\big(X_0^{(\varphi^\star)}(T) + X_1^{(\varphi^\star)}(T) - \varepsilon|X_1^{(\varphi^\star)}(T)|\big)\Big) \end{aligned}$$
   for $\widetilde{S}$ from statement 2 if $E(\widetilde{u}(Z_0(T))) < \infty$. For $(Z_0, Z_1) = (Y_0, Y_1)$ equality holds in (1.112, 1.113) by (1.103) and $\widetilde{S} = S^Z$.
4. This follows from the proof of statement 3. □

Remark 1.78 Consider a frictionless price process $\widetilde{S}$ without transaction costs and satisfying $(1-\varepsilon)S(t) \leq \widetilde{S}(t) \leq (1+\varepsilon)S(t)$ at any time t. Since this asset can be bought cheaper and sold more expensively than the original S, trading $\widetilde{S}$ should lead to higher terminal wealth and hence expected utility. Suppose now that the utility-maximising investment $\varphi^\star$ in the market $\widetilde{S}$ only buys shares when $\widetilde{S} = (1+\varepsilon)S$ and sells when $\widetilde{S} = (1-\varepsilon)S$. Put differently, its terminal liquidation wealth in both markets coincides; it amounts to $v_0 + \varphi^\star \bullet \widetilde{S}$. Then $\varphi^\star$ must *a fortiori* be optimal in the market S because the maximal expected utility in the latter is certainly bounded above by that in the frictionless alternative $\widetilde{S}$.

This is precisely the economical interpretation of the process in the previous theorem. It represents a fictitious frictionless asset such that the optimal portfolio, its terminal wealth and hence expected utility coincide with that of the original market with transaction costs. Indeed, by Theorem 1.72 the martingale property of Y and $Y\widetilde{S}$ together with (1.103) yield optimality of $\varphi^\star$ for the frictionless terminal wealth problem of Example 1.71 corresponding to $\widetilde{S}$ instead of S.

As a concrete example for Theorems 1.69 and 1.72 we reconsider the portfolio problem of Example 1.48. The related problems in Examples 1.49, 1.64, 1.65 can be treated similarly.

Example 1.79 (Logarithmic Utility of Terminal Wealth) We study Example 1.71 for $u = \log$ as in Example 1.48. In order to apply Theorem 1.72 we need a candidate optimal control $\varphi^\star$ and a candidate process Y with terminal value $Y(T) = u'(V_{\varphi^\star}(T)) = 1/V_{\varphi^\star}(T)$. We boldly guess that

$$Y(t) := 1/V_{\varphi^\star}(t), \quad t = 0, \dots, T \tag{1.114}$$

may do the job for some $\varphi^\star \in \mathscr{A}$ that is yet to be determined. In order to simplify computations, we write

$$\varphi^\star(t) = \pi^\star(t)\frac{V_\varphi(t-1)}{S(t-1)}$$

similarly as in (1.46). Observe that the increments of the compensators of Y and YS equal

$$\begin{aligned}
a^Y(t) &= E(Y(t) - Y(t-1)|\mathscr{F}_{t-1})\\
&= E\left(\frac{1}{V_{\varphi^\star}(t-1)(1+\pi^\star(t)\Delta X(t))} - Y(t-1)\middle|\mathscr{F}_{t-1}\right)\\
&= -Y(t-1)\pi^\star(t)E\left(\frac{\Delta X(t)}{1+\pi^\star(t)\Delta X(t)}\middle|\mathscr{F}_{t-1}\right)
\end{aligned} \tag{1.115}$$

and

$$
\begin{aligned}
a^{YS}(t) &= E(Y(t)S(t) - Y(t-1)S(t-1)|\mathscr{F}_{t-1}) \\
&= Y(t-1)S(t-1)E\left(\frac{1+\Delta X(t)}{1+\pi^\star(t)\Delta X(t)} - 1\middle|\mathscr{F}_{t-1}\right) \\
&= Y(t-1)S(t-1)\big(1-\pi^\star(t)\big)E\left(\frac{\Delta X(t)}{1+\pi^\star(t)\Delta X(t)}\middle|\mathscr{F}_{t-1}\right).
\end{aligned} \tag{1.116}
$$

In order for Y and YS to be martingales, we want both (1.115) and (1.116) to vanish, which naturally leads to the condition

$$
E\left(\frac{\Delta X(t)}{1+\pi^\star(t)\Delta X(t)}\middle|\mathscr{F}_{t-1}\right) = 0 \tag{1.117}
$$

for the optimal fraction $\pi^\star(t)$ of wealth. In view of Theorem 1.72 and Remark 1.73(2), we have found the optimal investment strategy.

How does this relate to Example 1.48 where $\pi^\star(t)$ was assumed to be the maximiser of (1.48)? Unless the maximum is attained at the boundary, the first-order condition for the optimiser is precisely (1.117) if we may exchange differentiation and integration. If the maximiser of (1.48) is at the boundary of the set of fractions γ satisfying $1+\gamma\Delta X(t) \geq 0$ almost surely, (1.117) may fail to have a solution $\varphi^\star(t)$. This means that the present approach does not lead to the optimal portfolio in this case.

1.5.4.1 The Markovian Situation

How are the present approach and in particular the process Y related to dynamic programming? In general this may not be obvious but in a Markovian framework there is a natural candidate, namely the derivative of the value function evaluated at the optimal controlled process. More specifically, we will argue informally that one can often choose

$$
Y(t) = \nabla_x v(t, X^{(\alpha^\star)}(t)), \tag{1.118}
$$

where v denotes the value function of Sect. 1.5.3, $\nabla_x v(t, x_0)$ the gradient of $x \mapsto v(t,x)$ at any x_0, and $\alpha^\star \in \mathscr{A}$ an optimal control.

To this end, suppose that the following holds.

1. The objective function g is concave and the state space A of admissible controls is convex.
2. We are in the setup of Sect. 1.5.3 and Theorem 1.60. In particular, $v : \{0,\dots,T\} \times \mathbb{R}^d \to \mathbb{R}$ denotes the value function of the problem. Slightly more specifically than in Theorem 1.60, we assume that the supremum in (1.77)

is attained in some $a^\star(t,x)$ such that the optimal control satisfies $\alpha^\star(t) = a^\star(t, X^{(\alpha^\star)}(t))$ in line with (1.78).
3. The value function $v(t,x)$ is differentiable in x for $t = 0, \dots, T$.
4. There exist random variables $U(t)$, $t = 1, \dots, T$, with values in some space Γ, and a function $\delta : \{1, \dots, T\} \times \mathbb{R}^d \times A \times \Gamma \to \mathbb{R}^d$ such that

$$\Delta X^{(\alpha)}(t) = \delta(t, X^{(\alpha)}(t-1), \alpha(t), U(t)). \tag{1.119}$$

Moreover, $(x,a) \mapsto \delta(t,x,a,u)$ is assumed to be differentiable with partial derivatives denoted as $\nabla_x \delta(t,x,a,u)$ and $\nabla_a \delta(t,x,a,u)$, respectively.

Equation (1.119) means that, conditionally on $\mathscr{F}_{t-1}$, the increments $\Delta X^{(\alpha)}(t)$ of all controlled processes $X^{(\alpha)}$, $\alpha \in \mathscr{A}$ can be represented as a smooth function of the same random variable $U(t)$. This assumption is crucial for the derivation below.
5. Set

$$H(t,x,a) = \int \nabla_x v\Big(t, X^{(\alpha^\star)}(t-1) + \delta\big(t, X^{(\alpha^\star)}(t-1), \alpha^\star(t), u\big)\Big)^\top \\ \times \delta(t,x,a,u) P^{U(t)|\mathscr{F}_{t-1}}(du) + f(t,x,a) \tag{1.120}$$

and

$$h(t,x) = \sup_{a \in A} H(t,x,a).$$

We assume that $(x,a) \mapsto H(t,x,a)$ is concave. Alternatively, we assume that $x \mapsto h(t,x)$ is differentiable and concave, $a \mapsto H(t,x,a)$ is concave, and $\nabla_x H(t,x,a^\star(t,x)) = \nabla_x h(t,x)$ for $t = 1, \dots, T$ and any x. The concavity of $a \mapsto H(t,x,a)$ holds, for example, if $x \mapsto \nabla_x v(t,x)$ has nonnegative components and $a \mapsto \delta(t,x,a,u)$, $a \mapsto f(t,x,a)$ are concave.
6. Integrals are finite and derivatives and integrals are exchangeable whenever needed.

Define the process Y by (1.118). We show that it has the properties required in Theorem 1.69 resp. Remark 1.70(2).

Step 1: Observe that

$$\sum_{i=1}^d Y_i(t-1) a^{X_i^{(\alpha)}}(t) + \widetilde{c}^{Y_i, X_i^{(\alpha)}}(t) + f\big(t, X^{(\alpha)}(t-1), \alpha(t)\big)$$
$$= \nabla_x v\big(t-1, X^{(\alpha^\star)}(t-1)\big)^\top E\Big(X^{(\alpha)}(t) - X^{(\alpha)}(t-1)\Big|\mathscr{F}_{t-1}\Big)$$
$$+ E\Big(\Big(\nabla_x v\big(t, X^{(\alpha^\star)}(t)\big) - \nabla_x v\big(t-1, X^{(\alpha^\star)}(t-1)\big)\Big)^\top$$

$$
\begin{aligned}
&\times \Big(X^{(\alpha)}(t) - X^{(\alpha)}(t-1)\Big)\Big|\mathscr{F}_{t-1}\Big) \\
&+ f\big(t, X^{(\alpha)}(t-1), \alpha(t)\big) \\
&= E\Big(\nabla_x v\big(t, X^{(\alpha^\star)}(t)\big)^\top \big(X^{(\alpha)}(t) - X^{(\alpha)}(t-1)\big)\Big|\mathscr{F}_{t-1}\Big) \\
&+ f\big(t, X^{(\alpha)}(t-1), \alpha(t)\big) \\
&= \int \nabla_x v\Big(t, X^{(\alpha^\star)}(t-1) + \delta\big(t, X^{(\alpha^\star)}(t-1), \alpha^\star(t), u\big)\Big)^\top \\
&\times \delta\big(t, X^{(\alpha)}(t-1), \alpha(t), u\big) P^{U(t)|\mathscr{F}_{t-1}}(du) \\
&+ f\big(t, X^{(\alpha)}(t-1), \alpha(t)\big) \\
&= H(t, X^{(\alpha)}(t-1), \alpha(t))
\end{aligned}
$$

in line with (1.85). Differentiation of (1.120) yields

$$
\begin{aligned}
\nabla_x H(t,x,a) = \sum_{i=1}^{d} \int & \partial_{x_i} v\Big(t, X^{(\alpha^\star)}(t-1) + \delta\big(t, X^{(\alpha^\star)}(t-1), \alpha^\star(t), u\big)\Big) \\
& \times \nabla_x \delta_i(t,x,a,u) P^{U(t)|\mathscr{F}_{t-1}}(du) + \nabla_x f(t,x,a). \qquad (1.121)
\end{aligned}
$$

Step 2: Recall from (1.79) that

$$
\begin{aligned}
v(t-1,x) &= \sup_{a \in A}\Big(f(t,x,a) + \int v(t,x+y)\kappa(a,t,x,dy)\Big) \\
&= f(t,x,a^\star(t,x)) + \int v(t,x+y)\kappa(a^\star(t,x),t,x,dy),
\end{aligned}
$$

which reads as

$$
v(t-1,x) = \sup_{a \in A}\Big(f(t,x,a) + \int v(t,x+\delta(t,x,a,u))P^{U(t)|\mathscr{F}_{t-1}}(du)\Big) \qquad (1.122)
$$

$$
= f(t,x,a^\star(t,x)) + \int v\big(t, x + \delta(t,x,a^\star(t,x),u)\big)P^{U(t)|\mathscr{F}_{t-1}}(du) \qquad (1.123)
$$

in the present notation. We obtain

$$\begin{aligned}
a^Y(t) &= E\big(Y(t) - Y(t-1)\big|\mathscr{F}_{t-1}\big)\\
&= E\Big(\nabla_x v\Big(t, X^{(\alpha^\star)}(t-1) + \delta\big(t, X^{(\alpha^\star)}(t-1), \alpha^\star(t), U(t)\big)\Big)\Big|\mathscr{F}_{t-1}\Big)\\
&\quad - \nabla_x v\big(t-1, X^{(\alpha^\star)}(t-1)\big)\\
&= \int \nabla_x v\Big(t, X^{(\alpha^\star)}(t-1) + \delta\big(t, X^{(\alpha^\star)}(t-1), \alpha^\star(t), u\big)\Big) P^{U(t)|\mathscr{F}_{t-1}}(du)\\
&\quad - \nabla_x f(t, X^{(\alpha^\star)}(t-1), \alpha^\star(t))\\
&\quad - \sum_{i=1}^d \int \partial_{x_i} v\Big(t, X^{(\alpha^\star)}(t-1) + \delta\big(t, X^{(\alpha^\star)}(t-1), \alpha^\star(t), u\big)\Big)\\
&\quad \times \Big(e_i + \nabla_x \delta_i\big(t, X^{(\alpha^\star)}(t-1), \alpha^\star(t), u\big)\Big) P^{U(t)|\mathscr{F}_{t-1}}(du)\\
&= -\nabla_x f(t, X^{(\alpha^\star)}(t-1), \alpha^\star(t))\\
&\quad - \sum_{i=1}^d \int \partial_{x_i} v\Big(t, X^{(\alpha^\star)}(t-1) + \delta\big(t, X^{(\alpha^\star)}(t-1), \alpha^\star(t), u\big)\Big)\\
&\quad \times \nabla_x \delta_i\big(t, X^{(\alpha^\star)}(t-1), \alpha^\star(t), u\big) P^{U(t)|\mathscr{F}_{t-1}}(du)\\
&= -\nabla_x H\big(t, X^{(\alpha^\star)}(t-1), \alpha^\star(t)\big),
\end{aligned}$$

where we used the derivative of (1.123) for the third equality, (1.121) for the last, and $e_i \in \mathbb{R}^d$ denotes the ith unit vector. This yields (1.86) while (1.87) is obvious from the definition of Y.

Step 3: It remains to be shown that $\alpha^\star(t)$ maximises $a \mapsto H(t, X^{(\alpha^\star)}(t-1), a)$. Since this function is concave, it suffices to verify that its derivative vanishes in $\alpha^\star(t)$, i.e.

$$\begin{aligned}
0 = &\int \sum_{i=1}^d \partial_{x_i} v\Big(t, X^{(\alpha^\star)}(t-1) + \delta\big(t, X^{(\alpha^\star)}(t-1), \alpha^\star(t), u\big)\Big)\\
&\times \nabla_a \delta_i(t, X^{(\alpha^\star)}(t-1), \alpha^\star(t), u) P^{U(t)|\mathscr{F}_{t-1}}(du))\\
&+ \nabla_a f(t, X^{(\alpha^\star)}(t-1), \alpha^\star(t)).
\end{aligned} \tag{1.124}$$

On the other hand, the first-order condition for the maximiser $a^\star(t,x)$ in (1.122) is

$$0 = \nabla_a f(t,x,a^\star(t,x))$$
$$+ \int \sum_{i=1}^d \partial_{x_i} v\big(t, x + \delta(t,x,a^\star(t,x),u)\big) \nabla_a \delta_i\big(t,x,a^\star(t,x),u\big) P^{U(t)|\mathscr{F}_{t-1}}(du).$$

Since $\alpha^\star(t) = a^\star(t, X^{(\alpha^\star)}(t-1))$, we obtain (1.124) as desired. □

Example 1.80 In order to illustrate the above statement, let us reconsider Examples 1.48, 1.79. Comparing (1.49) and (1.114) shows that Y is indeed obtained by taking the derivative of the value function at the optimal control.

1.6 Deterministic Calculus

This section considers continuous deterministic rather than discrete stochastic calculus. It can be skipped altogether. It just serves as motivation and as a reference for the generalised notions and concepts of continuous-time stochastic calculus.

1.6.1 Constant Growth

The time-varying state of a system is often described as a real- or vector-valued continuous function $X(t)$ of time t. In relatively simple situations the state $X(t)$ is of *constant growth* in the sense that the increment $X(t) - X(s)$ depends only on $t - s$ but not explicitly on s and t. If we assume in addition $X(0) = 0$, this class of functions coincides of course with the family of *linear functions*. They are generally of the form $X(t) = bt$ with some real number or vector b which represents the growth rate of X. The parameter b suffices to characterise the entire function $X(t)$. The random counterpart of linear functions is treated in the next chapter on *Lévy processes*.

Sometimes the state of a system is of constant growth in a multiplicative rather than in an additive sense, meaning that the ratio $X(t)/X(s)$ (or $(X_1(t)/X_1(s), \dots, X_d(t)/X_d(s))$ for $\mathbb{R}^d$-valued processes) instead of the difference depends only on the length $t - s$ of the interval. If we assume $X(0) = 1$ (resp. $X(0) = (1,\dots,1)$), these functions match those of the form $X(t) = e^{bt}$ (or $X(t) = (e^{b_1 t}, \dots, e^{b_d t})$ in the vector-valued case). Put differently, *constant relative growth* leads to exponentials of linear functions. *Exponential Lévy processes* as a random analogue are considered in Sect. 3.7.

1.6.2 Integration

In many real-life situations the state grows with approximately constant rate on a small but not on a large scale. Analysis provides two concepts to study such phenomena which behave locally but not globally linear, namely integration and differentiation. Integration can be seen as a means to add small pieces of linear functions $b(t)dt$ in order to obtain a quite arbitrary function

$$X(t) = X(0) + \int_0^t b(s)ds. \tag{1.125}$$

This Riemann or more generally Lebesgue integral can be generalised to the Stieltjes integral $\int_0^t b(s)dY(s)$, which for differentiable functions Y means nothing else than $\int_0^t b(s)\dot{Y}(s)ds$ with $\dot{Y}(s) = \frac{dY(s)}{ds}$. In the sequel we occasionally use the shorthand notation

$$b \bullet Y(t) := \int_0^t b(s)dY(s).$$

The *stochastic integral* in Chap. 3 extends the Stieltjes integral in a rather non-trivial way to the random case.

The rule of *integration by parts* for differentiable functions

$$\int_0^t X(s)\dot{Y}(s)ds = X(t)Y(t) - X(0)Y(0) - \int_0^t \dot{X}(s)Y(s)ds$$

can be expressed conveniently with Stieltjes integrals as

$$\begin{aligned} X(t)Y(t) &= X(0)Y(0) + \int_0^t X(s)dY(s) + \int_0^t Y(s)dX(s) \\ &= X(0)Y(0) + X \bullet Y(t) + Y \bullet X(t). \end{aligned} \tag{1.126}$$

Its random generalisation is stated in Theorem 3.15.

Similarly, we may write the *chain rule*

$$\frac{df(X(t))}{dt} = f'(X(t))\dot{X}(t)$$

as

$$f(X(t)) - f(X(0)) = \int_0^t \frac{df(X(s))}{ds}ds = \int_0^t f'(X(s))\dot{X}(s)ds$$

and hence

$$\begin{aligned} f(X(t)) &= f(X(0)) + \int_0^t f'(X(s))dX(s) \\ &= f(X(0)) + f'(X) \bullet X(t) \end{aligned} \tag{1.127}$$

in terms of Stieltjes integrals. The random counterpart of (1.127) is called *Itō's formula*, cf. Theorem 3.16. If X is an $\mathbb{R}^d$-valued function and $f : \mathbb{R}^d \to \mathbb{R}$, we have accordingly

$$\frac{df(X(t))}{dt} = \sum_{i=1}^{d} D_i f(X(t))\dot{X}_i(t),$$

which leads to

$$\begin{aligned} f(X(t)) &= f(X(0)) + \sum_{i=1}^{d} \int_0^t D_i f(X(s))dX_i(s) \\ &= f(X(0)) + \sum_{i=1}^{d} D_i f(X) \bullet X_i(t). \end{aligned}$$

If the growth rate $b(s)$ in (1.125) depends on the state of the system itself, we obtain an integral equation of the form

$$X(t) = X(0) + \int_0^t \beta(X(s))ds \tag{1.128}$$

where $X(0)$ and the function β are given but X is unknown in the first place. Under regularity conditions on β, such an equation has a unique solution. If β is a linear or more generally an affine function, (1.128) can be solved explicitly, leading to

$$X(t) = X(0)e^{bt} \tag{1.129}$$

for $\beta(x) = bx$ and

$$X(t) = X(0)e^{b_1 t} + \frac{b_0}{b_1}\left(e^{b_1 t} - 1\right) \tag{1.130}$$

for $\beta(x) = b_0 + b_1 x$.

Integral equations could also be driven by some arbitrary function Y instead of time, i.e. we consider an equation of Stieltjes type

$$X(t) = X(0) + \int_0^t \beta(X(s))dY(s) \tag{1.131}$$

where $X(0)$ and the functions β, Y are given. These functions may also be $\mathbb{R}^d$-valued, in which case we interpret the integral as a sum of integrals

$$\int_0^t \beta(X(s))dY(s) = \sum_{i=1}^d \int_0^t \beta_i(X(s))dY_i(s).$$

In the univariate linear case $\beta(x) = x$, (1.131) is solved by

$$X(t) = X(0)e^{Y(t)-Y(0)}.$$

More generally, an affine equation

$$X(t) = \int_0^t X(s)dY(s) + Z(t)$$

with two drivers Y, Z is solved by

$$\begin{aligned} X(t) &= e^{Y(t)-Y(0)}\left(Z(0) + \int_0^t \frac{1}{\exp(Y(s)-Y(0))}dZ(s)\right) \\ &= e^{Y(t)-Y(0)}Z(0) + \int_0^t e^{Y(t)-Y(s)}dZ(s). \end{aligned}$$

Stochastic extensions of such integral equations are treated thoroughly in Chap. 3.

1.6.3 *Differentiation*

Integration assembles a nonlinear function from many bits of linear functions. Differentiation works the other way around. It starts with a given nonlinear function and decomposes it into small approximately linear pieces. Specifically, the derivative

$$b^X(t) := \frac{dX(t)}{dt} \tag{1.132}$$

indicates that the function $X(t)$ behaves locally around t as a linear function with growth rate $b^X(t)$. Moreover, a function X is linear if and only if its derivative is constant, in which case the latter equals the growth rate b^X of X. These ideas are generalised to stochastic processes in Chap. 4.

Contrary to the deterministic case, differentiation in the sense of Chap. 4 and stochastic integration do *not* turn out to be simple converses of each other.

Nevertheless, we will encounter a rule corresponding to differentiation of Stieltjes integrals. The derivative of $Y(t) = \int_0^t \varphi(s)dX(s)$ is obviously given by

$$b^Y(t) = \varphi(t)b^X(t).$$

In our notation the chain rule for $Y(t) = f(X(t))$ reads as

$$b^Y(t) = f'(X(t))b^X(t)$$

or, for $\mathbb{R}^d$-valued X and $f : \mathbb{R}^d \to \mathbb{R}^n$, as

$$b^{Y_j}(t) = \sum_{i=1}^{d} D_i f_j(X(t))b^{X_i}(t). \tag{1.133}$$

In the situation $Z(t) = X(Y(t))$ with some smooth increasing function $Y : \mathbb{R}_+ \to \mathbb{R}_+$, we could also write the chain rule as

$$b^Z(t) = b^X(Y(t))\dot{Y}(t). \tag{1.134}$$

We strangely mix notations here in order to motivate a rule on random time changes in Sect. 4.2.

Functions in applications are often defined in terms of their derivative rather than explicitly. In our notation such *ordinary differential equations (ODEs)* read as

$$b^X(t) = \beta(X(t)), \quad X(0) = x_0 \tag{1.135}$$

with given β, x_0 and unknown X. The random counterparts of such equations are considered in Sects. 5.3 and 4.6. For constant β, (1.135) obviously leads to linear functions. In the linear case $\beta(x) = bx$ we have (1.129) whereas an affine function $\beta(x) = b_0 + b_1 x$ yields (1.130). Of course, linear (resp. affine) integral equations of the form (1.128) are just another way of writing linear (resp. affine) ODEs, which explains why the solutions coincide. Surprisingly, this connection ceases to hold in the random case. The stochastic generalisation of

$$X(t) = x + \int_0^t (b_0 + b_1 X(s))ds$$

in Sect. 3.9 leads to a different class of processes than that of affine ODEs

$$b^X(t) = b_0 + b_1 X(t), \quad X(0) = x.$$

The latter give rise to so-called affine processes, which are treated in Chap. 6.

Appendix 1: Problems

The exercises correspond to the section with the same number.

1.1 Denote by Z the density process of $Q \sim P$. Show that $1/Z$ is the density process of P relative to Q.

1.2 For adapted processes X, Y set $Z := \mathscr{E}(X)(Y(0) + \mathscr{E}(X)^{-1} \bullet Y)$. Show that Z solves the equation $Z = Y + Z_- \bullet X$.

1.3 For an adapted process X define $\widetilde{X} := \mathscr{L}(e^X)$. Show that $\widetilde{X}$ is a random walk if and only if X is a random walk. In this case determine the law of $\Delta\widetilde{X}(1)$ in terms of the law of $\Delta X(1)$.

1.4 Suppose that $M \in \mathbb{R}^{n\times n}$ is a **stochastic matrix**, i.e. $M_{ij} \geq 0$ and $\sum_{k=1}^n M_{ik} = 1$ for $i, j = 1, \dots, n$. Moreover, let X be a **Markov chain** with finite state space $E = \{x_1, \dots, x_n\}$ and transition matrix M, i.e. an E-valued process such that

1. $P(X(t+1) = y_{t+1}|X(0) = y_0, \dots, X(t) = y_t) = P(X(t+1) = y_{t+1}| X(t) = y_t)$ for any $t \in \mathbb{N}$ and any $y_0, \dots, y_{t+1} \in E$ such that $P(X(0) = y_0, \dots, X(t) = y_t) > 0$,
2. $P(X(t+1) = x_j|X(t) = x_i) = M_{ij}$ for any $t \in \mathbb{N}$, $i, j = 1, \dots, n$.

Show that

1. X is a Markov process relative to the filtration generated by X,
2. its transition function p_t and its generator G satisfy $p_t f = M^t f$ and $Gf = (M - 1)f$ if we identity functions $f : E \to \mathbb{R}$ with vectors $(f(x_1), \dots, f(x_n)) \in \mathbb{R}^n$ and $1 \in \mathbb{R}^{n\times n}$ denotes the identity matrix,
3. the adjoint operator A of the generator G satisfies $A\mu = G^\top\mu = (M-1)^\top\mu$ if we identify measures μ on E with vectors $(\mu(\{x_1\}), \dots, \mu(\{x_n\}) \in \mathbb{R}^n$ and likewise linear mappings $\mu : B(E) \to \mathbb{R}$ with $(\mu(1_{\{x_1\}}), \dots, \mu(1_{\{x_n\}})) \in \mathbb{R}^n$.

1.5 Elisabeth wants to sell her house within T days. She receives daily offers which are assumed to be independent random variables that are uniformly distributed on $[m, M]$. Elisabeth has the option of recalling earlier offers and consider them again but she must pay maintenance costs $c > 0$ for every day the house remains unsold. Determine the optimal time to sell, i.e. the stopping time maximising her expected reward from the sale.

Hint: Try the ansatz that the value function is of the form

$$v(t,x) = \begin{cases} x - c(t-1) & \text{for } x \geq \underline{x}, \\ \widetilde{v}(t,x) - ct & \text{for } x < \underline{x}, \end{cases}$$

where x represents the maximal offer so far, $\frac{\partial}{\partial x}\widetilde{v}(t,x) \leq 1$, and the threshold $\underline{x} \in [m, M]$ is chosen such that $x \mapsto v(t,x)$ is continuous in $\underline{x}$.

1.6 For $x, b_0 \in \mathbb{R}^2$ and $b_1 \in \mathbb{R}^{2\times 2}$ determine the function $X : \mathbb{R}_+ \to \mathbb{R}^2$ with $b^X(t) = b_0 + b_1 X(t)$, where b^X is defined as in (1.132).

Appendix 2: Notes and Comments

The material in this chapter is mostly classical. For an introduction to probability theory including martingales and discrete-time Markov processes see, for example, [153, 275]. Some results in Sects. 1.1–1.3 are discrete-time versions of statements from the general theory in [152, 154, 238]. Many additional references can be found in these texts. The presentation of Sect. 1.4 is based on the parallel more subtle results in Chap. 5. For stochastic optimal control in discrete time see [18, 271] and the references therein. However, we consider a non-Markovian framework similarly as in [96]. Moreover, the exposition here tries to mimic the continuous-time theory of Chap. 7 as much as possible. Section 1.6 presents standard results from calculus in stochastic process notation.

For early solutions to the portfolio problems in Examples 1.48, 1.49, 1.64, 1.65 see [222, 258]. A history on quadratic hedging in the martingale case of Example 1.50 and beyond can be found in [270]. For a nice short introduction to optimal stopping we refer to [203]. The perpetual American put is treated in [277]. For the background of Example 1.58 we refer to [102]. Proposition 1.59 is based on [135, 249]. The dual approach to optimal investment in Examples 1.71, 1.74 is inspired by more general characterisations in [188, 197] but the idea is already present in [27]. The extension of the dual approach underlying Example 1.76 goes back to [65, 161]. Example 1.79 is a special case of the results in [125]. The relationship (1.118) has been stated in [28] in a Brownian motion framework. Problem 1.5 is a slight modification of [271, Example 1.34].

Except for the few examples in Sect. 1.5, we do not discuss Mathematical Finance in discrete time. Classical references include [117, 237, 278].

Chapter 2
Lévy Processes

The continuous-time analogue of a random walk is called a Lévy process. One may also view Lévy processes as the stochastic counterpart of linear functions. Both viewpoints illustrate that these processes play a fundamental role. Before we look at them in detail in Sects. 2.2–2.5, we introduce some notions from the *general theory of stochastic processes* in Sect. 2.1.

2.1 Processes, Stopping Times, Martingales

This section corresponds directly to its discrete counterpart in Chap. 1. As in that case we work on a fixed **filtered probability space** $(\Omega, \mathscr{F}, \mathbf{F}, P)$. The **filtration** $\mathbf{F} = (\mathscr{F}_t)_{t\geq 0}$ is now indexed by continuous time $\mathbb{R}_+$ rather than discrete time $\mathbb{N}$. Since information grows with time, we assume $\mathscr{F}_s \subset \mathscr{F}_t$ for $s \leq t$ as before. Moreover, the filtration is assumed to be **right-continuous** in the sense that

$$\mathscr{F}_s = \bigcap_{t>s} \mathscr{F}_t, \quad s \geq 0.$$

Intuitively, this means that no new information arrives *immediately* after s. One imposes this rather technical condition in order to obtain a more streamlined theory.

Collections of real- or vector-valued random variables $X = (X(t))_{t\geq 0}$ are again called **stochastic processes**. Only the index set $\mathbb{N}$ is now replaced by $\mathbb{R}_+$. As before, we sometimes view them as real- or vector-valued functions $X(\omega, t)$ of two arguments.

Alternatively, they can be seen as single random variables whose values $X(\omega)$ are functions on the real line. The process X is called **continuous, left-continuous** etc. if the **paths** $X(\omega) : \mathbb{R}_+ \to \mathbb{R}$ (resp. $X(\omega) : \mathbb{R}_+ \to \mathbb{R}^d$) have this property. Later we consider mostly **càdlàg** processes, i.e. processes whose paths are right-continuous and have finite left-hand limits. The French acronym càdlàg stands for *continue à*

© Springer Nature Switzerland AG 2019

E. Eberlein, J. Kallsen, *Mathematical Finance*, Springer Finance,
https://doi.org/10.1007/978-3-030-26106-1_2

droite avec des limites à gauche. If X is a càdlàg process, its left-hand limits

$$X_-(t) := X(t-) := \begin{cases} \lim_{s \uparrow t} X(s) & \text{if } t > 0, \\ X(0) & \text{if } t = 0 \end{cases}$$

and its jumps

$$\Delta X(t) := X(t) - X(t-)$$

are well defined. Since they are defined for any t, we consider them as stochastic processes $X_- = (X_-(t))_{t \geq 0}$, $\Delta X = (\Delta X(t))_{t \geq 0}$ on their own. Observe that X_- is a **càglàd** process, i.e. it is left-continuous and has finite limits from the right.

Adaptedness (to the given filtration $\mathbf{F}$) is defined as in discrete time, i.e. $X(t)$ is supposed to be $\mathscr{F}_t$-measurable. The minimal filtration $\mathbf{F}$ making a given process X adapted is again called the **filtration generated by** X or **natural filtration of** X. Here, however, we must use a slightly different formula than (1.5) because the latter may not be right-continuous for some processes. Instead we have

$$\mathscr{F}_t = \bigcap_{u > t} \sigma(X(s) : s \leq u)$$

in continuous time.

For the purposes of stochastic integration we need the notion of *predictability* as well. As before we want to formalise that $X(t)$ is known slightly ahead of t. Coming from discrete time one is tempted to claim that $X(t)$ must be $\mathscr{F}_{t-}$-measurable for any t, where

$$\mathscr{F}_{t-} := \sigma\left(\bigcup_{s < t} \mathscr{F}_s\right)$$

stands for the information up to but excluding t. However, this definition turns out to be too weak for the theory of stochastic integration.

The "right" concept of predictability is more involved. One starts by defining the smallest σ-field $\mathscr{P}$ on $\Omega \times \mathbb{R}_+$ such that all left-continuous, adapted, real-valued processes are $\mathscr{P}$-measurable (as mappings $X : \Omega \times \mathbb{R}_+ \to \mathbb{R}$). In a second step one calls arbitrary processes X **predictable** if they are $\mathscr{P}$-measurable (again viewed as mappings $X : \Omega \times \mathbb{R}_+ \to \mathbb{R}$ or $\mathbb{R}^d$).

It follows immediately that left-continuous, adapted processes such as left-limits X_- of adapted càdlàg processes are predictable. In particular, any continuous, adapted process is predictable. This partly explains why one can do without this concept if one works only with continuous processes. On the other hand, general predictable processes may be very far from being left-continuous. For intuitive purposes it usually suffices to think of predictable processes as being known a "nanosecond in advance."

There are more concepts of measurability in the general theory of stochastic processes. For example, X is called **measurable** if it is measurable as a mapping $X : \Omega \times \mathbb{R}_+ \to \mathbb{R}$ (resp. $\mathbb{R}^d$) relative to the σ-field $\mathscr{F} \otimes \mathscr{B}$. It is **progressive(ly measurable)** if $X|_{[0,t]}$—viewed as a mapping $\Omega \times [0, t] \to \mathbb{R}$ (resp. $\mathbb{R}^d$)—is $\mathscr{F}_t \times \mathscr{B}([0, t])$-measurable for any $t \geq 0$. Finally, X is **optional** if it is measurable with respect to the *optional* σ-field $\mathscr{O}$ on $\Omega \times \mathbb{R}_+$, which is generated by all right-continuous, adapted, real-valued processes. One can show that predictable processes are optional, optional processes are progressive, and progressive processes are adapted and measurable, cf. [75, Theorem IV.67].

However, we will not need these concepts in the remainder of this book.

Random times τ that are consistent with the information structure are called **stopping times** as in discrete time. Here we have to use the formal definition

$$\{\tau \leq t\} \in \mathscr{F}_t, \quad t \geq 0$$

because the weaker requirement $\{\tau = t\} \in \mathscr{F}_t$ does not suffice for the theory in continuous time. In any case, one can think of stopping times as alarm clocks that are triggered by some observable information. As in discrete time a natural example is the first time when some adapted process enters a given set. We skip the difficult proof of this seemingly simple statement.

Proposition 2.1 *For any adapted càdlàg process X and any Borel set B,*

$$\tau := \inf\{t \geq 0 : X(t) \in B\}$$

is a stopping time.

Idea The statement is highly intuitive in spite of its nontrivial proof. If the objects correspond to their natural interpretation, the first time that an observable (i.e. adapted) process enters a given set should certainly be observable itself, i.e. a stopping time. For a proof, see [75, Theorem IV.50], and for related statements see [154, Section I.1.c]. □

For technical reasons it is often necessary to stop a process X at a stopping time τ. As in discrete time, *stopping* means to keep the process constant from τ onward, i.e. we define the **stopped process** X^τ as

$$X^\tau(t) := X(\tau \wedge t).$$

We occasionally need the σ-field representing the information up to some stopping time τ. As in discrete time it is defined as

$$\mathscr{F}_\tau := \big\{F \in \mathscr{F} : F \cap \{\tau \leq t\} \in \mathscr{F}_t \text{ for any } t \geq 0\big\}.$$

It has the same properties as in Proposition 1.3:

Proposition 2.2 *Let σ, τ, τ_n be stopping times.*

1. *$\mathscr{F}_\tau$ is a σ-field.*
2. *If $\tau = t$ is a constant stopping time, then $\mathscr{F}_\tau = \mathscr{F}_t$.*
3. *$\sigma \leq \tau$ implies that $\mathscr{F}_\sigma \subset \mathscr{F}_\tau$.*
4. *τ is $\mathscr{F}_\tau$-measurable.*
5. *Infima and suprema of finitely or countably many stopping times are again stopping times.*
6. *$\tau_n \downarrow \tau$ implies $\mathscr{F}_\tau = \bigcap_{n\in\mathbb{N}} \mathscr{F}_{\tau_n}$.*

Proof The proof of the first four statements is the same as for Proposition 1.3. For statement 5 see, for example, [186, Lemma 1.2.11 and Proposition 1.2.3]. The last statement is shown in [75, Theorem IV.56d]. □

The notation $X(\tau)$ resp. $X(\tau)1_{\{\tau<\infty\}}$ is used as in discrete time. We have the following counterpart of Proposition 1.4:

Proposition 2.3 *If X is an adapted càdlàg process and τ denotes a stopping time, $X(\tau)1_{\{\tau<\infty\}}$ is $\mathscr{F}_\tau$-measurable.*

Idea For stopping times with only countably many values this can be shown as in the proof of Lemma 1.4. The general statement follows from approximating τ from above by such stopping times, cf. [154, Proposition I.1.21]. □

Occasionally, we need a notation for the random interval between two stopping times. We set

$$]\!]\sigma, \tau]\!] := \{(\omega, t) \in \Omega \times \mathbb{R}_+ : \sigma(\omega) < t \leq \tau(\omega)\}$$

and accordingly $[\![\sigma, \tau]\!]$, $[\![\sigma, \tau[\![$, $]\!]\sigma, \tau[\![$.

Martingales play an even more important role than in discrete-time stochastic calculus. Similarly as in Chap. 1 we call an adapted càdlàg process X a **martingale** (resp. **submartingale**, **supermartingale**) if it is **integrable** (in the sense that $E(|X(t)|) < \infty$ for any t) and satisfies

$$E(X(t)|\mathscr{F}_s) = X(s) \quad (\text{resp. } \geq X(s), \leq X(s))$$

for $s \leq t$.

Remark 2.4 Above we assume the càdlàg property as part of the definition of martingales resp. sub-/supermartingales. If we omit this requirement, we can show that

1. any martingale X has a càdlàg *version*, i.e. there exists an adapted càdlàg process $\widetilde{X}$ such that $P(\widetilde{X}(t) = X(t)) = 1$ for any $t \geq 0$,
2. any sub-/supermartingale X has a càdlàg version if and only if $t \mapsto E(X(t))$ is right-continuous.

A proof can be found in [241, Theorem II.2.9]. It relies on martingale convergence theorems.

Any integrable random variable ξ naturally leads to a martingale via

$$X(t) = E(\xi|\mathscr{F}_t), \tag{2.1}$$

called again the **martingale generated by** ξ. Except for the càdlàg property this follows immediately from the law of iterated expectations. If the random variable generating the martingale is the density of some probability measure $Q \sim P$, we obtain again the **density process** of Q relative to P. We refer to Example 1.5, which extends without change to continuous time.

As in discrete time, the martingale property generalises to stopping times.

Theorem 2.5 (Doob's Stopping Theorem) *If X is a martingale, then $E(|X(\tau)|) < \infty$ and*

$$E(X(\tau)|\mathscr{F}_\sigma) = X(\sigma)$$

hold for any two bounded stopping times $\sigma \leq \tau$. If X is generated by some random variable $X(\infty)$, we need not require σ, τ to be bounded.

For a supermartingale X, we have accordingly $E(|X(\tau)|) < \infty$ and

$$E(X(\tau)|\mathscr{F}_\sigma) \leq X(\sigma)$$

for any two bounded stopping times $\sigma \leq \tau$.

Idea For stopping times with only finitely many values this can be reduced to Theorem 1.11. The general statement follows from approximation from above by such stopping times, cf. [76, Theorem VI.10]. □

The following technical lemma agrees verbatim with Lemma 1.7. It is needed only for the proof of Proposition 7.9.

Lemma 2.6 *Let X be a supermartingale, Y a martingale, $t \leq T$, and $F \in \mathscr{F}_t$ with $X(t) = Y(t)$ on F and $X(T) \geq Y(T)$. Then $X(s) = Y(s)$ on F for $t \leq s \leq T$. The statement continues to hold if we only require $X^T - X^t$, $Y^T - Y^t$ instead of X, Y to be a supermartingale resp. martingale.*

Proof This follows along the same lines as Lemma 1.7. □

As in discrete time, a fair/favourable/unfavourable game remains fair/favourable/ unfavourable if one stops playing at an observable time, cf. Proposition 1.9.

Proposition 2.7 *If X is a martingale (resp. submartingale, supermartingale), the same is true for X^τ if τ denotes an arbitrary stopping time.*

Proof We focus on the case when X is a supermartingale. The other cases follow by considering $-X$ or $\pm X$. Let $0 \leq s \leq t$. Since $X(\tau \wedge s)$ is $\mathscr{F}_s$-measurable by Propositions 2.2 and 2.3, we have

$$\begin{aligned} E\big(X^\tau(t)1_{\{\tau\leq s\}}\big|\mathscr{F}_s\big) &= E\big(X(\tau\wedge s)1_{\{\tau\leq s\}}\big|\mathscr{F}_s\big) \\ &= X(\tau\wedge s)1_{\{\tau\leq s\}} \\ &= X^\tau(s)1_{\{\tau\leq s\}}. \end{aligned}$$

Moreover, Theorem 2.5 yields

$$\begin{aligned} E\big(X^\tau(t)1_{\{\tau>s\}}\big|\mathscr{F}_s\big) &= E\big(X((\tau\wedge t)\vee s)1_{\{\tau>s\}}\big|\mathscr{F}_s\big) \\ &= E\big(X((\tau\wedge t)\vee s)\big|\mathscr{F}_s\big)1_{\{\tau>s\}} \\ &\leq X(s)1_{\{\tau>s\}} \\ &= X^\tau(s)1_{\{\tau>s\}}. \end{aligned}$$

The assertion follows from adding the two equations. □

Theorem 1.12 has a direct counterpart in continuous time.

Theorem 2.8 (Doob's Inequality) *We have*

$$E\Big(\sup_{s\leq t} X(s)^2\Big) \leq 4E\big(X(t)^2\big), \quad t\geq 0 \tag{2.2}$$

for any martingale X.

Idea This statement can be shown along the same lines as Theorem 1.12 or by discretising time and passing to the continuous-time limit, cf. [241, Theorem II.1.7]. □

For the purposes of stochastic calculus we often need a "local" counterpart of the martingale concept. There are in fact two such notions, namely *local martingales* defined below and *σ-martingales* in the next chapter. In discrete time they both reduce to *local martingales* in the sense of Chap. 1. However, neither of them relies on the intuitive equation (1.9), which does not make sense in continuous time. The way out is to introduce the concept of *localisation*, which was secretly applied in Proposition 1.10. Since it is used not only for martingales, we take a more general perspective.

A class $\mathscr{C}$ of processes is said to be **stable under stopping** if the stopped process X^τ belongs to $\mathscr{C}$ for any $X\in\mathscr{C}$. This property holds for most classes of processes considered in this book, such as martingales, sub-/supermartingales, bounded processes, processes of finite variation, etc.

We say that a process X belongs to the **localised class** $\mathscr{C}_{\text{loc}}$ if there exists a sequence of stopping times $(\tau_n)_{n\in\mathbb{N}}$, increasing to ∞ almost surely and such that for all n the stopped process X^{τ_n} belongs to $\mathscr{C}$.

If the set $\mathscr{C}$ is the class of martingales (or sub-/supermartingales, bounded processes, ...), we call the elements of $\mathscr{C}_{\text{loc}}$ **local martingales** (resp. **local sub-/supermartingales, locally bounded**, ...). Roughly speaking, one may say that a local martingale X coincides with a martingale up to an "arbitrarily large" stopping time τ_n. Nevertheless, X may behave quite differently from a martingale on a "global" scale. It needs some experience to get a good feeling for the difference between local and true martingales, which is best explained by a certain lack of uniform integrability. For the time being, it may suffice to think of a local martingale as almost being a martingale. The fact that not all local or, more generally, σ-martingales are martingales leads to many annoying problems in Mathematical Finance.

In Chap. 1 we observed the intuitive fact that any submartingale can be decomposed into a martingale and an increasing compensator representing the trend of the process. This statement has a continuous counterpart:

Theorem 2.9 (Doob–Meyer Decomposition) *Any submartingale (resp. supermartingale) X with $E(\sup_{s\in[0,t]}|X(s)|) < \infty$, $t \in \mathbb{R}_+$, can be uniquely decomposed as*

$$X = X(0) + M^X + A^X \tag{2.3}$$

*with some martingale M^X and some increasing (resp. decreasing) predictable càdlàg process A^X satisfying $M^X(0) = A^X(0) = 0$. The process A^X is called the **compensator** of X.*

Idea This statement can be proved by considering a sequence of discrete time sets $(t_n^{(k)})_{n\in\mathbb{N}}$, $k = 1, 2, \dots$ instead of $\mathbb{R}_+$, where the mesh size $\max\{|t_n^{(k)} - t_{n-1}^{(k)}| : n = 1, 2, \dots\}$ tends to 0 as $k \to \infty$. The hard part of the proof consists in showing that the discrete-time Doob decomposition corresponding to the grid $(t_n^{(k)})_{n\in\mathbb{N}}$ converges in an appropriate sense to a continuous-time decomposition of the required form, cf. [238, Theorem III.8]. □

Remark 2.10 The integrability condition can be weakened by requiring instead that X is of class (LD), see Sect. 4.5 for a definition.

Decompositions of this type play an important role in stochastic calculus, which is why one wants to extend them to more general classes of processes similar to Theorem 1.13. Contrary to the discrete case, predictability of the compensator does not suffice to ensure uniqueness of the decomposition. We need an additional path property, namely *finite variation*.

We say that a real- or vector-valued function α on $\mathbb{R}_+$ is of **finite variation** if its **total variation** on $[0, t]$, i.e.

$$\mathrm{Var}(\alpha)(t) := \sup\left\{\sum_{i=1}^{n} |\alpha(t_i) - \alpha(t_{i-1})| : n \in \mathbb{N}, 0 = t_0 < t_1 < \cdots < t_n = t\right\}, \tag{2.4}$$

is finite for any finite $t \geq 0$. It is not hard to show that a real-valued function α is of finite variation if and only if it can be written as a difference

$$\alpha = \beta - \gamma \tag{2.5}$$

of two increasing functions β, γ. Indeed, (2.5) implies

$$\mathrm{Var}(\alpha)(t) \leq \beta(t) + \gamma(t) < \infty.$$

Conversely, one may take $\beta := \mathrm{Var}(\alpha)$ if α is of finite variation.

In order to provide some intuition, suppose that $\alpha(t)$ denotes the location of a particle at time t. The total distance covered by the particle between 0 and t is finite if and only if α is of finite variation on $[0, t]$.

A process X is called of **finite variation** if its paths $X(\omega)$ have this property. In this case the **variation process** $\mathrm{Var}(X) = (\mathrm{Var}(X)(t))_{t\geq 0}$ is defined pathwise. If X is an adapted càdlàg process, so is $\mathrm{Var}(X)$. Note that we use the same notation for the total variation of a process X and for the variance of a random variable X. We hope that the respective meaning should be clear from the context.

Let us come back to extensions of Theorem 2.9. The following result ensures that there is at most one decomposition (2.3) of a given process X into a martingale and a predictable process of finite variation.

Proposition 2.11 *Any predictable local martingale X of finite variation is constant, i.e. $X(t) = X(0)$.*

Idea Since X is of finite variation, it can be written as a difference

$$X = X(0) + A - B$$

with increasing processes A, B satisfying $A(0) = 0 = B(0)$. One can show that A and B can be chosen to be predictable. Since it is increasing, B is a submartingale if it is integrable. Lacking integrability is taken care of by considering the stopped processes X^{τ_n} instead of X, where the stopping times τ_n are chosen such that X^{τ_n} is a sufficiently integrable martingale. Observe that B allows for two Doob–Meyer type decompositions

$$0 + B = B = (X - X(0)) + A.$$

By uniqueness of the Doob–Meyer decomposition we have $0 = X - X(0)$ and $B = A$, cf. Theorem 2.9. For a complete proof, see [154, Corollary I.3.16]. □

The previous proposition does not tell us which processes actually allow for a representation (2.3). The class of these processes—called *special semimartingales*—turns out to be quite large if we relax the martingale property of M^X slightly. More specifically, an adapted càdlàg process X is called a **special semimartingale** if it allows for a decomposition

$$X = X(0) + M^X + A^X \tag{2.6}$$

with some local martingale M^X and some predictable càdlàg process A^X of finite variation satisfying $M^X(0) = A^X(0) = 0$. The process A^X is again called the **compensator** of X. Equation (2.6) obviously generalises the Doob–Meyer decomposition. Since it is unique, it is called the **canonical decomposition** of X.

Physicists' Corner 2.12 On an intuitive level, (2.6) decomposes the increment

$$dX(t) := X(t) - X(t - dt)$$

in the infinitesimal time interval $(t - dt, t]$ into two parts, namely a predictable trend

$$dA^X(t) := E(dX(t)|\mathscr{F}_{t-dt}) \tag{2.7}$$

and a random deviation

$$dM^X(t) := dX(t) - E(dX(t)|\mathscr{F}_{t-dt})$$

from that trend. The equality

$$E(dM^X(t)|\mathscr{F}_{t-dt}) = 0 \tag{2.8}$$

means that M^X stays "constant on average in an infinitesimal period", which can be interpreted rigorously by saying that it is a local martingale. Note that (2.7–2.8) are only meant to provide some intuitive idea of (2.6). □

Let us briefly mention the local counterpart of Theorem 2.9.

Proposition 2.13 *A process X with integrable $X(0)$ is a local submartingale (resp. supermartingale) if and only if it is a special semimartingale with increasing (resp. decreasing) compensator A^X.*

Idea If (τ_n) denotes a localising sequence for the local submartingale X, Theorem 2.9 and Remark 2.10 yield $X^{\tau_n} = X^{\tau_n}(0) + M^{(n)} + A^{(n)}$ for some martingale $M^{(n)}$ and some increasing predictable process $A^{(n)}$. The uniqueness part actually allows us to show that $(M^{(n)})^{\tau_m} = M^{(m)}$ and $(A^{(n)})^{\tau_m} = A^{(m)}$ for $m \leq n$. Defining processes M^X, A^X via $M^X(t) = M^{(n)}(t)$, $A^X(t) = A^{(n)}(t)$ for $t \leq \tau_n$ we obtain the desired decomposition (2.6).

If, on the other hand, X is a special semimartingale with increasing A^X, choose a localising sequence (τ_n) for both M^X and A^X, which is possible by Propositions 2.2(5) and 2.7. Then

$$E\big(X^{\tau_n}(t) - X^{\tau_n}(s)\big|\mathscr{F}_s\big)$$
$$= E\Big((M^X)^{\tau_n}(t) - (M^X)^{\tau_n}(s)\Big|\mathscr{F}_s\Big) + E\Big((A^X)^{\tau_n}(t) - (A^X)^{\tau_n}(s)\Big|\mathscr{F}_s\Big) \geq 0$$

as desired.

For a more detailed argument we refer to [152, Théorème 2.19]. □

Special semimartingales in applications typically have a compensator which is absolutely continuous in time, i.e. it can be written as

$$A^X(t) = \int_0^t a^X(s)ds \tag{2.9}$$

with some predictable process a^X. The random variable $a^X(t)$ can be interpreted as the average **growth rate** of X at time t. This trend may vary randomly through time.

Most processes in concrete applications belong to the class of special semimartingales. However, this class is not closed under certain operations in stochastic calculus such as stochastic integration and application of smooth mappings, and change of measure. For mathematical reasons it is therefore desirable to base the theory on an even larger class of processes called *semimartingales*. At first glance, their definition looks very similar. A **semimartingale** is an adapted càdlàg process X allowing for a decomposition

$$X = X(0) + M + A \tag{2.10}$$

with some local martingale M and some adapted càdlàg process A of finite variation. Here the part of finite variation is no longer assumed to be predictable. Therefore uniqueness ceases to hold. The process A should not be called a trend component in this more general context because it may itself be a martingale. Whether or not a semimartingale is special can be interpreted as an integrability property of its big jumps. A sufficient condition is provided by the following

Proposition 2.14 *A semimartingale with bounded or at least locally bounded jumps is special.*

Idea Denote by $X = X(0) + M + A$ a semimartingale decomposition as in (2.10). Since A is of finite variation, it can be written as $A = B - C$ with increasing processes B, C. If these are integrable and hence submartingales, they allow for Doob–Meyer decompositions $B = M^B - A^B$ resp. $C = M^C - A^C$, and we obtain the special semimartingale decomposition

$$X = X(0) + (M + M^B - M^C) + (A^C - A^B)$$

of X. In general, however, B, C may fail to be integrable. But locally bounded jumps suffice to show that B, C are locally integrable, in which case the above argument works. For a complete proof, see [154, Lemma I.4.24 and Proposition I.4.25b]. □

Proposition 2.14 implies that any continuous semimartingale is special. It can be shown that the compensator A^X is also continuous in this case, cf. [154, Lemma I.4.24].

But as noted above, A^X is even continuous for most discontinuous processes in applications.

Discrete-time filtrations $(\mathscr{F}_t)_{t=0,1,2,...}$ and processes $(X(t))_{t=0,1,2,...}$ as in Chap. 1 can be naturally embedded in continuous time via

$$\mathscr{F}_t := \mathscr{F}_{[t]}, \quad X(t) := X([t])$$

for any $t \geq 0$, where $[t] := \max\{s \leq t : s \in \mathbb{N}\}$ denotes the integer part of t. In such continuous-time extensions of discrete models many concepts of stochastic calculus such as adaptedness, predictability, martingales, stopping times, stochastic integrals etc. reduce to their simple counterpart in Chap. 1.

In Mathematical Finance we often consider a finite time horizon $T \geq 0$, i.e. filtrations, processes etc. are defined only for $t \leq T$. Formally, this can be considered as a special case of the general framework in Sect. 2.1 by setting

$$\mathscr{F}_t := \mathscr{F}_{T\wedge t}, \quad X(t) := X(T \wedge t)$$

for $t \geq 0$, i.e. all filtrations and processes stay constant after the ultimate time horizon T. We will always identify finite horizon models with their extension on the whole positive real line. Consequently, there is no need to repeat the notions and results in this and later sections for finite time horizons.

2.2 Characterisation and Properties of Lévy Processes

If one wants to model concrete phenomena such as stock prices, one has to decide upon a reasonable class of stochastic processes. The driving forces behind such phenomena are often very complex or even unknown. Therefore theoretical considerations typically do not lead to a concrete process. A reasonable way out is to base models on simple processes having nice analytical properties and sharing the stylised features of real data. If the statistical evidence suggests, one may gradually move on to more complex models.

Lévy processes constitute a natural starting point because they represent constant growth. They can be viewed as the random counterpart to linear functions of time in deterministic calculus. Here, of course, *constant growth* has to be understood in a stochastic sense. In successive time intervals, the process does not necessarily change by the same amount but by the same kind of random experiment. This point of view leads to *random walks* in discrete time, where the increments $\Delta X(t), t = 1, 2, \dots$ are independent and identically distributed and hence "the same" in a stochastic sense. The independence makes sure that the random experiment really starts afresh in any period. In order to define the continuous-time analogue of a random walk, we fix a filtered probability space $(\Omega, \mathscr{F}, \mathbf{F}, P)$ as in the previous section.

A **Lévy process** or **process with independent, stationary increments** is an adapted càdlàg real- or vector-valued process X such that

1. $X(0) = 0$,
2. $X(t) - X(s)$ is independent of $\mathscr{F}_s$ for $s \leq t$ (*independent increments*),
3. the law of $X(t) - X(s)$ only depends on $t - s$ (*stationary increments*).

We call an adapted real- or vector-valued process X a **shifted Lévy process** if $X - X(0)$ is a Lévy process.

Properties 2 and 3 correspond directly to the i.i.d. increments of random walks, especially if the filtration is assumed to be generated by X. The càdlàg property of paths warrants the semimartingale property and rules out pathological cases. Even for deterministic functions $X(t)$, claiming that $X(t) - X(s)$ only depends on $t - s$ does not imply that X is linear unless, say, right-continuity is assumed.

Remark 2.15 The family $(P_t)_{t \geq 0}$ of laws $P_t := \mathscr{L}(X(t))$ of a Lévy process is called a **convolution semigroup** because it is a semigroup relative to the convolution product

$$(P_s * P_t)(B) = \int_B P_t(B - x) P_s(dx).$$

Indeed, we have

$$\begin{aligned} P_s * P_t &= \mathscr{L}(X(s)) * \mathscr{L}(X(s+t) - X(s)) \\ &= \mathscr{L}(X(s) + X(s+t) - X(s)) \\ &= P_{s+t} \end{aligned}$$

by stationarity and independence of the increments and because the convolution represents the law of the sum of independent random variables.

Linear functions as well as random walks are easily characterised by low- or at least lower-dimensional objects. Linear functions $X(t) = bt$ can be reduced to their constant growth rate b, which is determined by $X(1)$. Similarly, the law of $\Delta X(1)$ determines the distribution of a whole random walk X. Theorem 2.17 below provides a similar characterisation of an arbitrary Lévy process in terms of its *Lévy–Khintchine triplet*, which describes the characteristic function of the process uniquely. The characteristic function provides a particularly attractive representation of the law because it opens the door to studying random variables and processes from an analytical point of view.

Physicists' Corner 2.16 For readers who are inclined to informal intuitive derivations, we start with a very heuristic discussion of Lévy processes. Readers preferring rigorous statements should move on immediately to Theorem 2.17.

For simplicity we consider the univariate case. The idea is to derive the characteristic function

$$\varphi_{X(t)}(u) := E(\exp(iuX(t)))$$

by interpreting a Lévy process as a random walk on an infinitesimal time grid. To this end, we write $X(t)$ as a sum

$$X(t) = \sum_{n=1}^{\frac{t}{dt}} (X(n\,dt) - X((n-1)dt)$$

of increments in very short, virtually infinitesimal periods $((n-1)dt, n\,dt]$ of fixed length dt. By stationarity, these infinitesimal increments all share the same distribution, namely that of $X(dt) - X(0) = X(dt)$. Since the characteristic function of a sum of independent random variables is given by the product of the corresponding characteristic functions, we have

$$\varphi_{X(t)}(u) = \prod_{n=1}^{\frac{t}{dt}} \varphi_{X(n\,dt)-X((n-1)dt)}(u) = \left(\varphi_{X(dt)}(u)\right)^{\frac{t}{dt}}. \tag{2.11}$$

If the characteristic function of $X(dt)$ does not attain the value 0, we can write it as

$$\varphi_{X(dt)}(u) = \exp(\psi(u)dt)$$

with some complex-valued function ψ. Using (2.11) this leads to

$$\varphi_{X(t)}(u) = \exp(\psi(u)t).$$

We conclude that the logarithm of the characteristic function of $X(t)$ is linear in t if we liberally ignore the ambiguity of the complex logarithm. Let us have a closer look at $X(dt)$ in order to express the characteristic function more explicitly. Since $E(X(t)) = E(X(dt))\frac{t}{dt}$, the expected value of $X(dt)$ must have the order of magnitude dt, say $E(X(dt)) = b\,dt$. In the simplest case $X(dt)$ is deterministic and hence coincides with its expectation, i.e. we have $X(dt) = b\,dt$ and $\varphi_{X(dt)}(u) = \exp(iub\,dt)$. It follows that

$$\varphi_{X(t)}(u) = \exp(iubt)$$

or

$$X(t) = bt, \tag{2.12}$$

i.e. X is nothing else but a simple linear function, cf. Fig. 2.1.

Even if $X(dt)$ fails to be deterministic, we have

$$\mathrm{Var}(X(t)) = \sum_{n=1}^{\frac{t}{dt}} \mathrm{Var}\big(X(n\,dt) - X((n-1)dt)\big) = \mathrm{Var}(X(dt))\frac{t}{dt}$$

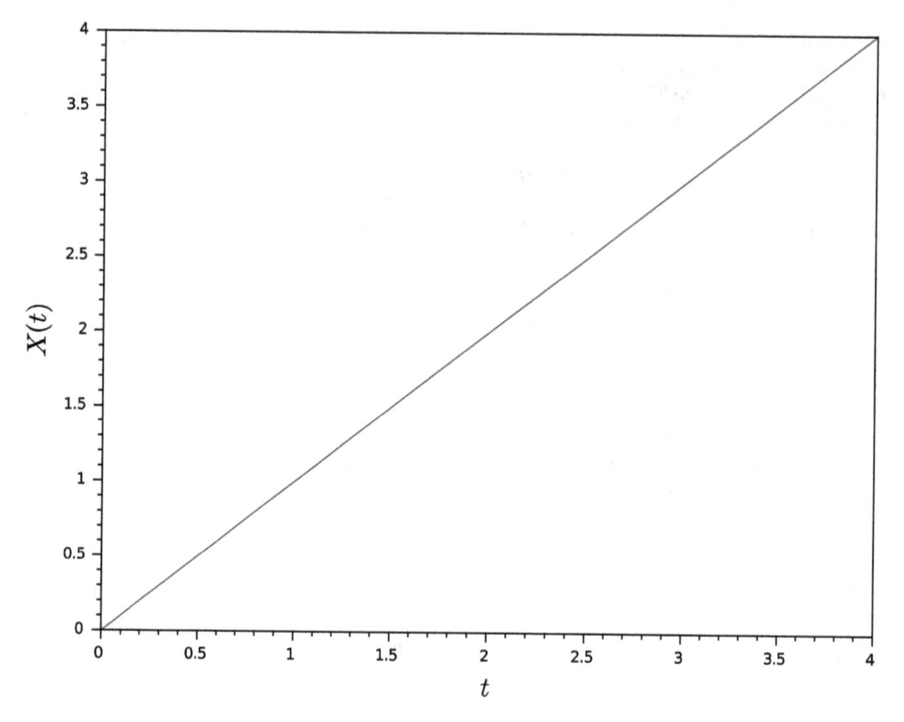

Fig. 2.1 Linear function

by independence of the increments. Therefore the variance of $X(dt)$ is of order dt as well. We will consider two different situations leading to a variance of this order. The starting point is a probability distribution Q with Fourier transform

$$\widehat{Q}(u) := \int e^{iux} Q(dx).$$

We want to use this measure as a candidate for the law of $X(dt)$. In order for the variance of $X(dt)$ to be of the right order, we have to adapt or rescale Q properly.

One may, for example, lower the probability of moving at all. We toss a coin which shows *heads* with small probability $\lambda\, dt$ for some $\lambda > 0$. Only if *heads* show up, X moves according to the law Q, otherwise it remains constant, cf. Fig. 2.2. The characteristic function of $X(dt)$ is given by

$$\varphi_{X(dt)}(u) = e^0(1 - \lambda\, dt) + \widehat{Q}(u)\lambda\, dt = 1 + (\widehat{Q}(u) - 1)\lambda\, dt$$

in this case. Since $e^x \approx 1 + x$ for small x, we can rephrase this as

$$\varphi_{X(dt)}(u) \approx \exp\big((\widehat{Q}(u) - 1)\lambda\, dt\big) = \exp\left(\int (e^{iux} - 1)\lambda Q(dx)\, dt\right),$$

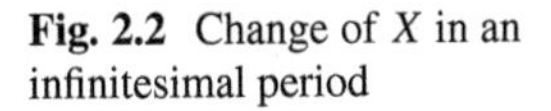

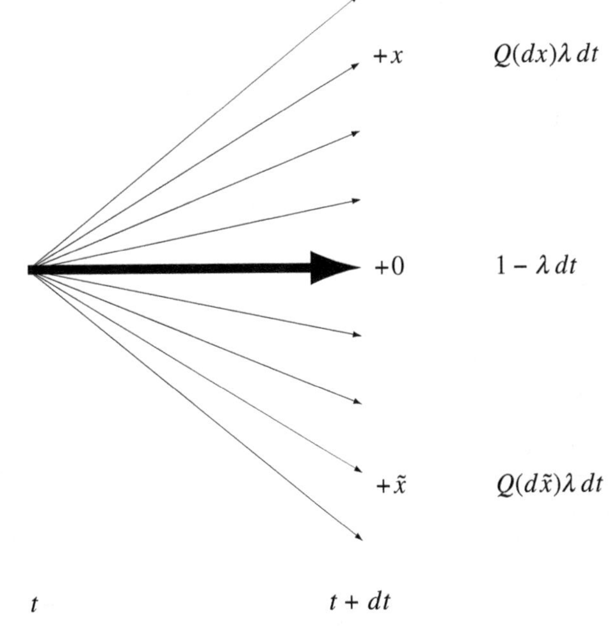

Fig. 2.2 Change of X in an infinitesimal period

which in turn leads to

$$\varphi_{X(t)}(u) = \exp\left(\int (e^{iux} - 1)\lambda Q(dx)t\right). \tag{2.13}$$

By construction, such a Lévy process jumps with probability $\lambda\, dt$ in small intervals of length dt, according to the *jump distribution* Q. Between jumps it remains constant. As a prime example consider a *Poisson process* counting, say, the arrival of customers or phone calls, cf. Fig. 2.3.

The *jump intensity* λ is given by the average number of calls in a unit time interval. Since the process moves only by jumps of size 1, we have $Q = \varepsilon_1$, i.e. the Dirac measure in 1. More generally, one calls Lévy processes satisfying (2.13) *compound Poisson processes*. Consider, for example, the cumulative claims put in at an insurance company, cf. Fig. 2.4. In this case Q stands for the distribution of the size of a randomly arriving claim and λ for the arrival rate of claims.

Another way of obtaining a variance of order dt for $X(dt)$ is to rescale Q, i.e. Q now stands for the distribution of $\frac{1}{\sqrt{dt}}X(dt)$, cf. Fig. 2.5. Rather than moving with *small probability*, we now reduce the *size of the movements*. This leads to

$$\varphi_{X(dt)}(u) = \widehat{Q}(u\sqrt{dt}).$$

For simplicity let us assume that the expectation of $X(dt)$ is 0, i.e. $\int xQ(dx) = 0$. Since $\widehat{Q}(0) = 1$ and $\widehat{Q}'(0) = i\int xQ(dx) = 0$, a second-order Taylor expansion yields

$$\varphi_{X(dt)}(u) \approx 1 + \frac{1}{2}\widehat{Q}''(0)u^2\, dt.$$

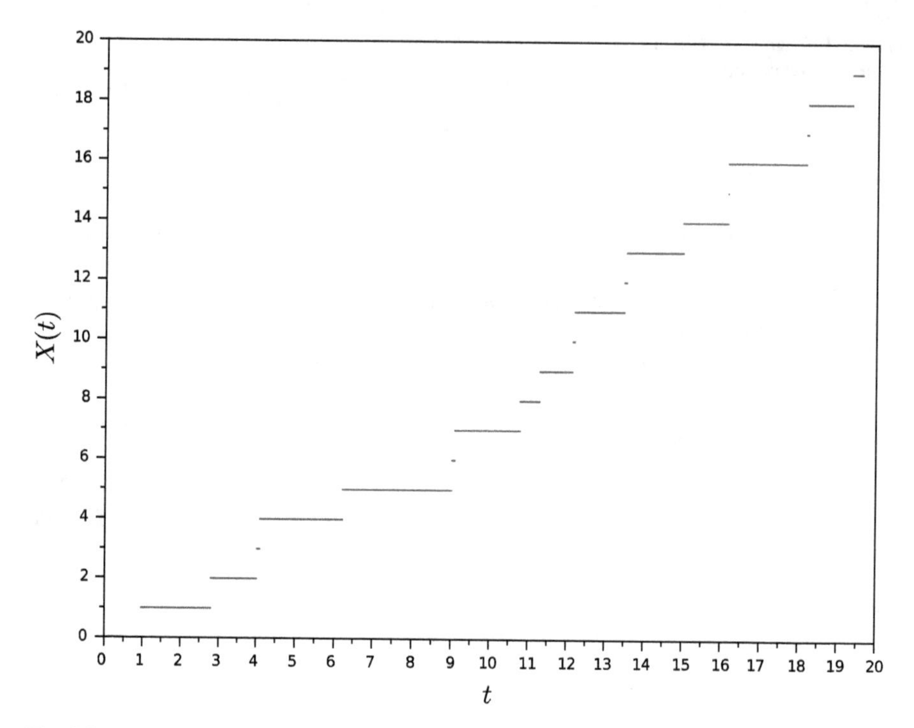

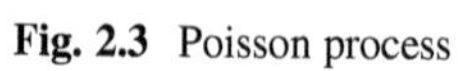
Fig. 2.3 Poisson process

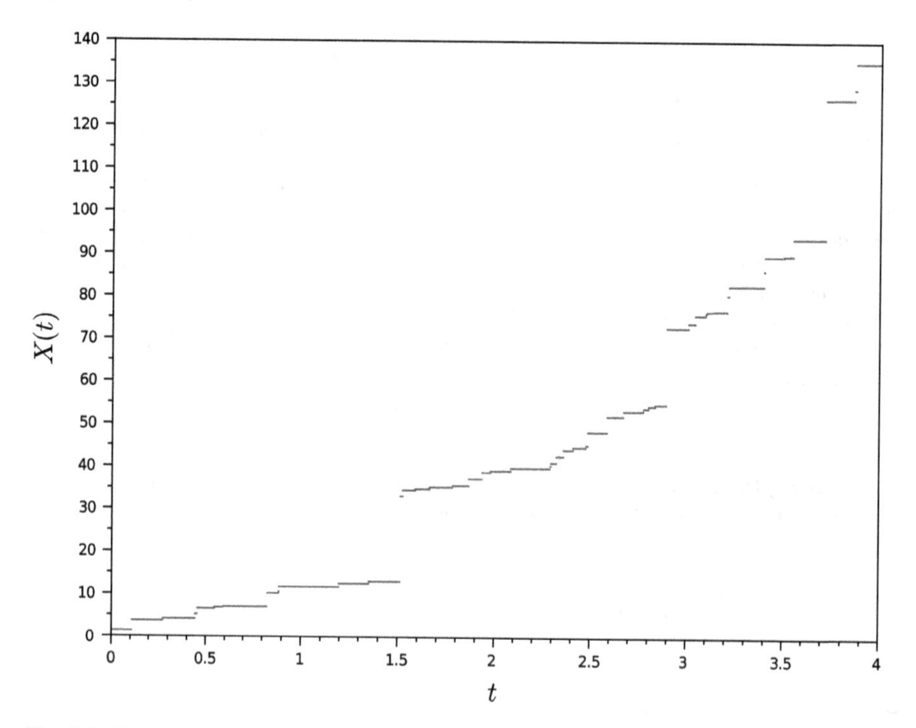

Fig. 2.4 Compound Poisson process

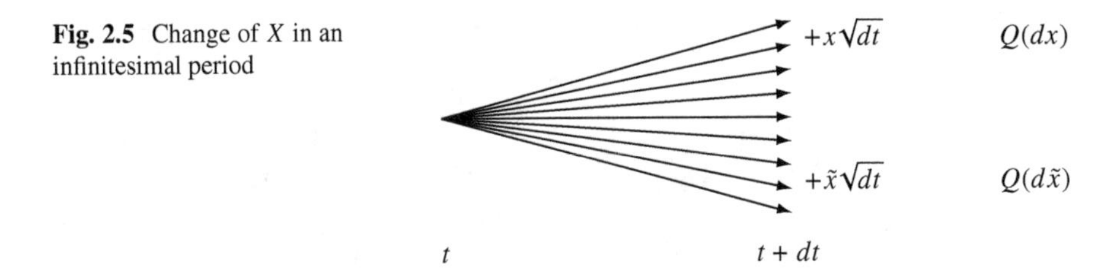

Fig. 2.5 Change of X in an infinitesimal period

Writing $c := -\widehat{Q}''(0)$ and using the approximation $e^x \approx 1 + x$, we obtain

$$\varphi_{X(dt)}(u) \approx \exp\Big(-\frac{1}{2}cu^2\,dt\Big)$$

or

$$\varphi_{X(t)}(u) = \exp\Big(-\frac{1}{2}cu^2 t\Big) \tag{2.14}$$

on a global scale. Observe that this characteristic function depends on Q only through its variance

$$\int x^2 Q(dx) = -Q''(0) = c.$$

The corresponding Lévy process behaves in a subtle way. By construction it does not change by rare jumps but by many small movements of order $\sqrt{dt}$. Since $\sqrt{dt}$ tends to 0 as $dt \to 0$, it has continuous paths, cf. Fig. 2.6. Differentiability, on the other hand, does not hold because $\sqrt{dt}$ vanishes more slowly than dt. As one can see from (2.14) or guess from Fig. 2.6, we have constructed *Brownian motion*, which is used to model all kinds of random phenomena such as particle movements, logarithmic asset prices etc.

Altogether, we have come across three types of Lévy processes, moving deterministically, by isolated jumps, or by continuous wiggling, respectively. Are there other types? The sum of independent Lévy processes as above is again a process of independent, stationary increments. We obtain its characteristic function by multiplying the respective characteristic functions (2.12, 2.13, 2.14):

$$\varphi_{X(t)}(u) = \exp\bigg(\Big(iub - \frac{1}{2}cu^2 + \int (e^{iux} - 1)\lambda Q(dx)\Big)t\bigg).$$

By construction, such a process can be viewed as the sum of a linear drift with slope b, a Brownian motion with variance coefficient c, and occasional jumps occurring with rate λ and distribution Q, cf. Fig. 2.7. Note that $b\,dt$ does not equal the expectation of $X(dt)$ any more because

$$E(X(dt)) = -i\varphi'_{X(dt)}(0) = \Big(b + \int x\lambda Q(dx)\Big)\,dt =: \widetilde{b}\,dt.$$

It turns out to be useful to introduce the finite measure $K := \lambda Q$. Using this new notation, we obtain the alternative representations

$$\varphi_{X(t)}(u) = \exp\bigg(\Big(iub - \frac{1}{2}cu^2 + \int (e^{iux} - 1)K(dx)\Big)t\bigg) \tag{2.15}$$

$$= \exp\bigg(\Big(iu\widetilde{b} - \frac{1}{2}cu^2 + \int (e^{iux} - 1 - iux)K(dx)\Big)t\bigg). \tag{2.16}$$

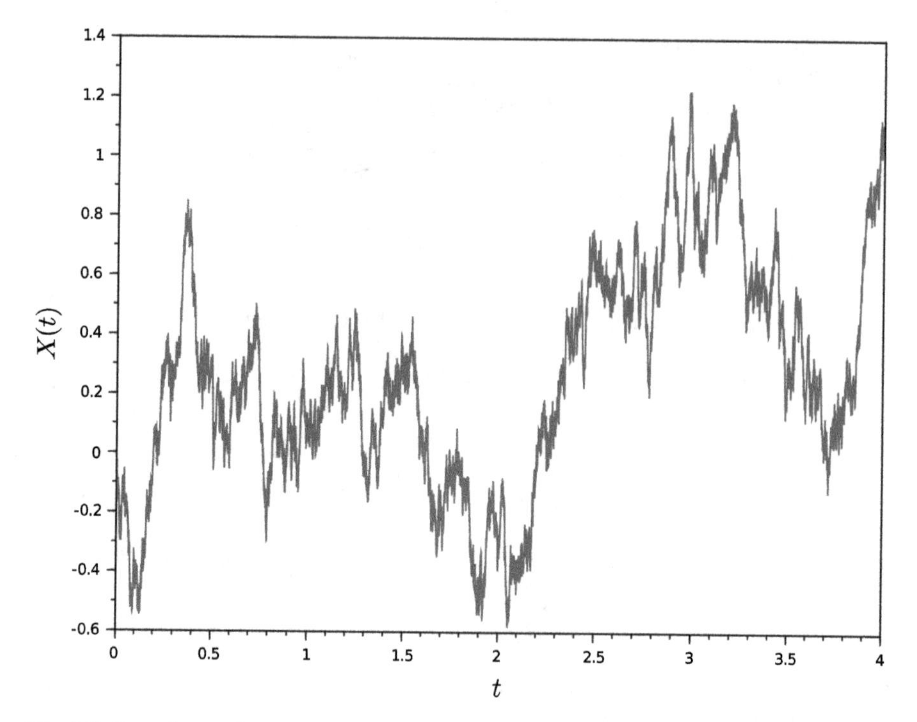

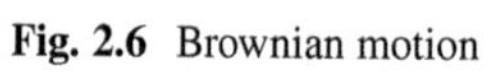
Fig. 2.6 Brownian motion

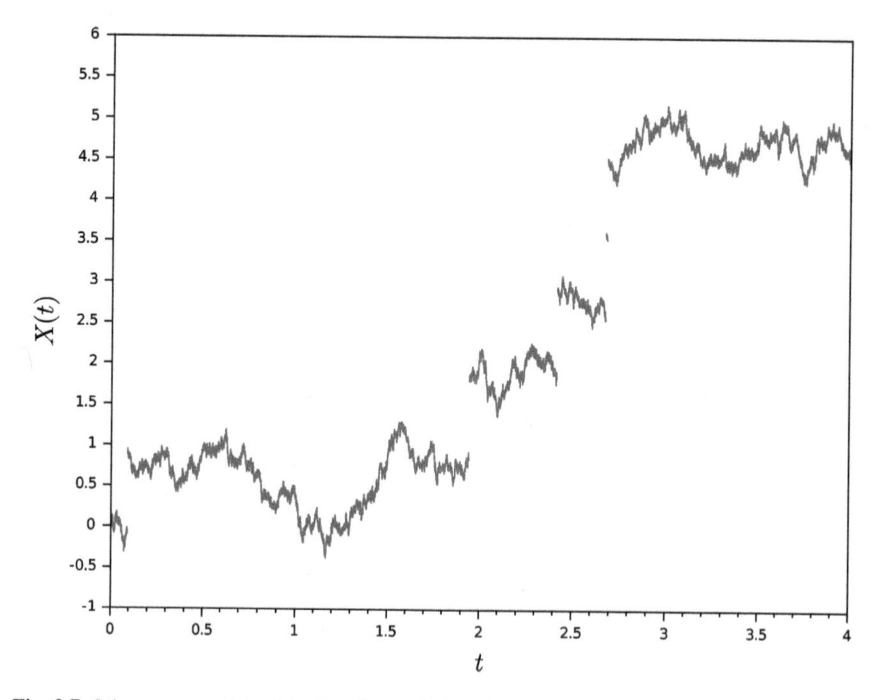

Fig. 2.7 Lévy process with triplet $b = 1$, $c = 1$, $\lambda = 1$, $Q = \varepsilon_1$

In order to cover all Lévy processes in one formula, we introduce yet another representation, which interpolates between (2.15) and (2.16). To this end, let h denote a *truncation function*, i.e. a bounded real-valued measurable function satisfying $h(x) = x$ for small x. A standard example is

$$h(x) = x1_{\{|x|\leq 1\}}$$

but the results are independent of its particular choice. If we set

$$b^h := b + \int h(x)K(dx),$$

(2.15) can be rewritten as

$$\varphi_{X(t)}(u) = \exp\Big(\Big(iub^h - \frac{1}{2}cu^2 + \int (e^{iux} - 1 - iuh(x))K(dx)\Big)t\Big). \tag{2.17}$$

One can show that any Lévy process allows for a representation of the form (2.17) if we extend the set of measures K. The minimal requirement for the integral to make sense for arbitrary u is that the *Lévy measure* K is a—possibly infinite—measure satisfying

$$\int (1 \wedge |x|^2)K(dx) < \infty.$$

How do Lévy processes corresponding to infinite Lévy measures behave? As one may guess, they exhibit infinitely many jumps in any finite time interval (cf. e.g. Fig. 2.13), which means that they interpolate in some sense between the compound Poisson and the Brownian motion case. We will discuss their properties in more detail later. Equation (2.17) is called the *Lévy–Khintchine formula*. We state it below in the general multivariate case. Since we motivated b as an expectation and c as a variance, it does not come as a surprise that they generalise to a vector and a covariance matrix, respectively. □

Theorem 2.17 (Lévy–Khintchine Formula) *Fix a* **truncation function** *h on $\mathbb{R}^d$, i.e. a bounded measurable function $h : \mathbb{R}^d \to \mathbb{R}^d$ satisfying $h(x) = x$ for small x. The characteristic function of any $\mathbb{R}^d$-valued Lévy process X is of the form*

$$\varphi_{X(t)}(u) = \exp(\psi(u)t), \quad u \in \mathbb{R}^d, \tag{2.18}$$

where the **characteristic exponent** *ψ allows for a unique representation*

$$\psi(u) = iu^\top b^h - \frac{1}{2}u^\top cu + \int \left(e^{iu^\top x} - 1 - iu^\top h(x)\right) K(dx) \tag{2.19}$$

with some **drift vector** *$b^h \in \mathbb{R}^d$, some symmetric, nonnegative definite* **diffusion matrix** *$c \in \mathbb{R}^{d\times d}$, and some* **Lévy measure** *K on $\mathbb{R}^d$ satisfying $\int(1\wedge|x|^2)K(dx) < \infty$ and $K(\{0\}) = 0$. We call the Lévy process X* **Brownian motion with drift** *if $K = 0$. The* **Lévy–Khintchine triplet** *(b^h, c, K) determines the law of X uniquely. Moreover, there exists a Lévy process for any such triplet.*

Idea Uniqueness of the law of $X(t)$ for fixed t holds because the characteristic function determines the law of a random variable uniquely. If we set $D_n := X(nt) - X((n-1)t)$, we have that $D_1, D_2, \ldots$ are independent and have the same law as $X(t)$. This means that the joint law of the sequence $D_1, D_2, \ldots$ is uniquely determined by ψ. Since the sequence $X(0), X(t), X(2t), \ldots$ is obtained as a cumulative sum of $D_1, D_2, \ldots$, its law is also determined by ψ. Since t can be chosen arbitrarily small, we have uniqueness of the law of the process on an arbitrarily fine grid, which can be used to conclude uniqueness of the law of the entire process.

In order to understand existence, we follow the informal reasoning of Physicists' Corner 2.16. The independent sum of a linear function, a Brownian motion, and a compound Poisson process yields a Lévy process having a characteristic function (2.18) with finite Lévy measure K. In order to obtain infinite jump intensity, one adds a whole sequence of independent compensated compound Poisson processes taking care of smaller and smaller jumps.

As a reference we mention [154, Corollary II.4.19 and Theorem II.5.2]. □

In particular, the law of X is determined by the law of $X(1)$.

Corollary 2.18 *Fix $t > 0$. The law of $X(t)$ determines the law of X uniquely (and vice versa). In the language of Remark 2.15 the convolution semigroup $(P_t)_{t\geq 0}$ is determined by P_1 or any other P_t.*

Examples are provided in Sect. 2.4. As discussed above, the parameter c represents the variance or covariance matrix of a Brownian motion component. The Lévy measure K represents jumps. Given a Borel set $B \subset \mathbb{R}^d \setminus \{0\}$ of possible jump sizes, $K(B)t$ coincides with the expected number of jumps with size in B during a period of length t, i.e.

$$K(B)t = E(|\{s \in [0, t] : \Delta X(s) \in B\}|). \tag{2.20}$$

The parameter b^h stands for a linear drift but it depends on the choice of h. As one can easily see from (2.19), changing the truncation function affects the drift vector via

$$b^{\widetilde{h}} = b^h + \int \big(\widetilde{h}(x) - h(x)\big)K(dx). \tag{2.21}$$

Truncation functions constitute an annoying technical feature in the theory of Lévy processes and more general semimartingales. They confuse the novice and lead to complicated expressions. Two particular choices lead to easier algebra and meaningful formulas, namely $h(x) = 0$ and $h(x) = x$. Unfortunately, both of them fail to be true truncation functions which can be used for arbitrary processes. $h(x) = 0$ cannot be chosen for processes whose absolute jumps do not sum up, i.e. if $\sum_{s\leq t} |\Delta X(s)| = \infty$. In this case the integrand $e^{iu^\top x} - 1$ in (2.19) fails to be integrable relative to K. The function $h(x) = x$, on the other hand, cannot be chosen for Lévy processes lacking first moments, i.e. if $E(|X(t)|) = \infty$. In this case $e^{iu^\top x} - 1 - iu^\top x$ in (2.19) is not K-integrable.

Most processes in Mathematical Finance do have finite expectation. Therefore we will often work with $h(x) = x$ even if this does not cover all processes. As a general rule one may say that using $h(x) = 0$ or $h(x) = x$ does not lead to wrong but at most to undefined results. By applying (2.21) one can typically turn them into correct and meaningful formulas.

Moments can easily be obtained from the characteristic function by differentiation. They can be expressed in terms of the Lévy–Khintchine triplet. Observe that the existence of moments is related only to the behaviour of the Lévy measure in the tails, i.e. by the occurrence of big jumps.

Theorem 2.19 (Moments) *Let X be a real- or vector-valued Lévy process with characteristic exponent and triplet as in Theorem 2.17.*

1. *X has moments of order $p \geq 1$ (i.e. we have $E(|X(t)|^p) < \infty$ for some or equivalently all $t > 0$) if and only if*

$$\int_{\{|x|>1\}} |x|^p K(dx) < \infty. \tag{2.22}$$

2. *Its expectation is given by*

$$E(X(t)) = \left(b^h + \int (x - h(x))K(dx)\right) t = -iD\psi(0)t \tag{2.23}$$

if it exists, i.e. if and only if (2.22) holds for $p = 1$. Here $D\psi$ denotes the derivative of ψ, i.e. ψ' in the univariate case.

3. *In the real-valued case, its variance is given by*

$$\mathrm{Var}(X(t)) = \left(c + \int x^2 K(dx)\right) t = -\psi''(0)t \tag{2.24}$$

if it exists, i.e. if and only if (2.22) holds for $p = 2$. More generally, the covariance of a vector-valued Lévy process X is given by

$$\mathrm{Cov}(X(t))_{ij} = \left(c_{ij} + \int x_i x_j K(dx)\right) t = -D_{ij}\psi(0)t \tag{2.25}$$

if (2.22) holds for $p = 2$.

4. *In the real-valued case, we have*

$$E\Big((X(t) - EX(t))^3\Big) = \int x^3 K(dx)t = i\psi'''(0)t \tag{2.26}$$

and

$$E\Big((X(t)-EX(t))^4\Big)=\int x^4K(dx)t+3\big(\mathrm{Var}(X(t))\big)^2$$
$$=\psi''''(0)t+3\big(\mathrm{Var}(X(t))\big)^2$$

if X has third or fourth moments, respectively.

Idea Moments of random variables are obtained by differentiating the characteristic function. Indeed, for integer p and real-valued Z we have

$$\frac{d^p}{du^p}E(e^{iuZ})=E\Big((iZ)^pe^{iuZ}\Big)$$

and hence

$$E(Z^p)=(-i)^p\frac{d^pE(e^{iuZ})}{du^p}\bigg|_{u=0}.$$

By $E(e^{iuX(t)})=\exp(t\psi(u))$ and $\psi(0)=0$ this yields

$$E(X(t))=-i\psi'(0)t=\left(b^h+\int(x-h(x))K(dx)\right)t.$$

For the integral to make sense we need that $\int|x-h(x)|K(dx)<\infty$, which is equivalent to $\int_{\{|x|>1\}}|x|K(dx)<\infty$.

For the second moment, we obtain $E(X(t)^2)=-\psi''(0)t-\psi'(0)^2t^2$ with $\psi''(0)=-c-\int x^2K(dx)$. This explains (2.24). For the integral to be finite we need that $\int_{\{|x|>1\}}x^2K(dx)<\infty$ because $\int_{\{|x|\le 1\}}x^2K(dx)<\infty$ holds for any Lévy measure K.

Considering higher resp. second-order partial derivatives we similarly obtain (2.25) and (2.26). The p-th derivative of ψ in 0 involves integrals $\int x^pK(dx)$. Since $\int_{\{|x|\le 1\}}|x|^pK(dx)<\infty$ for any Lévy measure, this illustrates that (2.22) is a natural necessary condition. One can show that it is sufficient even for non-integer p and vector-valued X.

For a complete proof apply [259, Theorem 25.3 and Example 25.12] together with Proposition A.5(2). □

By differentiating the characteristic function, one can also express higher order moments in terms of the characteristic exponent or the triplet, cf. Proposition A.5(4). Exponential moments are typically obtained by inserting complex arguments in (2.18, 2.19) as the following results shows.

Theorem 2.20 (Exponential Moments) *Let X be a real- or vector-valued Lévy process with characteristic exponent and triplet as in Theorem 2.17.*

1. *X has exponential moments of order $p \in \mathbb{R}^d$ (i.e. we have*

$$E\Big(\exp(p^\top X(t)\Big) < \infty \tag{2.27}$$

for some or equivalently all $t > 0$) if and only if

$$\int_{\{|x|>1\}} e^{p^\top x} K(dx) < \infty. \tag{2.28}$$

In this case they are given by

$$E\Big(e^{p^\top X(t)}\Big) = \exp\bigg(\Big(p^\top b^h + \frac{1}{2}p^\top cp + \int (e^{p^\top x} - 1 - p^\top h(x))K(dx)\Big)t\bigg). \tag{2.29}$$

This also holds for $p \in \mathbb{C}^d$ if we replace (2.27, 2.28) with

$$E\Big(\exp(\mathrm{Re}(p)^\top X(t)\Big) < \infty$$

and

$$\int_{\{|x|>1\}} e^{\mathrm{Re}(p)^\top x} K(dx) < \infty. \tag{2.30}$$

2. *Let $U \subset \mathbb{C}^d$ be an open convex neighbourhood of 0 such that the characteristic exponent $\psi : \mathbb{R}^d \to \mathbb{C}$ allows for an analytic extension to U, which we denote again by ψ. Then X has exponential moments of order p for any $p \in \mathbb{C}^d$ with $-i\mathrm{Re}(p) \in U$. If $-ip \in U$, it is given by*

$$E\Big(e^{p^\top X(t)}\Big) = \exp(\psi(-ip)t). \tag{2.31}$$

Idea On a heuristic level, (2.29) and (2.31) are immediately obtained by letting $u = -ip$ on both sides of (2.18) resp. (2.19). In order for the integral in (2.29) to be defined, we need that integrability condition (2.28) resp. (2.30) holds.

The formal proof of (2.29, 2.31) relies on the fact that both $u \mapsto E(e^{iu^\top X(t)})$ and the right-hand side of (2.18) are analytic functions of u on a subset of $\mathbb{C}^d$ whose size depends on the integrability of $X(t)$ resp. K. Therefore both sides of (2.18) coincide if this is the case for $u \in \mathbb{R}^d$.

For rigorous proofs see [259, Theorem 25.17] and Proposition A.5(4). □

Since martingales, submartingales etc. are defined in terms of conditional expectations and hence moments, it is not surprising that the presence of these properties is determined by the triplet as well.

Theorem 2.21 *Let X denote a real-valued Lévy process with Lévy–Khintchine triplet (b^h, c, K) and characteristic exponent ψ.*

1. *X is a semimartingale.*
2. *It is special if and only if*

$$\int |x - h(x)| K(dx) < \infty, \tag{2.32}$$

which holds if and only if the random variables $X(t)$ have finite expectation. In this case the compensator of X is given by $A^X(t) = a^X t$ with

$$a^X = b^{\mathrm{id}} = b^h + \int (x - h(x)) K(dx) = -i\psi'(0).$$

3. *It is a (local) martingale if and only if (2.32) and $b^{\mathrm{id}} = 0$ hold.*
4. *It is a (local) supermartingale if and only if (2.32) and $b^{\mathrm{id}} \leq 0$ hold.*
5. *It is a (local) submartingale if and only if (2.32) and $b^{\mathrm{id}} \geq 0$ hold.*

The equivalences hold with and without the limitation "local" in parentheses.

Proof By Theorem 2.19, $X(t)$ has finite expectation if and only if

$$\int_{\{|x|>1\}} |x| K(dx) < \infty$$

or, equivalently, (2.32) holds. In this case, we have $E(X(t)) = b^{\mathrm{id}} t$ by (2.23). This implies that

$$\begin{aligned} E(X(t)|\mathscr{F}_s) &= X(s) + E(X(t) - X(s)|\mathscr{F}_s) \\ &= X(s) + E(X(t) - X(s)) \\ &= X(s) + E(X(t-s)) \\ &= X(s) + b^{\mathrm{id}}(t-s) \end{aligned}$$

for $s \leq t$ by independence and stationarity of the increments. Therefore X is a martingale (resp. super-/submartingale) if and only if $b^{\mathrm{id}} = 0$ (resp. ≤ 0 or ≥ 0). In particular, it is a special semimartingale and hence a semimartingale. The "local" statements in 3–5 and the fact that (2.32) is also necessary in statement 2 are less obvious.

If (2.32) does not hold, one may consider the process $Y = X - J$ with

$$\begin{aligned} J(t) &:= \sum_{s \leq t} \Delta X(s) 1_{\{|\Delta X(s)|>1\}} \\ &= \sum_{s \leq t} \Delta X(s) 1_{\{\Delta X(s)>1\}} + \sum_{s \leq t} \Delta X(s) 1_{\{\Delta X(s)<-1\}}. \end{aligned}$$

It can be shown that Y is a Lévy process whose Lévy measure is concentrated on $[-1, 1]$. Hence it is a special semimartingale by statement 2. The process J is a semimartingale as well because it is adapted and of finite variation. Together, it follows that $X = Y + J$ is a semimartingale even if (2.32) does not hold.

The "local" assertions as well as statement 2 can be derived from Propositions 4.6, 4.21, 4.23 or [154, Proposition II.2.29a]. For statement 1 we quote [154, Corollary II.4.19] for completeness. □

Exponentials of Lévy processes play an even more important role in applications than Lévy processes themselves. They are considered more thoroughly in Sect. 3.7.

Theorem 2.22 *Let X denote a real-valued Lévy process with Lévy–Khintchine triplet (b^h, c, K).*

1. *e^X is a semimartingale.*
2. *e^X is a special semimartingale if and only if*

$$\int_{\{x>1\}} e^x K(dx) < \infty, \tag{2.33}$$

   which holds if and only if it has finite exponential moments of order 1.
3. *e^X is a (local) martingale if and only if (2.33) and*

$$\psi(-i) := \Big(b^h + \frac{1}{2}c + \int (e^x - 1 - h(x))K(dx)\Big) = 0$$

   hold.
4. *e^X is a (local) supermartingale if and only if (2.33) and $\psi(-i) \leq 0$ hold.*
5. *e^X is a (local) submartingale if and only if (2.33) and $\psi(-i) \geq 0$ hold.*

The equivalences hold with and without the limitation "local" in parentheses.

Proof We argue similarly as for Theorem 2.21. By Theorem 2.20, $E(e^{X(t)}) < \infty$ if and only if (2.33) holds. In this case, we have $E(e^{X(t)}) = \exp(\psi(-i)t)$ by (2.29). This implies that

$$\begin{aligned} E(e^{X(t)}|\mathscr{F}_s) &= e^{X(s)}E(\exp(X(t) - X(s))|\mathscr{F}_s) \\ &= e^{X(s)}E(\exp(X(t) - X(s))) \\ &= e^{X(s)}E(\exp(X(t - s))) \\ &= \exp(X(s) + \psi(-i)(t - s)) \end{aligned}$$

for $s \leq t$ by independence and stationarity of the increments. Therefore X is a martingale (resp. super-/submartingale) if and only if $\psi(-i) = 0$ (resp. ≤ 0 or ≥ 0). In particular, it is a special semimartingale and hence a semimartingale.

For the remaining statements we refer to tools from the subsequent chapters. Since X is a semimartingale by Theorem 2.20, e^X is also a semimartingale by Itō's formula, cf. Theorem 3.16 in Chap. 3. The missing "local" assertions as well as statement 2 can be derived from Propositions 4.6, 4.13, 4.21, 4.23. □

We have seen that Lévy processes show diverse path behaviour. They may or may not have jumps, may or may not be constant between jumps etc. The Lévy–Khintchine triplet also allows us to determine these sample path properties. In contrast to moments, a crucial role is played by the behaviour of the Lévy measure close to the origin, i.e. by the intensity of small jumps.

Theorem 2.23 (Path Properties) *Let X denote a real-valued Lévy process with Lévy–Khintchine triplet (b^h, c, K).*

1. *X has differentiable paths if and only if $c = 0$, $K = 0$, i.e. if it is a deterministic linear function.*
2. *It is continuous if and only if $K = 0$, i.e. if it is a Brownian motion with drift.*
3. *It has finitely many jumps in any interval of finite length if and only if $K(\mathbb{R}) < \infty$. Otherwise one observes almost surely infinitely many jumps in any interval of positive length.*
4. *It is piecewise constant between jumps if and only if $c = 0$, $K(\mathbb{R}) < \infty$ and*

$$b^0 := b^h - \int h(x)K(dx) = 0,$$

   i.e. if it is a compound Poisson process.
5. *It is of finite variation if and only if $c = 0$ and $\int |h(x)|K(dx) < \infty$. Otherwise almost all paths are of infinite variation on any interval of positive length.*

Idea As we will see in Sect. 2.5, a Lévy process with triplet (b^h, c, K) can be interpreted as an independent sum of a linear function, a Brownian motion and jumps whose intensity is characterised by the Lévy measure K. This explains why continuity holds if and only if $K = 0$. Moreover, we argued in Physicists' Corner 2.16 that differentiability does not hold unless $c = 0$, which means that linear functions $X(t) = bt$ are the only differentiable Lévy processes.

Recall that the total mass $K(\mathbb{R})$ can be interpreted as the arrival rate of jumps. In other words, $K(\mathbb{R})t$ is the expected number of jumps in an interval of length t. It can only be finite if the process has almost surely only finitely many jumps in this interval. It is less obvious that almost any path has infinitely many jumps on intervals of positive length if $K(\mathbb{R}) = \infty$.

In an interval of short length Δt, the probability of observing a jump of size in $(x, x + \Delta x]$ for small Δx is approximately $K((x, x + \Delta x])\Delta t$. This jump has a contribution of $|\Delta x|$ to the total variation of the path of X. Summing resp. integrating over Δt and Δx yields that the contribution of jumps of absolute size smaller than 1 to the expected total variation of X in an interval of length t amounts to $\int_{\{|x|\leq 1\}} |x|K(dx)t$. Hence, $\int_{\{|x|\leq 1\}} |x|K(dx) < \infty$ or equivalently $\int |h(x)|K(dx) < \infty$ implies that the absolute sum of such "small" jumps is finite on intervals of finite length. The absolute sum of "big" jumps is finite as well because càdlàg functions have only finitely many big jumps on intervals of finite length.

Again, it is less obvious that $\int |h(x)|K(dx) = \infty$ implies that the total variation from jumps is infinite on any path. With regards to the Brownian motion part, recall that the latter moves in the order of magnitude of $\sqrt{c\Delta t}$ on an interval of short length Δt. Hence its total variation on an interval of length t exceeds $t\sqrt{c/\Delta t}$, which diverges as $\Delta t \to 0$. This illustrates why paths of finite variation can only occur for $c = 0$.

Finally, recall from (2.13) that piecewise constant processes arise if the characteristic function is of Lévy–Khintchine form with $c = 0$ and $b^h = 0$ for the "truncation" function $h(x) = 0$. Since $c \neq 0$ adds a rapidly oscillating Brownian motion and $b^0 \neq 0$ a linear drift, this is the only case where piecewise constant paths occur.

For a rigorous proof see [259, Section 21]. □

Theorem 2.23 can be extended to vector-valued Lévy processes in a straightforward manner.

2.3 Constructing Lévy Processes from Simpler Building Blocks

In many real world applications such as the modelling of asset returns the hypothesis of stationary, independent increments appears to be natural or at least reasonable. For statistical estimation and concrete computations one may wish to restrict the large class of Lévy processes further. However, choosing a reasonable subset of models on purely theoretical grounds will often be impossible.

A way out is to construct sufficiently rich Lévy families with desirable analytical properties such as explicit representations for the density, the characteristic function, and the Lévy–Khintchine triplet. In concrete situations one examines whether one or more of these models suits the data under consideration. In this section we discuss some tools that help to construct such families from simple building blocks, namely subordination, convolution, and Esscher transforms.

Increasing Lévy processes are called **subordinators**. This terminology is motivated by the fact that they can be used to construct new processes by way of subordination.

Theorem 2.24 (Subordination) *Let $L = (L(t))_{t\geq 0}$ and $U = (U(\vartheta))_{\vartheta\geq 0}$ be independent processes on the same probability space. Suppose that both are Lévy processes with respect to their own natural filtrations, with triplets $(b^{L,h}, 0, K^L) = (b^{L,0} + \int h(\vartheta)K^L(d\vartheta), 0, K^L)$ and $(b^{U,h}, c^U, K^U)$, where the superscripts h and 0 refer to the truncation function as usual. Moreover, L is assumed to be a subordinator. Then $X = (X(t))_{t\geq 0}$ defined as*

$$X(t) := U(L(t))$$

is a Lévy process (relative to its own natural filtration) with triplet $(b^{X,h}, c^X, K^X)$ *given by*

$$\begin{aligned} b^{X,h} &= b^{U,h} b^{L,0} + \int E\big(h(U(\vartheta))\big) K^L(d\vartheta), \\ c^X &= b^{L,0} c^U, \\ K^X(B) &= b^{L,0} K^U(B) + \int P(U(\vartheta) \in B) K^L(d\vartheta). \end{aligned}$$

Consequently,

$$a^X = a^U a^L \tag{2.34}$$

if both U, L *are integrable Lévy processes. If the Laplace transform of* $L(t)$ *is written as*

$$\lambda_{L(t)}(z) := E(e^{zL(t)}), \quad \operatorname{Re}(z) \leq 0$$

and ψ_U *denotes the characteristic exponent of* U*, the characteristic function of* X *satisfies*

$$\varphi_{X(t)}(u) := E(e^{iuX(t)}) = \lambda_{L(t)}(\psi_U(u)), \quad u \in \mathbb{R}.$$

($\lambda_{L(t)}$ *is typically obtained from the characteristic function simply by inserting complex values in its analytic representation, cf. Theorem 2.20 and Proposition A.5(4).)*

If $L(t)$ *has density* $\ell_t(\vartheta)$ *and* $U(\vartheta)$ *has density* $u_\vartheta(x)$ *for any* $\vartheta > 0$*, then* $X(t)$ *has density*

$$\varrho_t(x) = \int u_\vartheta(x) \ell_t(\vartheta) d\vartheta. \tag{2.35}$$

The statements remain true if U *and hence* X *are vector-valued.*

Idea Let $0 \leq t_0 \leq t_1 \leq \cdots \leq t_n$ and $u_1, \ldots, u_n \in \mathbb{R}$. The Lévy property of U and L yield that

$$\begin{aligned} &E\left(\prod_{k=1}^n \exp\big(iu_k(X(t_k) - X(t_{k-1}))\big)\right) \\ &\quad = E\left(E\left(\prod_{k=1}^n \exp\big(iu_k(U(L(t_k)) - U(L(t_{k-1})))\big)\middle|\sigma(L)\right)\right) \\ &\quad = E\left(\prod_{k=1}^n \exp\big(\psi_U(u_k)(L(t_k) - L(t_{k-1}))\big)\right) \end{aligned}$$

$$= \prod_{k=1}^{n} E\Big(\exp\big(\psi_U(u_k)(L(t_k) - L(t_{k-1}))\big)\Big)$$

$$= \prod_{k=1}^{n} \lambda_{L(t_k - t_{k-1})}\big(\psi_U(u_k)\big).$$

By Proposition A.6 this implies that X has independent increments $X(t_k) - X(t_{k-1})$, $k = 1, \ldots, n$, whose characteristic functions are of the form

$$\varphi_{X(t_k)-X(t_{k-1})}(u) = \lambda_{L(t_k - t_{k-1})}\big(\psi_U(u)\big).$$

In particular, X has stationary increments. In view of Theorem 2.20 we have

$$\lambda_{L(t)}(z) = \exp\bigg(\Big(zb^{L,0} + \int (e^{z\vartheta} - 1)K^L(d\vartheta)\Big)t\bigg),$$

which implies that

$$\begin{aligned}
\varphi_{X(1)}(u) &= \exp\bigg(\psi_U(u)b^{L,0} + \int \big(e^{\psi_U(u)\vartheta} - 1\big)K^L(d\vartheta)\bigg)\\
&= \exp\bigg(iub^{U,h}b^{L,0} - \frac{u^2}{2}c^U b^{L,0} + \int (e^{iux} - 1 - iuh(x))K^U(dx)b^{L,0}\\
&\quad + \int E\big(e^{iuU(\vartheta)} - 1\big)K^L(d\vartheta)\bigg)\\
&= \exp\bigg(iub^{U,h}b^{L,0} + iu\int E\big(h(U(\vartheta))\big)K^L(d\vartheta) - \frac{u^2}{2}c^U b^{L,0}\\
&\quad + \int (e^{iux} - 1 - iuh(x))b^{L,0}K^U(dx)\\
&\quad + \iint \big(e^{iux} - 1 - iuh(x)\big)P^{U(\vartheta)}(dx)K^L(d\vartheta)\bigg).
\end{aligned}$$

Therefore the Lévy–Khintchine triplet $(b^{X,h}, c^X, K^X)$ of X is as claimed. Switching to the "truncation" function $h(x) = x$ for both U and L, we obtain (2.34).

Now, consider the Fourier transform of the right-hand side of (2.35), which equals

$$\begin{aligned}\int e^{iux}\int u_\vartheta(x)\ell_t(\vartheta)d\vartheta dx &= \iint e^{iux}u_\vartheta(x)dx\ell_t(\vartheta)d\vartheta\\ &= \int e^{\psi_U(u)\vartheta}\ell_t(\vartheta)d\vartheta\\ &= \lambda_{L(t)}(\psi_U(u))\\ &= \varphi_{X(t)}(u).\end{aligned}$$

Hence the right-hand side of (2.35) is indeed the density of $X(t)$.

For a complete proof see [259, Theorem 30.1].

Equation (2.35) follows from [259, equation (30.5)] by integrating the candidate on test sets and applying Tonelli's theorem. □

The Lévy process L in Theorem 2.24 plays the role of a time change. In finance such a time change may occur by passing from calendar time t to business or operational time ϑ measured, for example, in terms of transactions or trading volume. The process U is often chosen to be Brownian motion, which leads to a mixture of normals for the law of X, cf. Sect. 2.4 for examples.

Summing independent Lévy processes also leads to Lévy processes.

Proposition 2.25 (Convolution) *Suppose that U, V are independent Lévy processes with Lévy–Khintchine triplets $(b^{U,h}, c^U, K^U)$ and $(b^{V,h}, c^V, K^V)$, respectively. Then $X = U + V$ is a Lévy process with triplet*

$$\left(b^{X,h}, c^X, K^X\right) = \left(b^{U,h} + b^{V,h}, c^U + c^V, K^U + K^V\right) \tag{2.36}$$

(at least unless the filtration is chosen too large, in which case $X(t) - X(s)$ may cease to be independent of $\mathscr{F}_s$). The characteristic function of $X(t)$ equals

$$\varphi_{X(t)}(u) = \varphi_{U(t)}(u)\varphi_{V(t)}(u). \tag{2.37}$$

If $U(t)$ has a density $u_t(x)$, the density of $X(t)$ is given by

$$\varrho_t(x) = \int u_t(x-y)P^{V(t)}(dy), \tag{2.38}$$

where $P^{V(t)}$ denotes the law of $V(t)$. If $V(t)$ also has a density $v_t(x)$, this means

$$\varrho_t(x) = \int u_t(x-y)v_t(y)dy. \tag{2.39}$$

Proof Let $0 \leq t_0 \leq t_1 \leq \cdots \leq t_n$ and $u_1, \ldots, u_n \in \mathbb{R}$. Independence and the Lévy property of U and V yield that

$$\begin{aligned}
E\left(\prod_{k=1}^{n} \exp\left(iu_k(X(t_k) - X(t_{k-1}))\right)\right)
&= E\left(\prod_{k=1}^{n} \exp\left(iu_k(U(t_k) - U(t_{k-1}))\right) \prod_{\ell=1}^{n} \exp\left(iu_\ell(V(t_\ell) - V(t_{\ell-1}))\right)\right) \\
&= E\left(\prod_{k=1}^{n} \exp\left(iu_k(U(t_k) - U(t_{k-1}))\right)\right) E\left(\prod_{\ell=1}^{n} \exp\left(iu_\ell(V(t_\ell) - V(t_{\ell-1}))\right)\right) \\
&= \prod_{k=1}^{n} \varphi_{U(t_k)-U(t_{k-1})}(u_k) \prod_{\ell=1}^{n} \varphi_{V(t_\ell)-V(t_{\ell-1})}(u_\ell) \\
&= \prod_{k=1}^{n} \left(\varphi_{U(t_k-t_{k-1})}(u_k)\varphi_{V(t_k-t_{k-1})}(u_k)\right).
\end{aligned}$$

In view of Proposition A.6, this shows both stationarity and independence of the increments of X as well as (2.37). This equation in turn immediately yields (2.36). Equations (2.38, 2.39) are just convolution formulas which generally hold for sums of independent random variables. □

In the simplest case $V(t) = bt$ one just adds a linear drift to the process U, which means that b is added to the drift term in the triplet, the characteristic function of $U(t)$ is multiplied by e^{iubt}, and its density is shifted by bt.

An appropriate change of measure allows us to construct skewed Lévy processes from symmetric ones. The *Esscher transform* corresponds to an exponential density both on the level of marginal distributions and the Lévy measure, which makes it handy for explicit formulas.

Proposition 2.26 (Exponential Tilting) *Let X denote a Lévy process with Lévy–Khintchine triplet (b^h, c, K). For $\beta \in \mathbb{R}^d$ with $E(\exp(\beta^\top X(t))) < \infty$ we define a new probability measure P^β in terms of its density process*

$$Z(t) = \frac{dP^\beta|_{\mathscr{F}_t}}{dP|_{\mathscr{F}_t}} = \frac{e^{\beta^\top X(t)}}{E(e^{\beta^\top X(t)})} = \exp(\beta^\top X(t) - t\kappa(\beta)), \tag{2.40}$$

where $\kappa(\beta) := \log E(\exp(\beta^\top X(1)))$. *Then* X *is a Lévy process relative to* P^β *with triplet* $(b^{\beta,h}, c^\beta, K^\beta)$ *given by*

$$b^{\beta,h} = b^h + c\beta + \int h(x)\big(e^{\beta^\top x} - 1\big)K(dx), \tag{2.41}$$

$$c^\beta = c, \tag{2.42}$$

$$\frac{dK^\beta}{dK}(x) = e^{\beta^\top x}. \tag{2.43}$$

In particular,

$$a^\beta = a + c\beta + \int x\big(e^{\beta^\top x} - 1\big)K(dx) \tag{2.44}$$

if X *is integrable under both* P *and* P^β *with drift rates* a, a^β*, respectively. The characteristic function of* $X(t)$ *relative to* P^β *equals*

$$E^\beta\big(e^{iu^\top X(t)}\big) = \frac{E(e^{(iu+\beta)^\top X(t)})}{E(e^{\beta^\top X(t)})}. \tag{2.45}$$

Proof For ease of notation we consider $d = 1$, i.e. the univariate case. Z is a martingale by Theorems 2.20 and 2.22. Let $0 \le t_0 \le t_1 \le \cdots \le t_n$ and $u_1, \ldots, u_n \in \mathbb{R}$. The Lévy property of X yields that

$$\begin{aligned}
&E^\beta\left(\prod_{k=1}^n \exp\big(iu_k(X(t_k) - X(t_{k-1}))\big)\right) \\
&= E\left(\prod_{k=1}^n \exp\big(i(u_k - i\beta)(X(t_k) - X(t_{k-1})) - \kappa(\beta)(t_k - t_{k-1})\big)\right) \\
&= \prod_{k=1}^n \Big(\varphi_{X(t_k)-X(t_{k-1})}(u_k - i\beta)\exp\big(-\kappa(\beta)(t_k - t_{k-1})\big)\Big) \\
&= \prod_{k=1}^n \frac{\varphi_{X(t_k-t_{k-1})}(u_k - i\beta)}{\varphi_{X(t_k-t_{k-1})}(-i\beta)},
\end{aligned}$$

where we use the notation $\varphi_Y(u) := E(e^{iuY})$ for random variables Y and $u \in \mathbb{C}$ whenever possible. Using Proposition A.6 we conclude that X is a Lévy process relative to P^β, which satisfies (2.45). By Theorem 2.20 the right-hand side of (2.45) is of the form

$$\exp\left(iub^{\beta,h} - \frac{u^2}{2}c^\beta + \int \big(e^{iux} - 1 - iuh(x)\big)K^\beta(dx)\right)$$

with $b^{\beta,h}, c^\beta, K^\beta$ as in (2.41–2.43). Finally, (2.44) is obtained by choosing the "truncation" function $h(x) = x$ in (2.41). The multivariate case follows along the same lines. □

Recall that the exponential moments $E(e^{\beta^\top X(t)})$ and $E(e^{(iu+\beta)^\top X(t)})$ in (2.45) can usually be computed by inserting complex arguments into the analytic expression of the characteristic exponent, cf. Theorem 2.20 and Proposition A.5.

Let us illustrate equation (2.40) by its effect on densities: If $X(t)$ has density $\varrho_t(x)$ relative to P, its P^β-density is given by

$$\varrho^\beta(x) := \varrho_t(x)\exp(\beta^\top x - t\kappa(\beta)),$$

which is sometimes called *exponential tilting*.

2.4 Examples

We turn to particular classes of Lévy processes in this section. Linear functions, Brownian motions, stable processes, and Poisson processes naturally arise from limit considerations or from their path properties. The remaining classes are primarily motivated by the need to model real-world phenomena as in particular logarithmic asset prices. Since one typically ignores the mechanism driving real data, we consider families with nice statistical properties. They should be parsimonious with clear interpretation of the parameters, flexible enough to allow for the stylised features of real data, and analytically tractable with regard to explicit formulas for characteristic functions, densities, Lévy measures, and moments.

2.4.1 Linear Functions

To warm up we briefly return to the simplest of all Lévy processes, namely **linear functions** $X(t) = bt$, cf. Fig. 2.1. They are the only deterministic and the only differentiable processes with stationary, independent increments. Since the law of $X(t)$ is the Dirac measure ε_{bt} in bt, the characteristic exponent is

$$\psi(u) = iu^\top b,$$

which in turn yields the Lévy–Khintchine triplet $(b, 0, 0)$.

Although it may seem paradoxical at first glance, deterministic linear functions do play a role in modelling random phenomena. Lévy processes and random walks X typically grow linearly on a large scale by stationarity of their increments. If we rescale them as $X^{(n)}(t) := \frac{X(nt)}{n}$ in order to stay within the same order of magnitude, we have convergence

$$X^{(n)}(t) \overset{n\to\infty}{\longrightarrow} E(X(1))t$$

almost surely by the law of large numbers. This statement can be extended to the whole process:

Theorem 2.27 (Law of Large Numbers) *Let X be an integrable Lévy process (or an integrable random walk extended to continuous time in a piecewise constant way by $X(t) := X([t])$ for $t \geq 0$). Then we have*

$$\sup_{s\leq t} \left| X^{(n)}(s) - E(X(1))s \right| \overset{n\to\infty}{\longrightarrow} 0 \tag{2.46}$$

almost surely for fixed t.

Proof W.l.o.g. let $E(X(1)) = 0$ and $t = 1$. Fix $\omega \in \Omega$. The pointwise law of large numbers $\lim_{r\to\infty} X(\omega, r)/r = 0$ implies that there exists some finite function $K : (0, \infty) \to (0, \infty)$ satisfying $\lim_{r\to\infty} K(r) = 0$ and $\sup_{s\geq r} |X(\omega, s)|/s < K(r)$ for any r. Let $\varepsilon > 0$ and choose r_0 so large that $K(r_0) < \varepsilon$. For

$$n > \max\left\{ \frac{\sup_{r\in[0,1]} |X(\omega, r)|}{\varepsilon}, \frac{K(1)r_0}{\varepsilon} \right\}$$

we have

$$\left| \frac{X(\omega, ns)}{n} \right| \leq \frac{\sup_{r\in[0,1]} |X(\omega, r)|}{n} < \varepsilon$$

for $s < 1/n$, moreover

$$\left| \frac{X(\omega, ns)}{n} \right| = s\frac{|X(\omega, ns)|}{ns} < \frac{\varepsilon}{K(1)} K(1) = \varepsilon$$

for $1/n \leq s < \varepsilon/K(1)$, and finally

$$\left| \frac{X(\omega, ns)}{n} \right| = s\frac{|X(\omega, ns)|}{ns} \leq K(r_0) < \varepsilon$$

for $\varepsilon/K(1) \leq s \leq 1$, which implies $ns > r_0$. This yields $\sup_{s\leq 1} |X(\omega, ns)/n| < \varepsilon$.

Since

$$\sup_{s\leq t} \left| X^{(n)}(s) \right| \leq ([n] + 1) \sup_{s\leq t} \left| X^{([n]+1)}(s) \right|,$$

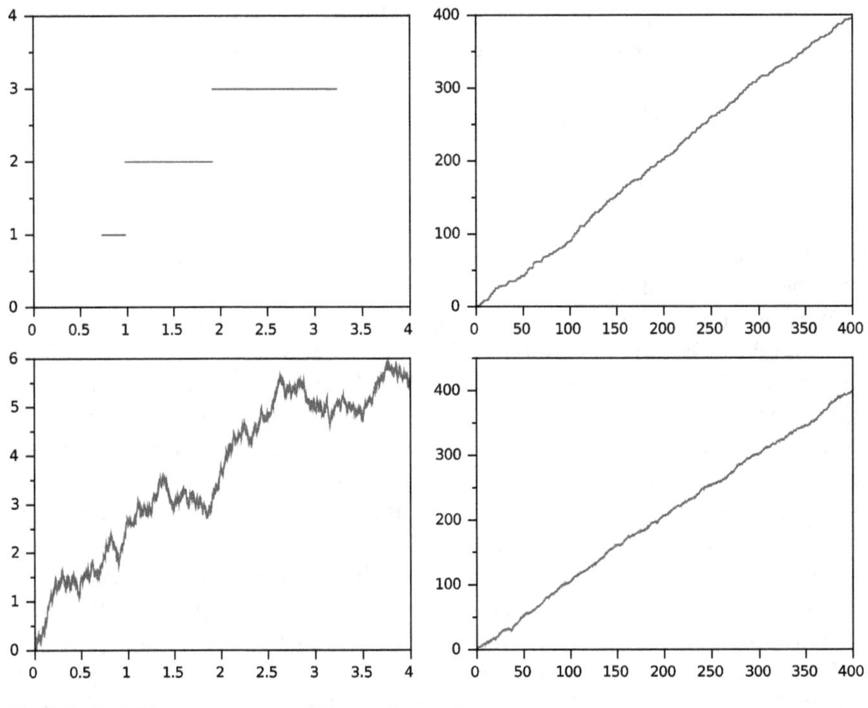

Fig. 2.8 Two Lévy processes on different time scales

convergence in (2.46) holds as well if we allow for real rather than integer n. The statement for random walks follows similarly. □

Put differently, a properly rescaled version of an arbitrary integrable Lévy process converges uniformly to a linear function, cf. Fig. 2.8. This explains why many truly random phenomena can ultimately be modelled deterministically. For example, insurance companies often work with expectations because their cumulative claim process as in Fig. 2.4 behaves linearly in a first-order approximation.

2.4.2 Brownian Motion with Drift

We move on to investigate the class of all continuous Lévy processes. From their characteristic function

$$\varphi_{X(t)}(u) = \exp\Big(\Big(iub - \frac{1}{2}u^2c\Big)t\Big)$$

we observe that $X(t)$ is a normally distributed random variable with mean bt and variance ct. The centred third and fourth moments are given by

$$E\Big((X(t) - EX(t))^3\Big) = 0,$$
$$E\Big((X(t) - EX(t))^4\Big) = 3\big(\mathrm{Var}(X(t))\big)^2 = 3c^2t^2.$$

Note that the Gaussianity of the increments need not be assumed since it follows from the continuity of paths.

Writing $X(t) = bt + \sqrt{c}W(t)$, we may also view X as an affine transformation of a **standard Brownian motion** or **Wiener process** W, which is the unique continuous Lévy process with $E(W(1)) = 0$ and $\mathrm{Var}(W(1)) = 1$. Hence the class of all continuous Lévy processes reduces essentially to a two-parameter family. This may seem particularly surprising in comparison to discrete-time random walks, where the law of $X(1) - X(0) = X(1)$ can be freely chosen.

In the multivariate case, **standard Brownian motion** refers accordingly to the continuous Lévy process W with $E(W(1)) = (0, \dots, 0)$ and $\mathrm{Cov}(W(1)) = 1$, where 1 here denotes the identity matrix. Again, affine transformations $X(t) = bt+\sqrt{c}W(t)$ lead to arbitrary continuous Lévy processes with triplet $(b, c, 0)$, where $\sqrt{c}$ now denotes the symmetric, nonnegative definite matrix with $\sqrt{c}\sqrt{c} = c$.

Brownian motion plays a predominant role in applications. Indeed, any phenomenon of constant growth which is continuous by default such as the location of a particle is naturally modelled by Brownian motion with drift. But it also enjoys much popularity in instances where continuity fails to hold or is at least not obvious such as for asset prices. This may be partly due to the fact that one needs only very little information about the mechanism driving the data. We observed already that only the first two moments of the small-scale randomness affect Brownian motion on a macroscopic scale, whereas the whole Lévy measure is needed in the general case.

Brownian motion also arises naturally from limit considerations. Above we argued that Lévy processes grow linearly in the long run. This does not hold if their expectation is close to 0 or if we subtract their drift in the first place. Since the variance of an integrable Lévy process X grows linearly in time by Theorem 2.19, we define

$$X^{(n)}(t) := \frac{X(nt) - E(X(1))nt}{\sqrt{n}}$$

as a properly rescaled version of the centred process $X(t) - E(X(1))t$. Its characteristic function equals

$$\varphi_{X^{(n)}(t)}(u) = \exp\bigg(\Big(-\frac{1}{2}cu^2 + \int n\big(e^{iux/\sqrt{n}} - 1 - iux/\sqrt{n}\big)K(dx)\Big)t\bigg), \tag{2.47}$$

where (b^h, c, K) denotes the Lévy–Khintchine triplet of X, cf. Proposition A.5(5). Since $e^{iux/\sqrt{n}} \approx 1 + iux/\sqrt{n} - \frac{1}{2}u^2x^2/n$ for large n, we can approximate this by

$$\varphi_{X^{(n)}(t)}(u) \approx \exp\left(-\frac{1}{2}\left(c + \int x^2 K(dx)\right)u^2 t\right),$$

which coincides with the characteristic function of a Gaussian random variable having mean 0 and variance $(c + \int x^2 K(dx))t$. This heuristic calculation suggests that $X^{(n)}$ tends to Brownian motion for large n. This is substantiated by

Theorem 2.28 (Donsker's Invariance Principle) *Let X be a Lévy process (or a random walk extended to continuous time as in Theorem 2.27) with finite second moments. Then $X^{(n)}$ as above converges in the sense of finite-dimensional distributions to a Brownian motion W with mean 0 and variance parameter $c + \int x^2 K(dx)$, i.e. we have*

$$E\big(f(X^{(n)}(t_1), \dots, X^{(n)}(t_k))\big) \stackrel{n\to\infty}{\longrightarrow} E\big(f(W(t_1), \dots, W(t_k))\big)$$

for any $t_1 \leq \dots \leq t_k$ and any bounded continuous $f : \mathbb{R}^k \to \mathbb{R}$.

Proof We have to show convergence in law of $(X^{(n)}(t_1), \dots, X^{(n)}(t_k))$ to $(W(t_1), \dots, W(t_k))$, which is equivalent to convergence in law of

$$(X^{(n)}(t_1), X^{(n)}(t_2) - X^{(n)}(t_1), \dots, X^{(n)}(t_k) - X^{(n)}(t_{k-1})) \tag{2.48}$$

to

$$(W(t_1), W(t_2) - W(t_1), \dots, W(t_k) - W(t_{k-1})). \tag{2.49}$$

By Lévy's continuity theorem it suffices to show pointwise convergence of the corresponding characteristic functions, cf. Proposition A.5(8). Independence and stationarity of the increments imply that the characteristic function of the random vector (2.48) is a product of factors $\varphi_{X^{(n)}(t_\ell - t_{\ell-1})}(u_\ell)$, $\ell = 1, \dots, k$ as in (2.47). By dominated convergence, they tend as $n \to \infty$ to

$$\varphi_{W(t_\ell - t_{\ell-1})}(u_\ell) = \exp\left(-\frac{1}{2}\left(c + \int x^2 K(dx)\right)u_\ell^2(t_\ell - t_{\ell-1})\right).$$

Independence and stationarity of the increments of W imply that the product of these functions coincides with the characteristic function of (2.49). Altogether, the claim follows.

For the related statement on random walks we refer to [154, Corollary VII.3.11] or Example 4.37. □

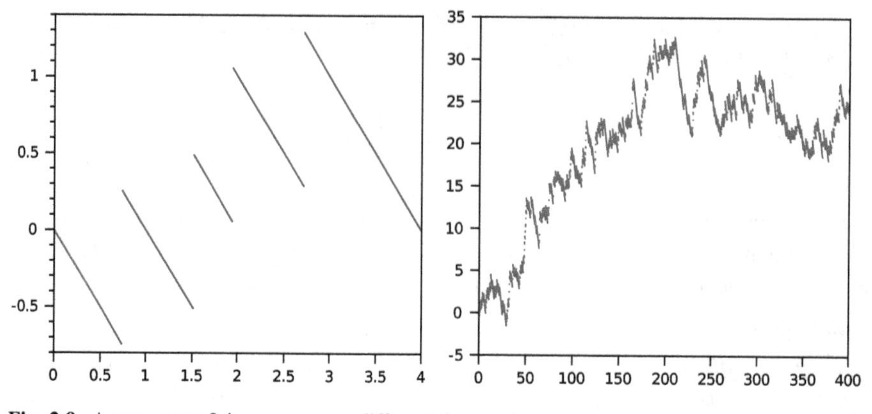

Fig. 2.9 A zero-mean Lévy process on different time scales

Convergence in law holds in fact in a more general sense, namely on the whole space of càdlàg paths relative to the Skorokhod topology, cf. [154, Corollaries VII.3.6 and VII.3.11].

Summing up, any Lévy process with finite variance resembles Brownian motion on a large scale or maybe on an intermediate scale where the linear drift $E(X(1))t$ does not dominate the whole path behaviour, cf. Fig. 2.9. This constitutes another justification for the ubiquitous use of Brownian motion.

2.4.3 *Poisson and Compound Poisson Processes*

Counting processes X are processes with values in the nonnegative integers which change only by jumps of height $+1$. As the name indicates, the random variable $X(t)$ stands for the number of random events which have happened up to time t. The Lévy processes among the counting processes are called *Poisson processes*, cf. Fig. 2.3 and Problem 2.3. They are used to model the number of phone calls, customer arrivals, radioactive decays and many other events that occur independently with more or less constant rate. Their Lévy–Khintchine triplet is of the form

$$(b^0, c, K) = (0, 0, \lambda\varepsilon_1)$$

for $h(x) = 0$ and with an intensity parameter λ that signifies the average number of jumps in a unit time interval. By Theorem 2.19 the process has moments and exponential moments of any order and we have

$$E(X(t)) = \mathrm{Var}(X(t)) = \lambda t.$$

The triplet also yields the characteristic exponent

$$\psi(u) = \lambda(e^{iu} - 1)$$

of X. From the characteristic function we conclude that $X(t)$ has a Poisson distribution with parameter λt, i.e.

$$P(X(t) = k) = e^{-\lambda t}\frac{(\lambda t)^k}{k!}.$$

Proposition 2.29 *The waiting times*

$$\tau_n := \inf\{t \geq 0 : X(t) = n\} - \inf\{t \geq 0 : X(t) = n - 1\}$$

between successive jumps are independent, with parameter λ exponentially distributed random variables.

Idea This is stated in [259, Theorem 21.3 and the following comment].

For a Lévy process X we have that $(X(t + s) - X(t))_{s\geq 0}$ is independent of the stopped process X^t and has the same law as $(X(s))_{s\geq 0}$. One can show that Lévy processes are *strong* Lévy processes, which means that this holds not only for fixed times but for stopping times t as well, cf. Chap. 5 in this respect, in particular Proposition 5.8 and Example 5.10. This implies that τ_n is independent of $(\tau_1, \ldots, \tau_{n-1})$ and has the same law as τ_1. Moreover, note that

$$P(\tau_1 > t) = P(X(t) = 0) = e^{-\lambda t}$$

for any $t \geq 0$, which means that τ_1 is exponentially distributed with parameter λ. □

If the jump size is random rather than fixed, we obtain the larger class of Lévy processes whose paths are piecewise constant, cf. Fig. 2.4. The Lévy–Khintchine triplet of these *compound Poisson processes* is given by

$$\left(b^0, c, K\right) = (0, 0, K)$$

with some finite measure K. The probability measure $K/K(\mathbb{R})$ signifies the law of single jumps whereas $K(\mathbb{R})$ stands for the total intensity of jumps parallel to λ above. Indeed, compound Poisson processes can be generated as

$$X(t) = \sum_{k=1}^{N(t)} Y_k,$$

where N denotes a Poisson process with parameter $\lambda = K(\mathbb{R}) > 0$ and $Y_1, Y_2, \dots$ a sequence of independent random variables with law $K/K(\mathbb{R})$, cf. Problem 2.4. Provided that they exist, the first four moments are given by

$$\begin{aligned} E(X(t)) &= -i\widehat{K}'(0)t, \\ \operatorname{Var}(X(t)) &= -\widehat{K}''(0)t, \\ E\Big((X(t) - EX(t))^3\Big) &= i\widehat{K}'''(0)t, \\ E\Big((X(t) - EX(t))^4\Big) &= \widehat{K}''''(0)t + 3\big(\operatorname{Var}(X(t))\big)^2, \end{aligned}$$

where $\widehat{K}(u) := \int e^{iux}K(dx)$ denotes the Fourier transform of K. Indeed, this follows from Theorem 2.19 and $\psi(u) = \widehat{K}(u) - \widehat{K}(0)$.

The extension to the multivariate case is straightforward. Compound Poisson processes are applied, for instance, to model cumulative insurance claims. Here $K/K(\mathbb{R})$ stands for the distribution of single claims and $K(\mathbb{R})$ for their arrival rate.

2.4.4 *Subordinators*

We discuss a number of increasing Lévy processes which ultimately lead to asset price processes that have been suggested in the literature. The first example is the multiple of a Poisson process with positive drift, i.e.

$$L(t) = bt + \gamma^2 N(t), \tag{2.50}$$

where $b, \gamma^2 \geq 0$ and N denotes a Poisson process with intensity $\lambda > 0$. The Lévy–Khintchine triplet of L equals

$$\big(b^0, c, K\big) = (b, 0, \lambda\varepsilon_{\gamma^2})$$

relative to $h(x) = 0$. Subordinated to Brownian motion, it leads to Merton's jump diffusion process in Sect. 2.4.5.

Another subordinator plays a role in Kou's model in Sect. 2.4.6. It is a compound Poisson process with exponentially distributed jumps, i.e. it can be written as

$$L(t) = \sum_{k=1}^{N(t)} Y_k, \tag{2.51}$$

where N denotes a Poisson process with parameter $\lambda > 0$ and $Y_1, Y_2, \dots$ a sequence of independent, exponentially distributed random variables with parameter $\alpha > 0$.

The Lévy–Khintchine triplet of L equals

$$\left(b^0, c, K(dx)\right) = \left(0, 0, \lambda\alpha e^{-\alpha x} 1_{\{x>0\}} dx\right).$$

The integral in the Lévy–Khintchine formula can be computed easily, which leads to the characteristic exponent

$$\psi(u) = \frac{iu\lambda}{\alpha - iu}. \tag{2.52}$$

The family of gamma distributions gives rise to a subordinator L with infinite activity but of finite variation, the *gamma process*. The law of $L(t)$ is $\Gamma(\lambda t, \gamma)$ with parameters $\lambda, \gamma > 0$, i.e. it has density

$$\varrho_t(x) = \frac{\gamma^{\lambda t}}{\Gamma(\lambda t)} x^{\lambda t - 1} e^{-\gamma x} 1_{\{x>0\}}$$

and characteristic exponent

$$\psi(u) = -\lambda \log\left(1 - \frac{iu}{\gamma}\right). \tag{2.53}$$

The branch of the complex logarithm is chosen such that the right-hand side is continuous and vanishes at 0. The Lévy–Khintchine triplet of L is given by

$$\left(b^0, c, K(dx)\right) = (0, 0, \lambda x^{-1} e^{-\gamma x} 1_{\{x>0\}} dx). \tag{2.54}$$

We will meet this process again in Sect. 2.4.7.

The family of *inverse-Gaussian distributions* $IG(\xi, \eta)$, $\xi, \eta > 0$ with density

$$\varrho(x) = \frac{\xi}{\sqrt{2\pi}} \exp(\xi\eta) x^{-3/2} \exp\left(-\frac{1}{2}(\xi^2 x^{-1} + \eta^2 x)\right) 1_{\{x>0\}} \tag{2.55}$$

and characteristic exponent

$$\psi(u) = -\xi\left(\sqrt{-2iu + \eta^2} - \eta\right) \tag{2.56}$$

also leads to an increasing Lévy process L if we choose $L(t)$ to be $IG(\xi t, \eta)$-distributed. The Lévy–Khintchine triplet of this *inverse-Gaussian Lévy process* equals

$$\left(b^0, c, K(dx)\right) = \left(0, 0, \frac{\xi}{\sqrt{2\pi}} x^{-3/2} \exp\left(-\tfrac{1}{2}\eta^2 x\right) 1_{\{x>0\}} dx\right). \tag{2.57}$$

Subordinated to Brownian motion it leads to the process in Sect. 2.4.8.

As final example we consider a subordinator whose characteristic function at time t equals

$$E(e^{iuL(t)}) = \left(\cosh\sqrt{-iu\alpha^2/2}\right)^{-2\delta t}, \tag{2.58}$$

where $\alpha, \delta > 0$ denote parameters and the possibly multi-valued power is chosen such that the right-hand side is continuous and equals 1 for $u = 0$. Subordinated to Brownian motion, L leads to the Lévy process in Sect. 2.4.10.

Except (2.51), the above subordinators are used in the following sections to construct new Lévy processes via Theorem 2.24. This procedure leads to diverse families which have been suggested in the literature for the modelling of asset returns. We consider univariate processes here but by subordination to multivariate Brownian motion one can similarly construct vector-valued Lévy processes with correlated components.

2.4.5 *The Merton Model*

If we subordinate the rescaled Poisson process with drift (2.50) to standard Brownian motion, we obtain a jump-diffusion Lévy process X whose Lévy–Khintchine triplet equals

$$\left(b^0, c, K\right) = \left(0, b, \lambda N(0, \gamma^2)\right)$$

by Theorem 2.24. Having real data in mind one may wish to allow for asymmetric jump distributions. Two natural approaches, namely either subordination to a Brownian motion *with drift*

$$U(\vartheta) = \beta\vartheta + W(\vartheta)$$

or exponential tilting along the lines of Proposition 2.26, lead to the same family of triplets

$$\left(b^0, c, K\right) = \left(b\beta, b, \lambda N(\beta\gamma^2, \gamma^2)\right)$$

with constants $\beta \in \mathbb{R}$, $\lambda, b, \gamma^2 \geq 0$. If we allow X to have an additional linear drift, we end up with Lévy processes X whose triplets are of the general form

$$\left(b^0, c, K\right) = \left(\mu, \sigma^2, \lambda N(\beta, \gamma^2)\right)$$

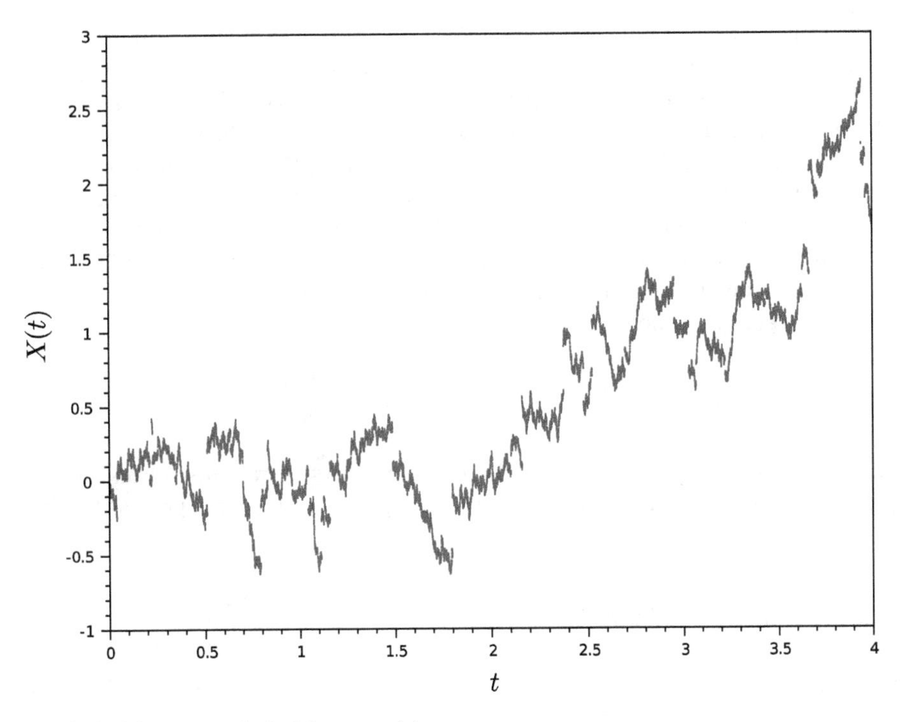

Fig. 2.10 Lévy process in the Merton model

with parameters $\mu, \beta \in \mathbb{R}$, $\lambda, \sigma^2, \gamma^2 \geq 0$, cf. Fig. 2.10. The characteristic exponent

$$\psi(u) = i\mu u - \frac{1}{2}\sigma^2 u^2 + \lambda(e^{i\beta u - \frac{1}{2}\gamma^2 u^2} - 1)$$

is easily obtained from (2.19). The first four moments are given by

$$\begin{aligned} E(X(t)) &= (\mu + \lambda\beta)t, \\ \text{Var}(X(t)) &= \left(\sigma^2 + \lambda\left(\gamma^2 + \beta^2\right)\right)t, \\ E\left((X(t) - EX(t))^3\right) &= \lambda\beta\left(3\gamma^2 + \beta^2\right)t, \\ E\left((X(t) - EX(t))^4\right) &= \lambda\left(3\gamma^4 + 6\beta^2\gamma^2 + \beta^4\right)t + 3\left(\text{Var}(X(t))\right)^2. \end{aligned}$$

From the shape of the triplet we see that X can be written as

$$X(t) = \mu t + \sigma W(t) + \sum_{k=1}^{N(t)} Y_k \tag{2.59}$$

with some standard Brownian motion W, some Poisson process N with parameter λ and a sequence of i.i.d. Gaussian random variables $Y_1, Y_2, \ldots$ with mean β and variance γ^2. Since the density of the Lévy measure decays rapidly in the tails, $X(t)$ has moments and even exponential moments of any order by Theorem 2.19.

The Lévy process (2.59) has been suggested by [220]. It extends Brownian motion by some jump component to account for the observed behaviour of asset returns. Unfortunately, a density for the law of $X(t)$ is only available in series expansion. Moreover, very different choices of parameters may lead to similar finite-dimensional distributions. For example, for $\mu = \sigma^2 = \beta = 0$, λ large, $\lambda\gamma^2 = 1$, we obtain the approximation

$$\psi(u) = \lambda(e^{-\frac{1}{2}\gamma^2 u^2} - 1) \approx -\frac{1}{2}u^2$$

for the characteristic exponent of X. In probabilistic terms this means that X resembles Brownian motion very closely although it is a pure jump process.

2.4.6 *The Kou Model*

The difference

$$X(t) = U(t) - V(t)$$

of two independent subordinators as in (2.51) yields a Lévy process with triplet

$$\left(b^0, c, K(dx)\right) = \left(0, 0, \left(\lambda_+\alpha_+ e^{-\alpha_+|x|}1_{\{x>0\}} + \lambda_-\alpha_- e^{-\alpha_-|x|}1_{\{x<0\}}\right) dx\right),$$

cf. Proposition 2.25. Allowing for an additional Brownian motion with drift, we obtain triplets of the form

$$\left(b^0, c, K\right) = \left(\mu, \sigma^2, K\right)$$

with K as before and parameters $\mu \in \mathbb{R}$, $\sigma^2, \lambda_+, \alpha_+, \lambda_-, \alpha_- \geq 0$, cf. Fig. 2.11 for a sample path. Equation (2.52) and Proposition 2.25 yield the characteristic exponent

$$\psi(u) = i\mu u - \frac{1}{2}\sigma^2 u^2 + \frac{iu\lambda_+}{\alpha_+ - iu} - \frac{iu\lambda_-}{\alpha_- + iu}.$$

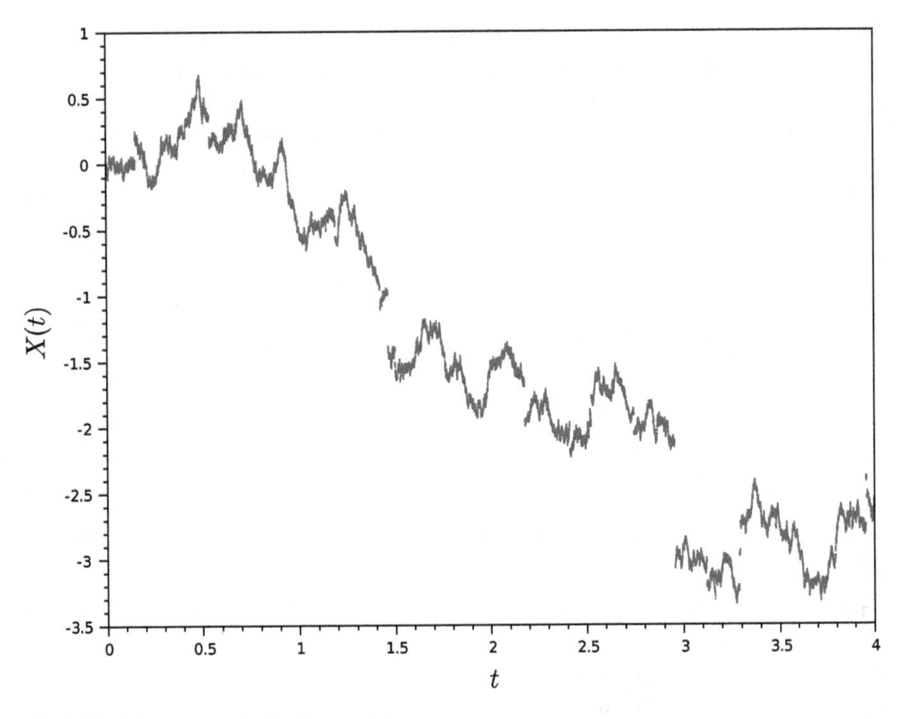

Fig. 2.11 Lévy process in the Kou model

The first four moments are given by

$$E(X(t)) = \left(\mu + \frac{\lambda_+}{\alpha_+} - \frac{\lambda_-}{\alpha_-}\right) t,$$

$$\mathrm{Var}(X(t)) = \left(\sigma^2 + 2\frac{\lambda_+}{\alpha_+^2} + 2\frac{\lambda_-}{\alpha_-^2}\right) t,$$

$$E\Big((X(t) - EX(t))^3\Big) = 6\left(\frac{\lambda_+}{\alpha_+^3} - \frac{\lambda_-}{\alpha_-^3}\right) t,$$

$$E\Big((X(t) - EX(t))^4\Big) = 24\left(\frac{\lambda_+}{\alpha_+^4} + \frac{\lambda_-}{\alpha_-^4}\right) t + 3\big(\mathrm{Var}(X(t))\big)^2.$$

One can also write the process in the form (2.59) with jump intensity $\lambda := \lambda_+ + \lambda_-$ if one considers jumps $Y_1, Y_2, \ldots$ of double exponential type, i.e. with density

$$\varrho(x) = \frac{\lambda_+}{\lambda_+ + \lambda_-}\alpha_+ e^{-\alpha_+ x} 1_{\{x>0\}} + \frac{\lambda_-}{\lambda_+ + \lambda_-}\alpha_- e^{\alpha_- x} 1_{\{x<0\}}.$$

The Lévy process X shares some properties of the previous one in Sect. 2.4.5: it is a Wiener process with additional discrete jumps, has moments of any order, resembles Brownian motion even for $\sigma = 0$ if the remaining parameters are chosen accordingly, and it allows for an analytic representation of the characteristic function but not of the density. However, it does not have exponential moments of any order. Applying Theorem 2.20, we see that exponential moments of order p exist for $-\alpha_- < p < \alpha_+$.

2.4.7 The Variance-Gamma Process and Extensions

If the gamma process (2.53, 2.54) is subordinated to Brownian motion, we obtain the Lévy–Khintchine triplet

$$\left(b^0, c, K(dx)\right) = \left(0, 0, \frac{\lambda}{|x|}\exp(-\alpha|x|)\,dx\right) \tag{2.60}$$

with $\alpha := \sqrt{2\gamma}$. Skewed extensions can be constructed by exponential tilting (cf. Proposition 2.26) or by subordination to Brownian motion with nonzero drift, both approaches leading to the same class of Lévy processes, namely *variance-gamma (VG) processes* with Lévy measure

$$K(dx) = \frac{\lambda}{|x|}\exp(\beta x - \alpha|x|)dx, \tag{2.61}$$

where $\alpha := \sqrt{\beta^2 + 2\gamma}$ if X is obtained by subordination to $U(t) = \beta t + W(t)$ for Brownian motion W. Equations (2.60, 2.61) follow from Theorem 2.24 and the fact that (2.55) integrates to 1. It makes sense to allow for an additional linear drift, leading to the triplet

$$\left(b^0, c, K\right) = (\mu, 0, K)$$

with K as before and parameters $\alpha, \lambda > 0$, $\beta \in (-\alpha, \alpha)$, $\mu \in \mathbb{R}$. A sample path is shown in Fig. 2.12. With the help of Theorem 2.24 we obtain the characteristic exponent

$$\psi(u) = i\mu u - \lambda \log \frac{\alpha^2 - (\beta + iu)^2}{\alpha^2 - \beta^2}, \tag{2.62}$$

where the branch of the complex logarithm is chosen such that the right-hand side is continuous in u and vanishes at 0. In contrast to the previous two examples, the variance-gamma process has no Brownian component. An examination of the Lévy measure close to the origin shows that X is of finite variation but has infinite activity, cf. Theorem 2.23. The decay of the Lévy density in the tails yields that $X(t)$ has

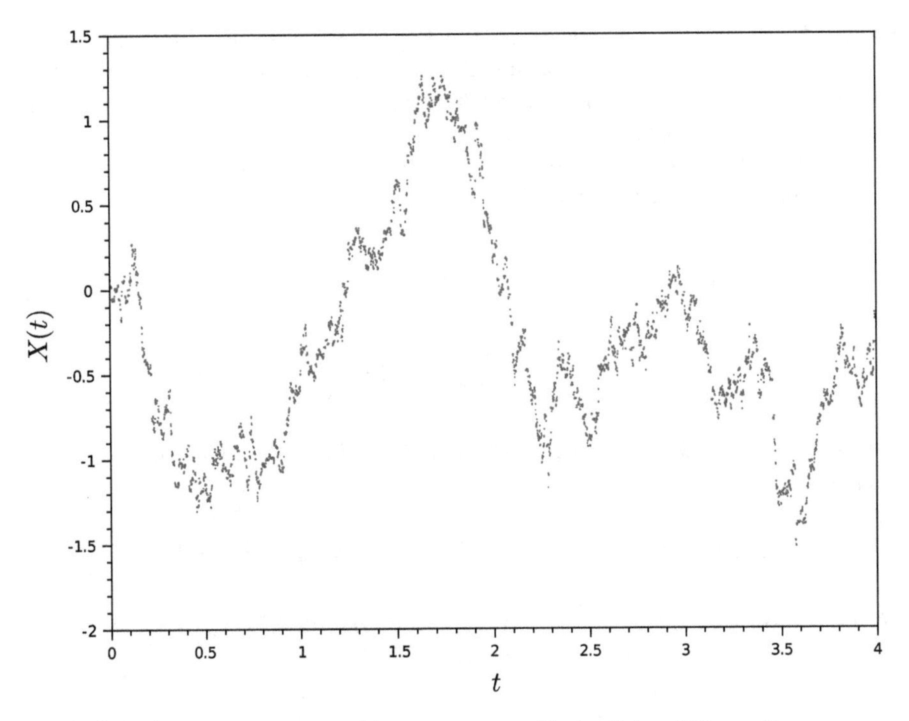

Fig. 2.12 Variance-gamma process with parameters $\alpha = 17$, $\beta = 0$, $\lambda = 150$, $\mu = 0$

finite exponential moments of order p for $-\alpha - \beta < p < \alpha - \beta$, cf. Theorem 2.19. Differentiation of the characteristic function yields moments of $X(t)$ such as

$$\begin{aligned}
E(X(t)) &= \left(\mu + \frac{2\beta\lambda}{\alpha^2 - \beta^2}\right) t, \\
\mathrm{Var}(X(t)) &= \left(1 + \frac{2\beta^2}{\alpha^2 - \beta^2}\right) \frac{2\lambda}{\alpha^2 - \beta^2} t, \\
E\Big((X(t) - EX(t))^3\Big) &= \left(3 + \frac{4\beta^2}{\alpha^2 - \beta^2}\right) \frac{4\beta\lambda}{(\alpha^2 - \beta^2)^2} t, \\
E\Big((X(t) - EX(t))^4\Big) &= \left(3 + \frac{24\beta^2}{\alpha^2 - \beta^2} + \frac{24\beta^4}{(\alpha^2 - \beta^2)^2}\right) \frac{4\lambda}{(\alpha^2 - \beta^2)^2} t \\
&\quad + 3\big(\mathrm{Var}(X(t))\big)^2.
\end{aligned}$$

The characteristic function can be computed in closed form, leading to the density $\varrho_t(x)$ of $X(t)$ given by

$$\varrho_t(x + \mu t) = \frac{\exp(\beta x)}{\sqrt{\pi}\Gamma(\lambda t)} (\alpha^2 - \beta^2)^{\lambda t} \left(\frac{|x|}{2\alpha}\right)^{\lambda t - \frac{1}{2}} K_{\lambda t - \frac{1}{2}}(\alpha |x|),$$

where K_ν denotes the modified Bessel function of the third kind with index ν.

We observe that the four parameters play quite distinctive roles: μ and β represent location and skewness parameters, respectively. The remaining values α and λ take care of both shape and scale.

The variance-gamma process with $\mu = 0$ has been extended to the so-called *CGMY process* from [51] with triplet

$$\left(b^{\mathrm{id}}, c, K(dx)\right) = \Bigg(C(M^{Y-1} - G^{Y-1})\Gamma(1-Y), 0,$$

$$\frac{C}{|x|^{1+Y}} \exp\left(\frac{G-M}{2}x - \frac{G+M}{2}|x|\right) dx\Bigg),$$

where $C, G, M > 0$, $Y \in (-\infty, 2)$ are parameters. For $Y \notin \{0, 1\}$ the characteristic exponent equals

$$\psi(u) = C\Gamma(-Y)\left((M - iu)^Y - M^Y + (G + iu)^Y - G^Y\right).$$

CGMY processes are themselves instances of *generalised tempered stable Lévy processes*, whose Lévy–Khintchine triplet equals

$$\left(b^{\mathrm{id}}, c, K(dx)\right) = \left(\mu, 0, \left(\frac{\lambda_-}{|x|^{1+\eta_-}} e^{-\alpha_-|x|} 1_{\{x<0\}} + \frac{\lambda_+}{|x|^{1+\eta_+}} e^{-\alpha_+|x|} 1_{\{x>0\}}\right) dx\right)$$

with parameters $\mu \in \mathbb{R}$, $\lambda_\pm, \alpha_\pm > 0$, $\eta_\pm \in (-\infty, 2)$. For $\eta_\pm \notin \{0, 1\}$ their characteristic exponent equals

$$\begin{aligned}\psi(u) = {} & i\mu u + \lambda_+\Gamma(-\eta_+)\left((\alpha_+ - iu)^{\eta_+} - \alpha_+^{\eta_+} + iu\alpha_+^{\eta_+-1}\eta_+\right) \\ & + \lambda_-\Gamma(-\eta_-)\left((\alpha_- + iu)^{\eta_-} - \alpha_-^{\eta_-} - iu\alpha_-^{\eta_--1}\eta_-\right).\end{aligned}$$

These classes allow for more flexible modelling but there is a price to be paid. Closed form expressions for the density cease to be available in the general case. Moreover, reliable estimation becomes increasingly difficult as the number of parameters grows.

A subclass of tempered stable Lévy processes with explicit density has been suggested under the name *bilateral gamma process*. It can be interpreted as the difference of two independent gamma processes with parameters $\lambda_+, \alpha_+ > 0$ and $\lambda_-, \alpha_- > 0$, respectively. Consequently, its Lévy–Khintchine triplet is given by

$$\left(b^0, c, K(dx)\right) = \left(0, 0, \left(\frac{\lambda_-}{|x|} e^{-\alpha_-|x|} 1_{\{x<0\}} + \frac{\lambda_+}{|x|} e^{-\alpha_+|x|} 1_{\{x>0\}}\right) dx\right).$$

One easily derives the characteristic exponent

$$\psi(u) = \lambda_+ \log \frac{\alpha_+}{\alpha_+ - iu} + \lambda_- \log \frac{\alpha_-}{\alpha_- + iu}$$

where the complex logarithm must be chosen such that the right-hand side is continuous and vanishes at 0. The density of the bilateral gamma process $X(t)$ can be expressed as

$$\begin{aligned}\varrho_t(x) = & \frac{\alpha_+^{\lambda_+ t} \alpha_-^{\lambda_- t}}{(\alpha_+ + \alpha_-)^{\frac{1}{2}(\lambda_+ + \lambda_-)t} \Gamma(\lambda_+ t)} x^{\frac{1}{2}(\lambda_+ + \lambda_-)t - 1} e^{-\frac{x}{2}(\alpha_+ - \alpha_-)} \\ & \times W_{\frac{1}{2}(\lambda_+ - \lambda_-)t, \frac{1}{2}(\lambda_+ + \lambda_-)t - \frac{1}{2}}(x(\alpha_+ + \alpha_-)),\end{aligned}$$

using the *Whittaker function*

$$W_{\lambda,\mu}(x) = \frac{x^\lambda e^{-x/2}}{\Gamma(\mu - \lambda + \frac{1}{2})} \int_0^\infty u^{\mu - \lambda - \frac{1}{2}} e^{-u} \left(1 + \frac{u\eta_+}{2}\right)^{\mu + \lambda - \frac{1}{2}} du.$$

From the triplet one can conclude that variance-gamma processes with $\mu = 0$ are particular instances of bilateral gamma processes, namely those with $\lambda_+ = \lambda_-$. Differentiation of the characteristic function yields

$$\begin{aligned}E(X(t)) &= \left(\frac{\lambda_+}{\alpha_+} - \frac{\lambda_-}{\alpha_-}\right) t, \\ \mathrm{Var}(X(t)) &= \left(\frac{\lambda_+}{\alpha_+^2} + \frac{\lambda_-}{\alpha_-^2}\right) t, \\ E\Big((X(t) - EX(t))^3\Big) &= 2\left(\frac{\lambda_+}{\alpha_+^3} - \frac{\lambda_-}{\alpha_-^3}\right) t, \\ E\Big((X(t) - EX(t))^4\Big) &= 6\left(\frac{\lambda_+}{\alpha_+^4} + \frac{\lambda_-}{\alpha_-^4}\right) t + 3\big(\mathrm{Var}(X(t))\big)^2.\end{aligned}$$

Finite exponential moments of order p exist for $-\alpha_- < p < \alpha_+$, which can be seen from the decay of the Lévy measure. Bigamma processes have finite variation and infinite activity because they are differences of Gamma processes with the same property.

The variance-gamma process allows for a multivariate extension. It is based on subordinating a gamma process to a multivariate rather than a univariate Brownian motion. More specifically, we consider

$$X(t) = \mu t + U(L(t))$$

with $\mu \in \mathbb{R}^d$ as well as a Brownian motion U in $\mathbb{R}^d$ with positive definite diffusion matrix $\Delta \in \mathbb{R}^{d\times d}$ and drift vector $\Delta\beta \in \mathbb{R}^d$. In order to avoid non-unique parameters one usually demands $\det(\Delta) = 1$. The subordinator L is assumed to be a gamma process as in (2.53, 2.54) with parameters $\lambda > 0$ and $\gamma := (\alpha^2 - \beta^\top\Delta\beta)/2$ for $0 \leq \sqrt{\beta^\top\Delta\beta} < \alpha$. The resulting *multivariate variance-gamma (VG) Lévy process* has Lévy–Khintchine triplet

$$\left(b^0, c, K\right) = (\mu, 0, K)$$

with

$$K(dx) = \frac{2\alpha^{d/2}\lambda e^{x^\top\beta}}{\left(2\pi\sqrt{x^\top\Delta^{-1}x}\right)^{d/2}} K_{d/2}\left(\alpha\sqrt{x^\top\Delta^{-1}x}\right)dx.$$

The characteristic exponent is of the form

$$\psi(u) = i\mu^\top u - \lambda\log\frac{\alpha^2 - (\beta+iu)^\top\Delta(\beta+iu)}{\alpha^2 - \beta^\top\Delta\beta},$$

where the complex logarithm must be chosen such that the right-hand side is continuous and vanishes at 0. The density $\varrho_t(x)$ of $X(t)$ satisfies

$$\varrho_t(x+\mu t) = \frac{e^{\beta^\top x}(\alpha^2 - \beta^\top\Delta\beta)^{\lambda t}}{(2\pi)^{d/2}2^{\lambda t-1}\Gamma(\lambda t)}\left(\frac{x^\top\Delta^{-1}x}{\alpha^2}\right)^{(\lambda t-\frac{d}{2})/2} K_{\lambda t-\frac{d}{2}}\left(\alpha\sqrt{x^\top\Delta^{-1}x}\right).$$

The first two moments of the process are given by

$$E(X(t)) = \left(\mu + \frac{2\lambda}{\alpha^2 - \beta^\top\Delta\beta}\Delta\beta\right)t,$$

$$\mathrm{Cov}(X(t)) = \left(\Delta + \frac{2}{\alpha^2 - \beta^\top\Delta\beta}\beta^\top\Delta^2\beta\right)\frac{2\lambda}{\alpha^2 - \beta^\top\Delta\beta}t.$$

There are more ways to generate multivariate distributions via subordination, cf. e.g. [209].

2.4.8 The Normal-Inverse-Gaussian Lévy Process

If we replace the gamma subordinator by an inverse-Gaussian process as given in (2.56, 2.57), we end up with a Lévy process X of similar merits. In line with Sect. 2.4.7 we subordinate the inverse-Gaussian process to Brownian motion with drift $U(t) = \beta t + W(t)$. Equivalently, one can also subordinate to standard Brownian motion and introduce skewness in a second step by exponential tilting. Finally, we allow for an additional linear trend.

Set $\delta = \xi$ and $\sqrt{\alpha^2 + \beta^2} = \eta$ if X is generated by subordination with respect to Brownian motion with drift or $\alpha = \eta$ if it is obtained by exponential tilting. The resulting *normal-inverse-Gaussian (NIG) Lévy process* has Lévy–Khintchine triplet

$$\left(b^{\mathrm{id}}, c, K(dx)\right) = \left(\mu + \delta \frac{\beta}{\sqrt{\alpha^2 - \beta^2}}, 0, \frac{\alpha\delta}{\pi} e^{\beta x} \frac{K_1(\alpha|x|)}{|x|} dx\right)$$

for $h(x) = x$, where $\alpha, \delta > 0$, $\beta \in (-\alpha, \alpha)$, $\mu \in \mathbb{R}$ are parameters and K_ν denotes the modified Bessel function of the third kind with index ν. A sample path is shown in Fig. 2.13. The characteristic exponent equals

$$\psi(u) = i\mu u + \delta\left(\sqrt{\alpha^2 - \beta^2} - \sqrt{\alpha^2 - (\beta + iu)^2}\right). \tag{2.63}$$

From the behaviour of the Lévy density at the origin we see that the process is of infinite variation. Moreover, Theorem 2.19 implies as above that $X(t)$ has finite

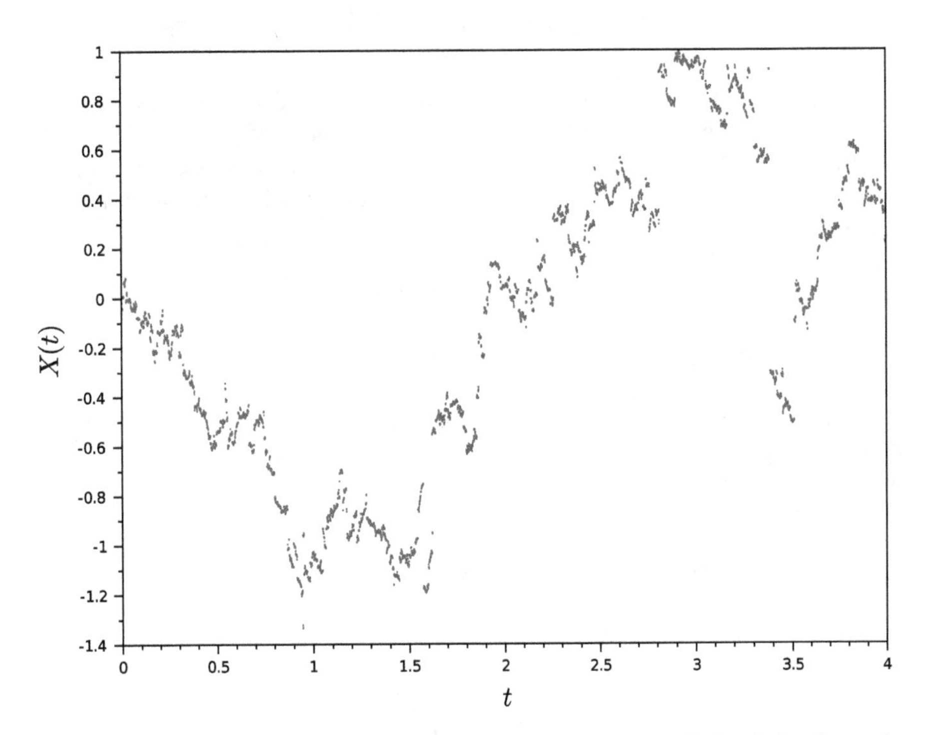

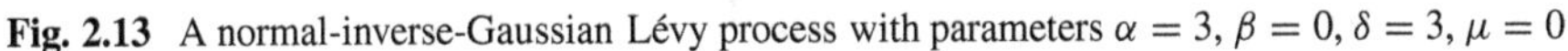
Fig. 2.13 A normal-inverse-Gaussian Lévy process with parameters $\alpha = 3, \beta = 0, \delta = 3, \mu = 0$

exponential moments of order p for $-\alpha - \beta < p < \alpha - \beta$. Differentiation yields moments such as

$$E(X(t)) = \left(\mu + \delta \frac{\beta}{\sqrt{\alpha^2 - \beta^2}}\right) t,$$

$$\mathrm{Var}(X(t)) = \frac{\alpha^2 \delta}{(\alpha^2 - \beta^2)^{3/2}} t,$$

$$E\left((X(t) - EX(t))^3\right) = \frac{3\beta\alpha^2\delta}{(\alpha^2 - \beta^2)^{5/2}} t,$$

$$E\left((X(t) - EX(t))^4\right) = \frac{3\alpha^2\delta(\alpha^2 + 4\beta^2)}{(\alpha^2 - \beta^2)^{7/2}} t + 3\left(\mathrm{Var}(X(t))\right)^2.$$

Similarly to above, the characteristic function can be computed explicitly, which yields the density ϱ_t of $X(t)$:

$$\varrho_t(x + \mu t) = \frac{\alpha\delta t}{\pi} \exp\left(\beta x + \delta t\sqrt{\alpha^2 - \beta^2}\right) \frac{K_1\left(\alpha\sqrt{(\delta t)^2 + x^2}\right)}{\sqrt{(\delta t)^2 + x^2}}.$$

Altogether, we observe that μ and β stand for location and skewness whereas the remaining parameters α, δ together take care of shape and scale.

As in Sect. 2.4.7 the process allows for a multivariate extension. Again, we consider subordination to a multivariate rather than univariate Brownian motion, namely

$$X(t) = \mu t + U(L(t))$$

with $\mu \in \mathbb{R}^d$ and a Brownian motion in $\mathbb{R}^d$ having positive definite diffusion matrix $\Delta \in \mathbb{R}^{d \times d}$ normalised by requiring $\det(\Delta) = 1$ and drift vector $\beta\Delta \in \mathbb{R}^d$. The subordinator L is assumed to be an inverse-Gaussian process as in (2.56, 2.57). The resulting *multivariate normal-inverse-Gaussian (NIG) Lévy process* has Lévy–Khintchine triplet

$$\left(b^{\mathrm{id}}, c, K\right) = \left(\mu + \frac{\delta}{\sqrt{\alpha^2 - \beta^\top \Delta\beta}} \Delta\beta, 0, K\right)$$

with

$$K(dx) = 2\delta e^{x^\top \beta} \left(\frac{\alpha}{2\pi\sqrt{x^\top \Delta^{-1} x}}\right)^{\frac{d+1}{2}} K_{\frac{d+1}{2}}\left(\alpha\sqrt{x^\top \Delta^{-1} x}\right) dx.$$

The characteristic exponent equals

$$\psi(u) = i\mu u + \delta\left(\sqrt{\alpha^2 - \beta^\top\Delta\beta} - \sqrt{\alpha^2 - (\beta + iu)^\top\Delta(\beta + iu)}\right)$$

and the density ϱ_t of $X(t)$ satisfies

$$\begin{aligned}
&\varrho_t(x + \mu t) \\
&= 2\delta t\left(\frac{\alpha}{2\pi}\right)^{\frac{d+1}{2}} \exp\left(\beta^\top x + \delta t\sqrt{\alpha^2 - \beta^\top\Delta\beta}\right) \frac{K_{\frac{d+1}{2}}\left(\alpha\sqrt{(\delta t)^2 + x^\top\Delta^{-1}x}\right)}{\left((\delta t)^2 + x^\top\Delta^{-1}x\right)^{\frac{d+1}{4}}}.
\end{aligned}$$

The first two moments of the process are given by

$$\begin{aligned}
E(X(t)) &= \left(\mu + \frac{\delta}{\sqrt{\alpha^2 - \beta^\top\Delta\beta}}\Delta\beta\right)t, \\
\mathrm{Cov}(X(t)) &= \frac{\delta t}{\sqrt{\alpha^2 - \beta^\top\Delta\beta}}\left(\Delta + \frac{1}{\alpha^2 - \beta^\top\Delta\beta}\beta^\top\Delta^2\beta\right).
\end{aligned}$$

2.4.9 Generalised Hyperbolic Lévy Processes

The NIG and VG Lévy processes in Sects. 2.4.8 and 2.4.7 can be viewed as special resp. limiting cases of a more general class of processes. To this end, we replace the subordinator of inverse-Gaussian resp. Gamma-type by a *generalised inverse-Gaussian (GIG) Lévy process* L whose law at $t = 1$ is *generalised inverse-Gaussian*, i.e. it has density

$$\varrho_{L(1)}(x) = \left(\frac{\gamma}{\delta}\right)^\lambda \frac{1}{2K_\lambda(\delta\gamma)} x^{\lambda-1} \exp\left(-\frac{1}{2}(\delta^2 x^{-1} + \gamma^2 x)\right) 1_{\{x>0\}}$$

and characteristic function

$$\varphi_{L(1)}(u) = \left(\frac{\gamma^2}{\gamma^2 - 2iu}\right)^{\frac{\lambda}{2}} \frac{K_\lambda\left(\delta\sqrt{\gamma^2 - 2iu}\right)}{K_\lambda(\delta\gamma)},$$

where λ, γ, δ denote real-valued parameters with $\gamma, \delta > 0$. Let $\alpha > 0$, $\beta \in (-\alpha, \alpha)$ and $\gamma := \sqrt{\alpha^2 - \beta^2}$. The Lévy process

$$X(t) = \mu t + U(L(t)) \tag{2.64}$$

with $\mu \in \mathbb{R}$ as well as Brownian motion U having drift rate $\beta \in \mathbb{R}$ and diffusion coefficient 1 is called the *generalised hyperbolic (GH) Lévy process* with parameters $\lambda, \alpha, \beta, \delta, \mu$. It has Lévy–Khintchine triplet

$$\left(b^{\mathrm{id}}, c, K\right) = \left(\mu + \frac{\delta\beta}{\sqrt{\alpha^2-\beta^2}} \frac{K_{\lambda+1}\left(\delta\sqrt{\alpha^2-\beta^2}\right)}{K_\lambda\left(\delta\sqrt{\alpha^2-\beta^2}\right)}, 0, K\right)$$

with Lévy measure

$$K(dx) = \frac{e^{\beta x}}{|x|}\left(\int_0^\infty \frac{\exp\left(-\sqrt{2y+\alpha^2}|x|\right)}{\pi^2 y\left(J^2_{|\lambda|}(\delta\sqrt{2y}) + Y^2_{|\lambda|}(\delta\sqrt{2y})\right)} dy + \lambda e^{-\alpha|x|} 1_{\{\lambda>0\}}\right) dx,$$

where $J_{|\lambda|}, Y_{|\lambda|}$ denote modified Bessel functions of the first resp. second kind with index $|\lambda|$. For $t = 1$ its law is *generalised hyperbolic*, i.e. $X(1)$ has density

$$\varrho_{X(1)}(x+\mu) = \frac{(\alpha^2-\beta^2)^{\lambda/2}}{\sqrt{2\pi}\alpha^{\lambda-\frac{1}{2}}\delta^\lambda K_\lambda\left(\delta\sqrt{\alpha^2-\beta^2}\right)} e^{\beta x}\left(\delta^2+x^2\right)^{(\lambda-\frac{1}{2})/2} K_{\lambda-\frac{1}{2}}\left(\alpha\sqrt{\delta^2+x^2}\right) \tag{2.65}$$

and characteristic function

$$\varphi_{X(1)}(u) = e^{i\mu u}\left(\frac{\alpha^2-\beta^2}{\alpha^2-(\beta+iu)^2}\right)^{\frac{\lambda}{2}} \frac{K_\lambda\left(\delta\sqrt{\alpha^2-(\beta+iu)^2}\right)}{K_\lambda\left(\delta\sqrt{\alpha^2-\beta^2}\right)}. \tag{2.66}$$

Since the family of generalised hyperbolic distributions is not stable under convolution, the law of $X(t)$ generally fails to be generalised hyperbolic except for $t = 1$. The first two moments of generalised hyperbolic Lévy processes are given by

$$E(X(t)) = \left(\mu + \frac{\delta\beta}{\sqrt{\alpha^2-\beta^2}} \frac{K_{\lambda+1}\left(\delta\sqrt{\alpha^2-\beta^2}\right)}{K_\lambda\left(\delta\sqrt{\alpha^2-\beta^2}\right)}\right) t$$

and

$$\begin{aligned}\operatorname{Var}(X(t)) = \Bigg(&\frac{K_{\lambda+1}\left(\delta\sqrt{\alpha^2-\beta^2}\right)}{\sqrt{\alpha^2-\beta^2}K_\lambda\left(\delta\sqrt{\alpha^2-\beta^2}\right)} \\ &+ \frac{\delta\beta^2}{\alpha^2-\beta^2}\left(\frac{K_{\lambda+2}\left(\delta\sqrt{\alpha^2-\beta^2}\right)}{K_\lambda\left(\delta\sqrt{\alpha^2-\beta^2}\right)} - \left(\frac{K_{\lambda+1}\left(\delta\sqrt{\alpha^2-\beta^2}\right)}{K_\lambda\left(\delta\sqrt{\alpha^2-\beta^2}\right)}\right)^2\right)\Bigg)\delta t.\end{aligned}$$

For $\lambda = -1/2$ we recover the NIG Lévy process as a special case. For positive λ the variance-gamma Lévy process can be considered as the limiting case $\delta \to 0$. For $\lambda = 1$ we obtain *hyperbolic Lévy processes*. In this case the law at time $t = 1$ is *hyperbolic*, i.e. the density

$$\varrho_{X(1)}(x+\mu) = \frac{\sqrt{\alpha^2-\beta^2}}{2\delta\alpha K_1\left(\delta\sqrt{\alpha^2-\beta^2}\right)}\exp\left(-\alpha\sqrt{\delta^2+x^2}+\beta x\right)$$

is the exponential of a hyperbola with asymptotic slopes $\alpha+\beta$, $-\alpha+\beta$.

The limiting case of (2.65) for $\alpha, \beta \to 0$ and $\lambda < 0$ is a scaled and shifted Student-t distribution with density

$$\varrho_{X(1)}(x+\mu) = \frac{\Gamma(-\lambda+\frac{1}{2})}{\sqrt{\pi}\delta\Gamma(-\lambda)}\left(1+\frac{x^2}{\delta^2}\right)^{\lambda-\frac{1}{2}}.$$

The corresponding characteristic function (2.66) is

$$\varphi_{X(1)}(u) = e^{i\mu u}\frac{2^{\lambda+1}K_\lambda(\delta|u|)}{\Gamma(-\lambda)(\delta|u|)^\lambda}.$$

Many other distributions are obtained as limiting cases of generalised hyperbolic laws, e.g. normal, Cauchy, exponential, Laplace, Gamma, reciprocal Gamma, and generalised inverse-Gaussian distributions, cf. [91] for details.

For a multivariate extension of the generalised hyperbolic Lévy process we replace the univariate Brownian motion in (2.64) by a multivariate one with positive definite diffusion matrix $\Delta \in \mathbb{R}^{d\times d}$ and drift vector $\Delta\beta \in \mathbb{R}^d$, where $\Delta \in \mathbb{R}^{d\times d}$ is normalised by $\det(\Delta) = 1$. Moreover, $\mu \in \mathbb{R}^d$. The resulting *multivariate generalised hyperbolic (GH) Lévy process* X has Lévy–Khintchine triplet

$$\left(b^{\mathrm{id}}, c, K(dx)\right) = \left(\mu + \frac{\delta}{\sqrt{\alpha^2-\beta^\top\Delta\beta}}\frac{K_{\lambda+1}\left(\delta\sqrt{\alpha^2-\beta^\top\Delta\beta}\right)}{K_\lambda\left(\delta\sqrt{\alpha^2-\beta^\top\Delta\beta}\right)}\Delta\beta, 0, K(dx)\right)$$

with Lévy measure

$$K(dx) = \frac{2e^{\beta^\top x}}{\left(2\pi\sqrt{x^\top\Delta^{-1}x}\right)^{d/2}}\left(\int_0^\infty \frac{(2y+\alpha^2)^{d/4}K_{d/2}\left(\sqrt{(2y+\alpha^2)x^\top\Delta^{-1}x}\right)}{\pi^2 y\left(J^2_{|\lambda|}(\delta\sqrt{2y})+Y^2_{|\lambda|}(\delta\sqrt{2y})\right)}dy\right.$$
$$\left.+\lambda\alpha^{d/2}K_{d/2}\left(\alpha\sqrt{x^\top\Delta^{-1}x}\right)1_{\{\lambda>0\}}\right)dx.$$

The first two moments satisfy

$$E(X(t)) = \left(\mu + \frac{\delta}{\sqrt{\alpha^2-\beta^\top\Delta\beta}}\frac{K_{\lambda+1}\left(\delta\sqrt{\alpha^2-\beta^\top\Delta\beta}\right)}{K_\lambda\left(\delta\sqrt{\alpha^2-\beta^\top\Delta\beta}\right)}\Delta\beta\right)t$$

and

$$\mathrm{Cov}(X(t)) = \Bigg(\frac{K_{\lambda+1}\big(\delta\sqrt{\alpha^2 - \beta^\top \Delta\beta}\big)}{\sqrt{\alpha^2 - \beta^\top \Delta\beta}\, K_\lambda\big(\delta\sqrt{\alpha^2 - \beta^\top \Delta\beta}\big)} \Delta + \frac{\delta}{\alpha^2 - \beta^\top \Delta\beta}$$

$$\times \left(\frac{K_{\lambda+2}\big(\delta\sqrt{\alpha^2 - \beta^\top \Delta\beta}\big)}{K_\lambda\big(\delta\sqrt{\alpha^2 - \beta^\top \Delta\beta}\big)} - \left(\frac{K_{\lambda+1}\big(\delta\sqrt{\alpha^2 - \beta^\top \Delta\beta}\big)}{K_\lambda\big(\delta\sqrt{\alpha^2 - \beta^\top \Delta\beta}\big)} \right)^2 \right) \beta^\top \Delta^2 \beta \Bigg) \delta t.$$

The density and characteristic function of $X(1)$ are given by

$$\varrho_{X(1)}(x+\mu) = \frac{(\alpha^2 - \beta^\top \Delta\beta)^{\lambda/2}}{(2\pi)^{d/2} \alpha^{\lambda - d/2} \delta^\lambda K_\lambda\big(\delta\sqrt{\alpha^2 - \beta^\top \Delta\beta}\big)}$$

$$\times\, e^{\beta^\top x} \big(\delta^2 + x^\top \Delta^{-1} x\big)^{(\lambda - \frac{d}{2})/2} K_{\lambda - \frac{d}{2}}\Big(\alpha\sqrt{\delta^2 + x^\top \Delta^{-1} x}\Big)$$

and

$$\varphi_{X(1)}(u) = e^{i\mu^\top u} \left(\frac{\alpha^2 - \beta^\top \Delta\beta}{\alpha^2 - (\beta + iu)^\top \Delta(\beta + iu)} \right)^{\frac{\lambda}{2}} \frac{K_\lambda\big(\delta\sqrt{\alpha^2 - (\beta + iu)^\top \Delta(\beta + iu)}\big)}{K_\lambda\big(\delta\sqrt{\alpha^2 - \beta^\top \Delta\beta}\big)},$$

respectively.

Again, the special case $\lambda = 1$ is called a hyperbolic Lévy process, in which case the density at $t = 1$ reduces to

$$\varrho_{X(1)}(x+\mu) = \frac{(2\pi)^{-\frac{d-1}{2}} (\alpha^2 - \beta^\top \Delta\beta)^{\frac{d+1}{4}}}{2\alpha\delta^{\frac{d+1}{2}} K_{\frac{d+1}{2}}\big(\delta\sqrt{\alpha^2 - \beta^\top \Delta\beta}\big)} e^{-\alpha\sqrt{\delta^2 + x^\top \Delta^{-1} x} + \beta^\top x}.$$

For $\lambda = -1/2$ we recover the multivariate NIG process and in the limit $\delta \to 0$ the multivariate VG process.

Another limiting case of (2.65), namely for $\alpha \to 0$, $\beta \to (0, \ldots, 0)$ and $\lambda < 0$, is a scaled and shifted multivariate Student-t distribution with density

$$\varrho_{X(1)}(x+\mu) = \frac{\Gamma(-\lambda + \frac{d}{2})}{(\delta^2\pi)^{d/2}\Gamma(-\lambda)} \left(1 + \frac{x^\top \Delta^{-1} x}{\delta^2}\right)^{\lambda - \frac{d}{2}}$$

and characteristic function

$$\varphi_{X(1)}(u) = e^{i\mu^\top u} \left(\frac{2}{\delta}\right)^\lambda \frac{2K_\lambda\big(\delta\sqrt{u^\top \Delta u}\big)}{\Gamma(-\lambda)\big(u^\top \Delta u\big)^{\lambda/2}}.$$

2.4.10 The Meixner Process

Finally, we take a look at the class of *Meixner processes* put forward by [263]. Parallel to variance-gamma and normal inverse-Gaussian Lévy processes they allow for explicit representations of the triplet, the characteristic function and also the density. Symmetric Meixner processes X are obtained by subordination of the increasing process (2.58) to a Wiener process. By Theorem 2.24 they have characteristic function

$$E(e^{iuX(t)}) = \left(\frac{1}{\cosh(\alpha u/2)}\right)^{2\delta t}.$$

Exponential tilting and an additional linear drift lead to the general case with characteristic function

$$E(e^{iuX(t)}) = e^{i\mu ut}\left(\frac{\cos(\beta/2)}{\cosh((\alpha u - i\beta)/2)}\right)^{2\delta t},$$

where $\alpha, \delta > 0$, $\beta \in (-\pi, \pi)$, $\mu \in \mathbb{R}$ are parameters and the power is chosen such that the right-hand side is continuous and 1 for $u = 0$, cf. Fig. 2.14 for a sample

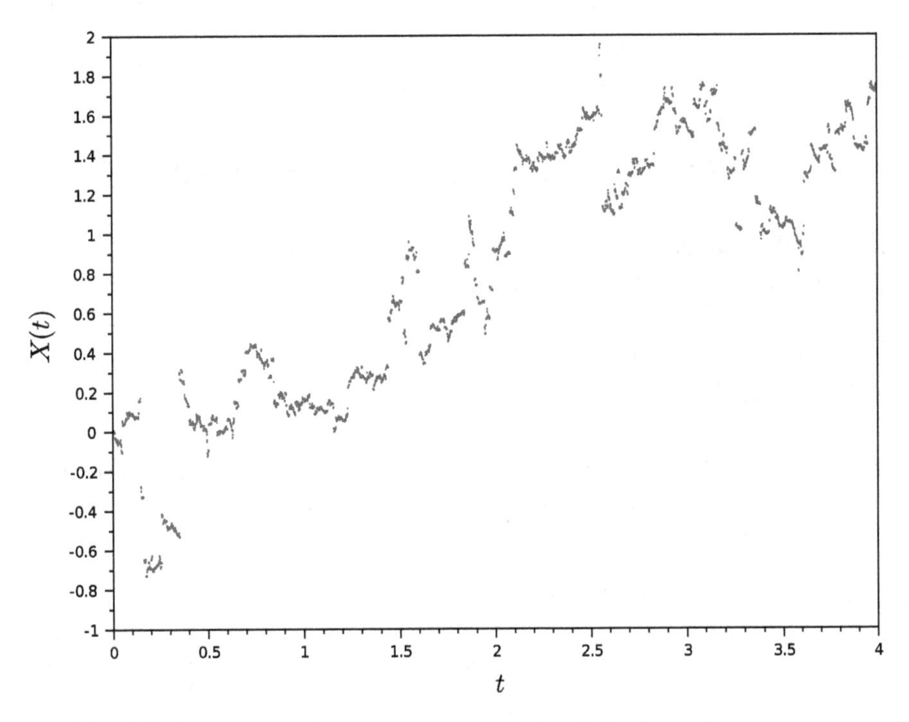

Fig. 2.14 A Meixner process with parameters $\alpha = 0.82$, $\beta = 0$, $\delta = 3$, $\mu = 0$

path. The corresponding Lévy–Khintchine triplet equals

$$(b^{\mathrm{id}}, c, K(dx)) = \left(\left(\mu + \frac{\alpha\delta \sin\beta}{1+\cos\beta}\right), 0, \frac{\delta \exp(\beta x/\alpha)}{x \sinh(\pi x/\alpha)} dx\right).$$

One can derive an explicit representation of the density ϱ_t of $X(t)$:

$$\varrho_t(x + \mu t) = \frac{(2\cos(\beta/2))^{2\delta t}}{2\pi\alpha\Gamma(2\delta t)} e^{\beta x/\alpha} \left(\Gamma\left(\delta t + \frac{ix}{\alpha}\right)\right)^2.$$

Examination of the Lévy measure near the origin yields that X is of infinite variation. Moreover, it has exponential moments of order

$$-\frac{\pi - \beta}{\alpha} < p < \frac{\pi + \beta}{\alpha}.$$

The first four moments are given by

$$E(X(t)) = \left(\mu + \frac{\alpha\delta \sin\beta}{1+\cos\beta}\right) t,$$

$$\mathrm{Var}(X(t)) = \frac{\alpha^2\delta}{1+\cos\beta} t,$$

$$E\left((X(t) - EX(t))^3\right) = \frac{\alpha^3\delta \sin\beta}{(1+\cos\beta)^2} t,$$

$$E\left((X(t) - EX(t))^4\right) = \frac{\alpha^4\delta(2-\cos\beta)}{(1+\cos\beta)^2} t + 3\big(\mathrm{Var}(X(t))\big)^2.$$

As in earlier classes above, μ and β stand for location and skewness whereas the remaining two parameters take care of shape and scale.

2.4.11 *Stable Lévy Motions*

What happens to Lévy processes which do not have the first or second moments that were assumed in Theorems 2.27 and 2.28? Do they converge to anything on a large scale? They may or may not. If they do, the limit belongs to the class of **strictly stable Lévy motions**.

These processes distinguish themselves by their scaling behaviour. They are *self-similar* in the sense that, for any $a > 0$, there is some $\gamma_a > 0$ such that the rescaled process $X^{(a)}(t) := \gamma_a X(at)$ has the same law as X. The slightly larger class of **stable Lévy motions** satisfies the relaxed self-similarity condition that, for any $a > 0$, there are $\gamma_a > 0$, $\delta_a > 0$ such that the process $X^{(a)}(t) := \gamma_a X(at) - \delta_a(t)$ has the same law as X. It turns out that $\gamma_a = a^{-1/\alpha}$, where the *index of stability* α varies between 0 and 2, cf. [256, Section 7.5] and [259, Proposition 13.5]. This inspires the name (strictly) α-*stable* Lévy motion. We have already come across strictly α-stable Lévy motions for $\alpha = 1$ and $\alpha = 2$, namely linear functions and Brownian motion, respectively.

Theorem 2.30 *A real-valued Lévy process is an α-stable Lévy motion if its Lévy–Khintchine triplet (b^h, c, K) satisfies*

$$\begin{aligned} c = 0, \quad & K(dx) = \left(M_+ 1_{\{x>0\}} + M_- 1_{\{x<0\}}\right) |x|^{-(1+\alpha)} dx && \text{if } 0 < \alpha < 2, \\ & K = 0 && \text{if } \alpha = 2 \end{aligned}$$

with some constants $M_+, M_- \geq 0$. The process is a strictly α-stable Lévy motion if we have in addition

$$\begin{aligned} b^0 &= 0 \quad \text{if } 0 < \alpha < 1, \\ M_+ &= M_- \ \text{if } \alpha = 1, \\ b^{\mathrm{id}} &= 0 \quad \text{if } 1 < \alpha < 2, \\ b^h &= 0 \quad \text{if } \alpha = 2, \end{aligned}$$

where the superscripts 0, id *refer to $h(x) = 0$ and $h(x) = x$, respectively.*

Proof The Lévy–Khintchine triplet $(b^{(a),h}, c^{(a)}, K^{(a)})$ of $X^{(a)}$ satisfies

$$\begin{aligned} b^{(a),h} &= a^{1-1/\alpha} b^h + \int \left(h(xa^{-1/\alpha}) - a^{-1/\alpha} h(x)\right) aK(dx), \\ c^{(a)} &= a^{1-2/\alpha} c, \quad K^{(a)}(B) = \int 1_B(xa^{-1/\alpha}) aK(dx). \end{aligned}$$

X is a stable Lévy process if and only if $c^{(a)} = c$, $K^{(a)} = K$. Consequently, X can only be self-similar if $c = 0$ or $\alpha = 2$.

Since $K^{(a)}([1, \infty)) = aK([a^{1/\alpha}, \infty))$ for $a > 0$, we have $K([x, \infty)) = x^{-\alpha} K([1, \infty))$ and hence

$$K(dx) = \alpha K([1, \infty)) x^{-(1+\alpha)} dx$$

for $x > 0$ if $K^{(a)} = K$. Similarly, we obtain

$$K(dx) = \alpha K((-\infty, -1]) |x|^{-(1+\alpha)} dx$$

for $x < 0$. Since Lévy measures satisfy the moment condition $\int (1 \wedge |x|^2) K(dx) < \infty$, the only possible indices satisfy $0 < \alpha < 2$. For strictly α-stable Lévy motions, $b^h = b^{(a),h}$ must hold as well, which leads to the drift condition in the assertion. □

A parallel statement holds in the vector-valued case: the diffusion coefficient c turns into a covariance matrix as usual. In order to understand the Lévy measure we identify $\mathbb{R}^d$ with the product space $\mathbb{R}_+ \times S^{d-1}$, where S^{d-1} denotes the unit sphere in $\mathbb{R}^d$. A pair (r, ϑ) in this product space stands for $r\vartheta \in \mathbb{R}^d$, i.e. $r = |x| \in \mathbb{R}_+$ represents the length and $\vartheta = x/|x| \in S^{d-1}$ the direction of a vector x. In this sense the Lévy measure of an α-stable Lévy motion satisfies

$$K(A) = \int_0^\infty \int_{S^{d-1}} 1_A(r\vartheta)\mu(d\vartheta)r^{-(1+\alpha)}dr$$

with some finite measure μ on S^{d-1}, i.e. it is the product of a power law for the radial part and an arbitrary angular part, cf. [259, Remark 14.4].

The integrals in the Lévy–Khintchine representation of the characteristic function can be evaluated in closed form. This leads to the following

Theorem 2.31 *The characteristic exponent of an α-stable Lévy motion equals*

$$\psi(u) = \begin{cases} iu\mu - \sigma^\alpha |u|^\alpha \left(1 - i\beta \operatorname{sgn} u \tan \frac{\pi\alpha}{2}\right) & \text{if } \alpha \neq 1, \\ iu\mu - \sigma|u|\left(1 + i\beta \frac{2}{\pi} \operatorname{sgn} u \log |u|\right) & \text{if } \alpha = 1 \end{cases} \tag{2.67}$$

with constants $\mu \in \mathbb{R}$, $\sigma > 0$, $\beta \in [-1, 1]$. The latter are related to the Lévy–Khintchine representation in Theorem 2.30 as follows:

$$c = \begin{cases} 2\sigma^2 & \text{if } \alpha = 2, \\ 0 & \text{otherwise,} \end{cases}$$

$$M_\pm = \begin{cases} (2\Gamma(-\alpha)\cos(\frac{\pi\alpha}{2}))^{-1}(1 \pm \beta)\sigma^\alpha & \text{if } \alpha \neq 1, \\ \frac{1}{\pi}(1 \mp \beta)\sigma & \text{if } \alpha = 1, \end{cases}$$

$$b^h = \begin{cases} \mu & \text{if } 0 < \alpha < 1, h(x) = 0, \\ \mu + (M_+ - M_-)(1 - \gamma) & \text{if } \alpha = 1, h(x) = x1_{\{|x| \leq 1\}}, \\ \mu & \text{if } 1 < \alpha \leq 2, h(x) = x, \end{cases}$$

where $\gamma \approx 0.577$ denotes the Euler–Mascheroni constant.

For strictly α-stable Lévy motions we have $\mu = 0$ if $\alpha \neq 1$ and $\beta = 0$ if $\alpha = 1$.

Idea This follows essentially by evaluating the integrals in Theorem 2.30.

Case $\alpha = 2$: This is evident.
Case $0 < \alpha < 2$: In the proof of [259, Theorem 14.10] it is shown that

$$\begin{aligned} \int_{-\infty}^0 (e^{-iux} - 1)|x|^{-(1+\alpha)}dx &= \int_0^\infty (e^{iux} - 1)|x|^{-(1+\alpha)}dx \\ &= \Gamma(-\alpha)\cos\left(\frac{\pi\alpha}{2}\right)|u|^\alpha\left(1 - i\tan\left(\frac{\pi\alpha}{2}\right)\operatorname{sgn} u\right). \end{aligned}$$

Collecting terms yields the first equation in (2.67).

Case $1 < \alpha < 2$: In this case the calculation in the proof of [259, Theorem 14.10] implies

$$\int_{-\infty}^{0} (e^{-iux} - 1 + iux)|x|^{-(1+\alpha)}dx = \int_{0}^{\infty} (e^{iux} - 1 - iux)|x|^{-(1+\alpha)}dx$$
$$= \Gamma(-\alpha)\cos\left(\frac{\pi\alpha}{2}\right)|u|^{\alpha}\left(1 - i\tan\left(\frac{\pi\alpha}{2}\right)\operatorname{sgn} u\right),$$

which once more yields the first equation in (2.67).

Case $\alpha = 1$: In this case the calculation in the proof of [259, Theorem 14.10] contains the equation

$$\int_{-\infty}^{0} (e^{-iux} - 1 + iux 1_{\{|x|\le 1\}})|x|^{-2}dx = \int_{0}^{\infty} (e^{iux} - 1 - iux 1_{\{|x|\le 1\}})|x|^{-2}dx$$
$$= -\frac{\pi}{2}|u| - iu\log|u| + i(1-\gamma)u$$

with

$$1 - \gamma = \int_0^1 \frac{\sin r - r}{r^2}dr + \int_1^\infty \frac{\sin r}{r^2}dr.$$

Integration by parts yields that this equals

$$1 - \sin(1) - \mathrm{Cin}(1) + \sin(1) - \mathrm{Ci}(1),$$

where

$$\mathrm{Ci}(x) := -\int_x^\infty \frac{\cos t}{t}dt$$

and

$$\mathrm{Cin}(x) := \int_0^x \frac{1-\cos t}{t}dt$$

denote cosine integrals. The identity

$$\mathrm{Cin}(x) = \gamma + \log x - \mathrm{Ci}(x)$$

yields the claim, where $\gamma \approx 0.577$ denotes the Euler–Mascheroni constant, cf. [1, 5.2.2] □

The distribution of $X(1)$ with characteristic function $\exp(\psi(u))$ given by (2.67) is called the **stable law** with parameters $\alpha, \sigma, \beta, \mu$ and written as $S_\alpha(\sigma, \beta, \mu)$. It is **strictly stable** if $\mu = 0$ for $\alpha \neq 1$ and $\beta = 0$ for $\alpha = 1$. Stable laws are characterised by the fact that

$$\mathscr{L}\big(X^{(1)} + \cdots + X^{(n)}\big) = \mathscr{L}(c_n X + d_n)$$

for some numbers $c_n > 0$ and d_n, where $X^{(1)}, \dots, X^{(n)}, X$ denote independent random variables with law $S_\alpha(\sigma, \beta, \mu)$, cf. [256, Definition 1.1.4].

In the strictly stable case we have $d_n = 0$.

Since $\varphi_{X(t)} = \exp(\psi(u)t)$, one easily verifies that $X(t)$ is distributed according to $S_\alpha(\sigma t^{1/\alpha}, \beta, \mu t)$. The stable law $S_1(\sigma, 0, \mu)$ is called the *Cauchy distribution*. It belongs to the few examples where the density is known in closed form, namely

$$\varrho(x) = \frac{\sigma}{\pi((x-\mu)^2 + \sigma^2)}.$$

For $\alpha = 2$ we obtain a Gaussian distribution or more precisely $S_2(\sigma, \beta, \mu) = N(\mu, 2\sigma^2)$. Note that the skewness parameter β has no effect in this case and that σ^2 differs from the variance by a factor of 2.

From Theorems 2.19 and 2.23 we can immediately derive the moments and path properties of α-stable Lévy motions. For $\alpha < 2$ any moment of order $p < \alpha$ exists, the absolute moment of order α itself being infinite. In particular, any α-stable Lévy motion has infinite variance unless it is Brownian motion with drift. The expectation exists for $\alpha > 1$. Any α-stable Lévy motion X except Brownian motion jumps infinitely often on any interval of positive length. X is of infinite variation if and only if $\alpha \geq 1$. Increasing α-stable Lévy motions exist for $\alpha < 1$, namely if $\mu \geq 0$ and the skewness parameter is chosen as $\beta = 1$. Simulated sample paths are shown in Figs. 2.15, 2.16, 2.17 for $\alpha = 0.5, 1, 1.5$. The closer α gets to 2, the more closely these paths resemble Brownian motion.

Finally we come back to the question of convergence. We start with an arbitrary Lévy process X. Suppose that L is the limit in law of the rescaled processes

$$X^{(n)}(t) := n^{-1/\alpha} X(nt) \tag{2.68}$$

as $n \to \infty$. Since X is a Lévy process, $X(knt)$ is distributed as the sum of k independent copies of $X(nt)$. Consequently,

$$\mathscr{L}\big(X^{(kn)}(t)\big) = \mathscr{L}\Big(k^{-1/\alpha}\big(X^{(n,1)}(t) + \cdots + X^{(n,k)}(t)\big)\Big), \tag{2.69}$$

where $X^{(n,1)}, \dots, X^{(n,k)}$ denote k independent copies of $X^{(n)}$. By assumption, $X^{(kn)}(t)$ converges in law to $L(t)$ and the right-hand side of (2.69) to

$$k^{-1/\alpha}\Big(L^{(1)}(t) + \cdots + L^{(k)}(t)\Big),$$

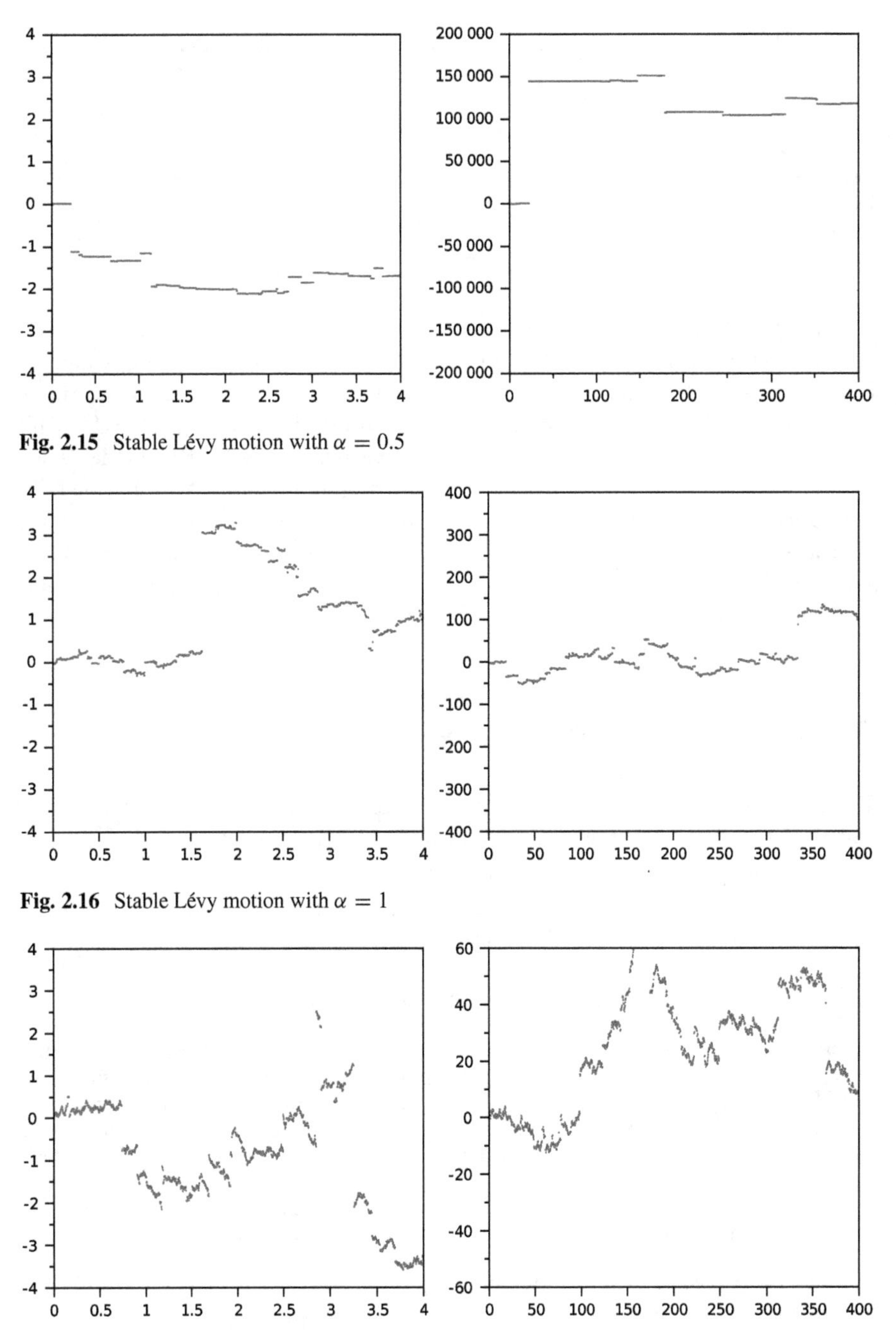

Fig. 2.15 Stable Lévy motion with $\alpha = 0.5$

Fig. 2.16 Stable Lévy motion with $\alpha = 1$

Fig. 2.17 Stable Lévy motion with $\alpha = 1.5$

where $L^{(1)}, \dots, L^{(k)}$ denote independent copies of L. It follows that

$$\mathscr{L}\big(L^{(1)}(t) + \cdots + L^{(k)}(t)\big) = \mathscr{L}\big(k^{1/\alpha} L(t)\big),$$

which means that $L(t)$ is strictly α-stable or, put differently, L is a strictly α-stable Lévy motion. If we allow for some drift $\delta_n t$ in (2.68), i.e.

$$X^{(n)}(t) := n^{-1/\alpha} X(nt) + \delta_n t, \tag{2.70}$$

we end up with an α-stable Lévy motion in the limit.

Conversely, it is easy to see that any α-stable resp. strictly α-stable Lévy motion L may occur as the limiting object of a rescaled Lévy process X in the sense of (2.70) resp. (2.68). By self-similarity we may, for example, choose $X = L$ itself.

Occasionally, stable laws and processes have been suggested to model asset returns because they allow for the leptocurtosis present in real data. Today the tails of stable laws are mostly considered to be heavier than those of real data. Moreover, their lacking higher moments make them difficult to use in Mathematical Finance. Therefore, we discuss them here at the very end in spite of their importance in the mathematical theory.

2.5 The Lévy–Itō Decomposition

The characteristic exponent of a general Lévy process decomposes into three terms, namely a linear one for the drift, a quadratic one for Brownian motion and an integral term taking care of jumps.

Physicists' Corner 2.32 The integral term, on the other hand, may liberally be viewed as an infinite sum of possibly "infinitesimal functions"

$$\psi_x(u) := \big(e^{iu^\top x} - 1 - iu^\top h(x)\big) K(dx). \tag{2.71}$$

Let us fix $h(x) = x 1_{\{|x| \leq 1\}}$ as the truncation function. Equation (2.71) coincides with the characteristic function of a rescaled and partly compensated Poisson process with jump size x and possibly infinitesimal jump intensity $K(dx)$. By *partly compensated* we mean that a linear drift is added which subtracts the average increase $xK(dx)$ of the original Poisson process if $|x| \leq 1$. Since a sum in the characteristic exponent corresponds to a sum of independent random variables, one is tempted to view the Lévy process as a "sum" of a linear drift, a Brownian motion, and a continuum of partly compensated Poisson processes with different jump sizes. The Lévy–Itō decomposition stated in equation (2.76) below makes this intuition precise. □

One may wonder whether an arbitrary Lévy process X cannot simply be decomposed into a linear function for the drift, a Brownian motion, and a compound Poisson process for the jumps. Since X may have infinitely many jumps or even be of infinite variation, such a representation does not exist—but see (2.78) below for the case where it does. Nevertheless, X can be approximated arbitrarily well by such processes. The idea is essentially to remove compensated jumps below some threshold.

Theorem 2.33 (Lévy–Itō Decomposition) *Let X be a Lévy process with Lévy–Khintchine triplet (b^h, c, K) relative to $h(x) = x1_{\{|x|\leq 1\}}$. Then X can be written as*

$$X(t) = b^h t + \sqrt{c}W(t) + \sum_{s\leq t} \Delta X(s) 1_{\{|\Delta X(s)|>1\}} \qquad (2.72)$$

$$+ \lim_{\varepsilon\to 0}\left(\sum_{s\leq t} \Delta X(s) 1_{\{\varepsilon\leq|\Delta X(s)|\leq 1\}} - \int x 1_{\{\varepsilon\leq|x|\leq 1\}} K(dx)t\right)$$

with some standard Wiener process W. The limit refers to convergence in probability, uniformly on intervals $[0, t]$. In the vector-valued case $\sqrt{c}$ is the square root of c, i.e. the unique symmetric, nonnegative definite matrix with $\sqrt{c}\sqrt{c} = c$, and W is a standard Brownian motion in $\mathbb{R}^d$.

Idea The Lévy–Itō decomposition is a special case of a more general decomposition in Theorem 4.19. We confine ourselves to giving a less formal argument at this point.

To this end, consider independent Lévy processes $L_1, L_2, L_{3,\varepsilon}$ with triplets $(b^h, c, 0)$, $(0, 0, K|_{[-1,1]^C})$, and $(-\int x 1_{\{\varepsilon\leq|x|\leq 1\}} K(dx), 0, K|_{[-1,1]\setminus[-\varepsilon,\varepsilon]})$ with some small $\varepsilon > 0$, all relative to the truncation function 0. As a Brownian motion with drift, L_1 can be written as $L_1(t) = b^h t + \sqrt{c}W(t)$ for some Wiener process W, cf. Sect. 2.4.2. L_2 is a compound Poisson process and hence of the form

$$L_2(t) = \sum_{s\leq t} \Delta L_2(s),$$

cf. Sect. 2.4.3. Finally, $L_{3,\varepsilon}$ is a compound Poisson process with an additional drift, i.e.

$$L_{3,\varepsilon}(t) = \sum_{s\leq t} \Delta L_{3,\varepsilon}(s) - \int x 1_{\{\varepsilon\leq|x|\leq 1\}} K(dx)t. \qquad (2.73)$$

According to (2.21), switching to the truncation function h yields triplets $(b^h, c, 0)$, $(0, 0, K|_{[-1,1]^C})$, and $(0, 0, K|_{[-1,1]\setminus[-\varepsilon,\varepsilon]})$, respectively. The limit $(0, 0, K|_{[-1,1]})$ of the third triplet for $\varepsilon \to 0$ is a Lévy–Khintchine triplet as well. It can be shown that the corresponding process $L_{3,0}$ allows for a representation similar to (2.73), namely

$$L_{3,0}(t) = \lim_{\varepsilon\to 0}\left(\sum_{s\leq t} \Delta L_{3,\varepsilon}(s) 1_{\{\varepsilon\leq|\Delta L_{3,\varepsilon}(s)|\leq 1\}} - \int x 1_{\{\varepsilon\leq|x|\leq 1\}} K(dx)t\right).$$

By Proposition 2.25 the sum

$$
\begin{aligned}
L(t) &:= L_1(t) + L_2(t) + L_{3,0}(t) \\
&= b^h t + \sqrt{c} W(t) + \sum_{s \le t} \Delta L(s) 1_{\{|\Delta L(s)| > 1\}} \\
&\quad + \lim_{\varepsilon \to 0} \left(\sum_{s \le t} \Delta L(s) 1_{\{\varepsilon \le |\Delta L(s)| \le 1\}} - \int x 1_{\{\varepsilon \le |x| \le 1\}} K(dx) t \right)
\end{aligned}
$$

is a Lévy process with triplet (b^h, c, K) relative to h. Since X has the same Lévy–Khintchine triplet, it coincides in law with L, which explains why X allows for a corresponding decomposition. □

Note that the two sums in (2.72) and also their sum represent compound Poisson processes. The integral yields a deterministic linear function. For convergence to hold the two terms in parentheses cannot generally be separated because the limit of either term may not exist. The four separate terms of the right-hand side of (2.72) are illustrated in Fig. 2.18 for a sample path of the Lévy process in Fig. 2.19 with Lévy–Khintchine triplet $(b^0, c, K(dx)) = (0.5, 1, e^{-x} 1_{\{x>0\}} dx)$ as in Sect. 2.4.6.

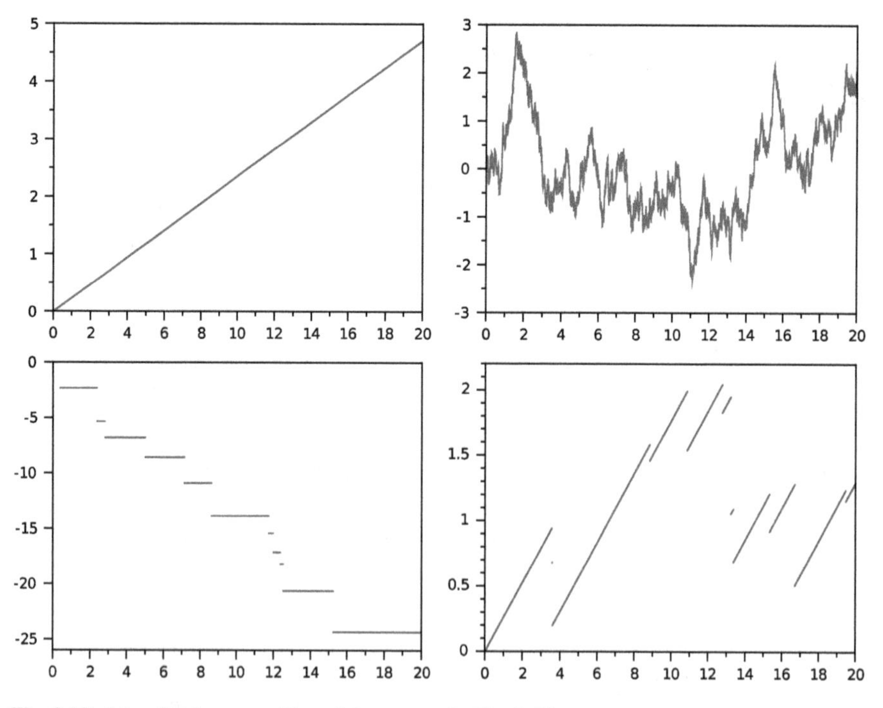

Fig. 2.18 Lévy–Itō decomposition of the process in Fig. 2.19

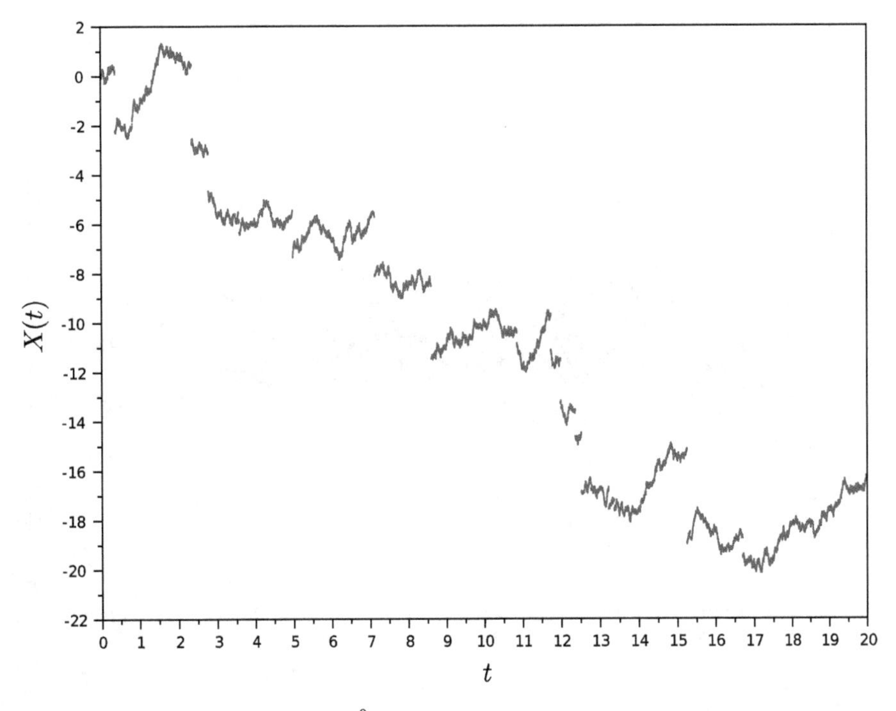

Fig. 2.19 Lévy process with triplet $(b^0, c, K(dx)) = (0.5, 1, e^{-x}1_{\{x>0\}}dx)$

In order to formulate the less obvious decomposition alluded to in Physicists' Corner 2.32, we introduce the random measure of jumps of a Lévy process, which also plays a key role in later chapters on stochastic calculus. The idea is to interpret the jumps of an $\mathbb{R}^d$-valued Lévy process X as points (t, x) or unit masses $\varepsilon_{(t,x)}$ in $\mathbb{R}_+ \times \mathbb{R}^d$. This is visualised in Fig. 2.20, which represents the jumps of X for a fixed path $(X(\omega, t))_{t\geq 0}$. Any point (t, x) in this diagram signifies that $\Delta X(\omega, t) = x$ for the random outcome ω under consideration. The **random measure of jumps** μ^X of X counts the points in the diagram, i.e.

$$\mu^X([0,t] \times B) := \left|\left\{(s,x) \in [0,t] \times B \setminus \{0\} : \Delta X(s) = x\right\}\right| \tag{2.74}$$

for sets $B \in \mathscr{B}^d$. It is random because $\Delta X(s)$ and hence also $\mu^X([0,t] \times B)$ depend on the particular path ω.

The random measure of jumps of a Lévy process is an example of a **homogeneous Poisson random measure** μ on $\mathbb{R}_+ \times \mathbb{R}^d$. This means that:

1. $\mu(\omega, \cdot)$ is an integer-valued measure on $\mathbb{R}_+ \times \mathbb{R}^d$ for fixed $\omega \in \Omega$. *Integer-valued* here means that it takes its values in $\mathbb{N} \cup \{\infty\}$.
2. $\mu(\omega, \{t\} \times \mathbb{R}^d) \leq 1$ for any $\omega \in \Omega$, $t \geq 0$.
3. μ is *adapted* in the sense that $\mu(\cdot, [0,t] \times B)$ is $\mathscr{F}_t$-measurable for any $t \geq 0$ and any $B \in \mathscr{B}^d$.

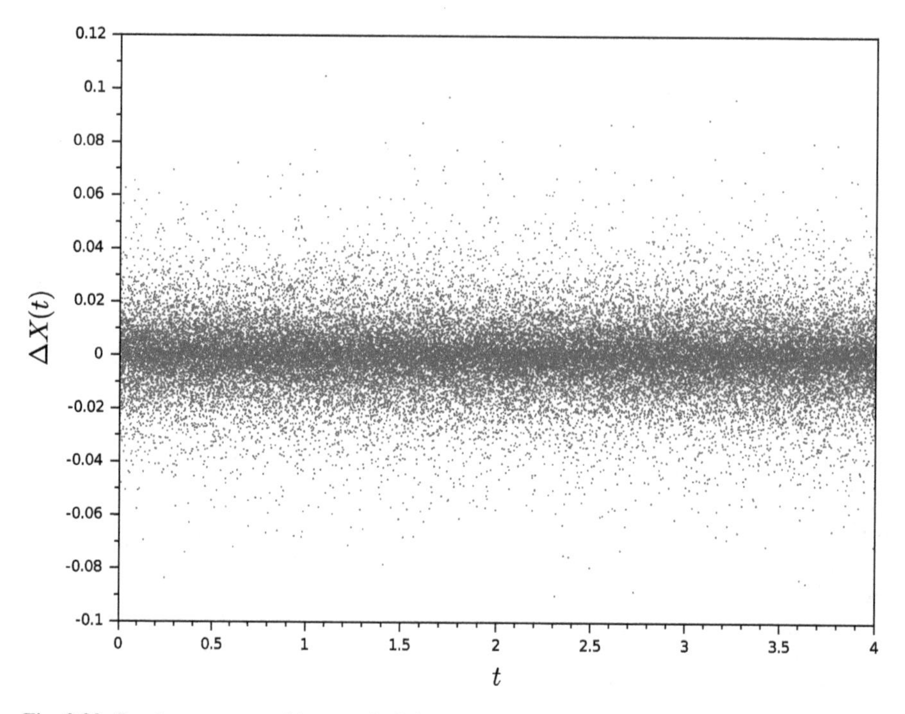

Fig. 2.20 Random measure of jumps of a Lévy process

4. The random variable $\mu(\cdot, (s, t] \times B)$ is independent of $\mathscr{F}_s$ for $s \leq t$, $B \in \mathscr{B}^d$.
5. $E(\mu(\cdot, [0, t] \times B)) = tK(B)$ for $t \geq 0$, $B \in \mathscr{B}^d$ with some (non-random) *intensity measure* K on $\mathbb{R}^d$.
6. μ satisfies some technical condition called *predictable σ-finiteness* which is not crucial for the following, cf. Sect. 3.3 for a precise definition.

A homogeneous Poisson random measure can be interpreted as a counting measure which counts random events in $\mathbb{R}^d$ occurring with constant rate. $\mu(\omega, [0, t] \times B)$ stands for the random number of events (e.g. jumps of a Lévy process) in B happening in the time interval $[0, t]$. The average number of these events is $tK(B)$ by condition 5. Condition 2 means that two events do not happen at the same time. Moreover, the axioms imply that $(\mu(\cdot, [0, t] \times B))_{t\geq 0}$ is an increasing Lévy process growing only by jumps of size 1. Consequently, it is a Poisson process with parameter $K(B)$. Subtracting its compensator yields a martingale, namely the compensated Poisson process $(\mu(\cdot, [0, t] \times B) - tK(B))_{t\geq 0}$.

Summing up jumps as in (2.72) means integrating with respect to the random measure of jumps. Indeed, we have

$$\sum_{s\leq t} f(\Delta X(s)) = \int_{[0,t]\times\mathbb{R}} f(x)\mu^X(d(s, x)) = \int_{\mathbb{R}} f(x)\mu^X([0, t] \times dx))$$

for any real-valued function f, where the last expression is just another notation for the second. Consequently, equation (2.72) can be rewritten as

$$\begin{aligned} X(t) &= b^h t + \sqrt{c}W(t) + \int_{\mathbb{R}} x 1_{\{|x|>1\}} \mu^X([0,t] \times dx) \\ &\quad + \lim_{\varepsilon \to 0} \left(\int_{\mathbb{R}} x 1_{\{\varepsilon \le |x| \le 1\}} \mu^X([0,t] \times dx) - \int_{\mathbb{R}} x 1_{\{\varepsilon \le |x| \le 1\}} K(dx)t \right) \\ &= b^h t + \sqrt{c}W(t) + \int_{\mathbb{R}} x 1_{\{|x|>1\}} \mu^X([0,t] \times dx) \\ &\quad + \lim_{\varepsilon \to 0} \int_{\mathbb{R}} x 1_{\{\varepsilon \le |x| \le 1\}} \left(\mu^X([0,t] \times dx) - K(dx)t \right). \end{aligned} \tag{2.75}$$

This suggests to simply write

$$\begin{aligned} X(t) &= b^h t + \sqrt{c}W(t) + \int_{\mathbb{R}} x 1_{\{|x|>1\}} \mu^X([0,t] \times dx) \\ &\quad + \int_{\mathbb{R}} x 1_{\{|x| \le 1\}} \left(\mu^X([0,t] \times dx) - K(dx)t \right) \end{aligned} \tag{2.76}$$

and to call it a **Lévy–Itō decomposition** as well. But since the integrals

$$\int_{\mathbb{R}} x 1_{\{|x| \le 1\}} \mu^X([0,t] \times dx) \quad \text{and} \quad \int_{\mathbb{R}} x 1_{\{|x| \le 1\}} K(dx)t$$

may not be defined separately, the right-hand side of (2.76) does not seem to make sense. For the time being we use it as a shorthand for (2.75) but in Sect. 3.3 we will get to know an alternative definition.

Physicists' Corner 2.34 Equation (2.76) represents the promised pathwise decomposition of an arbitrary Lévy process into a linear drift, a Brownian motion part and a superposition of uncountably many jump processes of fixed jump size x. For $|x| > 1$, the latter are rescaled Poisson processes

$$\left(x\mu^X([0,t] \times dx) \right)_{t \ge 0}$$

with possibly "infinitesimal" intensity $K(dx)$. For $|x| \le 1$, on the other hand, the corresponding jump processes are multiples of compensated Poisson processes

$$\left(x \left(\mu^X([0,t] \times dx) - K(dx)t \right) \right)_{t \ge 0}.$$

Of course, this admittedly somewhat doubtful interpretation is meant only to help understand (2.76). □

If X has finite expectation, (2.22) holds for $p = 1$, which means that $h(x) = x$ can be used as a "truncation function." Thus (2.76) can be written in the simpler form

$$X(t) = a^X t + \sqrt{c}W(t) + \int_{\mathbb{R}} x \left(\mu^X([0,t] \times dx) - K(dx)t\right) \tag{2.77}$$

with

$$a^X = b^{\text{id}} = b^h + \int (x - h(x))K(dx)$$

as usual. The three terms correspond to the drift part of the special semimartingale X, to a Brownian motion, and to a martingale which is determined entirely by its jumps. Figure 2.21 illustrates (2.77) for a Kou model.

If, on the other hand, the sum of jumps of X exists on any finite interval $[0, t]$, we can use $h(x) = 0$. This leads to

$$\begin{aligned} X(t) &= b^0 t + \sqrt{c}W(t) + \int x\mu^X([0,t] \times dx) \\ &= b^0 t + \sqrt{c}W(t) + \sum_{s \le t} \Delta X(s). \end{aligned} \tag{2.78}$$

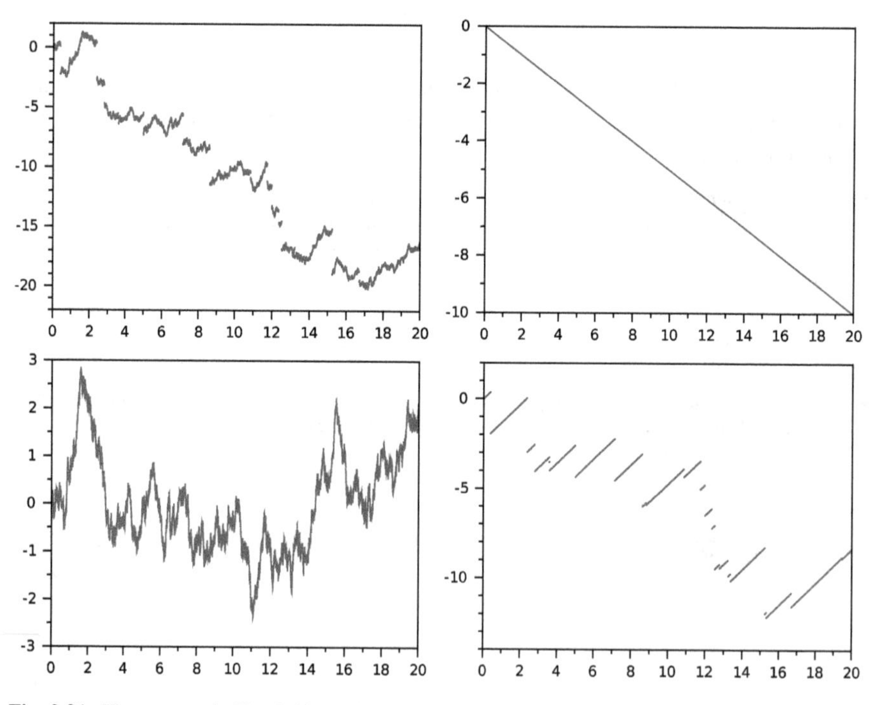

Fig. 2.21 The process in Fig. 2.19 and its decomposition (2.77)

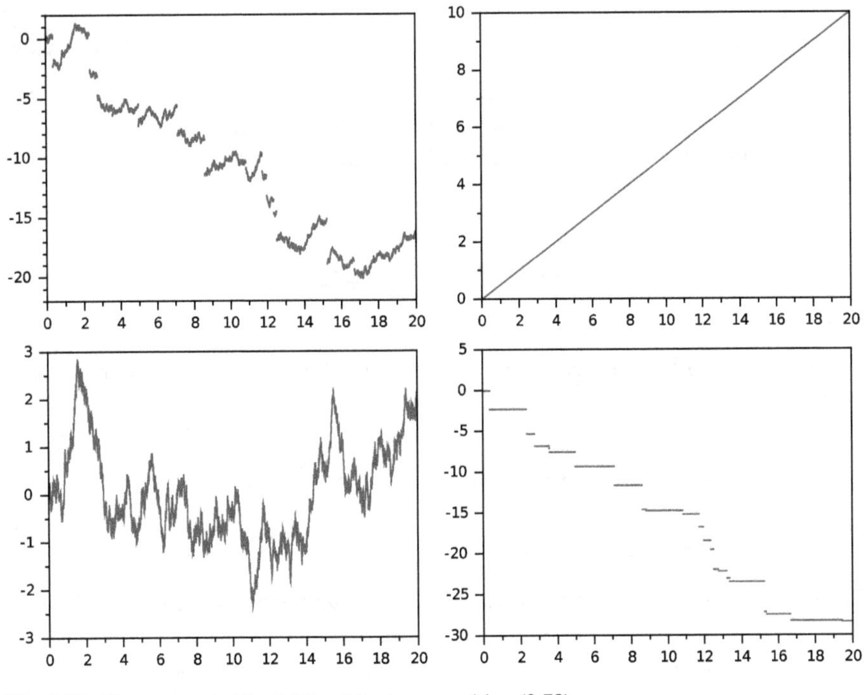

Fig. 2.22 The process in Fig. 2.19 and its decomposition (2.78)

Here the three terms correspond to a linear function, a Brownian motion, and a compound Poisson process for the jumps. Decomposition (2.78) is visualised in Fig. 2.22 for a Kou model.

Appendix 1: Problems

The exercises correspond to the section with the same number.

2.1 Show that an adapted càdlàg process X is a martingale if and only if $X(\tau)$ is integrable and $E(X(\tau)) = E(X(0))$ for any bounded stopping time τ.

2.2 Let X be a Lévy process and B a Borel set. Let $\tau := \inf\{t \geq 0 : \Delta X(t) \in B\}$, i.e. τ denotes the first time where X has a jump of size in B. Show that τ has exponential distribution unless it is almost surely 0 or ∞.

Hint: recall that the exponential distribution is characterised by having no memory, i.e. $P(\tau > t + s | \tau > t) = P(\tau > s)$ for any $s, t \geq 0$.

2.3 Show that if a counting process N is a Lévy process with $E(N(1)) = \lambda$, then

$$E(e^{iuN(1)}) = \exp((e^{iu} - 1)\lambda), \quad u \in \mathbb{R},$$

i.e. N is a Poisson process with parameter λ.

Hint: consider $N_n := |\{i \in \{1, \ldots, 2^n\} : N_{i/2^n} - N_{(i-1)/2^n} \geq 1\}|$ and show $N_n \to N(1)$ in law as $n \to \infty$.

2.4 Let N be a Poisson process with parameter λ and $(Y_n)_{n\in\mathbb{N}}$ a sequence of random variables with law Q such that N and the Y_n are all independent. Define the process X by

$$X(t) := \sum_{n=1}^{N(t)} Y_n, \quad t \in \mathbb{R}_+.$$

Show that X is a Lévy process relative to its own filtration and has characteristic function

$$E(e^{iuX(1)}) = \exp\left(\int (e^{iux} - 1)\lambda Q(dx)\right), \quad u \in \mathbb{R}.$$

Hint: condition on the value of N.

2.5 For an adapted càdlàg process X define its *random measure of jumps* μ^X by

$$\mu^X([0, t] \times B) := \left|\{(s, x) \in [0, t] \times B : \Delta X(s) = x\}\right|$$

for $t \geq 0$, $B \in \mathscr{B}$. Show that μ^X is an *integer-valued random measure* on $\mathbb{R}_+ \times \mathbb{R}$ in the sense that

1. $\mu^X(\omega, \cdot)$ is a measure on $\mathbb{R}_+ \times \mathbb{R}$ for fixed $\omega \in \Omega$,
2. $\mu^X(\omega, \{0\} \times \mathbb{R}) = 0$ for any $\omega \in \Omega$,
3. $\mu^X(\omega, \cdot)$ has values in $\mathbb{N} \cup \{\infty\}$ for any $\omega \in \Omega$,
4. $\mu^X(\omega, \{t\} \times \mathbb{R}) \leq 1$ for any $\omega \in \Omega, t \geq 0$,
5. μ^X is *adapted* in the sense that $\mu^X(\cdot, [0, t] \times B)$ is $\mathcal{F}_t$-measurable for fixed $t \geq 0$ and any $B \in \mathscr{B}$,
6. μ^X is *predictably σ-finite* in the sense that there exists some strictly positive $\mathcal{P} \otimes \mathscr{B}$-measurable function η on $\Omega \times \mathbb{R}_+ \times \mathbb{R}$ such that

$$\iint_{\mathbb{R}_+ \times \mathbb{R}} \eta(\omega, t, x)\mu^X(\omega, d(t, x))P(d\omega) < \infty.$$

Moreover, show that the random measure of jumps μ^X of a Lévy process satisfies in addition:

7. the random variable $\mu^X(\cdot, (s, t] \times B)$ is independent of $\mathcal{F}_s$ for $s \leq t$, $B \in \mathscr{B}$,
8. $E(\mu^X(\cdot, [0, t] \times B)) = tK(B)$, $t \geq 0$, $B \in \mathscr{B}$ for some (non-random) *intensity measure* K on $\mathbb{R}$. (One can show that K is in fact the Lévy measure of X.)

Hint: You may use without proof that $\tau := \inf\{t \geq 0 : \Delta X(t) \in B\}$ is a stopping time for any adapted càdlàg process X and any Borel set B. Strictly speaking, this can be guaranteed only if the filtration is *complete*, i.e. if $\mathcal{F}_0$ contains all $\mathcal{F}$-null sets.

Appendix 2: Notes and Comments

For Sect. 2.1, see for example [154]. General references on Lévy processes include [2, 238, 259]. References [23, 202] treat the fluctuation theory which is not discussed here, whereas [38, 60, 263] consider Lévy processes from the point of view of applications in finance.

For Theorem 2.24 we refer to [259, Theorem 30.1]. Proposition 2.25 can be viewed as a special case of [259, Proposition 11.10], cf. also [60, Example 4.1]. Proposition 2.26 can be obtained as an application of [182, Corollary 2.21] and [154, Corollary II.4.19]. A law of large numbers related to Theorem 2.27 is stated in [259, Theorem 36.5].

Most of the Lévy processes in Sect. 2.4 and their properties are discussed in [60, Chapter 4] and [263, Section 5.3]. For specific processes see also [9, 82, 88, 90, 91, 195, 210, 220]. The excellent fit of hyperbolic distributions to empirical return distributions of German stock price data led to the introduction of hyperbolic Lévy motions in finance, cf. [82]. The bilateral gamma process is introduced in [200], and for multivariate generalised hyperbolic processes we refer to [288, 289]. Stable Lévy motions are discussed in detail in [256, 259], cf. also [60, Section 3.7]. For the results in Sect. 2.4.11 we refer in particular to [256, Sections 1.1, 2.3, 3.1, 7.5]. The Lévy–Itô decomposition of Theorem 2.33 can be found in [259, Theorem 19.2], even stated with almost sure convergence.

Chapter 3
Stochastic Integration

Traditional stochastic calculus is based on stochastic integration. It constitutes the basis of modern Mathematical Finance. The most important notions and results from the theory are presented in this chapter.

3.1 More on General Semimartingale Theory

Before we define the stochastic integral in the subsequent sections, we review some notions from the general theory of stochastic processes. They are mostly of interest in their own right.

3.1.1 Quadratic Variation

Rules from stochastic calculus such as Itō's formula involve a concept which has no counterpart in ordinary calculus, namely the *quadratic variation* or *covariation* process. We encountered it already in Sect. 1.2 where we discussed stochastic "integration" in discrete time. What is the analogue of equation (1.13) in continuous time? A reasonable idea is to consider semimartingales artificially as discrete-time processes on a fine grid $0 = t_0 < t_1 < t_2 < \dots$ and let the mesh size $\sup_{i\geq 1} |t_i - t_{i-1}|$ tend to 0 afterwards. This leads to

Definition 3.1 If X is a semimartingale, the semimartingale $[X, X]$ defined by

$$[X, X](t) := \lim_{n\to\infty} \sum_{i\geq 1} \left(X(t_i^{(n)} \wedge t) - X(t_{i-1}^{(n)} \wedge t) \right)^2 \tag{3.1}$$

© Springer Nature Switzerland AG 2019
E. Eberlein, J. Kallsen, *Mathematical Finance*, Springer Finance,
https://doi.org/10.1007/978-3-030-26106-1_3

is called the **quadratic variation** of X. Here, $(t_i^{(n)})_{i\geq 1}$, $n \in \mathbb{N}$ denote sequences of numbers with $0 = t_0^{(n)} < t_1^{(n)} < t_2^{(n)} < \dots$ with $t_i^{(n)} \uparrow \infty$ for $i \to \infty$ and such that the mesh size $\sup_{i\geq 1} |t_i^{(n)} - t_{i-1}^{(n)}|$ tends to 0 as $n \to \infty$.

The **covariation** $[X, Y]$ of semimartingales X, Y is defined accordingly as

$$[X, Y](t) := \lim_{n\to\infty} \sum_{i\geq 1} \left(X(t_i^{(n)} \wedge t) - X(t_{i-1}^{(n)} \wedge t)\right) \left(Y(t_i^{(n)} \wedge t) - Y(t_{i-1}^{(n)} \wedge t)\right). \tag{3.2}$$

The limits are to be understood in measure, uniformly on compact intervals $[0, t]$.

It may seem far from evident why the limits in the above definition should exist, let alone be semimartingales. One can bypass this problem by using (3.17) below as a definition of the covariation, cf. [154, Definition I.4.45].

In this case the above representation is derived only afterwards as a theorem, cf. [154, Theorem I.4.47]. We prefer to introduce $[X, X]$ and $[X, Y]$ by (3.1) resp. (3.2) because these equations motivate the denomination and give a feeling for what these processes look like. However, for concrete calculations we will hardly need the representation of quadratic variation as a limit.

Example 3.2 (Lévy Process) Suppose that X is a real-valued Lévy process with Lévy–Khintchine triplet (b^h, c, K). Then its quadratic variation equals

$$\begin{aligned}[X, X](t) &= ct + \sum_{s\leq t} (\Delta X(s))^2 \\ &= ct + \int x^2 \mu^X([0, t] \times dx),\end{aligned} \tag{3.3}$$

where μ^X denotes the random measure of jumps defined in Sect. 2.5. $[X, X]$ is again a Lévy process and its Lévy–Khintchine triplet equals $(\widetilde{b}^0, \widetilde{c}, \widetilde{K}) = (c, 0, \widetilde{K})$ with

$$\widetilde{K}(B) = \int 1_B(x^2) K(dx), \quad B \in \mathscr{B}. \tag{3.4}$$

Idea From (3.1) one observes that $[X, X]$ is increasing and has jumps of height $\Delta[X, X](t) = (\Delta X(t))^2$. Moreover, as $[X, X](t) - [X, X](s)$ depends only on $(X(r) - X(s))_{r\in[s,t]}$, we see that $[X, X]$ is a Lévy process as well. Since it is of finite variation, it cannot have a Brownian motion part, i.e. we have

$$[X, X](t) = b^{[X,X]}t + \sum_{s\leq t} (\Delta[X, X](s))^2 = b^{[X,X]}t + \sum_{s\leq t} (\Delta X(s))^2$$

with some constant $b^{[X,X]} \geq 0$. From (3.1) and

$$
\begin{aligned}
E\big((X(t)-X(s))^2\big) &= E\big((X(t-s))^2\big) \\
&= \mathrm{Var}(X(t-s)) + E(X(t-s))^2 \\
&= \left(c + \int x^2 K(dx)\right)(t-s) + E(X(1))^2(t-s)^2 \qquad (3.5)
\end{aligned}
$$

(cf. Theorem 2.19) we conclude that $E([X,X](t)) = ct + \int x^2 K(dx)t$ because the contribution in (3.1) of the second term in (3.5) vanishes as the mesh size tends to zero. Since

$$
E\Big(\sum_{s\leq t}(\Delta X(s))^2\Big) = E\Big(\int x^2 \mu^X([0,t]\times dx)\Big) = E\Big(\int x^2 K(dx)\Big)t,
$$

we obtain $b^{[X,X]} = c$. Equation (3.4) for the Lévy measure is a consequence of the fact that $[X,X]$ jumps by x^2 whenever X jumps by x.

A short rigorous proof of the Lévy property of $[X,X]$ and its triplet is obtained by using the notions of Chap. 4, namely combining Propositions 4.11(1), 4.6, 2.25. □

As a side remark we note that any continuous local martingale X starting at 0 and satisfying $[X,X](t) = t$ is a standard Brownian motion. This follows from Example 3.28 below.

If X is a vector-valued Lévy process, the covariation between its components is given by

$$
\begin{aligned}
[X_i, X_j](t) &= c_{ij}t + \sum_{s\leq t}\Delta X_i(s)\Delta X_j(s) \\
&= c_{ij}t + \int x_i x_j \mu^X([0,t]\times dx),
\end{aligned}
$$

which naturally generalises (3.3).

If $[X,X]$ is a *special* semimartingale, it has a decomposition

$$
[X,X] = M^{[X,X]} + A^{[X,X]}
$$

as in (2.6). The compensator $A^{[X,X]}$ in this decomposition is called **predictable quadratic variation** and it is written $\langle X,X\rangle$. Likewise, the compensator of $[X,Y]$ is called **predictable covariation** and denoted $\langle X,Y\rangle$ if it exists. Figures 3.1 and 3.2 illustrate both quadratic variation and predictable quadratic variation for Brownian motion and a normal-inverse-Gaussian Lévy process, respectively.

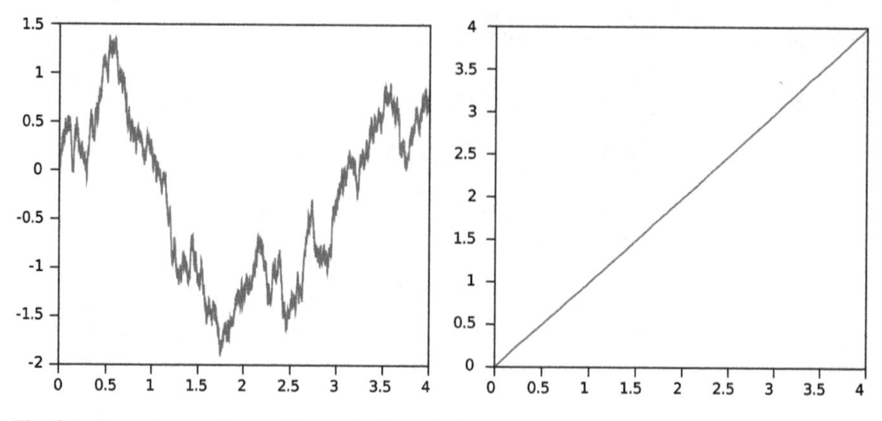

Fig. 3.1 Brownian motion and its quadratic variation

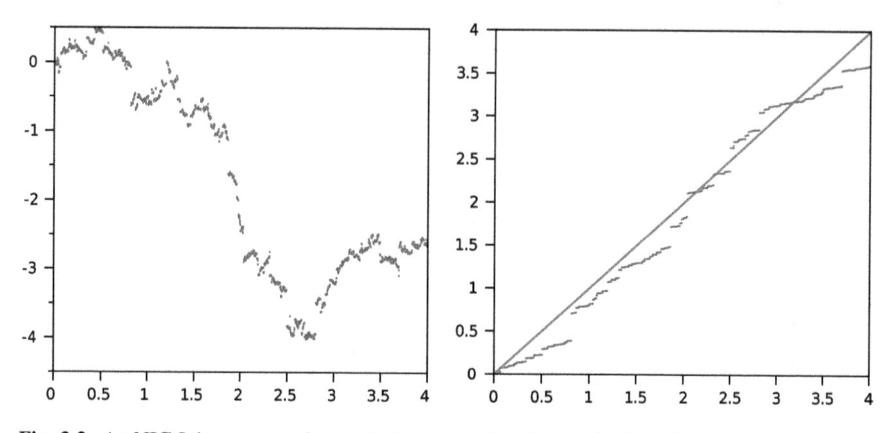

Fig. 3.2 An NIG Lévy process, its quadratic and predictable quadratic variation

Example 3.3 (Lévy Process) As in Example 3.2 we consider a Lévy process with triplet (b^h, c, K). In the univariate case, its predictable quadratic variation is given by

$$\langle X, X\rangle(t) = \left(c + \int x^2 K(dx)\right) t = \text{Var}(X(t)) \tag{3.6}$$

provided that the integral is finite. Indeed, by Theorems 2.21 and 2.19 the compensator of the Lévy process $[X, X]$ equals

$$A^{[X,X]}(t) = \left(\widetilde{b}^0 + \int x \widetilde{K}(dx)\right)t = \left(c + \int x^2 K(dx)\right)t = \text{Var}(X(t)),$$

where we use the notation of Example 3.2.

In the multivariate case, we obtain accordingly

$$\langle X_i, X_j \rangle(t) = \left(c_{ij} + \int x_i x_j K(dx)\right) t = \mathrm{Cov}\big(X_i(t), X_j(t)\big)$$

for the predictable covariation between the components of X.

Let us state a few properties of covariation processes:

Proposition 3.4 *For semimartingales X, Y the following properties hold.*

1. *$[X, Y]$ is symmetric and linear in X and Y. The same holds for $\langle X, Y \rangle$ if it exists.*
2. *$[X, X]$ (and $\langle X, X \rangle$ if it exists) is an increasing process.*
3. *$[X, Y]$ (and $\langle X, Y \rangle$ if it exists) is a process of finite variation.*
4. *$\langle X, Y \rangle = [X, Y]$ if X or Y is continuous.*
5. *$[X, Y] = 0$ if X is continuous and X or Y is of finite variation.*

Proof

1. This follows directly from the definition.
2. This is also immediate from the definition.
3. This follows from statement 2 and the decomposition

$$[X, Y] = \frac{1}{4}\Big([X + Y, X + Y] - [X - Y, X - Y]\Big)$$

and accordingly for $\langle X, Y \rangle$.

4. From the definition it follows that $\Delta[X, Y](t) = \Delta X(t)\Delta Y(t)$. If X or Y is continuous, this implies that $[X, Y]$ is continuous and hence predictable. Since predictable processes of finite variation coincide with their compensator, we are done.
5. Fix $\varepsilon > 0$, $t \in \mathbb{R}_+$ and consider a fixed path. Let $0 = t_0 < t_1 < t_2 < \ldots \uparrow \infty$. For sufficiently small $t_i - t_{i-1}$ we have $|X(t_i \wedge t) - X(t_{i-1} \wedge t)| < \varepsilon$ by continuity of X.

Case 1: Y is of finite variation. If the variation process of Y is denoted by $\mathrm{Var}(Y)$, we have

$$\sum_{i \geq 1} |X(t_i \wedge t) - X(t_{i-1} \wedge t)|\,|Y(t_i \wedge t) - Y(t_{i-1} \wedge t)| \leq \varepsilon \mathrm{Var}(Y)(t),$$

which yields $[X, Y](t) = 0$.

Case 2: X is of finite variation. Decompose Y as $Y = Y' + Y''$ with $Y'(t) := \sum_{s \leq t} \Delta Y(s) 1_{\{|\Delta Y(s)| > \varepsilon\}}$. Since Y is càdlàg, it has only finitely many "big" jumps on $[0, t]$, which implies that Y' is of finite variation. Hence case 1 applies and yields $[X, Y'](t) = 0$. Since the jumps of Y'' are bounded by ε, we have $|Y''(t_i \wedge t) - Y''(t_{i-1} \wedge t)| < 2\varepsilon$ for sufficiently small $t_i - t_{i-1}$.

This yields

$$\sum_{i\geq 1} |X(t_i \wedge t) - X(t_{i-1} \wedge t)|\, |Y(t_i \wedge t) - Y(t_{i-1} \wedge t)| \leq \mathrm{Var}(X)(t)2\varepsilon$$

and hence $[X, Y](t) = [X, Y'](t) + [X, Y''](t) < 2\varepsilon \mathrm{Var}(X)(t)$.

For a different proof see [154, Proposition I.4.49d]. □

We mentioned already in Sect. 2.1 that compensators are typically absolutely continuous in time, i.e. the predictable covariation is often of the form

$$\langle X, Y\rangle(t) = \int_0^t \widetilde{c}^{X,Y}(s)ds \tag{3.7}$$

with some predictable process $\widetilde{c}^{X,Y}$. For Lévy processes this integrand reduces to a deterministic constant, as the previous example shows. We come back to this issue in Theorem 4.26.

3.1.2 Square-Integrable and Purely Discontinuous Martingales

Martingales M may or may not converge to an integrable limit random variable $M(\infty)$ such that the martingale condition

$$E(M(t)|\mathscr{F}_s) = M(s), \quad s \leq t$$

also holds for $t = \infty$. Such martingales are called **uniformly integrable martingales**, cf. Sect. 4.5. Since the whole martingale M is generated from its limit $M(\infty)$ by conditional expectation, we may identify the set of uniformly integrable martingales with the set $L^1(\Omega, \mathscr{F}_\infty, P)$ of integrable random variables.

For purposes of stochastic integration, the subset $\mathscr{H}^2$ of **square-integrable martingales** plays a more important role. Here one requires in addition that the limit has finite second moment $E(M(\infty)^2)$. Equivalently $E([M, M](\infty)) < \infty$, which holds if $[M, M]$ is special and $E(\langle M, M\rangle(\infty)) < \infty$, cf. [154, Theorem I.4.50].

By way of the above identification, these processes correspond to the Hilbert space $L^2(\Omega, \mathscr{F}_\infty, P)$ of square-integrable random variables. The elements of the corresponding localised class $\mathscr{H}^2_{\mathrm{loc}}$ in the sense of Sect. 2.1 are called **locally square-integrable martingale** s. For such processes there is an alternative interpretation of predictable covariation processes:

Proposition 3.5 *If M, N are locally square-integrable martingales, $\langle M, N\rangle$ is the compensator of MN. If M, N are square-integrable martingales, the difference $MN - \langle M, N\rangle$ is even a uniformly integrable martingale.*

Idea $[M, N]$ is defined as a limit of discrete-time covariation if the mesh size of the time grid tends to zero. Leaving integrability issues aside, Proposition 3.5 corresponds to Proposition 1.14(10) in discrete time, which was proved using integration by parts. The above result can thus be understood and obtained as a limiting case.

For a rigorous argument, note that the properties in the proposition are taken as a definition of $\langle M, N\rangle$ in [154, Theorem I.4.2].

Reference [154, Theorem I.4.50b] yields that it coincides with the compensator of $[M, N]$. □

The space of square-integrable random variables is a Hilbert space, i.e. it is endowed with a scalar product defined as $E(UV)$ for $U, V \in L^2(\Omega, \mathscr{F}_\infty, P)$. In view of the above identification, this naturally induces a scalar product on $\mathscr{H}^2$, namely $E(M(\infty)N(\infty))$ for $M, N \in \mathscr{H}^2$. This turns $\mathscr{H}^2$ into a Hilbert space relative to the norm

$$\|M\|_{\mathscr{H}^2} := E\big(M(\infty)^2\big).$$

This scalar product can be used to define the concept of a **purely discontinuous (square-integrable) martingale**. It refers to any square-integrable martingale M that is orthogonal to all continuous square-integrable martingales, i.e.

$$E(M(\infty)N(\infty)) = 0$$

for any continuous $N \in \mathscr{H}^2$. As a result we can decompose any square-integrable martingale M uniquely as an orthogonal sum

$$M = M(0) + M^c + M^d \tag{3.8}$$

of a continuous martingale M^c with $M^c(0) = 0$ and a purely discontinuous martingale M^d, respectively.

Equation (3.8) can be extended to local martingales. To this end, we first observe the following connection between orthogonality and predictable covariation.

Proposition 3.6 *If M, N are square-integrable martingales with $M(0) = 0 = N(0)$ and $\langle M, N\rangle = 0$, then M and N are orthogonal in the Hilbert space sense, i.e. $E(M(\infty)N(\infty)) = 0$.*

Proof Since $MN - \langle M, N\rangle$ is a uniformly integrable martingale, we have

$$E(M(\infty)N(\infty) - \langle M, N\rangle(\infty)) = E(M(0)N(0) - \langle M, N\rangle(0)) = 0. \quad □$$

The converse is generally not true. Inspired by Proposition 3.6 one calls local martingales M, N **(strongly) orthogonal** if $\langle M, N\rangle = 0$, i.e. if MN is a local martingale as well. Moreover, local martingales M with $M(0) = 0$ are called **purely discontinuous** if they are strongly orthogonal to all continuous local martingales. For square-integrable martingales this coincides with the concept above, cf. Problem 3.1. One can show that any local martingale can be uniquely decomposed as in (3.8), where M^c, M^d now denote a continuous and a purely discontinuous local martingale, respectively, cf. [154, Theorem I.4.18].

For square-integrable martingales the two decompositions coincide.

From its definition one hardly gets an intuition for what a purely discontinuous local martingale looks like. Such processes occur, for example, as last terms in the Lévy–Itō decompositions (2.76) and (2.77), which are illustrated in Figs. 2.18 and 2.21. Loosely speaking, the randomness of purely discontinuous local martingales is entirely due to jumps. This intuition is supported by the following

Proposition 3.7 *Purely discontinuous local martingales M, N with identical jumps coincide, i.e. $\Delta M = \Delta N$ implies $M = N$.*

Proof Since $M - N$ is both a continuous and a purely discontinuous local martingale, it must be zero by uniqueness of its decomposition into a continuous and a purely discontinuous part, respectively. □

Moreover, we have

Proposition 3.8 *Any local martingale M of finite variation is purely discontinuous if $M(0) = 0$. In particular, any continuous local martingale of finite variation is constant.*

Proof This follows immediately from statements 4 and 5 of Proposition 3.4. □

If X is a semimartingale with decomposition $X = X(0) + M + A$ as in (2.10), the continuous part M^c of the local martingale is called the **continuous martingale partof** X and it is denoted by X^c. One can show that it does not depend on the choice of M, which is generally not unique, cf. [154, Proposition I.4.27] and Sect. 2.1. This notion allows us to decompose the covariation of two semimartingales into two parts, one coming from the continuous martingale part and the other from jumps.

Proposition 3.9 *For any two semimartingales X, Y we have*

$$[X, Y](t) = \langle X^c, Y^c\rangle(t) + \sum_{s\le t} \Delta X(s)\Delta Y(s). \tag{3.9}$$

Idea In this illustration we consider only the case where $\sum_{s\le t} |\Delta X(s)| < \infty$ for any $t \geq 0$ and likewise for Y.

Step 1: We start with $X = Y$. Denote the sum of jumps of X as $J(t) = \sum_{s\le t} \Delta X(s)$. By (3.1) we have

$$[J, J](t) = \sum_{s\le t} (\Delta X(s))^2. \tag{3.10}$$

Since $X - J$ is continuous and of finite variation, Proposition 3.4(5) implies that $[X - J, J]$ vanishes. Observe that the special semimartingale decomposition of $X - J$ is $X - J = X^c + A$ with some continuous process A of finite variation. Indeed, $X - J$ and X have the same continuous martingale part because they differ only by a process of finite variation. Again by Proposition 3.4(5) we have $[X^c, A] = 0$ and $[A, A] = 0$. Altogether we obtain

$$\begin{aligned}[X, X] &= [X - J, X - J] + 2[X - J, J] + [J, J] \\ &= [X^c, X^c] + 2[X^c, A] + [A, A] + 0 + [J, J] \\ &= [X^c, X^c] + [J, J] \\ &= \langle X^c, X^c\rangle + \sum_{s\leq t}(\Delta X(s))^2,\end{aligned}$$

where the last line follows from (3.10) and Proposition 3.4(4).

Step 2: The general case is obtained from step 1 using

$$[X, Y] = \frac{1}{4}\big([X + Y, X + Y] - [X - Y, X - Y]\big).$$

For a rigorous proof see [154, Theorem I.4.52]. □

Corollary 3.10 *For any semimartingales X, Y, Z we have*

$$[X, [Y, Z]](t) = [[X, Y], Z](t) = \sum_{s\leq t}\Delta X(s)\Delta Y(s)\Delta Z(s).$$

Remark 3.11 (Finite Time Horizon) Let us briefly discuss how the theory changes if we work with a compact time interval $[0, T]$ rather than $\mathbb{R}_+$. As noted at the end of Sect. 2.1, this can be considered as a special case of the infinite time horizon setup by setting $\mathscr{F}_t := \mathscr{F}_{T\wedge t}$, $X(t) := X(T \wedge t)$ etc. for $t \geq 0$. In this sense any martingale converges as $t \to \infty$ because it is constant after T. Therefore there is no difference between martingales and uniformly integrable martingales. Moreover, a martingale X is square-integrable if and only if $X(T)$ has finite second moment $E(X(T)^2)$.

Example 3.12 (Lévy Process) Recall the Lévy–Itō decomposition

$$\begin{aligned}X(t) = b^h t + \sqrt{c}W(t) &+ \int x 1_{\{|x|>1\}}\mu^X([0, t] \times dx) \\ &+ \int x 1_{\{|x|\leq 1\}}\Big(\mu^X([0, t] \times dx) - K(dx)t\Big)\end{aligned} \tag{3.11}$$

of a Lévy process with triplet (b^h, c, K) relative to the truncation function $h(x) = x1_{\{|x|\leq 1\}}$. Let us show that

$$X^c = \sqrt{c}W, \tag{3.12}$$

i.e. the continuous martingale part of any Lévy process is a Brownian motion unless it vanishes. Indeed, the first and third term on the right-hand side of (3.11) are of finite variation. Hence, they do not contribute to the continuous martingale part of X. The last integral process in (3.11) is a limit of martingales of finite variation, cf. (2.75). These martingales are purely discontinuous by Proposition 3.8. It can be shown that the limit is also a purely discontinuous local martingale. Hence, the remaining continuous martingale $\sqrt{c}W(t)$ is in fact the continuous local martingale part of X.

Recall that X is a martingale if it satisfies the drift condition

$$b^h + \int x1_{\{|x|>1\}}K(dx) = 0, \tag{3.13}$$

cf. Theorem 2.21. In this case (3.11) yields

$$\begin{aligned} X(t) &= -\int x1_{\{|x|>1\}}K(dx)t + \sqrt{c}W(t) + \int x1_{\{|x|>1\}}\mu^X([0,t]\times dx) \\ &\quad + \int x1_{\{|x|\leq 1\}}\left(\mu^X([0,t]\times dx) - K(dx)t\right) \\ &= \sqrt{c}W(t) + \int x\left(\mu^X([0,t]\times dx) - K(dx)t\right). \end{aligned} \tag{3.14}$$

Since $X(t) = X^c(t) + X^d(t)$ and $X^c = \sqrt{c}W$, the integral term in (3.14) constitutes the purely discontinuous martingale part of X.

Nonzero Lévy processes do not converge as $t \to \infty$. Therefore they cannot be uniformly integrable martingales unless we work with a finite time horizon as in the previous remark. But X is a locally square-integrable martingale if second moments exist, i.e. if (2.22) for $p = 2$ holds in addition to (3.13). In this case the difference $X^2 - \langle X, X\rangle$ is a martingale, cf. (3.6) and Proposition 3.5.

3.2 The Stochastic Integral for Processes

Stochastic integration forms an indispensable basis of stochastic calculus and its applications to finance—much more than the concepts of "differentiation" that are discussed in Chap. 5. From a calculus point of view integration allows us to proceed from the linear to the nonlinear world.

Physicists' Corner 3.13 Linear functions increase by $b\,dt$ in a small time interval $(t-dt,t]$. Their growth rate b is constant. Arbitrary functions may have a growth rate which depends on time, i.e. they increase by $b(t)dt$. Through integration

$$X(t) = X(0) + \int_0^t b(s)ds \tag{3.15}$$

this intuition is given a precise meaning.

Lévy processes $L(t)$ stand for constant growth and can be viewed as a stochastic counterpart of linear functions. But as in the deterministic case, constant growth is an idealisation which holds only on a small scale in many applications. Similarly as above we may consider arbitrary processes which grow as $b(t)dL(t)$ in a small time interval $(t-dt,t]$, i.e. the increment $dL(t) := L(t) - L(t-dt)$ of the Lévy process is multiplied by some time-dependent and generally also random coefficient $b(t)$. This intuition motivates the desire to define the integral in the expression

$$X(t) = X(0) + \int_0^t b(s)dL(s),$$

which generalises (3.15).

From the finance point of view one may even wish to integrate relative to processes more general than just Lévy processes. Recall from the discrete-time discussion in Chap. 1 that stochastic integration proves useful to formalise financial gains of a dynamic portfolio. For an asset price process $S(t)$ we denote by $dS(t)$ its increment in an infinitesimal time interval $(t-dt,t]$. If $\varphi(t)$ stands for the number of shares in your portfolio at time t, your wealth increases by $\varphi(t)dS(t)$ during this short period. Again, the wish arises to define an integral

$$\varphi \bullet S(t) = \int_0^t \varphi(s)dS(s)$$

in order to give a precise meaning to the gains from trade between times 0 and t. □

3.2.1 A Careless Approach

Let us start with a natural albeit not entirely correct definition of the stochastic integral with respect to a semimartingale X. This hand-waving approach probably suffices for the technically less interested reader. A proper mathematical construction turns out to be surprisingly difficult. We sketch its main steps in Sect. 3.2.3 below.

Integrals can typically be viewed as limits of sums. Consider a fine grid of times $0 = t_0 < t_1 < t_2 < \dots \uparrow \infty$ as in the definition of quadratic variation above. In view of Sect. 1.2 it seems natural to define the integral

$$\varphi \bullet X(t) = \int_0^t \varphi(s)dX(s)$$

as a limit of sums

$$\sum_{i\geq 1}\varphi(t_{i-1})\big(X(t_i\wedge t)-X(t_{i-1}\wedge t)\big), \tag{3.16}$$

where we let the mesh size $\sup_{i\geq 1}|t_i - t_{i-1}|$ tend to 0. However, it is not obvious whether these sums converge to anything. They may in fact fail to do so, which is why a more subtle path is taken in the actual definition. Still, the limit of (3.16) leads to the correct result for sufficiently regular φ, and it helps to understand the idea behind stochastic integration.

If X is a vector-valued process, we view it as the price process of a number of assets, cf. Sect. 1.2. In this case the dynamic portfolios φ should be vector-valued as well and the gains of the individual assets sum up to yield the total gains. This leads us to define

$$\varphi\bullet X := \sum_{i=1}^{d}\varphi_i\bullet X_i$$

for $\mathbb{R}^d$-valued φ, X.

We will primarily consider the case where φ is a predictable process. We motivated in Sect. 1.2 that trading strategies φ should be predictable from a finance point of view because decisions can only be based on information from the past. Moreover, predictability is needed for the technical construction of $\varphi\bullet X$ unless X is of finite variation, in which case adaptedness of φ suffices, for example.

Let us state a number of properties of the stochastic integral. We remain somewhat vague about our assumptions. The technically inclined reader may find precise statements in Sect. 3.2.3.

Rule 3.14 *For semimartingales X, Y, predictable processes φ, ψ, and any stopping time τ we have:*

1. *$\varphi\bullet X$ is linear in φ and X.*
2. *$[X, Y]$ and $\langle X, Y\rangle$ are symmetric and linear in X and Y.*
3. *$\psi\bullet(\varphi\bullet X) = (\psi\varphi)\bullet X$.*
4. *$[\varphi\bullet X, Y] = \varphi\bullet[X, Y]$.*
5. *$\langle\varphi\bullet X, Y\rangle = \varphi\bullet\langle X, Y\rangle$.*
6. *If X is a local martingale, so is $\varphi\bullet X$.*
7. *If $\varphi\geq 0$ and X is a local sub-/supermartingale, $\varphi\bullet X$ is a local sub-/supermartingale as well.*
8. *$A^{\varphi\bullet X} = \varphi\bullet A^X$*
9. *If X, Y are local martingales, the process $XY - \langle X, Y\rangle$ is a local martingale, i.e. $\langle X, Y\rangle$ is the compensator of XY.*
10. *$\Delta(\varphi\bullet X) = \varphi\Delta X$*
11. *$(\varphi\bullet X)^\tau = \varphi\bullet X^\tau = (\varphi 1_{[\![0,\tau]\!]})\bullet X$*

12. *If X is the sum of its jumps (i.e. $X(t) = \sum_{s \leq t} \Delta X(s)$, $t \geq 0$), we have $\varphi \bullet X(t) = \sum_{s \leq t} \varphi(s) \Delta X(s)$.*
13. $(\varphi \bullet X)^c = \varphi \bullet X^c$
14. *If X is of finite variation, so is $\varphi \bullet X$.*

Idea Statements 1–12 hold for the discrete-time integral of Sect. 1.2. For 10–12 this is obvious. Properties 1–9 are stated in Proposition 1.14. Together, this naturally suggests 1–12 above because the stochastic integral here constitutes both a counterpart and a limit of the one in Sect. 1.2. Statement 13 may at least seem plausible because both sides are continuous martingales involving φ and X. The last claim is discussed later.

Rigorous versions of these assertions can be found in Propositions 3.20, 3.23, 3.22 and Corollary 3.24. Note, however, that statements 6, 7, 14 are generally false. More precisely, they require stronger assumptions than just existence of the involved objects. □

From the discrete-time case one would expect $\varphi \bullet X$ to be a martingale whenever this holds for X, implying that you cannot turn a fair game into a favourable one by ingenious trading. Although such a statement is true for many "decent" integrands φ, it fails to hold in full generality. It may happen that $\varphi \bullet X$ is only a local martingale or even a little less, namely a *σ-martingale*. We refer to Sect. 3.2.3 and in particular Proposition 3.23 as well as Example 3.26 for details.

The integration by parts formula of Proposition 1.14 transfers directly to continuous time. Note that (3.17) differs from its deterministic counterpart (1.126) by the covariation term. This extra term constitutes a major difference between deterministic and stochastic calculus.

Theorem 3.15 (Integration by Parts) *For semimartingales X, Y we have*

$$XY = X(0)Y(0) + X_- \bullet Y + Y_- \bullet X + [X, Y]. \tag{3.17}$$

If X is of finite variation or Y is predictable and of finite variation, this equals

$$XY = X(0)Y(0) + X_- \bullet Y + Y \bullet X. \tag{3.18}$$

Idea Theorem 3.15 is the counterpart of the corresponding statement in Proposition 1.14(6) for discrete time.

For rigorous proofs see [154, Definition I.4.45 and Theorems I.4.47, I.4.49]. □

The chain rule of stochastic calculus is called *Itō's formula*. Its terms should seem vaguely familiar from (1.15) and (1.16) in discrete time. Even for continuous processes it does not reduce to its deterministic counterpart (1.127). It is again a quadratic variation term that makes the difference.

Theorem 3.16 (Itō's Formula) *If X is an $\mathbb{R}^d$-valued semimartingale and $f : \mathbb{R}^d \to \mathbb{R}$ a twice continuously differentiable function, then $f(X)$ is a semimartingale of the form*

$$f(X(t)) = f(X(0)) + Df(X_-) \bullet X(t) + \frac{1}{2}\sum_{i,j=1}^{d} D_{ij} f(X_-) \bullet \langle X_i^c, X_j^c\rangle(t)$$

$$+ \sum_{s\leq t}\Big(f(X(s)) - f(X(s-)) - Df(X(s-))^\top \Delta X(s)\Big), \qquad (3.19)$$

where $Df(x)$ denotes the derivative or gradient of f in x and $D_{ij}f(x)$ partial derivatives of second order. In the univariate case ($d = 1$) this reduces to

$$f(X(t)) = f(X(0)) + f'(X_-) \bullet X(t) + \frac{1}{2} f''(X_-) \bullet \langle X^c, X^c\rangle(t)$$

$$+ \sum_{s\leq t}\big(f(X(s)) - f(X(s-)) - f'(X(s-))\Delta X(s)\big). \qquad (3.20)$$

More generally, the statement holds for twice continuously differentiable $f : U \to \mathbb{R}$ with some open set $U \subset \mathbb{R}^d$ such that X, X_- are U-valued.

Idea Equation (3.20) is the counterpart of (1.15) in discrete time. This formula, however, does not explain the third term on the right-hand side.

Equation (3.20) can be shown by considering polynomials and passing to the limit. Let us focus on the case $d = 1$. By linearity it is enough to consider monomials $f(x) = x^n$. These in turn can be treated by induction. For $n = 0$ there is nothing to be shown. Integration by parts yields

$$\begin{aligned} f(X(t)) &= X(t)^n \\ &= X(t)^{n-1}X(t) \\ &= X(0)^n + X_-^{n-1} \bullet X(t) + X_- \bullet X^{n-1}(t) + [X^{n-1}, X](t). \qquad (3.21) \end{aligned}$$

Provided that Itō's formula holds for $x \mapsto x^{n-1}$, we have

$$\begin{aligned} X(t)^{n-1} &= X(0)^{n-1} + (n-1)X_-^{n-2} \bullet X(t) \\ &\quad + \frac{1}{2}(n-1)(n-2)X_-^{n-3} \bullet \langle X^c, X^c\rangle(t) \\ &\quad + \sum_{s\leq t}\big(X(s)^{n-1} - X(s-)^{n-1} - (n-1)X(s-)^{n-2}\Delta X(s)\big), \end{aligned}$$

which implies $(X^{n-1})^c = (n-1)X_-^{n-2} \bullet X^c(t)$ by Rule 3.14(13) and

$$\begin{aligned}X_- \bullet X^{n-1}(t) &= (n-1)X_-^{n-1} \bullet X(t) \\ &\quad + \frac{1}{2}(n-1)(n-2)X_-^{n-2} \bullet \langle X^c, X^c\rangle(t) \\ &\quad + \sum_{s\leq t}\big(X(s-)X(s)^{n-1} - X(s-)^n - (n-1)X(s-)^{n-1}\Delta X(s)\big).\end{aligned}$$

Using (3.9) we have

$$\begin{aligned}[X^{n-1}, X](t) &= \langle (X^{n-1})^c, X^c\rangle + \sum_{s\leq t}\Delta X^{n-1}(s)\Delta X(s) \\ &= (n-1)X_-^{n-2} \bullet \langle X^c, X^c\rangle(t) + \sum_{s\leq t}\big(X(s)^{n-1} - X(s-)^{n-1}\big)\Delta X(s)\end{aligned}$$

by Rule 3.14(4). Summing the terms in (3.21), we obtain

$$\begin{aligned}X(t)^n &= X(0)^n + nX_-^{n-1} \bullet X(t) + \frac{1}{2}n(n-1)X_-^{n-2} \bullet \langle X^c, X^c\rangle(t) \\ &\quad + \sum_{s\leq t}\big(X(s)^n - X(s-)^n - nX(s-)^{n-1}\Delta X(s)\big),\end{aligned}$$

which equals (3.20) for $f(x) = x^n$.

For a complete proof see [154, Theorem I.4.57]. □

By taking care of the real and the imaginary part separately, stochastic integration and even results such as Theorem 3.16 naturally extend to the complex-valued case. We will occasionally use this extension without further notice.

3.2.2 *Differential Notation*

Let us stress that both $\varphi \bullet X(t)$ and $\int_0^t \varphi(s)dX(s)$ are used to denote the same stochastic integral. Moreover, an intuitive differential notation enjoys a certain popularity in the literature. An equation such as

$$dX(t) = b(t)dt + \sigma(t)dW(t) \tag{3.22}$$

is given precise meaning by adding integral signs:

$$\int_0^s dX(t) = \int_0^s b(t)dt + \int_0^s \sigma(t)dW(t), \quad s \geq 0,$$

which in turn means

$$X(s) = X(0) + \int_0^s b(t)dt + \int_0^s \sigma(t)dW(t), \quad s \geq 0.$$

In other words, differential equations such as (3.22) should be viewed as a convenient shorthand for corresponding integral equations where the integral signs have been omitted.

Physicists' Corner 3.17 Occasionally, the differential notation is used even more liberally. In this spirit, a process X increases by

$$X(t) - X(t - dt) = dX(t) \tag{3.23}$$

in an infinitesimal time interval $(t - dt, t]$, the stochastic integral by

$$d(\varphi \bullet X(t)) = \varphi(t)dX(t),$$

the covariation by

$$d[X, Y](t) = dX(t)dY(t), \tag{3.24}$$

the compensator of a special semimartingale X by

$$dA^X(t) = E(dX(t)|\mathscr{F}_{t-}),$$

and the predictable covariation by

$$d\langle X, Y\rangle(t) = E(d[X, Y](t)|\mathscr{F}_{t-}) = E(dX(t)dY(t)|\mathscr{F}_{t-}). \tag{3.25}$$

If X is a local martingale, then its average increase in an infinitesimal period vanishes, i.e.

$$E(dX(t)|\mathscr{F}_{t-}) = 0,$$

which implies that the predictable covariation can also be written as

$$d\langle X, Y\rangle(t) = \text{Cov}(dX(t), dY(t)|\mathscr{F}_{t-})$$

and in particular

$$d\langle X, X\rangle(t) = \text{Var}(dX(t)|\mathscr{F}_{t-})$$

in the local martingale case.

Such notation can be motivated by the discrete-time formulas in Chap. 1, where we have, for example,

$$\Delta\langle X, Y\rangle(t) = E(\Delta X(t)\Delta Y(t)|\mathscr{F}_{t-1}),$$

which resembles (3.25). Literally, the above differential notation seems to make sense only in a vague intuitive sense. But expressions containing differentials can be given a precise meaning if they are related to well-defined concepts from stochastic calculus. As an example consider the equation

$$d(XY)(t) = X(t-)dY(t) + Y(t-)dX(t) + dX(t)dY(t),$$

which follows from (3.23) if we identify $X(t-)$ with $X(t-dt)$ etc. What is it supposed to mean? Using (3.24), one may write it more accurately as

$$d(XY)(t) = X(t-)dY(t) + Y(t-)dX(t) + d[X,Y](t).$$

As noted above, "differential" equations of such type turn into proper statements if an integral sign is added on both sides, which yields

$$\int_0^s d(XY)(t) = \int_0^s X(t-)dY(t) + \int_0^s Y(t-)dX(t) + \int_0^s d[X,Y](t)$$

or

$$X(s)Y(s) - X(0)Y(0) = \int_0^s X(t-)dY(t) + \int_0^s Y(t-)dX(t) + [X,Y](s) - [X,Y](0),$$

which is precisely the integration by parts formula because $[X,Y](0) = 0$.

Intuitive manipulation of differentials often leads to correct results. Consider, for instance,

$$\begin{aligned}
E\big(dM^2(t)\big|\mathscr{F}_{t-}\big) &= E\big((M(t-dt)+dM(t))^2 - M(t-dt)^2\big|\mathscr{F}_{t-}\big)\\
&= E\big((dM(t))^2\big|\mathscr{F}_{t-}\big) + 2M(t-dt)E(dM(t)|\mathscr{F}_{t-})\\
&= E\big((dM(t))^2\big|\mathscr{F}_{t-}\big)\\
&= d\langle M,M\rangle(t)
\end{aligned}$$

for a local martingale M. This suggests that the compensator of M^2 equals $\langle M,M\rangle$, which is indeed the case, cf. Proposition 3.5. □

3.2.3 A Careful Approach

We start by considering the stochastic integral $\varphi \bullet X$ for a general semimartingale X and locally bounded, predictable integrands φ, turning to more general integrands later. Let us interpret X as a price process and φ as a dynamic trading strategy. In Sect. 1.2 we motivated why predictability is a reasonable requirement for trading strategies. For predictable processes local boundedness is essentially equivalent to claiming that $\varphi(s)$ is bounded on any finite interval $[0,t]$ for fixed $\omega \in \Omega$, cf. [163, Lemma A.1]. This weak condition is met, for example, by continuous or more generally càglàd processes. Nevertheless, the integral must be extended even further for certain representation theorems to hold.

Firstly, consider an integrand of the form

$$\varphi(t) = Y 1_{]t_1,t_2]}(t) \tag{3.26}$$

with real numbers $t_1 \leq t_2$ and some $\mathscr{F}_{t_1}$-measurable random variable Y. This corresponds to an investor who buys Y shares of stock at time t_1 and sells them again at t_2. Obviously, the gains from this portfolio up to time t are given by

$$\varphi \bullet X(t) := Y\big(X(t_2 \wedge t) - X(t_1 \wedge t)\big). \tag{3.27}$$

Secondly, stochastic integration should be linear in φ, i.e.

$$(\varphi + \psi) \bullet X = \varphi \bullet X + \psi \bullet X,$$
$$(c\varphi) \bullet X = c(\varphi \bullet X)$$

should hold for any locally bounded, predictable φ, ψ and any constant c. If the investor buys, say, twice the number of shares, her gains from trade obviously double. A third reasonable requirement is some degree of continuity, not only for mathematical reasons. Real trading strategies are piecewise constant; more general ones occur only in the limit $\varphi^{(n)} \to \varphi$ as a mathematical idealisation. This should be reflected by stochastic integration. If the integral of the approximating strategies $\varphi^{(n)}$ did not converge to the integral of φ, the financial interpretation of the integral $\varphi \bullet X$ would be questionable. More precisely, we require the following continuity:

> If $\varphi^{(n)}$, φ, ψ are locally bounded, predictable strategies with $|\varphi^{(n)}| \leq \psi$ and $\varphi^{(n)}(\omega, t) \to \varphi(\omega, t)$ for any fixed $\omega \in \Omega$, $t \geq 0$, then $\varphi^{(n)} \bullet X(t)$ converges to $\varphi \bullet X(t)$ in probability for any fixed t.

Finally, we want $\varphi \bullet X$ to be an adapted, càdlàg process. It turns out that these properties, namely simple integrals (3.27), linearity, continuity in the above sense, adaptedness and càdlàg paths suffice to determine the stochastic integral uniquely, cf. [154, Theorem I.4.31]. There are several ways to actually construct the integral satisfying the above properties. We proceed by considering different subclasses of semimartingales separately.

If X is an increasing process, one can easily use the **Lebesgue–Stieltjes integral**. It is based on the idea that any increasing, right-continuous function $f : \mathbb{R}_+ \to \mathbb{R}$ can be identified with a unique measure on $\mathbb{R}_+$, namely the one assigning mass

$$\mu((a, b]) := f(b) - f(a)$$

to an arbitrary interval $(a, b]$. The Stieltjes integral of another function $g : \mathbb{R}_+ \to \mathbb{R}$ relative to f is now simply defined as the integral relative to the *Lebesgue–Stieltjes measure* μ of f:

$$\int_0^t g(s)df(s) := \int_{(0,t]} g(s)\mu(ds). \tag{3.28}$$

Since $X(\omega, t)$ is an increasing function in t for fixed ω, we can use this immediately to define

$$\varphi \bullet X(t) = \int_0^t \varphi(s)dX(s)$$

path by path as a Lebesgue–Stieltjes integral in the sense of (3.28).

If X is an adapted process of finite variation, it can be written as a difference $X = Y - Z$ of two increasing semimartingales, cf. [154, Proposition I.3.3]. In this case, we set

$$\varphi \bullet X(t) := \varphi \bullet Y(t) - \varphi \bullet Z(t).$$

It remains to verify that this process is a semimartingale having the desired properties. Note that we do not need predictability for this approach. If we only assume φ to be adapted, the definition of $\varphi \bullet X$ as a Lebesgue–Stieltjes integral still yields a well-defined semimartingale. We will occasionally use this stochastic interval for non-predictable integrands, e.g. in the integration by parts formula (3.18) where Y in the second integral may only be adapted. As far as integrability is concerned, we could replace local boundedness of φ by the weaker requirement

$$\int_0^t |\varphi(s)| d(\mathrm{Var}(X))(s) < \infty \quad \forall t \geq 0, \tag{3.29}$$

where $\mathrm{Var}(X)$ denotes the total variation of X as in (2.4). We denote by $L_s(X)$ the set of all predictable processes φ satisfying (3.29). For the identity process $I(t) = t$ this condition simplifies to

$$\int_0^t |\varphi(s)| ds < \infty \quad \forall t \geq 0$$

and the Lebesgue–Stieltjes integral is just the ordinary Lebesgue integral

$$\varphi \bullet I(t) = \int_0^t \varphi(s) ds.$$

For a general semimartingale X the above construction does not work. Suppose now that X is a square-integrable martingale. In this case the Hilbert space structure of $\mathscr{H}^2$ leads to an integral. To this end we view predictable processes φ as real-valued functions defined on the product space $\Omega \times \mathbb{R}_+$, which is endowed with the predictable σ-field $\mathscr{P}$. On this space we define the so-called *Doléans measure* m by

$$m(A) := E(1_A \bullet \langle X, X \rangle(\infty))$$

for $A \in \mathscr{P}$. Here $1_A \bullet \langle X, X \rangle(\infty)$ denotes the Lebesgue–Stieltjes integral of the process 1_A. Rather than locally bounded φ we consider the space $L^2(X) := L^2(\Omega \times \mathbb{R}_+, \mathscr{P}, m)$ of integrands having finite second moment relative to m, i.e. with

$$\|\varphi\|^2_{L^2(X)} := E(\varphi^2 \bullet \langle X, X \rangle(\infty)) < \infty.$$

We have now introduced two L^2- and hence Hilbert spaces, namely the space $\mathscr{H}^2$ of square-integrable martingales and the space $L^2(X)$ of integrands. Recall that we want the integral $\varphi \bullet X$ to be defined by (3.27) for integrands of the form (3.26). By linearity, this immediately yields an expression for linear combinations of such processes, which we call *simple integrands*. It is easy to show that

$$\|\varphi \bullet X\|_{\mathscr{H}^2} = \|\varphi\|_{L^2(X)} \tag{3.30}$$

for simple integrands

$$\varphi(t) = \sum_{i=1}^{n} Y_i 1_{]t_{i-1},t_i]}(t)$$

with $0 \leq t_0 < \cdots < t_n$ and $\mathscr{F}_{t_{i-1}}$-measurable Y_i. Indeed, we have

$$\|\varphi \bullet X\|^2_{\mathscr{H}^2} = E\left(\left(\sum_{i=1}^{n} (Y_i 1_{]t_{i-1},t_i]}) \bullet X(\infty)\right)^2\right). \tag{3.31}$$

For $j > i$ the martingale property of X yields

$$\begin{aligned}
&E\Big((Y_i 1_{]t_{i-1},t_i]}) \bullet X(\infty)(Y_j 1_{]t_{j-1},t_j]}) \bullet X(\infty)\Big)\\
&= E\Big(E\big(Y_i(X(t_i) - X(t_{i-1}))Y_j(X(t_j) - X(t_{j-1}))\big|\mathscr{F}_{t_{j-1}}\big)\Big)\\
&= E\Big(Y_i(X(t_i) - X(t_{i-1}))Y_j\big(E(X(t_j)|\mathscr{F}_{t_{j-1}}) - X(t_{j-1})\big)\Big)\\
&= 0.
\end{aligned}$$

Hence (3.31) equals

$$\sum_{i=1}^{n} E\Big(\big(Y_i(X(t_i) - X(t_{i-1}))\big)^2\Big) = \sum_{i=1}^{n} E\Big(Y_i^2 E\big((X(t_i) - X(t_{i-1}))^2\big|\mathscr{F}_{t_{i-1}}\big)\Big).$$

Since X and $X^2 - \langle X, X\rangle$ are martingales, we have

$$\begin{aligned}
E\big((X(t_i) - X(t_{i-1}))^2\big|\mathscr{F}_{t_{i-1}}\big) &= E\big(X(t_i)^2\big|\mathscr{F}_{t_{i-1}}\big) - X(t_{i-1})^2\\
&= E\big(\langle X, X\rangle(t_i) - \langle X, X\rangle(t_{i-1})\big).
\end{aligned}$$

Consequently, (3.31) equals

$$\sum_{i=1}^{n} E\Big(E\big(Y_i^2(\langle X,X\rangle(t_i)-\langle X,X\rangle(t_{i-1}))\big|\mathscr{F}_{t_{i-1}}\big)\Big)$$

$$= \sum_{i=1}^{n} E\Big((Y_i^2 1_{]t_{i-1},t_i]})\bullet\langle X,X\rangle(\infty)\Big)$$

$$= E\big(\varphi^2\bullet\langle X,X\rangle(\infty)\big)$$

$$= \|\varphi\|^2_{L^2(X)},$$

which yields (3.30).

In other words we have shown that the mapping $\varphi\mapsto\varphi\bullet X$ is an isometry on the subspace $\mathscr{E}\subset L^2(X)$ of simple integrands. This subspace $\mathscr{E}$ can be shown to be dense in $L^2(X)$, i.e. any $\varphi\in L^2(X)$ can be approximated by some sequence $(\varphi^{(n)})$ in $\mathscr{E}$. The corresponding sequence of integrals $\varphi^{(n)}\bullet X$ is by isometry a Cauchy sequence in $\mathscr{H}^2$, which implies that it converges to some square-integrable martingale which we denote by $\varphi\bullet X$ and call the *stochastic integral* of φ relative to X. It remains to verify that this definition does not depend on the chosen approximating sequence $(\varphi^{(n)})$ and that it satisfies the desired properties. Moreover, one can show that the isometry (3.30) extends to arbitrary $\varphi\in L^2(X)$, cf. [154, Section III.6a].

Example 3.18 (Lévy Process) Let X be a Lévy process with triplet (b^h,c,K), which satisfies $b^{\mathrm{id}}:=b^h+\int(x-h(x))K(dx)=0$ and (2.22) for $p=2$, i.e. it is a locally square-integrable martingale. In order to make it a true square-integrable martingale, we consider a finite time interval $[0,T]$ with $T\in\mathbb{R}_+$ or, equivalently, we consider X to be stopped at T. In view of (3.6), the space $L^2(X)$ contains the predictable processes φ satisfying

$$E\bigg(\int_0^T\varphi(t)^2dt\bigg)<\infty.$$

Since the integral is a martingale, we have obviously

$$E(\varphi\bullet X(T))=0$$

for these φ. The isometry (3.30) and (3.6) yield

$$\mathrm{Var}(\varphi\bullet X_T)=E(\varphi^2\bullet\langle X,X\rangle_T)=\bigg(c+\int x^2K(dx)\bigg)E\bigg(\int_0^T\varphi(t)^2dt\bigg).$$

This generalises the *Itō isometry* for standard Brownian motion, in which case we have $c=1$ and $K=0$.

The previous definition can easily be extended to the case where X is only a locally square-integrable martingale and φ is only *locally in* $L^2(X)$ (written $\varphi \in L^2_{\text{loc}}(X)$), which here means that

$$E\big(\varphi^2 \bullet \langle X, X\rangle(\tau_n)\big) < \infty$$

holds for a sequence of stopping times satisfying $\tau_n \uparrow \infty$. If (τ_n) denotes a localising sequence for both $X \in \mathscr{H}^2_{\text{loc}}$ and $\varphi \in L^2_{\text{loc}}(X)$, we have $X^{\tau_n} \in \mathscr{H}^2$ and $\varphi \in L^2(X^{\tau_n})$. Therefore we can set

$$\varphi \bullet X := \varphi \bullet (X^{\tau_n})$$

on the interval $[\![0, \tau_n]\!]$. It remains to verify that this procedure leads to a well-defined semimartingale satisfying the desired properties, cf. [154, Section III.6a].

Moreover, one can easily see that any locally bounded predictable process φ belongs to $L^2_{\text{loc}}(X)$.

Example 3.19 (Lévy Process) We consider once more the Lévy process from Example 3.18. Since it is locally square-integrable, there is no need to consider a finite time horizon. By

$$\varphi^2 \bullet \langle X, X\rangle(t) = \left(c + \int x^2 K(dx)\right) \int_0^t \varphi(s)^2 ds$$

the integrability condition $\varphi \in L^2_{\text{loc}}(X)$ holds if and only if

$$\int_0^t \varphi(s)^2 ds < \infty$$

almost surely for any $0 \leq t < \infty$.

The case of an arbitrary semimartingale X can now be put together from the two integrals above. Recall that any semimartingale can be decomposed as

$$X = X(0) + M + A \tag{3.32}$$

with a local martingale M and a process of finite variation A. It is even possible to choose M as a locally square-integrable martingale, cf. [154, Proposition I.4.17]. For any locally bounded predictable φ, we naturally define

$$\varphi \bullet X := \varphi \bullet M + \varphi \bullet A. \tag{3.33}$$

Of course, it remains to verify that this definition does not depend on the decomposition of X and that it satisfies the desired properties, cf. [154, Section I.4d]. If both

X and φ are $\mathbb{R}^d$-valued processes, we simply set

$$\varphi \bullet X := \sum_{i=1}^{d} \varphi_i \bullet X_i \tag{3.34}$$

as in Sects. 1.2 resp. 3.2.1.

We have noted earlier that the class of locally bounded φ turns out to be too small for some applications. Let us turn things around and wonder how far we can reasonably extend the class of integrands. In any case we want $\varphi \bullet X$ to be a semimartingale satisfying

$$\psi \bullet (\varphi \bullet X) = (\psi\varphi) \bullet X \tag{3.35}$$

for processes ψ such that the expressions on both sides are defined. Observe that both ψ and $\psi\varphi$ are bounded predictable processes for $\psi := 1_{\{|\varphi|\leq n\}}$, which implies that the right-hand side of (3.35) makes sense. We say that a predictable process φ is **X-integrable** (written $X \in L(X)$) if there is a semimartingale Z (the candidate for $\varphi \bullet X$) such that

$$1_{\{|\varphi|\leq n\}} \bullet Z = (1_{\{|\varphi|\leq n\}}\varphi) \bullet X$$

for any $n \geq 0$ in accordance with (3.35). Moreover, $\varphi \bullet X := Z$ is called the **stochastic integral** of φ relative to X. This definition works for both univariate and vector-valued φ, X. Surprisingly, it may happen that φ is X-integrable but the components φ_i fail to be X_i-integrable in the multivariate case, which implies that the integral $\varphi \bullet X$ cannot generally be written as in (3.34). But as noted above, (3.33, 3.34) generally hold for locally bounded integrands φ, regardless of the chosen decomposition (3.32). In particular, locally bounded predictable processes are integrable with respect to any semimartingale X of the same dimension.

For real-valued processes the vector space $L(X)$ of most general integrands can be related to the sets $L_s(X)$ and $L^2_{\text{loc}}(X)$ above. One can show that φ is X-integrable if and only if there exists a decomposition

$$X = X(0) + M + A$$

with a locally square-integrable martingale M and a process of finite variation A such that $\varphi \in L^2_{\text{loc}}(M) \cap L_s(A)$. In this case, the general integral is obtained as in (3.33), cf. [154, Section III.6c].

With a little extra work one can generalise $L^2_{\text{loc}}(M)$ and $L_s(A)$ to the vector-valued case, which leads to the multivariate analogue of this statement. Note that we cannot take just any decomposition of X. In particular, the canonical decomposition of a special semimartingale may not be the one that actually allows for integrability of the components.

Let us come back to the properties of stochastic integration and under what conditions they actually hold. A more accurate version of Rule 3.14 reads as follows.

Proposition 3.20 *For semimartingales X, Y, predictable processes φ, ψ with $\varphi \in L(X)$, and any stopping time τ we have:*

1. *$L(X)$ is a vector space and $\varphi \bullet X$ is linear in φ.*
2. *$[X, Y]$ is symmetric and linear in X and Y.*
3. *$\langle X, Y\rangle$ is symmetric and linear in X and Y if it exists.*
4. *$\varphi \bullet X$ is linear in X (provided that the integrals exist).*
5. *$\psi \in L(\varphi \bullet X)$ holds if and only if $\psi\varphi \in L(X)$. In this case*
$$\psi \bullet (\varphi \bullet X) = (\psi\varphi) \bullet X.$$
6. *$[\varphi \bullet X, Y] = \varphi \bullet [X, Y]$.*
7. *If X, Y are locally square-integrable martingales and $\varphi \in L^2_{\mathrm{loc}}(X)$, we have*
$$\langle \varphi \bullet X, Y\rangle = \varphi \bullet \langle X, Y\rangle.$$
8. *If $[X, Y]$ is a special semimartingale and $\varphi \in L_s(X)$, we have*
$$\langle \varphi \bullet X, Y\rangle = \varphi \bullet \langle X, Y\rangle$$
as well.
9. *$\Delta(\varphi \bullet X) = \varphi\Delta X$.*
10. *$(\varphi \bullet X)^\tau = \varphi \bullet X^\tau = (\varphi 1_{[\![0,\tau]\!]}) \bullet X$.*
11. *If $X = \sum_{t\leq\cdot} \Delta X(t)$ and $\varphi \in L_s(X)$, then $\varphi \bullet X = \sum_{t\leq\cdot} \varphi(t)\Delta X(t)$.*
12. *If X is a continuous local martingale, so is $\varphi \bullet X$.*
13. *$(\varphi \bullet X)^c = \varphi \bullet X^c$.*

Idea These statements correspond directly to their informal versions in Rule 3.14. For rigorous proofs we refer to [154, Theorems III.6.19, III.6.4, III.6.22] and [125, Propositions A.1, A.2]. □

If they make sense, the assertions from the previous proposition also hold for vector-valued processes.

Theorems 3.15 and 3.16 hold literally as they are stated above. There are a number of dominated convergence theorems for stochastic integrals. The following version refines the statement from the beginning of Sect. 3.2.3.

Proposition 3.21 *If $\varphi^{(n)}, \varphi, \psi$ are locally bounded, predictable processes with $|\varphi^{(n)}| \leq \psi$ and $\varphi^{(n)}(\omega, t) \to \varphi(\omega, t)$ for any fixed $\omega \in \Omega$, $t \geq 0$, then*
$$\sup_{s\leq t} \left|\varphi^{(n)} \bullet X(s) - \varphi \bullet X(s)\right| \to 0$$
in probability for any fixed t.

Proof See [154, Theorem I.4.31]. □

One could guess that $\varphi \bullet X$ is of finite variation if this holds for X. However, this does not hold for general integrands, cf. [154, III.6.20].

However, local boundedness suffices for that purpose:

Proposition 3.22 *If X is of finite variation and φ is a locally bounded, predictable process, then φ is X-integrable and $\varphi \bullet X$ is of finite variation as well.*

Proof Locally bounded processes are actually in $L_s(X)$ and hence the stochastic integral is a pathwise Lebesgue–Stieltjes integral. Since this integral yields paths of finite variation, we are done. □

From discrete time and also from the heuristic "physicists' style" derivation

$$E(d(\varphi \bullet X)(t)|\mathscr{F}_{t-dt}) = E(\varphi(t)dX(t)|\mathscr{F}_{t-dt}) = \varphi(t)E(dX(t)|\mathscr{F}_{t-dt})$$

(cf. Physicists' Corner 3.17) one may expect that $\varphi \bullet X$ is a local martingale if this holds for X. This fails to be true in general, cf. [154, III.6.21]. But we have the

Proposition 3.23

1. *If X is a local martingale and φ a locally bounded, predictable process, then $\varphi \bullet X$ is a local martingale as well. In this case $\varphi \bullet X$ is purely discontinuous if this holds for X.*
2. *If X is a locally square-integrable martingale and φ is predictable with*

$$E(\varphi^2 \bullet \langle X, X\rangle(\infty)) < \infty,$$

then $\varphi \bullet X$ is a square-integrable martingale.

3. *If X is a special semimartingale and φ a locally bounded, predictable process, then $\varphi \bullet X$ is special as well with compensator $A^{\varphi \bullet X} = \varphi \bullet A^X$.*
4. *If X is a local sub-/supermartingale and φ a nonnegative locally bounded, predictable process, then $\varphi \bullet X$ is a local sub-/supermartingale as well.*

Proof

1. See [154, I.4.34b and Corollary I.4.55d].
2. This follows from [154, Theorem III.6.4d].
3. Linearity of the integral yields $\varphi \bullet X = \varphi \bullet M^X + \varphi \bullet A^X$, where $\varphi \bullet M^X$ is a local martingale by statement 1. Since the Stieltjes integral $\varphi \bullet A^X$ is of finite variation, it coincides with the compensator of $\varphi \bullet X$.
4. If X is a local submartingale, it is special and A^X is increasing by Proposition 2.13. In view of the previous statements, $A^{\varphi \bullet X} = \varphi \bullet A^X$ is also increasing and $\varphi \bullet M^X$ is a local martingale, which in turn implies that $\varphi \bullet X$ is a local submartingale, cf. Proposition 2.13. For local supermartingales, the assertion follows along the same lines. □

This allows us to verify the statement in Rule 3.14(9).

Corollary 3.24 *If X, Y are local martingales such that XY is special, $\langle X, Y\rangle$ is the compensator of XY.*

Proof Since the integral terms in (3.17) are local martingales by Proposition 3.23, XY has compensator A if and only if $[X, Y]$ has compensator A. By definition of $\langle X, Y\rangle$ this yields the claim. □

Using the notation (2.9) and (3.7), we can rephrase the integration by parts rule in terms of compensators.

Proposition 3.25 *We have*

$$a^{XY}(t) = X(t-)a^Y(t) + Y(t-)a^X(t) + \widetilde{c}^{XY}(t)$$

provided that X, Y and XY are special semimartingales whose compensators are absolutely continuous.

Proof Computing the compensators of

$$XY = X(0)Y(0) + X_- \bullet Y + Y_- \bullet X + [X, Y]$$

yields

$$a^{XY} \bullet I = (X_- a^Y) \bullet I + (Y_- a^X) \bullet I + \widetilde{c}^{XY} \bullet I$$

by Propositions 3.23(3) and 3.20(5). □

For arbitrary integrands φ the integral $\varphi \bullet X$ relative to a local martingale is in general only a *σ-martingale*. Such processes are obtained by a more general type of localisation from the set of martingales. To this end, observe that any stopped process X^τ satisfies

$$X^\tau = X(0) + 1_{[\![0,\tau]\!]} \bullet X. \tag{3.36}$$

Inspired by (3.36) we define

$$X^D := (X1_D)(0) + 1_D \bullet X$$

for arbitrary predictable sets D, which implies $X^\tau = X^{[\![0,\tau]\!]}$ for stopping times τ. This leads to the following generalisation of localisation. Let $\mathscr{C}$ denote a class of processes that is stable under stopping. We say that a process X belongs to the **σ-localised class** $\mathscr{C}_\sigma$ if there exists some sequence $(D_n)_{n\in\mathbb{N}}$ of predictable sets with $D_n \uparrow \Omega \times \mathbb{R}_+$ and such that $X^{D_n} \in \mathscr{C}$ for any n. The elements of the σ-localised class of martingales are called **σ-martingales**. We will see later that uniformly integrable martingales, martingales, local martingales, and σ-martingales can all be viewed as processes with vanishing drift rate that differ only in their degree of integrability, cf. Proposition 4.22. In the same vein, σ-martingales are local martingales if and only if they are special semimartingales.

However, this does not imply that σ- or local martingales behave similarly as martingales on a global scale. The following example shows that we may have $X(0) = 0 < 1 = X(1)$ with probability 1 for a local martingale.

Example 3.26 Consider

$$Y(t) := \begin{cases} W(\frac{1}{1-t}) & \text{for } t < 1, \\ 0 & \text{for } t \geq 1, \end{cases}$$

where W denotes standard Brownian motion. If we set $\tau := \inf\{s \geq 0 : W(s) = 1\}$ and $\widetilde{\tau} := \inf\{t \geq 0 : Y(t) = 1\}$, then $\widetilde{\tau} < 1$ because τ is almost surely finite, cf. e.g. [204, Proposition 3.3.6]. The stopped process $X := Y^{\widetilde{\tau}}$ satisfies $X(0) = 0 < 1 = X(1)$ with probability 1. In order to see that X is a local martingale in the filtration generated by Y, define stopping times

$$\tau_n := \begin{cases} 1 - \frac{1}{n} & \text{if } 1 - \frac{1}{n} < \widetilde{\tau}, \\ \infty & \text{otherwise.} \end{cases}$$

Note that $X^{\tau_n}(t) = W^{\tau}(n \wedge (1-(1 \wedge t))^{-1})$. The martingale property of X^{τ_n} follows now from the fact that W^{τ} is a martingale in its own filtration.

The local martingale in the previous example is of the form $X = \varphi \bullet S$ for some locally bounded φ and geometric Brownian motion S, cf. Example 11.46. In the language of finance this means that one can produce sure profits by trading in a martingale S. In Part II of this book we rule out such strategies that lead to paradoxical conclusions and are practically unfeasible anyway. A useful fact in this context is the following

Proposition 3.27 *Any nonnegative local martingale (or more generally nonnegative σ-martingale with $E(X(0)) < \infty$) is a supermartingale. Any bounded σ-martingale is a martingale.*

Idea We prove the first statement only for local martingales X. If $(\tau_n)_n$ denotes a localising sequence for X, Fatou's lemma and the martingale property of X^{τ_n} yield

$$\begin{aligned} E(X(t)|\mathscr{F}_s) &= E\Big(\lim_{n\to\infty} X^{\tau_n}(t)\Big|\mathscr{F}_s\Big) \\ &\leq \liminf_{n\to\infty} E\big(X^{\tau_n}(t)\big|\mathscr{F}_s\big) \\ &= \liminf_{n\to\infty} X^{\tau_n}(s) = X(s), \quad s \leq t. \end{aligned}$$

The second statement follows from the first by considering both X and $-X$, possibly shifted in order to become nonnegative.

The proof for the statements on σ-martingales can be found in [168, Proposition 3.1]. □

As an application of the previous results on stochastic integration we come back to a remark in Sect. 3.1.1.

Example 3.28 (Lévy Characterisation of Brownian Motion) Suppose that X is a continuous local martingale starting at $X(0) = 0$ and with quadratic variation $[X, X](t) = t$. Fix $u \in \mathbb{R}$. Itō's formula applied to $f(x,t) = \exp(iux + \frac{u^2}{2}t)$ yields for $M = f(X, I)$:

$$\begin{aligned} M(t) &= M(0) + iuM \bullet X(t) + \frac{1}{2}u^2 M \bullet I - \frac{1}{2}u^2 M \bullet [X, X](t) \\ &= M(0) + iuM \bullet X(t). \end{aligned}$$

M is a local martingale because it is an integral relative to the local martingale X. Since it is bounded on any interval $[0, T]$, it is in fact a martingale, cf. Proposition 3.27. This implies

$$\begin{aligned} E\left(e^{iu(X(t)-X(s))}\middle|\mathscr{F}_s\right) &= \frac{E(M(t)|\mathscr{F}_s)}{M(s)} \exp\left(-\frac{u^2}{2}(t-s)\right) \\ &= \exp\left(-\frac{u^2}{2}(t-s)\right) \end{aligned}$$

for $s \leq t$. This means that the characteristic function of the increment $X(t) - X(s)$ given $\mathscr{F}_s$ coincides with the Fourier transform of the $N(0, t-s)$-distribution. Since the Fourier transform characterises probability measures uniquely, we conclude that the conditional law of $X(t) - X(s)$ given $\mathscr{F}_s$ is $N(0, t-s)$. In particular, it is deterministic, which means that X has independent increments. Together, we obtain that X is standard Brownian motion.

The order of integration can typically be exchanged for multiple integrands because integrands are limits of sums where this property naturally holds. We state here a Fubini-type theorem for stochastic and ordinary integrals, which is used in particular in interest rate theory.

Theorem 3.29 (Fubini's Theorem for Stochastic Integrals) *Let X be a semimartingale and $\varphi(\omega, t, x)$ a process which depends on an additional parameter $x \in [a, b]$. If φ is $\mathscr{P} \otimes \mathscr{B}$-measurable and*

$$\psi(t) := \sqrt{\int_a^b \varphi(t,x)^2 dx}$$

is an X-integrable process, one may exchange the order of stochastic and ordinary integration:

$$\int_0^t \int_a^b \varphi(s, x) dx dX(s) = \int_a^b \int_0^t \varphi(s, x) dX(s) dx.$$

Proof See [238, Theorem IV.65]. □

What happens if f in Itō's formula fails to be twice continuously differentiable? In general we may leave the realm of semimartingales. It can, for example, be shown that $\sqrt{|W(t)|}$ is not a semimartingale for standard Brownian motion W, cf. [58, Theorem 5.5 and Remark 5.7]. Nevertheless, Itō's formula can be extended to some functions which are not C^2, the absolute value $|x|$ being a prominent example. However, some care has to be taken. Naive application of Itō's formula for standard Brownian motion W would yield

$$|W(t)| = \operatorname{sgn}(W) \bullet W(t) \tag{3.37}$$

because $f(x) := |x|$ implies $f'(x) = \operatorname{sgn}(x)$ and $f''(x) = 0$ up to the single point $x = 0$. But (3.37) cannot hold because the right-hand side is a martingale whereas the left-hand side is certainly not. The difference of both sides is called the *local time* of W at 0. More generally, for an arbitrary semimartingale X the process

$$\begin{aligned} L^a(t) := {} & |X(t) - a| - |X(0) - a| - \operatorname{sgn}(X_- - a) \bullet X(t) \\ & - \sum_{s \leq t} \big(|X(s) - a| - |X(s-) - a| - \operatorname{sgn}(X(s-) - a) \Delta X(s) \big) \end{aligned} \tag{3.38}$$

is called the **local time** of X at $a \in \mathbb{R}$. Here, the sign function is defined as

$$\operatorname{sgn}(x) := \begin{cases} 1 & \text{for } x > 0, \\ 0 & \text{for } x = 0, \\ -1 & \text{for } x < 0. \end{cases}$$

As in the Brownian motion case, the right-hand side of (3.38) would vanish if naive application of Itō's formula were allowed. It is easy to see that L^a is a continuous, increasing, adapted process starting at 0, cf. [152, Définition 5.47]. The name *local time* can be motivated from its representation

$$\int_B L^a(t) da = \int_0^t 1_B(X(s)) d\langle X^c, X^c \rangle (s), \quad B \in \mathscr{B},$$

cf. [152, (5.53)]. So $L^a(t)$ stands for the Lebesgue density of the time X spends in a up to t. However, time is measured according to the "clock" $\langle X^c, X^c \rangle$ rather

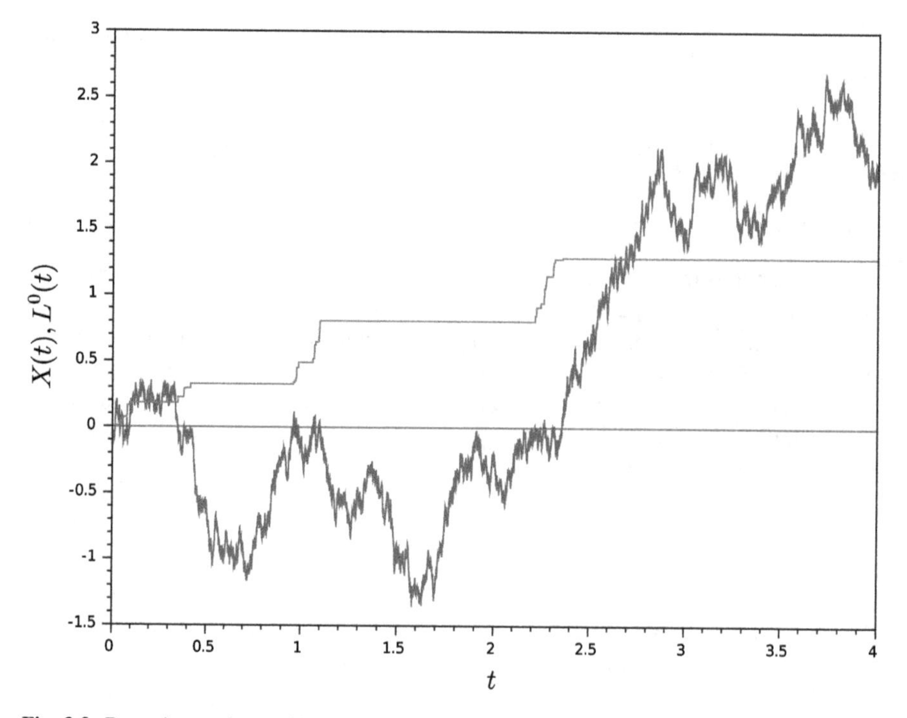

Fig. 3.3 Brownian motion and its local time at 0

than calendar time t. But note that the two clocks coincide for standard Brownian motion.

Brownian motion and its local time at 0 are shown in Fig. 3.3. With the help of local times one can formulate the following generalisation of Itō's formula.

Theorem 3.30 (Itō–Meyer Formula) *If X is a semimartingale and $f : \mathbb{R} \to \mathbb{R}$ the difference of two convex functions, $f(X)$ is a semimartingale satisfying*

$$f(X(t)) = f(X(0)) + f'(X_-) \bullet X(t) + \frac{1}{2}\int_{\mathbb{R}} L^a(t)\varrho(da)$$
$$+ \sum_{s \leq t} \big(f(X(s)) - f(X(s-)) - f'(X(s-))\Delta X(s)\big). \tag{3.39}$$

Here, $f' := (f'_l + f'_r)/2$ denotes the average of the left- and right-hand derivatives of f. Moreover, the signed measure ϱ stands for the second derivative of f in the sense of distributions.

Idea By linearity it suffices to consider convex functions f, which we write as limits of linear combinations of functions of the form $x \mapsto a$ resp. $x \mapsto |x-a|$ with $a \in \mathbb{R}$. Again by linearity, we can even assume $f(x) = |x - a|$. In this case, (3.39) reduces to (3.38) because ϱ is twice the Dirac measure in a.

For a rigorous proof see [152, Théorème 5.52]. □

One can show that the local time L^0 of Brownian motion at 0 grows only on a set of Lebesgue measure 0, cf. [152, Proposition 5.51] and the fact that

$$E\big(\lambda(\{t \geq 0 : W(t) = 0\})\big) = \int E\big(1_{\{0\}}(W(t))\big)dt = 0.$$

Consequently,

$$|W(t)| = \operatorname{sgn}(W) \bullet W(t) + L^0(t) = M^{|W|}(t) + A^{|W|}(t) \tag{3.40}$$

does not allow for a drift rate in the sense of (2.9). Fortunately, the situation simplifies considerably for pure jump processes, which are a main focus of this book.

Proposition 3.31 *If we have $X^c = 0$ for the continuous martingale part of a semimartingale X, its local time at any $a \in \mathbb{R}$ satisfies $L^a = 0$. In particular, Itō's formula (3.20) holds for any f that is a difference of two convex functions and in particular for any continuously differentiable function f.*

Idea Comparing (3.39) with (3.20) for twice continuously differentiable functions f yields

$$f''(X_-) \bullet \langle X^c, X^c\rangle(t) = \int_{\mathbb{R}} L^a(t)\varrho(da) = \int_{\mathbb{R}} L^a(t) f''(a)da.$$

This equality can be extended to arbitrary bounded measurable functions g instead of f''. Choosing $g = 1_B$ for some subset $B \in \mathscr{B}$, we obtain

$$1_B(X_-) \bullet \langle X^c, X^c\rangle(t) = \int_{\mathbb{R}} L^a(t) 1_B(a)da = \int_A L^a(t)da. \tag{3.41}$$

If $X^c = 0$, (3.41) vanishes for any $B \in \mathscr{B}$, which yields the claim.

For a rigorous proof see [152, Corollaire 5.56]. □

3.3 The Stochastic Integral for Random Measures

In the Lévy–Itō decomposition the jump part of a Lévy process is written as an integral with respect to the random measure of jumps. These concepts—random measures and their stochastic integration—turn out to be very useful tools for studying jump processes. They are introduced in this section. In particular, we will generalise the Lévy–Itō decomposition to general semimartingales.

The starting point is the **random measure of jumps** of an $\mathbb{R}^d$-valued semimartingale X, which is defined as for Lévy processes by

$$\mu^X([0,t] \times B) := \left|\{(s,x) \in [0,t] \times B : \Delta X(s) = x\}\right|$$

for $t \geq 0$, $B \in \mathscr{B}^d$, cf. (2.74). For general semimartingales, $\mu = \mu^X$ will typically not be a homogeneous Poisson random measure, but still an **integer-valued random measure** on $\mathbb{R}_+ \times \mathbb{R}^d$ in the sense that

1. $\mu(\omega, \cdot)$ is a measure on $\mathbb{R}_+ \times \mathbb{R}^d$ for fixed $\omega \in \Omega$,
2. $\mu(\omega, \{0\} \times \mathbb{R}^d) = 0$ for any $\omega \in \Omega$,
3. $\mu(\omega, \cdot)$ has values in $\mathbb{N} \cup \{\infty\}$ for any $\omega \in \Omega$,
4. $\mu(\omega, \{t\} \times \mathbb{R}^d) \leq 1$ for any $\omega \in \Omega, t \geq 0$,
5. μ is *adapted* in the sense that $\mu(\cdot, [0,t] \times B)$ is $\mathscr{F}_t$-measurable for fixed $t \geq 0$ and any Borel-measurable $B \in \mathscr{B}^d$,
6. μ is *predictably σ-finite* in the sense that there exists some strictly positive $\mathscr{P} \otimes \mathscr{B}^d$-measurable function η on $\Omega \times \mathbb{R}_+ \times \mathbb{R}^d$ such that

$$\int \int_{\mathbb{R}_+ \times \mathbb{R}^d} \eta(\omega, t, x) \mu(\omega, d(t,x)) P(d\omega) < \infty.$$

As for processes we usually omit the argument ω. Conditions 1, 3 tell us that μ is a counting measure, i.e. $\mu([0,t] \times B)$ counts the number of events belonging to B during the period $[0,t]$. For $\mu = \mu^X$ these events are the jumps of X. Condition 2 signifies that no event (e.g. jump) happens at time 0. Condition 4 tells us that at most one event happens at a time. Adaptedness means that the events are observable; it corresponds directly to adaptedness of stochastic processes. The technical integrability condition 6 is introduced to rule out pathological cases. It does not play any role in the following. If μ satisfies 1, 2, 5, 6 but not necessarily 3, 4, we refer more generally to a **random measure**.

Keeping ω as a fixed parameter, we can integrate deterministic or even random functions relative to these measures. To this end, let $\xi(\omega, t, x)$ denote a real- or possibly vector-valued function on $\Omega \times \mathbb{R}_+ \times \mathbb{R}^d$. For the purposes of integration we consider only **predictable functions** in the sense that ξ is measurable with respect to the product-σ-field $\widetilde{\mathscr{P}} := \mathscr{P} \otimes \mathscr{B}^d$ on $\Omega \times \mathbb{R}_+ \times \mathbb{R}^d$. The **integral process** $\xi * \mu$ is defined by pathwise integration of ξ relative to μ, namely

$$\xi * \mu(\omega, t) := \int_{[0,t] \times \mathbb{R}^d} \xi(\omega, s, x) \mu(\omega, d(s,x)),$$

whenever the right-hand side makes sense, i.e. if

$$\int_{[0,t] \times \mathbb{R}^d} |\xi(\omega, s, x)| \mu(\omega, d(s,x)) < \infty, \quad t \geq 0.$$

We write $L(\mu)$ for the set of such **μ-integrable** predictable functions ξ. The integral process $\xi * \mu$ is an adapted càdlàg process of finite variation and in particular a semimartingale. For integer-valued random measures it reduces to a sum or more precisely a series. For the random measure of jumps μ^X of a semimartingale X we have

$$\xi * \mu^X(t) = \sum_{s \leq t} \xi(s, \Delta X(s)) 1_{\{\Delta X(s) \neq 0\}}. \tag{3.42}$$

With this notion of an integral process we can formulate Itō's formula (3.20) equivalently as

$$\begin{aligned} f(X(t)) = f(X(0)) + f'(X_-) \bullet X(t) + \frac{1}{2} f''(X_-) \bullet \langle X^c, X^c \rangle(t) \\ + \big(f(X_- + x) - f(X_-) - f'(X_-)x\big) * \mu^X(t) \end{aligned} \tag{3.43}$$

and accordingly for vector-valued X. In view of Proposition 3.9, we can write the quadratic variation of a semimartingale X as

$$[X, X] = \langle X^c, X^c \rangle + x^2 * \mu^X. \tag{3.44}$$

In the multivariate case we have

$$[X_i, X_j] = \langle X_i^c, X_j^c \rangle + (x_i x_j) * \mu^X \tag{3.45}$$

for the covariation of the components of X. In the preceding and also in later expressions, x refers implicitly to the argument of the function $\xi(x)$ that is integrated. For example, $x^2 * \mu^X$ means $\xi * \mu^X$ for $\xi(x) = x^2$.

Proposition 3.32 *Let μ denote some random measure and ξ some μ-integrable function.*

1. *The integral $\xi * \mu$ is linear in ξ.*
2. *We have $\varphi \bullet (\xi * \mu) = (\varphi\xi) * \mu$ if $\varphi \in L_s(\xi * \mu)$ or equivalently $\varphi\xi \in L(\mu)$.*

Proof The first statement follows from linearity of Lebesgue integration. For ease of notation, we assume $d = 1$ for the second assertion. By standard arguments from measure theory (cf. the proof of Proposition 1.38), it suffices to consider $\xi = 1_A$ for $A \in \widetilde{\mathscr{P}}$ and $\varphi = 1_B$ for $B \in \mathscr{P}$. In this case

$$(\varphi\xi) * \mu(\omega, t) = \mu\big(\omega, ([0, t] \times \mathbb{R}) \cap A_\omega \cap (B_\omega \times \mathbb{R})\big)$$

with $A_\omega := \{(t, x) \in \mathbb{R}_+ \times \mathbb{R} : (\omega, t, x) \in A\}$ and $B_\omega := \{t \in \mathbb{R}_+ : (\omega, t) \in B\}$. The process $\xi * \mu$, on the other hand, has pathwise Lebesgue–Stieltjes measure

$$\widetilde{\mu}(\omega, C) = \mu(\omega, (C \times \mathbb{R}) \cap A_\omega), \quad C \in \mathscr{B}(\mathbb{R}_+)$$

on $\mathbb{R}_+$. Consequently,

$$\begin{aligned}(\varphi \bullet (\xi * \mu))(\omega, t) &= \int_{(0,t]} \varphi(\omega, s)\widetilde{\mu}(\omega, ds) \\ &= \mu\big(\omega, ((B_\omega \cap (0, t]) \times \mathbb{R}) \cap A_\omega\big) \\ &= (\varphi\xi) * \mu(\omega, t)\end{aligned}$$

as desired. □

For concrete calculations one often needs to determine the martingale and the drift part of a special semimartingale. For jump processes they can be expressed in terms of the *predictable compensator of a random measure.* A random measure μ is called **predictable** if the integral process $\xi * \mu$ is predictable for any predictable integrand $\xi \in L(\mu)$. The **(predictable) compensator** of an integer-valued random measure μ is the unique predictable random measure ν with the following property:

> ξ is ν-integrable and $\xi * \nu$ is the compensator of $\xi * \mu$ for any μ-integrable ξ such that $\xi * \mu$ is a special semimartingale

or, equivalently, $E(\xi * \nu(\infty)) = E(\xi * \mu(\infty))$ for any nonnegative predictable function ξ, cf. [154, Theorem II.1.8]. Using this notion, we immediately obtain the canonical decomposition of an integral process $X = \xi * \mu$ if it can be decomposed at all, i.e. if it is a special semimartingale. It is given by $X = M^X + A^X$ with martingale part $M^X = \xi * \mu - \xi * \nu$ and drift $A^X = \xi * \nu$. In the sequel we use the shorthand

$$\xi * (\mu - \nu) := \xi * \mu - \xi * \nu$$

for the local martingale part.

From now on we focus on the measure of jumps μ^X of a given semimartingale X. We denote its compensator generally by ν^X. It occurs, for instance, in the predictable quadratic variation of a semimartingale X. From (3.44) we conclude that the latter is given by

$$\langle X, X\rangle = \langle X^c, X^c\rangle + x^2 * \nu^X$$

if it exists. Accordingly, we have

$$\langle X_i, X_j\rangle = \langle X_i^c, X_j^c\rangle + (x_i x_j) * \nu^X$$

in the multivariate case.

Figure 3.4 depicts the stochastic exponential X of a Lévy process with bounded jumps, its random measure of jumps μ^X, and the density of its predictable compensator ν^X.

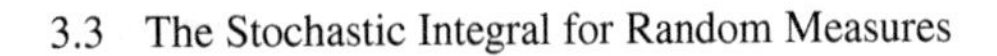

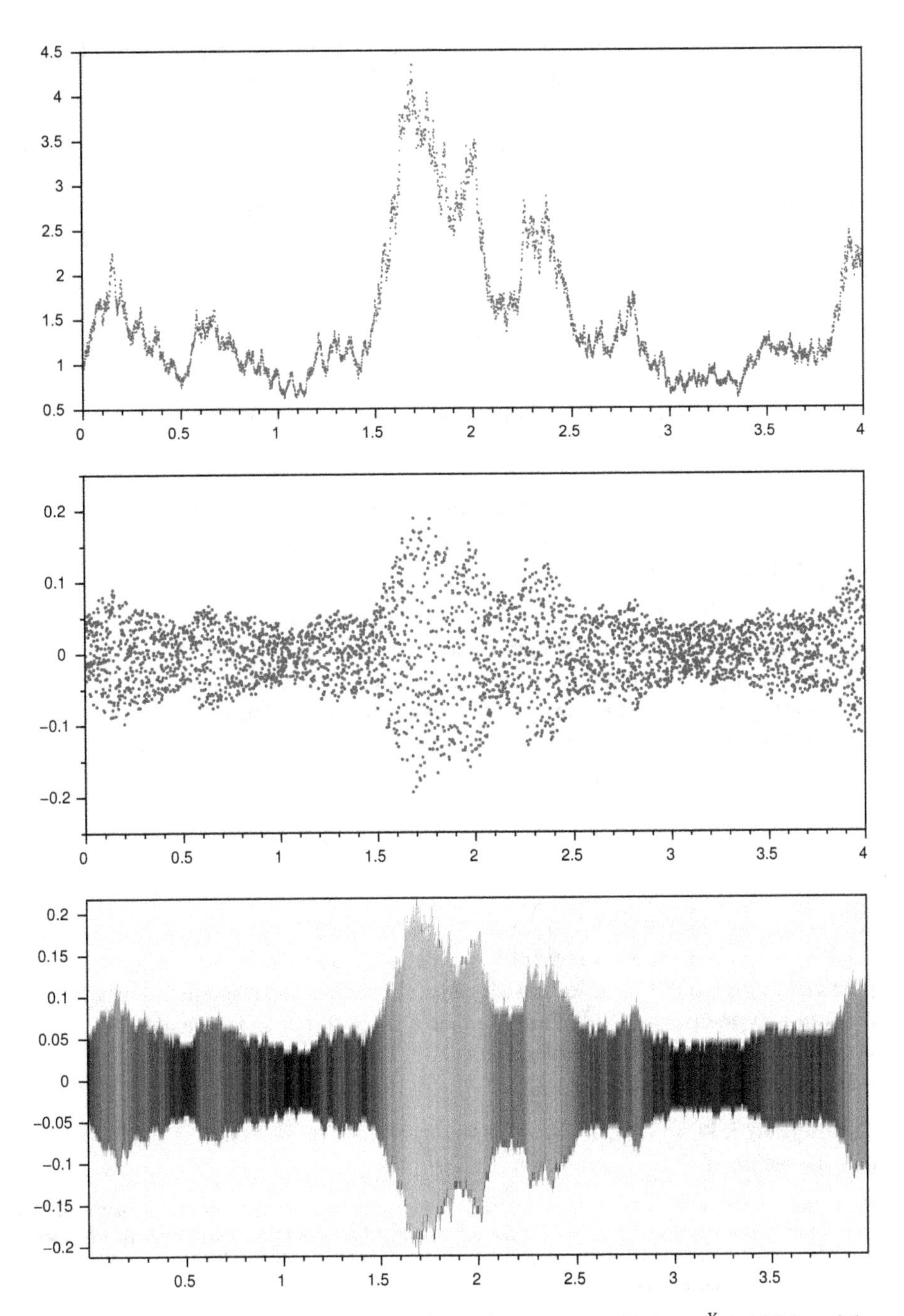

Fig. 3.4 An exponential Lévy process (top), its random measure of jumps μ^X (middle), and the density of its compensator ν^X (bottom)

Example 3.33 (Lévy Process) In the case of a Lévy process X with Lévy–Khintchine triplet (b^h, c, K), the compensator ν^X of μ^X is given by

$$\nu^X([0,t] \times B) = K(B)t. \tag{3.46}$$

Indeed, ν^X in (3.46) is a predictable random measure. Using standard arguments from measure theory, it suffices to consider $\xi = 1_{F\times(s,t]\times B}$ with $s \leq t$, $F \in \mathscr{F}_s$, $B \in \mathscr{B}^d$ such that $K(B) < \infty$. In this case

$$\begin{aligned}&\xi * \mu^X(v) - \xi * \nu^X(v) - (\xi * \mu^X(u) - \xi * \nu^X(u))\\&= 1_F\left(\sum_{r\in(u\vee s,v\wedge t]} 1_B(\Delta X(r)) - K(B)((v \wedge t) - (u \vee s))\right)\end{aligned}$$

for $u \leq v$. Since X has independent increments, this random variable is independent of $\mathscr{F}_u$. Moreover, it has expectation 0 by (2.20). Consequently, $\xi * \mu^X - \xi * \nu^X$ is a martingale as desired.

Note that $(\nu^X([0,t]\times B))_{t\geq 0}$ is a deterministic linear function for Lévy processes. For general semimartingales it may be neither deterministic nor linear and it can even have jumps. But similarly as we noted in Sect. 2.1 for the drift part of special semimartingales, the vast majority of processes in applications have a compensator ν^X which is absolutely continuous with respect to time. This means that it can be written as

$$\nu^X([0,t] \times B) = \int_0^t K(s,B)ds \tag{3.47}$$

with some transition kernel K from $(\Omega \times \mathbb{R}_+, \mathscr{P})$ to $\mathbb{R}^d$, i.e. $B \mapsto K(\omega,t,B) =: K(t,B)(\omega)$ is a measure on $\mathbb{R}^d$ for fixed (ω,t) and $(\omega,t) \mapsto K(\omega,t,B)$ is predictable for fixed B. The kernel K stands for a local jump intensity. It is called the **intensity measure** of μ^X. Observe that $K(t,B)(\omega)$ does not depend on ω and not even on t in the situation of Example 3.33. For the remainder of this section we make the following standing

Assumption 3.34 ν^X *allows for a factorisation (3.47). This implies* $\Delta A^X(t) = 0$ *if* X *is special, cf. [154, Proposition II.2.29].*

Physicists' Corner 3.35 What is the idea behind the compensator of a random measure? The quantity $\mu^X((t-dt,t] \times B)$ stands for the random number of jumps of size in B that happen during the infinitesimal period from $t-dt$ to t. We do not know in advance how many this will be. On an intuitive level, we may write

$$\nu^X((t-dt,t] \times B) = E(\mu^X((t-dt,t] \times B)|\mathscr{F}_{t-dt}),$$

i.e. the compensator represents the expected number of jumps given the information up to time $t-dt$. This way of thinking corresponds directly to the compensator of a special semimartingale in Physicists' Corner 2.12. As regards the intensity measure, we expect $K(t, B)dt$ jumps of size in B in the infinitesimal period $(t - dt, t]$ if we are given the information up to $t - dt$. □

Let us come back to integration.

Proposition 3.36

1. *A predictable function ξ is ν^X-integrable if*

$$\int_0^t \int |\xi(s, x)| K(s, dx) ds < \infty, \quad t \geq 0,$$

which in turn holds if and only if ξ is μ^X-integrable and $\xi * \mu^X$ is a special semimartingale.

2. *If ξ is a ν^X-integrable function, then*

$$\xi * \nu^X(t) = \int_0^t \int \xi(s, x) K(s, dx) ds. \tag{3.48}$$

3. *If ξ is a ν^X-integrable function, then $\xi * (\mu^X - \nu^X)$ is a purely discontinuous local martingale whose jumps are given by*

$$\Delta(\xi * (\mu^X - \nu^X))(t) = \xi(t, \Delta X(t)) 1_{\{\Delta X(t) \neq 0\}}.$$

Idea

1. The first statement is obvious. For the remaining ones see [152, Lemme 3.67a].
2. For $\xi(\omega, t, x) = 1_B(x)$ this follows from (3.47). The extension to the general case is obtained by the standard procedure of approximating measurable functions, cf. the proof of Proposition 1.38.
3. $\xi * (\mu^X - \nu^X)$ is a local martingale because $\xi * \nu^X$ is the compensator of $\xi * \mu^X$. It is purely discontinuous by Proposition 3.8. Since $\xi * \nu^X$ is continuous by (3.48), the jumps of $\xi * (\mu^X - \nu^X)$ equal those of $\xi * \mu^X$, which are immediately obtained from (3.42). □

In some instances we need to define an integral $\xi * (\mu^X - \nu^X)$ even though ξ is not μ^X-integrable. We encountered this situation in the Lévy–Itō decomposition (2.76), where the jumps of X may not sum to a finite value. The way out is to use the properties of $\xi * (\mu^X - \nu^X)$ in statement 3 of Proposition 3.36 in order to extend its definition beyond μ^X-integrable ξ.

We say that a predictable function ξ is **$(\mu^X - \nu^X)$-integrable** (written $\xi \in G_{\text{loc}}(\mu^X)$) if

$$E\left(\sqrt{\sum_{t \leq \tau_n} \xi(t, \Delta X(t))^2 1_{\{\Delta X(t) \neq 0\}}}\right) < \infty, \quad n \in \mathbb{N}, \tag{3.49}$$

i.e., using different notation, $E(\sqrt{\xi^2 * \mu^X(\tau_n)}) < \infty$, $n \in \mathbb{N}$ for a localising sequence of stopping times $\tau_n \uparrow \infty$. The unique purely discontinuous local martingale M with

$$\Delta M(t) = \xi(t, \Delta X(t)) 1_{\{\Delta X(t) \neq 0\}} \tag{3.50}$$

is called the **stochastic integral** of ξ relative to $\mu^X - \nu^X$ and it is written $\xi * (\mu^X - \nu^X)$ or occasionally

$$\left(\int_{[0,t] \times \mathbb{R}^d} \xi(s, x)(\mu^X - \nu^X)(d(s, x)) \right)_{t \geq 0}.$$

This integral in (2.76) can be interpreted in the sense of the previous definition. Before we extend this Lévy–Itō decomposition to more general processes, we mention some properties of stochastic integration relative to compensated random measures. In particular, statements 1 and 2 contain sufficient conditions for integrability which are easier to verify than the somewhat terrifying definition (3.49).

Proposition 3.37

1. *If ξ is ν^X-integrable, then it is both μ^X- and $(\mu^X - \nu^X)$-integrable and we have*

$$\xi * (\mu^X - \nu^X) = \xi * \mu^X - \xi * \nu^X.$$

   This is the case if and only if ξ is $(\mu^X - \nu^X)$-integrable and $\xi * (\mu^X - \nu^X)$ is of finite variation.
2. *A predictable function ξ satisfies $\xi^2 * \nu^X(t) < \infty$, $t \geq 0$ (written $\xi^2 \in L(\nu^X)$ or $\xi \in G^2_{\text{loc}}(\mu^X)$) if and only if it is $(\mu^X - \nu^X)$-integrable and if the stochastic integral $\xi * (\mu^X - \nu^X)$ is a locally square-integrable martingale. The integral is a square-integrable martingale if and only if $E(\xi^2 * \nu^X(\infty)) < \infty$ (written $\xi \in G^2(\mu^X)$).*
3. *The integral $\xi * (\mu^X - \nu^X)$ is linear in ξ.*
4. *If φ is a predictable, locally bounded process and ξ is $(\mu^X - \nu^X)$-integrable, then $\varphi\xi$ is $(\mu^X - \nu^X)$-integrable as well and*

$$\varphi \bullet \big(\xi * (\mu^X - \nu^X)\big) = (\varphi\xi) * (\mu^X - \nu^X). \tag{3.51}$$

5. *We have*

$$\big\langle \eta * (\mu^X - \nu^X), \xi * (\mu^X - \nu^X) \big\rangle = (\eta\xi) * \nu^X$$

   if the predictable covariation exists, which happens if and only if both ξ and η are $(\mu^X - \nu^X)$-integrable and $\xi\eta$ is ν^X-integrable.

6. *If Y denotes a semimartingale whose jumps satisfy $\Delta Y(t) = \zeta(t, \Delta X)$ for some predictable function ζ on $\Omega \times \mathbb{R}_+ \times \mathbb{R}^d$, we have*

$$\begin{aligned} \xi * \mu^Y &= (\xi \circ \zeta) * \mu^X, \\ \xi * \nu^Y &= (\xi \circ \zeta) * \nu^X, \\ \xi * (\mu^Y - \nu^Y) &= (\xi \circ \zeta) * (\mu^X - \nu^X) \end{aligned}$$

for any predictable function ξ such that either of the integrals in the respective equations exists. Here, $\xi \circ \zeta$ denotes the predictable function defined as

$$(\xi \circ \zeta)(\omega, t, x) := \xi(\omega, t, \zeta(\omega, t, x)).$$

Proof

1. The equality follows from Proposition 3.36(3). The statements concerning integrability can be derived from [154, Proposition II.1.28, Theorem II.1.33b, Lemma I.3.11].
2. This is stated in [154, Theorem II.1.33a].
3. This follows from the definition.
4. This is stated in [125, Proposition A.1(5)]. We focus here on (3.51). By Propositions 3.23(1) and 3.20(9), the left-hand side is a purely discontinuous local martingale whose jumps satisfy

$$\begin{aligned} \Delta\big(\varphi \bullet (\xi * (\mu^X - \nu^X))\big)(t) &= \varphi(t)\Delta(\xi * (\mu^X - \nu^X))(t) \\ &= \varphi(t)\xi(t, \Delta X(t)). \end{aligned}$$

   Consequently, it has the defining properties of the process on the right.
5. By Proposition 3.9 we have

$$\begin{aligned} &\big[\eta * (\mu^X - \nu^X), \xi * (\mu^X - \nu^X)\big](t) \\ &\quad = \sum_{s \le t} \Delta(\eta * (\mu^X - \nu^X))(s)\Delta(\xi * (\mu^X - \nu^X))(s) \\ &\quad = \sum_{s \le t} \eta(s, \Delta X(s))\xi(s, \Delta X(s))1_{\{\Delta X(s) \neq 0\}} \\ &\quad = (\eta\xi) * \mu^X(t), \quad t \ge 0 \end{aligned}$$

   because the integrals are purely discontinuous by definition. By definition of ν^X its compensator exists if and only if $\eta\xi$ is ν^X-integrable. In this case the compensator equals $(\eta\xi) * \nu^X$.

6. The first equation follows from

$$\begin{aligned}\xi * \mu^Y(t) &= \sum_{s\leq t} \xi(s, \Delta Y(s)) \\ &= \sum_{s\leq t} \xi(s, \zeta(s, \Delta X(s))) \\ &= (\xi \circ \zeta) * \mu^X(t).\end{aligned}$$

The second is obtained from the first by taking compensators. The third equation holds because both sides are purely discontinuous local martingales with the same jumps, whence they coincide by Proposition 3.7. □

Remark 3.38 If h is a truncation function, we have $h \in G_{\text{loc}}(\mu^X)$. Indeed, $h^2 * \mu^X$ has bounded jumps, which implies that it is special by Proposition 2.14. Proposition 3.36(1) yields that h^2 is ν^X-integrable and hence $h \in G_{\text{loc}}(\mu^X)$ by Proposition 3.37(2).

Dominated convergence holds for this integral as well.

Proposition 3.39 (Dominated Convergence) *Suppose that ξ^n, η are predictable functions with pointwise limit $\xi = \lim_{n\to\infty} \xi^n$ and $|\xi^n| \leq \eta$, $n \geq 1$. If η is $(\mu^X - \nu^X)$-integrable, then ξ^n, ξ are $(\mu^X - \nu^X)$-integrable as well and we have*

$$\xi^n * (\mu^X - \nu^X) \to \xi * (\mu^X - \nu^X)$$

in probability uniformly on compact intervals $[0, t]$, $t \geq 0$. Analogous statements hold for

$$\begin{aligned}\xi^n * \mu^X &\to \xi * \mu^X, \\ \xi^n * \nu^X &\to \xi * \nu^X.\end{aligned}$$

Proof The assertion for μ^X and ν^X follows from dominated convergence. If $\eta \in L(\nu^X)$, we have $\xi^n * (\mu^X - \nu^X) = \xi^n * \mu^X - \xi^n * \nu^X$, whence the first convergence is implied by the second and third. If $\eta^2 \in L(\nu^X)$, Proposition 3.37(5) yields

$$\langle (\xi^n - \xi) * (\mu^X - \nu^X), (\xi^n - \xi) * (\mu^X - \nu^X) \rangle (\tau) = (\xi^n - \xi)^2 * \nu^X(\tau) \to 0$$

as $n \to \infty$ and any bounded stopping time τ, which, by Proposition 3.5, implies

$$\begin{aligned}&E\Big(\big((\xi^n - \xi) * (\mu^X - \nu^X)(\tau_k)\big)^2\Big) \\ &\quad = E\Big(\langle (\xi^n - \xi) * (\mu^X - \nu^X), (\xi^n - \xi) * (\mu^X - \nu^X) \rangle (\tau_k)\Big) \overset{n\to\infty}{\longrightarrow} 0\end{aligned}$$

for some properly chosen localising sequence $(\tau_k)_{k=0,1,...}$ of stopping times. From Doob's inequality (2.2) we even obtain

$$E\left(\sup_{t\leq\tau_k}\left((\xi^n-\xi)*(\mu^X-\nu^X)(t)\right)^2\right)\stackrel{n\to\infty}{\longrightarrow}0,$$

which implies the claimed convergence.

The case of general η can be reduced to the above cases. Indeed, for a continuous, symmetric truncation function h and $\overline{h}(x) := x - h(x)$ we have $h(\xi^n) \to h(\xi)$, $\overline{h}(\xi^n) \to \overline{h}(\xi)$, $|h(\xi^n)| \leq |h(\eta)|$, $|\overline{h}(\xi^n)| \leq |\overline{h}(\eta)|$. Since $h(\xi) * (\mu^X - \nu^X)$ has bounded jumps, it is locally square-integrable. By Proposition 3.37(2) we have $h(\eta)^2 \in L(\nu^X)$. Since $\eta * (\mu^X - \nu^X)$ has only finitely many "big" jumps, $\overline{h}(\eta)$ is μ^X-integrable. Note that $B = \overline{h}(\eta) * (\mu^X - \nu^X) - \overline{h}(\eta) * \mu^X$ satisfies $B^c = 0$ and $\Delta B = 0$, which implies that it equals the predictable part of finite variation in its canonical decomposition. Consequently $\overline{h}(\eta) * (\mu^X - \nu^X)$ is of finite variation and hence $\overline{h}(\eta) \in L(\nu^X)$. Together, we obtain

$$\begin{aligned}\xi^n * (\mu^X-\nu^X) &= h(\xi^n) * (\mu^X-\nu^X) + \overline{h}(\xi^n) * (\mu^X-\nu^X)\\ &\to h(\xi) * (\mu^X-\nu^X) + \overline{h}(\xi) * (\mu^X-\nu^X) = \xi * (\mu^X-\nu^X). \quad \square\end{aligned}$$

Special semimartingales allow for a canonical representation

$$X = X(0) + M^X + A^X,$$

which can be decomposed further as

$$X = X(0) + X^c + M^{X,d} + A^X.$$

If we express the purely discontinuous martingale part as a stochastic integral, we end up with a generalisation of the Lévy–Itō decomposition (2.77), namely

$$X = X(0) + A^X + X^c + x * (\mu^X - \nu^X). \tag{3.52}$$

Indeed, both $M^{X,d}$ and $x * (\mu^X - \nu^X)$ are purely discontinuous local martingales with jumps ΔX, cf. Assumption 3.34. Here we secretly apply Proposition 4.21 for the existence of $x * (\mu^X - \nu^X)$.

As an example, consider $f(X)$ in Itō's formula. Provided that both X and $f(X)$ are special, decomposition (3.52) of $f(X)$ in (3.20) and (3.43) reads as

$$
\begin{aligned}
f(X(t)) = f(X(0)) &+ \Big(f'(X_-) \bullet A^X(t) + \frac{1}{2} f''(X_-) \bullet \langle X^c, X^c \rangle(t) \\
&+ \big(f(X_- + x) - f(X_-) - f'(X_-)x \big) * \nu^X(t) \Big) \\
&+ f'(X_-) \bullet X^c(t) + \big(f(X_- + x) - f(X_-) \big) * (\mu^X - \nu^X)(t),
\end{aligned} \tag{3.53}
$$

cf. [125, Proposition A.3].

However, arbitrary semimartingales cannot be written in the form (3.52) because x is not $(\mu^X - \nu^X)$-integrable. A way out is to subtract the "big" jumps

$$J^X(t) := \sum_{s \leq t} \Delta X(s) 1_{\{|\Delta X(s)|>1\}}.$$

The remainder $\widetilde{X} := X - J^X$ is a special semimartingale by Proposition 2.14. Hence it can be written as

$$
\begin{aligned}
\widetilde{X} &= \widetilde{X}(0) + A^{\widetilde{X}} + \widetilde{X}^c + x * (\mu^{\widetilde{X}} - \nu^{\widetilde{X}}) \\
&= X(0) + A^{\widetilde{X}} + X^c + x 1_{\{|x| \leq 1\}} * (\mu^X - \nu^X).
\end{aligned}
$$

Together we obtain

$$X = X(0) + A^{\widetilde{X}} + x 1_{\{|x|>1\}} * \mu^X + X^c + x 1_{\{|x| \leq 1\}} * (\mu^X - \nu^X), \tag{3.54}$$

which generalises (2.76) to general (i.e. not necessarily special) semimartingales. We will come back to the issue of decompositions in Sect. 4.3.

So far, we have considered primarily random measures of jumps and their compensators. Stochastic integration can easily be generalised to more general random measures. For the definition of $\xi * (\mu - \nu)$ it suffices to assume that μ is an integer-valued random measure whose compensator ν is absolutely continuous in time in the sense that

$$\nu([0, t] \times B) = \int_0^t K(s, B) ds \tag{3.55}$$

for some random intensity measure as in (3.47). The integrability condition (3.49) is replaced with

$$E\left(\sqrt{\xi^2 * \mu(\tau_n)} \right) < \infty$$

and the jumps of $\xi * (\mu - \nu)$ are assumed to equal

$$\Delta M(t) := \begin{cases} \xi(t, x) & \text{if } \mu(\{(t, x)\}) = 1, \\ 0 & \text{if } \mu(\{(t\} \times \mathbb{R}^d\}) = 0, \end{cases}$$

which generalises (3.49) and (3.50) to arbitrary integer-valued random measures. The applicable parts of Propositions 3.37 and 3.39 carry over to this slightly more general integral, cf. [154, Section II.1d].

One could even do without the existence of intensity measures as in (3.47) resp. (3.55) for the definition of stochastic integrals $\xi * (\mu^X - \nu^X)$ resp. $\xi * (\mu - \nu)$. This allows us to extend (3.52–3.54) literally to general semimartingales. However, we chose not to do so because the general formulas for $\Delta(\xi * (\mu^X - \nu^X))$ and $\langle \xi * (\mu^X - \nu^X), \xi * (\mu^X - \nu^X) \rangle$, for instance, look confusing at first glance and we hardly need such generality in applications.

3.4 Itō Semimartingales

In the beginning of Sect. 3.2 we briefly discussed processes that resemble Lévy processes on a local scale. Let us return to this issue from a more general perspective. For example, by Propositions 4.6, 4.12, any process of the form

$$X(t) = b^{\mathrm{id}} t + \sigma W(t) + \delta(x) * (\mu - \nu)(t)$$

or, more generally,

$$X(t) = b^h t + \sigma W(t) + h(\delta(x)) * (\mu - \nu)(t) + \overline{h}(\delta(x)) * \mu(t) \tag{3.56}$$

is a Lévy process, where $b^{\mathrm{id}}, b^h, \sigma$ are constants, W denotes a Wiener process, μ a homogeneous Poisson random measure with intensity measure K and compensator $\nu(dt, dx) = K(dx)dt$, and δ a deterministic function such that the integrals exist. h in (3.56) stands for some truncation function and

$$\overline{h}(x) := x - h(x).$$

In view of the Lévy–Itō decomposition (3.54), any Lévy process with Lévy–Khintchine triplet (b^h, c, K) can be represented in this form, e.g. letting $\sigma = \sqrt{c}$, $W = X^c/\sqrt{c}$ (unless $c = 0$), $\mu = \mu^X$, $\delta(x) = x$. Using differential notation, we rewrite (3.56) as

$$dX(t) = b^h dt + \sigma dW(t) + h(\delta(x)) * (\mu - \nu)(dt) + \overline{h}(\delta(x)) * \mu(dt).$$

As in Sect. 3.3 we use the implicit convention that the argument x refers to the running variable of integration relative to $\mu - \nu$ resp. μ.

We pass from the linear to the non-linear case by letting b^h, σ, δ depend on (ω, t):

$$dX(t) = b^h(t)dt + \sigma(t)dW(t) + h(\delta(t,x)) * (\mu - \nu)(dt) + \overline{h}(\delta(t,x)) * \mu(dt). \tag{3.57}$$

This means that X resembles a Lévy process on a local scale but not globally. The terms are to be interpreted as stochastic integrals in the sense of Sects. 3.2 and 3.3. In analogy to **Itō processes** where the jump terms are missing, we call such processes of the form

$$\begin{aligned} X(t) &= X(0) + \int_0^t b^h(s)ds + \int_0^t \sigma(s)dW(s) \\ &\quad + \int_0^t \int h(\delta(s,x))(\mu - \nu)(ds,dx) + \int_0^t \int \overline{h}(\delta(s,x))\mu(ds,dx) \\ &= X(0) + b^h \bullet I(t) + \sigma \bullet W(t) + (h \circ \delta) * (\mu - \nu)(t) + (\overline{h} \circ \delta) * \mu(t) \end{aligned} \tag{3.58}$$

Lévy–Itō processes or **Itō semimartingales**. As above W denotes a Wiener process and μ a homogeneous Poisson random measure with intensity measure K and compensator $\nu(dt,dx) = K(dx)dt$. Moreover, b^h, σ are I- resp. W-integrable processes and δ some predictable function such that $h \circ \delta, \overline{h} \circ \delta$ are $(\mu - \nu)$- resp. μ-integrable. The Brownian motion part stands for continuous wiggling whereas the random measure contributes jumps. μ can be interpreted as a collection of random events. Each one is located at some (t,x), which leads to a jump of size $\delta(t,x)$ at time t.

An Itō semimartingale (3.58) is a special semimartingale if and only if $\overline{h} \circ \delta$ is ν-integrable, i.e. if

$$\int_0^t \int |\overline{h}(\delta(s,x))| K(s,dx)ds < \infty, \quad t \geq 0.$$

In this case, the canonical decomposition of X can be written as

$$\begin{aligned} dX(t) &= dA^X(t) + dM^{X,c}(t) + dM^{X,d}(t) \\ &= \left(b^h(t) + \int \overline{h}(\delta(t,x))K(dx)\right)dt + \sigma(t)dW(t) + \delta(t,x) * (\mu - \nu)(dt) \\ &= a^X(t)dt + \sigma(t)dW(t) + \delta(t,x) * (\mu - \nu)(dt) \end{aligned} \tag{3.59}$$

with $a^X(t) := b^{\mathrm{id}}(t) := b^h(t) + \int \overline{h}(\delta(t,x))K(dx)$. Hence there is no need for an integral relative to μ in (3.58). Put differently, we can choose $h(x) = x$ as the truncation function.

The general form (3.57) is chosen if there are too many big jumps, which happens, for example, for stable Lévy motions where only the general form (2.76) of the Lévy–Itō decomposition makes sense. But in applications it often suffices to consider a predictable drift term, a Brownian part, and a compensated jump martingale as in (3.59).

The generalisation of Itō semimartingales to the multivariate case is straightforward. In this context W denotes a multivariate standard Brownian motion.

3.5 Stochastic Differential Equations

In Sect. 3.2 we motivated stochastic integration by the fact that many processes resemble Lévy processes on a small scale but not globally, leading to an equation

$$dX(t) = b(t)dL(t), \tag{3.60}$$

which generalises

$$dX(t) = b(t)dt \tag{3.61}$$

to the random case. In deterministic applications the integrand is often a function of the current state $X(t)$, i.e. we have an equation of the form

$$dX(t) = \beta(X(t))dt \tag{3.62}$$

instead of (3.61). This is typically interpreted as an ordinary differential equation (ODE)

$$\frac{dX(t)}{dt} = \beta(X(t)).$$

The counterpart of ODE (3.62) in the random case reads as

$$dX(t) = \beta(X(t-))dL(t) \tag{3.63}$$

with some deterministic function β. Here the left-hand limit $X(t-)$ is chosen as the current state in order for the integrand to be predictable. In contrast to (3.62) a differential notation

$$\frac{dX(t)}{dL(t)} = \beta(X(t-))$$

does not make sense because derivatives relative to processes are not defined. Equation (3.63) refers properly to

$$\int_0^t dX(s) = \int_0^t \beta(X(s-))dL(s), \quad t \geq 0$$

or equivalently

$$X(t) = X(0) + \int_0^t \beta(X(s-))dL(s) = X(0) + \beta(X_-) \bullet L(t), \quad t \geq 0.$$

It should therefore be called a *stochastic integral equation* rather than a *stochastic differential equation* but the latter term is used more frequently. As for ordinary differential equations, unique solutions exist under Lipschitz and growth conditions. We state a corresponding result for general semimartingale drivers L rather than just Lévy processes. Moreover, we allow for β to depend explicitly on time.

Theorem 3.40 *Suppose that L is an $\mathbb{R}^n$-valued semimartingale and $\beta : \mathbb{R}_+ \times \mathbb{R}^n \to \mathbb{R}^{d\times n}$ some Borel-measurable function such that the following conditions hold.*

1. *(Lipschitz condition) For any $m \in \mathbb{N}$ there exists a constant ϑ_m such that*

$$|\beta(t, x) - \beta(t, y)| \leq \vartheta_m |x - y|$$

   holds for $t \leq m$ and $x, y \in \mathbb{R}^d$.
2. *(Continuity condition) $t \mapsto \beta(t, x)$ is continuous for any $x \in \mathbb{R}^d$.*

Then there exists a unique $\mathbb{R}^d$-valued semimartingale X which solves the **stochastic differential equation (SDE)**

$$dX(t) = \beta(t, X(t-))dL(t) \tag{3.64}$$

for any given initial value $X(0)$ (in general an $\mathscr{F}_0$-measurable random variable).

Idea This is shown as in the classical Picard–Lindelöf theorem for ordinary differential equations. X is a solution to the SDE (3.64) if and only if it is a fixed point of the operator Φ defined by $\Phi(X) = X(0) + \beta(I, X_-) \bullet L$. The idea is now to choose an appropriate Banach space of processes such that Φ is a contraction. Banach's fixed point theorem then yields the existence of a solution. In order to show uniqueness even beyond the chosen space, we may use Grönwall's lemma, again relative to some appropriate norm. The Lipschitz condition yields that, for any two solutions $X, \widetilde{X}$ to (3.64), the process $|X - \widetilde{X}|$ does not grow faster than exponentially. Since $X(0) = \widetilde{X}(0)$, uniqueness follows.

A complete proof may be found in [238, Theorem V.7]. □

The previous result can be generalised to functions β which depend on the whole past $(X(s))_{s<t}$ rather than only $X(t-)$.

Itō semimartingales are generally written in the form

$$dX(t) = b^h(t)dt + \sigma(t)dW(t) + h(\delta(t, x)) * (\mu - \nu)(dt) + \overline{h}(\delta(t, x)) * \mu(dt)$$

rather than (3.60), cf. Sect. 3.4. If the coefficients b^h, σ, δ depend only on t and the current value $X(t-)$ (resp. t, $X(t-)$ and x) but not explicitly on ω, we end up with an SDE of the form

$$dX(t) = \beta^h(t, X(t-))dt + \sigma(t, X(t-))dW(t) \qquad (3.65)$$
$$+ h(\delta(t, X(t-), x)) * (\mu - \nu)(dt) + \overline{h}(\delta(t, X(t-), x)) * \mu(dt),$$

which involves stochastic integrals relative to random measures. The theory of such equations resembles that of the more common SDEs in Theorem 3.40. Under Lipschitz and growth conditions on β^h, σ, δ, unique solutions to (3.65) exist.

We consider here the multivariate case. Fix an $\mathbb{R}^d$-valued Wiener process W and a homogeneous Poisson random measure μ on $\mathbb{R}_+ \times \mathbb{R}^d$.

Theorem 3.41 *Suppose that* $\beta^h : \mathbb{R}_+ \times \mathbb{R}^d \to \mathbb{R}^d$, $\sigma : \mathbb{R}_+ \times \mathbb{R}^d \to \mathbb{R}^{d \times n}$, $\delta : \mathbb{R}_+ \times \mathbb{R}^d \times \mathbb{R}^d \to \mathbb{R}^d$ *are Borel-measurable functions such that the following conditions hold:*

1. *(Local Lipschitz conditions) For any* $m \in \mathbb{N}$ *there exist a constant* ϑ_m *and a function* $\varrho_m : \mathbb{R}^d \to \mathbb{R}_+$ *with* $\int \varrho_m(x)^2 K(dx) < \infty$ *such that*

$$|\beta^h(t,x) - \beta^h(t,y)| \le \vartheta_m |x-y|,$$
$$|\sigma(t,x) - \sigma(t,y)| \le \vartheta_m |x-y|,$$
$$|h(\delta(t,x,z)) - h(\delta(t,y,z))| \le \varrho_m(z)|x-y|,$$
$$|\overline{h}(\delta(t,x,z)) - \overline{h}(\delta(t,y,z))| \le \varrho_m(z)^2|x-y|$$

hold for $t \le m$, $|x| \le m$, $|y| \le m$, $z \in \mathbb{R}^d$.

2. *(Linear growth conditions) For any* $m \in \mathbb{N}$ *there exist* ϑ_m *and* ϱ_m *as above such that*

$$|\beta^h(t,x)| \le \vartheta_m(1+|x|),$$
$$|\sigma(t,x)| \le \vartheta_m(1+|x|),$$
$$|h(\delta(t,x,z))| \le \varrho_m(z)(1+|x|),$$
$$|\overline{h}(\delta(t,x,z))| \le \left(\varrho_m(z)^2 \wedge \varrho_m(z)^4\right)(1+|x|)$$

hold for $t \le m$, $x \in \mathbb{R}^d$.

Then there exists a unique $\mathbb{R}^d$*-valued semimartingale* X *which solves the SDE (3.65) for any given initial value* $X(0)$ *(in general a* $\mathscr{F}_0$*-measurable random variable).*

Proof This theorem can be found in [154, Theorem III.2.32], cf. also [152, Théorème 14.95]. Its proof relies on the same ideas as Theorem 3.40. □

If one allows for an enlargement of the probability space, solutions exist under weaker conditions. The solution processes—or rather their laws—are called *weak solutions*. We briefly return to this notion in Sect. 5.7.3.

The SDEs in this section move *forward* in time in the sense that they start from a given initial value. Another kind of SDE with given terminal condition is discussed in Sect. 3.11.

3.6 The Doléans Exponential

Linear differential equations constitute at the same time the simplest and most frequently applied differential equations. In the deterministic case they lead to exponential functions, cf. Sect. 1.6.2 in this context. This motivates us to call the solutions to linear SDEs *stochastic exponentials*.

Definition 3.42 The solution Z to the SDE

$$dZ(t) = Z(t-)dX(t), \quad Z(0) = 1 \tag{3.66}$$

for an arbitrary semimartingale X is called the **Doléans exponential** or **stochastic exponential** of X. It is denoted by $\mathscr{E}(X) := Z$.

In view of Theorem 3.40, SDE (3.66) has a unique solution. Moreover, one easily verifies that $Z = c\mathscr{E}(X)$ solves $dZ(t) = Z(t-)dX(t)$ for $Z(0) = c$ with general deterministic or $\mathscr{F}_0$-measurable c.

The stochastic exponential can be expressed explicitly.

Proposition 3.43 *We have*

$$\mathscr{E}(X)(t) = \exp\Big(X(t) - X(0) - \frac{1}{2}\langle X^c, X^c\rangle(t)\Big)\prod_{s\leq t} e^{-\Delta X(s)}(1 + \Delta X(s)) \tag{3.67}$$

for any semimartingale X. If $\Delta X > -1$, this equals

$$\begin{aligned}\mathscr{E}(X)(t) &= \exp\Big(X(t) - X(0) - \frac{1}{2}\langle X^c, X^c\rangle(t) + \sum_{s\leq t}\big(\log(1 + \Delta X(s)) - \Delta X(s)\big)\Big)\\ &= \exp\Big(X(t) - X(0) - \frac{1}{2}\langle X^c, X^c\rangle(t) + \big(\log(1 + x) - x\big) * \mu^X(t)\Big).\end{aligned}$$

$\mathscr{E}(X)$ attains the value 0 only by jumps, namely if $\Delta X(t) = -1$. Afterwards it stays at 0.

Idea The statement can be found in [154, Theorem I.4.61]. We provide a proof for $\Delta X > -1$. Define $Y(t) = X(t) - X(0) - \frac{1}{2}\langle X^c, X^c\rangle(t) + \sum_{s\leq t}(\log(1+\Delta X(s)) - \Delta X(s))$. Note that $Y^c = X^c$ and $\Delta Y(s) = \log(1 + \Delta X(s))$. Itō's formula applied to Y yields $\exp(Y) = 1 + \exp(Y_-) \bullet U$ for

$$U(t) = Y(t) + \frac{1}{2}\langle Y^c, Y^c\rangle(t) + \sum_{s\leq t}\left(e^{\Delta Y(t)} - 1 - \Delta Y(t)\right) = X(t) - X(0).$$

We conclude that $\exp(Y) = \mathscr{E}(X - X(0)) = \mathscr{E}(X)$ as desired. □

Doléans exponentials of Lévy processes are discussed in the next section.

The product of two Doléans exponentials can be represented as Doléans exponential as well.

Proposition 3.44 (Yor's Formula) *We have*

$$\mathscr{E}(X)\mathscr{E}(Y) = \mathscr{E}(X + Y + [X, Y])$$

for any two semimartingales X, Y.

Proof Integration by parts and (3.66) yields for $Z = \mathscr{E}(X)\mathscr{E}(Y)$:

$$\begin{aligned} Z &= 1 + \mathscr{E}(X)_- \bullet \mathscr{E}(Y) + \mathscr{E}(Y)_- \bullet \mathscr{E}(X) + [\mathscr{E}(X), \mathscr{E}(Y)] \\ &= 1 + (\mathscr{E}(X)_-\mathscr{E}(Y)_-) \bullet Y + (\mathscr{E}(Y)_-\mathscr{E}(X)_-) \bullet X \\ &+ (\mathscr{E}(X)_-\mathscr{E}(Y)_-) \bullet [X, Y] \\ &= 1 + Z_- \bullet (Y + X + [X, Y]) \end{aligned}$$

and hence $Z = \mathscr{E}(X + Y + [X, Y])$ as desired. □

Up to the starting value $X(0)$, the original process X can be recovered from its Doléans exponential $Z = \mathscr{E}(X)$. Indeed, (3.66) implies $dX(t) = \frac{1}{Z(t-)}dZ(t)$, which leads to the following

Definition 3.45 Let Z be a semimartingale such that Z, Z_- do not attain the value 0. The semimartingale

$$\mathscr{L}(Z) := \frac{1}{Z_-} \bullet Z$$

is called the **stochastic logarithm** of Z.

It is easy to see that we have $\mathscr{L}(\mathscr{E}(X)) = X - X(0)$ if X does not jump by -1 and $\mathscr{E}(\mathscr{L}(Z)) = Z/Z(0)$ if Z, Z_- do not attain the value 0. In other words, the stochastic logarithm is the inverse of the Doléans exponential up to the initial value. The stochastic logarithm allows for a representation similar to (3.67).

Proposition 3.46 *For Z as in Definition 3.45 we have*

$$\mathscr{L}(Z)(t) = \log\left|\frac{Z(t)}{Z(0)}\right| + \frac{1}{2Z_-^2} \bullet \langle Z^c, Z^c\rangle(t) - \sum_{s\le t}\left(\log\left|\frac{Z(s)}{Z(s-)}\right| + 1 - \frac{Z(s)}{Z(s-)}\right). \tag{3.68}$$

Proof If we denote the right-hand side of (3.68) by $U(t)$, we have $U^c = \frac{1}{Z_-} \bullet Z^c$ by Itō's formula and

$$\Delta U(t) = \log\left|\frac{Z(t)}{Z(t-)}\right| - \log\left|\frac{Z(t)}{Z(t-)}\right| - 1 + \frac{Z(t)}{Z(t-)} = -1 + \frac{Z(t)}{Z(t-)}.$$

Inserting U in (3.67) yields $\mathscr{E}(U)(t) = Z(t)/Z(0)$ after a straightforward computation. □

Note that we use the notation $\mathscr{L}(Z)$ for both the stochastic logarithm of a process and for the law of a random variable Z. The meaning should be clear from the context.

Finally we consider affine SDEs. From the deterministic case in Sect. 1.6.2 one may expect them to allow for an explicit solution as well. This is indeed the case.

Definition 3.47 For any two semimartingales X, Y, we call the solution Z to the SDE

$$dZ(t) = Z(t-)dX(t) + dY(t), \quad Z(0) = Y(0) \tag{3.69}$$

the **generalised stochastic exponential** and denote it by $\mathscr{E}_Y(X) := Z$.

In particular we have $\mathscr{E}_1(X) = \mathscr{E}(X)$. From Theorem 3.40 we know that (3.69) allows for a unique solution. Unless X jumps by -1 it can be represented as follows.

Proposition 3.48 *If X, Y are semimartingales with $\Delta X(t) \neq -1$ for $t \ge 0$, we have*

$$\mathscr{E}_Y(X) = \mathscr{E}(X)\left(Y(0) + \frac{1}{\mathscr{E}(X)_-} \bullet Y - \frac{1}{\mathscr{E}(X)} \bullet [X, Y]\right).$$

Proof Let $Z := Y(0) + \frac{1}{\mathscr{E}(X)_-} \bullet Y - \frac{1}{\mathscr{E}(X)} \bullet [X, Y]$. Observe that

$$\begin{aligned}[X, Z](t) &= \frac{1}{\mathscr{E}(X)_-} \bullet [X, Y](t) - \sum_{s\le t} \frac{1}{\mathscr{E}(X)(s-)(1+\Delta X(s))}\Delta X(s)^2 \Delta Y(s) \\ &= \frac{1}{\mathscr{E}(X)_-} \bullet \left([X, Y] - \frac{\Delta X}{1+\Delta X} \bullet [X, Y]\right)(t) \\ &= \frac{1}{\mathscr{E}(X)} \bullet [X, Y](t).\end{aligned}$$

Integration by parts yields

$$\begin{aligned}\mathscr{E}(X)Z &= Y(0) + \mathscr{E}(X)_- \bullet Z + Z_- \bullet \mathscr{E}(X) + [\mathscr{E}(X), Z] \\ &= Y(0) + Y - Y(0) - \frac{\mathscr{E}(X)_-}{\mathscr{E}(X)} \bullet [X, Y] + (Z_- \mathscr{E}(X)_-) \bullet X + \mathscr{E}(X)_- \bullet [X, Z] \\ &= Y + (\mathscr{E}(X)Z)_- \bullet X,\end{aligned}$$

which implies $\mathscr{E}(X)Z = \mathscr{E}_Y(X)$ as desired. □

3.7 Exponential Lévy Processes

We motivated Lévy processes by their constant growth in a stochastic sense. This means that increments over periods of fixed length have the same law. In applications, however, it is often the *relative* growth that stays constant over time. This leads to the concept of an exponential Lévy process. An **exponential** (or **geometric**) **Lévy process** is a nonvanishing semimartingale X starting at $X(0) = 1$ such that the relative increments $(X(t) - X(s))/X(s)$ are independent of $\mathscr{F}_s$ for $s \leq t$ and such that their law depends only on $t - s$. Such processes are widely applied for price processes of stocks, currencies etc. Constant relative growth appears natural from an intuitive and economic point of view, at least much more natural than constant absolute growth as in the definition of Lévy processes.

As a particularly simple example consider a deterministic savings account growing with a fixed continuously compounded interest rate. It is easy to see that any such deterministic exponential Lévy process has a representation $X(t) = e^{bt}$ with some constant b. This in turn means that it solves the linear ordinary differential equation

$$dX(t) = X(t)bdt, \quad X(0) = 1 \tag{3.70}$$

driven by the deterministic Lévy process $L(t) = bt$. These statements have a natural stochastic counterpart.

Theorem 3.49 *For any positive semimartingale X we have equivalence between*

1. *X is an exponential Lévy process,*
2. *$X = e^L$ for some Lévy process L,*
3. *$X = \mathscr{E}(\widetilde{L})$ for some Lévy process $\widetilde{L}$.*

We have

$$\widetilde{L} = L + \frac{1}{2}\langle L^c, L^c\rangle + (e^x - 1 - x) * \mu^L \tag{3.71}$$

and

$$L = \widetilde{L} - \frac{1}{2}\langle \widetilde{L}^c, \widetilde{L}^c \rangle + (\log(1+x) - x) * \mu^{\widetilde{L}}. \tag{3.72}$$

The Lévy–Khintchine triplets $(b^{L,h}, c^L, K^L)$ *resp.* $(b^{\widetilde{L},h}, c^{\widetilde{L}}, K^{\widetilde{L}})$ *of* L *and* $\widetilde{L}$ *are related to each other via*

$$b^{\widetilde{L},h} = b^{L,h} + \frac{1}{2}c^L + \int \big(h(e^x - 1) - h(x)\big)K^L(dx), \tag{3.73}$$

$$c^{\widetilde{L}} = c^L, \tag{3.74}$$

$$K^{\widetilde{L}}(B) = \int 1_B(e^x - 1)K^L(dx), \quad B \in \mathscr{B} \tag{3.75}$$

and

$$b^{L,h} = b^{\widetilde{L},h} - \frac{1}{2}c^{\widetilde{L}} + \int \big(h(\log(1+x)) - h(x)\big)K^{\widetilde{L}}(dx), \tag{3.76}$$

$$c^L = c^{\widetilde{L}}, \tag{3.77}$$

$$K^L(B) = \int 1_B(\log(1+x))K^{\widetilde{L}}(dx), \quad B \in \mathscr{B}^d. \tag{3.78}$$

If both $L, \widetilde{L}$ *have finite first moments, we have*

$$a^{\widetilde{L}} = a^L + \frac{1}{2}c^L + \int \big(e^x - 1 - x\big)\, K^L(dx), \tag{3.79}$$

$$a^L = a^{\widetilde{L}} - \frac{1}{2}c^{\widetilde{L}} + \int (\log(1+x) - x)K^{\widetilde{L}}(dx) \tag{3.80}$$

for the drift rates.

Proof If $e^L = \mathscr{E}(\widetilde{L})$, then (3.71) and (3.72) follow from Itō's formula (3.43).

1⇔2: Set $L(t) := \log X(t)$. First note that $L(t) - L(s) = \log(X(t)/X(s))$ is independent of $\mathscr{F}_s$ if and only if this is the case for $X(t)/X(s)$. Moreover, the law of $L(t) - L(s)$ depends only on $\mathscr{F}_s$ if and only if this holds for the law of $X(t)/X(s)$.

2⇔3: This is shown along with (3.73–3.78) in Example 4.17. □

In other words, an exponential Lévy process solves a linear SDE

$$dX(t) = X(t-)d\widetilde{L}(t), \quad X(0) = 1 \tag{3.81}$$

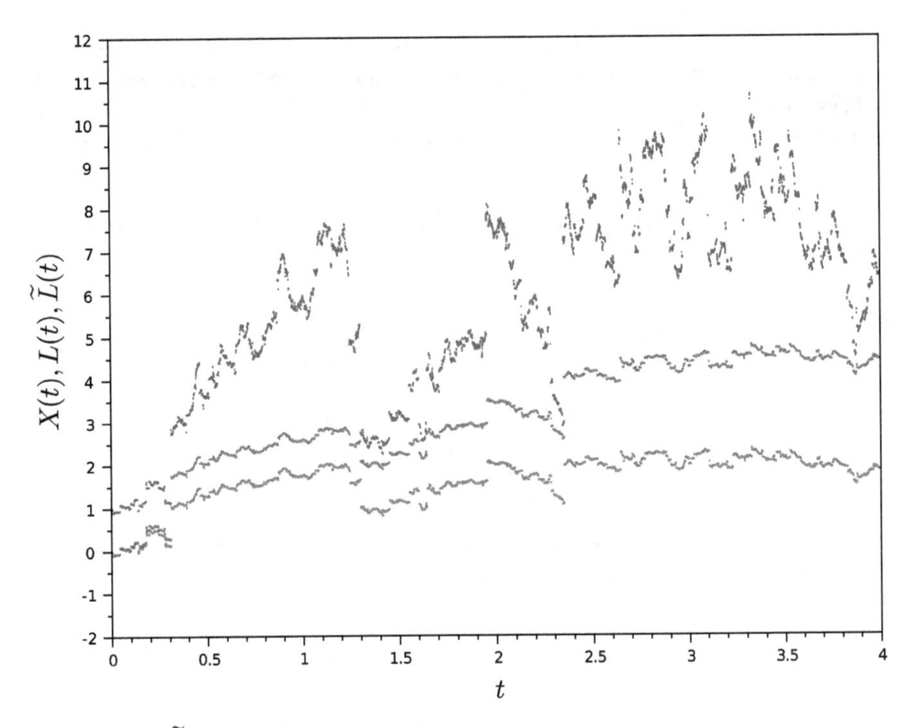

Fig. 3.5 $X, L, \widetilde{L}$ for an NIG Lévy process L

driven by a Lévy process $\widetilde{L}$. This also indicates that such processes are stochastic counterparts of exponential functions e^{bt}, which solve the deterministic restriction (3.70) of (3.81). In financial terms the constant interest $b\,dt$ in an infinitesimal period $(t - dt, t]$ is replaced with some random return $d\widetilde{L}(t)$. Figure 3.5 shows $L, \widetilde{L}, X$ in the case of a normal-inverse-Gaussian Lévy process L.

The exponential moments of a Lévy process can easily be derived from its Lévy–Khintchine triplet—or even better—from its characteristic exponent if it is available in closed form. This immediately yields formulas for the expectation, the variance and higher moments of an exponential Lévy process. Moreover, recall that positive exponential Lévy processes are martingales if and only if they are local martingales, cf. Theorem 2.22.

Theorem 3.49 shows that the stochastic exponential $X = \mathscr{E}(\widetilde{L})$ of a Lévy process can be written as the ordinary exponential e^L of another Lévy process and vice versa, provided that X is positive. This correspondence can be generalised to arbitrary $\mathbb{R} \setminus \{0\}$-valued exponential Lévy processes X if we allow for complex-valued L.

In the continuous case of **geometric Brownian motion**, L and $\widetilde{L}$ differ only by a linear drift. We have $L = \widetilde{L}$ if and only if L is deterministic, i.e. if it is a linear function.

We can extend Theorem 3.49 to the multivariate case. We call an $\mathbb{R}^d$-valued semimartingale with nonvanishing components an **exponential** or **geometric Lévy process** if it starts at $X(0) = (1, \dots, 1)$ and if $(X_1(t)/X_1(s), \dots, X_d(t)/X_d(s))$ is independent of $\mathscr{F}_s$ and has a law which depends only on $t - s$. Similarly to the above we obtain

Theorem 3.50 *For any $\mathbb{R}^d$-valued semimartingale X we have equivalence between*

1. *X is an exponential Lévy process,*
2. *$X = (e^{L_1}, \dots, e^{L_d})$ for some $\mathbb{R}^d$-valued Lévy process $L = (L_1, \dots, L_d)$,*
3. *$X = (\mathscr{E}(\widetilde{L}_1), \dots, \mathscr{E}(\widetilde{L}_d))$ for some $\mathbb{R}^d$-valued Lévy process $\widetilde{L} = (\widetilde{L}_1, \dots, \widetilde{L}_d)$.*

Suppose that the truncation function h on $\mathbb{R}^d$ is of the form

$$h(x_1, \dots, x_d) = \big(\widetilde{h}(x_1), \dots, \widetilde{h}(x_d)\big)$$

for some univariate truncation function $\widetilde{h} : \mathbb{R} \to \mathbb{R}$. The Lévy–Khintchine triplets of L and $\widetilde{L}$ are related to each other via

$$\begin{aligned} b_i^{\widetilde{L},h} &= b_i^{L,h} + \frac{1}{2} c_{ii}^L + \int \big(\widetilde{h}(e^{x_i} - 1) - \widetilde{h}(x_i)\big) K^L(dx), \\ c^{\widetilde{L}} &= c^L, \\ K^{\widetilde{L}}(B) &= \int 1_B(e^{x_1} - 1, \dots, e^{x_d} - 1) K^L(dx), \quad B \in \mathscr{B}^d \end{aligned} \tag{3.82}$$

and

$$\begin{aligned} b_i^{L,h} &= b_i^{\widetilde{L},h} - \frac{1}{2} c_{ii}^{\widetilde{L}} + \int \big(\widetilde{h}(\log(1 + x_i)) - \widetilde{h}(x_i)\big) K^{\widetilde{L}}(dx), \\ c^L &= c^{\widetilde{L}}, \\ K^L(B) &= \int 1_B(\log(1 + x_1), \dots, \log(1 + x_d)) K^{\widetilde{L}}(dx), \quad B \in \mathscr{B}^d. \end{aligned} \tag{3.83}$$

In the case of finite first moments, equations corresponding to (3.79, 3.80) can be derived from (3.82, 3.83) by setting $h(x) = x$.

Proof The proof is essentially the same as for Theorem 3.49. □

In Part II of this book we use the term **exponential** or **geometric Lévy process** liberally if the initial value $X(0)$ is positive rather than exactly 1 resp. $(1, \dots, 1)$. Similarly, **geometric Brownian motion** refers to the exponential of a shifted Brownian motion with drift.

3.8 Time-Inhomogeneous Lévy Processes

Lévy processes are characterised by stationary and independent increments. Many results for this class of processes still hold in a similar form if we do not require the stationarity of increments. For ease of exposition we focus on the case where the processes have no *fixed time of discontinuity*, i.e. they do not jump with positive probability at any fixed time.

A **time-inhomogeneous Lévy process** or **additive process** is an adapted càdlàg real- or vector-valued process X such that

1. $X(0) = 0$,
2. $X(t) - X(s)$ is independent of $\mathscr{F}_s$ for $s \leq t$ (*independent increments*),
3. $P(\Delta X(t) \neq 0) = 0$ for any $t > 0$.

The counterpart of Theorem 2.17 reads as follows.

Theorem 3.51 *Fix a truncation function h on $\mathbb{R}^d$. The characteristic function of any $\mathbb{R}^d$-valued time-inhomogeneous Lévy process X satisfies*

$$\varphi_{X(t)-X(s)}(u) = \exp(\Psi(t,u) - \Psi(s,u)), \quad u \in \mathbb{R}^d,$$

where $\Psi : \mathbb{R}_+ \times \mathbb{R}^d \to \mathbb{C}$ is of the form

$$\Psi(t,u) = iu^\top B^h(t) - \frac{1}{2}u^\top C(t)u + \int_{[0,t]\times\mathbb{R}^d} \left(e^{iu^\top x} - 1 - iu^\top h(x)\right) \nu(d(s,x))$$

with some continuous $\mathbb{R}^d$-valued function B^h on $\mathbb{R}_+$, some continuous $\mathbb{R}^{d\times d}$-valued function C on $\mathbb{R}_+$, and some measure ν on $\mathbb{R}_+ \times \mathbb{R}^d$. Moreover, the components of B^h are increasing with $B^h(0) = 0$, the increments $C(t) - C(s)$ are symmetric, nonnegative matrices for $s \leq t$ with $C(0) = 0$, and $\int_{[0,t]\times\mathbb{R}^d}(1 \wedge |x|^2)\nu(d(t,x)) < \infty$ as well as $\nu(\{t\} \times \mathbb{R}^d) = 0$ for any $t \geq 0$.

Conversely, for any triplet (B^h, C, ν) with the above properties there exists a corresponding time-homogeneous Lévy process.

Idea This is shown similarly as for Lévy processes. For a rigorous proof see, for example, [154, Theorem II.5.2] □

Moments and exponential moments can be computed by differentiation as in Theorem 2.19 resp. by inserting imaginary arguments in the characteristic function as in Theorem 2.20.

In applications one mostly considers **time-inhomogeneous Lévy processes with absolutely continuous characteristics** (also called **processes with independent increments and absolutely continuous characteristics, PIIAC**), for which B^h, C, ν are absolutely continuous in time in the sense that

$$B^h(t) = \int_0^t b^h(s)ds, \quad C(t) = \int_0^t c(s)ds, \quad \nu([0,t] \times dx) = \int_0^t K(s,dx)ds$$

holds for some functions b^h, c and some transition kernel K from $\mathbb{R}_+$ into $\mathbb{R}^d$. In this case we call

$$\psi(t,u) = iu^\top b^h(t) - \frac{1}{2}u^\top c(t)u + \int \left(e^{iu^\top x} - 1 - iu^\top h(x)\right) K(s,dx)$$

the **characteristic exponent** of X.

3.9 (Generalised) Ornstein–Uhlenbeck Processes

In this section we consider generalised stochastic exponentials

$$X = \mathscr{E}_{X(0)+L}(M)$$

which solve an equation

$$dX(t) = dL(t) + X(t-)dM(t)$$

driven by a bivariate Lévy process (L, M). We call such processes **generalised Ornstein–Uhlenbeck (OU) processes**. If the Lévy process M is just a linear function $M(t) = -\lambda t$, we call X simply a **(Lévy-driven) Ornstein–Uhlenbeck (OU) process**. It is not easy to say much about the law of generalised Ornstein–Uhlenbeck processes. At least, we can derive their first two moments.

Proposition 3.52 *Let (L, M) denote an $\mathbb{R}^2$-valued Lévy process. We write*

$$E\begin{pmatrix} L(1) \\ M(1) \end{pmatrix} = \begin{pmatrix} a^L \\ a^M \end{pmatrix}, \quad \mathrm{Cov}\begin{pmatrix} L(1) \\ M(1) \end{pmatrix} = \begin{pmatrix} \widetilde{c}^L & \widetilde{c}^{LM} \\ \widetilde{c}^{ML} & \widetilde{c}^M \end{pmatrix}$$

provided that first resp. second moments exist. If L, M have finite first moments, then

$$E(X(t)) = \left(E(X(0)) + \frac{a^L}{a^M}\right)e^{a^M t} - \frac{a^L}{a^M}. \tag{3.84}$$

If second moments are finite as well, then

$$E\big(X(t)^2\big) = \alpha_0 + \alpha_1 e^{a^M t} + \alpha_2 e^{(2a^M + \widetilde{c}^M)t} \tag{3.85}$$

and

$$\operatorname{Cov}\big(X(t), X(t+h)\big) = \Big(E\big(X(t)^2\big) - \big(E(X(t))\big)^2\Big)e^{a^M h}$$

for $t, h \geq 0$, *where*

$$\begin{aligned}
\alpha_0 &:= \frac{2(a^L + \widetilde{c}^{LM})a^L/a^M - \widetilde{c}^L}{2a^M + \widetilde{c}^M},\\
\alpha_1 &:= -2\frac{a^L + \widetilde{c}^{LM}}{a^M + \widetilde{c}^M}\left(E(X(0)) + \frac{a^L}{a^M}\right),\\
\alpha_2 &:= E(X(0)^2) - \alpha_0 - \alpha_1.
\end{aligned}$$

Proof We write

$$dX(t) = (a^L + X(t-)a^M)dt + dN(t)$$

with some local martingale N. If this local martingale is a true martingale, we obtain

$$E(X(t)) = E(X(0)) + a^L t + \int_0^t E(X(s))a^M ds,$$

i.e. $f(t) = E(X(t))$ solves the ODE $f'(t) = a^L + a^M f(t)$. Solving this ODE leads to (3.84). Similarly, we conclude that

$$E(X(t)|\mathscr{F}_s) = \left(X(s) + \frac{a^L}{a^M}\right)e^{a^M(t-s)} - \frac{a^L}{a^M} \tag{3.86}$$

for $s \leq t$. Itō's formula yields

$$\begin{aligned}
dX(t)^2 &= 2X(t-)dL(t) + 2X(t-)^2 dM(t) + d[L,L](t)\\
&\quad + 2X(t-)d[L,M](t) + X(t-)^2 d[M,M](t)\\
&= (2X(t-)a^L + 2X(t-)^2 a^M + c^L + 2X(t-)c^{LM} + X(t-)^2 c^M)dt + d\widetilde{N}
\end{aligned}$$

for some local martingale $\widetilde{N}$. Again assuming that the local martingale is a martingale, we obtain the ODE

$$g'(t) = 2f(t)a^L + 2g(t)a^M + c^L + 2f(t)c^{LM} + g(t)c^M$$

for the second moments $g(t) = E(X(t)^2)$. The solution to this ODE is (3.85). From (3.86) we obtain

$$
\begin{aligned}
E(X(t)E(X(t+h)) &= E\big(X(t)E(X(t+h)|\mathscr{F}_t)\big) \\
&= \left(E\big(X(t)^2\big) + \frac{a^L}{a^M}E(X(t))\right)e^{a^M h} - \frac{a^L}{a^M}E(X(t))
\end{aligned}
$$

which, together with (3.84) and (3.85), yields the formula for the covariance.

On a rigorous level, the statements follow from Theorem 6.15. □

More can be said about Ornstein–Uhlenbeck processes which solve an SDE of the form

$$
dX(t) = dL(t) - \lambda X(t-)dt \tag{3.87}
$$

with some Lévy process L and some constant λ. Such processes are applied for phenomena that show a mean-reverting behaviour such as interest rates, volatility, or default rates. For such applications one must use or may prefer positive Ornstein–Uhlenbeck processes, which in turn correspond to subordinators L, i.e. increasing Lévy processes.

The previous proposition yields

$$
E(X(t)) = \left(E(X(0)) - \frac{a^L}{\lambda}\right)e^{-\lambda t} + \frac{a^L}{\lambda}, \tag{3.88}
$$

$$
\operatorname{Var}(X(t)) = \left(\operatorname{Var}(X(0)) - \frac{\widetilde{c}^L}{2\lambda}\right)e^{-2\lambda t} + \frac{\widetilde{c}^L}{2\lambda}, \tag{3.89}
$$

$$
\operatorname{Cov}\big(X(t), X(t+h)\big) = \operatorname{Var}(X(t))e^{-\lambda h}. \tag{3.90}
$$

Proposition 3.53 *SDE (3.87) is solved by*

$$
X(t) = e^{-\lambda t}X(0) + \int_0^t e^{-\lambda(t-s)}dL(s). \tag{3.91}
$$

Proof This follows immediately from the general formula in Proposition 3.48. □

Representation (3.91) of the solution to (3.87) holds in fact for general semimartingales L.

The characteristic function or more generally the conditional characteristic function of $X(t)$ given the information up to $s \leq t$ can be represented in closed form. For its derivation the following proposition turns out to be useful.

Proposition 3.54 *If L is a Lévy process with characteristic exponent ψ and $f : \mathbb{R}_+ \to \mathbb{R}$ denotes a continuous function, the conditional characteristic function of $X(t) := X(0) + \int_0^t f(s)dL(s)$ is given by*

$$E\big(e^{iuX(t)}\big|\mathscr{F}_s\big) = \exp\bigg(\int_s^t \psi(uf(r))dr + iuX(s)\bigg). \tag{3.92}$$

Proof Suppose first that $f(s) = \sum_{j=1}^n c_j 1_{(t_{j-1},t_j]}(s)$ for some $c_1, \dots, c_n$ and $0 \leq t_0 \leq t_1 \leq \dots \leq t_n = t$. Then $X(t) = X(0) + \sum_{j=1}^n c_j(L(t_j) - L(t_{j-1}))$. W.l.o.g. $s = t_k$ for some k. By independence and stationarity of the increments of L we have

$$\begin{aligned}
E\big(e^{iuX(t)}\big|\mathscr{F}_s\big) &= e^{iuX(s)} E\bigg(\exp\bigg(\sum_{j=k+1}^n iuc_j\big(L(t_j) - L(t_{j-1})\big)\bigg)\bigg|\mathscr{F}_s\bigg) \\
&= e^{iuX(s)} \prod_{j=k+1}^n E\Big(\exp\big(iuc_j(L(t_j) - L(t_{j-1}))\big)\Big) \\
&= e^{iuX(s)} \prod_{j=k+1}^n \exp\big(\psi(uc_j)(t_j - t_{j-1})\big) \\
&= \exp\bigg(iuX(s) + \int_s^t \psi(uf(r))dr\bigg).
\end{aligned}$$

For general f one can choose a sequence of piecewise constant functions f_n such that $\int_0^t f_n(s)dL(s) \to \int_0^t f(s)dL(s)$ in probability and hence in law. The statement for f follows now from the one for f_n, passing to the limit, and dominated convergence.

For an alternative proof under weaker conditions on f see Example 4.18. □

With the help of the previous proposition we obtain the characteristic function of the Ornstein–Uhlenbeck process X from (3.87).

Proposition 3.55 *We have*

$$E\big(e^{iuX(t)}\big|\mathscr{F}_s\big) = \exp\Big(\Theta(u) - \Theta(ue^{-\lambda(t-s)}) + iue^{-\lambda(t-s)}X(s)\Big)$$

for $s \leq t$, $u \in \mathbb{R}$, where ψ denotes the characteristic exponent of L and

$$\Theta(u) := \frac{1}{\lambda}\int_0^u \frac{\psi(x)}{x}dx.$$

Proof This follows from the previous proposition applied to $f(s) = e^{-\lambda(t-s)}$. Indeed, we have

$$\begin{aligned}\int_s^t \psi(ue^{-\lambda(t-r)})dr &= \int_0^{t-s} \psi(ue^{-\lambda r})dr \\ &= -\int_u^{ue^{-\lambda(t-s)}} \frac{\psi(x)}{\lambda x}dx \\ &= \Theta(u) - \Theta(ue^{-\lambda(t-s)})\end{aligned}$$

by substitution $x = ue^{-\lambda r}$. □

If $\lambda > 0$, then $X(t)$ converges in law as $t \to \infty$ to some random variable Y with characteristic function

$$E(e^{iuY}) = \exp(\Theta(u)),$$

cf. Proposition A.5(8). If we choose the law of Y as the initial distribution $\mathscr{L}(X(0))$, we end up with a stationary process, i.e. the law of $X(t)$ does not depend on t or more generally, $(X(t+h))_{h\geq 0}$ has the same law for all $h \geq 0$. The marginal laws of stationary Ornstein–Uhlenbeck processes are called **self-decomposable distributions**, cf. [259, Theorem 17.5]. One can show that they are a subset of the class of **infinitely divisible distributions**. The latter are the laws of $L(t)$ (or equivalently $L(1)$) for arbitrary Lévy processes L.

Let us consider some examples which may be used for modelling stochastic volatility, default intensities, or interest rates.

Example 3.56 (Gaussian Ornstein–Uhlenbeck Process) If L in (3.87) is a Brownian motion with drift, i.e. $L(t) = \mu t + \sigma W(t)$ for some $\mu, \sigma \in \mathbb{R}$, we end up with a Gaussian process. Applying Proposition 3.55 we conclude that the conditional law of $X(t)$ given $\mathscr{F}_s$ for $s \leq t$ is Gaussian with conditional mean

$$\frac{\mu}{\lambda} + \left(X(s) - \frac{\mu}{\lambda}\right)e^{-\lambda(t-s)}$$

and conditional variance

$$\frac{1-e^{-2\lambda(t-s)}}{2\lambda}\sigma^2.$$

An example with finite jump activity is the following:

Example 3.57 (Gamma-OU Process) If we start with a compound Poisson process with exponential jump distribution, i.e. with Lévy–Khintchine triplet

$$\left(b^0, c, K(dx)\right) = \left(0, 0, \beta\alpha e^{-\alpha x}1_{\{x>0\}}dx\right)$$

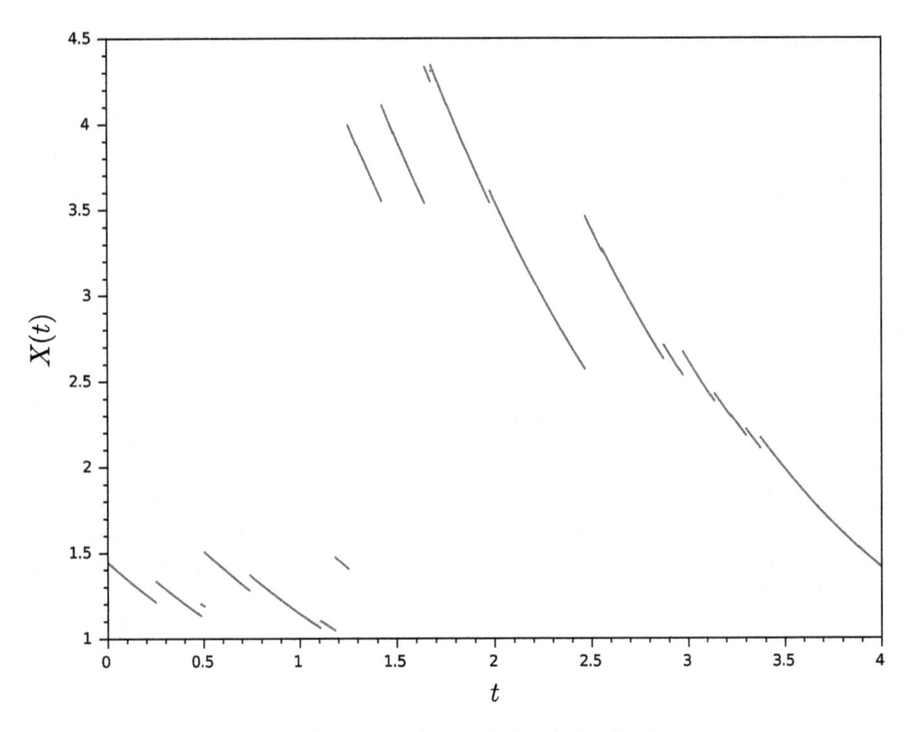

Fig. 3.6 Gamma-OU process with parameters $\alpha = 1, \beta = 1, \lambda = \log 2$

and characteristic exponent

$$\psi(u) = \frac{iu\beta}{\alpha - iu} \tag{3.93}$$

for parameters $\alpha, \beta > 0$, the corresponding Ornstein–Uhlenbeck process (3.87) is called a Gamma-OU process. A sample path is shown in Fig. 3.6. The parameters needed for the computation of its moments are given by

$$a^L = \frac{\beta}{\alpha}, \quad \tilde{c}^L = \frac{2\beta}{\alpha^2}.$$

The name of the process is motivated by the fact that its stationary distribution is Gamma. Indeed, the characteristic exponent in Proposition 3.55 is easily computed as

$$\Theta(u) = -\frac{\beta}{\lambda} \log\left(1 - \frac{iu}{\alpha}\right),$$

which corresponds to $\Gamma(\beta/\lambda, \alpha)$.

Alternatively, we consider an OU process with infinitely many jumps in finite periods:

Example 3.58 (Inverse-Gaussian OU Process) One can also construct an Ornstein–Uhlenbeck process whose stationary distribution is inverse-Gaussian. To this end, consider a Lévy process L with Lévy–Khintchine triplet

$$(b^0, c, K(dx)) = \left(0, 0, \frac{\beta}{2\sqrt{2\pi}}(x^{-3/2} + \alpha^2 x^{-1/2}) \exp\left(-\tfrac{1}{2}\alpha^2 x\right) 1_{\{x>0\}} dx\right)$$

and characteristic exponent

$$\psi(u) = \frac{iu\beta}{\sqrt{\alpha^2 - 2iu}}$$

for parameters $\alpha, \beta > 0$. It can be viewed as the sum of an inverse-Gaussian subordinator and an independent compound Poisson process whose jumps are squares of zero-mean Gaussian random variables. For the computation of first and second moments we need

$$a^L = \frac{\beta}{\alpha}, \quad \tilde{c}^L = \frac{2\beta}{\alpha^3}.$$

The characteristic exponent in Proposition 3.55 is derived as

$$\Theta(u) = \frac{\beta}{\lambda}\left(\alpha - \sqrt{\alpha^2 - 2iu}\right),$$

which corresponds to an inverse-Gaussian law with parameters $\beta/\lambda, \alpha$. A sample path of this OU process is shown in Fig. 3.7.

Occasionally, a formula for the characteristic function of integrated Ornstein–Uhlenbeck processes turns out to be useful.

Proposition 3.59 *The conditional characteristic function of the integrated Ornstein–Uhlenbeck process*

$$Y(t) := \int_0^t X(s) ds$$

is given by

$$E\left(e^{iuY(t)} \middle| \mathscr{F}_s\right) = \exp\left(\Upsilon(u, t-s) + iu\left(Y(s) + g(t-s)X(s)\right)\right) \tag{3.94}$$

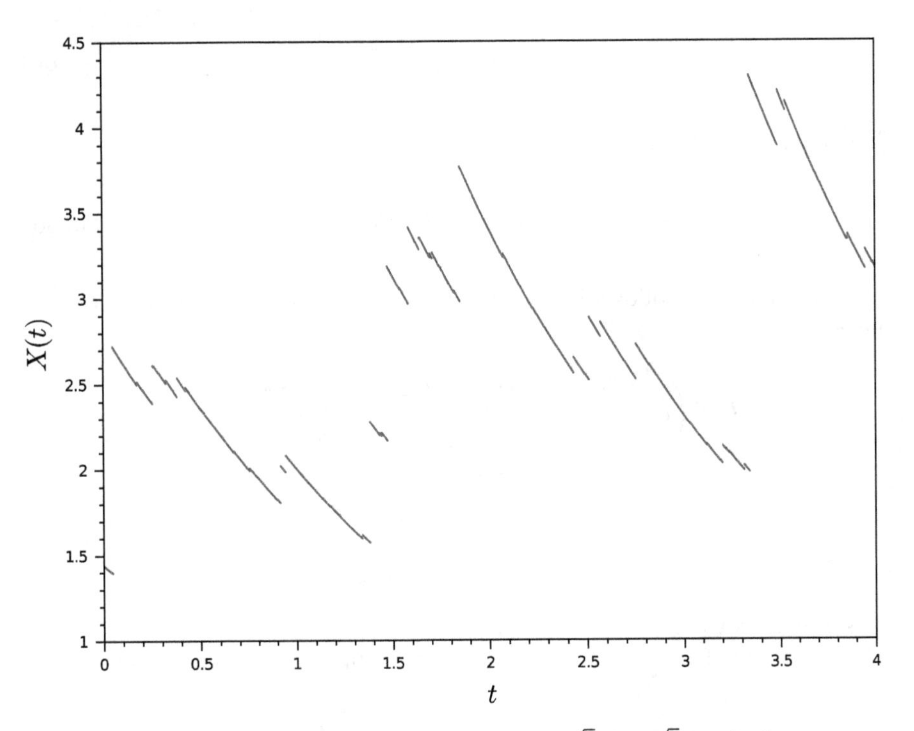

Fig. 3.7 Inverse-Gaussian OU process with parameters $\alpha = \sqrt{2}$, $\beta = \sqrt{2}$, $\lambda = \log 2$

for $s \leq t$, where

$$\Upsilon(u,t) := \int_0^{ug(t)} \frac{\psi(x)}{u - \lambda x} dx$$

and $g(t) := (1 - e^{-\lambda t})/\lambda$.

More generally, the joint conditional characteristic function $E\big(e^{iuY(t)+ivX(t)}\big| \mathscr{F}_s\big)$ is of the form (3.94) as well, but with $g(t) = (1 - (1 - \lambda v/u)e^{-\lambda t})/\lambda$ and

$$\Upsilon(u,t) := \int_v^{ug(t)} \frac{\psi(x)}{u - \lambda x} dx$$

instead of g, Υ above.

Proof For ease of notation let $s = 0$. The stochastic Fubini theorem 3.29 yields

$$\begin{aligned} Y(t) &= \int_0^t \left(e^{-\lambda s} X(0) + \int_0^s e^{-\lambda(s-r)} dL(r) \right) ds \\ &= \frac{1 - e^{-\lambda t}}{\lambda} X(0) + \int_0^t \int_r^t e^{-\lambda(s-r)} ds dL(r) \end{aligned}$$

$$= \frac{1-e^{-\lambda t}}{\lambda} X(0) + \int_0^t \frac{1-e^{-\lambda(t-r)}}{\lambda} dL(r). \tag{3.95}$$

In view of (3.91) we obtain

$$Y(t) + \frac{v}{u} X(t) = g(t)X(0) + \int_0^t g(t-r)dL(r). \tag{3.96}$$

The statement now follows from Proposition 3.54 applied to $f(s) := g(t-s)$. Indeed, we have

$$\begin{aligned}\int_0^t \psi\Big(uf(r)\Big)dr &= \int_0^t \psi\left(u\frac{1-(1-\frac{\lambda v}{u})e^{-\lambda(t-r)}}{\lambda}\right)dr \\ &= \int_{ug(t)}^{v} \frac{\psi(x)}{\lambda x - u} dx \\ &= \Upsilon(u,t)\end{aligned}$$

by substitution $x = ug(t-r)$.

The assertion for $s > 0$ follows along the same lines by observing that

$$Y(t) + \frac{v}{u} X(t) = Y(s) + g(t-s)X(s) + \int_s^t g(t-r)dL(r)$$

similar to (3.96). □

For later use we reconsider the processes in Examples 3.56–3.58.

Example 3.60

1. *(Gaussian OU process)* In the previous proposition we have

$$\Upsilon(u,t) = iu\mu\left(\frac{t}{\lambda} - \frac{1-e^{-\lambda t}}{\lambda^2}\right) - \frac{u^2\sigma^2}{4}\left(\frac{2t}{\lambda^2} - \frac{3-4e^{-\lambda t}+e^{-2\lambda t}}{\lambda^3}\right)$$

for the Gaussian Ornstein–Uhlenbeck process in Example 3.56, which means that the conditional law of $Y(t)$ given $\mathscr{F}_s$ is Gaussian with mean

$$\mu\left(\frac{t-s}{\lambda} - \frac{1-e^{-\lambda(t-s)}}{\lambda^2}\right) + \frac{1-e^{-\lambda(t-s)}}{\lambda} X(s) + Y(s)$$

and variance

$$\frac{\sigma^2}{2}\left(\frac{2(t-s)}{\lambda^2} - \frac{3-4e^{-\lambda(t-s)}+e^{-2\lambda(t-s)}}{\lambda^3}\right).$$

2. *(Gamma-OU process)* Slightly generalising the Gamma-OU process let

$$\psi(u) = \frac{iu\beta}{\alpha - iu} + iu\kappa$$

with $\kappa \in \mathbb{R}$. The corresponding Ornstein–Uhlenbeck process X differs from the one in Example 3.57 by an additional constant κ/λ. Indeed, $X - \kappa/\lambda$ solves the SDE (3.87) for L with exponent (3.93). The function Υ in Proposition 3.59 for fixed general v reads as

$$\begin{aligned}\Upsilon(u,t) &= \frac{\beta}{\lambda\alpha - iu}\left(\alpha \log \frac{\alpha - iug(t)}{\alpha - iv} + iut\right) \\ &\quad + \frac{\kappa}{\lambda}\left(iut + (1 - iv\lambda)\frac{1 - e^{-\lambda t}}{\lambda}\right) \end{aligned} \tag{3.97}$$

with

$$g(t) = \frac{1 - (1 - \frac{\lambda v}{u})e^{-\lambda t})}{\lambda}.$$

3. *(Inverse-Gaussian OU process)* The parallel extension of the OU process in Example 3.58 relies on the exponent

$$\psi(u) = \frac{iu\beta}{\sqrt{\alpha^2 - 2iu}} + iu\kappa$$

with $\kappa \in \mathbb{R}$. The function Υ in Proposition 3.59 for fixed general v now reads as

$$\begin{aligned}\Upsilon(u,t) &= \frac{2iu\beta}{\lambda^2}\Bigg(\frac{\sqrt{\alpha^2 - 2iv} - \sqrt{1 + \gamma\lambda g(t)}}{\gamma} \\ &\quad + \frac{1}{\sqrt{\alpha^2+\gamma}}\left(\operatorname{artanh}\sqrt{\frac{\alpha^2 + \gamma\lambda g(t)}{\alpha^2+\gamma}} - \operatorname{artanh}\sqrt{\frac{\alpha^2 - 2iv}{\alpha^2+\gamma}}\right)\Bigg) \\ &\quad + \frac{\kappa}{\lambda}\left(iut + (1 - iv\lambda)\frac{1 - e^{-\lambda t}}{\lambda}\right) \end{aligned} \tag{3.98}$$

with $\gamma := -2iu/\lambda$ and $g(t)$ as in 2.

One can also define **multivariate Ornstein–Uhlenbeck processes** of the form

$$dX(t) = dL(t) - \Lambda X(t-)dt \tag{3.99}$$

where X denotes an $\mathbb{R}^d$-valued process, Λ a fixed $d \times d$-matrix and L an $\mathbb{R}^d$-valued Lévy process. The solution is given by

$$X(t) = e^{-\Lambda t} X(0) + \int_0^t e^{-\Lambda(t-s)} dL(s)$$

in extension of Proposition 3.53. As an example consider

$$\Lambda = \begin{pmatrix} 0 & 1 & 0 & \dots & 0 \\ 0 & 0 & 1 & \dots & 0 \\ \vdots & \vdots & & \ddots & \vdots \\ 0 & 0 & 0 & \dots & 1 \\ a_d & a_{d-1} & a_{d-2} & \dots & a_1 \end{pmatrix}$$

and $L = (0, \dots, 0, L_d)$ with constants $a_1, \dots, a_d$ and some univariate Lévy process L_d. For constants $b_1, \dots, b_d$ the process

$$Y(t) := \sum_{i=1}^{d} b_i X_i(t) \tag{3.100}$$

can be viewed as a continuous-time generalisation of an ARMA time series. Equations (3.99–3.100) constitute the state-space representation of a Lévy-driven **CARMA process** as put forward by [43].

3.10 Martingale Representation

Under certain conditions any martingale may be written as a stochastic integral with respect to some given martingale. Such results have important implications for the existence of hedging strategies in Mathematical Finance. For Brownian motion the following classical statement holds.

Theorem 3.61 *Suppose that the filtration is generated by an $\mathbb{R}^d$-valued standard Brownian motion W. Then any local martingale M can be written in the form*

$$M = M(0) + \varphi \bullet W$$

for some W-integrable process φ. The integrand φ satisfies the equation

$$d\langle M, W_i \rangle(t) = \varphi_i(t)dt, \quad i = 1, \dots, d,$$

i.e. $\varphi_i(t) = d\langle M, W_i \rangle(t)/dt$.

Proof This is a special case of Theorem 3.62 below. On an intuitive level it can be viewed as the continuous-time counterpart of the discrete-time representation in Proposition 1.21. □

In financial applications the left-hand side refers to some discounted option price process, whereas the right-hand side can be interpreted as the evolution of a dynamic hedging portfolio. Theorem 3.61 is used to show the existence of perfect hedging strategies in Black–Scholes-type models.

The previous theorem has an analogue for general semimartingales with jumps. We state here the version for Lévy processes.

Theorem 3.62 *Suppose that the filtration is generated by an $\mathbb{R}^d$-valued Lévy process L. Then any local martingale M can be written in the form*

$$M(t) = M(0) + \varphi \bullet L^c + \psi * (\mu^L - \nu^L) \tag{3.101}$$

for some L^c-integrable process φ and some $(\mu^L - \nu^L)$-integrable function ψ.

Idea The proof can be found in [154, Theorem III.4.34].

The idea is to show that M can be written as

$$M(t) = M(0) + \varphi \bullet L^c + \psi * (\mu^L - \nu^L) + N$$

for some local martingale N which is in some sense orthogonal to L^c and μ^L. If N does not vanish, it can be used to construct a probability measure $Q \sim P$ such that L is a Lévy process with the same Lévy–Khintchine triplet as under P. Since the law of L is determined by its triplet and since the filtration is generated by L, the measures P and Q must coincide, which yields a contradiction.

Let us illustrate the last term on an informal level for a purely discontinuous local martingale M. Since $\Delta M(t)$ is measurable with respect to the σ-field generated by $(L(s))_{s\leq t}$, it must be of the form

$$\Delta M(t) = f\big((L(\omega, s))_{s<t}, \Delta L(\omega, t))\big) = \xi(\omega, t, \Delta L(\omega, t))$$

for some functions f, ξ. The function ξ turns out to be predictable because it depends on L only up to $t-$, i.e. excluding t. We obtain $M - M(0) = \xi * (\mu^L - \nu^L)$ because both sides are purely discontinuous local martingales with jumps $\xi(\omega, t, \Delta L(\omega, t))$ if they jump at all. □

Formally, this result extends Theorem 3.61. However, it does not lead to the existence of hedging strategies in a financial context. The reason is that the integral relative to the measure of jumps does not refer to gains from trade in the sense of a dynamic portfolio. The complicated form of (3.101) rather indicates that options cannot typically be hedged perfectly in jump-type models. This does not necessarily imply a weakness of such models but rather corresponds to the situation in real markets.

For applications to finance we need representations in terms of stochastic integrals relative to processes rather than random measures. Such a statement is provided below. It is applied in the context of quadratic hedging in Chap. 12.

Theorem 3.63 (Galtchouk–Kunita–Watanabe (GKW) Decomposition) *Suppose that M and X are real- resp. $\mathbb{R}^d$-valued locally square-integrable martingales. Then M can be written in the form*

$$M = M(0) + \varphi \bullet X + N$$

with some $\varphi \in L^2_{\mathrm{loc}}(X)$ and some local martingale N that is strongly orthogonal to X in the sense of Sect. 3.1.2. The processes $\varphi \bullet X$ and N (but not necessarily φ) are unique. The integrand φ satisfies the equation

$$\langle M, X \rangle = \varphi \bullet \langle X, X \rangle \tag{3.102}$$

in the sense that $\langle M, X_i \rangle = \varphi \bullet (\langle X_j, X_i \rangle)_{j=1,\dots,d}$ for $i = 1, \dots, d$.

If M is a square-integrable martingale, then $\varphi \in L^2(X)$, which implies that both $\varphi \bullet X$ and N are square-integrable martingales as well.

Idea A rigorous proof can be found in [152, Théorème 4.27].

For an illustration assume that X and M are univariate square-integrable martingales, which implies that they are of the form $X(t) = E(X(\infty)|\mathscr{F}_t)$, $M(t) = E(M(\infty)|\mathscr{F}_t)$ for some square-integrable random variables $X(\infty)$, $M(\infty)$. Denote by $Y = m + \varphi \bullet X(\infty)$ the orthogonal projection of $M(\infty)$ on the closed subspace

$$\{c + \xi \bullet X(\infty) : c \in \mathbb{R}, \xi \in \mathscr{L}^2(X)\} \subset L^2(\Omega, \mathscr{F}, P).$$

The orthogonal remainder $M(\infty) - Y$ generates a martingale N with

$$N(t) = E(M(\infty) - Y|\mathscr{F}_t) = M(t) - m - \varphi \bullet X(t).$$

Orthogonality in $L^2(\Omega, \mathscr{F}, P)$ yields

$$\begin{aligned} 0 &= E(N(\infty)1_{F\times(s,t]} \bullet X(\infty)) \\ &= E(1_F(N(\infty)X(t) - N(\infty)X(s))) \\ &= E(1_F(N(t)X(t))) - E(1_F N(s)X(s))) \\ &= E(1_F(N(t)X(t) - N(s)X(s))) \end{aligned}$$

for any $F \in \mathscr{F}_s$. Hence NX is a martingale, which in turns means that N is orthogonal to X in the sense of Sect. 3.1.2. Moreover, the orthogonality also yields $0 = E(N(\infty)c) = cE(N(\infty))$ for any $c \in \mathbb{R}$, whence $m = M(0)$.

The second statement follows easily from

$$\langle M, X\rangle = \langle \varphi \bullet X, X\rangle + \langle N, X\rangle \;= \varphi \bullet \langle X, X\rangle$$

for $d = 1$ and accordingly for higher dimensions. □

Equation (3.102) can be written as

$$\varphi(t) = \frac{d\langle M, X\rangle(t)}{d\langle X, X\rangle(t)}$$

for univariate X. Typically, we have $d\langle M, X\rangle(t) = \tilde{c}^{MX}(t)dt$ and $d\langle X, X\rangle(t) = \tilde{c}^{X}(t)dt$ with some predictable processes $\tilde{c}^{MX}, \tilde{c}^{X}$. In this case (3.102) means $\varphi(t) = \tilde{c}^{MX}/\tilde{c}^{X}$. In financial applications, the processes $\varphi \bullet X$ and N refer to the hedgeable and the unhedgeable part of a contingent claim with price process M.

The following decomposition plays an important role for the existence of superhedging strategies in Mathematical Finance.

Theorem 3.64 (Optional Decomposition) *Let S denote an $\mathbb{R}^d$-valued process and $\mathscr{Q}$ the set of all $Q \sim P$ such that S is a Q-local martingale. Moreover, let X be a process which is a Q-local supermartingale relative to all $Q \in \mathscr{Q}$. If $\mathscr{Q}$ is not empty, there exists a predictable $\mathbb{R}^d$-valued process φ such that $C := X(0) + \varphi \bullet S - X$ is increasing.*

Proof This is stated and proved in [114, Theorem 1]. It generalises the discrete-time analogue in Theorem 1.22. □

Theorem 3.64 is applied later to processes of the form

$$X(t) = \operatorname*{ess\,sup}_{Q\in\mathscr{Q}} E_Q(H|\mathscr{F}_t), \quad t \geq 0, \tag{3.103}$$

where H denotes a bounded random variable.

Proposition 3.65 *X in (3.103) is a supermartingale relative to all $Q \in \mathscr{Q}$. More precisely, there is a version of X with this property, i.e. X can be chosen with càdlàg paths.*

Idea The reasoning is the same as for Proposition 1.23. For a full proof including the càdlàg property see [196, Proposition 4.2]. □

3.11 Backward Stochastic Differential Equations

The martingale representation theorem 3.61 can be rephrased as follows. Given $T \in \mathbb{R}_+$ and a square-integrable, $\mathscr{F}_T$-measurable random variable H there exist unique processes X and $\varphi \in L^2(W)$ with $\varphi(t) = 0$ for $t > T$ and

$$X(T) = H,$$
$$dX(t) = 0dt + \varphi(t)dW(t). \tag{3.104}$$

Uniqueness for φ here means that $\varphi(t) = \widetilde{\varphi}(t)$ almost surely for Lebesgue-almost all $t \in [0, T]$ and any such $\varphi, \widetilde{\varphi}$. The uniqueness of X follows from the fact that it is a martingale by (3.104) and hence $X(t) = E(H|\mathscr{F}_t)$ for any $t \geq 0$.

If we replace the vanishing drift in (3.104) by some deterministic or even random function of X and φ, we can still obtain existence and uniqueness of such a pair (X, φ). To this end, suppose that the filtration is generated by an $\mathbb{R}^d$-valued standard Brownian motion W. Fix $T \geq 0$ and an $\mathscr{F}_T$-measurable random variable H with $E(H^2) < \infty$. Let $f : \Omega \times [0, T] \times \mathbb{R} \times \mathbb{R}^d \to \mathbb{R}$ denote a function that is $\mathscr{P} \otimes \mathscr{B} \otimes \mathscr{B}^d$-measurable. We call a pair (X, φ) of an adapted process X with $E(\sup_{t\in[0,T]} X_t^2) < \infty$ and $\varphi \in L^2(W)$ a solution to the **backward stochastic differential equation (BSDE)**

$$X(T) = H, \quad dX(t) = -f(t, X(t), \varphi(t))dt + \varphi(t)dW(t) \tag{3.105}$$

if (3.105) holds, i.e. if $X(T) = H$ and

$$X(t) = X(0) - \int_0^t f(s, X(s), \varphi(s))ds + \int_0^t \varphi(s)dW(s), \quad t \in [0, T].$$

Equivalently, (3.105) can be written as

$$X(t) = H + \int_t^T f(s, X(s), \varphi(s))ds - \int_t^T \varphi(s)dW(s).$$

The function f is called the **generator** or **driver** of the BSDE (3.105). Similarly to usual forward SDEs in Sect. 3.5, existence and uniqueness of solutions holds under Lipschitz conditions.

Theorem 3.66 *Suppose that* $E(\int_0^T f(t, 0, 0)^2 dt) < \infty$ *and*

$$|f(\omega, t, x, y) - f(\omega, t, \widetilde{x}, \widetilde{y})| \leq c(|x - \widetilde{x}| + |y - \widetilde{y}|), \ \omega \in \Omega, \ x, \widetilde{x} \in \mathbb{R}, \ y, \widetilde{y} \in \mathbb{R}^d$$

holds for some constant $c < \infty$. *Then the BSDE (3.105) has a solution. It is unique in the sense that, for any two solutions* (X, φ) *and* $(\widetilde{X}, \widetilde{\varphi})$, *we have* $X(t) = \widetilde{X}(t)$, $t \in [0, T]$ *outside some* P*-null set and*

$$(P \otimes \lambda)\big(\{(\omega, t) \in \Omega \times [0, T] : \varphi(\omega, t) \neq \widetilde{\varphi}(\omega, t)\}\big) = 0.$$

Idea A proof can be found in [230, Theorem 3.1]. The idea is to proceed similarly as for SDEs in Sect. 3.5, namely by relating solutions of BSDEs to fixed points of certain operators. In the present context, define an operator Φ mapping a pair of processes (X, φ) to $(\widetilde{X}, \widetilde{\varphi})$, where $\widetilde{X}(t) = E(H + \int_t^T f(s, X(s), \varphi(s))ds|\mathscr{F}_t)$ and $\widetilde{\varphi}$ denotes the integrand in the martingale representation $M^{\widetilde{X}} = \widetilde{\varphi} \bullet W$ of the martingale part $M^{\widetilde{X}}$ in the canonical decomposition of $\widetilde{X}$, cf. Theorem 3.61. Observe that fixed points of Φ are solutions to (3.105). Indeed,

$$\begin{aligned}
X(t) &= E\left(H + \int_t^T f(s, X(s), \varphi(s))ds\middle|\mathscr{F}_t\right) \\
&= E\left(H + \int_0^T f(s, X(s), \varphi(s))ds\middle|\mathscr{F}_t\right) - \int_0^t f(s, X(s), \varphi(s))ds
\end{aligned}$$

and $M^X = \varphi \bullet W$ imply $A^X(t) = -\int_0^t f(s, X(s), \varphi(s))ds$ and

$$\begin{aligned}
\int_0^t \varphi(s)dW(s) &= M^X(t) \\
&= X(t) - X(0) - A^X(t) \\
&= X(t) - X(0) + \int_0^t f(s, X(s), \varphi(s))ds
\end{aligned}$$

as desired. This fixed point property of the solution now opens the door to proceeding similarly as indicated in the proof of Theorem 3.40. □

Both the concept of a BSDE and the above existence and uniqueness result 3.66 can be extended to equations driven by Brownian motion and Poisson random measures. We take here Theorem 3.62 as a starting point. Similarly as above it implies that given a Lévy process L with Lévy–Khintchine triplet $(b^{L,h}, c^L, K^L)$, a time $T \geq 0$ and an $\mathscr{F}_T$-measurable square-integrable random variable H, there are essentially unique processes resp. integrands X, φ, ψ such that

$$\begin{aligned}
X(T) &= H, \\
dX(t) &= 0dt + \varphi(t)dL^c(t) + \psi(t, x) * (\mu^L - \nu^L)(dt). \qquad (3.106)
\end{aligned}$$

As before, a backward stochastic differential equation means considering (3.106) for non-vanishing drift terms. To this end, suppose that the filtration is generated by some $\mathbb{R}^d$-valued Lévy process L. Fix $T \geq 0$ and an $\mathscr{F}_T$-measurable random variable H with $E(H^2) < \infty$. Let $f : \Omega \times [0, T] \times \mathbb{R} \times \mathbb{R}^d \times L^2(\mathbb{R}^d, \mathscr{B}^d, K^L) \to \mathbb{R}$ denote a function that is $\mathscr{P} \otimes \mathscr{B} \otimes \mathscr{B}^d \otimes \mathscr{B}(L^2(\mathbb{R}^d, \mathscr{B}^d, K^L))$-measurable. Recall that $L^2(\mathbb{R}^d, \mathscr{B}^d, K^L)$ denotes the set of all measurable functions $g : \mathbb{R}^d \to \mathbb{R}$ with

$$\|g\|_2 := \sqrt{\int g(x)^2 K^L(dx)} < \infty.$$

Since $g \mapsto \|g\|_2$ is a norm, the Borel-σ-field $\mathscr{B}(L^2(\mathbb{R}^d, \mathscr{B}^d, K^L))$ of this space is well defined.

We call a triple (X, φ, ψ) of an adapted process X with $E(\sup_{t\in[0,T]} X_t^2) < \infty$, a process $\varphi \in L^2(L^c)$, and a function $\psi \in G^2(\mu^L)$ a solution to the **backward stochastic differential equation (BSDE)**

$$\begin{aligned} X(T) &= H, \\ dX(t) &= -f(t, X(t-), \varphi(t), \psi(t, \cdot))dt \\ &\quad + \varphi(t)dL^c(t) + \psi(t, x) * (\mu^L - \nu^L)(dt) \end{aligned} \tag{3.107}$$

if $X(T) = H$ and (3.107) holds on $[0, T]$. This means that $X(T) = H$ and

$$\begin{aligned} X(t) = X(0) &- \int_0^t f(s, X(s-), \varphi(s), \psi(s, \cdot))ds + \int_0^t \varphi(s)dL^c(s) \\ &+ \int_{[0,t]\times\mathbb{R}^d} \psi(s, x)(\mu^L - \nu^L)(d(s, x)) \end{aligned}$$

or, equivalently,

$$\begin{aligned} X(t) = H &+ \int_t^T f(s, X(s-), \varphi(s), \psi(s, \cdot))ds - \int_t^T \varphi(s)dL^c(s) \\ &- \int_{(t,T]\times\mathbb{R}^d} \psi(s, x)(\mu^L - \nu^L)(d(s, x)). \end{aligned}$$

The counterpart of Theorem 3.66 reads as follows.

Theorem 3.67 *Suppose that* $E(\int_0^T f(t, 0, 0, 0)^2 dt) < \infty$ *and*

$$|f(\omega, t, x, y, z) - f(\omega, t, \widetilde{x}, \widetilde{y}, \widetilde{z})| \leq c(|x - \widetilde{x}| + |y - \widetilde{y}| + \|z - \widetilde{z}\|_2)$$

for any $\omega \in \Omega, x, \widetilde{x} \in \mathbb{R}, y, \widetilde{y} \in \mathbb{R}^d, z, \widetilde{z} \in L^2(\mathbb{R}^d, \mathscr{B}^d, K^L)$ *holds for some constant* $c < \infty$. *Then the BSDE (3.107) has a solution. It is unique in the sense that, for any two solutions* (X, φ, ψ) *and* $(\widetilde{X}, \widetilde{\varphi}, \widetilde{\psi})$, *we have* $X(t) = \widetilde{X}(t)$, $t \in [0, T]$ *outside some* P*-null set,* $c^L(\varphi - \widetilde{\varphi})(\omega, t) = 0$ *and* $\|(\psi - \widetilde{\psi})(\omega, t)\|_2 = 0$ *for all* $(\omega, t) \in \Omega \times [0, T]$ *outside some* $P \otimes \lambda$*-null set.*

Idea The assertion is proved in [284, Lemma 2.4], cf. also [8, Theorem 2.1]. The idea is basically as indicated in Theorem 3.66, now relying on the more general martingale representation theorem 3.62 instead of Theorem 3.61. □

Remark 3.68 The BSDEs (3.105, 3.107) and Theorems 3.66, 3.67 allow for straightforward extensions to the case of $\mathbb{R}^n$-valued processes. In this case the uniqueness statement in Theorem 3.67 reads as $((\varphi - \widetilde{\varphi})^\top c^L(\varphi - \widetilde{\varphi}))(\omega, t) = 0$ outside some $P \otimes \lambda$-null set.

3.12 Change of Measure

Changes of the probability measure play an important role in statistics and finance. In statistics the true law underlying the data is typically unknown. Therefore one works with a whole family of candidate measures at the same time. In finance measure changes occur in the context of derivative pricing. The fundamental theorem of asset pricing tells us that option prices are typically obtained as expected values relative to so-called *equivalent martingale measures*. This is discussed in detail in Part II. At this point we consider how certain properties behave under an equivalent change of the probability measure.

Recall that probability measures P, Q are *equivalent* (written $P \sim Q$) if the events of probability 0 are the same under both measures. More generally, Q is called **absolutely continuous** with respect to P (written $Q \ll P$) if events of P-probability 0 have Q-probability 0 as well. In both cases, the P-density dQ/dP of Q exists, which in turn leads to the density process Z of Q relative to P. This process is a nonnegative P-martingale with $E(Z(t)) = 1$, $t \geq 0$. If Z attains the value 0, it stays at 0 afterwards. This does not happen for $Q \sim P$, in which case Z and Z_- are positive processes. A density process Z of Q relative to P still exists if we only require $Q|_{\mathscr{F}_t} \ll P|_{\mathscr{F}_t}$ for any $t \in \mathbb{R}_+$, which is weaker than $Q \ll P$. This property is denoted $Q \overset{\text{loc}}{\ll} P$. Accordingly, $P \overset{\text{loc}}{\sim} Q$ means that both $Q \overset{\text{loc}}{\ll} P$ and $P \overset{\text{loc}}{\ll} Q$ hold.

For the rest of this section we assume $Q \sim P$ with density process Z. The martingale property depends on the probability measure because it involves expectations. We have the following

Proposition 3.69 *An adapted càdlàg process M is a Q-martingale if and only if MZ is a P-martingale. Corresponding statements hold for local and σ-martingales, respectively.*

Idea This is shown in [154, Proposition III.3.8] and [168, Proposition 5.1]. We sketch here the proof for martingales and local martingales.

M is a Q-martingale if and only if

$$E_Q(M(t)|\mathscr{F}_s) = M(s), \quad s \leq t. \tag{3.108}$$

By Bayes' rule (1.8) the left-hand side equals $E_P(M(t)Z(t)|\mathscr{F}_s)/Z(s)$. Hence (3.108) is equivalent to $E_P(M(t)Z(t)|\mathscr{F}_s) = M(s)Z(s)$, which is the martingale property of MZ relative to P. The integrability property is shown similarly.

The local martingale case follows by localisation. □

In view of the previous proposition it may not seem evident that the class of semimartingales is invariant under equivalent measure changes. However, this can be shown to hold, cf. [154, Proposition III.6.24]. Similarly, the covariation $[X, Y]$ (but not $\langle X, Y\rangle$), the set $L(X)$ of X-integrable processes and stochastic integrals

$\varphi \bullet X$ themselves remain the same if P is replaced with $Q \sim P$. As noted above, P-martingales typically cease to be martingales under Q, i.e. they acquire a trend. This drift is quantified in the following

Proposition 3.70 (Girsanov) *Suppose that M is a P-local martingale such that $[M, Z]$ is a P-special semimartingale. Then*

$$M' := M - \langle M, \mathscr{L}(Z)\rangle^P$$

is a Q-local martingale. Put differently, M is a Q-special semimartingale with compensator $\langle M, \mathscr{L}(Z)\rangle^P$. The superscript indicates that the predictable covariation refers to the measure P.

Proof Since $A := \langle \mathscr{L}(Z), M\rangle = \frac{1}{Z_-} \bullet \langle Z, M\rangle$ is a predictable process of finite variation and arguing as in the proof of Theorem 1.13, it suffices to show that $M - M(0) - A$ is a Q-local martingale. By Proposition 3.69 this amounts to proving that $Z(M - M(0) - A)$ is a P-local martingale. Integration by parts yields

$$\begin{aligned} Z(M - M(0) - A) &= Z_- \bullet M + (M - M(0))_- \bullet Z + [Z, M - M(0)] \\ &\quad - Z_- \bullet A - A \bullet Z. \end{aligned} \tag{3.109}$$

The integrals relative to M and Z are local martingales. Moreover,

$$Z_- \bullet A = \frac{Z_-}{Z_-} \bullet \langle Z, M\rangle = \langle Z, M\rangle$$

is the compensator of $[Z, M] = [Z, M - M(0)]$, which implies that the difference is a local martingale. Altogether the right-hand side of (3.109) is indeed a local martingale. □

Corollary 3.71 *Suppose that $X = X(0) + M^X + A^X$ is a P-special semimartingale such that $[M^X, Z]$ is a P-special semimartingale as well. Then X is a Q-special semimartingale whose Q-canonical decomposition $X = X(0) + M^{X,Q} + A^{X,Q}$ is given by*

$$\begin{aligned} M^{X,Q} &= M^X - \langle M^X, \mathscr{L}(Z)\rangle^P, \\ A^{X,Q} &= A^X + \langle M^X, \mathscr{L}(Z)\rangle^P. \end{aligned}$$

Proof This follows by applying Proposition 3.70 to M^X. □

Remark 3.72 If M is a P-standard Brownian motion, M' in Proposition 3.70 is a Q-standard Brownian motion. Indeed, one easily verifies that M' is a continuous Q-local martingale satisfying $\langle M', M'\rangle = I$. These properties characterise Wiener processes by Example 3.28. This result extends to multivariate Wiener processes but it has no counterpart for Lévy processes with jumps. The Q-martingale part of a P-Lévy process may fail to be a Q-Lévy process.

The compensator ν of a random measure μ depends on the probability measure as well.

Proposition 3.73 *Suppose that the density process of $Q \sim P$ is of the form*

$$Z = \mathscr{E}(\varphi \bullet W + \psi * (\mu - \nu)),$$

where W denotes some continuous P-local martingale and μ some integer-valued random measure with P-compensator $\nu(dt, dx) = K(t, dx)dt$. Then the Q-compensator $\nu'(d(t, x)) = K'(t, dx)dt$ of μ satisfies

$$\frac{dK'(t, dx)}{dK(t, dx)} = 1 + \psi(t, x). \tag{3.110}$$

Proof Set $\nu'(d(t, x)) = K'(t, dx)dt$ with K' as in (3.110). We need to show

$$E_Q(\xi * \mu(\infty)) = E_Q(\xi * \nu(\infty)) \tag{3.111}$$

for any predictable $\xi \geq 0$. By monotone convergence it suffices to consider $\xi \leq n\eta/(1 + |\psi|)$, where η is as in the definition of predictable σ-finiteness in Sect. 3.3. But using Proposition 3.37(5) we obtain

$$\begin{aligned}
\xi * \mu - \xi * \nu' &= \xi * \mu - \xi(1 + \psi) * \nu \\
&= \xi * \mu - \xi * \nu - \xi\psi * \nu \\
&= \xi * \mu - \xi * \nu - \langle \xi * (\mu - \nu), \psi * (\mu - \nu) \rangle \\
&= \xi * (\mu - \nu) - \langle \xi * (\mu - \nu), \mathscr{L}(Z) \rangle,
\end{aligned}$$

which is a Q-local martingale by Proposition 3.70. Hence we get $E_Q(\xi * \mu(\tau_n)) = E_Q(\xi * \nu(\tau_n))$ for some increasing sequence (τ_n) of stopping times. Dominated convergence now yields (3.111). □

Appendix 1: Problems

The exercises correspond to the section with the same number.

3.1 Suppose that M, N are square-integrable martingales. Show that M, N are strongly orthogonal if and only if M^τ and $N - N(0)$ are orthogonal in the Hilbert space sense for any stopping time τ.

3.2 For a semimartingale X and $z \in \mathbb{R}$ set $Z(t) := e^{-\lambda t}z + \int_0^t e^{-\lambda(t-s)}dX(s)$. Show that $Z(t) = z - \lambda Z_- \bullet I + X$.

3.3 Show that a predictable function ξ is $(\mu^X - \nu^X)$-integrable and $\xi * (\mu^X - \nu^X)$ is a locally square-integrable martingale if $\xi^2 * \nu^X(t) < \infty$ for any $t \geq 0$.

3.4 Let X be a Lévy–Itō process as in (3.57) and $f : \mathbb{R} \to \mathbb{R}$ twice continuously differentiable. Show that $f(X)$ is a Lévy–Itō process and determine its coefficients.

3.5 Let $W = (W_1, W_2)$ be a standard Brownian motion in $\mathbb{R}^2$. Show that $X := W_1^2 + W_2^2$ solves the stochastic differential equation $dX(t) = 2dt + 2\sqrt{X(t)}d\widetilde{W}(t)$ with some univariate Wiener process $\widetilde{W}$.

Hint: you may use Lévy's characterisation of Brownian motion in Example 3.28.

3.6 Suppose that X is a semimartingale which does not have jumps of size -1. Show that $1/\mathscr{E}(X) = \mathscr{E}(Y)$ for

$$Y := -X + [X^c, X^c] + \frac{x^2}{1+x} * \mu^X.$$

3.7 Let S be an $\mathbb{R}^d$-valued exponential Lévy process and $X = 1 + \varphi \bullet S$ with $\varphi_i(t) = \frac{X(t-)}{S_i(t-)}\gamma_i, i = 1, \dots, d$ for some $\gamma_1, \dots, \gamma_d \in \mathbb{R}_+$ satisfying $\gamma_1 + \dots + \gamma_d < 1$. Show that X is an exponential Lévy process. In Mathematical Finance, X corresponds to the wealth of a trader who invests constant fractions $\gamma_1, \dots, \gamma_d$ of her wealth in assets $S_1, \dots, S_d$.

3.8 Let X be a time-inhomogeneous Lévy process with $d = 1$ and $\nu = 0$ in Theorem 3.51. Show that X is Gaussian and determine the autocovariance $\mathrm{Cov}(X(s), X(t))$ for any $s, t \geq 0$. (From Proposition 4.7 for PIIACs or [154, Theorem II.4.15] for general time-inhomogeneous Lévy process it follows that $\nu = 0$ holds if and only if X is continuous.)

3.9 Determine the solution to the system of SDEs

$$\begin{aligned} dX_1(t) &= -\lambda_1 X_2(t)dt + dW_1(t), \quad X_1(0) = x_1, \\ dX_2(t) &= -\lambda_2 X_1(t)dt + dW_2(t), \quad X_2(0) = x_2 \end{aligned}$$

for $x_1, x_2, \lambda_1, \lambda_2 \in \mathbb{R}$ and some two-dimensional Wiener process $W = (W_1, W_2)$.

3.10 Let $W = (W_1, W_2)$ be a Brownian motion in $\mathbb{R}^2$ with $E(W(t)) = 0$ and

$$\mathrm{Cov}(W(t)) = \begin{pmatrix} \sigma_1^2 & \varrho\sigma_1\sigma_2 \\ \varrho\sigma_1\sigma_2 & \sigma_2^2 \end{pmatrix}$$

for some $\sigma_1, \sigma_2 > 0$, $\varrho \in [-1, 1]$. Determine the Galtchouk–Kunita–Watanabe decomposition of $M := \mathscr{E}(W_2)$ relative to $X := \mathscr{E}(W_1)$.

3.11 (Linear BSDE) Consider the BSDE (3.107) with

$$f(\omega, t, x, y, z) = \alpha(t) + \beta(t)x + \gamma(t)c^L y + \int \delta(t, \xi)z(\xi)K^L(d\xi)$$

for some bounded measurable functions $\alpha, \beta, \gamma : [0, T] \to \mathbb{R}$ and a function $\delta : [0, T] \times \mathbb{R} \to (-1, \infty]$ such that $t \mapsto \int \delta(t, \xi)^2 K^L(d\xi)$ is bounded on $[0, T]$. Define $Z := \mathscr{E}(\beta \bullet I + \gamma \bullet L^c + \delta * (\mu^L - \nu^L))$ and

$$X(t) := E\left(H\frac{Z(T)}{Z(t)} + \int_t^T \alpha(s)\frac{Z(s-)}{Z(t)}ds \middle| \mathscr{F}_t \right).$$

Verify that there are integrands φ, ψ such that (X, φ, ψ) solves (3.107). In particular, we have $\varphi(t) = d\langle X^c, L^c\rangle(t)/(c^L dt)$.

3.12 Suppose that the $\mathbb{R}^d$-valued Itō process X is of the form $X(t) = X(0) + \mu \bullet I + \sigma \bullet W$ in the sense

$$X_i(t) = X_i(0) + \mu_i \bullet I + \sigma_{i\cdot} \bullet W, \quad i = 1, \dots, d$$

for some $\mathbb{R}^d$-valued standard Brownian motion W, some $\mathbb{R}^d$-valued process $\mu \in L(I)$ and an $\mathbb{R}^{d\times d}$-valued process σ such that $\sigma_1, \dots, \sigma_d \in L(W)$. Moreover, assume that $\sigma^{-1}\mu \in L(W)$ and that $Q \sim P$ is a probability measure with density process $Z := \mathscr{E}((\sigma^{-1}\mu) \bullet W)$. Show that $X(t) = X(0) + \sigma \bullet W^Q(t)$ for some Q-standard Brownian motion W^Q in $\mathbb{R}^d$.

Hint: you my use the multivariate extension of Example 3.28, namely that W is an $\mathbb{R}^d$-valued standard Brownian motion if it is a continuous local martingale with $W(0) = 0$ and $[W_i, W_j](t) = t$ as well as $[W_i, W_j](t) = t$ for $i, j = 1, \dots, d$ with $i \neq j$.

Appendix 2: Notes and Comments

Stochastic integration is covered in [2, 76, 154, 186, 238, 241, 252] for stochastic processes. For integration relative to random measures we refer to [152, 154].

Example 3.28 follows [154, Theorem II.4.4]. Example 3.33 is part of [154, Corollary II.4.19]. Proposition 3.36 is based on [154, Proposition II.1.28]. For the integral relative to compensated random measures we refer to [154, Definition II.1.27]. Most of Proposition 3.37 can be found in [154, Section II.1d]. Proposition 3.44 is stated in [152, Proposition 6.4]. For Proposition 3.46 see [182, Lemma 2.4]. Proposition 3.48 is a special case of [152, Théorème 6.8]. Theorem 3.49 is based on [125, Lemma A.8]. Equations (3.88–3.90) can be found, for example, in [60, Section

15.3.1]. Proposition 3.54 is stated in [60, Lemma 15.1]. For Proposition 3.55, Examples 3.57, 3.58, Proposition 3.59, (3.97, 3.98) we refer to [263, Sections 5.2.2, 5.3.3, 5.3.4, 5.5.1, 5.5.2]. Proposition 3.70 is similar to [154, Theorem III.3.11]. Proposition 3.73 is related to [169, Proposition 2.6]. Problem 3.1 can be found in [154, Proposition I.4.15]. For Problem 3.11 we refer to [239, Theorem 3.3].

Chapter 4
Semimartingale Characteristics

The stochastic calculus in Chap. 3 is based on integration. Small Lévy-like bits of processes are pieced together to yield something that behaves differently from any Lévy process on a global scale. In this chapter we take the point of view of differentiation. How can the local behaviour of a given stochastic process be quantified? There are two different approaches which lead to related notions. *Semimartingale characteristics* play the role of a derivative in general semimartingale theory. They give rise to a whole calculus which makes them handy for concrete calculations. *Infinitesimal generators*, on the other hand, are defined for Markov processes. They are discussed in the next chapter. It is useful to know both concepts. In particular, both lead to *martingale problems* as stochastic counterparts to ordinary differential equations.

4.1 Definition

Recall that we view Lévy processes as functions of constant random growth. An arbitrary semimartingale X does not in general exhibit this constant growth behaviour, but it often resembles some Lévy process *locally* around t. The law of a Lévy process is characterised by its Lévy–Khintchine triplet (b^h, c, K), but since X does not behave as a Lévy process on a global scale, this triplet may now vary randomly through time. Put differently, we must deal with *local* triplets $(b^h, c, K)(\omega, t)$. This parallels the time-varying derivative $b(t) = dX(t)/dt$ of a deterministic function. In both cases a general object (differentiable function, semimartingale) is viewed on a local scale as an object of constant growth (linear function, Lévy process).

It may not seem obvious how to define a local triplet for arbitrary processes. As our starting point we consider the Lévy–Khintchine formula. For the whole section we fix a truncation h as in Theorem 2.17. If (b^h, c, K) denotes the Lévy–

© Springer Nature Switzerland AG 2019

E. Eberlein, J. Kallsen, *Mathematical Finance*, Springer Finance,
https://doi.org/10.1007/978-3-030-26106-1_4

Khintchine triplet of a Lévy process X and ψ its characteristic exponent, let us define its *(integral) characteristics* by

$$B^h(t) := b^h t, \quad C(t) := ct, \quad \nu([0,t] \times dx) := K(dx)t$$

and its *integral characteristic exponent* by

$$\begin{aligned}\Psi(t,u) &:= \psi(u)t \\ &= iu^\top B^h(t) - \frac{1}{2}u^\top C(t)u + \int_{[0,t]\times\mathbb{R}^d} \big(e^{iu^\top x} - 1 - iu^\top h(x)\big)\nu(d(s,x))\end{aligned}$$

for $t \geq 0$, $u \in \mathbb{R}^d$. Here, B^h, C are deterministic functions and ν denotes a measure on $\mathbb{R}_+ \times \mathbb{R}^d$. An application of Itō's formula yields that

$$M(t) := e^{iu^\top X(t)} - \int_0^t e^{iu^\top X(s-)}\Psi(ds,u) \tag{4.1}$$

is a local martingale for any $u \in \mathbb{R}^d$. Indeed, recall that in the case $d = 1$ and if we may choose $h(x) = x$, we have $A^X = b^{\mathrm{id}}t$, $\langle X^c, X^c\rangle(t) = ct$, and $\nu^X = \nu$ by Theorem 2.21 and (3.12, 3.46). Itō's formula as in (3.53) yields

$$\begin{aligned}e^{iuX(t)} &= e^{iuX(0)} + iue^{iuX_-} \bullet A^X(t) - \frac{1}{2}u^2 e^{iuX_-} \bullet \langle X^c, X^c\rangle(t) \\ &\quad + (e^{iu(X_-+x)} - e^{iuX_-} - iue^{iuX_-}x) * \nu^X(t) \\ &\quad + iue^{iuX_-} \bullet X^c(t) \\ &\quad + (e^{iu(X_-+x)} - e^{iuX_-}) * (\mu^X - \nu^X)(t) \\ &= e^{iuX(0)} + \int_0^t e^{iuX(s-)}\Psi(ds,u) \\ &\quad + iue^{iuX_-} \bullet X^c(t) + (e^{iu(X_-+x)} - e^{iuX_-}) * (\mu^X - \nu^X)(t),\end{aligned} \tag{4.2}$$

which implies the local martingale property of M. For a rigorous proof see [154, Theorem II.2.42].

The local martingale property of M in (4.1) can be extended to arbitrary semimartingales if we allow for more general triplets.

Definition 4.1 Let X denote an $\mathbb{R}^d$-valued semimartingale. Suppose that B^h is a predictable $\mathbb{R}^d$-valued process, C a predictable $\mathbb{R}^{d\times d}$-valued process whose values are non-negative symmetric matrices, both with components of finite variation, and ν a predictable random measure on $\mathbb{R}_+ \times \mathbb{R}^d$, cf. Sect. 3.3. The triplet (B^h, C, ν) is called **(integral) characteristics** of X if and only if

$$M(t) := e^{iu^\top X(t)} - \int_0^t e^{iu^\top X(s-)}\Psi(ds,u) \tag{4.3}$$

is a local martingale for any $u \in \mathbb{R}^d$, where the predictable process $(\Psi(t, u))_{t \geq 0}$ is defined as

$$\Psi(t, u) := iu^\top B^h(t) - \frac{1}{2}u^\top C(t)u + \left(e^{iu^\top x} - 1 - iu^\top h(x)\right) * \nu(t). \tag{4.4}$$

Observe that B^h depends on the choice of the truncation function h but C, ν, Ψ do not. If we pass from h to another truncation function $\widetilde{h}$, we must replace B^h by

$$B^{\widetilde{h}} := B^h + \left(\widetilde{h} - h\right) * \nu, \tag{4.5}$$

which parallels (2.21).

It is not obvious from the definition that such a triplet of characteristics generally exists. It can be constructed quite easily. To this end, let

$$X^h := X - (x - h(x)) * \mu^X, \tag{4.6}$$

$$B^h := A^{X^h}, \tag{4.7}$$

$$C_{ij} := \langle X_i^c, X_j^c \rangle, \tag{4.8}$$

$$\nu := \nu^X, \tag{4.9}$$

where A^{X^h} denotes the compensator of X^h in the sense of Sect. 2.1, X^c stands for the continuous local martingale part, μ^X for the random measure of jumps of X and ν^X for its compensator. An application of Itō's formula parallel to (4.2) shows that M in (4.3) is a local martingale, cf. [154, Theorem II.2.42].

X^h differs from X by the missing or partially missing big jumps, e.g. of size > 1 if $h(x) = x1_{\{|x| \leq 1\}}$. Since X^h has bounded jumps, it is a special semimartingale and hence its compensator A^{X^h} exists. If X itself is special, we may use $h(x) = x$ as the "truncation" function, in which case $B^h = B^{\text{id}}$ coincides with the compensator A^X of X. Another convenient choice is $h(x) = 0$, which works for processes whose jump part is of finite variation. In this case $X^h = X^0$ equals X without its jumps and $B^h = B^0$ represents the compensator of this "continuous part" of X. Since A^X and X^0 do not exist in general, we work with true truncation functions in the sense of Theorem 2.17, which however implies that B^h is slightly less intuitive.

If $(\widetilde{B}, \widetilde{C}, \widetilde{\nu})$ is another version of the integral characteristics leading to $\widetilde{\Psi}$ via (4.4), then

$$\int_0^t e^{iu^\top X(s-)}(\widetilde{\Psi} - \Psi)(ds, u)$$

is a local martingale for fixed u. By integrating the process $e^{-iu^\top X_-}$ one can conclude that $((\widetilde{\Psi}-\Psi)(t,u))_{t\geq 0}$ is a predictable local martingale of bounded variation as well and hence 0, cf. Proposition 2.11. Since Ψ determines the coefficients (B^h, C, ν) in its Lévy–Khintchine representation (4.4) uniquely (cf. [154, Lemma II.2.44]), we have uniqueness of the integral characteristics. Consequently, we have shown

Proposition 4.2 *The integral characteristics of an arbitrary semimartingale are unique. They are given by (4.6–4.9).*

The integral characteristics of X are not the object we have been looking for in the first place. They involve the dynamics of the whole process between 0 and t and not just in a neighbourhood of t. For instance, we have $X = A^X$ and hence $(B^h, C, \nu) = (X, 0, 0)$ for differentiable functions X. Often, however, the triplet (B^h, C, ν) is absolutely continuous in time, i.e. we have

$$B^h(t) = \int_0^t b^h(s)ds, \quad C(t) = \int_0^t c(s)ds, \quad \nu([0,t]\times dx) := \int_0^t K(s,dx)ds \tag{4.10}$$

for some predictable processes b^h, c and some transition kernel K from $(\Omega \times \mathbb{R}_+, \mathscr{P})$ to $\mathbb{R}^d$. In this case we call the triplet (b^h, c, K) the *local characteristics* of X. In view of the above results, the following definition makes sense.

Definition 4.3 Let X be an $\mathbb{R}^d$-valued semimartingale. We call a triplet (b^h, c, K) the **local** (or **differential**) **characteristics** of X if b^h is a predictable $\mathbb{R}^d$-valued process, c a predictable $\mathbb{R}^{d\times d}$-valued process whose values are non-negative symmetric matrices, and K a transition kernel from $(\Omega \times \mathbb{R}_+, \mathscr{P})$ to $\mathbb{R}^d$ with $K(t,\{0\}) = 0$, if $b^h, c, \int(1\wedge |x|^2)K(t,dx) \in L(I)$, and if one of the following equivalent conditions is satisfied.

1.

$$M(t) := e^{iu^\top X(t)} - \int_0^t e^{iu^\top X(s-)}\psi(s,u)ds \tag{4.11}$$

is a local martingale for any $u \in \mathbb{R}^d$, where

$$\psi(t,u) := iu^\top b^h(t) - \frac{1}{2}u^\top c(t)u + \int_{\mathbb{R}^d}\left(e^{iu^\top x} - 1 - iu^\top h(x)\right)K(t,dx) \tag{4.12}$$

denotes the **characteristic exponent** of $(b^h, c, K)(\omega, t)$.

2.

$$A^{X^h}(t) = \int_0^t b^h(s)ds,$$
$$\langle X_i^c, X_j^c\rangle(t) = \int_0^t c_{ij}(s)ds,$$
$$\nu^X([0,t] \times dx) = \int_0^t K(s, dx)ds$$

for $t \geq 0, i, j = 1, \dots, d$.

The equivalence of the two conditions follows from Proposition 4.2.

Remark 4.4 It can be shown that the semimartingale property of an adapted càdlàg process X already follows from the local martingale property of (4.3) resp. (4.11), cf. [154, Proposition II.2.42].

Differential characteristics enjoy a uniqueness property similarly as their integral counterpart.

Proposition 4.5 *The differential characteristics and the characteristic exponent are unique in the sense that any two versions coincide outside some $P \otimes \lambda$-null set of $(\omega, t) \in \Omega \times \mathbb{R}_+$.*

Proof Equation (4.11) yields uniqueness of the exponent, cf. the proof of Proposition 4.2. Since the coefficients (b^h, c, K) in the Lévy–Khintchine representation (4.12) are uniquely determined by the exponent ψ (cf. [154, Lemma II.2.44]), we obtain uniqueness of the differential characteristics as well. □

It may not be obvious from the definition but the local characteristics can be interpreted as a local Lévy–Khintchine triplet. Put differently, the process X resembles locally after t a Lévy process with triplet $(b^h, c, K)(\omega, t)$. Since this local behaviour may depend on the history up to t, the local characteristics may be random albeit predictable. As noted above, the connection between Lévy processes and local characteristics parallels the one between linear functions and derivatives of deterministic functions. In fact, b^h equals the ordinary derivative if X has absolutely continuous paths (and we have $c = 0$, $K = 0$ in this case).

As above and as in the Lévy case, only the first characteristic b^h depends on the choice of the truncation function h, whereas c, K and the characteristic exponent ψ do not. If h is replaced by another truncation function $\widetilde{h}$, the first characteristic changes according to

$$b^{\widetilde{h}}(t) = b^h(t) + \int \big(\widetilde{h}(x) - h(x)\big)K(t, dx), \tag{4.13}$$

which corresponds directly to (2.21). For special semimartingales we can choose $h(x) = x$ because A^X exists. In this case $b^h = b^{\mathrm{id}}$ equals the drift rate a^X in the sense of (2.9). Since most processes in applications are special and since concrete calculations become simpler and more intuitive, we recommend to use $h(x) = x$ whenever this is possible. Even in the general case, using $h(x) = x$ will not lead to wrong but at most to undefined results. By applying (4.13) one easily returns to well-defined objects, cf. Example 4.9 below.

4.2 Rules

For derivatives of deterministic functions we have the well-known differentiation rules which are summarised in Sect. 1.6.3. If one views local characteristics as a stochastic process analogue of derivatives, one naturally expects counterparts of these rules at this more general level. Such results are presented in this section. We begin with the statement that constant growth corresponds to a constant derivative. On the level of processes constant growth refers as before to Lévy processes.

Proposition 4.6 (Lévy Process) *An $\mathbb{R}^d$-valued semimartingale X with $X_0 = 0$ is a Lévy process if and only if it has a version (b^h, c, K) of the local characteristics which does not depend on (ω, t). In this case, (b^h, c, K) equals the Lévy–Khintchine triplet.*

Proof It follows from the considerations in Sect. 4.1 that a Lévy process with Lévy–Khintchine triplet (b^h, c, K) has local characteristics (b^h, c, K).

Conversely, suppose that X is an $\mathbb{R}^d$-valued semimartingale with deterministic and constant local characteristics (b^h, c, K). The local martingale M in (4.11) is actually a martingale because it is bounded on any interval $[0, t]$. Hence

$$\begin{aligned} E\big(e^{iu^\top X(t)}\big|\mathscr{F}_s\big) &= e^{iu^\top X(s)} + E\bigg(\int_s^t e^{iu^\top X(r)}\psi(u)dr\bigg|\mathscr{F}_s\bigg) \\ &= e^{iu^\top X(s)} + \int_s^t E(e^{iu^\top X(r)}|\mathscr{F}_s)\psi(u)dr \end{aligned}$$

for $s \le t$. Setting $f(r) := E(e^{iu^\top X(r)}|\mathscr{F}_s)$ we obtain $f(s) = e^{iu^\top X(s)}$ and $f'(r) = f(r)\psi(u)$ for $r \ge s$. Solving this linear ODE yields

$$E(e^{iu^\top X(t)}|\mathscr{F}_s) = f(t) = \exp\big(iu^\top X(s) + \psi(u)(t-s)\big).$$

This shape of the conditional characteristic function tells us that X is a Lévy process with Lévy–Khintchine triplet (b^h, c, K). □

Differential characteristics generally depend on both randomness ω and time t. If we only allow for dependence on time, we end up with time-inhomogeneous Lévy processes as in Sect. 3.8.

Proposition 4.7 (Time-Inhomogeneous Lévy Process) *A real- or vector-valued semimartingale X with $X(0) = 0$ is a time-inhomogeneous Lévy process with absolutely continuous characteristics if and only if it has a version $(b^h, c, K)(\omega, t)$ of the local characteristics which does not depend on ω. In this case, the characteristic function of its increments is given by*

$$\varphi_{X(t)-X(s)}(u) = \exp\left(\int_s^t \psi(r,u)dr\right)$$

for $s \leq t$, where $\psi(t, u)$ denotes the characteristic exponent of $(b^h, c, K)(t)$.

Proof This follows similarly as Proposition 4.6. □

The local characteristics of a semimartingale X immediately yield the characteristics of stochastic integrals relative to X.

Proposition 4.8 (Stochastic Integration I) *Let X be an $\mathbb{R}^d$-valued semimartingale and φ an $\mathbb{R}^{n\times d}$-valued predictable process whose columns are integrable with respect to X. If X has local characteristics (b^h, c, K), the $\mathbb{R}^n$-valued integral process $\varphi \bullet X := (\varphi_{j\cdot} \bullet X)_{j=1,\dots,n}$ has characteristics $(\widetilde{b}^{\widetilde{h}}, \widetilde{c}, \widetilde{K})$ (relative to some truncation function $\widetilde{h}$ on $\mathbb{R}^n$) given by*

$$\widetilde{b}^{\widetilde{h}}(t) = \varphi(t)b^h(t) + \int \big(\widetilde{h}(\varphi(t)x) - \varphi(t)h(x)\big)K(t, dx), \tag{4.14}$$

$$\widetilde{c}(t) = \varphi(t)c(t)\varphi(t)^\top,$$

$$\widetilde{K}(t, B) = \int 1_B(\varphi(t)x)K(t, dx)$$

for $t \geq 0$ and $B \in \mathscr{B}^n$ not containing 0.

If ψ denotes the characteristic exponent of (b^h, c, K), the corresponding exponent $\widetilde{\psi}$ of $(\widetilde{b}^{\widetilde{h}}, \widetilde{c}, \widetilde{K})$ is given by

$$\widetilde{\psi}(t,u) = \psi(t, \varphi(t)^\top u).$$

Proof This is shown in [181, Lemma 3]. The statement for the exponent follows from the triplet version and from the definition of the exponent.

We sketch the idea for $d = 1 = n$ by letting $h = \mathrm{id}$, $\widetilde{h} = \mathrm{id}$. Observe that

$$\begin{aligned}\varphi \bullet X &= \varphi \bullet A^X + \varphi \bullet X^c + \varphi \bullet (x * (\mu^X - \nu^X)) \\ &= (\varphi b^{\mathrm{id}}) \bullet I + \varphi \bullet X^c + (\varphi x) * (\mu^X - \nu^X).\end{aligned}$$

The first term is the compensator $A^{\varphi\bullet X}$ of this process because the others are local martingales. The second term is the continuous martingale part because the others are of finite variation resp. a purely discontinuous local martingale. We conclude that

$$\begin{aligned}\langle(\varphi\bullet X)^c,(\varphi\bullet X)^c\rangle &= \langle\varphi\bullet X^c,\varphi\bullet X^c\rangle\\ &= \varphi^2\bullet\langle X^c,X^c\rangle\\ &= (\varphi^2 c)\bullet I\end{aligned}$$

as claimed. Finally, the jumps of $\varphi\bullet X$ are

$$\Delta(\varphi\bullet X)(t)=\varphi(t)\Delta X(t)=\xi(t,\Delta X(t))$$

for $\xi(t,x)=\varphi(t)x$. Proposition 3.37(6) yields the statement on $\widetilde{K}$. □

If both X and $\varphi\bullet X$ are special semimartingales, we may use $h(x)=x$, $\widetilde{h}(x)=x$. Equation (4.14) reduces to

$$a^{\varphi\bullet X}(t)=\widetilde{b}^{\mathrm{id}}(t)=\varphi(t)b^{\mathrm{id}}(t)=\varphi(t)a^X(t), \tag{4.15}$$

where we use the notation of (2.9) and the superscripts id refer to the identity "truncation" function as usual.

Example 4.9 Let us see what happens if we wrongly use (4.15) for non-special semimartingales, where the drift rates a^X, $a^{\varphi\bullet X}$ do not exist. In order to return to well-defined objects, we apply (4.13), which here reads as

$$\begin{aligned}b^h(t)&=b^{\mathrm{id}}(t)+\int\big(h(x)-x\big)K(t,dx),\\ \widetilde{b}^{\widetilde{h}}(t)&=\widetilde{b}^{\mathrm{id}}(t)+\int\big(\widetilde{h}(x)-x\big)\widetilde{K}(t,dx).\end{aligned}$$

The superscript refers to the truncation function under consideration. This leads to

$$\begin{aligned}\widetilde{b}^{\widetilde{h}}(t)&=\widetilde{b}^{\mathrm{id}}(t)+\int\big(\widetilde{h}(x)-x\big)\widetilde{K}(t,dx)\\ &=\varphi(t)b^{\mathrm{id}}(t)+\int\big(\widetilde{h}(\varphi(t)x)-\varphi(t)x\big)\,K(t,dx)\\ &=\varphi(t)b^h(t)+\int\big(\varphi(t)(x-h(x))+\widetilde{h}(\varphi(t)x)-\varphi(t)x\big)\,K(t,dx)\\ &=\varphi(t)b^h(t)+\int\big(\widetilde{h}(\varphi(t)x)-\varphi(t)h(x)\big)\,K(t,dx)\end{aligned}$$

and hence (4.14). Note that the result is correct even though the intermediate steps do not make sense for arbitrary semimartingales.

Stopping can be viewed as a special case of integration. This leads to the following rule.

Proposition 4.10 (Stopping) *Let X be an $\mathbb{R}^d$-valued semimartingale and τ a stopping time. If X has local characteristics (b^h, c, K), the stopped process X^τ has local characteristics $(\widetilde{b}^h, \widetilde{c}, \widetilde{K})$ given by*

$$\widetilde{b}^h(t) = b^h(t)1_{[\![0,\tau]\!]}(t),$$
$$\widetilde{c}(t) = c(t)1_{[\![0,\tau]\!]}(t),$$
$$\widetilde{K}(t, dx) = K(t, dx)1_{[\![0,\tau]\!]}(t).$$

Proof This follows from Proposition 4.8 for $\varphi := 1_{[\![0,\tau]\!]}$. □

The definition of Lévy–Itō processes involves stochastic integrals relative to random measures. What can be said about the characteristics of such integrals?

Proposition 4.11 (Stochastic Integration II) *Let μ denote an integer-valued random measure with intensity measure K as in (3.55).*

1. *If ξ is a μ-integrable function, $X = \xi * \mu$ has local characteristics $(\widetilde{b}^h, \widetilde{c}, \widetilde{K})$ given by*

$$\widetilde{b}^h(t) = \int h(\xi(t, x))K(t, dx),$$
$$\widetilde{c}(t) = 0,$$
$$\widetilde{K}(t, B) = \int 1_B(\xi(t, x))K(t, dx)$$

for $t \geq 0$ and $B \in \mathscr{B}$ not containing 0. If X is special, we have

$$a^X(t) = \widetilde{b}^{\mathrm{id}}(t) = \int \xi(t, x)K(t, dx)$$

for the identity truncation function.

2. *If ξ is a $(\mu - \nu)$-integrable function, the integral process $X := \xi * (\mu - \nu)$ has local characteristics $(\widetilde{b}^h, \widetilde{c}, \widetilde{K})$ given by*

$$\widetilde{b}^h(t) = \int \big(h(\xi(t, x)) - \xi(t, x)\big)K(t, dx),$$
$$\widetilde{c}(t) = 0,$$
$$\widetilde{K}(t, B) = \int 1_B(\xi(t, x))K(t, dx)$$

for $t \geq 0$ and $B \in \mathscr{B}$ not containing 0. Since X is special, the identity can be used as truncation function, and we have $a^X(t) = \widetilde{b}^{\mathrm{id}}(t) = 0$.

Proof For ease of notation we suppose that $h = \mathrm{id}$ and that $\mu = \mu^Y$ for some semimartingale Y.

1. Since $\xi * \mu = \xi * \nu + \xi * (\mu - \nu)$, we have that

$$A^X(t) = \xi * \nu(t) = \int_0^t \int \xi(s, x) K(s, dx) ds$$

   as desired. The second characteristic vanishes because the continuous martingale part is absent. Since $\Delta(\xi * \mu)(t) = \xi(t, \Delta Y(t))$, Proposition 3.37(6) yields the third characteristic.
2. Since $\xi * (\mu - \nu)$ is a purely discontinuous local martingale, the first and second characteristic vanish. Since $\Delta(\xi * (\mu - \nu))(t) = \xi(t, \Delta Y(t))$, Proposition 3.37(6) yields the third characteristic as in the previous statement. □

The combination of the previous two results yields the characteristics of general Lévy–Itō processes. We state here the univariate version but the extension to the $\mathbb{R}^d$-valued process is straightforward.

Proposition 4.12 (Lévy–Itō Process) *Let*

$$dX(t) = \beta(t)dt + \sigma(t)dW(t) + (h \circ \delta) * (\mu - \nu)(dt) + (\overline{h} \circ \delta) * \mu(dt)$$

denote a Lévy–Itō process as in (3.58), where in particular μ denotes a homogeneous Poisson random measure with intensity measure K. Then X has local characteristics $(\widetilde{b}^h, \widetilde{c}, \widetilde{K})$ given by

$$\widetilde{b}^h(t) = \beta(t), \quad \widetilde{c}(t) = \sigma(t)^2, \quad \widetilde{K}(t, B) = \int 1_B(\delta(t, x)) K(t, dx)$$

for $t \geq 0$ and $B \in \mathscr{B}$ not containing 0.

Proof For ease of exposition we suppose that $\mu = \mu^Y$ for some semimartingale Y. Since the continuous martingale part is $\sigma \bullet W$, we have that $\langle X^c, X^c \rangle = \sigma^2 \bullet \langle W, W \rangle = \sigma^2 \bullet I$ and hence $\widetilde{c} = \sigma^2$ as desired. The statement for the jump characteristic follows from

$$\Delta X(t) = h(\delta(t, \Delta Y(t))) + \overline{h}(\delta(t, \Delta Y(t))) = \delta(t, \Delta Y(t))$$

and Proposition 3.37(6). Since

$$\begin{aligned} X^h(t) &= X(t) - \sum_{s \leq t} \overline{h}(\Delta X(s)) \\ &= X(t) - (\overline{h} \circ \delta) * \mu^Y(t) \\ &= \beta \bullet I + \sigma \bullet W + (h \circ \delta) * (\mu^Y - \nu^Y)(t), \end{aligned}$$

we have $A^{X^h} = \beta \bullet I$, which yields the statement on $\widetilde{b}^h$. □

Itō's formula can be used to compute the characteristics of a smooth function of a semimartingale. The corresponding proposition generalises the chain rule (1.133) to the random case.

Proposition 4.13 **(C^2-Function)** *Let X be an $\mathbb{R}^d$-valued semimartingale with local characteristics (b^h, c, K). Suppose that $f : U \to \mathbb{R}^n$ is twice continuously differentiable on some open subset $U \subset \mathbb{R}^d$ such that X, X_- are U-valued. Then the $\mathbb{R}^n$-valued semimartingale $f(X)$ has local characteristics $(\widetilde{b^{\widetilde{h}}}, \widetilde{c}, \widetilde{K})$ (relative to some truncation function $\widetilde{h}$ on $\mathbb{R}^n$), where*

$$\widetilde{b}_i^{\widetilde{h}}(t) = \sum_{k=1}^{d} D_k f_i(X(t-))b_k^h(t) + \frac{1}{2}\sum_{k,l=1}^{d} D_{kl} f_i(X(t-))c_{kl}(t)$$
$$+ \int \Bigg(\widetilde{h}_i\Big(f(X(t-)+x) - f(X(t-))\Big)$$
$$- \sum_{k=1}^{d} D_k f_i(X(t-))h_k(x)\Bigg)K(t, dx),$$

$$\widetilde{c}_{ij}(t) = \sum_{k,l=1}^{d} D_k f_i(X(t-))c_{kl}(t)D_l f_j(X(t-)),$$

$$\widetilde{K}(t, B) = \int 1_B\big(f(X(t-)+x) - f(X(t-))\big)K(t, dx)$$

for $t \geq 0$ and $B \in \mathscr{B}^n$ not containing 0. If both X and $f(X)$ are special, we have

$$a_i^{f(X)}(t) = \sum_{k=1}^{d} D_k f_i(X(t-))a_k^X(t) + \frac{1}{2}\sum_{k,l=1}^{d} D_{kl} f_i(X(t-))c_{kl}(t)$$
$$+ \int \left(f_i(X(t-)+x) - f_i(X(t-)) - \sum_{k=1}^{d} D_k f_i(X(t-))x_k\right)K(t, dx),$$
$$(4.16)$$

where $a^X = b^{\text{id}}$ and $a^{f(X)} = \widetilde{b}^{\text{id}}$ represent the drift rates of $X, f(X)$.

Proof For ease of notation we consider $h = \text{id}$ and $d = 1$. The general statement follows along the same lines, cf. [125, Corollary A.6]. Itō's formula (3.53) yields that $(f(X))^c = f'(X_-) \bullet X^c$ because the other terms are of finite variation or a purely discontinuous local martingale, respectively. This implies

$$\langle f(X)^c, f(X)^c\rangle = f'(X_-)^2 \bullet \langle X^c, X^c\rangle = (f'(X_-)^2 c) \bullet I$$

and hence the statement on $\widetilde{c}$. Also from (3.53) we observe that $A^{f(X)} = a^{f(X)} \bullet I$ for $a^{f(X)}$ as in (4.16). Finally, $\Delta f(X)(t) = \zeta(t, \Delta X(t))$ for $\zeta(t, x) = f(X(t-) + x) - f(X(t-))$ and Proposition 3.37(6) yield the statement on $\widetilde{K}$. □

Another kind of chain rule is obtained if we consider *time changes*. We encountered such transformations of the time scale in the context of Lévy processes in Sect. 2.3. A **finite time change** is a right-continuous, increasing process $(\tau(\vartheta))_{\vartheta \geq 0}$ such that $\tau(\vartheta)$ is a finite stopping time for any ϑ. It often helps to think of the time change as a kind of operational time. For example, $\tau(\vartheta)$ may stand for the number of trades or the trading volume up to calendar time ϑ. A time change naturally gives rise to a time-changed filtration and to time-changed processes. By $\widetilde{\mathscr{F}}_\vartheta := \mathscr{F}_{\tau(\vartheta)}$ we define the **time-changed filtration** $\widetilde{\mathbf{F}} = (\widetilde{\mathscr{F}}_\vartheta)_{\vartheta \geq 0}$. Similarly, a process $(X(t))_{t \geq 0}$ generates a **time-changed process** $(\widetilde{X}(\vartheta))_{\vartheta \geq 0}$ via $\widetilde{X}(\vartheta) := X(\tau(\vartheta))$. $\widetilde{X}$ can be interpreted as the process X expressed on a random time scale. It can be shown that the time-changed version $\widetilde{X}$ of a semimartingale X is a semimartingale relative to the new filtration $\widetilde{\mathbf{F}}$, cf. [152, Corollaire 10.12].

However, its characteristics may change dramatically. Even time-changed Brownian motion may have jumps if the time change jumps, as can be seen in Sects. 2.4.7–2.4.10. To avoid the involved complications we focus here on sufficiently smooth time changes. Specifically, $\tau(\vartheta)$ is supposed to be *absolutely continuous* in the sense that $\tau(\vartheta) = \int_0^\vartheta \dot{\tau}(\varrho)d\varrho$ for $\varrho \geq 0$. In this case we obtain the characteristics of $\widetilde{X}$ by a rule which resembles (1.134).

Proposition 4.14 (Absolutely Continuous Time Change) *Let X be an $\mathbb{R}^d$-valued semimartingale with local characteristics (b^h, c, K). Suppose that $(\tau(\vartheta))_{\vartheta \geq 0}$ is an absolutely continuous finite time change with derivative $\dot{\tau}(\varrho)$. Then the time-changed process $\widetilde{X}$ is a semimartingale relative to the time-changed filtration $\widetilde{\mathbf{F}}$ with local characteristics $(\widetilde{b}^h, \widetilde{c}, \widetilde{K})$ given by*

$$\widetilde{b}^h(\vartheta) = b^h(\tau(\vartheta))\dot{\tau}(\vartheta),$$
$$\widetilde{c}(\vartheta) = c(\tau(\vartheta))\dot{\tau}(\vartheta),$$
$$\widetilde{K}(\vartheta, dx) = K(\tau(\vartheta), dx)\dot{\tau}(\vartheta).$$

Proof By Definition 4.3 and Remark 4.4 it suffices to show that for any $u \in \mathbb{R}^d$

$$\widetilde{M}(\vartheta) := e^{iu^\top \widetilde{X}(\vartheta)} - \int_0^\vartheta e^{iu^\top \widetilde{X}(\varrho-)} \widetilde{\psi}(\varrho, u)d\varrho$$

is a local martingale in ϑ-time for $\widetilde{\psi}(\vartheta, u) = \psi(\tau(\vartheta), u)\dot{\tau}(\vartheta)$ and ψ as in (4.12). This follows from the local martingale property of M in (4.11) and Doob's stopping theorem 2.5. □

Semimartingale characteristics depend on the probability measure because they involve the concept of a local martingale, which itself depends on expectations.

For applications in statistics as well as in finance one needs to know how the characteristics change if the probability measure is replaced with an equivalent one.

Theorem 4.15 (Change of the Probability Measure) *Let X be an $\mathbb{R}^d$-valued semimartingale and $Q \ll P$ with density process $Z = Z(0)\mathscr{E}(N)$.*

1. *Suppose that X has local characteristics (b^h, c, K) and*

$$N = \varphi \bullet X^c + \eta * (\mu^X - \nu^X)$$

for some X^c-integrable process φ and some $(\mu^X - \nu^X)$-integrable function η. Then the local characteristics $(\widetilde{b}^h, \widetilde{c}, \widetilde{K})$ of X relative to Q are given by

$$\widetilde{b}^h(t) = b^h(t) + \varphi(t)^\top c(t) + \int h(x)\eta(t,x)K(t,dx), \tag{4.17}$$

$$\widetilde{c}(t) = c(t),$$

$$\widetilde{K}(t,B) = \int 1_B(x)(1+\eta(t,x))K(t,dx)$$

for $t \geq 0$, $B \in \mathscr{B}^d$.

2. *By b^h we denote the first local characteristic of X. Moreover, we suppose that the $\mathbb{R}^{d+1}$-valued semimartingale (X, N) has local characteristics $(*, c, K)$. (The star stands for the first characteristic, which depends on b^h and on the truncation function on $\mathbb{R}^{d+1}$ and which is not needed in the sequel.) Then the local characteristics $(\widetilde{b}^h, \widetilde{c}, \widetilde{K})$ of X relative to Q are given by*

$$\widetilde{b}^h_i(t) = b^h_i(t) + c_{i,d+1}(t) + \int h_i(x)x_{d+1}K(t,dx), \tag{4.18}$$

$$\widetilde{c}_{ij}(t) = c_{ij}(t),$$

$$\widetilde{K}(t,B) = \int 1_B(x_1,\dots,x_d)(1+x_{d+1})K(t,dx)$$

for $i, j = 1, \dots, d$, $t \geq 0$, $B \in \mathscr{B}^d$.

Proof

1. Define the local martingale M as in (4.11). Itō's formula (3.53) resp. (3.43) yields that

$$e^{iu^\top X(t)} = e^{iu^\top X(0)} + e^{iu^\top X_-} iu \bullet X^c + e^{iu^\top X_-}(e^{iu^\top x} - 1) * (\mu^X - \nu^X) + A,$$

where A is predictable and of finite variation. This implies that

$$\begin{aligned}\langle M, N\rangle(t) &= e^{iu^\top X_-} \bullet \left(\varphi \bullet \langle iu^\top X^c, X^c\rangle + ((e^{iu^\top x} - 1)\eta) * \nu^X\right)(t)\\ &= \int_0^t e^{iu^\top X(s-)}\left(iu^\top c(s)\varphi(s) + \int (e^{iu^\top x} - 1)\eta(s, x)K(s, dx)\right)ds.\end{aligned} \tag{4.19}$$

From Proposition 3.70 we conclude that

$$\begin{aligned}M(t) - \langle M, N\rangle(t) &= e^{iu^\top X(t)} - \int_0^t e^{iu^\top X(s-)}\psi(s, u)ds\\ &\quad - \int_0^t e^{iu^\top X(s-)}\left(iu^\top c(s)\varphi(s) + \int (e^{iu^\top x} - 1)\eta(s, x)K(s, dx)\right)ds\\ &= e^{iu^\top X(t)} - \int_0^t e^{iu^\top X(s-)}\widetilde{\psi}(s, u)ds\end{aligned}$$

is a Q-local martingale with ψ as in (4.12) and accordingly $\widetilde{\psi}$ for the triplet $(\widetilde{b}^h, \widetilde{c}, \widetilde{K})$. By Definition 4.3 this implies that $(\widetilde{b}^h, \widetilde{c}, \widetilde{K})$ are local characteristics of X relative to Q.

2. This follows along the same lines as statement 1, but with

$$\begin{aligned}\langle M, N\rangle(t) &= e^{iu^\top X_-} \bullet \Big(\langle iu^\top X^c, N^c\rangle\\ &\quad + ((e^{iu^\top (x_1,\dots,x_d)} - 1)x_{d+1}) * \nu^{(X,N)}\Big)(t)\\ &= \int_0^t e^{iu^\top X(s-)}\Big(iu^\top c_{\cdot,d+1}(s)\\ &\quad + \int (e^{iu^\top (x_1,\dots,x_d)} - 1)x_{d+1}K(s, dx)\Big)ds\end{aligned}$$

instead of (4.19). □

As usual, a statement for drift rates $a^{X,P} = b^{\text{id}}$ resp. $\widetilde{a}^{X,Q} = \widetilde{b}^{\text{id}}$ relative to P and Q is derived from (4.17, 4.18) for $h(x) = x$.

Often the ordinary rather than the stochastic exponential of the density process is known in closed form. Therefore we also state the following alternative version.

Proposition 4.16 (Change of the Probability Measure) *Let X be an $\mathbb{R}^d$-valued semimartingale. Moreover, let $Q \ll P$ with density process $Z = \exp(N)$ which satisfies*

$$N^c = \varphi \bullet X^c, \quad \Delta N(t) = \eta(t, \Delta X(t))$$

for some X^c-integrable process φ and some predictable function η. If X has local characteristics (b^h, c, K), the local characteristics $(\widetilde{b}^h, \widetilde{c}, \widetilde{K})$ of X relative to Q are given by

$$\begin{aligned}\widetilde{b}^h(t) &= b^h(t) + \varphi(t)^\top c(t) + \int h(x)\big(e^{\eta(t,x)} - 1\big)K(t, dx),\\ \widetilde{c}(t) &= c(t),\\ \widetilde{K}(t, B) &= \int 1_B(x)e^{\eta(t,x)}K(t, dx)\end{aligned}$$

for $t \geq 0$, $B \in \mathscr{B}^d$.

Proof Itō's formula (3.53) resp. (3.43) and Proposition 3.37(6) yield that

$$\begin{aligned}Z &= Z(0) + Z_- \bullet N^c + (Z_- e^x - Z_-) * (\mu^N - \nu^N) + B\\ &= 1 + Z_- \bullet (\varphi \bullet X^c + (e^\eta - 1) * (\mu^X - \nu^X)) + B\end{aligned}$$

for some predictable process B of finite variation. The latter process must actually vanish because Z is a martingale, i.e. $Z = \mathscr{E}(\widetilde{N})$ with $\widetilde{N} = \varphi \bullet X^c + (e^\eta - 1) * (\mu^X - \nu^X)$. Theorem 4.15(1) now yields the claim. □

Example 4.17 As an application of the rules in this section we prove the equivalence 2⇔3 and equations (3.73–3.78) in Theorem 3.49. The multivariate case in Theorem 3.50 follows along the same lines.

If L is a Lévy process with triplet $(b^{L,h}, c^L, K^L)$, Propositions 4.13, 4.8 and a straightforward calculation yield the characteristics of $\widetilde{L} = \mathscr{L}(e^L)$, which are of the form $(b^{\widetilde{L},h}, c^{\widetilde{L}}, K^{\widetilde{L}})$ as in (3.76–3.78). In view of Proposition 4.6, $\widetilde{L}$ is a Lévy process with this triplet.

Conversely, if $\widetilde{L}$ is a Lévy process with triplet $(b^{\widetilde{L},h}, c^{\widetilde{L}}, K^{\widetilde{L}})$, the differential characteristics $(b^{L,h}, c^L, K^L)$ of $L = \log \mathscr{E}(\widetilde{L})$ are again obtained from Propositions 4.8, 4.13 and a straightforward calculation. As before, Proposition 4.6 yields that L is a Lévy process with the desired triplet.

Example 4.18 In this example we reconsider Proposition 3.54 and give an alternative proof. If L is a Lévy process with exponent ψ, Proposition 4.8 yields that the semimartingale $X(t) = \int_0^t f(s)dL(s)$ has local exponent $\widetilde{\psi}(t, u) := \psi(f(t)u)$. Since the latter is deterministic, X is a time-inhomogeneous Lévy process with absolutely continuous characteristics by Proposition 4.7. From that proposition, we also obtain the conditional characteristic function of the form (3.92).

4.3 Canonical Decomposition of a Semimartingale

The Lévy–Itō decomposition allows us to write Lévy processes as a sum of a linear drift, a Brownian motion and one or two purely discontinuous processes in integral form. A generalisation of this representation exists for general semimartingales.

Theorem 4.19 *Let X be an $\mathbb{R}^d$-valued semimartingale with integral characteristics (B^h, C, ν). Then X can be written as*

$$X = X(0) + B^h + X^c + h(x) * (\mu^X - \nu^X) + (x - h(x)) * \mu^X. \tag{4.20}$$

If X is special, we have the simpler decomposition

$$X = X(0) + A^X + X^c + x * (\mu^X - \nu^X) \tag{4.21}$$

in the initial value, the compensator, the continuous and the purely discontinuous martingale part, respectively.

If X is instead of finite variation, we obtain

$$X = X(0) + B^0 + x * \mu^X, \tag{4.22}$$

where B^0 is continuous and $x * \mu^X$ equals the sum of jumps of X.

Proof The statement can be found in [154, Theorem II.2.34 and Corollary II.2.38]. We give the proof here only in our setup of characteristics which are absolutely continuous in time, cf. Assumption 3.34. In this case B^h does not jump, cf. [154, equation II.2.14].

By definition of B^h and X^c, we have $X^h = X(0) + B^h + X^c + M$ for some purely discontinuous local martingale M. Since $\Delta M(t) = \Delta X^h(t) = h(\Delta X(t)) = \Delta(h*(\mu^X - \nu^X))(t)$, we conclude $M = h*(\mu^X - \nu^X)$. Finally, we have $X - X^h(t) = \sum_{s\leq t} \overline{h}(\Delta X(s)) = \overline{h} * \mu^X$ for $\overline{h}(x) := x - h(x)$, which yields (4.20). The second statement follows for $h = \mathrm{id}$, where we secretly use Proposition 4.21 below.

For the third statement observe that $B^0(t) := X(t) - X(0) - \sum_{s\leq t} \Delta X(s) = X(t) - X(0) - x * \mu^X(t)$ is continuous and of finite variation. By Proposition 3.8 we have $X^c = 0$, which means that (4.22) directly corresponds to (4.20) for $h = 0$. □

Decompositions (4.20–4.22) reduce to (2.76–2.78) for Lévy processes.

4.4 Extensions

Some semimartingales do not allow for local characteristics. As a simple example consider discrete-time processes which are embedded in continuous time in the usual way, cf. the end of Sect. 2.1. On the other hand, integral characteristics are defined for any semimartingale. If one allows for absolute continuity relative to more general increasing processes instead of linear time, the concept of local characteristics can be extended to arbitrary semimartingales X. Specifically, one can show that there exist an increasing predictable process A and predictable triplets (b^h, c, K) similar as in Definition 4.3 such that the integral characteristics of X satisfy

$$B^h(t) = \int_0^t b^h(s)dA(s),$$

$$C(t) = \int_0^t c(s)dA(s),$$

$$\nu([0,t] \times dx) := \int_0^t K(s,dx)dA(s),$$

cf. [154, Proposition II.2.9]. Put differently, the process $A(t)$ plays the role of time t in (4.10). These more general characteristics are not unique: instead of (A, b^h, c, K) one can also choose, say, $(\frac{1}{2}A, 2b^h, 2c, 2K)$. The possible choices of A depend on X. Moreover, it is debatable whether (b^h, c, K) should be viewed as local Lévy–Khintchine triplets for general A. For general mathematical reasons, however, it is sometimes helpful to have a tool available which exists for arbitrary semimartingales. The rules in the previous section can be extended to cover the general case. From an engineering point of view, the vast majority of continuous-time processes in applications do possess local characteristics in the sense of Definition 4.3. Hence there is usually no need to deal with their less intuitive generalisation.

4.5 Applications

In Mathematical Finance one often works with two equivalent probability measures at the same time. The *objective measure* concerns "real" probabilities driving the data. By contrast, derivative prices are obtained as expectations under some equivalent *risk-neutral measure*. Let us focus on the case that some asset price process S follows an exponential Lévy process under the objective measure, i.e. $X(t) = \log S(t) - \log S(0)$ is a Lévy process. It is not obvious whether X is also a Lévy process under the risk-neutral measure because the property of stationary, independent increments is generally not invariant under equivalent measure changes. But for computational ease, one may want to limit modelling to pairs of probability measures such that X is a Lévy process under both measures, say with Lévy–Khintchine triplets (b^h, c, K) and $(\widetilde{b}^h, \widetilde{c}, \widetilde{K})$. This in turn raises the question whether a given pair of triplets does in fact correspond to the same process under two equivalent probability measures. The answer is given by the following theorem. For this particular result we consider a finite time interval $[0, T]$ rather than $\mathbb{R}_+$, where T denotes a positive real number. One can identify Lévy processes on $[0, T]$ with Lévy processes on $\mathbb{R}_+$ that are stopped at T.

Theorem 4.20 *Suppose that $(X(t))_{t\in[0,T]}$ is an $\mathbb{R}^d$-valued Lévy process with Lévy–Khintchine triplet (b^h, c, K). Let $(\widetilde{b}^h, \widetilde{c}, \widetilde{K})$ denote another Lévy–Khintchine triplet. There exists some probability measure $Q \ll P$ such that X is a Lévy process with triplet $(\widetilde{b}^h, \widetilde{c}, \widetilde{K})$ under Q if and only if there are $\beta \in \mathbb{R}^d$ and a measurable function $\kappa : \mathbb{R}^d \to \mathbb{R}_+$ satisfying*

1. $\frac{d\widetilde{K}}{dK} = \kappa$,
2. $\widetilde{b}^h = b^h + c\beta + \int h(x)(\kappa(x) - 1)K(dx)$,
3. $\widetilde{c} = c$,
4. $\int \left(1 - \sqrt{\kappa(x)}\right)^2 K(dx) < \infty$.

We have $Q \sim P$ if and only if κ above can be chosen strictly positive.

Proof Below we consider only the case $Q \sim P$ resp. $\kappa > 0$. A full proof can be found in [154, Theorem IV.4.39] with the additional assumption

$$\int |h(x)(1 - \kappa(x))|K(dx) < \infty. \tag{4.23}$$

However, the latter is implied by condition 4. Indeed, for $\varrho(x) := 1 - \sqrt{\kappa(x)}$ condition 4 reads as $\int \varrho(x)^2 K(dx) < \infty$. Moreover, $\int |h(x)|^2 K(dx) < \infty$ holds because K is a Lévy measure. Since

$$\int |h(x)(1 - \kappa(x))|K(dx) = \int |h(x)||2\varrho(x) - \varrho(x)^2|K(dx),$$

we have

$$\int |h(x)(1 - \kappa(x))|K(dx) \le \int h(x)^2 K(dx) + \int (1 + |h(x)|)\varrho(x)^2 K(dx)$$

because $|2ab| \le a^2 + b^2$. Since h is bounded, (4.23) follows.

$\Leftarrow$: Condition 4 and $\kappa > 0$ warrant that $N = \beta^\top X^c + (\kappa - 1) * (\mu^X - \nu^X)$ is a well-defined local martingale with $\Delta N > -1$. Since it is a Lévy process and a local martingale, $Z = \mathscr{E}(N)$ is a martingale by Theorems 3.49 and 2.22, whence it is the density process of some probability measure $Q \sim P$. Theorem 4.15 yields that the characteristics of X relative to Q are given by $(\widetilde{b}^h, \widetilde{c}, \widetilde{K})$. Since they are deterministic and constant, X is a Q-Lévy process with this triplet, cf. Proposition 4.6.

$\Rightarrow$: W.l.o.g. we may assume that the filtration is generated by X. Denote by $Z > 0$ the density process of Q relative to P and set $N := \mathscr{L}(Z)$. The martingale representation theorem 3.62 yields that it is of the form $N = \beta \bullet X^c + (\kappa - 1) * (\mu^X - \nu^X)$ for some integrands β and κ. The positivity of Z implies $\Delta N > -1$ and hence $\kappa > 0$. Theorem 4.15 shows that the characteristics of X relative to

Q are given by $(\widetilde{b}^h, \widetilde{c}, \widetilde{K})$, which implies that β, κ are or can at least be chosen to be deterministic and constant in time. This yields conditions 1–3. The last one follows from the integrability of $\kappa - 1$ relative to $\mu^X - \nu^X$, cf. [154, Theorem II.1.33d]. □

Note that the triplet may change arbitrarily in some respects (e.g. from b^h to $\widetilde{b}^h$) but not at all in others (from c to $\widetilde{c}$). This is explained by the different nature of b^h, c, K. If a coefficient involves expectations (such as b^h), it may change because expectations depend on the underlying probability measure. If, on the other hand, it refers to a path property such as continuity, finite variation etc., it cannot change because path properties can be read off each single realisation. The diffusion coefficient c is fixed because it refers to the quadratic variation of the continuous part, which is a path property. The drift coefficient b^h may change arbitrarily if c is non-degenerate. For $c = 0$ it is determined by the density of $\widetilde{K}$ relative to K. This is easy to understand for compound Poisson processes, in which case we have drift coefficient $b^0 = 0$ for $h(x) = 0$ and hence $b^h = \int h(x)K(dx)$ for general truncation functions h. Condition $b^0 = 0$ means that the process remains constant between jumps, which holds under Q as well. Consequently, we have $\widetilde{b}^0 = 0$ for $h(x) = 0$ and hence

$$\widetilde{b}^h = \int h(x)\widetilde{K}(dx) = \int h(x)\kappa(x)K(dx) = b^h + \int h(x)(\kappa(x) - 1)K(dx)$$

for arbitrary h. This illustrates Theorem 4.20(2) in the compound Poisson case.

The Lévy measure may change because it involves the expected number of jumps at certain times. But impossible jump sizes under P remain impossible under Q. Similarly as in the Radon–Nikodym theorem this leads to the existence of a density of $\widetilde{K}$ relative to K. If X has infinitely many jumps, the Lévy measure has a singularity at 0. In this case, the asymptotics of $K(dx)$ for small x can be read off each single path. By invariance of path properties under measure changes, K and $\widetilde{K}$ must behave similarly near 0. This loose statement is made precise in condition 4 above.

Some objects and properties in stochastic calculus can be expressed in terms of characteristics. We start with special semimartingales.

Proposition 4.21 *A semimartingale X is special if and only if x is integrable with respect to $\mu^X - \nu^X$ or, equivalently, if*

$$\int_0^t \int |x - h(x)|K(s, dx)ds < \infty, \quad t \geq 0, \tag{4.24}$$

where K denotes the third component of the local characteristics (b^h, c, K) of X.

Proof Since X^h in (4.6) is special, X is special if and only if this is the case for $(x - h(x)) * \mu^X$. By Proposition 3.36(1) this holds if and only if we have (4.24).

Equation (4.24) in turn implies $x - h(x) \in G_{\text{loc}}(\mu^X)$ by Proposition 3.37(1). Since $h(x) \in G_{\text{loc}}(\mu^X)$ by Remark 3.38, we obtain that $x \in G_{\text{loc}}(\mu^X)$. If, on the other hand, $x \in G_{\text{loc}}(\mu^X)$, the process $X - x * (\mu^X - \nu^X)$ is a semimartingale with jumps $\Delta X - \Delta X = 0$. By Proposition 2.14 it is special, which yields that X is special as well. □

We have already observed that the compensator and the drift rate of a special semimartingale are related to the integral and local characteristics (B^h, C, ν), (b^h, c, K) via $A^X = B^{\text{id}}$, $a^X = b^{\text{id}}$ if we choose the identity as the "truncation" function. If B^h, b^h refer to a true truncation function h, we have instead

$$\begin{aligned} A^X &= B^h + (x - h(x)) * \nu, \\ a^X &= b^h + \int (x - h(x))K(t, dx) \end{aligned} \tag{4.25}$$

by (4.5, 4.13).

We know that a special semimartingale is a local martingale if its compensator vanishes, i.e. if

$$a^X(t) = b^h(t) + \int (x - h(x))K(t, dx) = 0, \quad t \geq 0.$$

Even for some non-special semimartingales we may define a "drift rate" via (4.25). In order for a^X to be defined, we only need

$$\int |x - h(x)|K(t, dx) < \infty, \quad t \geq 0 \tag{4.26}$$

but not the slightly stronger integrability condition (4.24). A vanishing drift rate in this generalised sense corresponds to a σ-martingale. Related notions such as local martingales, martingales etc. differ from these only in their degree of integrability.

Proposition 4.22 *Let X be a semimartingale with local characteristics (b^h, c, K) such that (4.26) holds.*

1. *X is a σ-martingale if and only if $a^X = 0$ for the drift rate from (4.25).*
2. *X is a local martingale if and only if $E(|X(0)|) < \infty$ and if X is a σ-martingale as well as a special semimartingale, i.e. if its drift rate vanishes and if it satisfies (4.24).*
3. *X is a locally square-integrable martingale if and only if it is a σ-martingale with $E(X(0)^2) < \infty$ and*

$$\int_0^t \int x^2 K(s, dx)ds < \infty, \quad t \geq 0.$$

4. *X is a square-integrable martingale if and only if it is a σ-martingale satisfying* $E(X(0)^2) < \infty$ *and*

$$E\Big(\int_0^\infty \Big(c(t) + \int x^2 K(t, dx)\Big)dt\Big) < \infty. \tag{4.27}$$

Proof This is shown in resp. follows from [168, Lemma 3.1 and Corollary 3.1] and [154, Proposition II.2.29 and Theorems I.4.2, I.4.40b]. □

For the sake of completeness, we briefly discuss the corresponding statement for martingales. To this end, recall that a family of random variables $(X(i))_{i\in I}$ is called **uniformly integrable** if

$$\lim_{n\to\infty} \sup_{i\in I} E\big(|X(i)| 1_{\{|X(i)|>n\}}\big) = 0.$$

This property implies $\sup_{i\in I} E(|X(i)|) < \infty$ but is itself implied by the stronger requirement $E(\sup_{i\in I} |X(i)|) < \infty$. A process $(X(t))_{t\geq 0}$ belongs to **class (D)** if the family $\{X(\tau) : \tau \text{ finite stopping time}\}$ is uniformly integrable. Under the weaker requirement that the stopped process X^t is of class (D) for any fixed time $t \in \mathbb{R}_+$, one says that X is of **class (LD)**. With these notions one can complement Proposition 4.22 as follows.

5. *X is a martingale if and only if it is a σ-martingale of class (LD).*
6. *X is a* **uniformly integrable martingale** *(i.e. a martingale such that* $(X(t))_{t\geq 0}$ *is uniformly integrable) if and only if it is a σ-martingale of class (D).*

Proof This is shown in [168, Corollary 3.1]. □

Let us note in passing that the family of uniformly integrable martingales coincides with the class of martingales that are generated in the sense of (2.1) by some integrable random variable, cf. [154, Theorem I.1.42]. We will hardly need the previous characterisations because of the involved definition of class (D). Nevertheless, they immediately yield that any bounded σ-martingale is a uniformly integrable martingale.

In the context of stochastic control, we need the more precise refinement of Propositions 4.22(1) and 3.27. It specifies the intuitive statement that supermartingales (resp. submartingales) are characterised by nonnegative (resp. nonpositive) drift rate a^X.

Proposition 4.23 *Let X be a semimartingale with local characteristics* (b^h, c, K) *such that (4.26) holds. Then X is a σ-supermartingale if and only if* $a^X \leq 0$ *for the drift rate from (4.25). Moreover, a σ-supermartingale X is a supermartingale if its negative part* X^- *is of class (LD).*

Proof If X is a σ-martingale with σ-localising sequence $(D_n)_{n\in\mathbb{N}}$, we have $1_{D_n} a^X = a^{(X^{D_n})} \leq 0$ and hence $a^X = 0$, cf. Proposition 4.8. For the converse statement let $D_n := \{\int |x - h(x)| K(\cdot, dx) < n\}$, $n \in \mathbb{N}$. Condition (4.26) implies

$D_n \uparrow \Omega \times \mathbb{R}_+$. Moreover, X^{D_n} is special with $a^{(X^{D_n})} = 1_{D_n} a^X \leq 0$, which implies that it is a local supermartingale. Hence X is a σ-supermartingale.

For the second statement in Proposition 4.23 let X be σ-supermartingale whose negative part is of class (LD).

Step 1: Suppose first that X is a local supermartingale with localising sequence $(\tau_n)_n$. Since $X^{\tau_n}(t)^- \to X(t)^-$ as $n \to \infty$ and X^- is of class (LD), we have $\lim_{n\to\infty} E(X^{\tau_n}(t)^-|\mathscr{F}_s) = E(X(t)^-|\mathscr{F}_s)$ as $n \to \infty$ and $s \leq t$. Similarly, Fatou's lemma yields $\liminf_{n\to\infty} E(X^{\tau_n}(t)^+|\mathscr{F}_s) \geq E(X(t)^+|\mathscr{F}_s)$ as $n \to \infty$ and $s \leq t$. Using the supermartingale property of X^{τ_n} we obtain

$$
\begin{aligned}
E(X(t)|\mathscr{F}_s) &= E(X(t)^+|\mathscr{F}_s) - E(X(t)^-|\mathscr{F}_s) \\
&\leq \liminf_{n\to\infty} E\big(X^{\tau_n}(t)^+\big|\mathscr{F}_s\big) - \lim_{n\to\infty} E\big(X^{\tau_n}(t)^-\big|\mathscr{F}_s\big) \\
&= \liminf_{n\to\infty} E\big(X^{\tau_n}(t)\big|\mathscr{F}_s\big) \\
&\leq \liminf_{n\to\infty} X^{\tau_n}(s) = X(s)
\end{aligned}
$$

for $s \leq t$.

Step 2: Now let X be a σ-supermartingale. Set

$$
A(t) := x^- 1_{\{|x|>1\}} * \mu^X(t) = \sum_{s\leq t} \Delta X(s)^- 1_{\{|\Delta X(s)|>1\}}.
$$

For $\tau_n := \{t \geq 0 : A(t) \geq n\}$ we observe

$$
A^{\tau_n}(t) \leq n + (\Delta X(\tau_n \wedge t))^- \leq 2n + (X^-(\tau_n \wedge t)).
$$

The right-hand side has finite expectation because X^- is of class (LD). By [154, Proposition I.4.23] this implies that $x^- 1_{\{|x|>1\}} * \mu^X(t)$ is a special semimartingale. This in turn yields that $\int_0^t \int x^- 1_{\{|x|>1\}} K(s, dx) ds < \infty, t \geq 0$ by Proposition 3.36(1). The claim follows now as in the proof of [168, Proposition 3.1]. □

For processes of stochastic exponential type $\mathscr{E}(X)$ neither class (D) nor the sufficient criterion (4.27) may be easy to verify. However, these processes play an important role in applications, e.g. in stochastic control or as prospective density processes. Therefore the issue of deriving sufficient conditions for uniform integrability has received a lot of attention in the literature. A famous criterion for processes of exponential type is *Novikov's condition*, which has been generalised to processes with jumps by Lepingle and Mémin. It is far from necessary but quite useful because it depends only on the characteristics of X.

Theorem 4.24 (Novikov–Lepingle–Mémin Condition) *Suppose that X is a σ-martingale with $\Delta X \geq -1$ and*

$$E\left(\exp\left(\int_0^\tau \left(\frac{1}{2}c(t) + \int ((1+x)\log(1+x) - x)K(t,dx)\right)dt\right)\right) < \infty,$$

where $\tau := \inf\{t \geq 0 : \Delta X(t) = -1\}$, the triplet (b^h, c, K) denotes the local characteristics of X, and $0 \log 0 := 0$. Then $\mathscr{E}(X)$ is a uniformly integrable martingale.

Proof $\mathscr{E}(X)$ is a non-negative σ-martingale and hence a local martingale by Propositions 4.8, 4.22(1), 4.23, 4.22(2). The statement follows now from [207, Theorem IV.3]. □

In Sect. 3.2 we considered stochastic integration in various degrees of generality, leading to classes $L^2(X)$, $L^2_{\mathrm{loc}}(X)$ and $L(X)$. These families of X-integrable processes can be characterised in terms of the local characteristics of X.

Theorem 4.25 *Let X be a semimartingale with local characteristics (b^h, c, K) and φ a predictable process.*

1. *If X is a square-integrable martingale, φ belongs to $L^2(X)$ if and only if*

$$E\left(\int_0^\infty \varphi(t)^2\left(c(t) + \int x^2 K(t,dx)\right)dt\right) < \infty.$$

2. *If X is a locally square-integrable martingale, φ belongs to $L^2_{\mathrm{loc}}(X)$ if and only if*

$$\int_0^t \varphi(s)^2\left(c(s) + \int x^2 K(s,dx)\right)ds < \infty, \quad t \geq 0.$$

3. *X belongs to $L(X)$ if and only if*

$$\int_0^t \left(\varphi(s)b^h(s) + \varphi(s)^2\left(c(s) + \int x^2 1_{\{|x|\leq 1, |\varphi(s)x|\leq 1\}} K(s,dx)\right)\right)ds < \infty$$

for all $t \geq 0$. If both X and φ are $\mathbb{R}^d$-valued, we have accordingly $\varphi \in L(X)$ if and only if

$$\int_0^t \left(\left|\varphi(s)^\top b^h(s) + \int \left(\varphi(s)^\top x 1_{\{|x|\leq 1, |\varphi(s)^\top x|\leq 1\}} - \varphi(s)^\top h(x)\right)K(s,dx)\right|\right.$$
$$\left. + \varphi(s)^\top c(s)\varphi(s) + \int (1 \wedge |\varphi(s)^\top x|^2) 1_{\{|x|\leq 1\}} K(s,dx)\right)ds < \infty, \quad t \geq 0.$$

Proof The first two statements follow from [154, Theorem II.2.29] or Theorem 4.26 below, the third from [154, Theorem III.6.30]. □

We turn now to the predictable quadratic variation of a semimartingale.

Theorem 4.26 *Let X be a semimartingale with local characteristics (b^h, c, K). Its quadratic variation $[X, X]$ is special if and only if*

$$\int_0^t \int x^2 K(s, dx)ds < \infty, \quad t \geq 0.$$

In this case the predictable quadratic variation of X is given by

$$\langle X, X \rangle(t) = \int_0^t \left(c(s) + \int x^2 K(s, dx) \right) ds.$$

If X is instead $\mathbb{R}^d$-valued, we have that $[X_i, X_j]$ is special if and only if

$$\int_0^t \int |x_i x_j| K(s, dx)ds < \infty, \quad t \geq 0. \tag{4.28}$$

In this case the predictable covariation of components X_i, X_j is given by

$$\langle X_i, X_j \rangle(t) = \int_0^t \left(c_{ij}(s) + \int x_i x_j K(s, dx) \right) ds. \tag{4.29}$$

Proof Recall from (3.45) that

$$[X_i, X_j] = \langle X_i^c, X_j^c \rangle + (x_i x_j) * \mu^X = c_{ij} \bullet I + (x_i x_j) * \mu^X, \tag{4.30}$$

which is special if and only if $(x_i x_j) * \mu^X$ is special. By Proposition 3.36(1) this holds if and only if (4.28) is satisfied. Moreover, the compensator of (4.30) is

$$\langle X_i, X_j \rangle = c_{ij} \bullet I + (x_i x_j) * \nu^X,$$

as desired. □

In contrast to the quadratic variation $[X, X]$, the predictable quadratic variation depends on the probability measure because K does, cf. Theorem 4.15.

For an $\mathbb{R}^d$-valued semimartingale X with local characteristics (b^h, c, K), we call the $\mathbb{R}^{d \times d}$-valued predictable process

$$\widetilde{c} := c + \int x x^\top K(dx) \tag{4.31}$$

the **modified second characteristic** if it exists, i.e. if $\int_0^t \int |x|^2 K(dx)dt < \infty, t \geq 0$. Consequently, we have

$$\langle X_i, X_j \rangle = \widetilde{c}_{ij} \bullet I, \quad i, j = 1, \ldots, d.$$

Observe that the values of $\widetilde{c}$ are nonnegative symmetric matrices because this holds for c.

As an application of modified characteristics we consider the following result.

Proposition 4.27 *Let N denote a local martingale with jumps $\Delta N > -1$ or, more generally, $\neq -1$. Moreover, let S be a special semimartingale such that NS is special as well and (N, S) allows for local characteristics. Then $S\mathscr{E}(N)$ is a local martingale if and only if $a^S + \widetilde{c}^{N,S} = 0$, where we use the notation $\langle N, S \rangle = \widetilde{c}^{N,S} \bullet I$.*

Proof Integration by parts yields

$$\begin{aligned} S\mathscr{E}(N) &= S(0) + S_- \bullet \mathscr{E}(N) + \mathscr{E}(N)_- \bullet S + [S, \mathscr{E}(N)] \\ &= S(0) + \mathscr{E}(N)_- \bullet (S_- \bullet N + S + [S, N]), \end{aligned}$$

which implies that its drift rate equals

$$a^{S\mathscr{E}(N)} = \mathscr{E}(N)_- \big(a^S + \widetilde{c}^{N,S}\big).$$

By Proposition 4.22(2) the claim follows. □

Many theorems and arguments rely on integrability conditions. In order to verify such requirements, the general theory of stochastic processes provides a large number of inequalities such as Doob's inequality, which we encountered in Theorem 2.8. In this text we confine ourselves to mentioning just two others, which involve the following S^p- and H^p-norms.

For $p \in [1, \infty]$ the possibly infinite S^p**-norm** of any measurable process X is defined as

$$\|X\|_{S^p} := \big\| \sup\{|X(t)| : t \in \mathbb{R}_+\} \big\|_{L^p},$$

where L^p denotes the usual L^p-norm for random variables. Moreover, the possibly infinite H^p**-norm** of any semimartingale X is defined as

$$\begin{aligned} \|X\|_{H^p} := \inf\Big\{ &\Big\| |X(0)| + \sqrt{[M, M](\infty)} + \mathrm{Var}(A)(\infty) \Big\|_{L^p} : \\ &X = X(0) + M + A, M \text{ local martingale}, A \text{ of finite variation} \Big\}, \end{aligned}$$

where $\mathrm{Var}(A)$ denotes the total variation process of A as in (2.4).

The S^p-norm is dominated by the H^p-norm in the following sense.

Theorem 4.28 *Let $p \in [1, \infty)$. There is a constant $c_p < \infty$ such that*

$$\|X\|_{S^p} \leq c_p \|X\|_{H^p}$$

for any semimartingale X.

Proof See [238, Theorem V.2]. □

This inequality is related to the *Burkholder–Davis–Gundy inequalities*, cf. e.g. [241, Section IV.4]. Moreover, Doob's inequality (2.2) together with $E(X(t)^2) = E([X,X](t))$ tells us that we can choose $c_2 = 2$ if X is a square-integrable martingale which starts at 0.

The H^p-norm proves to be particularly handy for estimating stochastic integrals:

Theorem 4.29 (Emery's Inequality) *Let $p, q, r \in [1, \infty]$ with $\frac{1}{p} + \frac{1}{q} = \frac{1}{r}$ (where we set $\frac{1}{\infty} := 0$). For any semimartingale X and any X-integrable process φ we have*

$$\|\varphi \bullet X\|_{H^r} \leq \|\varphi\|_{S^p} \|X - X(0)\|_{H^q}.$$

Proof See [238, Theorem V.3], where left-continuity of φ is assumed but not needed in the proof. □

It does not seem obvious how to compute or bound the H^p-norm in concrete applications because it involves the quadratic variation which may not be known well enough for processes with jumps. For $p = 2$ and $p = 1$, however, we can find upper bounds which depend only on the local characteristics of X.

Proposition 4.30 *Let X be a special semimartingale with differential characteristics (b^X, c, K), modified second characteristic $\widetilde{c}$, and growth rate $a := a^X$. Then we have*

$$\|X\|_{H^2} \leq \sqrt{E\left(\int_0^\infty \widetilde{c}(t)dt\right)} + \sqrt{E\left(\left(|X(0)| + \int_0^\infty |a(t)|dt\right)^2\right)} \tag{4.32}$$

and

$$\|X\|_{H^1} \leq \sqrt{E\left(\int_0^\infty \left(c(t) + 2\int h(x)^2 K(t,dx)\right)dt\right)}$$
$$+ \sqrt{2}E\left(\int_0^\infty \int |x - h(x)|K(t,dx)dt + |X(0)| + \int_0^\infty |a(t)|dt\right)$$

for any measurable function $h : \mathbb{R} \to \mathbb{R}$, e.g. $h(x) = x1_{\{|x|\leq 1\}}$ or $h(x) = 0$.

Proof In both cases we consider the canonical decomposition $X = X(0) + M + A$ with predictable A. Since $E([M,M](\infty)) = E(\langle M, M\rangle(\infty)) = E(\widetilde{c} \bullet I(\infty))$ and $\mathrm{Var}(A)(\infty) = |a| \bullet I(\infty)$, we obtain (4.32).

For the second estimate note that

$$\begin{aligned}[M,M](\infty) &= [M^c,M^c](\infty) + x^2 * \mu^X(\infty)\\ &\leq [M^c,M^c](\infty) + 2h(x)^2 * \mu^X(\infty) + 2(x-h(x))^2 * \mu^X(\infty)\\ &= [M^c,M^c](\infty) + 2h(x)^2 * \mu^X(\infty) + 2\sum_{s\leq t}(\Delta X(s) - h(\Delta X(s)))^2\end{aligned}$$

and hence

$$\begin{aligned}\sqrt{[M,M](\infty)} &\leq \sqrt{[M^c,M^c](\infty) + 2h(x)^2 * \mu^X(\infty)} + \sum_{t\geq 0}\sqrt{2(\Delta X(t) - h(\Delta X(t)))^2}\\ &= \sqrt{[M^c,M^c](\infty) + 2h(x)^2 * \mu^X(\infty)} + \sqrt{2}\sum_{t\geq 0}|\Delta X(t) - h(\Delta X(t))|\\ &= \sqrt{[M^c,M^c](\infty) + 2h(x)^2 * \mu^X(\infty)} + \sqrt{2}|x - h(x)| * \mu^X(\infty).\end{aligned}$$

Since the L^1-norm is dominated by the L^2-norm, we obtain

$$\begin{aligned}&\left\|\sqrt{[M^c,M^c](\infty) + 2h(x)^2 * \mu^X(\infty)}\right\|_{L^1}\\ &\quad\leq \left\|\sqrt{[M^c,M^c](\infty) + 2h(x)^2 * \mu^X(\infty)}\right\|_{L^2}\\ &\quad= \sqrt{E\big([M^c,M^c](\infty) + 2h(x)^2 * \mu^X(\infty)\big)}\\ &\quad= \sqrt{E\big([M^c,M^c](\infty) + 2h(x)^2 * \nu^X(\infty)\big)}\\ &\quad= \sqrt{E\left(\int_0^\infty c(s)ds + 2\int_0^\infty\int h(x)^2 K(t,dx)dt\right)}.\end{aligned}$$

Collecting terms yields the second statement. □

4.6 Martingale Problems

We turn now to the random counterpart of ODEs in the form (1.135). As in the deterministic case we want to express the local dynamics of X as a function of X itself, but now *local dynamics* refers to local characteristics rather than just the derivative.

Suppose that P^0 is a distribution on $\mathbb{R}^d$ and that measurable functions $\beta^h : \mathbb{R}_+ \times \mathbb{R}^d \to \mathbb{R}^d$, $\gamma : \mathbb{R}_+ \times \mathbb{R}^d \to \mathbb{R}^{d\times d}$ and a kernel $\kappa : \mathbb{R}_+ \times \mathbb{R}^d \times \mathscr{B}^d \to \mathbb{R}_+$

satisfying $\int(1 \wedge |y|^2)\kappa(t,x,dy) < \infty$, $(t,x) \in \mathbb{R}_+ \times \mathbb{R}^d$ are given. We call X a **solution to the martingale problem** related to P^0 and $(\beta^h, \gamma, \kappa)$ if X is an $\mathbb{R}^d$-valued semimartingale on some filtered space $(\Omega, \mathscr{F}, \mathbf{F}, P)$ such that the law of $X(0)$ is P^0 and X has local characteristics (b^h, c, K) given by

$$b^h(t) = \beta^h(t, X(t-)), \tag{4.33}$$

$$c(t) = \gamma(t, X(t-)), \tag{4.34}$$

$$K(t, dx) = \kappa(t, X(t-), dx). \tag{4.35}$$

The name *martingale problem* can be motivated from the following characterisations in terms of martingales.

Proposition 4.31 *X is a solution to the martingale problem related to P^0 and $(\beta^h, \gamma, \kappa)$ if and only if the law of $X(0)$ is P^0 and one of the following equivalent conditions holds:*

1.

$$M(t) = e^{iu^\top X(t)} - \int_0^t e^{iu^\top X(s-)}\psi(s, X(s-), u)ds$$

is a local martingale for any $u \in \mathbb{R}^d$, where

$$\psi(t,x,u) := iu^\top \beta^h(t,x) - \frac{1}{2}u^\top \gamma(t,x)u + \int \left(e^{iu^\top y} - 1 - iu^\top h(y)\right)\kappa(t,x,dy)$$

denotes the characteristic exponent of $(\beta^h, \gamma, \kappa)(t,x)$.

2.

$$M(t) := f(X(t)) - \int_0^t \widetilde{G}_s f(X(s-))ds$$

is a local martingale for any infinitely often differentiable function $f : \mathbb{R}^d \to \mathbb{R}$ with compact support, where

$$\widetilde{G}_t f(x) := \beta^h(t,x)^\top Df(x) + \frac{1}{2}\sum_{i,j=1}^d \gamma_{ij}(t,x)D_{ij}f(x) + \int \left(f(x+y) - f(x) - h(y)^\top Df(x)\right)\kappa(t,x,dy).$$

3.

$$M(t) := f(t, X(t)) - \int_0^t \left(D_1 f(s, X(s-)) + \widetilde{G}_s f(s, X(s-))\right) ds$$

is a local martingale for any bounded twice continuously differentiable function $f : \mathbb{R}^{1+d} \to \mathbb{R}$, *where* $\widetilde{G}_t f(t, \cdot)$ *here refers to* $\widetilde{G}_t$ *as in 2 applied to* $f(t, \cdot)$ *and* $D_1 f(t, x)$ *indicates the derivative of* $t \mapsto f(t, x)$.

Proof If X is a solution to the martingale problem, statement 3 follows from Itō's formula (3.53) or, more generally, (3.43), applied to the process (I, X). To this end, note that $\Delta(I, X)(t) = (0, \Delta X(t))$ and hence $K^{(I,X)}(t, B) = \int 1_B(0, x) K^X(t, dx)$ for $B \in \mathscr{B}^d$ not containing 0 by Proposition 3.37(6).

Statement 3 trivially implies statement 2. If statement 1 holds, X solves the martingale problem by the definition of local characteristics.

It remains to show that statement 2 implies statement 1. To this end it suffices to verify that statement 2 can be extended to bounded infinitely differentiable functions $f : \mathbb{R}^d \to \mathbb{R}$ without compact support. Indeed, statement 1 then follows from application to $x \mapsto e^{iu^\top x}$ or, more precisely, its real and imaginary parts. Writing $K_n := \{x \in \mathbb{R}^d : |x| \leq n\}$, choose infinitely differentiable functions $f_n : \mathbb{R}^d \to \mathbb{R}$ with compact support which coincide on K_{2n} with f. Define stopping times $\tau_n := \inf\{t \geq 0 : X(t) \notin K_n \text{ or } |\Delta X(t)| > n\}$. Then

$$
\begin{aligned}
M^{\tau_n}(t) &= f(X^{\tau_n}(t)) - \int_0^{\tau_n \wedge t} \widetilde{G}_s f(X(s-))ds \\
&= f_n(X^{\tau_n}(t)) - \int_0^{\tau_n \wedge t} \widetilde{G}_s f_n(X(s-))ds \\
&\quad + \Big(\big(f(X_- + x) - f_n(X_- + x)\big)1_{\{|x|>n\}} * \mu^X\Big)^{\tau_n}(t) \\
&\quad - \Big(\big(f(X_- + x) - f_n(X_- + x)\big)1_{\{|x|>n\}} * \nu^X\Big)^{\tau_n}(t) \\
&= f_n(X^{\tau_n}(t)) - \int_0^{\tau_n \wedge t} \widetilde{G}_s f_n(X(s-))ds \\
&\quad + \Big(\big(f(X_- + x) - f_n(X_- + x)\big)1_{\{|x|>n\}} * \big(\mu^X - \nu^X\big)\Big)^{\tau_n}(t)
\end{aligned}
$$

is a local martingale for any $n \in \mathbb{N}$. Therefore M is a local martingale as desired. □

More in line with the common language on martingale problems, one may also call the distribution P^X of X a **solution** to the martingale problem. Accordingly, we say that **uniqueness holds for the martingale problem** if any two solutions X, Y have the same law. We do not discuss existence and uniqueness of solutions to martingale problems here but refer instead to Sect. 5.7 for general results and to Chap. 6 for a particularly important class of martingale problems.

4.7 Limit Theorems

We turn now to an important application of characteristics which is not discussed in Sect. 4.5. In probability theory one often studies the limiting behaviour of a sequence of random variables or, more generally, stochastic processes. One may think of the central limit theorem and Donsker's invariance principle as prime examples.

Lévy's continuity theorem in Proposition A.5(8) provides a powerful tool for showing convergence in law of a sequence of real- or vector-valued random variables to some limiting random variable. The situation seems more involved for sequences of stochastic processes $X^{(n)}(t)$. In principle, one could prove convergence of the finite-dimensional distributions $\mathscr{L}(X^{(n)}(t_1), \dots, X^{(n)}(t_k))$ towards the law $\mathscr{L}(X(t_1), \dots, X(t_k))$ of some limiting process X by considering the corresponding characteristic functions and applying Lévy's theorem. However, in concrete applications the law and hence the characteristic functions of $(X^{(n)}(t_1), \dots, X^{(n)}(t_k))$ and $(X(t_1), \dots, X(t_k))$ often fail to be known explicitly.

As a way out, one may study the local dynamics of the processes. Intuitively, it seems natural that processes with similar local characteristics should have a similar law. This vague idea can in fact be made precise: the monograph [154] is devoted to this approach. In fact, the application to limit theorems served as a main motivation to study semimartingale characteristics in the first place.

In this section we cite a powerful general result of [154], slightly simplified to fit the present setup and notation. To this end, suppose that $X^{(n)}$, $n \in \mathbb{N}$ is a sequence of $\mathbb{R}^d$-valued stochastic processes with integral characteristics $(B^{h,(n)}, C^{(n)}, \nu^{(n)})$. The latter are assumed to be functions of the path of the process, i.e. for any $(\omega, t) \in \Omega \times \mathbb{R}_+$, $G \in \mathscr{B}^d$ we have

$$B^{h,(n)}(\omega, t) = \underline{B}^{h,(n)}(X^{(n)}(\omega, \cdot), t), \tag{4.36}$$

$$C^{(n)}(\omega, t) = \underline{C}^{(n)}(X^{(n)}(\omega, \cdot), t), \tag{4.37}$$

$$\nu^{(n)}(\omega, [0, t] \times G) = \underline{\nu}^{(n)}(X^{(n)}(\omega, \cdot), [0, t] \times G) \tag{4.38}$$

with some deterministic mappings $\underline{B}^{h,(n)} : \mathbb{D}^d \times \mathbb{R}_+ \to \mathbb{R}^d$, $\underline{C}^{(n)} : \mathbb{D}^d \times \mathbb{R}_+ \to \mathbb{R}^{d\times d}$, $\underline{\nu}^{(n)} : \mathbb{D}^d \times \mathbb{R}_+ \times \mathscr{B}^d \to [0, \infty]$ that satisfy some measurability properties specified in the following remark. Similar to the modified second characteristics in (4.31), we set

$$\widetilde{\underline{C}}^{(n)}(\alpha, t) := \underline{C}^{(n)}(\alpha, t) + \int_{[0,t]\times\mathbb{R}^d} h(x)h(x)^\top \underline{\nu}^{(n)}(\alpha, d(s, x))$$

for any $\alpha \in \mathbb{D}^d$, $t \geq 0$.

Remark 4.32

1. Above, $\mathbb{D}^d$ denotes the Skorokhod space of càdlàg functions $\mathbb{R}_+ \to \mathbb{R}^d$. By $\pi(t) : \mathbb{D}^d \to \mathbb{R}^d, \alpha \mapsto \alpha(t)$ we denote the projection mapping for $t \in \mathbb{R}_+$, which returns the value at time t of any càdlàg function. These mappings can be used to define a σ-field and even a filtration on $\mathbb{D}^d$. Indeed, $\mathscr{D}^d := \sigma(\pi(t) : t \in \mathbb{R}_+)$ is the σ-field generated by all projections. Since $(\pi(t))_{t\geq 0}$ can be viewed as a stochastic process on $\mathbb{D}^d$, it generates a filtration, which we denote by $\mathbf{D}^d = (\mathscr{D}^d_t)_{t\geq 0}$. For limit theorems it turns out to be useful to endow this Skorokhod space $(\mathbb{D}^d, \mathscr{D}^d, \mathbf{D}^d)$ with a topology and even a metric. We refer to [154, Section VI.1b] for its rather involved definition. According to this metric, two functions $\alpha, \beta \in \mathbb{D}^d$ are close if, for any $t \in \mathbb{R}_+$, the quantity $\sup_{s\leq t} |\alpha(\lambda(s)) - \beta(s)|$ is small for some continuous one-to-one mapping $\lambda : \mathbb{R}_+ \to \mathbb{R}_+$ such that $\sup_{t\in\mathbb{R}_+} |\lambda(t) - t|$ is likewise small. The point of this metric is that it turns $\mathbb{D}^d$ into a *Polish space*, i.e. a separable topological space which is complete relative to this metric. Moreover, $\mathscr{D}^d$ coincides with the Borel σ-field on $\mathbb{D}^d$ relative to this topology. Such Polish spaces have many desirable measure-theoretic properties. In order for (4.36–4.38) to make sense, we assume that $\underline{B}^{h,(n)}, \underline{C}^{(n)}$ are predictable (viewed as $\mathbb{R}^d$-valued resp. $\mathbb{R}^{d\times d}$-valued processes on $(\mathbb{D}^d, \mathscr{D}^d, \mathbf{D}^d)$) and that $\underline{\nu}^{(n)} : \mathscr{P} \otimes \mathscr{B}^d \to \mathbb{R}_+$ is a kernel, where $\mathscr{P}$ here denotes the predictable σ-field on $\mathbb{D}^d \times \mathbb{R}_+$.
2. The processes $X^{(n)}$ may be defined on different filtered spaces $(\Omega^{(n)}, \mathscr{F}^{(n)}, \mathbf{F}^{(n)}, P^{(n)})$. For ease of notation we do not indicate this possible dependence in the following.
3. We do not assume $X^{(n)}$ to have differential characteristics but in applications this is often the case, cf. Corollary 4.38 in this context. Skipping the superscript n in the following discussion for ease of notation, the second relevant case concerns the situation of discrete-time processes $(X(t_k))_{k\in\mathbb{N}}$ with $t_k = k\delta$, i.e. defined on a grid $0, \delta, 2\delta, \dots$ of times. Similarly as in Sect. 2.1, such processes can naturally be embedded in continuous time by setting $X(t) := X(t_k)$ for $t \in [t_k, t_{k+1})$. The continuous-time extension X is piecewise constant and hence its integral characteristics (B^h, C, ν) are of the form

$$B^h(t) = \sum_{t_k\leq t} \Delta B^h(t_k) = \sum_{t_k\leq t} E\big(h(\Delta X)(t_k)\big|\mathscr{F}_{t_k-}\big),$$

$$C(t) = 0,$$

$$\nu([0,t] \times G) = \sum_{t_k\leq t} \nu(\{t_k\} \times G) = \sum_{t_k\leq t} E\big(1_G(\Delta X(t_k))\big|\mathscr{F}_{t_k-}\big),$$

i.e. B^h jumps at t_k by the conditional mean of $h(\Delta X(t_k))$, the kernel ν by the conditional law of $\Delta X(t_k)$, and the continuous martingale part vanishes, see [154, Theorem II.3.11] for details. In a Markovian situation in the spirit of

Sect. 1.3, the conditional law of the jumps depends only on the present state of the process, i.e.

$$B^h(t) = \sum_{t_k \leq t} \beta^h(t_k, X(t_{k-1}))\delta,$$

$$\nu([0, t] \times G) = \sum_{t_k \leq t} \kappa(t_k, X(t_{k-1}), G)\delta,$$

for some deterministic functions β^h, κ. Since X is piecewise constant, we can rewrite this as

$$B^h(t) = \int_0^{[t/\delta]\delta} \beta^h([s/\delta]\delta, X(s-))ds,$$

$$\nu([0, t] \times G) = \int_0^{[t/\delta]\delta} \kappa([s/\delta]\delta, X(s-), G)ds,$$

which indicates that β^h, κ play a similar role as differential characteristics. Let us stress once more that these considerations concern the processes $X^{(n)}$ rather than the prospective limiting process X, i.e. $X, t_k, \delta, B^h, C, \nu, \beta^h, \kappa$ above should ideally all carry a superscript (n) in order to avoid confusion.

As a potential limiting object of the sequence $(X^{(n)})$ of processes with characteristics (4.36–4.38) we consider an $\mathbb{R}^d$-valued semimartingale X with differential characteristics as in (4.33–4.35), but with a triplet $(\beta^h, \gamma, \kappa)$ of functions that do not depend on the time parameter t. More specifically, we make the

Assumption 4.33 (Uniqueness in Law of the Limiting Process) *The martingale problem associated to the triplet $(\beta^h, \gamma, \kappa)$ allows for a unique solution for any deterministic starting value $x \in \mathbb{R}^d$ of the solution process.*

We refer to Sect. 5.7 for a discussion of this assumption. Denoting the integral characteristics of X by (B^h, C, ν), we have

$$B^h(t) = \int_0^t \beta^h(X(s-))ds,$$

$$C(t) = \int_0^t \gamma(X(s-))ds,$$

$$\nu([0, t] \times G) = \int_0^t \kappa(X(s-), G)ds.$$

More in line with (4.36–4.38), we can write

$$\begin{aligned} B^h(t) &= \underline{B}^h(X(\cdot), t), \\ C(t) &= \underline{C}(X(\cdot), t), \\ \nu([0, t] \times G) &= \underline{\nu}(X(\cdot), [0, t] \times G), \end{aligned}$$

where the mappings $\underline{B}^h$, $\underline{C}$, $\underline{\nu}$ are defined as

$$\begin{aligned} \underline{B}^h(\alpha, t) &:= \int_0^t \beta^h(\alpha(s-))ds, \\ \underline{C}(\alpha, t) &:= \int_0^t \gamma(\alpha(s-))ds, \\ \underline{\nu}(\alpha, [0, t] \times G) &:= \int_0^t \kappa(\alpha(s-), G)ds \end{aligned}$$

for any $\alpha \in \mathbb{D}^d$, $t \geq 0$, $G \in \mathscr{B}^d$.

The key message of the limit theorem 4.36 below is that the law of $X^{(n)}$ is close to the law of X if

$$\big(\underline{B}^{h,(n)}(\alpha, t), \underline{C}^{(n)}(\alpha, t), \underline{\nu}^{(n)}(\alpha, t, \cdot)\big) \approx (\underline{B}^h(\alpha, t), \underline{C}(\alpha, t), \underline{\nu}(\alpha, t, \cdot)),$$

i.e. if the local behaviour of $X^{(n)}$ resembles that of X. In order for this to hold, some additional regularity of the limiting triplet $(\beta^h, \gamma, \kappa)$ is needed.

Assumption 4.34 (Continuity of the Triplet) *The mappings $x \mapsto \beta^h(x)$,*

$$x \mapsto \tilde{\gamma}^h(x) := \gamma(x) + \int h(y)h(y)^\top \kappa(x, dy), \tag{4.39}$$

and $x \mapsto \int f(y)\kappa(x, dy)$ are continuous for any bounded continuous function $f : \mathbb{R}^d \to \mathbb{R}$ that vanishes in a neighbourhood of 0 (or, equivalently, any bounded continuous function $f : \mathbb{R}^d \to \mathbb{R}$ that is $o(|x|^2)$ for $x \to 0$, cf. [154, Theorem VII.2.9]).

This assumption will reappear in the context of existence of solutions to the martingale problem, cf. Theorems 5.36 and 5.38. Finally, the probability of big jumps should go to 0 *uniformly* in the present state:

Assumption 4.35 (Condition on Big Jumps)

$$\lim_{m\to\infty} \sup_{x \in \mathbb{R}^d \text{ with } |x| \leq a} \kappa\big(x, \{y \in \mathbb{R}^d : |y| > m\}\big) = 0$$

for any finite real number a.

We are now ready to state the limit theorem:

Theorem 4.36 *Suppose that Assumptions 4.33–4.35 hold and*

1. $X^{(n)}(0) \to X(0)$ *in law as* $n \to \infty$,
2. *for any* $t \in \mathbb{R}_+$, *any* $a \in \mathbb{R}_+$, *and any bounded continuous function* $f : \mathbb{R}^d \to \mathbb{R}$ *that vanishes in a neighbourhood of* 0 *we have*

$$\sup_{\alpha} \sup_{s \le t} \left| \underline{B}^{h,(n)}(\alpha, s) - \int_0^s \beta^h(\alpha(u-))du \right| \to 0,$$

$$\sup_{\alpha} \left| \underline{\widetilde{C}}^{(n)}(\alpha, t) - \int_0^t \widetilde{\gamma}^h(\alpha(s-))ds \right| \to 0,$$

$$\sup_{\alpha} \left| \int_{[0,t]\times\mathbb{R}^d} f(x)\underline{\nu}^{(n)}(\alpha, d(s,x)) - \int_0^t \int f(x)\kappa(\alpha(s-), dx)ds \right| \to 0$$

as $n \to \infty$, *where the supremum refers to all* $\alpha \in \mathbb{D}^d$ *with* $\sup_{s\ge 0} |\alpha(s)| \le a$.

Then $(X^{(n)})$ *converges to* X *in the sense of finite-dimensional distributions, i.e. for any* $k \in \mathbb{N}$, $t_1, \dots, t_k \in \mathbb{R}_+$, *and any bounded continuous function* $f : (\mathbb{R}^d)^k \to \mathbb{R}$ *we have*

$$E\big(f(X^{(n)}(t_1), \dots, X^{(n)}(t_k))\big) \to E\big(f(X(t_1), \dots, X(t_k))\big)$$

as $n \to \infty$.

Proof This follows from [154, Theorem IX.3.39]. Indeed, conditions i)-v) in this result are verified in [154, Theorem IX.4.8], based on our assumption here. An exception is the measurability assumption in [154, IX.4.3ii)], which, however, follows from Theorem 5.30 and [100, Proposition 4.1.2]. The remaining condition vi) in [154, Theorem IX.4.8] is implied by the convergence of the triplets as required in the assertion. □

Let us stress that Assumptions 4.33–4.35 in Theorem 4.36 concern only the limiting process X. Convergence holds in fact in a stronger sense, namely $X^{(n)} \to X$ in law as $\mathbb{D}^d$-valued random variables relative to the Skorokhod topology. This type of convergence has its advantages because it allows us to make statements on the convergence of suprema and integrals of the process. We refer to [154, Section VI.3] for details.

As an example we have a look at the convergence of autoregressive time series to an Ornstein–Uhlenbeck process.

Example 4.37 (AR(1) Time Series) We consider an autoregressive time series of order 1 in the sense that $X_0 = x$ and

$$X_k = (1 + \lambda)X_{k-1} + \sigma Z_k + \mu, \quad k = 1, 2, \dots$$

with some parameters $x \in \mathbb{R}$, $\alpha \in \mathbb{R}$, $\beta > 0$, $\gamma \in \mathbb{R}$ and a sequence $(Z_k)_{k\in\mathbb{N}}$ of independent, identically distributed random variables with mean 0 and variance 1. For simplicity we assume the Z_k to be bounded. We let the coefficients $\lambda_n := \lambda$, $\sigma_n := \sigma$, $\mu_n := \mu$ depend on n and suppose that the limits

$$\lambda := \lim_{n\to\infty} \lambda_n n, \tag{4.40}$$

$$\sigma := \lim_{n\to\infty} \sigma_n \sqrt{n}, \tag{4.41}$$

$$\mu := \lim_{n\to\infty} \mu_n n \tag{4.42}$$

exist.

We extend the model to continuous time by introducing the process $X^{(n)}$ with

$$X^{(n)}(t_k) := \alpha_n X^{(n)}(t_{k-1}) + \beta_n Z_k + \gamma_n, \quad k = 1, 2, \ldots$$

for $t_k := k/n$ and

$$X^{(n)}(t) := X^{(n)}(t_k)$$

for $t \in [t_k, t_{k+1})$.

We use Theorem 4.36 in order to show that $X^{(n)}$ converges in law as $n \to \infty$ to some process X with dynamics

$$dX(t) = (\mu + \lambda X(t))dt + \sigma dW(t), \quad X(0) = x,$$

where W denotes a Wiener process. Observe that X is an Ornstein–Uhlenbeck process as in Example 3.56.

In order to verify the conditions of Theorem 4.36, note that

$$\underline{B}^{h,(n)}(\alpha, t) = \int_0^t 1_{[0,[nt]/n]}(s)\beta^{h,(n)}([ns]/n, \alpha(s-))ds,$$

$$\underline{C}^{(n)}(\alpha, t) = 0,$$

$$\int_{[0,t]\times\mathbb{R}^d} f(x)\underline{\nu}^{(n)}(\alpha, d(x,s)) = \int_0^t \int 1_{[0,[nt]/n]}(s) f(y)\kappa^{(n)}([ns]/n, \alpha(s-), dy)ds$$

for any $\alpha \in \mathbb{D}$, $t \geq 0$, and any reasonable $f : \mathbb{R} \to \mathbb{R}$, where Q denotes the law of Z_1 and

$$\beta^{h,(n)}(k/n, x) = n \int h(\lambda_n x + \sigma_n + \mu_n)Q(dy),$$

$$\int f(y)\kappa^{(n)}(k/n, x, dy) = n \int f(\lambda_n x + \sigma_n + \mu_n)Q(dy)$$

for $k = 1, 2, \ldots$ and any $x \in \mathbb{R}$. Moreover, we have $\beta^h(x) = \mu + \lambda x$, $\gamma = \sigma^2$, $\kappa = 0$. For bounded measurable functions $f : \mathbb{R} \to \mathbb{R}$ that vanish in a neighbourhood of 0 we have

$$\begin{aligned}
&\sup_\alpha \left| \int_{[0,t]\times\mathbb{R}^d} f(x)\underline{\nu}^{(n)}(\alpha, d(s,x)) \right| \\
&\leq \int_0^t \int n \sup_x |\|f\|_\infty 1_{\{|\lambda_n x+\sigma_n+\mu_n|\geq \text{cst.}\}} Q(dy)ds \\
&\leq tnQ(\{y \in \mathbb{R} : |y| \geq \sqrt{n}\text{cst.}\}) \stackrel{n\to\infty}{\longrightarrow} 0,
\end{aligned}$$

where the suprema refer to α and x that are bounded by some $a < \infty$ and cst. denotes a constant which may vary from line to line. In view of (4.40, 4.42) we have

$$\begin{aligned}
&\sup_\alpha \sup_{s\leq t} \left| \underline{B}^{h,(n)}(\alpha, s) - \int_0^s \beta^h(\alpha(u-))du \right| \\
&= \sup_\alpha \sup_{s\leq t} \left| \underline{B}^{\text{id},(n)}(\alpha, s) - \int_0^s \beta^{\text{id}}(\alpha(u-))du \right| \\
&\leq \int_0^t \sup_x \left| n(\lambda_n x + \mu_n)1_{[0,[nt]/n]}(u) - \lambda x - \mu \right| du \stackrel{n\to\infty}{\longrightarrow} 0,
\end{aligned}$$

where we used $\int yQ(dy) = E(Z_1) = 0$ and the suprema refer to α and x that are bounded by some $a < \infty$. Applying (4.40–4.42) we conclude

$$\begin{aligned}
&\sup_\alpha \left| \underline{\widetilde{C}}^{(n)}(\alpha, t) - \int_0^t \widetilde{\gamma}^h(\alpha(s-))ds \right| \\
&= \sup_x \left| \int_0^t \left(\int n(\lambda_n x + \sigma_n y + \mu_n)^2 Q(dy)1_{[0,[nt]/n]}(s) - \sigma^2 \right) ds \right| \\
&\stackrel{n\to\infty}{\longrightarrow} 0,
\end{aligned}$$

where we used $\int y^2 Q(dy) = \text{Var}(Z_1) = 1$.

Assumption 4.33 follows from Theorem 3.41 and Sect. 5.7.3.

For the special case $\lambda_n = 0$, $\sigma_n = 1/\sqrt{n}$, $\mu_n = 0$, we obtain Donsker's invariance principle for random walks, cf. Theorem 2.28.

If the processes $X^{(n)}$ have differential characteristics of Markovian type as well, the statement of Theorem 4.36 can be slightly simplified. More specifically, assume that $X^{(n)}$ have characteristics as in (4.33–4.35), with triplets $(\beta^{h,(n)}, \gamma^{(n)}, \kappa^{(n)})$ that do not depend on the time parameter t, i.e.

$$B^{h,(n)}(t) = \int_0^t \beta^{h,(n)}(X^{(n)}(s-))ds,$$

$$C^{(n)}(t) = \int_0^t \gamma^{(n)}(X^{(n)}(s-))ds,$$

$$\nu^{(n)}([0,t] \times G) = \int_0^t \kappa^{(n)}(X^{(n)}(s-), G)ds.$$

In this case a condition of the form

$$\left(\beta^{h,(n)}(x), \gamma^{(n)}(x), \kappa^{(n)}(x,\cdot)\right) \approx \left(\beta^h(x), \gamma(x), \kappa(x,\cdot)\right)$$

should warrant that the law of $X^{(n)}$ is close to the law of X. Letting

$$\widetilde{\gamma}^{h,(n)}(x) := \gamma^{(n)}(x) + \int h(y)h(y)^\top \kappa^{(n)}(x, dy)$$

in line with (4.39), this can be made rigorous as follows:

Corollary 4.38 *Suppose that Assumptions 4.33–4.35 hold and*

1. $X^{(n)}(0) \to X(0)$ *in law as* $n \to \infty$,
2. *for any* $t \in \mathbb{R}_+$, *any* $a \in \mathbb{R}_+$, *and any bounded continuous function* $f : \mathbb{R}^d \to \mathbb{R}$ *that vanishes in a neighbourhood of* 0 *we have*

$$\sup_x |\beta^{h,(n)}(x) - \beta^h(x)| \to 0, \tag{4.43}$$

$$\sup_x |\widetilde{\gamma}^{h,(n)}(x) - \widetilde{\gamma}^h(x)| \to 0, \tag{4.44}$$

$$\sup_x \left| \int f(y)\kappa^{(n)}(x, dy) - \int f(y)\kappa(x, dy) \right| \to 0 \tag{4.45}$$

as $n \to \infty$, *where the supremum refers to all* $x \in \mathbb{R}^d$ *with* $|x| \le a$.

Then $(X^{(n)})$ *converges to* X *in the sense of finite-dimensional distributions.*

Proof This follows from [154, Theorem IX.4.8]. Indeed, all conditions in this theorem are explicitly stated above, except for the measurability assumption in [154, IX.4.3ii)], The latter, however, follows from Theorem 5.30 and [100, Proposition 4.1.2]. □

Convergence holds once more in the stronger sense that $X^{(n)} \to X$ in law relative to the Skorokhod topology.

The following example concerns a model that has been considered for financial data:

Example 4.39 (Hawkes Process) Consider a counting process N with intensity

$$\lambda(t) = \mu + \int_0^t a\eta e^{-\eta(t-s)} dN(s)$$

with $\mu \in \mathbb{R}$, $a, \eta \in (0, \infty)$. This means that N starts at $N(0) = 0$, has values only in the nonnegative integers, and changes only by positive jumps of size 1 which occur with rate $\lambda(t)$.

We now let the parameter $a_n := a$ and hence also the processes $N^{(n)} := N$ as well as $\lambda^{(n)} := \lambda$ depend on n such that $(1 - a_n)n \to \xi$ for some finite positive constant ξ. Define a sequence of processes $(X^{(n)}, V^{(n)})$ by

$$(X^{(n)}(t), V^{(n)}(t)) := \left((1 - a_n)\lambda^{(n)}(nt), \frac{1 - a_n}{n} N^{(n)}(nt)\right).$$

We use Corollary 4.38 in order to show that $(X^{(n)}, V^{(n)})$ converges in law to the bivariate process (X, V) satisfying the stochastic differential equation

$$\begin{aligned} dX(t) &= (\mu - X(t))\xi\eta dt + \sqrt{\xi X(t)} dW(t), \quad X(0) = \mu, \\ dV(t) &= X(t)dt, \quad V(0) = 0. \end{aligned}$$

In the language of Chap. 6, the limiting process (X, V) is an affine Markov process. Its first component is discussed in Example 6.9.

Assumption 4.33 holds by Theorem 6.6, whereas Assumptions 4.34, 4.35 are obvious. In order to verify (4.43–4.45), observe that the triplet $(\beta^{h,(n)}, \gamma^{(n)}, \kappa^{(n)})$ for $(X^{(n)}, V^{(n)})$ is given by

$$\beta^{h,(n)}(x, v) = \begin{pmatrix} (1 - a_n)n\eta(\mu - x) \\ x \end{pmatrix}, \tag{4.46}$$

$$\gamma^{(n)}(x, v) = 0, \tag{4.47}$$

$$\int f(\widetilde{x}, \widetilde{v})\kappa^{(n)}((x, v), d(\widetilde{x}, \widetilde{v})) = f\left((1 - a_n)a_n, \frac{1 - a_n}{n}\right)\frac{n}{1 - a_n}x \tag{4.48}$$

for sufficiently large n. Indeed, the local characteristics of (I, N) equal

$$\big(b^0(t), c(t), K(t, dx)\big) = \big((1, 0), 0, \lambda(t)\varepsilon_{(0,1)}(dx)\big).$$

Since

$$d\lambda(t) = -\eta(\lambda(t) - \mu)dt + a dN(t), \tag{4.49}$$

(4.46–4.48) are obtained by Propositions 4.8, 4.14.

For the triplet $(\beta^h, \gamma, \kappa)$ of (X, V) we have

$$\beta^h(x, v) = \begin{pmatrix} \xi\eta(\mu - x) \\ x \end{pmatrix},$$

$$\gamma(x, v) = \begin{pmatrix} \xi x & 0 \\ 0 & 0 \end{pmatrix},$$

$$\kappa((x, v), \cdot) = 0.$$

Convergence of the triplets now follows easily.

4.8 The Černý and Černý–Ruf Representations

We have looked at stochastic calculus from different angles. For example, Theorems 3.16 and Proposition 4.13 both state versions of Itō's formula. It may be hard to memorise the numerous rules which we have encountered so far. Moreover, they can lead to quite involved computations in practice if a number of operations are applied in a row. In this section we state the rules of stochastic calculus in yet another representation which is often applicable, easy to keep in mind, and above all makes concrete calculations relatively simple.

4.8.1 Černý-Representable Processes

We fix an $\mathbb{R}^d$-valued reference semimartingale L with local characteristics (b^h, c, K) relative to some truncation function h on $\mathbb{R}^d$. In applications L will typically be a Lévy process or, for example, an affine process in the sense of Chap. 6. We denote by μ^L the random measure of jumps of L as usual. We call an $\mathbb{R}^n$-valued semimartingale X L**-representable** if it can be written as

$$dX(t) = \beta(t)dt + \gamma(t)dL(t) + (\delta(t, x) - \gamma(t)x) * \mu^L(dt) \tag{4.50}$$

for some predictable processes β, γ and some predictable function δ such that the stochastic integrals exist, i.e. $\beta_i \in L(I)$, $\gamma_{i\cdot} \in L(L)$, and $\delta_i(t, x) - \gamma_{i\cdot}(t)x$ is μ^L-integrable for $i = 1, \dots, n$. The triplet (β, γ, δ) determines the process X entirely if L is fixed. We call it the **Černý triplet** of X. Note that it does not depend on the truncation function h and that it is invariant under equivalent changes of the probability measure. Obviously, the process L itself is L-representable with triplet $(0, 1, x)$, where 1 here denotes the identity matrix.

If the jump part of L is of finite variation, representation (4.50) can be simplified to

$$dX(t) = \beta(t)dt + \gamma(t)dL^0(t) + \delta(t,x) * \mu^L(dt), \tag{4.51}$$

where

$$L^0(t) := L(t) - \sum_{s\leq t} \Delta L(s)$$

denotes a continuous semimartingale, namely L without its jumps. If the discontinuous part of L is not of finite variation, i.e. if the infinitely many jumps of L in any finite interval $[0, t]$ cannot be summed, then L^0 and hence representation (4.51) do not make sense. But even in this more general case, we use the latter as a convenient shorthand for (4.50). Formal manipulations of (4.51) are much easier than those of (4.50) and—surprisingly or not—always lead to the correct answer.

We will hardly use the slightly intimidating definition (4.50) of representability. The point of L-representable processes is that they stay invariant under operations such as stochastic integration, application of smooth functions, compositions and measure changes. Moreover, the new Černý triplet can be computed easily.

Proposition 4.40 (Stochastic Integration) *Let X be an $\mathbb{R}^n$-valued L-representable process with Černý triplet (β, γ, δ), i.e.*

$$dX(t) = \beta(t)dt + \gamma(t)dL^0(t) + \delta(t,x) * \mu^L(dt)$$

in formal notation. If φ is a locally bounded $\mathbb{R}^{p\times n}$-valued process, then

$$\varphi(t)dX(t) = \varphi(t)\beta(t)dt + \varphi(t)\gamma(t)dL^0(t) + \varphi(t)\delta(t,x) * \mu^L(dt), \tag{4.52}$$

i.e. $\varphi \bullet X$ is L-representable as well with Černý triplet $(\varphi\beta, \varphi\gamma, \varphi\delta)$.

Proof Equation (4.50) yields

$$\varphi(t)dX(t) = \varphi(t)\beta(t)dt + \varphi(t)\gamma(t)dL(t) + \varphi(t)(\delta(t,x) - \gamma(t)x) * \mu^L(dt),$$

which, in formal notation, reads as the right hand-side of (4.52). □

Proposition 4.41 (Covariation) *Let real-valued processes $X, \widetilde{X}$ be L-representable with Černý triplets (β, γ, δ), $(\widetilde{\beta}, \widetilde{\gamma}, \widetilde{\delta})$. Then*

$$d[X, \widetilde{X}](t) = \gamma(t)c(t)\widetilde{\gamma}(t)^\top dt + \delta(t,x)\widetilde{\delta}(t,x) * \mu^L(dt), \tag{4.53}$$

i.e. $[X, \widetilde{X}]$ is L-representable with Černý triplet $(\gamma c\widetilde{\gamma}^\top, 0, \delta\widetilde{\delta})$.

Proof Observe that $X^c = (\gamma \bullet L)^c = \gamma \bullet L^c$ and likewise for $\widetilde{X}$. This implies

$$\langle X^c, \widetilde{X}^c \rangle = (\gamma c \widetilde{\gamma}) \bullet I. \tag{4.54}$$

Moreover,

$$\Delta X(t) = \gamma(t)\Delta L(t) + \delta(t, \Delta L(t)) - \gamma(t)\Delta L(t) = \delta(t, \Delta L(t)) \tag{4.55}$$

and the parallel statement for $\widetilde{X}$ yield

$$\sum_{s\leq t} \Delta X(s)\widetilde{\Delta} X(s) = \sum_{s\leq t} \delta(s, \Delta L(s))\widetilde{\delta}(s, \Delta L(s)) = \delta\widetilde{\delta} * \mu^L(t).$$

From representation (3.9) of the covariation we obtain (4.53). □

Proposition 4.42 (Itō's Formula) *If X is an $\mathbb{R}^n$-valued L-representable process with Černý triplet (β, γ, δ) and if $f : \mathbb{R}_+ \times \mathbb{R}^n \to \mathbb{R}^p$ is twice continuously differentiable, we have*

$$df(t, X(t)) = \widetilde{\beta}(t)dt + \widetilde{\gamma}(t)dL^0(t) + \widetilde{\delta}(t, x) * \mu^L(dt)$$

with

$$\begin{aligned}
\widetilde{\beta}_i(t) &= D_1 f_i(t, X(t-)) + \sum_{j=1}^{n} D_{1+j} f_i(t, X(t-))\beta_j(t) \\
&\quad + \frac{1}{2} \sum_{j,k=1}^{n} D_{1+j,1+k} f_i(t, X(t-)) \sum_{\ell,m=1}^{d} \gamma_{j\ell}(t)c_{\ell m}(t)\gamma_{km}(t), \\
\widetilde{\gamma}_{i\ell}(t) &= \sum_{j=1}^{n} D_{1+j} f_i(t, X(t-))\gamma_{j\ell}(t), \\
\widetilde{\delta}(t, x) &= f\big(t, X(t-) + \delta(t, x)\big) - f\big(t, X(t-)\big)
\end{aligned}$$

for $i = 1, \ldots, p$ and $\ell = 1, \ldots, d$. Hence $f(t, X(t))$ is L-representable as well with Černý triplet $(\widetilde{\beta}, \widetilde{\gamma}, \widetilde{\delta})$. If $p = n = 1$, this can be written as

$$\begin{aligned}
df(t, X(t)) &= \left(\dot{f}(t, X(t-)) + f'(t, X(t-))\beta(t) + \frac{1}{2} f''(t, X(t-))\gamma(t)c(t)\gamma(t)^\top \right) dt \\
&\quad + f'(t, X(t-))\gamma(t)dL^0(t) \\
&\quad + \big(f(t, X(t-) + \delta(t, x)) - f(t, X(t-))\big) * \mu^L(dt),
\end{aligned} \tag{4.56}$$

where $\dot{f}$ and f' refer to the derivatives relative to the first and second variable, respectively.

Proof For ease of notation, we consider the case $p = n = d = 1$, the general assertion following along the same lines. Since $X^c = \gamma \bullet L^c$ by the previous proof, Itō's formula (3.19) yields

$$
\begin{aligned}
f(t, X(t)) = f(0, X(0)) & \\
&+ \int_0^t \left(\dot{f}(s, X(s-)) + f'(s, X(s-))\beta(s) + \frac{1}{2} f''(s, X(s-))\gamma(s)^2 c(s)\right) ds \\
&+ \int_0^t f'(s, X(s-))\gamma(s) dL(s) \\
&+ \int_{[0,t]\times\mathbb{R}} f'(s, X(s-))(\delta(s, x) - \gamma(s)x)\mu^L(d(s, x)) \\
&+ \sum_{s\leq t} \Big(f(s, X(s-) + \Delta X(s)) - f(s, X(s-)) - f'(s, X(s-))\Delta X(s)\Big).
\end{aligned}
$$

By (4.55) the sum can be rewritten as

$$
\begin{aligned}
&\sum_{s\leq t} \Big(f(s, X(s-) + \delta(s, \Delta L(s))) - f(s, X(s-)) - f'(s, X(s-))\delta(s, \Delta L(s))\Big) \\
&= \int_{[0,t]\times\mathbb{R}} \Big(f\big(s, X(s-) + \delta(s, x)\big) - f(s, X(s-)) - f'(s, X(s-))\delta(s, x)\Big)\mu^L(d(s, x)).
\end{aligned}
$$

Summing up, we obtain

$$
\begin{aligned}
f(t, X(t)) = f(0, X(0)) & \\
&+ \int_0^t \left(\dot{f}(s, X(s-)) + f'(s, X(s-))\beta(s) + \frac{1}{2} f''(s, X(s-))\gamma(s)^2 c(s)\right) ds \\
&+ \int_0^t f'(s, X(s-))\gamma(s) dL(s) \\
&+ \int_{[0,t]\times\mathbb{R}} \Big(f\big(s, X(s-) + \delta(s, x)\big) - f(s, X(s-)) \\
&- f'(s, X(s-))\gamma(s)x\Big)\mu^L(d(s, x)),
\end{aligned}
$$

which can be written as (4.56) in formal notation. □

Proposition 4.43 (Composition) *If X is an $\mathbb{R}^n$-valued L-representable process with Černý triplet (β, γ, δ) and Y an $\mathbb{R}^p$-valued X-representable process with corresponding triplet $(\widetilde{\beta}, \widetilde{\gamma}, \widetilde{\delta})$ and locally bounded $\widetilde{\gamma}$, then*

$$
dY(t) = \big(\widetilde{\beta}(t) + \widetilde{\gamma}(t)\beta(t)\big)dt + \widetilde{\gamma}(t)\gamma(t)dL^0(t) + \widetilde{\delta}(t, \delta(t, x)) * \mu^L(dt),
$$

i.e. Y is L-representable as well with Černý triplet $(\overline{\beta}, \overline{\gamma}, \overline{\delta})$ given by

$$\begin{aligned}\overline{\beta}(t) &= \widetilde{\beta}(x) + \widetilde{\gamma}(t)\beta(t),\\ \overline{\gamma}(t) &= \widetilde{\gamma}(t)\gamma(t),\\ \overline{\delta}(t,x) &= \widetilde{\delta}(t,\delta(t,x)).\end{aligned}$$

Proof Equations (4.50, 4.55) yield

$$\begin{aligned}dY(t) &= \widetilde{\beta}(t)dt + \widetilde{\gamma}(t)dW(t) + \big(\widetilde{\delta}(t,x) - \widetilde{\gamma}(t)x\big) * \mu^X(dt)\\ &= \big(\widetilde{\beta}(t) + \widetilde{\gamma}(t)\beta(t)\big)dt + \widetilde{\gamma}(t)\gamma(t)dL(t)\\ &\quad + \Big(\widetilde{\gamma}(t)\delta(t,x) - \widetilde{\gamma}(t)\gamma(t)x\\ &\quad + \widetilde{\delta}(t,\delta(t,x)) - \widetilde{\gamma}(t)\delta(t,x)\Big) * \mu^L(d(t,x))\\ &= \big(\widetilde{\beta}(t) + \widetilde{\gamma}(t)\beta(t)\big)dt + \widetilde{\gamma}(t)\gamma(t)dL(t)\\ &\quad + \Big(\widetilde{\delta}(t,\delta(t,x)) - \widetilde{\gamma}(t)\gamma(t)x\Big) * \mu^L(d(t,x))\end{aligned}$$

and hence the claim. □

Since neither the definition nor the triplet of an L-representable process depends on the underlying probability measure, there is no need for a theorem on measure changes. From the Černý triplet we can derive the local characteristics of an L-representable process:

Proposition 4.44 (Characteristics) *If X is a real-valued L-representable process with Černý triplet (β, γ, δ), it has local characteristics $(\widetilde{b^h}, \widetilde{c}, \widetilde{K})$ given by*

$$\widetilde{b^h}(t) = \beta(t) + \gamma(t)b^h(t) + \int \big(\widetilde{h}(\delta(t,x)) - \gamma(t)h(x)\big)\, K(t,dx), \tag{4.57}$$

$$\widetilde{c}(t) = \gamma(t)c(t)\gamma(t)^\top, \tag{4.58}$$

$$\widetilde{K}(t,B) = \int 1_B(\delta(t,x))K(t,dx), \quad B \in \mathscr{B} \text{ with } 0 \notin B. \tag{4.59}$$

If X is $\mathbb{R}^n$-valued, this generalises to

$$\begin{aligned}\widetilde{b_i^h}(t) &= \beta_i(t) + \sum_{k=1}^d \gamma_{ik}(t)b_k^h(t)\\ &\quad + \int \Big(\widetilde{h}_i(\delta(t,x)) - \sum_{k=1}^d \gamma_{ik}(t)h_k(x)\Big)K(t,dx), \quad i = 1,\ldots,n,\end{aligned}$$

$$\widetilde{c}_{ij}(t) = \sum_{k,l=1}^{d} \gamma_{ik}(t) c_{kl}(t) \gamma_{jl}(t), \quad i, j = 1, \dots, n,$$

$$\widetilde{K}(t, B) = \int 1_B(\delta(t, x)) K(t, dx), \quad B \in \mathscr{B}^n \text{ with } 0 \notin B.$$

Proof For ease of notation, we consider again the case $n = 1$, the general case following along the same lines. Equation (4.58) is derived from (4.54) for $\widetilde{X} = X$. Since $\Delta X(t) = \delta(t, \Delta L(t))$ by (4.55), (4.59) is obtained from Proposition 3.37(6). In order to verify (4.57) observe that (4.50) yields

$$\begin{aligned} X(t) - \sum_{s \le t} \left(\Delta X(s) - \widetilde{h}(\Delta X(s))\right) \quad (4.60) \\ = \beta \bullet I(t) + \gamma \bullet L(t) + \left(\delta - \gamma x - \delta + \widetilde{h}(\delta)\right) * \mu^L(t) \\ = \beta \bullet I(t) + \gamma \bullet L^h(t) + \left(\widetilde{h}(\delta) - \gamma h(x)\right) * \mu^L(t) \end{aligned}$$

for $L^h = L - (x - h(x)) * \mu^L$. Since $b^h \bullet I$ is the compensator of L^h, the compensator of (4.60) equals

$$\left(\beta + \gamma b^h\right) \bullet I(t) + \left(\widetilde{h}(\delta) - \gamma h(x)\right) * \nu^L(t),$$

which is the integrated right-hand side of (4.57) as desired. □

In contrast to the natural equations (4.58) and (4.59), the drift rate (4.57) may look complicated at first glance. It can be deduced formally by simple manipulation of (4.51). If b^0 denotes the drift rate of L^0,

$$\widetilde{b}^0(t) = \beta(t) + \gamma(t) b^0(t) \quad (4.61)$$

because the jumps do not contribute to the drift for the truncation function 0. Changing the truncation function for X leads to

$$\begin{aligned} \widetilde{b}^h(t) &= \widetilde{b}^0(t) + \int \widetilde{h}(x) \widetilde{K}(t, dx) \\ &= \widetilde{b}^0(t) + \int \widetilde{h}(\delta(t, x)) K(t, dx) \end{aligned}$$

by (4.13) and (4.59). Since $b^h(t) = b^0(t) + \int h(x) K(t, dx)$, we conclude

$$\begin{aligned} \widetilde{b}^{\widetilde{h}} &= \beta(t) + \gamma(t) b^0(t) + \int \widetilde{h}(\delta(t, x)) K(t, dx) \\ &= \beta(t) + \gamma(t) b^h(t) + \int \left(\widetilde{h}(\delta(t, x)) - \gamma(t) h(x)\right) K(t, dx), \quad (4.62) \end{aligned}$$

which makes sense even if the terms in the formal derivation do not.

In contrast to Černý triplets, local characteristics of an L-representable process depend on the underlying probability measure P. This becomes apparent in (4.57–4.59) where b^h and K depend on P. The following result states how these objects are affected by a measure change with L-representable density process.

Proposition 4.45 (Change of Measure) *Let $Q \sim P$ be a probability measure whose density process Z is L-representable with Černý triplet $(\beta_Z, \gamma_Z, \delta_Z)$. Then the Q-local characteristics $(\widetilde{b}^h, \widetilde{c}, \widetilde{K})$ of L are given by*

$$\begin{aligned}\widetilde{b}^h(t) &= b^h(t) + c(t)\frac{\gamma_Z(t)}{Z(t-)} + \int h(x)\frac{\delta_Z(t,x)}{Z(t-)}K(t,dx),\\ \widetilde{c} &= c, \\ \frac{\widetilde{K}(t,dx)}{K(t,dx)} &= 1 + \frac{\delta_Z(t,x)}{Z(t-)}.\end{aligned} \tag{4.63}$$

Proof In view of Proposition 4.44, this follows from Theorem 4.15(2). □

If one prefers to derive the Q-drift rate of an L-representable process X starting from b^0 instead of b^h, one may replace (4.45) with the simpler equation

$$\widetilde{b}^0(t) = b^0(t) + c(t)\frac{\gamma_Z(t)}{Z(t-)} \tag{4.64}$$

for the Q-drift rate of L^0 if the latter exists. But even in general (4.64) leads to the correct expression for the Q-drift rate of X if one proceeds as in (4.61–4.62) with $\widetilde{K}, \widetilde{b}^0$ instead of K, b.

Example 4.46 We want to compute the Černý triplet of $\widetilde{X} := \mathscr{L}(e^X)$ for an L-representable process X of the form

$$dX(t) = \beta(t)dt + \gamma(t)dL^0(t) + \delta(t,x) * \mu^L(dt).$$

Recall that $\widetilde{X}$ satisfies $\mathscr{E}(\widetilde{X}) = e^{X-X(0)}$. Itō's formula in Proposition 4.42 yields

$$\begin{aligned}de^{X(t)} &= e^{X(t-)}\left(\beta(t) + \frac{1}{2}\gamma(t)^2c(t)\right)dt + e^{X(t-)}\gamma(t)dL^0(t)\\ &\quad + e^{X(t-)}\left(e^{\delta(t,x)} - 1\right) * \mu^L(dt).\end{aligned}$$

Multiplying by $e^{-X(t-)}$ leads to

$$\begin{aligned}d\widetilde{X}(t) &= d\mathscr{L}\left(e^{X(t)}\right)\\ &= \left(\beta(t) + \frac{1}{2}\gamma(t)^2c(t)\right)dt + \gamma(t)dL^0(t) + \left(e^{\delta(t,x)} - 1\right) * \mu^L(dt).\end{aligned}$$

In particular, $\widetilde{X}$ is L-representable with Černý triplet $(\widetilde{\beta}, \widetilde{\gamma}, \widetilde{\delta})$ given by

$$\widetilde{\beta}(t) = \beta(t) + \frac{1}{2}\gamma(t)^2 c(t), \tag{4.65}$$

$$\widetilde{\gamma}(t) = \gamma(t), \tag{4.66}$$

$$\widetilde{\delta}(t) = e^{\delta(t,x)} - 1. \tag{4.67}$$

For $X = L$ or $(\beta, \gamma, \delta) = (0, 1, x)$ together with Proposition 4.45 we obtain the implication 2⇒3 in Theorem 3.49 and in particular the Lévy property of $\widetilde{X}$.

Conversely, let us consider $X := \log \mathscr{E}(\widetilde{X})$ for an L-representable process $\widetilde{X}$ of the form

$$d\widetilde{X}(t) = \widetilde{\beta}(t)dt + \widetilde{\gamma}(t)dL^0(t) + \widetilde{\delta}(t, x) * \mu^L(dt),$$

where $\widetilde{\delta}(t, x) > -1$. This leads to

$$\begin{aligned} d\mathscr{E}(\widetilde{X})(t) &= \mathscr{E}(\widetilde{X})(t-)d\widetilde{X}(t) \\ &= \mathscr{E}(\widetilde{X})(t-)\widetilde{\beta}(t)dt + \mathscr{E}(\widetilde{X})(t-)\widetilde{\gamma}(t)dL^0(t) \\ &\quad + \mathscr{E}(\widetilde{X})(t-)\widetilde{\delta}(t, x) * \mu^L(dt). \end{aligned}$$

Itō's formula (4.56) now yields

$$\begin{aligned} dX(t) &= d\log(\mathscr{E}(\widetilde{X}))(t) \\ &= \left(\frac{1}{\mathscr{E}(\widetilde{X})(t-)}\mathscr{E}(\widetilde{X})(t-)\widetilde{\beta}(t) + \frac{1}{2}\frac{-1}{\left(\mathscr{E}(\widetilde{X})(t-)\right)^2}\left(\mathscr{E}(\widetilde{X})(t-)\widetilde{\gamma}(t)\right)^2 c(t)\right) dt \\ &\quad + \frac{1}{\mathscr{E}(\widetilde{X})(t-)}\mathscr{E}(\widetilde{X})(t-)\widetilde{\gamma}(t)dL^0(t) \\ &\quad + \left(\log\bigl(\mathscr{E}(\widetilde{X})(t-) + \mathscr{E}(\widetilde{X})(t-)\widetilde{\delta}(t, x)\bigr) - \log \mathscr{E}(\widetilde{X})(t-)\right) * \mu^L(dt) \\ &= \left(\widetilde{\beta}(t) - \frac{1}{2}\widetilde{\gamma}(t)^2 c(t)\right) dt + \widetilde{\gamma}(t)dL^0(t) + \log\left(1 + \widetilde{\delta}(t, x)\right) * \mu^L(dt). \end{aligned}$$

This implies that X is L-representable as well with Černý triplet (β, γ, δ) given by

$$\begin{aligned} \beta(t) &= \widetilde{\beta}(t) - \frac{1}{2}\widetilde{\gamma}(t)^2 c(t), \\ \gamma(t) &= \widetilde{\gamma}(t), \\ \delta(t) &= \log\left(1 + \widetilde{\delta}(t, x)\right), \end{aligned}$$

which is equivalent to (4.65–4.67).

The practical use of Černý triplets stems from the fact that the rules are easy to memorise in formal notation, especially if one is used to Itō calculus for continuous processes. However, often one can use an even simpler representation, namely that of the following Sect. 4.8.2.

4.8.2 Černý–Ruf Representable Processes

We use the notation from the previous section. In applications one frequently faces the situation that $x \mapsto \delta(t, x)$ is twice differentiable in 0 and

$$\gamma_{ij}(t) = \gamma_{ij}^{(\delta)}(t) := D_{1+j}\delta_i(t, 0, \dots, 0),$$

$$\beta_i(t) = \beta_i^{(\delta)}(t) := \frac{1}{2}\sum_{j,k=1}^{d} D_{1+j,1+k}\delta(t, 0, \dots, 0)c_{jk}(t)$$

for $i = 1, \dots, n$ and $j = 1, \dots, d$. Put differently, we are considering L-representable processes X with Černý triplet $(\beta^{(\delta)}, \gamma^{(\delta)}, \delta)$. Since $\beta^{(\delta)}$ and $\gamma^{(\delta)}$ are determined by δ, we call X **L-representable (in the sense of Černý–Ruf)** with **Černý–Ruf coefficient** δ and we write $X = X(0) + \delta \circ L$ with

$$\delta \circ L := \beta^{(\delta)} \bullet I + \gamma^{(\delta)} \bullet L + \left(\delta - \gamma^{(\delta)}x\right) * \mu^L.$$

Such processes occur often because the operations in Propositions 4.40–4.43 leave the class of Černý–Ruf representable processes invariant. More specifically, one easily derives the following statements.

1. L is L-representable with Černý–Ruf coefficient $\delta(t, x) = x$. In other words,

$$L = L(0) + x \circ L.$$

2. *(Covariation)* $[L_i, L_j]$ is L-representable with Černý–Ruf coefficient $\delta(t, x) = x_i x_j$. We write this as

$$[L_i, L_j] = (x_i x_j) \circ L.$$

3. *(Stochastic integration)* For locally bounded $\varphi : \mathbb{R}^d \to \mathbb{R}^n$ the stochastic integral $\varphi \bullet L$ is L-representable with Černý–Ruf coefficient $\delta(t, x) = \varphi(t)x$, i.e.

$$\varphi \bullet L = (\varphi x) \circ L.$$

4. *(Itō's formula)* If $f : \mathbb{R}^d \to \mathbb{R}^n$ is twice continuously differentiable, the process $f(L)$ is L-representable with Černý–Ruf coefficient $\delta(t,x) = f(L(t-)+x) - f(L(t-))$. This means

$$f(L) = f(L(0)) + (f(L_- + x) - f(L_-)) \circ L.$$

5. *(Composition)* Let X be L-representable with Černý–Ruf coefficient δ. Moreover, suppose that Y is an $\mathbb{R}^p$-valued X-representable process with coefficient $\widetilde{\delta}$ such that $(D_{1+j}\delta_i(t,0,\dots,0))_{t\in\mathbb{R}_+}$ and $(D_{1+j,1+k}\delta_i(t,0,\dots,0))_{t\in\mathbb{R}_+}$ are locally bounded for $i = 1,\dots,p$ and $j,k = 1,\dots,n$. Then Y is L-representable with Černý–Ruf coefficient $\overline{\delta}(t,x) = \widetilde{\delta}(t,\delta(t,x))$. In other words,

$$Y = Y(0) + \widetilde{\delta}(\delta) \circ L.$$

Consequently, the Černý–Ruf representation gives rise to a particularly simple calculus because only these intuitive rules for δ have to be kept in mind.

Appendix 1: Problems

The exercises correspond to the section with the same number.

4.1 Show that X is a Poisson process with intensity 1 if and only if it is a counting process with $X(0) = 0$ and such that $(X(t) - t)_{t\geq 0}$ is a local martingale.

4.2 Let X be an $\mathbb{R}^d$-valued semimartingale with characteristic exponent $\psi(t,u)$. Show that

$$\langle e^{iu^\top X}, e^{iv^\top X}\rangle(t) = \int_0^t e^{i(u+v)^\top X(s)}\big(\psi(s,u+v) - \psi(s,u) - \psi(s,v)\big)ds \tag{4.68}$$

for any $u,v \in \mathbb{R}^d$.

Remark: Provided that $e^{z^\top X}$ is a special semimartingale for $z = \mathrm{Re}(u/i)$, $\mathrm{Re}(v/i)$, $\mathrm{Re}((u+v)/i)$, both (4.68) and the local martingale property of (4.11) can be extended to complex $u,v \in \mathbb{C}^d$.

4.3 Let $h(x) = x1_{\{|x|\leq 1\}}$. Show that the terms on the right-hand sides of (4.20), (4.21), and (4.22) are independent Lévy processes if X is a Lévy process.

4.4 Let W be a Wiener process. Show that $X = |W|$ is a semimartingale which does not allow for absolutely continuous characteristics. Determine its integral characteristics B^h, C, ν.

Hint: Show first that $\{t \geq 0 : W(t) = 0\}$ is almost surely a Lebesgue-null set. Moreover, recall (3.40).

4.5 Suppose that X is a variance-gamma Lévy process under both P and $Q \sim P$ with parameters $\alpha, \lambda, \beta, \mu$ and $\widetilde{\alpha}, \widetilde{\lambda}, \widetilde{\beta}, \widetilde{\mu}$, respectively. What can be said about the possible values of $\widetilde{\alpha}, \widetilde{\lambda}, \widetilde{\beta}, \widetilde{\mu}$ in terms of $\alpha, \lambda, \beta, \mu$?

4.6 Let $x \in \mathbb{R}$ and suppose that $(\beta^h, \gamma, \kappa) = (\beta^h, 1, 0)$ for some bounded function $\beta^h : \mathbb{R}_+ \times \mathbb{R} \to \mathbb{R}$. Show that uniqueness holds for the martingale problem related to $P^0 = \varepsilon_x$ and $(\beta^h, \gamma, \kappa) = (\beta^h, 1, 0)$.

Hint: show that X is a Wiener process after an appropriate change of measure.

4.7 (Heston Limit of Hawkes Processes) For fixed n consider two counting processes $N_+^{(n)}, N_-^{(n)}$, not jumping at the same time, with intensities $\lambda_+^{(n)}, \lambda_-^{(n)}$ of the form

$$\begin{pmatrix} \lambda_+^{(n)} \\ \lambda_-^{(n)} \end{pmatrix} = \begin{pmatrix} \mu \\ \mu \end{pmatrix} + \int_0^t a_n \begin{pmatrix} \alpha_1 e^{-\eta(t-s)} & \alpha_2 e^{-\eta(t-s)} \\ \alpha_2 e^{-\eta(t-s)} & \alpha_1 e^{-\eta(t-s)} \end{pmatrix} d\begin{pmatrix} N_+^{(n)} \\ N_-^{(n)} \end{pmatrix}(s)$$

with constants $\mu, a_n, \alpha_1, \alpha_2, \eta \in \mathbb{R}_+$ such that $\xi := \lim_{n\to\infty}(1 - a_n)n$ exists, is positive, and $\alpha_1 + \alpha_2 = \eta$. Define bivariate processes $(X^{(n)}, V^{(n)})$ by

$$\left(X^{(n)}(t), V^{(n)}(t)\right) = \left(\frac{\lambda_+^{(n)}(nt) + \lambda_-^{(n)}(nt)}{n}, \frac{N_+^{(n)}(nt) - N_-^{(n)}(nt)}{n}\right).$$

Show that $(X^{(n)}, V^{(n)})$ converges in law as $n \to \infty$ to some bivariate process (X, V) solving the SDE

$$dX(t) = \eta(2\mu - \xi X(t))dt + \eta\sqrt{X(t)}dW_1(t),$$
$$dV(t) = \sqrt{X(t)}dW_2(t)$$

with two independent Wiener processes W_1, W_2.

In Mathematical Finance the events in $N_+^{(n)}, N_-^{(n)}$ can be interpreted as price moves. Since

$$d\lambda_+^{(n)}(t) = -\eta\big(\lambda_+^{(n)}(t) - \mu\big)dt + a_n\alpha_1 dN_+(t) - a_n\alpha_2 dN_-(t), \tag{4.69}$$

$$d\lambda_-^{(n)}(t) = -\eta\big(\lambda_-^{(n)}(t) - \mu\big)dt + a_n\alpha_2 dN_+(t) - a_n\alpha_1 dN_-(t), \tag{4.70}$$

each price change increases the probability of subsequent moves but this increased intensity fades after some time. The limiting process is an instance of the popular Heston model, which is discussed in Sect. 8.2.4.

Hint: Use Theorem 4.36. To this end, express the compensator of $V^{(n)}$ as an integral

$$A^{V^{(n)}}(t) = \int_0^t f^{(n)}(s)dV^{(n)}(s)$$
$$= f^{(n)}(t)V^{(n)}(t) - f^{(n)}(0)V^{(n)}(0) - \int_0^t V^{(n)}(s)(f^{(n)})'(s)ds$$

of the past values of $V^{(n)}$. Show that $f^{(n)}$, $(f^{(n)})'$ and hence $A^{(n)}$ converge to 0 as $n \to \infty$. The uniqueness of the law of (X, V) need not to be shown. It follows from Theorem 6.6.

4.8 Consider $S = (S_1, \ldots, S_d)$ of the form $S_i(t) = S_i(0)e^{X_i}$, $i = 1, \ldots, d$ for $S_1(0), \ldots, S_d(0) > 0$ and some $\mathbb{R}^d$-valued Lévy process or, more generally, semimartingale X. For $\varphi \in L(S)$ write $V_\varphi(t) := 1 + \varphi \bullet S(t)$. Determine φ such that $1/V_\varphi$ and S_i/V_φ, $i = 1, \ldots, d$ are local martingales—provided that it exists. In Mathematical Finance, φ is called the *numeraire portfolio* for the asset price process S.

Appendix 2: Notes and Comments

Semimartingale characteristics are covered in [138, 152, 154]. The equivalence in Definition 4.3 follows from [154, Theorem II.2.42]. Proposition 4.6 is stated in [154, Corollary II.4.19]. For Proposition 4.7 we refer to [154, Theorem II.4.15]. Proposition 4.14 is stated in [181, Lemma 5]. For Theorem 4.15 and Proposition 4.16 see [154, Theorem III.3.24], [169, Proposition 2.6], [168, Lemma 5.1]. Theorem 4.24 goes back to [207]. For Proposition 4.31 we refer to [154, Theorem II.2.42]. Convergence of Hawkes processes as in Example 4.39 and Problem 4.7 is studied in [155]. The general approach of Sect. 4.8 is due to Aleš Černý. The Černý–Ruf representation is introduced in [57], where a thorough treatment can be found. For a more general statement of Problem 4.6 see [241, Corollary IX.1.14]. Problem 4.1 goes back to [291].

Chapter 5
Markov Processes

Most processes in applications are Markov processes or can be viewed as components of multivariate Markov processes. As in discrete time the term *Markov* refers to a certain lack of memory. The theory of these processes provides an alternative way to express the local behaviour of a stochastic process. In this chapter we give only a brief account of Markov processes, concentrating on definitions and results that are needed in the sequel.

5.1 The Transition Function and Semigroup

Fix a closed subset $E \subset \mathbb{R}^d$. We call an adapted E-valued process X a **(homogeneous) Markov process** with state space E if

$$P(X(s+t) \in B|\mathscr{F}_s) = P(t, X(s), B), \quad s, t \geq 0, B \in \mathscr{B}(E) \tag{5.1}$$

holds for some family $(P(t, x, \cdot))_{t \geq 0, x \in E}$ of probability measures on E such that $(t, x) \mapsto P(t, x, B)$ is measurable for fixed B. The function $(t, x, B) \mapsto P(t, x, B)$ is called the *transition function* of X. Equation (5.1) implies that the future evolution $(X(s+t))_{t \geq 0}$ of a Markov process given the past up to s depends on this past only through the present value $X(s)$. In this sense it has *no memory*. Moreover, its dynamics do not depend explicitly on time: given the past up to s, the law of $X(s+t) - X(s)$ may depend on $X(s)$ but not on s itself. This second property is relaxed in Sect. 5.2.

The transition function "almost" satisfies the **Chapman–Kolmogorov equation**

$$P(s+t, x, B) = \int P(t, y, B) P(s, x, dy) \tag{5.2}$$

© Springer Nature Switzerland AG 2019

E. Eberlein, J. Kallsen, *Mathematical Finance*, Springer Finance,
https://doi.org/10.1007/978-3-030-26106-1_5

for $s, t \geq 0$, $B \in \mathscr{B}(E)$. Indeed, for arbitrary $r \geq 0$ we have

$$\begin{aligned} P(s+t, X(r), B) &= P(X(r+s+t) \in B | \mathscr{F}_r) \\ &= E\big(P(X(r+s+t) \in B | \mathscr{F}_{r+s}) \big| \mathscr{F}_r\big) \\ &= E\big(P(t, X(r+s), B) \big| \mathscr{F}_r\big) \\ &= \int P(t, y, B) P(s, X(r), dy), \end{aligned} \tag{5.3}$$

where the last line follows because $P(s, X(r), \cdot)$ is the law of $X(r+s)$ given the information up to time r. The restriction *almost* above refers to the fact that (5.3) yields (5.2) only up to some set of points x that is never or almost never visited by X. To be more precise, we call $(P(t, x, \cdot))_{t \geq 0, x \in E}$ a **(homogeneous) transition function** only if (5.2) actually holds for any x. Moreover, we generally require the family in the above definition of a **Markov process** to be a transition function in this sense. Of course, it is unique only up to some set of points x that are never visited by the process.

Example 5.1 (Lévy Processes) Shifted Lévy processes are Markov processes. Indeed, for a Lévy process X set $P(t, x, B) := P^{X(t)+x}(B)$. By stationarity and independence of increments we have (5.1). Again by stationarity and independence of increments we obtain

$$\begin{aligned} \int P(t, y, B) P(s, x, dy) &= \int P^{X(t)}(B - y) P^{X(s)+x}(dy) \\ &= \big(P^{X(t)} * P^{X(s)+x}\big)(B) \\ &= \big(P^{X(s+t)-X(s)} * P^{X(s)+x}\big)(B) \\ &= P^{X(s+t)+x}(B) \\ &= P(s+t, x, B), \end{aligned}$$

i.e. (5.2) holds.

One-to-one functions of Markov processes are Markov processes as well.

Proposition 5.2 *Let $\widetilde{E} \subset \mathbb{R}^d$ and $\widetilde{X} = f(X)$ for some E-valued Markov process and some one-to-one mapping $f : E \to \widetilde{E}$ such that f, f^{-1} are measurable. Then $\widetilde{X}$ is a Markov process with transition function $\widetilde{P}(t, \widetilde{x}, \widetilde{B}) := P(t, f^{-1}(\widetilde{x}), f^{-1}(\widetilde{B}))$.*

Proof First note that

$$\begin{aligned} P\big(\widetilde{X}(s+t) \in \widetilde{B} \big| \mathscr{F}_s\big) &= P\big(X(s+t) \in f^{-1}(\widetilde{B}) \big| \mathscr{F}_s\big) \\ &= P\big(t, X(s), f^{-1}(\widetilde{B})\big) \end{aligned}$$

$$= \widetilde{P}\big(t, f(X(s)), \widetilde{B}\big)$$
$$= \widetilde{P}\big(t, \widetilde{X}(s), \widetilde{B}\big)$$

as desired for (5.1). The Chapman–Kolmogorov equation for $\widetilde{X}$ follows from the Chapman–Kolmogorov equation for X by a similar calculation. □

Example 5.3 (Exponential Lévy Processes) Geometric Lévy processes are Markov processes. Indeed, for both $(0, \infty)$- and $(0, \infty)^d$-valued exponential Lévy processes this follows from Theorems 3.49 resp. 3.50 and Proposition 5.2.

The following result implies that the finite-dimensional distributions and hence the law of a Markov process are determined by its transition function together with its initial distribution $P^{X(0)} = \mathscr{L}(X(0))$, i.e. the law of $X(0)$.

Proposition 5.4 *Let X be an E-valued Markov process. For any bounded function or nonnegative measurable function $f : E^n \to \mathbb{R}$, we have*

$$E\big(f(X(t_1), \dots, X(t_n))\big) = \int \cdots \int f(x_1, \dots, x_n) P(t_n - t_{n-1}, x_{n-1}, dx_n)$$
$$\cdots P(t_2 - t_1, x_1, dx_2) P(t_1, x_0, dx_1) P^{X(0)}(dx_0).$$

Proof The proof is the same as for Proposition 1.37. □

If the transition function $(P(t, x, \cdot))_{t \geq 0, x \in E}$ is fixed, the distribution of the whole process is determined by the law of $X(0)$. Conversely, any homogeneous transition function gives rise to some Markov process.

Proposition 5.5 *If $(P(t, x, \cdot))_{t \geq 0, x \in E}$ is a transition function and μ some law on $E \subset \mathbb{R}^d$, there exists an E-valued Markov process X with initial distribution μ and transition function $(P(t, x, \cdot))_{t \geq 0, x \in E}$. It is defined on some filtered space $(\Omega, \mathscr{F}, \mathbf{F}, P_\mu)$. For $x \in E$, the notation P_x and E_x is typically used for P_{ε_x} and the expectation under P_{ε_x}. In particular, we have $P_x(X(0) = x) = 1$.*

Idea This is shown in [100, Theorem 4.1.1]. The idea is to define a law on the canonical path space $\Omega := E^{\mathbb{R}_+}$ such that the canonical process X with $X(\omega, t) := \omega(t)$ has the desired properties. To this end, the finite-dimensional marginals of P_μ are chosen as

$$P_\mu\big(\pi_{\{t_1, \dots, t_n\}} \in B\big) = \int \cdots \int 1_B(x_1, \dots, x_n) P(t_n - t_{n-1}, x_{n-1}, dx_n)$$
$$\cdots P(t_2 - t_1, x_1, dx_2) P(t_1, x_0, dx_1) \mu(dx_0)$$

for $B \in \mathscr{B}(E^n)$ and $\pi_{\{t_1, \dots, t_n\}} : \Omega \to E^n$, $\omega \mapsto (\omega(t_1), \dots, \omega(t_n))$. By Proposition 5.4 this is the only possible choice if P_μ is meant to be the law of X. □

The Markov property implies that the whole process starts afresh at any time s.

Proposition 5.6 *Let X be a Markov process with state space E and $(P_x)_{x\in E}$ as in Proposition 5.5. Then*

$$P\big((X(s+t))_{t\geq 0}\in B\big|\mathscr{F}_s\big)=P_{X(s)}\big((X(t))_{t\geq 0}\in B\big),\quad s\geq 0,\, B\in\mathscr{B}(E)^{\mathbb{R}_+}. \tag{5.4}$$

By slight abuse of notation, X in this equation stands for both the original Markov process and the one constructed in Proposition 5.5 on the canonical path space.

Proof It suffices to show that

$$P\big((X(s+t_1),\dots,X(s+t_n))\in B\big|\mathscr{F}_s\big)=P_{X(s)}\big((X(t_1),\dots,X(t_n))\in B\big)$$

for $B\in\mathscr{B}(E^n)$. For $n=1$ we have

$$P\big(X(s+t_1)\in B\big|\mathscr{F}_s\big)=P(t_1,X(s),B)=P_{X(s)}(X(t_1)\in B)$$

by (5.1). For $n\geq 1$ the induction hypothesis implies

$$\begin{aligned}
&P\big((X(s+t_1),\dots,X(s+t_n))\in B\big|\mathscr{F}_s\big)\\
&=E\Big(P\big((X(s+t_1),\dots,X(s+t_n))\in B\big|\mathscr{F}_{s+t_1}\big)\Big|\mathscr{F}_s\Big)\\
&=E\Big(P_{X(s+t_1)}\big((X(0),\dots,X(t_n-t_1))\in B\big)\Big|\mathscr{F}_s\Big)\\
&=\int g(x_1)P(t_1,X(s),dx_1)
\end{aligned}$$

with

$$\begin{aligned}
g(x_1)&=P_x\big((X(0),\dots,X(t_n-t_1))\in B\big)\\
&=\int\dots\int 1_B(x_1,\dots,x_n)P(t_n-t_{n-1},x_{n-1},dx_n)\\
&\qquad\cdots P(t_3-t_2,x_2,dx_3)P(t_2-t_1,x_1,dx_2).
\end{aligned}$$

Proposition 5.4 yields

$$\begin{aligned}
\int g(x_1)P(t_1,x_0,dx_1)&=\int\dots\int 1_B(x_1,\dots,x_n)P(t_n-t_{n-1},x_{n-1},dx_n)\\
&\qquad\cdots P(t_3-t_2,x_2,dx_3)P(t_2-t_1,x_1,dx_2)P(t_1,x_0,dx_1)\\
&=E_{x_0}\big(1_B(X(t_1),\dots,X(t_n))\big)
\end{aligned}$$

for any x_0 and hence the claim. □

The transition function leads naturally to a family $(p_t)_{t\geq 0}$ of operators on bounded measurable functions $f : E \to \mathbb{R}$, called the **transition semigroup**. These operators are defined via

$$p_t f(x) := \int f(y)P(t, x, dy)$$

for $t \geq 0$, bounded measurable functions $f : E \to \mathbb{R}$, and $x \in E$. In view of (5.1), $p_t f(x)$ is the expected value of $f(X(s+t))$ given that $X(s) = x$. Since

$$p_t 1_B(x) = P(t, x, B),$$

the transition function can be recovered from $(p_t)_{t\geq 0}$. The family $(p_t)_{t\geq 0}$ is called *semigroup* because

$$p_{s+t} = p_s \circ p_t, \tag{5.5}$$

which means that $p_{s+t} f = p_s(p_t f)$ for any bounded function f. Moreover, we have $p_0 = \text{id}$. Equation (5.5) follows easily from the Chapman–Kolmogorov equality.

Markov processes in applications are usually **strong** in the sense that (5.1) and consequently (5.4) also holds for finite stopping times τ instead of fixed $s \geq 0$. Typically, they are even **Feller**, i.e. the transition function $(t, x, B) \mapsto P(t, x, B)$ of X is continuous in x and t in the sense that $P(t, x_n, \cdot) \to P(t, x, \cdot)$ weakly for $x_n \to x$ and $P(t, x, \cdot) \to \varepsilon_x$ weakly for $t \to 0$. Moreover, one requires $\lim_{|x|\to\infty} p_t f(x) = 0$ for any $t \geq 0$ and any $f \in C_0(E)$, where $C_0(E)$ denotes the set of continuous functions on E which vanish at infinity.

Feller processes have some nice properties.

Proposition 5.7 *If X is a Feller process, the transition semigroup satisfies $p_t f \in C_0(E)$ for any $t \geq 0$ and any $f \in C_0(E)$. Moreover, $(p_t)_{t\geq 0}$ is strongly continuous in $C_0(E)$ in the sense that for any $f \in C_0(E)$ we have $\lim_{t\to 0} p_t f = f$ with convergence in $C_0(E)$, i.e. uniformly on E.*

Proof The first statement follows from the definition of weak convergence, for the second see [241, Theorem III.2.4]. □

By the previous proposition, $(p_t)_{t\geq 0}$ can also be viewed as a family or semigroup of operators acting on $C_0(E)$. We call it the **Feller semigroup** of X.

Proposition 5.8 *Feller processes are strong Markov processes.*

Proof See [241, Theorem III.3.1]. □

Proposition 5.9 *Feller processes have a càdlàg version.*

Proof See [241, Theorem III.2.7]. □

Example 5.10 (Lévy and Exponential Lévy Processes) Lévy processes as in Example 5.1 are Feller. Indeed, since $X(t) + x_n \to X(t) + x$ almost surely and in law if $x_n \to x$, we have $P(t, x_n, \cdot) \to P(t, x, \cdot)$ weakly for Lévy processes. Similarly, $X(t) \to X(0)$ holds almost surely and in law for $t \to 0$. This implies $P(t, x, \cdot) \to \varepsilon_x$ weakly for $t \to 0$. Finally, $p_t f(x) = E_x(f(X(t))) = E_0(f(X(t) + x)) \to 0$ for $|x| \to \infty$ by dominated convergence because $f(x) \to 0$ for $|x| \to \infty$.

Exponential Lévy processes as in Example 5.3 are Feller if they are $(0, \infty)$- resp. $(0, \infty)^d$-valued. This follows along the same lines as for Lévy processes. However, see Example 5.44 for a counterexample if the process may attain the value 0.

In fact, many processes resembling Lévy processes locally are Feller as well, cf. Sect. 5.8. In particular, processes as in Theorem 3.41 are often Feller if the coefficients β^h, σ, δ do not depend explicitly on the time argument t, cf. Example 5.44 below.

5.2 Time-Inhomogeneous Markov Processes

For Markov processes in the sense of Sect. 5.1 we have that the law of $X(s+t)$ given $\mathscr{F}_s$ does not depend explicitly on s. Here we abandon this time homogeneity while we keep the property that the future evolution $(X(s+t))_{t\geq 0}$ of a process depends on the past up to s only through the present value $X(s)$. An **inhomogeneous Markov process** with state space $E \subset \mathbb{R}^d$ is a process satisfying

$$P(X(t) \in B|\mathscr{F}_s) = P(s, t, X(s), B), \quad 0 \leq s \leq t, B \in \mathscr{B}(E)$$

for some family of probability measures $(P(s, t, x, \cdot))_{0\leq s\leq t, x\in E}$ satisfying the *Chapman–Kolmogorov equation*

$$P(r, t, x, B) = \int P(s, t, y, B)P(r, s, x, dy)$$

for $0 \leq r \leq s \leq t, x \in E, B \in \mathscr{B}(E)$. Proposition 5.4 now reads as

$$E\big(f(X(t_1), \dots, X(t_n))\big) = \int \cdots \int f(x_1, \dots, x_n)P(t_{n-1}, t_n, x_{n-1}, dx_n) \cdots P(t_1, t_2, x_1, dx_2)P(0, t_1, x_0, dx_1)P^{X(0)}(dx_0).$$

As before the *transition function* $(s, t, x, B) \mapsto P(s, t, x, B)$ induces a family of operators $(p_{s,t})_{0\leq s\leq t}$, namely

$$p_{s,t}f(x) := \int f(y)P(s, t, x, dy)$$

for $0 \leq s \leq t$, bounded measurable $f : E \to \mathbb{R}$, and $x \in E$. Equation (5.5) generalises to

$$p_{r,t} = p_{r,s} \circ p_{s,t}, \quad 0 \leq r \leq s \leq t$$

in the inhomogeneous case.

Time-inhomogeneous Markov processes can be reduced to homogeneous Markov processes by considering the $(\mathbb{R}_+ \times E)$-valued *space-time process*

$$\overline{X}(t) := (t, X(t)).$$

One easily verifies that X is a time-inhomogeneous Markov process if and only if $\overline{X}$ is a time-homogeneous Markov process. Therefore we confine ourselves to discussing the time-homogeneous case in this book. Let us indicate how the transition functions and operators of X and $\overline{X}$ are related to each other. If $\overline{P}(t, x, B)$ denotes the transition function of $\overline{X}$, we have

$$\overline{P}(t, (s, x), B) = P\big(s, s+t, x, \{y \in E : (s+t, y) \in B\}\big)$$

for $s, t \geq 0$, $x \in E$, $B \in \mathscr{B}(\mathbb{R}_+ \times E)$. This implies

$$\overline{p}_t f(s, x) = (p_{s,s+t} f(s+t, \cdot))(x)$$

for $s, t \geq 0$, $x \in E$ and bounded measurable $f : \mathbb{R}_+ \times E \to \mathbb{R}$, where $(\overline{p}_t)_{t \geq 0}$ denotes the transition semigroup of $\overline{X}$.

5.3 The Generator and Extended Generator

We observed in discrete time that the whole transition semigroup can be recovered from its generator. The latter is defined in Sect. 1.4 in terms of the transition operator for the smallest nontrivial time increment $t = 1$. Such a minimal period does not exist in continuous time; but by an appropriate limiting procedure one can again define a generator which typically allows us to recover the whole transition semigroup. In the sequel we focus on the particularly important case of Feller processes.

The **(infinitesimal) generator** of a Feller process X (or, more precisely, of its Feller semigroup $(p_t)_{t \geq 0}$) is defined as a kind of right-hand derivative of $t \mapsto p_t$ at 0, namely as the linear operator G defined via

$$Gf := \lim_{t \to 0} \frac{p_t f - p_0 f}{t} = \lim_{t \to 0} \frac{p_t f - f}{t}$$

for any function $f \in C_0(E)$ such that the limit exists in $C_0(E)$. This in turn means that $((p_t f)(x) - f(x))/t$ converges uniformly in x to some function in $C_0(E)$. The subset of functions f with this property is called the **domain** $\mathscr{D}_G$ of G. Note that Gf is again an element of $C_0(E)$ and hence a function from E to $\mathbb{R}$. On an intuitive level the generator describes the dynamics of the process in an infinitesimal period.

Proposition 5.11 *The generator of a Feller process determines the corresponding transition semigroup uniquely.*

Proof This is shown in [100, Proposition 1.2.9]. □

The transition semigroup satisfies a kind of differential equation as the following result shows.

Proposition 5.12 *If f belongs to the domain $\mathscr{D}_G$ of the generator G of a Feller process X, so does $p_t f$ for any $t \geq 0$ and we have*

$$\frac{d}{dt}(p_t f) = G(p_t f) = p_t(Gf), \tag{5.6}$$

where the derivative is to be understood relative to the uniform norm in $C_0(E)$, i.e. $\|f\| = \sup_{x \in E} |f(x)|$.

Idea On the formal level, we have

$$\begin{aligned}\frac{d}{dt}(p_t f) &= \lim_{h \to 0} \frac{p_{t+h} f - p_t f}{h} \\ &= \lim_{h \to 0} \frac{p_h(p_t f) - p_t f}{h} \\ &= G(p_t f)\end{aligned}$$

and

$$\begin{aligned}\frac{d}{dt}(p_t f) &= \lim_{h \to 0} \frac{p_{t+h} f - p_t f}{h} \\ &= \lim_{h \to 0} \frac{p_t(p_h f) - p_t f}{h} \\ &= p_t(Gf).\end{aligned}$$

For a rigorous proof see [100, Proposition 1.1.5b]. □

Remark 5.13

1. Formally, we can interpret (5.6) as a linear differential equation $\frac{d}{dt} p_t = G p_t$, which in turn suggests that the transition operator is given by $p_t = e^{tG}$ with an operator exponential of the form $e^{tG} = \sum_{n=0}^{\infty} \frac{1}{n!} t^n G^n$. However, since G is typically not continuous, the series on the right-hand side is not defined.

2. For $\lambda > 0$, define

$$R_\lambda := \int_0^\infty e^{-\lambda t} p_t dt$$

or, more precisely, the mapping $R_\lambda : C_0(E) \to C_0(E)$ with $R_\lambda f := \int_0^\infty e^{-\lambda t} p_t f dt$. Put differently, $(R_\lambda)_{\lambda>0}$ is the Laplace transform of the semigroup $(p_t)_{t\geq 0}$. It turns out that

$$R_\lambda = (\lambda - G)^{-1} \tag{5.7}$$

for any $\lambda > 0$ or, more precisely, $\lambda - G : \mathscr{D}_G \to C_0(E)$ is one-to-one with inverse $R_\lambda : C_0(E) \to \mathscr{D}_G$. In the language of functional analysis, R_λ is the **resolvent** of the generator G. It plays an important role in semigroup and Markov process theory but we do not need it in the sequel.

Formally, (5.7) follows from

$$\begin{aligned} GR_\lambda &= \int_0^\infty e^{-\lambda t} G p_t dt \\ &= \int_0^\infty e^{-\lambda t} \frac{d}{dt} p_t dt \\ &= \left[e^{-\lambda t} p_t\right]_0^\infty - \int_0^\infty \frac{de^{-\lambda t}}{dt} p_t dt \\ &= 0 - \mathrm{id} + \lambda \int_0^\infty e^{-\lambda t} p_t dt \\ &= \lambda R_\lambda - \mathrm{id}, \end{aligned}$$

where we used (5.6) in the second equality and integration by parts in the third.

The generator can be characterised in terms of a martingale property.

Proposition 5.14

1. *If f belongs to the domain $\mathscr{D}_G$ of the generator G of a Feller process X,*

$$M(t) := f(X(t)) - \int_0^t Gf(X(s))ds, \quad t \geq 0 \tag{5.8}$$

is a martingale.

2. *For the converse statement suppose that any $x \in E$ can be reached by X in the sense that $P(\inf\{t \geq 0 : X(t) = x\} < \infty) > 0$. If f and g are functions in*

$C_0(E)$ such that

$$M(t) := f(X(t)) - \int_0^t g(X(s))ds, \quad t \geq 0 \tag{5.9}$$

is a martingale, then $f \in \mathscr{D}_G$ and $Gf = g$.

Idea

1. Formally, we have

$$\begin{aligned}
E(M(s+t)|\mathscr{F}_s) &= p_t f(X(s)) - \int_0^s Gf(X(r))dr - \int_s^{s+t} p_{r-s}(Gf)(X(s))dr \\
&= p_t f(X(s)) - \int_0^s Gf(X(r))dr - \int_0^t p_u(Gf)(X(s))du \\
&= p_t f(X(s)) - \int_0^s Gf(X(r))dr - p_t f(X(s)) + p_0 f(X(s)) \\
&= M(s)
\end{aligned}$$

by substitution $u = r - s$ and Proposition 5.12. For a rigorous proof see [241, Proposition VII.1.6].

2. Formally, observe that

$$\begin{aligned}
\frac{p_t f(X(s)) - f(X(s))}{t} &= \frac{1}{t}\big(E(f(X(s+t))|\mathscr{F}_s) - f(X(s))\big) \\
&= E\Big(\frac{1}{t}\int_s^{s+t} g(X(r))dr \Big| \mathscr{F}_s\Big) \\
&\overset{t\to 0}{\longrightarrow} E(g(X(s)|\mathscr{F}_s) \\
&= g(X(s))
\end{aligned}$$

and hence $Gf = g$ as desired. For a complete proof see [241, Proposition VII.1.7]. □

Remark 5.15 The assumption in statement 2 on the finiteness of the hitting time of x can be replaced by the martingale property of M under any measure P_x, where we use the notation of Proposition 5.5.

In the previous proposition *martingale* can be replaced by *local martingale* without altering the statement because the processes in (5.8) and (5.9) are bounded. Proposition 5.14 leads to the notion of an **extended generator** $\widetilde{G}$ of X. A measurable function $f : E \to \mathbb{R}$ belongs to its **domain** $\mathscr{D}_{\widetilde{G}}$ if there exists some measurable function $\widetilde{G}f : E \to \mathbb{R}$ such that

$$M(t) := f(X(t)) - \int_0^t \widetilde{G}f(X(s))ds, \quad t \geq 0 \tag{5.10}$$

is a local martingale. Obviously, we have $\mathscr{D}_G \subset \mathscr{D}_{\widetilde{G}}$ and $\widetilde{G}f = Gf$ (up to sets that are visited with probability 0 by X) for $f \in \mathscr{D}_G$. Note that the extended generator can be defined for quite arbitrary adapted processes, which need not even be Markov. If X is a semimartingale allowing for differential characteristics, the extended generator can be obtained from Proposition 4.31 as is illustrated in the following examples.

Example 5.16 (Lévy Process) Any bounded twice continuously differentiable function $f : \mathbb{R}^d \to \mathbb{R}$ belongs to the domain $\mathscr{D}_{\widetilde{G}}$ of the extended generator $\widetilde{G}$ of a Lévy process X with Lévy–Khintchine triplet (b^h, c, K) and we have

$$\begin{aligned}\widetilde{G}f(x) = (b^h)^\top Df(x) + \frac{1}{2}\sum_{i,j=1}^d c_{ij}D_{ij}f(x)\\ + \int \Big(f(x+y) - f(x) - h(y)^\top Df(x)\Big)K(dy).\end{aligned}$$

If f is twice continuously differentiable, vanishes at infinity, and its first two derivatives vanish at infinity as well, then it belongs to the domain of the generator as well. Indeed, this follows from applying Itō's formula (3.43) to $f(X(t))$ and from Proposition 5.14(2).

Example 5.17 (Geometric Lévy Process) The extended generator of a geometric Lévy process X as in Theorem 3.50 can be written as

$$\begin{aligned}\widetilde{G}f(x) = \beta^h(x)^\top Df(x) + \frac{1}{2}\sum_{i,j=1}^d \gamma_{ij}(x)D_{ij}f(x)\\ + \int \Big(f(x+y) - f(x) - h(y)^\top Df(x)\Big)\kappa(x, dy)\end{aligned}$$

for any bounded twice continuously differentiable function $f : \mathbb{R}^d \to \mathbb{R}$. The functions β^h, γ and the kernel κ are given by

$$\begin{aligned}\beta_i^h(x) &= x_i b_i^{\widetilde{L},h},\\ \gamma_{ij}(x) &= x_i x_j c_{ij}^{\widetilde{L}},\\ \kappa(x, B) &= \int 1_B(x_1y_1, \ldots, x_dy_d)K^{\widetilde{L}}(dy).\end{aligned}$$

If f has compact support, it belongs to the domain of the generator as well. Indeed, this follows again by Itō's formula (3.43) and Proposition 5.14(2).

Example 5.18 Let us consider the Feller process X of Example 5.44 below, i.e. the solution to an SDE as in Theorem 3.41 such that the coefficients $\beta^h : \mathbb{R}^d \to \mathbb{R}^d$, $\sigma : \mathbb{R}^d \to \mathbb{R}^d$, $\delta : \mathbb{R}^d \times \mathbb{R}^d \to \mathbb{R}^d$ do not depend explicitly on the time argument t. Any bounded twice continuously differentiable function $f : \mathbb{R}^d \to \mathbb{R}$ belongs to the domain $\mathscr{D}_{\widetilde{G}}$ of the extended generator $\widetilde{G}$ of X and we have

$$\begin{aligned}\widetilde{G}f(x) &= \beta^h(x)^\top Df(x) + \frac{1}{2}\sum_{i,j=1}^d (\sigma\sigma^\top)_{ij}(x)D_{ij}f(x)\\ &\quad + \int \Big(f(x+\delta(x,y)) - f(x) - h(\delta(x,y))^\top Df(x)\Big) K^\nu((dy)\\ &= \beta^h(x)^\top Df(x) + \frac{1}{2}\sum_{i,j=1}^d \gamma_{ij}(x)D_{ij}f(x)\\ &\quad + \int \Big(f(x+y) - f(x) - h(y)^\top Df(x)\Big)\kappa(x,dy).\end{aligned}$$

Here K^ν denotes the measure in the representation $\nu(dx,dt) = K^\nu(dx)dt$ of the measure ν in (3.65) and

$$\begin{aligned}\gamma(x) &= \sigma(x)\sigma^\top(x),\\ \kappa(x,B) &= \int 1_B(\delta(x,y))K^\nu((dy), \quad B \in \mathscr{B}^d.\end{aligned}$$

If f has in addition compact support, it belongs to the domain $\mathscr{D}_G$ of the generator G of X. Indeed, as in Examples 5.17, 5.18 this follows from applying Itō's formula and from Proposition 5.14(2) for the generator. For the proof of $\widetilde{G}f \in C_0(\mathbb{R}^d)$ note that (5.42) implies Proposition 5.34(4).

What do generators of Markov processes generally look like? Proposition 4.31 and Example 5.18 suggest that the extended generator is typically of the form

$$\begin{aligned}\widetilde{G}f(x) &= \beta^h(x)^\top Df(x) + \frac{1}{2}\sum_{i,j=1}^d \gamma_{ij}(x)D_{ij}f(x)\\ &\quad + \int \Big(f(x+y) - f(x) - h(y)^\top Df(x)\Big)\kappa(x,dy)\end{aligned} \tag{5.11}$$

for sufficiently regular f, where h denotes a truncation function and β^h, γ, κ deterministic functions. This is indeed the case, as the following result shows.

Theorem 5.19 *Suppose that X is an $\mathbb{R}^d$-valued process which is right-continuous and strong Markov relative to any initial law ε_x, cf. Proposition 5.5. Moreover, let the domain $\mathscr{D}_{\widetilde{G}}$ of its extended generator contain all functions with compact support that are infinitely often differentiable or, alternatively, all functions of the form $f(x) = e^{iu^\top x}$, $u \in \mathbb{R}^d$. Then there are functions β^h, γ, κ such that the extended generator satisfies (5.11) for any bounded twice continuously differentiable function $f : \mathbb{R}^d \to \mathbb{R}$. Moreover, X is a semimartingale with local characteristics (b^h, c, K) of the form*

$$b^h(t) = \beta^h(X(t-)), \tag{5.12}$$

$$c(t) = \gamma(X(t-)), \tag{5.13}$$

$$K(t, dx) = \kappa(X(t-), dx). \tag{5.14}$$

Proof This follows from [58, Theorem 7.16]. □

We call a process as in the previous theorem a **Lévy–Itō diffusion.**

In view of (5.12–5.14), β^h, γ, κ can be interpreted as a local Lévy–Khintchine triplet. More specifically, one expects X to resemble a Lévy process with triplet $(\beta^h(x), \gamma(x), \kappa(x, \cdot))$ as long as $X(t)$ is close to x.

For purposes of quadratic hedging in Chap. 12 we also need the *carré du champ* operator of a Lévy–Itō diffusion X. It helps to determine the predictable covariation of a Markov process. We say that a pair (f, g) of measurable functions $f, g : E \to \mathbb{R}$ belongs to the domain $\mathscr{D}_\Gamma$ of the **carré du champ operator** Γ if there exists some measurable function $\Gamma(f, g) : E \to \mathbb{R}$ such that

$$[f(X), g(X)](t) - \int_0^t \Gamma(f, g)(X(s))ds, \quad t \geq 0$$

is a local martingale or, equivalently,

$$\langle f(X), g(X) \rangle = \int_0^t \Gamma(f, g)(X(s))ds, \quad t \geq 0. \tag{5.15}$$

$\Gamma(f, g)$ can be expressed in terms of the extended generator if f, g are sufficiently regular.

Proposition 5.20 *If f, g, fg belong to the domain $\mathscr{D}_{\widetilde{G}}$ of the extended generator $\widetilde{G}$ of X, then (f, g) belongs to the domain $\mathscr{D}_\Gamma$ of the carré du champ operator and we have*

$$\Gamma(f, g) = \widetilde{G}(fg) - f\widetilde{G}g - g\widetilde{G}f.$$

Proof Integration by parts and the definition of the extended generator yield

$$
\begin{aligned}
[f(X), g(X)](t) &= (fg)(X(t)) - (fg)(X(0)) - f(X_-) \bullet g(X)(t) - g(X_-) \bullet f(X)(t) \\
&\sim \int_0^t \Big(\widetilde{G}(fg)(X(s)) - f(X(s))\widetilde{G}g(X(s)) - g(X(s))\widetilde{G}f(X(s))\Big)ds \\
&= \int_0^t \Gamma(f, g)(X(s))ds,
\end{aligned}
$$

where $\sim$ denotes equality up to a local martingale. Hence the right-hand side is the compensator of the left as desired. □

Interestingly, (5.15) holds even if f and g depend explicitly on time.

Proposition 5.21 *Suppose that $f, g : \mathbb{R}_+ \times E \to \mathbb{R}$ are bounded and twice continuously differentiable. Then we have*

$$
\langle f(I, X), g(I, X)\rangle(t) = \int_0^t \Gamma(f(s, \cdot), g(s, \cdot))(X(s))ds, \quad t \geq 0.
$$

Proof Integration by parts and Itō's formula yield that

$$
\begin{aligned}
&[f(I, X), g(I, X)](t) \\
&\quad = (fg)(t, X(t)) - (fg)(0, X(0)) - f(I, X_-) \bullet g(I, X)(t) - g(I, X_-) \bullet f(I, X)(t) \\
&\quad \sim \int_0^t \Big(\widetilde{G}(fg)(s, \cdot)(X(s)) - f(s, X(s))\widetilde{G}g(s, \cdot)(X(s)) - g(s, X(s))\widetilde{G}f(s, \cdot)(X(s)) \\
&\qquad - (\dot{fg})(s, X(s)) - f\dot{g}(s, X(s)) - g\dot{f}(s, X(s))\Big)ds \\
&\quad = \int_0^t \Gamma(f(s, \cdot), g(s, \cdot))(X(s))ds,
\end{aligned}
$$

where $\sim$ denotes equality up to a local martingale and the dot stands for the derivative with respect to the time variable. Hence the right-hand side is the compensator of the left as desired. □

Remark 5.22 In the literature one can occasionally find extended generators of the form

$$
\begin{aligned}
\widetilde{G}f(x) = \alpha(x)f(x) + \beta^h(x)^\top Df(x) + \frac{1}{2}\sum_{i,j=1}^d \gamma_{ij}(x)D_{ij}f(x) \\
+ \int \Big(f(x + y) - f(x) - h(y)^\top Df(x)\Big)\kappa(x, dy)
\end{aligned}
$$

with an additional function $\alpha(x) \geq 0$ compared to (5.11). From a probabilistic point of view this corresponds to a process as in (5.11) which vanishes with intensity $\alpha(X(t))$, i.e. it jumps to some graveyard state Δ from which it does not escape. $\alpha(X(t))$ is called the **killing rate** of the process. Roughly speaking, this means that $\alpha(X(t))dt$ is the probability that the process vanishes in an interval of length dt around t.

For this to make sense we need to add the graveyard state Δ to E. To this end, E is replaced by its one-point compactification E^Δ. In order for results such as the backward equation to still hold, we set $f(\Delta) = 0$ for any $f \in C_0(E)$, which extends such f to a continuous function on E^Δ.

Even if killing is not intended, it sometimes makes sense to work with E^Δ instead of E. Indeed, the existence of processes with generator (5.11) holds under less restrictive conditions if the state space is compact. Specifically, for exploding processes $X(t) \in \mathbb{R}^d$ is not defined for all $t \geq 0$. But by letting $X(t) = \Delta$ after explosion, it can be defined as a process with values in $(\mathbb{R}^d)^\Delta$. Growth conditions as in Sect. 5.8 are often imposed to warrant that a process is **conservative**, i.e. it does not reach the graveyard state unless it started there. Hence there is no need to extend the state space for such processes.

We end this section with a comment on limit theorems for stochastic processes.

Remark 5.23 (Limit Theorems) Section 4.7 dealt with convergence in law of a sequence (X_n) of processes to a limiting process X. Under some additional regularity it could be derived from convergence of the corresponding sequences of semimartingale characteristics and of the initial distributions. Semimartingale characteristics, on the other hand, are closely linked to the extended generator of a Markov process, as we have seen in Theorem 5.19. So one may wonder whether limit theorems in the spirit of Theorem 4.36 and Corollary 4.38 can be phrased directly in terms of convergence of the sequence of extended generators of X_n. Such statements can be found in [100, Section 4.8], in particular Theorem 4.8.10 and the following corollaries.

5.4 The Backward Equation and Feynman–Kac Formula

What is the use of Markov processes in the context of Mathematical Finance? We will see in Part II that option prices are often of the form $E(f(X(t)))$ or $E(f(X(t))\exp(-\int_0^t g(X(s))ds))$ for some functions f and g. The generator of a Markov process allows us to compute such expectations by solving partial differential equations or, more generally, *partial integro-differential equations (PIDEs)*. This opens the door to sophisticated numerical methods for the computation of option prices.

Theorem 5.24 (Backward Equation) *Suppose that X is a Feller process with generator G. If f belongs to the domain $\mathscr{D}_G$ of G, then $u(t,\cdot) \in \mathscr{D}_G$, $t \geq 0$ for*

$$u(t,x) = \int f(y)P(t,x,dy). \tag{5.16}$$

Moreover, u is the unique solution to the Cauchy problem $u(0,x) = f(x)$ and

$$\frac{\partial}{\partial t}u = Gu. \tag{5.17}$$

Here, Gu means that G is applied to $u(t,\cdot)$ for fixed t. Moreover,

$$\frac{\partial}{\partial t}u(t,x) = \lim_{\tau \to 0} \frac{u(t+\tau,x) - u(t,x)}{\tau}$$

refers to a limit in $C_0(E)$, i.e. uniform in x.

Proof Equation (5.17) follows immediately from Proposition 5.12. Uniqueness of the solution to this Cauchy problem is stated, for example, in [292, Satz VII.4.7]. □

Equation (5.16) can be rephrased as

$$u(t,x) = E_x(f(X(t))),$$

where E_x denotes the expectation for $X(0) = x$ as in Proposition 5.5, or as

$$u(t,x) = E\big(f(X(s+t))\big|X(s) = x\big),$$

which means

$$u(t,X(s)) = E\big(f(X(s+t))\big|\mathscr{F}_s\big) \tag{5.18}$$

in terms of conditional expectations as in Sect. 1.1.

In order to compute expectations of the form $E(f(X(t))\exp(-\int_0^t g(X(s))ds))$, we need a generalised version of Theorem 5.24. Gu and $\partial u/\partial t$ are to be interpreted as above.

Theorem 5.25 (Feynman–Kac Formula) *Suppose that X is a Feller process with generator G. If f belongs to the domain $\mathscr{D}_G$ of G and g is a continuous bounded nonnegative function, then $u(t,\cdot) \in \mathscr{D}_G$, $t \geq 0$ holds for*

$$u(t,x) = E_x\left(f(X(t))\exp\left(-\int_0^t g(X(r))dr\right)\right),$$

which essentially means

$$u(t, X(s)) = E\left(f(X(s+t)) \exp\left(-\int_s^{s+t} g(X(r))dr \right) \middle| \mathscr{F}_s \right), \tag{5.19}$$

cf. Proposition 5.5. Moreover, u is the unique solution to $u(0, x) = f(x)$ and

$$\frac{\partial}{\partial t} u = Gu - gu. \tag{5.20}$$

Proof Define a transition function

$$Q(t, x, B) := E_x\left(1_B(X(t)) \exp\left(-\int_0^t g(X(r))dr \right) \right).$$

It can be shown that it generates a Feller semigroup $(q_t)_{t\geq 0}$ with generator $\tilde{G} := G - g$, i.e. for $f \in \mathscr{D}_G$ we have $f \in \mathscr{D}_{\tilde{G}}$ and $\tilde{G}f(x) = Gf(x) - g(x)f(x)$, cf. [251, Section III.19]. The assertion now follows by applying the semigroup version of Theorem 5.24 to this semigroup $(q_t)_{t\geq 0}$, cf. e.g. [292, Satz VII.4.7]. The transition function $Q(\cdot, \cdot, \cdot)$ corresponds to the process X *killed* at rate $g(X(t))$, i.e. it jumps to a graveyard state at some random time, from which it cannot escape. □

Remark 5.26 If f does not vanish at infinity and hence $f \notin \mathscr{D}_G$, one can show a sufficient version of the above results for Lévy–Itō diffusions X and bounded C^2-functions $u : \mathbb{R}_+ \times \mathbb{R}^d \to \mathbb{R}^d$. If such a u solves $u(0, x) = f(x)$ and

$$\frac{\partial}{\partial t} u = \tilde{G}u \tag{5.21}$$

(resp. $\frac{\partial}{\partial t} u = \tilde{G}u - gu$), it satisfies (5.18) (resp. (5.19)). Indeed, applying Itō's formula (3.43) and (5.11) yield that $(u(s + t - r, X(r))_{r\in[0,s+t]}$ is a bounded local martingale and hence a martingale. In particular, we have

$$u(t, X(s)) = E(u(0, X(s + t))|\mathscr{F}_s) = E(f(X(s + t))|\mathscr{F}_s)$$

as desired. The statement with (5.19) follows along the same lines. Note that this derivation does not really require X to be Markov; it only relies on the extended generator (5.11).

In view of (5.11), the computation of expectations essentially boils down to solving partial integro-differential equations (PIDEs). In general, however, we do not know whether the solution u is sufficiently smooth for the right-hand side of (5.11) to make sense. But according to Theorems 5.24 and 5.25, (5.17) and (5.20) hold for general $f \in \mathscr{D}_G$—even if u is not twice continuously differentiable.

The introduction of *viscosity solutions* provides another way to give meaning to such PIDEs in cases where u lacks the smoothness that is required for the right-hand side of (5.11) to make sense.

5.5 The Forward Equation

The extended generator or more precisely its adjoint allows us to express the evolution of the law $P^{X(t)}$ of $X(t)$ over time. To this end, suppose that X is a Lévy–Itō diffusion with state space $\mathbb{R}^d$ and extended generator $\widetilde{G}$. More specifically, we assume that $\widetilde{G}f$ is a bounded function for any infinitely differentiable function f with compact support. In view of (5.11), this is a mild boundedness condition on the coefficients β^h, γ, κ. Moreover, we suppose that X is the unique solution to its martingale problem in the sense of Sects. 4.6 or 5.7.1 below.

Let $C_c^\infty(\mathbb{R}^d)$ denote the set of all infinitely differentiable functions $f : \mathbb{R}^d \to \mathbb{R}$ with compact support. Any probability measure μ on the state space $\mathbb{R}^d$ can be viewed as a linear mapping $C_c^\infty(\mathbb{R}^d) \to \mathbb{R}$ by setting $\mu f := \int f(x)\mu(dx)$ for $f \in C_c^\infty(\mathbb{R}^d)$. We denote by $\mathscr{M}(\mathbb{R}^d)$, $C_c^\infty(\mathbb{R}^d)'$ the set of probability measures on $\mathbb{R}^d$ and the set of linear mappings $C_c^\infty(\mathbb{R}^d) \to \mathbb{R}$, respectively. Moreover, we define the **adjoint operator** $A : \mathscr{M}(\mathbb{R}^d) \to C_c^\infty(\mathbb{R}^d)'$ of $\widetilde{G}$ by setting $(A\mu)f := \mu(\widetilde{G}f)$.

Theorem 5.27 (Forward Equation) *$\mu_t := P^{X(t)}$, $t \geq 0$ is the unique solution to the integral equation*

$$\mu_t = \mu_0 + \int_0^t A\mu_s ds, \quad t \geq 0$$

in the sense that

$$\mu_t f = \mu_0 f + \int_0^t (A\mu_s) f ds, \quad t \geq 0 \tag{5.22}$$

for any $f \in C_c^\infty(\mathbb{R}^d)$.

Proof The assertion follows from [100, Proposition 4.9.19]. □

Under certain smoothness conditions, we can express the adjoint operator more explicitly. For ease of notation we consider the univariate case, i.e. with $d = 1$ and state space $E = \mathbb{R}$. As before, we assume that X is the unique solution to its martingale problem and that its extended generator $\widetilde{G}$ maps $C_c^\infty(\mathbb{R})$ to bounded functions. Moreover, suppose that $\widetilde{G}$ is of the form

$$\widetilde{G}f(x) = \beta^h(x)f'(x) + \frac{1}{2}\gamma(x)f''(x) + \int \big(f(x+y) - f(x) - h(y)f'(x)\big)\kappa(x,y)dy \tag{5.23}$$

with truncation function h, continuously differentiable $\beta^h : \mathbb{R} \to \mathbb{R}$, twice continuously differentiable $\gamma : \mathbb{R} \to \mathbb{R}$, and

$$\sup_{x\in\mathbb{R}} \int (|\kappa(x,y)| + |D_1\kappa(x,y)| + |D_{11}\kappa(x,y)|)\, y^2 dy < \infty, \tag{5.24}$$

where D_1, D_{11} denote the derivatives of $x \mapsto \kappa(x,y)$, which we assume to exist. If now $\mu \in \mathscr{M}(\mathbb{R})$ has twice continuously differentiable density ϱ, integration by parts yields that $A\mu$ has density

$$\begin{aligned}(A\varrho)(x) :=\; & -(\beta^h\varrho)'(x) + \frac{1}{2}(\gamma\varrho)''(x) \\ & + \int \Big(\kappa(x-y,y)\varrho(x-y) - \kappa(x,y)\varrho(x) + (\kappa(\cdot,y)\varrho)'(x)h(y)\Big)dy,\end{aligned} \tag{5.25}$$

i.e.

$$\begin{aligned}(A\mu)f = \int \Big(& -(\beta^h\varrho)'(x) + \frac{1}{2}(\gamma\varrho)''(x) \\ & + \int \Big(\kappa(x-y,y)\varrho(x-y) - \kappa(x,y)\varrho(x) + (\kappa(\cdot,y)\varrho)'(x)h(y)\Big)dy\Big) f(x)dx\end{aligned}$$

for $f \in C_c^\infty(\mathbb{R})$. This leads to

Theorem 5.28 (Fokker–Planck Equation) *Suppose that $\varrho_t : \mathbb{R} \to \mathbb{R}_+$, $t \geq 0$ are twice continuously differentiable probability density functions that solve*

$$\frac{\partial \varrho_t(x)}{\partial t} = (A\varrho_t)(x), \quad t \geq 0.$$

Moreover, we assume that $(t,x) \mapsto (A\varrho_t)(x)$ is bounded on compact subsets of $\mathbb{R}_+ \times \mathbb{R}$. If ϱ_0 is the density of $P^{X(0)}$, then ϱ_t is the density of $P^{X(t)}$ for any $t \geq 0$.

Proof For A as in (5.25), integration by parts yields

$$(A\mu_t)f = \mu_t(\widetilde{G}f) = \int (A\varrho_t)(x)f(x)dx$$

if μ_t denotes the probability measure with probability density function ϱ_t and f is a smooth function with compact support. Therefore

$$\begin{aligned}\int_0^t (A\mu_s)f ds &= \int_0^t \mu_s(\widetilde{G}f)ds \\ &= \int_0^t \int (A\varrho_s)(x)f(x)dxds\end{aligned}$$

$$= \int \int_0^t (A\varrho_s)(x) ds f(x) dx$$

$$= \int \int_0^t \frac{\partial \varrho_s(x)}{\partial s} ds f(x) dx$$

$$= \int (\varrho_t(x) - \varrho_0(x)) f(x) dx$$

$$= \mu_t f - \mu_0 f,$$

which means that the family $(\mu_t)_{t \geq 0}$ solves (5.22). By the uniqueness statement of Theorem 5.27, we are done. □

Example 5.29 (Lévy Process) If X is a univariate Lévy process with Lévy–Khintchine triplet (b^h, c, K), we have

$$(A\varrho)(x) = -b^h \varrho'(x) + \frac{1}{2} c \varrho''(x) + \int \Big(\varrho(x-y) - \varrho(x) + \varrho'(x) h(y) \Big) K(dy)$$

for twice continuously differentiable probability density functions ϱ. In this case, we need not assume that the Lévy measure K has a density as we implicitly did in (5.23, 5.24). Indeed, as for the proof of Theorem 5.28, one verifies using integration by parts that $\mu_t(\widetilde{G} f) = \int (A\varrho_t)(x) f(x) dx$ if μ_t denotes the probability measure with probability density function ϱ_t and f a smooth function with compact support. Therefore we can argue as above.

5.6 Markov Processes and Semimartingales

We have considered two important classes of stochastic processes, namely semimartingales and Markov processes. In both families there are tools to characterise the local dynamics of a process, namely semimartingale characteristics and generators, both leading to martingale problems in the end. This naturally raises the question of how these classes and concepts are related. Not surprisingly, there are many semimartingales which are not Markov processes. Indeed, the future evolution of an arbitrary semimartingale may well depend on its whole past and not just its present value. The converse is less obvious. But one can show that there are Markov processes which are not semimartingales, e.g. $X(t) = \sqrt{|W(t)|}$ for standard Brownian motion W, cf. e.g. [58, Remark 5.7]. Nevertheless, most processes in applications belong to both classes.

But even if a Markov process is a semimartingale, it does not necessarily allow for local characteristics. As an example consider $X(t) = |W(t)|$ with standard Brownian motion W. The Itō–Tanaka formula yields its canonical decomposition

$$X(t) = M^X(t) + A^X(t) = \int_0^t \operatorname{sgn} X(t) dW(t) + L^0(t),$$

where the sign function is defined here as

$$\operatorname{sgn}(x) := \begin{cases} 1 & \text{for } x \geq 0, \\ -1 & \text{for } x < 0. \end{cases} \tag{5.26}$$

The compensator $A^X = L^0$ of X coincides with the local time of W in 0, which is not absolutely continuous in time. Therefore a drift rate and hence local characteristics of X do not exist in this case. Moreover, the extended generator is not of the form (5.11) for arbitrary C^2-functions. However, recall from Theorem 5.19 that if the domain of the extended generator of a Markov process is sufficiently large, it is a semimartingale allowing for local characteristics. Furthermore, its extended generator is of the form (5.11) and it is linked naturally to the local characteristics (b^h, c, K) via (5.12–5.14).

What about a converse to Theorem 5.19? If the characteristics of a semimartingale are of the form (5.12–5.14) for some deterministic functions β^h, γ, κ, the local dynamics depend on ω and t only through the current value $X(t-)$. It is thus natural to conjecture that X is a Markov process whose generator is of the form (5.11). However, this does not hold in general. The point is that the characteristics may not determine the law of X uniquely.

As a simple example consider the deterministic process

$$X(t) := \begin{cases} 0 & \text{for } t \leq 1, \\ (t-1)^2 & \text{for } t > 1, \end{cases}$$

which has characteristics of the form (5.12–5.14) with $\beta^h(x) = 2\sqrt{|x|}$, $\gamma(x) = 0$, $\kappa(x, \cdot) = 0$. One easily verifies that X is not a Markov process. Indeed, the law of $X(2)$ given $X(1)$ is entirely different from the law of $X(1)$ given $X(0)$.

The following theorem shows that this phenomenon does not occur if X is the unique solution to its martingale problem. To this end, suppose that X is an $\mathbb{R}^d$-valued semimartingale whose local characteristics (b^h, c, K) are of the form (5.12–5.14) for some triplet $(\beta^h, \gamma, \kappa)$ that is continuous in the sense of Proposition 5.33.

Theorem 5.30 *If the martingale problem related to $(\beta^h, \gamma, \kappa)$ has a unique solution for any initial distribution on $\mathbb{R}^d$, then X is a strong Markov process whose extended generator is given by (5.11) for any bounded twice continuously differentiable function $f : \mathbb{R}^d \to \mathbb{R}$. In particular, X is a Lévy–Itō diffusion in the sense of Sect. 5.3.*

Proof This follows from [100, Theorems 4.4.2 and 4.4.6]. For the application of Theorem 4.4.6 one may consider the set of functions $x \mapsto \chi(nx)p(x)$ with a fixed cutoff function χ, a polynomial p with rational coefficients, and $n \in \mathbb{N}$. This set is dense in C_c^∞. □

Remark 5.31 In the setup of Theorem 5.30 it suffices to verify that the univariate marginals $\mathscr{L}(X(t))$, $\mathscr{L}(\widetilde{X}(t))$ of any two solutions $X, \widetilde{X}$ to the martingale problem coincide. This already implies that the laws $\mathscr{L}(X), \mathscr{L}(\widetilde{X})$ of the entire processes coincide, cf. [100, Theorems 4.4.2].

5.7 Existence and Uniqueness

Here we come back to questions raised in Sect. 4.6, namely whether there exists a unique process with a given triplet.

Remark 5.32 The generator is akin to the derivative of a deterministic function in that it quantifies the local behaviour of a process by comparing it to an object of constant growth, namely a linear function in analysis and a Lévy process in (5.11). But the local triplet $(\beta^h(x), \gamma(x), \kappa(x, \cdot))$ in (5.11) is expressed as a function of the current state $X(t-) = x$ of the process. This situation is reminiscent of ordinary differential equations (ODEs), where the derivative is not given explicitly but in terms of the function's current value. Indeed, let us consider a deterministic function $X(t)$ satisfying the ODE

$$\frac{dX}{dt}(t) = \beta(X(t)) \tag{5.27}$$

with some function β. If we allow for a general starting value $X(0) = x$, the function X can be interpreted as a deterministic Markov process with generator

$$Gf(x) = \lim_{t\to 0} \frac{f(X(t)) - f(X(0))}{t} = f'(X(0))\frac{dX}{dt}(0) = f'(x)\beta(x).$$

The operator equation

$$Gf(x) = f'(x)\beta(x)$$

can be viewed as another way of writing the ODE (5.27). It is obtained as a special case of (5.11) for $\gamma = 0, \kappa = 0$. To sum up, equations such as (5.11) transfer the idea of ordinary differential equations to the level of processes. This should be contrasted with stochastic differential equations which relate rather with *integral* equations.

This correspondence with ODEs immediately raises the question of existence and uniqueness of a Feller or more general Markov process X whose generator or extended generator satisfies (5.11). In line with Sect. 4.6 we say that such a process X or its law solves the *martingale problem* related to the triplet $(\beta^h, \gamma, \kappa)$ and initial distribution $\mathscr{L}(X(0))$. This term is motivated by the fact that it boils down to the martingale property of processes M as in (5.10). Obviously, uniqueness can only refer to the law of X. Indeed, any Lévy process with triplet (b^h, c, K) has the same generator.

In the special case of ODEs existence and uniqueness holds for sufficiently regular coefficients, with local existence holding under less restrictive assumptions than uniqueness. Very roughly speaking, the same is true for martingale problems. We discuss existence in Sect. 5.7.1 and several instances of uniqueness in the subsequent sections. In Chap. 6 we consider processes whose triplet $(\beta^h(x), \gamma(x), \kappa(x, \cdot))$ is an affine function of x. Thanks to their flexibility and mathematically tractability, such *affine* processes play an important role in financial modelling.

5.7.1 Martingale Problems

Fix a truncation function h on the state space $\mathbb{R}^d$. In the following we consider a triplet $(\beta^h, \gamma, \kappa)$ of measurable functions $\beta^h : \mathbb{R}^d \to \mathbb{R}^d$, $\gamma : \mathbb{R}^d \to \mathbb{R}^{d\times d}$, and $\kappa : \mathbb{R}^d \times \mathscr{B}^d \to \mathbb{R}$ such that $\gamma(x)$ is a covariance matrix and $\kappa(x, \cdot)$ is a Lévy measure. We call the mapping $\psi : \mathbb{R}^d \times \mathbb{R}^d \to \mathbb{C}$ with

$$\psi(x, u) = iu^\top \beta^h(x) - \frac{1}{2}u^\top \gamma(x)u + \int \left(e^{iu^\top y} - 1 - iu^\top h(y)\right) \kappa(x, dy)$$

the corresponding **symbol**. In other words, $(\beta^h(x), \gamma(x), \kappa(x, \cdot))$ is a Lévy–Khintchine triplet for fixed x and $\psi(x, \cdot)$ its characteristic exponent. Finally, we define the **operator** $\widetilde{G} : C_b^2(\mathbb{R}^d) \to M(\mathbb{R}^d)$ related to the triplet $(\beta^h, \gamma, \kappa)$ by (5.11), where $C_b^2(\mathbb{R}^d)$, $M(\mathbb{R}^d)$ denote the sets of bounded, twice continuously differentiable resp. measurable functions from $\mathbb{R}^d$ to $\mathbb{R}$. Observe that the symbol is obtained from the operator via

$$\psi(x, u) = e^{-iu^\top x}\widetilde{G}f(x), \quad x \in \mathbb{R}^d$$

for $f(x) = e^{iu^\top x}$. Moreover, the triplet $(\beta^h, \gamma, \kappa)$ is determined by the restriction of $\widetilde{G}$ to the set $C_c^\infty(\mathbb{R}^d)$ of infinitely differentiable functions from $\mathbb{R}^d$ to $\mathbb{R}$ with compact support, cf. [58, (7.8)].

In line with Sect. 4.6 we call an adapted $\mathbb{R}^d$-valued càdlàg process X a **solution to the martingale problem** related to the initial law P^0 on $\mathbb{R}^d$ and the triplet $(\beta^h, \gamma, \kappa)$, symbol ψ, or operator $\widetilde{G}$ if the law of $X(0)$ is P^0 and one of the following equivalent conditions holds:

1.

$$M(t) := e^{iu^\top X(t)} - \int_0^t e^{iu^\top X(s-)}\psi(X(s-), u)ds$$

is a local martingale for any $u \in \mathbb{R}^d$,

2.
$$M(t) := f(X(t)) - \int_0^t \widetilde{G} f(X(s))ds$$

is a local martingale for any $f \in C_c^\infty(\mathbb{R}^d)$,

3.
$$M(t) := f(t, X(t)) - \int_0^t \left(D_1 f(s, X(s)) + \widetilde{G} f(X(s))\right) ds$$

is a local martingale for any bounded twice continuously differentiable function $f : \mathbb{R}_+ \times \mathbb{R}^d \to \mathbb{R}$, where $\widetilde{G} f(t, \cdot)$ refers to $\widetilde{G}$ applied to $f(t, \cdot)$ and $D_1 f(t, x)$ indicates the derivative of $t \mapsto f(t, x)$.

The equivalence of these conditions is shown in Proposition 4.31. According to [154, Theorem II.2.42] any solution X is a semimartingale. We say that **uniqueness holds for the martingale problem** related to P^0 and triplet $(\beta^h, \gamma, \kappa)$, symbol ψ, or operator $\widetilde{G}$ if any two solutions $X, \widetilde{X}$, possibly defined on different filtered probability spaces, have the same law.

Both the triplet and the symbol may be defined on some closed subset $E \subset \mathbb{R}^d$ instead of $\mathbb{R}^d$ if we are interested in processes X with state space E, cf. e.g. Sect. 6.1 on affine processes. In this case we define $\widetilde{G}$ on $C_b^2(E) := \{f|_E : f \in C_b^2(\mathbb{R}^d)\}$. Sets such as $C_c^\infty(E) := \{f|_E : f \in C_c^\infty(\mathbb{R}^d)\}$ are defined accordingly.

5.7.2 Existence

Existence is known to hold under relatively weak continuity assumptions. Let $(\beta^h, \gamma, \kappa)$ be a triplet with corresponding symbol ψ and operator $\widetilde{G}$ as in Sect. 5.7.1.

Proposition 5.33 (Continuity) *The following conditions are equivalent.*

1. *The triplet $(\beta^h, \gamma, \kappa)$ is continuous in the sense that, for some continuous truncation function h, the mappings $x \mapsto \beta^h(x)$, $x \mapsto \widetilde{\gamma}^h(x) := \gamma(x) + \int h(y)h(y)^\top \kappa(x, dy)$, and $x \mapsto \int f(y)\kappa(x, dy)$ are continuous for any bounded continuous function $f : \mathbb{R}^d \to \mathbb{R}$ that vanishes in a neighbourhood of 0 (or, equivalently, any bounded continuous function $f : \mathbb{R}^d \to \mathbb{R}$ that is $o(|x|^2)$ for $x \to 0$).*
2. *The symbol ψ is continuous.*
3. *$\widetilde{G} f$ is continuous for any uniformly continuous $f \in C_b^\infty(\mathbb{R}^d)$, which denotes the set of bounded, infinitely differentiable functions on $\mathbb{R}^d$.*

If any of these conditions holds, we say that the coefficients of the martingale problem are ***continuous****.*

Proof $1 \Rightarrow 3$: We have

$$\widetilde{G}f(x) = \beta^h(x)^\top Df(x) + \frac{1}{2}\sum_{i,j=1}^{d} \widetilde{\gamma}_{ij}^h(x)D_{ij}f(x) + \int g(x,y)\kappa(x,dy) \tag{5.28}$$

with

$$\begin{aligned} g(x,y) &:= f(x+y) - f(x) - h(y)^\top Df(x) - \frac{1}{2}\sum_{i,j=1}^{d} h_i(y)h_j(y)D_{ij}f(x) \\ &= f(x+y) - f(x+h(y)) \\ &\quad + \frac{1}{2}\int_0^1 \sum_{i,j,k=1}^{d} (1-t)^2 h_i(y)h_j(y)h_k(y)D_{ijk}f(x+th(y))dt, \end{aligned}$$

where we used Taylor's theorem with integral remainder term. The first two terms on the right-hand side of (5.28) are continuous in x. Let $x_n \to x$. Then

$$|g(x,y) - g(x_n,y)| \leq \varepsilon_n(1 \wedge |y|^3)$$

with $\varepsilon_n \to 0$, which implies $\int |g(x_n,y) - g(x,y)|\kappa(x_n,dy) \to 0$ as $n \to \infty$ by continuity of κ in x. Once more applying this continuity, we have that

$$\begin{aligned} &\left|\int g(x_n,y)\kappa(x_n,dy) - \int g(x,y)\kappa(x,dy)\right| \\ &= \int |g(x_n,y) - g(x,y)|\kappa(x_n,dy) + \left|\int g(x,y)\kappa(x_n,dy) - \int g(x,y)\kappa(x,dy)\right| \end{aligned}$$

converges to 0 as well.

$3 \Rightarrow 2$: Since $\psi(x,u) = e^{-iu^\top x}\widetilde{G}f(x)$ for the bounded, uniformly continuous, smooth function $f(x) = e^{iu^\top x}$, we have that $\psi(x_n,u) \to \psi(x,u)$ pointwise in u for $x_n \to x$. Since $\exp(\psi(x_n,\cdot))$ and $\exp(\psi(x,\cdot))$ are characteristic functions, Lévy's continuity theorem yields $\exp(\psi(x_n,\cdot)) \to \exp(\psi(x,\cdot))$ and hence $\psi(x_n,\cdot) \to \psi(x,\cdot)$ uniformly on compact sets as $n \to \infty$, cf. Proposition A.5(8). This implies that both terms on the right-hand side of

$$|\psi(x_n,u_n) - \psi(x,u)| \leq |\psi(x_n,u_n) - \psi(x,u_n)| + |\psi(x,u_n) - \psi(x,u)|$$

converge to 0 for $(x_n,u_n) \to (x,u)$, which in turn yields continuity of the symbol.

$2 \Rightarrow 1$: If $\psi(x_n,u) \to \psi(x,u)$ pointwise in u for $x_n \to x$, the infinitely divisible laws corresponding to $\psi(x_n,\cdot)$ converge weakly to the law corresponding to $\psi(x,\cdot)$ by Lévy's continuity theorem, cf. Proposition A.5(8). Reference [154,

Theorem VII.2.9] states that this implies convergence $\beta^h(x_n) \to \beta^h(x)$ and likewise for $\widetilde{\gamma}^h$ and κ in the above sense. □

Note from the proof that in 2. it suffices to verify that $x \mapsto \psi(x, u)$ is continuous for any u.

Additionally, we need some assumption on the behaviour near infinity.

Proposition 5.34 (Vanishing at Infinity) *For continuous coefficients we consider the following conditions.*

1.

$$\lim_{|x|\to\infty} \kappa(x, B(-x, r)) = 0, \quad r \in \mathbb{R}_+, \tag{5.29}$$

where $B(-x, r) := \{y \in \mathbb{R}^d : |x + y| \leq r\}$.

2. $\kappa(x, dy) \leq (1 + |x|)K(dy)$ *for some measure* K *on* $\mathbb{R}^d$ *with* $\int_{[1,\infty)} |y|K(dy) < \infty$.

3.

$$\lim_{|x|\to\infty} \sup_{|u|\leq 1/|x|} \operatorname{Re}(\psi(x, u)) = 0.$$

4. $\lim_{|x|\to\infty} \widetilde{G}f(x) = 0$ *for any* $f \in C_c^\infty(\mathbb{R}^d)$.

We have $3 \Rightarrow 4 \Leftrightarrow 1 \Leftarrow 2$. *If any of these conditions hold, we say that the coefficients of the martingale problem* ***vanish at infinity***.

Proof $3 \Rightarrow 4 \Leftrightarrow 1$: This is shown in [36, Lemma 3.26].
$2 \Rightarrow 1$: The easy proof is left to the reader. □

A third property concerns the growth of the coefficients.

Proposition 5.35 (Linear Growth) *For continuous coefficients we consider the following conditions.*

1. There exists some $c < \infty$ *such that for any* $x \in \mathbb{R}^d$ *we have*

$$|\beta^{h_x}(x)| \leq c(1 + |x|), \tag{5.30}$$

$$\operatorname{tr}\left(\gamma(x) + \int h_x(x)h_x(x)^\top \kappa(x, dy)\right) \leq c(1 + |x|^2), \tag{5.31}$$

$$\kappa\big(x, \{y \in \mathbb{R}^d : |y| \geq 1 \vee |x|/2\}\big) \leq c, \tag{5.32}$$

where $h_x : \mathbb{R}^d \to \mathbb{R}^d$ *is defined as* $h_x(y) := y1_{\{|y|\leq 1\vee|x|/2\}}$.

2. There exists some $c < \infty$ *such that for any* $x \in \mathbb{R}^d$ *we have*

$$\left(|\beta^{\mathrm{id}}(x)|^2 + \operatorname{tr}(\gamma(x)) + \int |y|^2\kappa(x, dy)\right) < c(1 + |x|^2).$$

3.

$$\limsup_{|x|\to\infty} \sup_{|u|\le 1/|x|} |\psi(x,u)| < \infty. \tag{5.33}$$

4. There exists a sequence $(f_n)_{n\in\mathbb{N}}$ *in* $C_c^\infty(\mathbb{R}^d)$ *such that*

$$\sup_{n\in\mathbb{N}} \sup_{x\in\mathbb{R}^d} (|f_n(x)| + |\widetilde{G} f_n(x)|) < \infty$$

and $\lim_{n\to\infty} f_n(x) = 1$, $\lim_{n\to\infty} \widetilde{G} f_n(x) = 0$ *for any* $x \in \mathbb{R}^d$.

We have $2 \Rightarrow 1 \Leftrightarrow 3$. *If*

$$\kappa\big(x, \{y \in \mathbb{R}^d : |y| \ge 2 \vee |x|\}\big) = 0, \quad x \in \mathbb{R}^d, \tag{5.34}$$

we obtain the implication $1 \Rightarrow 4$ *as well. If any of 1.–4. hold, we say that the coefficients of the martingale problem satisfy the* **growth condition**.

Proof $1 \Leftrightarrow 3$: This follows from [201, Lemma 3.1]. To this end, note that local boundedness in the sense of [201] is implied by continuity of the coefficients.

$2 \Rightarrow 1$: This is obvious.

$1 \Rightarrow 4$ for $c = 0$ in (5.32): Let $g \in C_c^\infty(\mathbb{R}^d)$ such that $g(x) = 1$ for $|x| \le 1$ and $g(x) = 0$ for $|x| \ge 2$. For $n \in \mathbb{N}$ define $f_n \in C_c^\infty(\mathbb{R}^d)$ by $f_n(x) := g(x/n)$. One easily verifies that (5.30, 5.31) warrant that f_n has the desired properties. □

We are now ready to state an existence result:

Theorem 5.36 *Suppose that the coefficients of the martingale problem are continuous, vanish at infinity, and satisfy the growth condition. Then there exists a solution X to the martingale problem.*

If the growth conditions 1. or 3. holds, we need not assume that the coefficients vanish at infinity.

Proof

Step 1: If the growth condition 4. holds, this follows from [100, Theorems 4.5.4 and 4.3.8].

Step 2: Suppose that condition 1. in Proposition 5.35 holds. For some continuous function $\lambda : \mathbb{R}_+ \to [0, 1]$ with

$$\lambda(y) = \begin{cases} 0 \text{ on } [0, 1], \\ 1 \text{ on } [2, \infty) \end{cases}$$

define the kernels κ_b, κ_s by

$$\kappa_b(x, dy) := \lambda\left(\frac{|y|}{1 \vee |x|/2}\right)\kappa(x, dy),$$

$$\kappa_s(x, dy) := \left(1 - \lambda\left(\frac{|y|}{1 \vee |x|/2}\right)\right)\kappa(x, dy) = \kappa(x, dy) - \kappa_b(x, dy).$$

Let h be a continuous truncation function whose support is contained in $\{y \in \mathbb{R}^d : |y| \leq 1\}$. The continuity of λ implies that the triplet $(\beta^h, \gamma, \kappa_s)$ is continuous in the sense of Proposition 5.33. Moreover, it satisfies the growth condition of Proposition 5.35(1) as well as (5.34). Since κ_s satisfies (5.29) as well, step 1 yields the existence of the solution to the martingale problem related to $(\beta^h, \gamma, \kappa_s)$ and any initial law for $X(0)$.

The operator corresponding to the triplet $(0, 0, \kappa_b)$ for the truncation function h is of the form $\widetilde{G}f(x) = \int (f(x + y) - f(x))\kappa_b(x, dy)$. Since $\kappa_b(x, \mathbb{R}^d) \leq c$ for any $x \in \mathbb{R}^d$ by (5.32), [100, Proposition 4.10.2] yields that the martingale problem related to $(\beta^h, \gamma, \kappa)$ has a solution as well. □

It is not clear yet whether the process X in Theorem 5.36 is a Markov process, cf. Sect. 5.6 for a discussion. But typically, one can at least choose X to be Markovian, cf. [170, Theorem 2.7 and Corollary 2.9].

Let us briefly discuss how the situation changes if the state space of the desired process X is a closed subset E of $\mathbb{R}^d$. In this case the triplet $(\beta^h, \gamma, \kappa)(x)$ and the corresponding symbol $\psi(x, \cdot)$ may only be defined for $x \in E$. Propositions 5.33–5.35 hold accordingly if $\mathbb{R}^d$ is replaced by E. In Propositions 5.34 and 5.35 $\lim_{|x|\to\infty}$ refers only to sequences in E. If E is compact, any continuous coefficients vanish at infinity and satisfy the growth condition. In order to obtain the existence of a solution to the martingale problem, we must make sure that the process does not want to leave the state space E. It may not be evident how to express this property in terms of the triplet or even the symbol. The *positive maximum principle* turns out to clarify the situation in this context.

Definition 5.37 We say that $\widetilde{G}$ satisfies the **positive maximum principle** if $\widetilde{G}f(x) \leq 0$ for any $x \in E$ and any $f \in C_c^\infty(E)$ with $f(x) = \max_{y\in E} f(y)$.

Morally speaking, the positive maximum principle should hold if X is an E-valued solution to the martingale problem. Indeed, if $a^{f(X)}(t) = \widetilde{G}f(X(t)) > 0$ holds for the drift rate of $f(X)$, the process may go up with positive probability. However, this cannot happen if f is maximal in $X(t) = x$. Therefore, $\widetilde{G}f(x) \leq 0$ must hold at these points. Surprisingly, the positive maximum principle turns out to be sufficient for the existence of an E-valued solution to the martingale problem.

Theorem 5.38 *Suppose that the coefficients of the martingale problem are continuous, vanish at infinity, and satisfy the growth condition. Moreover, suppose that $\widetilde{G}$ fulfils the positive maximum principle. Then there exists a solution X to the martingale problem.*

If the growth conditions 1. or 3. hold, we need not assume that the coefficients vanish at infinity.

Proof This follows along the same lines as Theorem 5.36. To this end, note that the operators $\widetilde{G}_s$, $\widetilde{G}_b$ corresponding to the triplets $(\beta^h, \gamma, \kappa_s)$ resp. $(0, 0, \kappa_b)$ in the proof of Theorem 5.36 both satisfy the positive maximum principle.

Indeed, this is evident for $\widetilde{G}_b$. For $f \in C_c^\infty(E)$ with $f(x) = \max_{y \in E} f(y)$ define $g(y) := f(y) + (f(x) - f(y))m(y - x) \in [f(y), f(x)]$ for some cutoff function $m : \mathbb{R}^d \to [0, 1]$ which vanishes in a neighbourhood of 0 and equals 1 for $|y| \geq 1 \vee |x|/2$. From the explicit representation of $\widetilde{G}$, $\widetilde{G}_s$ we obtain $\widetilde{G}_s f(x) \leq \widetilde{G}_s g(x) = \widetilde{G} g(x) \leq 0$ as desired. □

Example 5.39 (Brownian Motion on the Unit Circle) Let $E := \{x \in \mathbb{R}^2 : |x| = 1\}$ be the unit circle and consider the coefficients (β, γ, κ) given by

$$\beta(x) := -\frac{1}{2}\begin{pmatrix} x_1 \\ x_2 \end{pmatrix}, \quad \gamma(x) := \begin{pmatrix} x_2^2 & -x_1 x_2 \\ -x_1 x_2 & x_1^2 \end{pmatrix}, \quad \kappa := 0.$$

Is there an E-valued process $X = (X_1, X_2)$ with $(X_1, X_2)(0) = (1, 0)$ that solves the martingale problem related to (β, γ, κ)?

To answer the question, note that

$$\begin{aligned} \widetilde{G} f(x) = &-\frac{1}{2} D_1 f(x) x_1 - \frac{1}{2} D_2 f(x) x_2 \\ &+ \frac{1}{2} D_{11} f(x) x_2^2 + D_{12} f(x) x_1 x_2 + \frac{1}{2} D_{22} f(x) x_2^2 \end{aligned}$$

for $f \in C_b^2(E)$. For $f : E \to \mathbb{R}$ define $g : \mathbb{R} \to \mathbb{R}$ by $g(\alpha) := f(\cos(\alpha)), \sin(\alpha))$. If $f \in C_b^2(E)$ is maximal in $x = (x_1, x_2) = (\cos(\alpha), \sin(\alpha))$, then g is maximal in α. This implies $0 = g'(\alpha)$ and

$$\begin{aligned} 0 &\geq g''(\alpha) \\ &= D_{11} f(x) x_2^2 - 2D_{12} f(x) x_1 x_2 + \frac{1}{2} D_{22} f(x) x_2^2 - D_1 f(x) x_2 - D_2 f(x) x_1 \\ &= \widetilde{G} f(x), \end{aligned}$$

whence the positive maximum principle holds. Theorem 5.38 yields that an E-valued solution exists.

In this example we could have verified this directly by observing that $X(t) = (\cos(W(t)), \sin W(t))$ has the desired properties for a real-valued Wiener process W. As a side remark, X happens to be a polynomial process in the sense of Sect. 6.6.

5.7.3 Stochastic Differential Equations

Stochastic differential equations driven by Brownian motion and homogeneous Poisson random measures constitute a prime way to construct Lévy–Itō diffusions. Recall from Proposition 4.12 that solutions to

$$\begin{aligned} dX(t) = \beta^h(X(t-))dt + \sigma(X(t-))dW(t) \\ + h(\delta(X(t-), x)) * (\mu - \nu)(dt) + \overline{h}(\delta(X(t-), x)) * \mu(dt) \end{aligned} \tag{5.35}$$

have local characteristics of the form

$$b^h(t) = \beta^h(X(t-)), \tag{5.36}$$

$$c(t) = \sigma(X(t-))\sigma(X(t-))^\top, \tag{5.37}$$

$$K(t, B) = \int 1_B(\delta(X(t-), x))K^\nu(dx), \quad B \in \mathscr{B} \text{ with } 0 \notin B, \tag{5.38}$$

where K^ν here denotes the intensity measure of μ. Hence they are Lévy–Itō diffusions with coefficient functions β^h as above, $\gamma := \sigma\sigma^\top$, and κ given by

$$\kappa(t, B) := \int 1_B(\delta(X(t-), y))K^\nu(dy), \quad B \in \mathscr{B} \text{ with } 0 \notin B. \tag{5.39}$$

An existence and uniqueness result for such SDEs is stated in Theorem 3.41. Although this is not immediately obvious, uniqueness in the sense of this result leads to uniqueness of the martingale problem related to $(\beta^h, \gamma, \kappa)$ as it is needed in Theorem 5.30, cf. the discussion below. Hence, X is really a Lévy–Itō diffusion in the sense of Sect. 5.3 if the conditions of Theorem 3.41 hold. It often turns out to be Feller as we will see in Example 5.44.

Note that we are not interested in W and μ in the first place. They only appear as auxiliary objects in (5.35) in order to construct a Lévy–Itō diffusion with coefficient functions β^h, γ, κ of a given form. The concept of *weak* solutions to (5.35) corresponds to this point of view. A process X on some filtered probability space $(\Omega, \mathscr{F}, \mathbf{F}, P)$ carrying an $\mathbb{R}^d$-valued standard Brownian motion W and a homogeneous Poisson random measure μ with compensator $\nu(d(t, x)) = K^\nu(dx)dt$ is called a **weak solution** to (5.35) if this equation holds for this particular choice of X, W, μ. Contrary to the situation in Theorem 3.41, W, μ and the underlying space are part of the solution and not given in the first place. But note that any weak solution in this sense has characteristics of the form (5.36–5.38), regardless of the particular choice of Ω, W and μ. Conversely, any semimartingale X with local characteristics of the form (5.36–5.38) can be written as a solution to SDE (5.35) for some W, μ, possibly after extending the probability space, cf. [152, Théorème 14.80] and Problem 5.6. **Weak uniqueness** of the solution measure means that any two weak solutions $X, \widetilde{X}$ to SDE (5.35) have the same law if they start in the same initial distribution $\mathscr{L}(X(0)) = \mathscr{L}(\widetilde{X}(0))$. A pathwise statement as in Theorem 3.41 would not make sense because $X, \widetilde{X}$ may not even be defined on the same probability space. One can show that weak uniqueness implies uniqueness of the solution to the martingale problem as it is required in Theorem 5.30, cf. [152, Théorème 14.80]. Hence, any weak solution is a Lévy–Itō diffusion if weak uniqueness holds for (5.35).

Surprisingly, this extended solution concept leads to existence and uniqueness theorems for SDEs which go far beyond the Lipschitz case in Theorem 3.41. Although hard to believe, equation (5.35) may have a solution on a sufficiently rich probability space but fail to do so on others. As an example consider the *Tanaka equation*

$$dX(t) = \operatorname{sgn} X(t) dW(t), \quad X(0) = 0, \tag{5.40}$$

where the sign function is defined here as in (5.26). It is easy to construct a weak solution to this equation. If we start from a standard Brownian motion X on some space $(\Omega, \mathscr{F}, \mathbf{F}, P)$,

$$W(t) = \int_0^t \frac{1}{\operatorname{sgn} X(s)} dX(s) = \int_0^t \operatorname{sgn} X(s) dX(s)$$

is a standard Brownian motion as well because W is a continuous local martingale satisfying

$$d[W, W](t) = (\operatorname{sgn} X(t))^2 d[X, X](t) = dt,$$

cf. Example 3.2. Consequently, X together with $(\Omega, \mathscr{F}, \mathbf{F}, P)$ and W constitute a weak solution to (5.40). We do in fact have weak uniqueness because any solution to (5.40) is a continuous local martingale with

$$d[X, X](t) = (\operatorname{sgn} X(t))^2 d[W, W](t) = dt$$

and hence a standard Brownian motion. Nevertheless one may show that the filtration generated by any solution X to (5.40) is strictly larger than the filtration generated by W, cf. e.g. [186, Example 5.3.5], where it is shown that the filtration of W coincides with the filtration of $|X|$, which does not contain any information about the sign of X. Therefore (5.40) cannot have a solution in the classical sense on a probability space where the filtration is generated by the driving Brownian motion W. This leads to the counterintuitive situation that the randomness of the driving Brownian motion W is not enough to define a solution to (5.40). For some reason X cannot be constructed without an additional "random number generator".

One can in fact define a pathwise concept of uniqueness for weak solutions to (5.35) if one only considers solutions $X, \widetilde{X}$ defined on the same probability space. Specifically, we say that **pathwise uniqueness** holds for (5.35) if any two solutions to $X, \widetilde{X}$ defined on the same probability space, referring to the same drivers W, μ, and having the same initial value $X(0) = \widetilde{X}(0)$ coincide up to a set of probability 0. One can show that pathwise uniqueness implies weak uniqueness, cf. [152, Théorème 14.94]. This implication is non-trivial because weak uniqueness also refers to solutions which are defined on different probability spaces and/or relative to different drivers W, μ. Moreover, pathwise uniqueness implies that any solution X to (5.35) is **strong**, which means that it is adapted to the filtration generated by the drivers W, μ, cf. once more [152, Théorème 14.94]. In particular, the counterintuitive situation of the Tanaka equation above does not occur. Combining these statements yields that pathwise uniqueness must fail for Tanaka's equation. And it does: if X denotes a weak solution to (5.40) on some sufficiently large probability space, then $\widetilde{X} := -X$ is a solution as well because

$$d\widetilde{X}(t) = -dX(t) = -\operatorname{sgn} X(t)dW(t) = \operatorname{sgn} \widetilde{X}(t)dW(t).$$

The last equation follows from the fact that $\{t \geq 0 : X(t) = 0\}$ is a set of Lebesgue measure 0, which implies that it is "not seen" by the integral relative to W, i.e. $1_{\{X=0\}} \bullet W = 0$ and hence $1_{\{X\geq 0\}} \bullet W = 1_{\{X>0\}} \bullet W$.

We do not discuss weak solutions in any detail in this book but refer instead to the literature, e.g. [152, 186, 241]. In particular, we do not state any weak existence and uniqueness theorems. One could derive such results from Stroock's existence and uniqueness theorem in Sect. 5.7.5. Alternatively, Theorem 3.41 immediately yields such a statement as well. Indeed, besides existence this theorem implies pathwise uniqueness, which entails weak uniqueness. Weak uniqueness in turn implies that uniqueness holds for the martingale problem related to $(\beta^h, \gamma, \kappa)$. Consequently, X from (5.35) is a Lévy–Itō diffusion under the conditions of Theorem 3.41, which is what we claimed above.

Let us briefly consider Lévy–Itō diffusions which solve SDEs driven by Lévy processes. Suppose that X solves an SDE of the form

$$dX(t) = \beta(X(t-))dL(t) \tag{5.41}$$

with some Lévy process L having Lévy–Khintchine triplet $(b^{h,L}, c^L, K^L)$. From Proposition 4.8 we conclude that its local characteristics (b^h, c, K) are of the form (5.12–5.14) with

$$\begin{aligned}
\beta^h(x) &= \beta(x)b^{h,L} + \int \big(h(\beta(x)y) - \beta(x)h(y)\big)K^L(dy), \\
\gamma(x) &= \beta(x)^2 c^L, \\
\kappa(x, B) &= \int 1_B(\beta(x)y)K^L(dy), \quad B \in \mathscr{B} \text{ with } 0 \notin B.
\end{aligned}$$

If these coefficient functions determine the law of X uniquely, we know that X is a Lévy–Itō diffusion by Theorem 5.30. But how do we know whether this is the case? The key is to write (5.41) in the form (5.35):

$$\begin{aligned}
dX(t) &= \beta(X(t-))b^{h,L}dt + \beta(X(t-))dL^c(t)\\
&\quad + \beta(X(t-))h(x) * (\mu^L - \nu^L)(dt) + \beta(X(t-))(x - h(x)) * \mu^L(dt),\\
&= \left(\beta(X(t-))b^{h,L} + \int \Big(h\big(\beta(X(t-))x\big) - \beta(X(t-))h(x)\Big)K^L(dx)\right)dt\\
&\quad + \beta(X(t-))\sqrt{c^L}dW(t)\\
&\quad + h\big(\beta(X(t-))x\big) * (\mu^L - \nu^L)(dt) + \overline{h}\big(\beta(X(t-))x\big) * \mu^L(dt)\\
&= \beta^h(X(t-))dt + \sigma(X(t-))dW(t)\\
&\quad + h\big(\delta(X(t-), x)\big) * (\mu^L - \nu^L)(dt) + \overline{h}\big(\delta(X(t-), x)\big) * \mu^L(dt),
\end{aligned}$$

where W denotes standard Brownian motion, μ^L is a homogeneous Poisson random measure with intensity measure K^L, function $\overline{h}(x) = x - h(x)$, and

$$\begin{aligned}
\beta^h(x) &= \beta(x)b^{h,L} + \int \big(h\big(\beta(x)y\big) - \beta(x)h(y)\big)K^L(dy),\\
\sigma(x) &= \beta(x)\sqrt{c^L},\\
\delta(x, z) &= \beta(x)z.
\end{aligned}$$

If β satisfies Lipschitz conditions as in Theorem 3.40, then β^h, σ, δ meet the conditions in Theorem 3.41 for a Lipschitz truncation function h. As argued above, this implies that the martingale problem related to $\mathscr{L}(X(0))$ and $(\beta^h, \gamma, \kappa)$ has a unique solution.

Example 5.40 (Geometric Lévy Processes) The stochastic exponential $X = \mathscr{E}(L)$ of a Lévy process L with Lévy–Khintchine triplet (b^h, c, K) is a Lévy–Itō diffusion with coefficient functions

$$\begin{aligned}
\beta^h(x) &= xb^h + \int \big(h(xy) - xh(y)\big)K(dy),\\
\gamma(x) &= x^2c,\\
\kappa(x, B) &= \int 1_B(xy)K(dy), \quad B \in \mathscr{B}.
\end{aligned}$$

So far we have considered homogeneous Lévy–Itō diffusions in the sense that the local characteristics depend on time only through $X(t-)$ but not explicitly. As for

Markov processes one can extend the above results and concepts easily to explicitly time-dependent functions satisfying

$$
\begin{aligned}
b^h(t) &= \beta^h(t, X(t-)),\\
c(t) &= \gamma(t, X(t-)),\\
K(t, dx) &= \kappa(t, X(t-), dx),
\end{aligned}
$$

which generalises (5.12–5.14). As a simple example consider a semimartingale where β^h, γ, κ depend on t only but not on x. In other words, X has deterministic local characteristics, which corresponds to a time-inhomogeneous Lévy process by Proposition 4.7.

Formally, we can interpret an **inhomogeneous** $\mathbb{R}^d$-valued **Lévy–Itō diffusion** X as the last d components of the homogeneous Lévy–Itō diffusion $\widetilde{X}(t) = (t, X(t))$ where time is added as the first component. The coefficient functions $\widetilde{\beta}^h, \widetilde{\gamma}, \widetilde{\kappa}$ of $\widetilde{X}$ are given by

$$
\begin{aligned}
\widetilde{\beta}^h(\widetilde{x}) &= \beta^h(t, x),\\
\widetilde{\gamma}_{ij}(\widetilde{x}) &= \begin{cases} \gamma_{i+1,j+1}(t, x) & \text{if } i, j \geq 2\\ 0 & \text{if } i = 1 \text{ or } j = 1, \end{cases}\\
\widetilde{\kappa}(\widetilde{x}, B) &= \kappa(t, x, \{x \in \mathbb{R}^d : (0, x) \in B\}), \quad B \in \mathscr{B}^{1+d}
\end{aligned}
$$

with $\widetilde{x} = (t, x) \in \mathbb{R}^d \times \mathbb{R}_+$.

5.7.4 Explicit Construction

A classical way to obtain existence and uniqueness is to construct the transition function or the corresponding semigroup explicitly. This works, for example, for the class of affine processes which is discussed in the next chapter. Uniqueness of the solution to the martingale problem can be obtained from Remark 5.26. Indeed, if existence of a solution to the Cauchy problem (5.21) holds for sufficiently many functions f such as those of the form $f(x) = \exp(iu^\top x)$, the univariate marginals $\mathscr{L}(X(t))$ are uniquely determined by the triplet and $\mathscr{L}(X(0))$. Reference [100, Theorem 4.4.2] in turn states that uniqueness of the univariate marginals yields uniqueness of the law of the process and the strong Markov property. According to Theorem 5.43 below, this almost implies that the solution is even Feller. For an illustration of this general approach we refer to Example 6.2.

5.7.5 Stroock's Existence and Uniqueness Theorem

Stroock proved a general existence and uniqueness result for Lévy–Itō diffusions based on martingale methods from [283]. We content ourselves with stating this result here:

Theorem 5.41 *Suppose that P^0 and $(\beta^h, \gamma, \kappa)$ as in Sect. 4.6 satisfy the following conditions.*

1. *$\beta^h(t, x)$, $\gamma(t, x)$, $\int_B (1 \wedge |y|^2)\kappa(t, x, dy)$ are bounded continuous functions of (t, x) for any $B \in \mathscr{B}^d$.*
2. *$\gamma(t, x)$ is everywhere invertible.*

Then the martingale problem related to P^0 and $(\beta^h, \gamma, \kappa)$ has a unique solution.

Proof This follows from [282, Theorem 4.3], cf. also [154, Theorem III.2.34]. □

Note that no Lipschitz condition is imposed in the previous theorem. The boundedness of the coefficients prevents explosion of the solution and could be relaxed. The only severe condition constitutes the regularity of γ, which rules out pure jump processes without diffusion part. But under the condition on γ we obtain uniqueness even in cases where it does not hold for deterministic equations. For example,

$$dX(t) = \sqrt{|X(t)|}dt, \quad X(0) = 0$$

is an ODE which is known to have many solutions as for example $X(t) = t^2/2$ and $X(t) = -t^2/2$. But by Theorem 5.41 the slightly perturbed equation

$$dX(t) = \sqrt{|X(t)|}dt + \varepsilon dW(t), \quad X(0) = 0$$

with $\varepsilon > 0$ has a unique weak solution no matter how small ε is chosen. In that sense Brownian motion has a stabilising effect.

5.7.6 Existence and Uniqueness for Smooth Symbols

From Picard–Lindelöf type theorems for ordinary and stochastic differential equations one may conjecture that a sufficiently smooth dependence of the characteristics on the state implies uniqueness of the solution to the corresponding martingale problem. However, it is not entirely obvious what kind of *smoothness* is needed in this context. Below, we cite a result stating that sufficient regularity of the symbol yields the desired uniqueness.

Theorem 5.42 *Let ψ be a continuous symbol with triplet $(\beta^h, \gamma, \kappa)$. Assume that*

$$\sup_{x \in \mathbb{R}^d} \left(|\beta^h(x)| + \operatorname{tr}(\gamma(x)) + \int |y|^2 \kappa(x, dy) \right) < \infty$$

and that $x \mapsto \psi(x,u)$ is $[d/2]+3$ times continuously differentiable for any $u \in \mathbb{R}^d$. Moreover, suppose that there exists a characteristic exponent $\varphi : \mathbb{R}^d \to \mathbb{C}$ such that

$$|\mathrm{Re}(\psi(x,u))| \geq g_1(x)|\varphi(u)|,$$
$$|D_x^\alpha \psi(x,u)| \leq g_2(x)|\varphi(u)|$$

for some bounded functions $g_1, g_2 : \mathbb{R}^d \to (0,\infty)$ and any $\alpha \in \mathbb{N}^d$ with $|\alpha| \leq [d/2]+3$. Then existence and uniqueness holds for the martingale problem related to the symbol ψ and any initial law P^0.

Proof This is shown in [170, Theorem 4.4]. □

5.8 The Feller Property

Interestingly, uniqueness of the solution to the martingale problem is not far from the Feller property. We consider the setup of Sects. 5.7.1, 5.7.2 with state space $E = \mathbb{R}^d$.

Theorem 5.43 *Let X be a solution to the martingale problem related to the triplet $(\beta^h, \gamma, \kappa)$, the symbol ψ, or the operator $\widetilde{G}$. Suppose that*

1. *the coefficients are continuous,*
2. *the coefficients vanish at infinity,*
3. *the coefficients satisfy the growth conditions 1., 2., or 3. in Proposition 5.35,*
4. *uniqueness holds for the martingale problem.*

Then X is a Feller process.

Proof This follows from [201, Theorem 1.1]. □

As an example, let us consider solutions to stochastic differential equations under Lipschitz conditions.

Example 5.44 Processes as in Theorem 3.41 are Feller if the coefficients β^h, σ, δ do not depend explicitly on the time argument t and

$$\lim_{|x|\to\infty} K^\nu\big(\{|\delta(x,z)+x| \leq r\}\big) = 0, \quad r \in \mathbb{R}_+ \tag{5.42}$$

for the intensity measure K^ν of μ.

Indeed, from the shape of the corresponding triplet $(\beta^h, \gamma, \kappa)$ in Sect. 5.7.3 we obtain that the symbol equals

$$\psi(x,u) = iu^\top \beta^h(x) - \frac{1}{2}u^\top \sigma(x)\sigma(x)^\top u + \int \big(e^{iu^\top \delta(x,y)} - 1 - iu^\top h(\delta(x,y))\big)K^\nu(dy).$$

The Lipschitz conditions on β^h, σ, δ imply that the coefficients are continuous. Equation (5.42) is a reformulation of (5.32) in the present setup. Moreover, (5.39) and the growth conditions on β^h, σ, δ warrant that (5.33) in Proposition 5.34 is satisfied. Since uniqueness of the solution to the martingale problem holds by Sect. 5.7.3, the assertion follows from Theorem 5.43.

Condition (5.42) cannot simply be removed. If N is a Poisson process with intensity $\lambda > 0$, the process $X = X(0)\mathscr{E}(-N)$ solves the SDE

$$dX(t) = -X(t)dN(t) = -X(t) * \mu^N(dt).$$

If we choose a truncation function h with $h(1) = 0$, its coefficients $\beta^h = 0, \sigma = 0, \delta(x, z) = -x$ satisfy the conditions of Theorem 3.41 but the process X is not Feller. Indeed, for nonnegative $f \in C_0(\mathbb{R}^d)$ with $f(0) = 1$ we have

$$p_1 f(x) := E_x(f(X(1))) \geq P_x(X(1) = 0) = P_x(N(1) \geq 1) = 1 - e^{-\lambda},$$

which does not tend to 0 as $x \to \infty$.

But even without satisfying (5.42) the processes of Theorem 3.41 are strong Markov if the coefficients β^h, σ, δ do not depend explicitly on the time argument t. This follows from the discussion in Sect. 5.7.3.

Appendix 1: Problems

The exercises correspond to the section with the same number.

5.1 (Reflection Principle) For a Wiener process W set $S(t) := \sup_{s \leq t} W(s)$. Use the strong Markov property of W in order to show that $P(W(t) \leq a - x, S(t) \geq a) = P(W(t) \geq a + x)$ for any $a > 0, x \geq 0, t \geq 0$. Deduce that the joint law of $(W(t), S(t))$ has density

$$\varrho(x, y) = \sqrt{\frac{2}{\pi t^3}}(2y - x)\exp\left(-\frac{(2y - x)^2}{2t}\right)1_{\mathbb{R}_+}(y - x)1_{\mathbb{R}_+}(y).$$

Hint: Show that

$$\widetilde{W}(t) := \begin{cases} W(t) & \text{for } t \leq \tau, \\ 2a - W(t) & \text{for } t > \tau \end{cases}$$

is a Wiener process for $a > 0$ and the stopping time $\tau := \inf\{t \geq 0 : W(t) \geq a\}$.

5.2 Show that X is a time-inhomogeneous Markov process if and only if the space-time process $\overline{X}(t) = (t, X(t))$ is a Markov Process.

5.3 Show that the resolvent mapping R_λ of standard Brownian motion satisfies

$$R_\lambda f(x) = \int_{-\infty}^{\infty} \frac{1}{\sqrt{2\lambda}} e^{-\sqrt{2\lambda}|x-y|} f(y)dy$$

for $f \in C_0(\mathbb{R})$.

5.4 Derive the characteristic function of $X(t)$ for the square-root process solving $X(t) = x + \sqrt{X} \bullet W(t)$ with initial value $x > 0$ and some Wiener process W.

Hint: Try the ansatz $u(t, x) = \exp(\Psi_0(t, v) + \Psi_1(t, v)x)$ with $\Psi_0, \Psi_1 : \mathbb{R}_+ \to \mathbb{R}$ for the backward equation, where $v \in \mathbb{R}$ denotes the argument of the characteristic function.

5.5 Derive the probability density function of $X(t)$ for an Ornstein–Uhlenbeck process as in (3.87) with $L(t) = \mu t + \sigma W(t)$ for $\mu \in \mathbb{R}, \sigma > 0$ and some Wiener process W. Suppose that the initial distribution $P^{X(0)}$ is Gaussian with mean μ_0 and variance σ_0^2.

Hint: Try the ansatz that $X(t)$ is Gaussian and derive ODEs for its mean and variance.

5.6 Let X denote a solution process to the martingale problem related to $(\beta, \gamma, 0)$ for some bounded continuous functions $\beta, \gamma : \mathbb{R} \to \mathbb{R}$ such that $\gamma > 0$. Define the process $W = \gamma(X)^{-1/2} \bullet X^c$. Show that W is a Wiener process and that X solves the SDE $X = X(0) + \beta(X) \bullet I + \sqrt{\gamma(X)} \bullet W$.

5.7 Show that $|W|^{1/4}$ is not a semimartingale for a Wiener process W.

Hint: If $X = |W|^{1/4}$ is a semimartingale, apply Itōs formula to $X^4 = |W|$ in order to show that the local time of W in 0 equals $L^0 = 1_{\{W=0\}} \bullet L^0 = 0$, which contradicts the fact that $|W|$ is not a local martingale.

5.8 Verify the implication $2 \Rightarrow 1$ in Proposition 5.34 and the growth condition in Example 5.44.

Appendix 2: Notes and Comments

For a thorough treatment of Markov process theory one may consult [100, 241, 251, 252]. For Example 5.10 see also [2, Theorem 3.1.9]. Concerning Sect. 5.2 we refer to [241, Exercise III.1.10]. Details on Remark 5.13(2) can be found in [241, Propositions III.2.4 and VII.1.4]. Our definition of the extended generator is close to the one in [241, Definition VII.1.8] and [152, Remarque 13.45]. For Proposition 5.20 see also [241, Definition VIII.3.2 and Proposition VIII.3.3] as well as [152, Remarque 13.46]. A result in the spirit of Remark 5.26 is stated in [226, Theorem 8.2.1]. For a version of Theorem 5.28 for continuous processes see [226, Exercise 8.3]. The approach to studying processes by their symbol is advocated in [36, 151]. Theorem 5.43 and the counterexample in Example 5.44 can be found in [201]. For Problem 5.1 see also [279, Section 3.7.3]. Problem 5.7 is based on [238, Theorem IV.71].

Chapter 6
Affine and Polynomial Processes

Affine processes appear here for two reasons. They are treated in this part on stochastic calculus because they solve linear or more generally affine martingale problems. If one views martingale problems as a stochastic analogue of ODEs, affine processes correspond to exponential functions because the latter solve linear ODEs. But crucially, affine processes appear in this *book* because they play an important role in financial modelling. They are applied to such diverse phenomena as interest rates, stochastic volatility, default intensity. This stems from their flexibility, accompanied by mathematical tractability.

Integer moments of affine processes can be computed elegantly by solving systems of linear ordinary differential equations. This property extends in fact to the more general family of *polynomial processes*, which also includes, for instance, the generalised Ornstein–Uhlenbeck processes from Sect. 3.9. We take a look at this family in the last section of this chapter.

6.1 Affine Markov Processes

Affine processes are introduced in [80] as Markov processes whose characteristic function is of exponentially affine form. We view them here as solutions to affine martingale problems, i.e. as semimartingales whose characteristics depend affine-linearly on the process itself.

Let us start with the linear univariate case. Specifically, we consider a semimartingale X with characteristics

$$b^h(t) = b^h_{(1)} X(t-), \tag{6.1}$$

$$c(t) = c_{(1)} X(t-), \tag{6.2}$$

$$K(t, dx) = K_{(1)}(dx) X(t-), \tag{6.3}$$

© Springer Nature Switzerland AG 2019

E. Eberlein, J. Kallsen, *Mathematical Finance*, Springer Finance,
https://doi.org/10.1007/978-3-030-26106-1_6

where $(b^h_{(1)}, c_{(1)}, K_{(1)})$ is a fixed Lévy–Khintchine triplet, i.e. $b^h_{(1)}$ and $c_{(1)} \geq 0$ are numbers and $K_{(1)}$ denotes a Lévy measure on $\mathbb{R}$. The subscript is introduced to avoid confusion between the local characteristics of X and the Lévy–Khintchine triplet in their representation on the right. Obviously, (b^h, c, K) can only be the characteristics of a semimartingale X if X is nonnegative or if $c_{(1)} = 0$, $K_{(1)} = 0$. Otherwise the positivity of c or K would be violated. In the deterministic case $c_{(1)} = 0$, $K_{(1)} = 0$, the above martingale problem reduces to the ODE

$$dX(t) = b_{(1)}X(t-)dt. \tag{6.4}$$

Since $K = 0$, the drift coefficient does not depend on the truncation function h here. Equation (6.4) is solved by $X(t) = X(0)\exp(b_{(1)}t)$. Consequently, solutions to (6.1–6.3) deserve the name "stochastic exponential" in some sense. We avoid this terminology because it could be confused with stochastic exponentials in the sense of Definition 3.42. Maybe surprisingly, the two kinds of "exponentials" lead to quite different processes in the random case. Whereas both the linear SDE

$$dX(t) = X(t-)dL(t) \tag{6.5}$$

and the linear martingale problem (6.1–6.3) are solved by the same "process"

$$X(t) = X(0)\exp(b_{(1)}t)$$

if the Lévy process L is a deterministic linear function $L(t) = b_{(1)}t$, this concordance ceases to hold for more general Lévy processes. For a Lévy process L with triplet $(b^h_{(1)}, c_{(1)}, K_{(1)})$, the solution to (6.5) has characteristics

$$\begin{aligned}
b^h(t) &= b^h_{(1)}X(t-),\\
c(t) &= c_{(1)}X(t-)^2,\\
K(t,G) &= \int 1_G(X(t-)x)K_{(1)}(dx), \quad G \in \mathscr{B},
\end{aligned}$$

which are generally non-linear in $X(t-)$. This is even true for standard Brownian motion $L = W$. Consequently, the stochastic exponential

$$X(t) = \mathscr{E}(W(t)) = \exp\left(W(t) - \frac{1}{2}t\right)$$

of Brownian motion does not solve a linear martingale problem. In order to obtain a process having linear characteristics (6.1–6.3) with $K_{(1)} = 0$, we must consider the SDE

$$dX(t) = b_{(1)}X(t)dt + \sqrt{c_{(1)}X(t)}dW(t).$$

Such *square-root processes* appear, for example, as special cases of the popular Cox–Ingersoll–Ross model in interest rate theory. For more general Lévy processes there is no natural way to express the solution to (6.1–6.3) as the solution to an SDE which is driven by a Lévy process.

Both the processes with characteristics as in (6.1–6.3) and those solving (6.5) deserve the name "exponential" in certain respects because their change depends linearly on their current value. However, they differ in how the loose term *linear dependence* is interpreted mathematically. Whereas the *size* of the increments depends linearly on the state $X(t-)$ for the stochastic exponential (6.5), the *speed* or *intensity* depends linearly on $X(t-)$ in the present context (6.1–6.3) of affine processes. As noted above, both concepts reduce to the same exponential function in the purely deterministic case.

Let us try to derive the characteristic function of a semimartingale X solving the above linear martingale problem. More precisely, we consider the conditional characteristic function

$$u \mapsto E(\exp(iuX(s+t))|\mathscr{F}_s) \tag{6.6}$$

for $s, t \geq 0$. Since the form of the characteristics suggests a Markov process by Sect. 5.6, we can write (6.6) as a function

$$f(t, X(s)) := E(\exp(iuX(s+t))|X(s)) = E(\exp(iuX(s+t))|\mathscr{F}_s). \tag{6.7}$$

Kolmogorov's backward equation (5.17) resp. (5.21) tells us that f should solve the PIDE

$$\begin{aligned}
\dot{f}(t,x) &= x\Bigg(f'(t,x)b^h_{(1)} + \frac{1}{2}f''(t,x)c_{(1)} \\
&\qquad + \int \Big(f(t,x+y) - f(t,x) - f'(t,x)^\top h(y)\Big) K_{(1)}(dy)\Bigg), \\
f(0,x) &= \exp(iux),
\end{aligned}$$

where the dot and prime refer to differentiation in the first and second variable, respectively. Trying an ansatz

$$f(t,x) = \exp(\Psi(t,u)x) \tag{6.8}$$

leads to the ODE

$$\begin{aligned}
\frac{d}{dt}\Psi(t,u) &= \Psi(t,u)b^h_{(1)} + \frac{1}{2}\Psi(t,u)^2 c_{(1)} + \int \Big(e^{\Psi(t,u)y} - 1 - \Psi(t,u)h(y)\Big) K_{(1)}(dy) \\
&= \psi_1(-i\Psi(t,u))
\end{aligned} \tag{6.9}$$

with initial condition

$$\Psi(0, u) = iu,$$

where

$$\psi_1(u) := iub^h_{(1)} - \frac{1}{2}u^2 c_{(1)} + \int \left(e^{iuy} - 1 - iuh(y)\right) K_{(1)}(dy)$$

denotes the characteristic exponent of $(b^h_{(1)}, c_{(1)}, K_{(1)})$ as in (2.19). Equation (6.9) is called a *generalised Riccati equation* because it reduces to a Riccati equation in the no-jump case $K_{(1)} = 0$. The conditional characteristic function (6.7) determines the transition function and hence the law of the whole process, cf. Propositions 5.4 and 6.8 below.

Under what conditions do we have existence of a semimartingale satisfying (6.1–6.3)? The Lévy measure $K_{(1)}$ must be concentrated on the positive real line to prevent X from jumping to negative values. This is enough to warrant existence of a unique Markov process whose generator corresponds to (6.1–6.3). This Markov process, however, may explode to $+\infty$ in finite time, which means that it is not a semimartingale in the usual sense with finite values on the whole positive real line. A sufficient—but not necessary—condition to prevent explosion is given by

$$\int_{\{x>1\}} x K_{(1)}(dx) < \infty. \tag{6.10}$$

Summing up, we have motivated the

Proposition 6.1 *Given $X(0) \geq 0$ and a Lévy–Khintchine triplet $(b^h_{(1)}, c_{(1)}, K_{(1)})$ with $K_{(1)}((-\infty, 0]) = 0$ and (6.10), there is a unique solution to the linear martingale problem (6.1–6.3), where uniqueness refers to a unique law as usual. Equivalently, there is a unique solution to the martingale problem related to the symbol q of the form*

$$q(x, u) := \psi_1(u)x, \quad (x, u) \in \mathbb{R}_+ \times \mathbb{R}.$$

The solution is a Feller process whose conditional characteristic function is given by

$$E(\exp(iuX(s + t))|\mathscr{F}_s) = \exp(\Psi(t, u)X(s)), \quad u \in \mathbb{R} \tag{6.11}$$

with $\Psi(\cdot, u)$ being the unique solution to the initial value problem

$$\frac{d}{dt}\Psi(t, u) = \psi_1(-i\Psi(t, u)), \quad \Psi(0, u) = iu. \tag{6.12}$$

Its extended generator is of the form (5.11) with

$$\beta^h(x) = b^h_{(1)}x,$$

$$\gamma(x) = c_{(1)}x,$$

$$\kappa(x, dy) = K_{(1)}(dy)x$$

for any bounded twice continuously differentiable function $f : \mathbb{R}_+ \to \mathbb{R}$. If f has compact support, it belongs to the domain of the generator of X.

Proof This result is a special case of Theorem 6.6 below. □

Equation (6.11) holds in fact for $u \in \mathbb{C}_- := -\mathbb{R}_+ + i\mathbb{R}$ if Ψ is defined accordingly.

Example 6.2 (Linear Birth and Death Process with Multiple Offsprings) Suppose that $X(t)$ denotes a number of individuals in some population. From time to time an individual splits into a random number of offsprings. Under idealised conditions, the local characteristics of X may be assumed to be of the form (6.1–6.3) with $b^0_{(1)} = 0$ (i.e. 0 for $h = 0$), $c_{(1)} = 0$, and $K_{(1)}(G) = \lambda Q(\{x \in \mathbb{R} : x - 1 \in G\})$. Here, λdt denotes the probability that a single individual splits during a short time interval of length dt. The probability measure Q on $\mathbb{N}$ represents the law of the number of offsprings. The drift and diffusion coefficients of X vanish because the process moves only on integers. We assume that the offspring distribution has finite expectation, i.e. $\sum_{n \in \mathbb{N}} nQ(\{n\}) < \infty$. In view of Proposition 6.1, the characteristic function of $X(t)$ is given by

$$E(\exp(iuX(t))) = \exp(\Psi(t, u)X(0)), \tag{6.13}$$

where $\Psi(\cdot, u)$ solves

$$\frac{d}{dt}\Psi(t, u) = \lambda \sum_{n \in \mathbb{N}} \left(e^{\Psi(t,u)(n-1)} - 1\right) Q(\{n\}), \quad \Psi(0, u) = iu \tag{6.14}$$

or, equivalently,

$$\frac{d}{dt}\Psi(t, u) = \lambda \left(g_Q(e^{\Psi(t,u)})e^{-\Psi(t,u)} - 1\right), \quad \Psi(0, u) = iu,$$

where $g_Q(s) := \sum_{n \in \mathbb{N}} s^n Q(\{n\})$ denotes the *probability generating function* of Q. Strictly speaking, Proposition 6.1 does not allow for the case $Q(\{0\}) > 0$ because it does not correspond to a Lévy measure $K_{(1)}$ on the positive real line. However, the concentration of $K_{(1)}$ on $\mathbb{R}_+$ was only imposed to prevent X from attaining negative values, which cannot happen in this example if $X(0) \in \mathbb{N}$. Indeed, one can show that the relevant statements of Proposition 6.1 still hold in this case. Specifically, let us verify that, for any $X(0) \in \mathbb{N}$, there exists a unique solution to the martingale problem (6.1–6.3). Moreover, the solution process is strong Markov with characteristic function of the form (6.13, 6.14) for $u \in \mathbb{R}$ or even $u \in \mathbb{C}_-$.

Existence of the solution is obtained from Theorem 5.38. To wit, linearity of the symbol implies continuity of the coefficients, cf. Proposition 5.33(2) and the following remark. Since Q has finite expectation $\sum_{n\in\mathbb{N}} nQ(\{n\})$, its characteristic function $\widehat{Q}$ is differentiable at 0, cf. Proposition A.5(2). Since the symbol corresponding to $(b^h_{(1)}, c_{(1)}, K_{(1)})$ is

$$q(x,u) = \psi_{(1)}(u)x = \int \left(e^{iu(x-1)} - 1\right)\lambda Q(dx) = \lambda\left(e^{-iu}\widehat{Q}(u) - 1\right)x,$$

one easily verifies that (5.33) holds. The positive maximum principle is obvious because

$$\widetilde{G}f(x) = \lambda \sum_{n\in\mathbb{N}} \big(f(x+n-1) - f(x)\big)Q(\{n\}).$$

Equation (6.14) is of the form $\frac{d}{dt}\Psi(t,u) = g(\Psi(t,u))$, where $g : \mathbb{C}_- \to \mathbb{C}_-$ is continuous and $g(x) \leq 2\lambda$ for $x \in \mathbb{C}_-$. Therefore (6.14) has a $\mathbb{C}_-$-valued solution for any $u \in \mathbb{C}_-$. Applying the backward equation in the sense of Remark 5.26 yields that

$$E\big(\exp(iuX(T))\big|\mathscr{F}_t\big) = \exp\big(\Psi(T-t,u)X(t)\big), \quad t \leq T. \tag{6.15}$$

By the proof of Proposition 5.4 this means that the finite-dimensional marginals and hence the law of X are uniquely determined by $X(0)$ and Q. Moreover, X is strong Markov by Theorem 5.30 or, more exactly, by [100, Theorems 4.4.2 and 4.4.6] because Theorem 5.30 only covers the state space $\mathbb{R}^d$.

Example 6.3 (Linear Birth and Death Process) Birth and death processes in the literature are usually restricted to single births, i.e. $K_{(1)}$ is of the form $K_{(1)} = \mu\varepsilon_{-1} + \lambda\varepsilon_1$ with birth rate $\lambda \geq 0$ and death rate $\mu \geq 0$, cf. e.g. [191]. The ODE (6.14) reduces to

$$\frac{d}{dt}\Psi(t,u) = \mu(e^{-\Psi(t,u)} - 1) + \lambda(e^{\Psi(t,u)} - 1), \quad \Psi(0,u) = iu$$

in this case. It is solved by

$$\Psi(t,u) = \log \frac{\mu(e^{\lambda t} - e^{\mu t}) - (\mu e^{\lambda t} - \lambda e^{\mu t})e^{iu}}{\lambda e^{\lambda t} - \mu e^{\mu t} - \lambda(e^{\lambda t} - e^{\mu t})e^{iu}},$$

which yields the characteristic or even conditional characteristic functions (6.13, 6.15) for $u \in \mathbb{C}_-$. As usual, the branch of the complex logarithm must be chosen such that the right-hand side is a continuous function of u. Differentiation yields

$$E(X(t)) = -i \left. \frac{dE(\exp(iuX(t)))}{du} \right|_{u=0} = X(0)e^{(\lambda-\mu)t}$$

as the expected number of individuals at time t, which is literally what one would expect.

Example 6.4 (Square-Root Process) The square-root process arises as a solution to an SDE of the form

$$dX(t) = -\lambda X(t)dt + \sigma\sqrt{X(t)}dW(t), \quad X(0) = x \tag{6.16}$$

with standard Brownian motion W and constants $\lambda, \sigma, x > 0$. It is a process with linear characteristics (6.1–6.3) of the form $(b^h_{(1)}, c_{(1)}, K_{(1)}) = (-\lambda, \sigma^2, 0)$. The initial value problem (6.12) is now truly of Riccati type and has the solution

$$\Psi(t, u) = \frac{iue^{-\lambda t}}{1 - \frac{\sigma^2 iu}{2\lambda}(1 - e^{-\lambda t})}.$$

The Cox–Ingersoll–Ross model in interest rate theory is based on the SDE

$$dX(t) = (\kappa - \lambda X(t))dt + \sigma\sqrt{X(t)}dW(t) \tag{6.17}$$

instead of (6.16) above. Its local characteristics

$$(b, c, K)(t) = (\kappa - \lambda X(t-), \sigma^2 X(t-), 0)$$

are affine rather than linear in $X(t-)$. The same holds for Lévy-driven Ornstein–Uhlenbeck processes solving

$$dX(t) = -\lambda X(t-)dt + dL(t).$$

If $(b^h_{(0)}, c_{(0)}, K_{(0)})$ denotes the Lévy–Khintchine triplet of L, the local characteristics of X are of affine form

$$\begin{aligned} b^h(t) &= b^h_{(0)} - \lambda X(t-), \\ c(t) &= c_{(0)}, \\ K(t, dx) &= K_{(0)}(dx). \end{aligned}$$

The general form of such an affine martingale problem is

$$b^h(t) = b^h_{(0)} + b^h_{(1)} X(t-), \tag{6.18}$$

$$c(t) = c_{(0)} + c_{(1)} X(t-), \tag{6.19}$$

$$K(t, dx) = K_{(0)}(dx) + K_{(1)}(dx) X(t-), \tag{6.20}$$

where $(b^h_{(0)}, c_{(0)}, K_{(0)})$, $(b^h_{(1)}, c_{(1)}, K_{(1)})$ denote two Lévy–Khintchine triplets on $\mathbb{R}$. As above, conditions on the triplets must be imposed in order to prevent the process from turning negative or exploding.

Proposition 6.5 (Univariate Affine Processes) *Given $X(0) > 0$ and Lévy–Khintchine triplets $(b^h_{(0)}, c_{(0)}, K_{(0)})$, $(b^h_{(1)}, c_{(1)}, K_{(1)})$ with*

$$c_{(1)} = 0, \quad K_{(1)} = 0$$

(real-valued case) or

$$\int h(x) K_{(0)}(dx) < \infty,$$

$$b^h_{(0)} - \int h(x) K_{(0)}(dx) \geq 0, \quad c_{(0)} = 0, \quad K_{(0)}((-\infty, 0]) = 0,$$

$$K_{(1)}((-\infty, 0]) = 0, \quad \int_{\{x>1\}} x K_{(1)}(dx) < \infty$$

(nonnegative case), there is a unique solution to the affine martingale problem (6.18–6.20). Equivalently, there is a unique solution to the martingale problem related to the symbol q of the form

$$q(x, u) := \psi_0(u) + \psi_1(u) x, \quad (x, u) \in \mathbb{R} \times \mathbb{R} \text{ resp. } \mathbb{R}_+ \times \mathbb{R},$$

where ψ_0 and ψ_1 denote the exponents of $(b^h_{(0)}, c_{(0)}, K_{(0)})$ and $(b^h_{(1)}, c_{(1)}, K_{(1)})$. The solution is a Feller process whose conditional characteristic function is given by

$$E(\exp(iuX(s+t)) | \mathscr{F}_s) = \exp(\Psi_0(t, u) + \Psi_1(t, u) X(s)), \quad u \in \mathbb{R} \tag{6.21}$$

with $\Psi_1(\cdot, u)$ being the unique solution to the initial value problem

$$\frac{d}{dt} \Psi_1(t, u) = \psi_1(-i\Psi_1(t, u)), \quad \Psi_1(0, u) = iu \tag{6.22}$$

and

$$\Psi_0(t,u) := \int_0^t \psi_0(-i\Psi_1(s,u))ds.$$

Its extended generator is of the form (5.11) with

$$\beta^h(x) = b^h_{(0)} + b^h_{(1)}x,$$
$$\gamma(x) = c_{(0)} + c_{(1)}x,$$
$$\kappa(x,dy) = K_{(0)}(dy) + K_{(1)}(dy)x$$

for any bounded twice continuously differentiable function $f : \mathbb{R} \to \mathbb{R}$ resp. $f : \mathbb{R}_+ \to \mathbb{R}$. If f has compact support, it belongs to the domain of the generator of X.

Idea On a formal level, the conditional characteristic function is obtained similarly as in the derivation leading to Proposition 6.1. The ansatz (6.8) is replaced by $f(t,x) = \exp(\Psi_0(t,u) + \Psi_1(t,u)x)$ in this case. This leads to the system of ODEs (6.22) and $\frac{d}{dt}\Psi_0(t,u) = \psi_0(-i\Psi_1(t,u))$, which is solved by Ψ_0 and Ψ_1 in the theorem.

Mathematically, the statement is a special case of Theorem 6.6 below. □

We turn now to multivariate affine processes. An $\mathbb{R}^d$-valued **affine process** is a process X whose local characteristics are affine functions of $X_1,\dots,X_d$. Specifically, it solves a martingale problem of the form

$$b^h(t) = b^h_{(0)} + \sum_{j=1}^d b^h_{(j)}X_j(t-), \tag{6.23}$$

$$c(t) = c_{(0)} + \sum_{j=1}^d c_{(j)}X_j(t-), \tag{6.24}$$

$$K(t,dx) = K_{(0)}(dx) + \sum_{j=1}^d K_{(j)}(dx)X_j(t-) \tag{6.25}$$

with Lévy–Khintchine triplets $(b^h_{(j)}, c_{(j)}, K_{(j)})$, $j = 0,\dots,d$.

In order for the characteristics (6.23–6.25) to make sense, some but not all of the components $X_1,\dots,X_d$ need to be nonnegative. If $c_{(j)} = 0$, $K_{(j)} = 0$ for some index j, a negative X_j does not lead to negative diffusion coefficients $c(t)$ or jump measures $K(t,\cdot)$. Without loss of generality we denote by $X_1,\dots,X_m$ components which need to be nonnegative and by $X_{m+1},\dots,X_d$ the remaining coordinates for which we have $c_{(j)} = 0$, $K_{(j)} = 0$. This leads to the following admissibility conditions for the Lévy–Khintchine triplets:

$$\left.\begin{aligned} b^h_{(j),k} - \int h_k(x)K_{(j)}(dx) &\geq 0 \\ K_{(j)}((\mathbb{R}^m_+ \times \mathbb{R}^{d-m})^C) &= 0 \\ \int h_k(x)K_{(j)}(dx) &< \infty \end{aligned}\right\} \text{if } 0 \leq j \leq m,\ 1 \leq k \leq m,\quad k \neq j;$$
$$\begin{aligned} c_{(j),k\ell} &= 0 \quad \text{if } 0 \leq j \leq m,\ 1 \leq k, \ell \leq m \text{ unless } k = \ell = j; \\ b^h_{(j),k} &= 0 \quad \text{if } j \geq m+1,\ 1 \leq k \leq m; \end{aligned}$$
$$\left.\begin{aligned} c_{(j)} &= 0 \\ K_{(j)} &= 0 \end{aligned}\right\} \text{if } j \geq m+1. \tag{6.26}$$

In particular, the "triplets" $(b^h_{(j)}, c_{(j)}, K_{(j)})$, $j = m+1, \ldots, d$ vanish except for $b^h_{(j),k}$ with $j, k = m+1, \ldots, d$. The sufficient condition preventing explosion now reads as

$$\int_{\{x_k > 1\}} x_k K_{(j)}(dx) < \infty, \quad j, k = 1, \ldots, m. \tag{6.27}$$

The symbol corresponding to the above characteristics is

$$q(x, u) := \psi_0(u) + \sum_{j=1}^{d} \psi_j(u)x_j, \quad (x, u) \in (\mathbb{R}^m_+ \times \mathbb{R}^{d-m}) \times \mathbb{R}^d, \tag{6.28}$$

where ψ_j, $j = 0, \ldots, d$ denote the characteristic exponents of $(b^h_{(j)}, c_{(j)}, K_{(j)})$ in the sense of (2.19).

Theorem 6.6 (Multivariate Affine Processes) *Given $X(0) \in \mathbb{R}^m_+ \times \mathbb{R}^{d-m}$ and Lévy–Khintchine triplets $(b^h_{(j)}, c_{(j)}, K_{(j)})$, $j = 0, \ldots, d$ satisfying (6.26) and (6.27), there is a unique solution to the affine martingale problem (6.23–6.25), where uniqueness refers to a unique law as usual. Equivalently, there is a unique solution to the martingale problem related to the symbol q in (6.28). The solution is a Feller process whose conditional characteristic function is given by*

$$E\Big(\exp\big(iu^\top X(s+t)\big)\Big|\mathscr{F}_s\Big) = \exp\Big(\Psi_0(t, u) + \sum_{j=1}^{d} \Psi_j(t, u)X_j(s)\Big), \quad u \in \mathbb{R}^d, \tag{6.29}$$

where $\Psi_{1,\ldots,d} = (\Psi_1, \ldots, \Psi_d)^\top$ is the unique solution to the following system of generalised Riccati equations

$$\frac{d}{dt}\Psi_j(t, u) = \psi_j(-i\Psi_{1,\ldots,d}(t, u)), \quad \Psi_{1,\ldots,d}(0, u) = iu \tag{6.30}$$

and

$$\Psi_0(t,u) := \int_0^t \psi_0(-i\Psi_{1,\dots,d}(s,u))ds. \tag{6.31}$$

Its extended generator is of the form (5.11) with

$$\beta^h(x) = b^h_{(0)} + \sum_{j=1}^d b^h_{(j)}x_j,$$

$$\gamma(x) = c_{(0)} + \sum_{j=1}^d c_{(j)}x_j,$$

$$\kappa(x,dy) = K_{(0)}(dy) + \sum_{j=1}^d K_{(j)}(dy)x_j$$

for any bounded twice continuously differentiable function $f : \mathbb{R}^m_+ \times \mathbb{R}^{d-m} \to \mathbb{R}$. If f has compact support, it belongs to the domain of the generator of X.

Idea On a formal level, the conditional characteristic function is derived as in the previous special cases. The proof of Theorem 6.6 is due to and can be found in [80, Theorems 2.7, 2.12, Lemma 9.2], cf. also [169, Theorem 3.2]. The idea for the existence part is to construct the Feller semigroup explicitly, based on (6.29). Uniqueness of the solution to the martingale problem can be shown as sketched in Sect. 5.7.4. The statement concerning the extended generator follows from Proposition 4.31. □

Equation (6.29) holds in fact for $u \in \mathbb{C}^m_- \times \mathbb{R}^{d-m}$ if Ψ is defined accordingly.

Remark 6.7 Observe that the solution to the inhomogeneous linear system of ODEs for $\Psi_{m+1,\dots,d} = (\Psi_{m+1},\dots,\Psi_d)^\top$ is given by

$$\Psi_{m+1,\dots,d}(t,u) = e^{Bt}\left(i(u_{m+1},\dots,u_d)^\top + \int_0^t e^{-Bs}A(\Psi_1,\dots,\Psi_m)^\top(s,u)ds\right),$$

where the matrices $A \in \mathbb{R}^{(d-m)\times m}$, $B \in \mathbb{R}^{(d-m)\times(d-m)}$ are defined as

$$A_{jk} := b_{(j+m),k}, \quad j = 1,\dots,d-m, \quad k = 1,\dots,m,$$

$$B_{jk} := b_{(j+m),k+m}, \quad j,k = 1,\dots,d-m.$$

Consequently, $\Psi_{1,\dots,m} := (\Psi_1, \dots, \Psi_m)^\top$ is a solution to the system

$$\begin{aligned}\frac{d}{dt}\Psi_{1,\dots,m}(t,u) &= \psi_{1,\dots,m}\bigg(-i\Psi_{1,\dots,m}(t,u), e^{Bt}\Big(i(u_{m+1},\dots,u_d)^\top \\ &\quad + \int_0^t e^{-Bs}A(\psi_1,\dots,\Psi_m)^\top(s,u)ds\Big)\bigg), \\ \Psi_{1,\dots,m}(0,u) &= i(u_1,\dots,u_m)^\top,\end{aligned}$$

where we set $\psi_{1,\dots,m} := (\psi_1, \dots, \psi_m)^\top$.

By introducing a zeroth component $X_0(t) = 1$, it is easy to see that an affine process in $\mathbb{R}_+^m \times \mathbb{R}^{d-m} \subset \mathbb{R}^d$ may in fact be interpreted as a process with *linear* characteristics in the $d+1$-dimensional state space $\mathbb{R}_+^{1+m} \times \mathbb{R}^{d-m} \subset \mathbb{R}^{1+d}$. But since the additional zeroth component is trivial, we prefer to work with the above representation.

We have shown that affine processes are Markov processes whose conditional characteristic function is an affine function of the components of X. It can be shown that the converse holds as well, cf. [80, Theorem 2.12], but we do not need this result in the sequel.

The conditional characteristic function in Theorem 6.6 allows us to compute the joint characteristic function of $X(t_1), \dots, X(t_n)$ which turns out to be of affine form as well.

Proposition 6.8 (Joint Characteristic Function) *Suppose that the assumptions of Theorem 6.6 hold and let* $0 = t_0 \le t_1 \le \cdots \le t_n$. *The joint characteristic function of* $X(t_1), \dots, X(t_n)$ *is given by*

$$\begin{aligned}E\left(\exp\left(i\sum_{k=1}^n u_k^\top X(t_k)\right)\right) &= \exp\left(\sum_{j=1}^d \Psi_j^{(n)}(t_1-t_0,\dots,t_n-t_{n-1};u_1,\dots,u_n)X_j(0)\right) \\ &\quad\times\exp\left(\sum_{k=1}^n \Psi_0^{(k)}(t_{n-k+1}-t_{n-k},\dots,t_n-t_{n-1};u_{n-k+1},\dots,u_n)\right) \qquad (6.32)\end{aligned}$$

for any $u_1, \dots, u_n \in \mathbb{R}^d$, *where* $\Psi^{(k)} = (\Psi_0^{(k)}, \dots, \Psi_d^{(k)})$ *is defined recursively via*

$$\Psi^{(1)}(\tau_1;u_1) := \big(\Psi_0(\tau_1,u_1), \Psi_{1,\dots,d}(\tau_1,u_1)\big)$$

and

$$\begin{aligned}&\Psi^{(k)}(\tau_1,\dots,\tau_k;u_1,\dots,u_k) \\ &\quad := \Psi^{(k-1)}\big(\tau_1,\dots,\tau_{k-1};u_1,\dots,u_{k-2},u_{k-1}-i\Psi_{1,\dots,d}(\tau_k,u_k)\big)\end{aligned}$$

for $u_1, \dots, u_k \in \mathbb{R}^d$.

Proof We proceed by induction. For $n = 1$, (6.32) reduces to (6.29). For the step from $n - 1$ to n observe that

$$E\left(\exp\left(i\sum_{k=1}^{n} u_k^\top X(t_k)\right)\right) = E\left(E\left(\exp\left(i\sum_{k=1}^{n} u_k^\top X(t_k)\right)\middle|\mathscr{F}_{t_{n-1}}\right)\right)$$

$$= E\left(\exp\left(i\sum_{k=1}^{n-1} u_k^\top X(t_k)\right)E\left(\exp\left(iu_n^\top X(t_n)\right)\middle|\mathscr{F}_{t_{n-1}}\right)\right)$$

which, by (6.29), equals

$$E\Bigg(\exp\Bigg(\Psi_0(t_n - t_{n-1}, u_n) + i\sum_{k=1}^{n-2} u_k^\top X(t_k)$$

$$+ \left(iu_{n-1} + \Psi_{1,\dots,d}(t_n - t_{n-1}, u_n)\right)^\top X(t_{n-1})\Bigg)\Bigg).$$

From (6.32) for $n - 1$ we obtain (6.32) for n. □

We have already encountered a number of affine semimartingales, cf. Examples 6.2–6.4. If the triplets in the linear part of (6.23–6.25) vanish, we simply end up with a Lévy process with Lévy–Khintchine triplet $(b^h_{(0)}, c_{(0)}, K_{(0)})$. As mentioned above, Lévy-driven Ornstein–Uhlenbeck processes as in Sect. 3.9 constitute further examples of affine processes. The process (6.17) which motivated the extension to affine characteristics is easily treated with the help of Theorem 6.6:

Example 6.9 (Extended Square-Root Process) The process satisfying

$$dX(t) = (\kappa - \lambda X(t))dt + \sigma\sqrt{X(t)}dW(t), \quad X(0) = x \tag{6.33}$$

with standard Brownian motion W and constants $\lambda \in \mathbb{R}, \kappa, \sigma, x > 0$ has affine characteristics in the sense of (6.18–6.20) with triplets $(b^h_{(0)}, c_{(0)}, K_{(0)}) = (\kappa, 0, 0)$ and $(b^h_{(1)}, c_{(1)}, K_{(1)}) = (-\lambda, \sigma^2, 0)$. The functions occurring in the characteristic function (6.21) are given by

$$\Psi_0(t, u) = -\frac{2\kappa}{\sigma^2}\log\left(1 - \frac{\sigma^2 iu}{2\lambda}(1 - e^{-\lambda t})\right),$$

$$\Psi_1(t, u) = \frac{iue^{-\lambda t}}{1 - \frac{\sigma^2 iu}{2\lambda}(1 - e^{-\lambda t})}.$$

As a side remark, it can be shown that SDE (6.33) has a unique strong solution but this does not follow from Theorems 3.40 or 3.41, cf. [241, Theorem IX.3.5].

Further and in particular multivariate examples of affine processes are discussed in Part II in the context of stochastic volatility and interest rate models. The Hawkes processes of Example 4.39 and Problem 4.7 constitute affine processes as well. The affine triplets are easily obtained from (4.49) and (4.69, 4.70).

6.2 Moments

In applications one often needs ordinary and exponential moments of random variables or processes. Since the characteristic function consists of purely imaginary exponential moments, the latter are immediately obtained from Theorem 6.6 for affine processes. Moreover, one can show that the statement of Theorem 6.6 also holds for

$$u \in \Big\{(x_1, \dots, x_d) \in \mathbb{C}^d : \\ \mathrm{Im}(x_j) \leq 0 \text{ for } j = 1, \dots, m \text{ and } \mathrm{Im}(x_j) = 0 \text{ for } j = m+1, \dots, d\Big\},$$

cf. [80]. Further exponential moments are obtained by applying Proposition A.5(4): if the characteristic function $u \mapsto E(\exp(iu^\top X(t)))$ in Theorem 6.6 has an analytic extension to some open convex set $U \in \mathbb{C}^d$, the exponential moment $E(\exp(p^\top X(t)))$ for $p \in \mathbb{C}^d$ with $-ip \in U$ and $-i\mathrm{Re}(p) \in U$ is obtained by inserting $-ip$ into this extension.

Alternatively, exponential moments $E(\exp(p^\top X(t)))$ with $p \in \mathbb{R}^d$ can typically be obtained by solving the system of generalised Riccati equations (6.30) with initial value p instead of iu.

Theorem 6.10 (Exponential Moments) *Let $p \in \mathbb{R}^d$ and $t \geq 0$. Suppose that there exist solutions $\Phi_{1,\dots,d}$ and Φ_0 to (6.30, 6.31) for p instead of iu, i.e. there exists a continuously differentiable function $\Phi_{1,\dots,d} = (\Phi_1, \dots, \Phi_d) : \mathbb{R}_+ \to \mathbb{R}^d$ satisfying*

$$\int_{\{|x|>1\}} e^{\Phi_{1,\dots,d}(s)^\top x} K_{(j)}(dx) < \infty$$

and

$$\frac{d}{dt}\Phi_j(s) = \psi_j(-i\Phi_{1,\dots,d}(s)), \quad \Phi_{1,\dots,d}(0) = p$$

for $s \in [0, t]$ and $j = 1, \dots, d$, and we set

$$\Phi_0(s) := \int_0^s \psi_0(-i\Phi_{1,\dots,d}(r))dr$$

for $s \in [0,t]$. Moreover, we need the moment condition

$$\sup_{s\in[0,t]} \int_{\{x_k>1\}} x_k e^{\Phi_{1,\dots,d}(s)^\top x} K_{(j)}(dx) < \infty, \quad 1 \le j,k \le m.$$

Then we have

$$E\Big(\exp\big(p^\top X(t)\big)\Big) = \exp\Big(\Phi_0(t) + \Phi_{1,\dots,d}(t)^\top X(0)\Big). \tag{6.34}$$

Proof On a formal level, (6.34) can be derived as (6.29). A rigorous proof can be found in [172, Theorem 5.1]. □

Example 6.11 (Extended Square-Root Process) By straightforward insertion one expects exponential moments of the extended square-root process in Example 6.9 to be of the form

$$E(\exp(pX(t))) = \exp\left(-\frac{2\kappa}{\sigma^2}\log\left(1-\frac{\sigma^2 p}{2\lambda}(1-e^{-\lambda t})\right) + \frac{pe^{-\lambda t}}{1-\frac{\sigma^2 p}{2\lambda}(1-e^{-\lambda t})}X(0)\right).$$

An analysis of the singularities of the expressions for Ψ_0, Ψ_1 shows that this is indeed the case for $\mathrm{Re}(p) < \frac{2\lambda}{\sigma^2(1-e^{-\lambda t})}$. For real numbers p one can alternatively verify that

$$\Phi_1(t) = \frac{pe^{-\lambda t}}{1-\frac{\sigma^2 p}{2\lambda}(1-e^{-\lambda t})}$$

and

$$\Phi_0(t) = -\frac{2\kappa}{\sigma^2}\log\left(1-\frac{\sigma^2 p}{2\lambda}(1-e^{-\lambda t})\right)$$

satisfy the conditions in Theorem 6.10 for $p < \frac{2\lambda}{\sigma^2(1-e^{-\lambda t})}$. This leads to the same expression for $E(\exp(pX(t)))$ as above.

In applications such as in interest rate theory, one also needs exponential moments of the form

$$E\left(\exp\left(\int_0^t -X(s)ds\right)\right), \tag{6.35}$$

where X denotes some affine process. It does not come as a surprise that these moments are of exponentially affine form as well. One way to obtain them is by considering the Feynman–Kac formula instead of Kolmogorov's backward equation in Sect. 6.1. Alternatively, $(\int_0^t X(s)ds)_{t\ge 0}$ can be viewed as an additional component of a multivariate affine process, which is shown in Proposition 6.21 below. This allows us to compute (6.35) via Theorem 6.6.

Ordinary moments $E(X^k)$, $k = 1, 2, \dots$ are obtained as derivatives of the characteristic function in 0, cf. Proposition A.5(2). Except for the first two moments in the univariate case, we leave its computation to the reader.

Example 6.12 (Expectation and Variance of an Affine Process) In the setup of Proposition 6.5 we obtain

$$E(X(t)) = -i\Psi_0'(t,0) - i\Psi_1'(t,0)X(0)$$

for the expectation of an affine process if it exists. This follows by differentiating the characteristic function according to Proposition A.5(2). The derivative refers to the second argument of Ψ_0 resp. Ψ_1. Similarly, we obtain

$$\begin{aligned} \mathrm{Var}(X(t)) &= E(X(t)^2) - (E(X(t)))^2 \\ &= -\Psi_0''(t,0) - \Psi_1''(t,0)X(0) \end{aligned} \tag{6.36}$$

for the variance if it exists.

In order to determine moments in this manner, the solutions $\Psi_0, \dots, \Psi_d$ to the generalised Riccati equations should of course be known in closed form. But even if they are, their relatively involved structure may turn higher order differentiation into a painful task. An alternative approach is to apply the backward equation directly to polynomial functions. This leads to a system of linear ODEs for integer moments. We start with the univariate case.

Theorem 6.13 *Let X be a univariate affine process as in Proposition 6.5 and $m \in \mathbb{N}$ even with*

$$\int_{\{|x|>1\}} |x|^m K_{(0)}(dx) < \infty, \quad \int_{\{|x|>1\}} |x|^m K_{(1)}(dx) < \infty. \tag{6.37}$$

These conditions hold if ψ_0, ψ_1 are m-times differentiable in 0. Fix $n \in \{0, 1, \dots, m\}$. Denote the kth derivatives of ψ_0, ψ_1 in 0 by $\psi_0^{(k)}(0)$, $\psi_1^{(k)}(0)$ for $k = 0, \dots, n$. Define the lower triangular matrix $B = (B_{jk})_{j,k=0,\dots,n} \in \mathbb{R}^{(1+n)\times(1+n)}$ by

$$B_{jk} := \begin{cases} 0 & \text{if } j = 0, \\ i^{-j}\psi_0^{(j)}(0) & \text{if } j \geq 1, \quad j = 0, \\ \binom{j}{k} i^{k-j}\psi_0^{(j-k)}(0) + \binom{j}{k-1} i^{k-j-1}\psi_1^{(j-k+1)}(0) & \text{if } j \geq 1, \quad k \in \{1, \dots, j-1\}, \\ -ij\psi_1'(0) & \text{if } j \geq 1, \quad k = j, \\ 0 & \text{if } j \geq 1, \quad k > j. \end{cases}$$

Then we have

$$E\big(X(s+t)^n\big|\mathscr{F}_s\big)=\sum_{k=0}^{n}a_k(t)X(s)^k,$$

where $a=(a_0,\dots,a_n):\mathbb{R}_+\to\mathbb{R}^{1+n}$ *are defined by*

$$a(t)=1_n^\top e^{Bt}$$

and $1_n=(0,\dots,0,1)^\top\in\mathbb{R}^{1+n}$ *denotes the nth unit vector.*

Condition (6.37) holds if ψ_0,ψ_1 *are n-times differentiable in* 0 *if n is even resp.* $(n+1)$*-times differentiable in* 0 *if n is odd.*

Proof This follows from Theorem 6.26 and Example 6.30. □

Observe that it is relatively easy to compute the matrix exponential e^{Bt} because B is in triangular form. Moreover, note that the coefficients $\psi_0^{(k)}(0)$, $\psi_1^{(k)}(0)$ appearing in B are determined by the moments $E(L(t)^k)$, $k=1,\dots,n$ of a Lévy process L with exponent ψ_0 resp. ψ_1.

Applied to Example 6.12 we obtain the following alternative representation of the first two moments.

Example 6.14 (Expectation and Variance of an Affine Process) Applying Theorem 6.13 to the unconditional first moment (i.e. $n=1$, $s=0$) for deterministic $X(0)=0$ yields

$$E(X(t))=\frac{\psi_0'(0)}{\psi_1'(0)}\big(e^{-i\psi_1'(0)t}-1\big)+e^{-i\psi_1'(0)t}X(0)$$

if $\psi_1'(0)\neq 0$ and

$$E(X(t))=-i\psi_0'(0)t+X(0)$$

otherwise. For the variance we obtain similarly

$$\begin{aligned}\operatorname{Var}(X(t))=&-\frac{\psi_0'(0)\psi_1''(0)}{i\psi_1'(0)^2}\big(e^{-i\psi_1'(0)t}-1\big)\\&+\left(\frac{\psi_0'(0)\psi_1''(0)}{2i\psi_1'(0)^2}+\frac{\psi_0''(0)}{2i\psi_1'(0)}\right)\big(e^{-2i\psi_1'(0)t}-1\big)\\&+\frac{\psi_1''(0)}{i\psi_1'(0)}\left(e^{-2i\psi_1'(0)t}-e^{-i\psi_1'(0)t}\right)X(0)\end{aligned}\tag{6.38}$$

if $\psi_1'(0) \neq 0$ and

$$\mathrm{Var}(X(t)) = -\psi_0''(0)t + \frac{i}{2}\psi_0'(0)\psi_1''(0)t^2 - \psi_1''(0)tX(0) \tag{6.39}$$

otherwise. The previous formulas hold of course only if the derivatives exist. At first glance, (6.38, 6.39) look more involved than (6.36). But typically the derivatives of ψ_0, ψ_1 are easier to determine than those for Ψ_0, Ψ_1 as one may verify in the case of the extended square-root process in Example 6.9.

We now turn to the general case. We content ourselves with stating the linear ODEs leading to integer moments. We start with some notation. For a multiindex $\lambda = (\lambda_1, \dots, \lambda_d) \in \mathbb{N}^d$ we write $|\lambda| := \sum_{j=1}^d \lambda_j$ for the sum of its entries,

$$f^{(\lambda)}(x) := \frac{\partial^{\lambda_1}}{\partial x_1^{\lambda_1}} \cdots \frac{\partial^{\lambda_d}}{\partial x_d^{\lambda_d}} f(x_1, \dots, x_d)$$

for higher-order partial derivatives of a function $f : \mathbb{R}^d \to \mathbb{C}$ and $x^\lambda := \prod_{k=1}^d x_k^{\lambda_k}$ for $x \in \mathbb{R}^d$. We denote by $\mathbb{N}_n^d$ the set of multiindices λ with $|\lambda| \leq n$. For multiindices λ, μ the inequality $\lambda \leq \mu$ means that $\lambda - \mu$ is a multiindex as well, i.e. it has nonnegative components. In this case, we set

$$\binom{\lambda}{\mu} := \prod_{j=1}^d \binom{\lambda_j}{\mu_j}.$$

Finally, 1_j denotes the multiindex with $(1_j)_k := \delta_{jk} := 1_{\{j=k\}}$.

Theorem 6.15 (Moments of Affine Processes) *Let X be an $\mathbb{R}^d$-valued affine process as in Theorem 6.6. Fix a multiindex $\lambda \in \mathbb{N}_m^d$, where $m \in \mathbb{N}$ is even and satisfies*

$$\int_{\{|x|>1\}} |x|^m K_{(j)}(dx) < \infty, \quad j = 0, \dots, d.$$

This holds if the partial derivatives $\psi_j^{(m1_k)}(0)$ for $j = 0, \dots, d$ and $k = 1, \dots, d$ exist.

Define functions $g_\mu : \mathbb{R}_+ \to \mathbb{R}$ for multiindices $\mu \in \mathbb{N}_{|\lambda|}^d$ as the unique solution to the linear system

$$g_\mu'(t) = \sum_{\nu \in \mathbb{N}_{|\lambda|}^d \text{ with } \mu \leq \nu} \binom{\nu}{\mu} i^{-|\nu-\mu|} \psi_0^{(\nu-\mu)}(0) g_\nu(t)$$

$$+\sum_{j=1}^{d} \sum_{\nu\in\mathbb{N}^d_{|\lambda|} \text{ with } \mu\leq\nu+1_j} \binom{\nu}{\mu-1_j} i^{-|\nu+1_j-\mu|}\psi_j^{(\nu+1_j-\mu)}(0)g_\nu(t),$$

$$g_\mu(0) = \begin{cases} 1 & \text{if } \mu = \lambda, \\ 0 & \text{otherwise.} \end{cases}$$

Then we have

$$E\big(X(s+t)^\lambda \big| \mathscr{F}_s\big) = \sum_{\mu\in\mathbb{N}^d_{|\lambda|}} g_\mu(t)X(s)^\mu.$$

Proof This is the multivariate extension of Theorem 6.13. It follows from Theorem 6.26 and Example 6.30 as well. □

As a side remark, the idea to compute integer moments by solving linear ODEs can also be applied to generalised Ornstein–Uhlenbeck processes as in Sect. 3.9. For these processes the characteristic function is generally not known in closed form. We will come back to this issue in Sect. 6.6.

Finally, we consider exponentially affine martingales. Recall that local martingales coincide with martingales in the case of Lévy processes. This pleasant property ceases to hold for components of affine processes. Nevertheless it does hold under an easy to check integrability condition on the Lévy–Khintchine triplets underlying the affine process.

Proposition 6.16 (Exponentially Affine Martingales) *Let X be an $\mathbb{R}^d$-valued affine process as in Theorem 6.6 and let $1 \leq j \leq d$.*

1. *If*

   a) $K_{(k)}(\{x \in \mathbb{R}^d : x_j < -1\}) = 0$ *for* $k = 0, \ldots, m$,
   b) $\int_{\{x_j>1\}} x_j K_{(k)}(dx) < \infty$ *for* $k = 0, \ldots, m$,
   c) $b^h_{(k),j} + \int (x_j - h_j(x))K_{(k)}(dx) = 0$ *for* $k = 0, \ldots, d$,
   d) $\int_{\{x_\ell>1\}} x_\ell(1 + x_j)K_{(k)}(dx) < \infty$ *for* $k, \ell = 1, \ldots, m$,

   then $\mathscr{E}(X_j)$ is a martingale.
2. *If*

   a) $\int_{\{x_j>1\}} e^{x_j} K_{(k)}(dx) < \infty$, $k = 0, \ldots, d$,
   b) $b^h_{(k),j} + \frac{1}{2}c_{(k),jj} + \int (e^{x_j} - 1 - h_j(x))K_{(k)}(dx) = 0$, $k = 0, \ldots, d$,
   c) $\int_{\{x_\ell>1\}} x_\ell e^{x_j} K_{(k)}(dx) < \infty$, $k, \ell = 1, \ldots, m$,

   then e^{X_j} is a martingale.

Proof See [172, Corollaries 3.9 and 3.4]. □

Condition 1a) in the previous proposition warrants that $\mathscr{E}(X^j)$ does not jump to negative values. Condition 1b) is just needed for the integral in Condition 1c) to be finite. This condition in turn essentially means that $\mathscr{E}(X_j)$ is a local martingale. The fourth condition guarantees that this local martingale is a true martingale, which is not always the case for affine processes. Conditions 2a–c) have a similar meaning.

6.3 Structure-Preserving Operations

The affine structure of a process is generally not preserved under changes of the probability measure, stochastic integration, or application of mappings. Nevertheless, we do recover an affine process for some operations that play a role in Mathematical Finance. We start with linear combinations.

Proposition 6.17 (Linear Combination) *Suppose that X is an $\mathbb{R}^d$-valued affine process as in Theorem 6.6. Let $X_{d+1} := \sum_{j=1}^d a_j X_j$ for some real numbers $a_1, \dots, a_d$. Then $\widetilde{X} := (X_1, \dots, X_d, X_{d+1})$ is an $\mathbb{R}^{d+1}$-valued affine process as in Theorem 6.6 with the same parameter m. The corresponding characteristic exponents $\widetilde{\psi}_j$, $j = 1, \dots, d+1$ are given by*

$$\widetilde{\psi}_j(u_1, \dots, u_{d+1}) = \psi_j(u_1 + u_{d+1}a_1, \dots, u_d + u_{d+1}a_d), \quad j = 0, \dots, d, \tag{6.40}$$

$$\widetilde{\psi}_{d+1}(u_1, \dots, u_{d+1}) = 0 \tag{6.41}$$

for $(u_1, \dots, u_{d+1}) \in \mathbb{R}^{d+1}$.

Proof By Propositions 4.13 or 4.42, the characteristics of $\tilde{X}$ are affine with exponents (6.40, 6.41). It is easy to verify that the conditions in Theorem 6.6 are satisfied. □

The quadratic variation or more generally covariation of components of an affine process retains the affine structure although it is a bilinear operation.

Proposition 6.18 (Quadratic Variation and Covariation) *Let χ be a truncation function on $\mathbb{R}$. Denote by h the componentwise truncation function*

$$h(x_1, \dots, x_d) := (\chi(x_1), \dots, \chi(x_d))$$

on $\mathbb{R}^d$ and likewise $\widetilde{h}$ on $\mathbb{R}^{d+1}$. Suppose that X is an $\mathbb{R}^d$-valued affine process as in Theorem 6.6. Let $X_{d+1} := [X_j, X_k]$ for some indices $j, k \in \{1, \dots, d\}$. Then $\widetilde{X} := (X_1, \dots, X_d, X_{d+1})$ is an $\mathbb{R}^{d+1}$-valued affine process as in Theorem 6.6 with the same parameter m. The corresponding Lévy–Khintchine triplets $(\widetilde{b}^h_{(\ell)}, \widetilde{c}_{(\ell)}, \widetilde{K}_{(\ell)})$, $\ell = 0, \dots, d+1$ are given by

$$\widetilde{b}^h_{(\ell)} = \begin{pmatrix} b^h_{(\ell)} \\ c_{(\ell),jk} + \int \chi(x_j x_k) K_{(\ell)}(dx) \end{pmatrix}, \quad \widetilde{c}_{(\ell)} = \begin{pmatrix} c_{(\ell)} & 0 \\ 0 & 0 \end{pmatrix},$$

$$\widetilde{K}_{(\ell)}(G) = \int 1_G(x_1, \ldots, x_d, x_j x_k) K_{(\ell)}(dx), \quad G \in \mathscr{B}^{d+1}$$

for $\ell = 0, \ldots, d$ *and*

$$(\widetilde{b}^h_{(d+1)}, \widetilde{c}_{(d+1)}, \widetilde{K}_{(d+1)}) = (0, 0, 0).$$

Proof By $[X_j, X_k] = X_j X_k - X_j(0) X_k(0) - (X_j)_- \bullet X_k - (X_k)_- \bullet X_j$, Propositions 4.8, 4.13 and a tedious calculation, the characteristics of $\widetilde{X}$ are as stated in the assertion. Alternatively, one can use the simpler calculus of Sect. 4.8, cf. Propositions 4.41 and 4.44. Finally, one verifies that the conditions in Theorem 6.6 are satisfied. □

The process $\widetilde{X} := \mathscr{L}(\exp(X))$ is sometimes called the **exponential transform** of a semimartingale X. Conversely, $X := \log \mathscr{E}(\widetilde{X})$ is the **logarithmic transform** of a semimartingale $\widetilde{X}$ with $\Delta\widetilde{X} > -1$. These operations map Lévy processes on Lévy processes as is shown in Theorem 3.49. They preserve the affine structure as well.

Proposition 6.19 (Exponential and Logarithmic Transforms) *Choose truncation functions* $\chi, h, \widetilde{h}$ *as in Proposition 6.18. Suppose that* X *is an* $\mathbb{R}^d$*-valued affine process as in Theorem 6.6.*

1. *Let* $X_{d+1} := \mathscr{L}(\exp(X_n))$ *for some index* $n \in \{1, \ldots, d\}$*. Then* $\widetilde{X} := (X_1, \ldots, X_d, X_{d+1})$ *is an* $\mathbb{R}^{d+1}$*-valued affine process as in Theorem 6.6 with the same parameter* m*. The corresponding Lévy–Khintchine triplets* $(\widetilde{b}^h_{(j)}, \widetilde{c}_{(j)}, \widetilde{K}_{(j)})$*,* $j = 0, \ldots, d+1$ *are given by*

$$\widetilde{b}^h_{(j)} = \begin{pmatrix} b^h_{(j)} \\ b_{(j),n} + \frac{1}{2} c_{(j),nn} + \int (\chi(e^{x_n} - 1) - \chi(x_n)) K_{(j)}(dx) \end{pmatrix},$$

$$\widetilde{c}_{(j),kl} = \begin{cases} c_{(j),kl} & \text{for } k, \ell = 1, \ldots, d, \\ c_{(j),n\ell} & \text{for } k = d+1, \quad \ell = 1, \ldots, d, \\ c_{(j),kn} & \text{for } k = 1, \ldots, d, \quad \ell = d+1, \\ c_{(j),ni} & \text{for } k, \ell = d+1, \end{cases}$$

$$\widetilde{K}_{(j)}(G) = \int 1_G(x_1, \ldots, x_d, e^{x_n} - 1) K_{(j)}(dx), \quad G \in \mathscr{B}^{d+1}$$

for $j = 0, \ldots, d$ *and*

$$(\widetilde{b}^h_{(d+1)}, \widetilde{c}_{(d+1)}, \widetilde{K}_{(d+1)}) = (0, 0, 0).$$

2. *Similarly, let* $X_{d+1} := \log \mathscr{E}(X_n)$ *for some index* $n \in \{1, \ldots, d\}$ *such that* $K_{(j)}(\{x \in \mathbb{R}^d : x_n < -1\}) = 0$ *for* $j = 0, \ldots, m$*. Then* $\widetilde{X} := (X_1, \ldots, X_d, X_{d+1})$ *is an* $\mathbb{R}^{d+1}$*-valued affine process as in Theorem 6.6 with*

the same parameter m. The corresponding Lévy–Khintchine triplets $(\widetilde{b}^h_{(j)}, \widetilde{c}_{(j)}, \widetilde{K}_{(j)})$, $j = 0, \dots, d+1$ are given by

$$\widetilde{b}^h_{(j)} = \begin{pmatrix} b^h_{(j)} \\ b_{(j),n} - \frac{1}{2}c_{(j),nn} + \int (\chi(\log(1+x_n)) - \chi(x_n))K_{(j)}(dx) \end{pmatrix},$$

$$\widetilde{c}_{(j),k\ell} = \begin{cases} c_{(j),k\ell} & \text{for } k, \ell = 1, \dots, d, \\ c_{(j),n\ell} & \text{for } k = d+1, \quad \ell = 1, \dots, d, \\ c_{(j),kn} & \text{for } k = 1, \dots, d, \quad \ell = d+1, \\ c_{(j),nn} & \text{for } k, \ell = d+1, \end{cases}$$

$$\widetilde{K}_{(j)}(G) = \int 1_G(x_1, \dots, x_d, \log(1+x_n))K_{(j)}(dx), \quad G \in \mathscr{B}^{d+1}$$

for $j = 0, \dots, d$ and

$$\left(\widetilde{b}^h_{(d+1)}, \widetilde{c}_{(d+1)}, \widetilde{K}_{(d+1)}\right) = (0, 0, 0).$$

Proof Using Propositions 4.8 and 4.13 or alternatively Example 4.46 and Proposition 4.44 in Sect. 4.8 we conclude that the characteristics of $\widetilde{X}$ are of affine form with triplets as stated in the assertion. □

If the density process of a new probability measure is of exponentially affine form, the change of measure preserves the affine structure of a process.

Proposition 6.20 (Change of Measure) *Suppose that X is an $\mathbb{R}^d$-valued affine process as in Theorem 6.6 and let $j \in \{1, \dots, d\}$.*

1. *If X_j satisfies Conditions 1a–d) in Proposition 6.16 and $Q \overset{\text{loc}}{\sim} P$ is a probability measure with density process $Z = \mathscr{E}(X_j)$, then X is affine relative to Q as well. The corresponding Lévy–Khintchine triplets $(\widetilde{b}^h_{(k)}, \widetilde{c}_{(k)}, \widetilde{K}_{(k)})$, $k = 0, \dots, d$ are given by*

$$\widetilde{b}^h_{(k)} = b^h_{(k)} + c_{(k),j\cdot} + \int h(x)x_j K_{(k)}(dx), \quad \widetilde{c}_{(k)} = c_{(k)},$$

$$\widetilde{K}_{(k)}(G) = \int 1_G(x)(1+x_j)K_{(k)}(dx), \quad G \in \mathscr{B}^d.$$

2. *If X_j satisfies Conditions 2a–c) in Proposition 6.16 and $Q \overset{\text{loc}}{\sim} P$ is a probability measure with density process $Z = \exp(X_j)$, then X is affine relative to Q as well.*

The corresponding Lévy–Khintchine triplets $(\widetilde{b}^h_{(k)}, \widetilde{c}_{(k)}, \widetilde{K}_{(k)})$ *and exponents* $\widetilde{\psi}_k$, $k = 0, \dots, d$ *are given by*

$$\widetilde{b}^h_{(k)} = b^h_{(k)} + c_{(k),j\cdot} + \int h(x)(e^{x_j} - 1)K_{(k)}(dx), \quad \widetilde{c}_{(k)} = c_{(k)},$$

$$\widetilde{K}_{(k)}(G) = \int 1_G(x)e^{x_j}K_{(k)}(dx), \quad G \in \mathscr{B}^d,$$

$$\widetilde{\psi}_k(u) = \psi_k(u - ie_j) - \psi_k(-ie_j), \quad u \in \mathbb{R}^d,$$

where $e_j \in \mathbb{R}^d$ *denotes the* j*th unit vector.*

Proof The differential characteristics of X relative to Q can be derived from Theorem 4.15 or Sect. 4.8, cf. Proposition 4.45. We observe that they are of affine form with triplets as in the assertion. □

Ordinary integrals of components of affine processes lead again to an affine process.

Proposition 6.21 (Integration) *Suppose that* X *is an* $\mathbb{R}^d$*-valued affine process as in Theorem 6.6. Let* $j \in \{1, \dots, d\}$ *and* $X_{d+1}(t) := \int_0^t X_j(s)ds$, $t \geq 0$. *Then* $\widetilde{X} := (X_1, \dots, X_d, X_{d+1})$ *is an* $\mathbb{R}^{d+1}$*-valued affine process as in Theorem 6.6 with the same parameter* m. *The corresponding characteristic exponents* $\widetilde{\psi}_k$, $k = 0, \dots, d+1$ *are given by*

$$\begin{aligned}
\widetilde{\psi}_k(u_1, \dots, u_{d+1}) &= \psi_k(u_1, \dots, u_d), \quad k \in \{0, \dots, d\} \setminus \{j\}, \\
\widetilde{\psi}_j(u_1, \dots, u_{d+1}) &= \psi_j(u_1, \dots, u_d) + iu_{d+1}, \\
\widetilde{\psi}_{d+1}(u_1, \dots, u_{d+1}) &= 0
\end{aligned}$$

for $(u_1, \dots, u_{d+1}) \in \mathbb{R}^{d+1}$.

Proof Applying Proposition 4.8 or Sect. 4.8, we conclude that the characteristics of $\widetilde{X}$ are of affine form, corresponding to the exponents above. □

6.4 Change of Measure

As will be discussed in Sect. 11.2.3, we may want to model financial data similarly under the objective and some risk-neutral probability measure. An affine structure comes in handy in both settings because the characteristic function is known explicitly or at least as a solution to some ODE. Explicit knowledge of the characteristic function allows us to compute moment-based estimators under the objective measure. In addition, it opens the door to efficient calibration procedures in order to determine the risk-neutral measure, as we will see in Sect. 11.2.3. In this context, the question arises whether two sets of affine triplets in the sense of (6.23–6.25) correspond to an equivalent change of measure of one and the same process. A complete characterisation seems out of reach here. But a simple sufficient criterion reads as follows:

Proposition 6.22 *Let X and Y be $\mathbb{R}^d$-valued semimartingales as in Theorem 6.6 with affine local characteristics relative to Lévy–Khintchine triplets $(b^h_{(j)}, c_{(j)}, K_{(j)})$ and $(\widetilde{b}^h_{(j)}, \widetilde{c}_{(j)}, \widetilde{K}_{(j)})$, $j = 0, \dots, d$, respectively. Suppose that there exist $\varphi \in \mathbb{R}^d$ and a function $\psi : \mathbb{R}^d \to [0, \infty)$ such that*

1. $\int (1 - \sqrt{\psi(x)})^2 K_{(j)}(dx) < \infty$,
2. $\widetilde{K}_{(j)}(G) = \int 1_G(x)\psi(x)K_{(j)}(dx)$ *for all* $G \in \mathscr{B}^d$,
3. $\widetilde{b}^h_{(j)} = b^h_{(j)} + c_{(j)}\varphi + \int h(x)(\psi(x) - 1)K_{(j)}(dx)$,
4. $\widetilde{c}_{(j)} = c_{(j)}$

for $0 \leq j \leq d$. Then the law of $(Y(t))_{t\in[0,T]}$ is absolutely continuous with respect to the law of $(X(t))_{t\in[0,T]}$ for any finite $T \in \mathbb{R}_+$. If ψ is strictly positive, the laws are actually equivalent for any finite $T \in \mathbb{R}_+$.

Proof See [172, Corollary 4.2] and the proof of Theorem 4.20. □

Recall from Theorem 4.20 that the sufficient conditions in Proposition 6.22 are in fact necessary for Lévy processes.

One may wonder what change of measure induces the change of triplets in Proposition 6.22. This can be easily seen from the proof. If we define the martingale

$$Z := \mathscr{E}\Big(\varphi^\top X^c + (\psi - 1) * (\mu^X - \nu^X)\Big),$$

the density $Z(T)$ defines a probability measure $Q_T \ll P$ (resp. $Q_T \sim P$) such that, under Q_T, $(X(t))_{t\in[0,T]}$ has affine characteristics as in (6.23–6.25) but relative to $(\widetilde{b}^h_{(j)}, \widetilde{c}_{(j)}, \widetilde{K}_{(j)})$ instead of $(b^h_{(j)}, c_{(j)}, K_{(j)})$, cf. Theorem 4.15.

6.5 Time-Inhomogeneous Affine Markov Processes

So far, we have considered time-homogeneous affine processes in the sense that the Lévy–Khintchine triplets $(b^h_{(j)}, c_{(j)}, K_{(j)})$ and hence the characteristic exponents in the generalised Riccati equations (6.30, 6.31) do not depend on time. For applications, for instance, in interest rate theory we need to study time-inhomogeneous affine processes X as well. Their local characteristics satisfy

$$b^h(t) = b^h_{(0)}(t) + \sum_{j=1}^{d} b^h_{(j)}(t)X_j(t-), \tag{6.42}$$

$$c(t) = c_{(0)}(t) + \sum_{j=1}^{d} c_{(j)}(t) X_j(t-), \tag{6.43}$$

$$K(t, dx) = K_{(0)}(t, dx) + \sum_{j=1}^{d} K_{(j)}(t, dx) X_j(t-) \tag{6.44}$$

with time-dependent but deterministic Lévy–Khintchine triplets $(b^h_{(j)}(t), c_{(j)}(t), K_{(j)}(t, dx))$. Characteristic exponents are defined accordingly as

$$\psi_j(t, u) := iu^\top b^h_{(j)}(t) - \frac{1}{2} u^\top c_{(j)}(t) u + \int \left(e^{iu^\top x} - 1 - iu^\top h(x)\right) K_{(j)}(t, dx). \tag{6.45}$$

In the simplest case the triplets for $j \neq 0$ vanish, in which case X is a time-inhomogeneous Lévy process with time-varying triplet $(b^h_{(0)}(t), c_{(0)}(t), K_{(0)}(t, \cdot))$. In general, some regularity must be imposed in addition to admissibility conditions (6.26) in order to warrant existence and uniqueness of semimartingales with local characteristics (6.42–6.44). In [104], inhomogeneous affine processes have been shown to exist if the triplets are continuous in an appropriate sense.

Theorem 6.23 (Time-Inhomogeneous Affine Processes) *Fix $X(0) \in \mathbb{R}^m_+ \times \mathbb{R}^{d-m}$ and let $(b^h_{(j)}, c_{(j)}, K_{(j)})$, $j = 0, \ldots, d$ be time-dependent Lévy–Khintchine triplets satisfying the admissibility conditions (6.26) for any fixed $t \geq 0$ and which are continuous in t in the sense of Proposition 5.33, i.e. for some continuous truncation function h, the mappings*

$$t \mapsto b^h_{(j)}(t), \quad t \mapsto \tilde{c}^h_{(j)}(t) := c_{(j)}(t) + \int h(x)h(x)^\top K_{(j)}(t, dx),$$

and

$$t \mapsto \int f(y) K_{(j)}(t, dx)$$

are continuous for any bounded continuous function $f : \mathbb{R}^d \to \mathbb{R}$ that vanishes in a neighbourhood of 0. Moreover, suppose that the no-explosion condition

$$\sup_{t \in [0,T]} \int_{\{x_k > 1\}} x_k K_{(j)}(t, dx) < \infty, \quad j, k = 1, \ldots, m$$

holds for any $T \geq 0$. Then there is a unique solution to the affine martingale problem (6.42–6.44), where uniqueness refers to a unique law as usual. The solution *is a time-inhomogeneous Markov process whose conditional characteristic function is given by*

$$E\big(\exp(iu^\top X(T))\big|\mathscr{F}_t\big) = \exp\left(\Psi_0(t,T,u) + \sum_{j=1}^d \Psi_j(t,T,u)X_j(s)\right), \quad u \in \mathbb{R}^d,$$

where $\Psi_{1,\dots,d} = (\Psi_1,\dots,\Psi_d)^\top$ *is the unique solution to the following system of* generalised Riccati equations

$$\frac{d}{dt}\Psi_j(t,T,u) = -\psi_j(t, -i\Psi_{1,\dots,d}(t,T,u)) \quad \textit{for } t \leq T,$$

$$\Psi_{1,\dots,d}(T,T,u) = iu,$$

Ψ_0 *is given by*

$$\Psi_0(t,T,u) := \int_t^T \psi_0\big(s, -i\Psi_{1,\dots,d}(s,T,u)\big)ds,$$

and the characteristic exponents ψ_j *are defined in (6.45). More specifically, the space-time process* (I,X) *is a Feller process with extended generator*

$$\widetilde{G}f(t,x) = D_1 f(t,x) + \sum_{j=1}^d \beta_j^h(t,x)D_{1+j}f(t,x) + \frac{1}{2}\sum_{j,k=1}^d \gamma_{jk}(t,x)D_{1+j,1+k}f(t,x)$$

$$+ \int \left(f(t,x+y) - f(t,x) - \sum_{j=1}^d h_j(y)D_{1+j}f(t,x)\right)\kappa(t,x,dy)$$

for

$$\beta^h(t,x) := b^h_{(0)}(t) + \sum_{j=1}^d b^h_{(j)}(t)x_j,$$

$$\gamma(t,x) := c_{(0)}(t) + \sum_{j=1}^d c_{(j)}(t)x_j,$$

$$\kappa(t,x,dy) := K_{(0)}(t,dy) + \sum_{j=1}^d K_{(j)}(t,dy)x_j$$

and any bounded twice continuously differentiable function $f : \mathbb{R}_+ \times \mathbb{R}^m_+ \times \mathbb{R}^{d-m} \to \mathbb{R}$. *If* f *has compact support, it belongs to the domain of the generator of* X.

Idea On a formal level we obtain the conditional characteristic function essentially as Theorem 6.6. A rigorous proof is provided in [104, Theorems 2.13, 2.14] together with [172, Lemma A.1]. The continuity conditions in [104, Definition 2.5] are slightly stronger than those in Theorem 6.23 but only the latter are needed in the proof. The statement concerning the extended generator follows from Proposition 4.31. □

Time-inhomogeneous counterparts can also be stated for Propositions and Theorems 6.8, 6.10, 6.15–6.22. Except for the additional time variable in triplets and characteristic exponents, this may occasionally require some continuity or uniform moment conditions, cf. e.g. the time-inhomogeneous version of Propositions 6.16 and 6.22 in [172].

Remark 6.24 Of course, affine processes in the sense of Sect. 6.1 are time-inhomogeneous affine. The functions $\Psi(t, T, u) = (\Psi_0, \dots, \Psi_d)(t, T, u)$ in Theorem 6.23 are obtained from $\Psi(t, u) = (\Psi_0, \dots, \Psi_d)(t, u)$ in Theorem 6.6 via $\Psi(t, T, u) = \Psi(T - t, u)$ for $t \leq T$, $u \in \mathbb{R}^d$.

6.6 Polynomial Processes

In Sect. 6.2 we computed integer moments of affine processes by solving linear ordinary differential equations. This method works in fact for the more general class of *polynomial* or *polynomial preserving processes*. Roughly speaking, this term signifies Markov processes whose transition semigroup or equivalently extended generator maps polynomials to polynomials of at most the same degree.

For monomials, multiindices etc. we refer to the notation before Theorem 6.15. Polynomials of degree at most m on $\mathbb{R}^d$ are functions $f : \mathbb{R}^d \to \mathbb{R}$ of the form $f(x) = \sum_{\lambda \in \mathbb{N}_m^d} c_\lambda x^\lambda$ with real-valued coefficients c_λ. We write $\mathscr{P}_m(\mathbb{R}^d)$ for the set of such polynomials. By definition, a polynomial of degree at most m on E is the restriction to E of some $f \in \mathscr{P}_m(\mathbb{R}^d)$, i.e. we set $\mathscr{P}_m(E) := \{f|_E : f \in \mathscr{P}_m(\mathbb{R}^d)\}$.

Fix a closed set $E \subset \mathbb{R}^d$ and an E-valued special semimartingale X that solves the martingale problem related to triplet $(\beta^{\mathrm{id}}, \gamma, \kappa)$ or symbol ψ. We define the extended generator $\widetilde{G}$ and its domain $\mathscr{D}_{\widetilde{G}}$ as in Sect. 5.3. Recall from Sect. 5.7.1 that $\widetilde{G} f$ is of the form (5.11) for bounded twice continuously differentiable functions f.

Theorem 6.25 (Polynomial Process) *The following statements are equivalent for fixed $m \in \mathbb{N}$.*

1. *There is a matrix $A_m = (a_{\lambda,\mu})_{\lambda,\mu \in \mathbb{N}_m^d}$ such that the triplet $(\beta^{\mathrm{id}}, \gamma, \kappa)$ satisfies*

$$\beta_j^{\mathrm{id}}(x) = \sum_{\mu \in \mathbb{N}_1^d} a_{1_j,\mu} x^\mu, \tag{6.46}$$

$$\widetilde{\gamma}_{jk}^{\mathrm{id}} := \gamma_{jk} + \int \xi_j \xi_k K(x, d\xi) = \sum_{\mu \in \mathbb{N}_2^d} a_{1_j + 1_k, \mu} x^{\mu}, \tag{6.47}$$

$$\int \xi^{\lambda} K(x, d\xi) = \sum_{\mu \in \mathbb{N}_{|\lambda|}^d} a_{\lambda, \mu} x^{\mu} \tag{6.48}$$

for any $j, k \in \{1, \dots, d\}$ and any $\lambda \in \mathbb{N}_m^d$ with $|\lambda| \geq 3$. Here we assume implicitly that

$$\int_{\{|\xi_j|>1\}} |\xi_j|^m K(x, d\xi) < \infty, \quad j = 1, \dots, d \tag{6.49}$$

in order for the integrals to make sense. Moreover, we set $a_{0,0} := 0$ and $a_{\lambda,\mu} := 0$ for any $\lambda, \mu \in \mathbb{N}_m^d$ with $|\mu| > |\lambda|$.

2. *$\psi^{(\lambda)}(x, 0) \in \mathscr{P}_k(E)$ holds for $k = 0, \dots, m$ and any $\lambda \in \mathbb{N}_k^d$, where $\psi^{(\lambda)}(x, u)$ refers to the derivative of order λ of the function $u \mapsto \psi(x, u)$. Here we make the same implicit assumption as in 1., which warrants that the derivatives exist. In this case define the matrix $A_m = (a_{\lambda,\mu})_{\lambda,\mu \in \mathbb{N}_m^d}$ by*

$$\psi^{(\lambda)}(x, 0) = i^{|\lambda|} \sum_{\mu \in \mathbb{N}_m^d} a_{\lambda, \mu} x^{\mu}. \tag{6.50}$$

3. *$f \in \mathscr{D}_{\widetilde{G}}$ and $\widetilde{G} f \in \mathscr{P}_k(E)$ hold for $k = 0, \dots, m$ and any $f \in \mathscr{P}_k(E)$. In this case, denote by $B_m = (b_{\lambda,\mu})_{\lambda,\mu \in \mathbb{N}_m^d}$ the matrix representing the linear mapping $\widetilde{G}$ on $\mathscr{P}_m$, i.e. satisfying*

$$\widetilde{G} f(x) = \sum_{\mu \in \mathbb{N}_m^d} b_{\lambda, \mu} x^{\mu} \tag{6.51}$$

for $f(x) = x^{\lambda}$. Note that $b_{0,0} = 0$ and $b_{\lambda,\mu} = 0$ if $|\mu| > |\lambda|$.

If either of these statements hold, the matrices A_m in 1., 2. coincide. Moreover, B_m is related to A_m via

$$b_{\lambda, \mu} := \sum_{\nu \leq \mu} \binom{\lambda}{\nu} a_{\lambda - \nu, \mu - \nu}. \tag{6.52}$$

If—for small state spaces E—the polynomials in (6.46, 6.47, 6.48, 6.50, 6.51) have more than one representation $\sum_{\mu \in \mathbb{N}_m^d} a_{\lambda,\mu} x^{\mu}$, we take one with minimal degree in each case.

Proof 1⇔2: Since

$$\psi(x,u)=iu^\top\beta^{\mathrm{id}}(x)-\frac{1}{2}u^\top\gamma(x)u+\int\left(e^{iu^\top\xi}-1-iu^\top\xi\right)\kappa(x,d\xi),$$

it is easy to see that

$$\begin{aligned}\psi^{(1_j)}(x,0)&=i\beta_j^{\mathrm{id}}(x),\quad j=1,\dots,d,\\ \psi^{(1_j+1_k)}(x,0)&=-\widetilde{\gamma}_{jk}^{\mathrm{id}}(x),\quad j,k=1,\dots,d,\\ \psi^{(\lambda)}(x,0)&=i^{|\lambda|}\int\xi^\lambda\kappa(x,d\xi)\quad\text{if }|\lambda|\geq 3.\end{aligned}$$

This yields the stated equivalence.
1⇒3: This follows by inserting (6.46–6.48) into (5.11). The integrability condition (6.35) warrants that $f(x)=x^\lambda$ belongs to the domain of the extended generator of X for $|\lambda|\leq m$. More specifically, (5.11) implies

$$\begin{aligned}\widetilde{G}f(x)=&\sum_{j=1}^d\lambda_jx^{\lambda-1_j}\beta_j^{\mathrm{id}}(x)+\sum_{j<k}\lambda_j\lambda_kx^{\lambda-1_j-1_k}\gamma_{jk}(x)+\sum_{j=1}^d\lambda_j(\lambda_j-1)x^{\lambda-21_j}\gamma_{jj}(x)\\&+\int\left((x+\xi)^\lambda-x^\lambda-\sum_{j=1}^d\lambda_jx^{\lambda-1_j}\xi_j\right)\kappa(x,d\xi)\end{aligned}$$

for $f(x)=x^\lambda$, where we set $x^\mu:=0$ if $\mu=(\mu_1,\dots,\mu_d)\notin\mathbb{N}^d$. Since

$$(x+\xi)^\lambda-x^\lambda-\sum_{j=1}^d\lambda_jx^{\lambda-1_j}\xi_j=\sum_{\substack{\mu\leq\lambda\\|\mu|<|\lambda|-1}}\binom{\lambda}{\mu}x^\mu\xi^{\lambda-\mu},$$

using (6.46–6.48) and sorting terms according to powers x^μ yields (6.52) after a straightforward but slightly tedious calculation.
3⇒1: By induction on $|\lambda|=1,2,\dots,m$ it follows from (6.51) that (6.46–6.48) holds for some index A_m. The implication 1⇒3 implies that A and B are related as claimed. □

In the situation of Theorem 6.25, we call X an m-**polynomial process** (or m-**polynomial preserving process**). Moreover, X is a **polynomial** or **polynomial preserving process** if it is m-polynomial for any $m\in\mathbb{N}$.

The following result explains why polynomial processes have been introduced in the first place.

Theorem 6.26 (Moment Formula) *For even $m \in \mathbb{N}$ let X be m-polynomial with corresponding matrices $A_m = (a_{\lambda,\mu})_{\lambda,\mu\in\mathbb{N}^d_m}$ and $B_m = (b_{\lambda,\mu})_{\lambda,\mu\in\mathbb{N}^d_m}$. For any $s,t \geq 0$ and any polynomial $f \in \mathscr{P}_m(E)$ we have*

$$E\big(f(X(s+t))\big|\mathscr{F}_s\big) = \sum_{\lambda,\mu\in\mathbb{N}^d_m} f_\lambda (e^{B_m t})_{\lambda,\mu} X(s)^\mu$$

if $(f_\lambda)_{\lambda\in\mathbb{N}^d_m}$ denote the coefficients of f in the sense that $f(x) = \sum_{\lambda\in\mathbb{N}^d_m} f_\lambda x^\lambda$.

Proof On the formal level we make the ansatz that

$$E\big(X(s+t)^\lambda\big|\mathscr{F}_s\big) = \sum_{\mu\in\mathbb{N}^d_{|\lambda|}} c_{\lambda,\mu}(t) X(s)^\mu =: u_\lambda(t, X(s))$$

with functions $c_{\lambda,\mu} : \mathbb{R}_+ \to \mathbb{R}$. Inserting this ansatz into the backward equation (5.21) yields the system of ODEs

$$\frac{d}{dt} c_{\lambda,\mu}(t) = \sum_{\nu\in\mathbb{N}^d_{|\lambda|}} c_{\lambda,\nu}(t)(B_m)_{\nu,\mu},$$

$$c_{\lambda,\mu}(0) = \begin{cases} 1 & \text{if } \lambda = \mu, \\ 0 & \text{otherwise.} \end{cases}$$

This is solved by $c_{\lambda,\mu}(t) = (e^{B_m t})_{\lambda,\mu}$, which yields the claim. In order to make the argument rigorous, one needs that the local martingale $(u_\lambda(s+t-r, X(r)))_{r\in[0,s+t]}$ is a true martingale, which holds by [63, Theorem 2.10] or more precisely its proof. □

In particular, we have

$$E(X(t)^\lambda) = \sum_{\mu\in\mathbb{N}^d_m} (e^{B_m t})_{\lambda,\mu} X(0)^\mu$$

for $f(x) = x^\lambda$ if $X(0)$ is deterministic. Therefore the previous theorem helps to express integer moments in terms of matrix exponentials or equivalently solutions to systems of linear ordinary differential equations. In the univariate case $d = 1$, the matrices A_m and hence B_m as well as $e^{B_m t}$ are triangular, which implies that the matrix exponential can be computed relatively easily.

Observe that the matrices in Theorem 6.25 do not really depend on m:

Proposition 6.27 *If X is polynomial, there are matrices $A = (a_{\lambda,\mu})_{\lambda,\mu\in\mathbb{N}^d}$, $B = (b_{\lambda,\mu})_{\lambda,\mu\in\mathbb{N}^d}$ such that the matrices in Theorem 6.25 are of the form $A_m = (a_{\lambda,\mu})_{\lambda,\mu\in\mathbb{N}^d_m}$, $B_m = (b_{\lambda,\mu})_{\lambda,\mu\in\mathbb{N}^d_m}$ for any $m \in \mathbb{N}$. We call A the* ***generating matrix*** *of the polynomial process X.*

Proof This is evident from how A and B are defined. □

In Theorem 6.25 we assumed (6.49). A sufficient condition in terms of the symbol reads as follows.

Proposition 6.28 *If $\psi^{(m1_j)}(x, 0)$, $j = 1, \dots, d$ exist for even m or $\psi^{((m+1)1_j)}(x, 0)$, $j = 1, \dots, d$ exist for odd m, then (6.49) holds.*

Proof Fix $j \in \{1, \dots, d\}$. Let m be even, the argument for odd m being essentially the same. Recall from Theorem 2.19 that (6.49) holds if and only if the infinitely divisible law corresponding to the exponent $u \mapsto \psi(x, u1_j)$ has moments of order m. By Proposition A.5(2) this holds if characteristic function $u \mapsto \varphi(u) := \exp(\psi(x, u1_j))$ has derivatives of order m in 0. This yields the claim. □

Theorem 6.25 allows us to compute joint moments as well.

Corollary 6.29 *Let X be an m-polynomial process as in Theorem 6.25. Moreover, fix $0 \leq t_0 \leq \cdots \leq t_n$ and $\lambda_{(1)}, \dots, \lambda_{(n)} \in \mathbb{N}^d$ with $|\lambda_{(1)} + \cdots + \lambda_{(n)}| \leq m$. Then*

$$E\left(\prod_{j=1}^{n} X(t_j)^{\lambda_{(j)}} \middle| \mathscr{F}_{t_0}\right) = \sum_{\mu \in \mathbb{N}_m^d} g_{(0),\mu} X(t_0)^{\mu},$$

where $(g_{(0),\mu})_{\mu \in \mathbb{N}_m^d}$ is obtained recursively via

$$g_{(n),\mu} = \delta_{\lambda_{(n)},\mu} := \begin{cases} 1 & \text{if } \lambda_{(n)} = \mu, \\ 0 & \text{otherwise,} \end{cases}$$

$$g_{(k-1),\mu} = \begin{cases} \sum_{\lambda \in \mathbb{N}_m^d} g_{(k),\lambda} (e^{B_m(t_k - t_{k-1})})_{\lambda, \mu - \lambda_{(k-1)}} & \text{if } \lambda_{(k-1)} \leq \mu, \\ 0 & \text{otherwise,} \end{cases}$$

$$g_{(0),\mu} = \sum_{\lambda \in \mathbb{N}_m^d} g_{(1),\lambda} \left(e^{B_m(t_1 - t_0)}\right)_{\lambda,\mu}$$

for $k = n, \dots, 2$ and any $\mu \in \mathbb{N}_m^d$.

Proof By iterating Theorem 6.25 it follows that

$$E\left(\prod_{j=1}^{n} X(t_j)^{\lambda_{(j)}} \middle| \mathscr{F}_{t_k}\right) = \prod_{j=1}^{k-1} X(t_j)^{\lambda_{(j)}} \sum_{\mu \in \mathbb{N}_m^d} g_{(k),\mu} X(t_k)^{\mu}$$

for $k = n, \dots, 0$. □

We turn now to concrete examples. As alluded to above, affine processes turn out to be polynomial if the corresponding moments exist.

Example 6.30 (Affine Processes) Let X be an $\mathbb{R}^d$-valued affine process as in Theorem 6.6 and such that $\int_{\{|x|>1\}} |x|^m K_{(j)}(dx) < \infty$ for some even $m \in \mathbb{N}$ and $j = 0, \dots, d$. Then X is m-polynomial with corresponding matrix

$$a_{\lambda,\mu} = \begin{cases} i^{-|\lambda|}\psi_0^{(\lambda)}(0) & \text{if } |\mu| = 0, \\ i^{-|\lambda|}\psi_j^{(\lambda)}(0) & \text{if } \mu = 1_j, \\ 0 & \text{otherwise.} \end{cases}$$

This implies that the matrix B_m has coefficients

$$b_{\lambda,\mu} = \binom{\lambda}{\mu} i^{-|\lambda-\mu|}\psi_0^{(\lambda-\mu)}(0) + \sum_{j=0}^{d} \binom{\lambda}{\mu - 1_j} i^{-|\lambda+1_j-\mu|}\psi_j^{(\lambda+1_j-\mu)}(0).$$

As a corollary we obtain the moment formulas in Theorems 6.13 and 6.15.

A second class of examples has been studied in Sect. 3.9.

Example 6.31 (Generalised Ornstein–Uhlenbeck Processes) Let X be a generalised Ornstein–Uhlenbeck process as in Sect. 3.9 and $m \in \mathbb{N}$ even such that

$$\int (|x_1|^m + |x_2|^m) K^{(L,M)}(dx) < \infty,$$

where $(b^{(L,M),\mathrm{id}}, c^{(L,M)}, K^{(L,M)})$ denotes the Lévy–Khintchine triplet of (L, M) relative to $h = \mathrm{id}$ and $\psi^{(L,M)}$ the corresponding exponent. Then X is m-polynomial with corresponding matrix

$$a_{\lambda,\mu} = \binom{\lambda}{\mu} \left(\psi^{(L,M)}\right)^{(\lambda,\lambda-\mu)}(0)$$

for $\lambda, \mu = 0, \dots, m$ with $\mu \leq \lambda$. This can be verified by computing the local characteristics of X and verifying (6.46–6.48). We conclude that the matrix B_m has coefficients

$$b_{\lambda,\mu} = \sum_{\nu=0}^{\mu} \binom{\lambda}{\nu} \binom{\lambda-\nu}{\mu-\nu} i^{\nu-\lambda} \left(\psi^{(L,M)}\right)^{(\lambda-\nu,\lambda-\mu)}(0).$$

Note that Theorem 6.26 extends Proposition 3.52 to higher-order moments.

Given a potential generating matrix $A \in \mathbb{R}^{\mathbb{N}^d \times \mathbb{N}^d}$, one may wonder whether a corresponding polynomial process X exists and whether its law is uniquely determined. More precisely, we say that **existence of an E-valued polynomial process** with generating matrix A holds if, for any law P^0 on E, there exists an E-valued polynomial process E with initial law $P^{X(0)} = P^0$ and generating matrix A. By **uniqueness of an E-valued polynomial process** with generating matrix A

we refer to the property that, for any initial distribution P^0 on E, any two such processes have the same law.

With regards to existence of solutions, recall from Sect. 5.7.2 that it holds under weak continuity and growth conditions on the triplet or symbol if $E = \mathbb{R}^d$. But note that only integer moments of κ resp. the derivatives of ψ in 0 are known in the first place, which does not necessarily imply that κ and ψ themselves are determined by A. If, however, the derivatives of $u \mapsto \psi(x, u)$ in 0 do not grow too quickly, it equals its power series in a neighbourhood of 0 and hence ψ resp. κ are determined by and can be recovered from $(a_{\lambda,\mu})_{\lambda,\mu\in\mathbb{N}^d}$.

In order to verify existence for some state space $E \subset \mathbb{R}^d$ one needs to warrant that the solution process does not want to leave E. This is expressed in Theorem 5.38 in terms of the positive maximum principle. Instead of a general statement we refer to Examples 5.39 and 6.33 for illustration.

Uniqueness of polynomial processes is a more delicate issue. In view of Theorem 6.26, all integer moments of X are uniquely determined by A and $X(0)$. According to [100, Theorem 4.4.2a], the martingale problem related to $(\beta^{\mathrm{id}}, \gamma, \kappa)$ has a unique solution if the marginal laws $\mathscr{L}(X(t)), t \in \mathbb{R}_+$ are uniquely determined by the triplet and $\mathscr{L}(X(0))$. Together, we conclude that uniqueness of the law of X in the desired sense holds if the moments $E(X(t)^\lambda)$, $\lambda \in \mathbb{N}^d$ determine the law of X uniquely. This does not hold in general. Indeed, there are instances of polynomial processes whose law is not determined by their coefficients and $X(0)$:

Example 6.32 Consider the symbol

$$\psi(t, x) := \begin{cases} \frac{\cos(ux)-1}{x^2} & \text{for } x \neq 0, \\ -u^2/2 & \text{for } x = 0, \end{cases}$$

which corresponds to the triplet

$$(\beta^{\mathrm{id}}, \gamma, \kappa)(x) := \begin{cases} (0, 0, \frac{\varepsilon_x+\varepsilon_{-x}}{2x^2}) & \text{for } x \neq 0, \\ (0, 1, 0) & \text{for } x = 0 \end{cases}$$

on $\mathbb{R}$. In [170, Example 3.1] it is shown that, for any $k > 0$, there exists a solution X to this martingale problem with initial value $X(0) = 0$ and values in $\{\pm k2^z : z \in \mathbb{Z}\} \cup \{0\}$. In particular, uniqueness in law does not hold. The process X is a strange animal: as long as it is in 0, the coefficient $\gamma(0) = 1$ indicates that it behaves as Brownian motion. But as soon as it leaves 0, it turns into a pure jump process.

The following second process shows a likewise odd behaviour. On the state space $E = \mathbb{R}_+$ consider the symbol

$$\psi(t, x) := \begin{cases} \frac{e^{iux}-1}{x} & \text{for } x > 0, \\ iu & \text{for } x = 0, \end{cases}$$

which corresponds to the triplet

$$(\beta^{\mathrm{id}}, \gamma, \kappa)(x) := \begin{cases} (1, 0, \frac{1}{x}\varepsilon_x) & \text{for } x > 0, \\ (1, 0, 0) & \text{for } x = 0, \end{cases}$$

cf. [170, Example 3.2]. Similarly as above, a solution X to this martingale problem with initial value $X(0) = 0$ and values in $\{k2^z : z \in \mathbb{Z}\} \cup \{0\}$ exists for any $k > 0$. The triplet $(1, 0, 0)$ indicates that it starts as a deterministic process with drift 1 in 0. But instantly, it turns into a piecewise constant pure jump process.

However, if for example $E(\exp(u^\top X)) < \infty$ for all u in a neighbourhood of $(0, \dots, 0)$, the law of an $\mathbb{R}^d$-valued random variable X is uniquely determined by its integer moments $E(X^\lambda)$, $\lambda \in \mathbb{N}^d$. This can sometimes be used to verify the desired uniqueness. The integrability condition certainly holds for random variables whose state space is bounded. Consequently, uniqueness holds for any polynomial process with bounded state space E. A popular example is the following:

Example 6.33 (Jacobi Process) Consider the SDE

$$dX(t) = (\kappa - \lambda X(t))dt + \sigma\sqrt{X(t)(1 - X(t))}dW(t) \tag{6.53}$$

with $\lambda, \sigma > 0$, $\kappa \in [0, \lambda]$, $X(0) \in E := [0, 1]$. The corresponding martingale problem has coefficients $(\beta, \gamma, 0)$ given by

$$\beta(x) = \kappa - \lambda x, \quad \gamma(x) = \sigma^2 x(1 - x)$$

for $x \in E$. It is easy to see that the positive maximum principle holds. By Theorem 5.38 there exists a solution to the martingale problem and hence a weak solution to SDE (6.53). Since X is a polynomial process on a compact state space, we have uniqueness in law as well. One can in fact show that pathwise uniqueness holds because the coefficients of the SDE are Hölder continuous of order 1/2, cf. [241, Theorem IX.3.5].

Appendix 1: Problems

The exercises correspond to the section with the same number.

6.1 The Gamma-OU stochastic volatility model is based on the bivariate affine process $X = (X_1, X_2)$ satisfying the SDE

$$dX_1(t) = -\lambda X_1(t-)dt + dZ(t),$$

$$dX_2(t) = (\mu + \delta X_1(t-))dt + \sqrt{X_1(t-)}dW(t) + \varrho Z(t),$$

with parameters $\mu, \delta, \varrho, \lambda$, a Wiener process W, and a subordinator Z with characteristic exponent $\psi^Z(u) = \frac{iu\beta}{\alpha - iu}$ for some $\alpha, \beta > 0$. Verify that the functions Ψ_0, Ψ_1, Ψ_2 in the characteristic function (6.29) are of the form

$$\Psi_0(t,u) = \int_0^t \psi^Z(-i\Psi_1(s,u) + \varrho u_2)ds + \mu t i u_2,$$

$$\Psi_1(t,u) = e^{-\lambda t} i u_1 + \frac{1 - e^{-\lambda t}}{\lambda} \psi^L(u_2),$$

$$\Psi_2(t,u) = i u_2,$$

where $\psi^L(u_2) := \delta i u_2 - \frac{1}{2} u_2^2$.

6.2 Compute the matrix B for $n = 4$ and the moments $E(X(t)^k)$, $k = 1, 2, 3, 4$ for the extended square-root process X of Example 6.9.

6.3 The Cox–Ingersoll–Ross model in interest rate theory is based on the extended square-root process X of Example 6.9. Compute the joint characteristic function $E(\exp(iu_1 X(t) + iu_2 \int_0^t X(s)ds))$, $u_1, u_2 \in \mathbb{R}$ of $X(t)$ and $\int_0^t X(s)ds$.

6.4 The Heston model in Mathematical Finance is based on the bivariate affine process $X = (X_1, X_2)$ solving the SDE

$$dX_1(t) = (\mu + \delta X_2(t))dt + \sqrt{X_2(t)}dW(t), \tag{6.54}$$

$$dX_2(t) = (\kappa - \lambda X_2(t))dt + \sigma\sqrt{X_2(t)}dZ(t), \tag{6.55}$$

where $\kappa \geq 0$, $\mu, \delta, \lambda, \sigma$ denote constants and W, Z Wiener processes with constant correlation ϱ. Determine all measure changes as in Proposition 6.22 such that X is affine and e^{X_1} is a local martingale under the new probability measure. Is X again of the form (6.54, 6.55) up to different parameter values?

6.5 The Hull–White model in interest rate theory is based on the solution to the SDE $dX(t) = (\kappa(t) - \lambda X_2(t))dt + \sigma dW(t)$ with parameters λ, σ, a continuous function $\kappa : \mathbb{R}_+ \to \mathbb{R}$, and some Wiener process W. Compute the joint characteristic function $E(\exp(iu_1 X(t) + iu_2 \int_0^t X(s)ds))$, $u_1, u_2 \in \mathbb{R}$ of $X(t)$ and $\int_0^t X(s)ds$.

6.6 (Pearson Diffusion) Show that continuous univariate polynomial processes are weak solutions to the SDE

$$dX(t) = \big(a_{1,0} + a_{1,1}X(t)\big)dt + \sqrt{a_{2,0} + a_{2,1}X(t) + a_{2,2}X(t)^2 dW(t)}$$

with some parameters $a_{1,0}, a_{1,1}, a_{2,0}, a_{2,1}, a_{2,2}$. Determine the matrices A, B of Proposition 6.27 in terms of these coefficients.

Appendix 2: Notes and Comments

Since textbook treatments of general affine processes are still missing, we refer to [80, 104] for results and references on affine processes and to [63, 106, 107] for the more general class of polynomial processes. Proposition 6.8 is stated in [169, Corollary 3.3], albeit with an incorrect expression for the joint characteristic function.

Theorems 6.13, 6.15, 6.26 are based on the observation of [63] that affine processes are polynomial and hence their moments can be obtained by solving linear ODEs. The first part of Proposition 6.19 is stated in [172, Lemma 2.7]. The models in Examples 6.4, 6.9 and in the problem section are revisited in Chaps. 8 and 14 in Part II of this book.

Concerning polynomial processes we follow here [107] in dispensing with the Markov property. Even if the latter holds in all relevant examples, the present approach allows us to apply the moment formula without verifying it in the first place. On the other hand, we stick to [63] by considering m-polynomial processes for finite m. This allows us to apply the moment formula even if not all moments exist.

As a final comment we remark that it does not seem easy to characterise all symbols or local characteristics that belong to polynomial processes. For results in this direction we refer to [64].

Chapter 7
Optimal Control

Dynamic stochastic optimisation problems play an important role in Mathematical Finance and other applications. In this chapter we provide basic tools for their mathematical treatment in continuous time. We proceed in close correspondence to Sect. 1.5. In particular, we distinguish two approaches, which are discussed in the following sections on dynamic programming and the stochastic maximum principle, respectively.

7.1 Dynamic Programming

As usual we work on a filtered probability space $(\Omega, \mathscr{F}, \mathbf{F}, P)$ with filtration $\mathbf{F} = (\mathscr{F}_t)_{t\geq 0}$. As in Sect. 1.5 we assume $\mathscr{F}_0$ to be trivial for simplicity, i.e. all $\mathscr{F}_0$-measurable random variables are deterministic.

The aim is to maximise the expected reward $E(u(\alpha))$ over all admissible controls $\alpha \in \mathscr{A}$. The set $\mathscr{A}$ of **admissible controls** contains $\mathbb{R}^m$-valued adapted processes and is assumed to be *stable under bifurcation*, i.e. for any stopping time τ, any event $F \in \mathscr{F}_\tau$, and any $\alpha, \widetilde{\alpha} \in \mathscr{A}$ with $\alpha^\tau = \widetilde{\alpha}^\tau$, the process $(\alpha|\tau_F|\widetilde{\alpha})$ defined by

$$(\alpha|\tau_F|\widetilde{\alpha})(t) := 1_{F^C}\alpha(t) + 1_F\widetilde{\alpha}(t)$$

is again an admissible control. Moreover, the initial values $\alpha(0)$ are assumed to coincide for all $\alpha \in \mathscr{A}$. By $u : \Omega \times (\mathbb{R}^m)^{[0,T]} \to \mathbb{R} \cup \{\infty\}$ we denote the **reward function**. The shorthand $u(\alpha)$ is used for the random variable $\omega \mapsto u(\omega, \alpha(\omega))$. The reward is meant to refer to some fixed time $T \in \mathbb{R}_+$, which motivates the assumption that $u(\alpha)$ is $\mathscr{F}_T$-measurable for any $\alpha \in \mathscr{A}$.

© Springer Nature Switzerland AG 2019
E. Eberlein, J. Kallsen, *Mathematical Finance*, Springer Finance,
https://doi.org/10.1007/978-3-030-26106-1_7

Example 7.1 The reward function is typically of the form

$$u(\alpha) = \int_0^T f(t, X^{(\alpha)}(t-), \alpha(t))dt + g(X^{(\alpha)}(T))$$

for some functions $f : [0, T] \times \mathbb{R}^d \times \mathbb{R}^m \to \mathbb{R} \cup \{-\infty\}$, $g : \mathbb{R}^d \to \mathbb{R} \cup \{-\infty\}$, and $\mathbb{R}^d$-valued adapted càdlàg *controlled processes* $X^{(\alpha)}$.

We call $\alpha^\star \in \mathscr{A}$ an **optimal control** if it maximises $E(u(\alpha))$ over all $\alpha \in \mathscr{A}$, where we set $E(u(\alpha)) := -\infty$ if $E(u(\alpha)^-) := -\infty$. Moreover, the **value process** of the optimisation problem is the family $(\mathscr{V}(\cdot, \alpha))_{\alpha \in \mathscr{A}}$ of adapted processes defined via

$$\mathscr{V}(t, \alpha) := \operatorname{ess\,sup}\left\{E(u(\widetilde{\alpha})|\mathscr{F}_t) : \widetilde{\alpha} \in \mathscr{A} \text{ with } \widetilde{\alpha}^t = \alpha^t\right\} \tag{7.1}$$

for $t \in [0, T]$, $\alpha \in \mathscr{A}$.

As in discrete time the value process concerns intermediate maximisation problems for all times $t \in [0, T]$. It is characterised by the following martingale/supermartingale property. The second statement is useful in order to prove that a given candidate control is optimal, cf. Example 7.5 below.

Theorem 7.2 *In the following statements—and only there—*supermartingale *is to be understood in the relaxed sense of Remark 2.4, i.e. without càdlàg property.*

1. *If $\mathscr{V}(0) := \sup_{\alpha \in \mathscr{A}} E(u(\alpha)) \neq \pm\infty$, the following holds.*
   a) *$\mathscr{V}(t, \widetilde{\alpha}) = \mathscr{V}(t, \alpha)$ if $\widetilde{\alpha}^t = \alpha^t$.*
   b) *$(\mathscr{V}(t, \alpha))_{t \in [0,T]}$ is a supermartingale with terminal value $\mathscr{V}(T, \alpha) = u(\alpha)$ for any admissible control α with $E(u(\alpha)) > -\infty$.*
   c) *If $\alpha^\star$ is an optimal control, $(\mathscr{V}(t, \alpha^\star))_{t \in [0,T]}$ is a martingale.*
2. *Suppose that $(\widetilde{\mathscr{V}}(\cdot, \alpha))_{\alpha \in \mathscr{A}}$ is a family of processes such that*
   a) *$\widetilde{\mathscr{V}}(t, \widetilde{\alpha}) = \widetilde{\mathscr{V}}(t, \alpha)$ if $\widetilde{\alpha}^t = \alpha^t$, the common value $\widetilde{\mathscr{V}}(0, \alpha)$ being denoted by $\widetilde{\mathscr{V}}(0)$,*
   b) *$(\widetilde{\mathscr{V}}(t, \alpha))_{t \in [0,T]}$ is a supermartingale with terminal value $\widetilde{\mathscr{V}}(T, \alpha) = u(\alpha)$ for any admissible control α with $E(u(\alpha)) > -\infty$,*
   c) *$(\widetilde{\mathscr{V}}(t, \alpha^\star))_{t \in [0,T]}$ is a submartingale—and hence a martingale—for some admissible control $\alpha^\star$ with $E(u(\alpha^\star)) > -\infty$.*

   Then $\alpha^\star$ is optimal and $\mathscr{V}(t, \alpha^\star) = \widetilde{\mathscr{V}}(t, \alpha^\star)$ for $t \in [0, T]$. In particular, $\mathscr{V}(0) = \widetilde{\mathscr{V}}(0)$.

Proof The proof is essentially the same as for Theorem 1.44. □

We state a simple continuous-time counterpart to Proposition 1.46.

Proposition 7.3 (Dynamic Programming Principle) *Suppose that $\mathscr{V}(0) \neq \pm\infty$. For $0 \leq s \leq t \leq T$ and any control $\alpha \in \mathscr{A}$ we have*

$$\mathscr{V}(s, \alpha) = \operatorname{ess\,sup}\left\{E(\mathscr{V}(t, \widetilde{\alpha}))|\mathscr{F}_s) : \widetilde{\alpha} \in \mathscr{A} \textit{ with } \widetilde{\alpha}^s = \alpha^s\right\}.$$

Proof The proof is essentially the same as for Proposition 1.46. □

A true *dynamic programming principle (DPP)* would allow for $[s, T]$-valued stopping times τ instead of deterministic t. In this form it plays a key role in stochastic control theory. In this introductory text, however, we need neither Proposition 7.3 nor the DPP.

As in discrete time the value function has a minimality property:

Proposition 7.4 *Let $\mathscr{V}(0) \neq \pm\infty$. For any family of processes $(\widetilde{\mathscr{V}}(\cdot, \alpha))_{\alpha\in\mathscr{A}}$ satisfying 2a,b) in Theorem 7.2, we have $\mathscr{V}(\cdot, \alpha) \leq \widetilde{\mathscr{V}}(\cdot, \alpha)$ for any $t \in [0, T]$ and any control $\alpha \in \mathscr{A}$.*

Proof The proof is the same as for Proposition 1.47. □

Remarks 1.45(1–3 and 5) concerning the existence of an optimal control, upper and lower bounds, resp. problems with infinite time horizon apply in the present continuous-time setup as well. In order to illustrate the previous results we consider the continuous-time extension of Example 1.48.

Example 7.5 (Logarithmic Utility of Terminal Wealth) Suppose that the price of a stock moves according to

$$S(t) = S(0)\mathscr{E}(X)(t)$$

for some semimartingale X with $\Delta X(t) > -1$. As in (1.45) the wealth of a trader with initial wealth $v_0 > 0$ and holding $\varphi(t)$ shares of stock at any time t can be expressed as

$$V_\varphi(t) := v_0 + \varphi \bullet S(t),$$

where we require φ to be predictable and *a fortiori* S-integrable for the stochastic integral to be defined. For the verification we need some further integrability condition. This leads to the set

$$\mathscr{A} := \left\{\varphi \in L(S) : V_\varphi > 0 \text{ and } \varphi(0) = 0 \text{ and } \log V_\varphi \text{ is of class (D)}\right\}$$

of admissible controls.

The investor aims at maximising the expected logarithmic utility $E(\log V_\varphi(T))$ of wealth at time T. As in discrete time we consider the *relative portfolio*

$$\pi(t) := \varphi(t)\frac{S(t-)}{V_\varphi(t-)}, \quad t \in [0, T],$$

i.e. the fraction of wealth invested in the stock. Starting with v_0, the stock holdings $\varphi(t)$ and the wealth process $V_\varphi(t)$ are obtained from π via

$$V_\varphi(t) = v_0\mathscr{E}(\pi \bullet X)(t) \tag{7.2}$$

and

$$\varphi(t) = \pi(t)\frac{V_\varphi(t-)}{S(t-)} = \pi(t)\frac{v_0\mathscr{E}(\pi \bullet X)(t-)}{S(t-)}.$$

Indeed, (7.2) follows from

$$dV_\varphi(t) = \varphi(t)dS(t) = \frac{V_\varphi(t-)\pi(t)}{S(t-)}dS(t) = V_\varphi(t-)d(\pi \bullet X)(t).$$

The natural candidate $\pi^\star(t)$ for the optimal fraction of wealth is the maximiser of the continuous-time counterpart of (1.48), namely of

$$\gamma \mapsto \gamma b^h(t) - \frac{1}{2}\gamma^2 c(t) + \int (\log(1+\gamma x) - 1 - \gamma h(x))K(t, dx),$$

where (b^h, c, K) denote the local characteristics of X. Indeed, Propositions 4.8 and 4.13 yield that the drift rate $a^{\log V_\varphi}$ of $\log V_\varphi$ in the sense of (2.9) equals

$$a^{\log V_\varphi}(t) = \pi(t)b^h(t) - \frac{1}{2}\pi(t)^2 c(t) + \int \big(\log(1+\pi(t)x) - \pi(t)h(x)\big)K(t, dx).$$

Note that $\pi^\star(t) = a^X(t)/c(t)$ if the stock price process does not jump. The analogue of the expression (1.49) for the candidate value process is

$$\begin{aligned}
\mathscr{V}(t,\varphi) &= E\left(\log\left(V_\varphi(t)\frac{\mathscr{E}(\pi^\star \bullet X)(T)}{\mathscr{E}(\pi^\star \bullet X)(t)}\right)\middle|\mathscr{F}_t\right)\\
&= \log V_\varphi(t) + E\Bigg(\int_t^T \Big(\pi^\star(s)b^h(s) - \frac{1}{2}\pi^\star(s)^2 c(s)\\
&\quad + \int \big(\log(1+\pi^\star(s)x) - \pi^\star(s)h(x)\big)K(s, dx)\Big)ds\,\Bigg|\,\mathscr{F}_t\Bigg),
\end{aligned}$$

where we used Propositions 4.8 and 4.13 along with Remark 2.10. Observe that

$$\begin{aligned}
a^{\mathscr{V}(\cdot,\varphi)}(t) &= a^{\log V_\varphi}(t) - \Big(\pi^\star(t)b^h(t) - \frac{1}{2}\pi^\star(t)^2 c(t)\\
&\quad + \int \big(\log(1+\pi^\star(t)x) - \pi^\star(t)h(x)\big)K(t, dx)\Big)\\
&= \pi(t)b^h(t) - \frac{1}{2}\pi(t)^2 c(t)\\
&\quad + \int \big(\log(1+\pi(t)x) - \pi(t)h(x)\big)K(t, dx)
\end{aligned}$$

$$
\begin{aligned}
&- \Big(\pi^\star(t) b^h(t) - \frac{1}{2}\pi^\star(t)^2 c(t) \\
&\quad + \int \big(\log(1+\pi^\star(t)x) - \pi^\star(t) h(x) \big) K(t,dx) \Big) \\
&\leq 0
\end{aligned}
$$

for any t and any $\varphi \in \mathscr{A}$, with equality for the candidate optimiser $\varphi^\star$ satisfying $\varphi^\star(t) = \pi^\star(t) V_{\varphi^\star}(t-)/S(t-)$. By Propositions 4.22, 4.23 and Theorem 7.2, we conclude that $\varphi^\star$ is indeed optimal if it is admissible at all, i.e. if $\log V_{\varphi^\star}$ is of class (D). As in discrete time the optimal fraction of wealth invested in stock depends only on the local dynamics of the stock, as expressed in terms of its semimartingale characteristics.

Example 1.49 can be transferred to continuous time along the same lines. We turn now to the counterpart of Example 1.50, which is discussed in Chap. 12 in a more general context.

Example 7.6 (Quadratic Hedging) As in Example 1.50 the goal is to approximate a square-integrable random variable X by a stochastic integral $v_0 + \varphi \bullet S(T)$, where $v_0 \in \mathbb{R}$ as well as the square-integrable martingale S are given and φ ranges over all admissible controls in the set

$$
\mathscr{A} := L^2(S) = \{\varphi \text{ predictable} : \varphi(0) = 0 \text{ and } E(\varphi^2 \bullet \langle S, S\rangle(T)) < \infty\}.
$$

More specifically, we want to minimise the expected squared hedging error

$$
E\big((V_\varphi(T) - X)^2\big), \tag{7.3}
$$

cf. Example 1.50 and in particular Chap. 12 for motivation.

As in discrete time we set $V_\varphi := v_0 + \varphi \bullet S$ and define the martingales $V(t) := E(X|\mathscr{F}_t)$ and $M_\varphi := V_\varphi - V$. This means that we have to maximise $E(u(\varphi))$ with $u(\varphi) = -M_\varphi(T)^2$. For simplicity we assume that (S, V) has local characteristics and we denote its modified second characteristics in the sense of (4.31) as

$$
\widetilde{c}^{(S,V)} = \begin{pmatrix} \widetilde{c}^S & \widetilde{c}^{S,V} \\ \widetilde{c}^{S,V} & \widetilde{c}^V \end{pmatrix}.
$$

Since this matrix is nonnegative, we have $(\widetilde{c}^{S,V})^2 \leq \widetilde{c}^S \widetilde{c}^V$. Also motivated by discrete time, we choose the candidate optimal control

$$
\varphi^\star(t) = \frac{d\langle S, V\rangle(t)}{d\langle S, S\rangle(t)},
$$

which means that $\varphi^\star$ is a predictable process satisfying

$$d\langle S, V\rangle(t) = \varphi^\star(t) d\langle S, S\rangle(t),$$

i.e.

$$\langle S, V\rangle = \varphi^\star \bullet \langle S, S\rangle.$$

In our setup it is obviously obtained as

$$\varphi^\star(t) := \frac{\widetilde{c}^{S,V}(t)}{\widetilde{c}^S(t)},$$

with an arbitrary definition such as $\varphi^\star(t) = 0$ if the denominator vanishes. Since

$$\begin{aligned}(\varphi^\star)^2 \bullet \langle S, S\rangle(T) &= \left(\left(\frac{\widetilde{c}^{S,V}}{\widetilde{c}^S}\right)^2 \widetilde{c}^S\right) \bullet I(T)\\ &= \frac{(\widetilde{c}^{S,V})^2}{\widetilde{c}^S} \bullet I(T)\\ &\leq \widetilde{c}^V \bullet I(T)\\ &= \langle V, V\rangle(T)\end{aligned}$$

and $E(\langle V, V\rangle(T)) = E(X^2) - E(X)^2 < \infty$, the process $\varphi^\star$ is actually an admissible control. In line with (1.58), a natural candidate value process is

$$\begin{aligned}\mathscr{V}(t, \varphi) :=& -M_\varphi(t)^2 - E\big(\langle M_{\varphi^\star}, M_{\varphi^\star}\rangle(T) - \langle M_{\varphi^\star}, M_{\varphi^\star}\rangle(t)\big|\mathscr{F}_t\big)\\ =& -M_\varphi(t)^2 - U(t) + \left(\widetilde{c}^V - \frac{(\widetilde{c}^{S,V})^2}{\widetilde{c}^S}\right) \bullet I(t)\end{aligned}$$

with a martingale U defined as

$$U(t) = E\big(\langle M_{\varphi^\star}, M_{\varphi^\star}\rangle(T)\big|\mathscr{F}_t\big) = E\left(\left(\widetilde{c}^V - \frac{(\widetilde{c}^{S,V})^2}{\widetilde{c}^S}\right) \bullet I(T)\middle|\mathscr{F}_t\right).$$

If $\sim$ denotes equality up to a local martingale, we have

$$\begin{aligned}d\mathscr{V}(t, \varphi) =& -dM_\varphi(t)^2 - dU(t) + \left(\widetilde{c}^V(t) - \frac{\widetilde{c}^{S,V}(t)^2}{\widetilde{c}^S(t)}\right) dt\\ \sim& -d\langle M_\varphi, M_\varphi\rangle(t) + \left(\widetilde{c}^V(t) - \frac{\widetilde{c}^{S,V}(t)^2}{\widetilde{c}^S(t)}\right) dt\end{aligned}$$

$$= -\Big(\widetilde{c}^V(t) - 2\varphi(t) \bullet \widetilde{c}^{S,V}(t) + \varphi(t)^2\widetilde{c}^S(t)$$
$$- \big(\widetilde{c}^V(t) - 2\varphi^\star(t) \bullet \widetilde{c}^{S,V}(t) + \varphi^\star(t)^2\widetilde{c}^S(t)\big)\Big)dt.$$

Similarly as in Example 1.50, we have that $\varphi^\star(t)$ minimises the mapping

$$\varphi(t) \mapsto \widetilde{c}^V(t) - 2\varphi(t) \bullet \widetilde{c}^{S,V}(t) + \varphi(t)^2\widetilde{c}^S(t).$$

Hence $\mathscr{V}(\cdot, \varphi)$ has nonpositive drift which vanishes for $\varphi = \varphi^\star$. The integrability conditions for $X, S, \varphi, \varphi^\star$ yield that $\mathscr{V}(\cdot, \varphi)$ is a true supermartingale resp. martingale. Since $\mathscr{V}(T, \varphi) = -(V_\varphi(T) - X)^2$, Theorem 7.2 yields the optimality of $\varphi^\star$ and the optimal value

$$-\mathscr{V}(\varphi^\star, 0) = (V_\varphi(0) - V(0))^2 + E(U(T))$$
$$= (v_0 - E(X))^2 + E\left(\left(\widetilde{c}^V - \frac{(\widetilde{c}^{S,V})^2}{\widetilde{c}^S}\right) \bullet I(T)\right)$$

for (7.3). As in discrete time the first term vanishes if we can minimise over v_0 as well, leading to the optimal choice $v_0 = E(X)$.

7.2 Optimal Stopping

We turn now to optimal stopping. Let X denote a semimartingale with continuous compensator A^X and $E(\sup_{t\in[0,T]} |X(t)|) < \infty$. Given a time horizon $T \geq 0$, the aim is to maximise the expected reward

$$\tau \mapsto E(X(\tau)) \tag{7.4}$$

over all stopping times τ with values in $[0, T]$.

Remark 7.7 Should the process X in an application fail to be adapted, we may want to replace it with $\widetilde{X}(t) := E(X(t)|\mathscr{F}_t)$ parallel to Remark 1.51(2). However, in continuous time it is less obvious whether this construction yields a well-defined adapted process with $E(\widetilde{X}(\tau)) = E(X(\tau))$ for any stopping time τ.

Here, we need the concept of an *optional projection*, cf. [76, Section VI.2]. If X is a measurable process with càdlàg paths and $E(\sup_{t\in[0,T]} |X(t)|) < \infty$, then there is a unique adapted càdlàg process $\widetilde{X}$ such that $E(\widetilde{X}(\tau)1_{\{\tau<\infty\}}) = E(X(\tau)1_{\{\tau<\infty\}})$ holds for any stopping time τ, cf. [76, Theorem VI.43 and Remarks VI.44(d,f)]. This process $\widetilde{X}$ is called the **optional projection** of X. Due to [115, Theorem 9], $\widetilde{X}$ is automatically a semimartingale if X is a semimartingale with respect to some filtration $(\mathscr{G}_t)_{t\in[0,T]}$ on $(\Omega, \mathscr{F})$ with $\mathscr{F}_t \subset \mathscr{G}_t$, $t \in [0, T]$.

As in discrete time the Snell envelope turns out to be the key to solving the stopping problem. Its existence is warranted by the following

Proposition 7.8 *There is a supermartingale V satisfying*

$$V(t) = \operatorname{ess\,sup}\big\{E(X(\tau)|\mathscr{F}_t) : \tau \text{ stopping time with values in } [t,T]\big\} \tag{7.5}$$

for any $t \in [0,T]$. Its compensator A^V is continuous. V is called the ***Snell envelope*** *of X.*

Proof

Step 1: We show that $\{E(X(\tau)|\mathscr{F}_t) : \tau \text{ stopping time}\}$ has the lattice property. Indeed, for stopping times τ_1, τ_2 set

$$\tau := \tau_1 1_{\{E(X(\tau_1)|\mathscr{F}_t) \geq E(X(\tau_2)|\mathscr{F}_t)\}} + \tau_2 1_{\{E(X(\tau_1)|\mathscr{F}_t) < E(X(\tau_2)|\mathscr{F}_t)\}}.$$

Then $E(X(\tau_1)|\mathscr{F}_t) \leq E(X(\tau)|\mathscr{F}_t)$ and likewise for τ_2.

Step 2: Define V by (7.5). By Lemma A.4 we have

$$\begin{aligned}
E(V(t)|\mathscr{F}_s) &= E\big(\operatorname{ess\,sup}\big\{E(X(\tau)|\mathscr{F}_t) : \tau \text{ stopping time with values in } [t,T]\big\}\big|\mathscr{F}_s\big)\\
&= \operatorname{ess\,sup}\big\{E(E(X(\tau)|\mathscr{F}_t)|\mathscr{F}_s) : \tau \text{ stopping time with values in } [t,T]\big\}\\
&= \operatorname{ess\,sup}\big\{E(X(\tau)|\mathscr{F}_s) : \tau \text{ stopping time with values in } [t,T]\big\}\\
&\leq V(s)
\end{aligned}$$

for $s \leq t$.

Step 3: We show that $t \mapsto E(V(t))$ is right continuous, which implies that V is a supermartingale by step 2 and Remark 2.4. To this end, observe that $E(V(t_n)) \leq E(V(t))$ for $t_n \geq t$ by step 2. Let $t_n \downarrow t$. For any stopping time $\tau \geq t$ set $\tau_n := \tau \vee t_n$. From $X(\tau_n) \to X(\tau)$ a.s. we conclude

$$E(X(\tau)) = \lim_{n\to\infty} E(X(\tau_n)) \leq \liminf_{n\to\infty} E(V(t_n))$$

by dominated convergence and hence

$$\begin{aligned}
E(V(t)) &= \sup\big\{E(X(\tau)) : \tau \text{ stopping time with values in } [t,T]\big\}\\
&\leq \liminf_{n\to\infty} E(V(t_n)).
\end{aligned}$$

Step 4: Since A^V is right continuous, it suffices to show that $A^V(t_n) \to A^V(t)$ for $t_n \uparrow t$. To this end, let τ_n be a stopping time with values in $[t_n, T]$ and such that $E(V(t_n)) \leq E(X(\tau_n)) + 1/n$. Then $E(V(t_n)) \leq E(X(0) + A^X(\tau_n)) + 1/n$.

Since $E(V(t)) \geq E(X(\tau_n \vee t))$, we have

$$E(V(t)) = E\big(X(0) + A^X(\tau_n \vee t)\big)$$
$$\geq E(V(t_n)) + E\big(A^X(\tau_n \vee t) - A^X(\tau_n)\big) - \frac{1}{n}.$$

By left-continuity of A^X we conclude $E(V(t)) \geq \limsup_{n\to\infty} E(V(t_n))$ and hence $E(A^V(t)) \geq \limsup_{n\to\infty} E(A^V(t_n))$. On the other hand, the supermartingale property of V yields $A^V(t_n) > A^V(t)$. Together, this implies $A^V(t_n) \to A^V(t)$ almost surely and we are done. □

As in discrete time we state two closely related verification theorems based on martingales. We start with the continuous-time version of Proposition 1.52, which is applied in Example 7.21 below.

Proposition 7.9

1. *Let τ be a stopping time with values in $[0, T]$. If V is an adapted process such that V^τ is a martingale, $V(\tau) = X(\tau)$, and $V(0) = M(0)$ for some martingale (or at least supermartingale) $M \geq X$, then τ is optimal for (7.4) and V coincides on $[\![0, \tau]\!]$ with the Snell envelope of Proposition 7.8.*
2. *More generally, let τ be a stopping time with values in $[t, T]$ and $F \in \mathscr{F}_t$. If V is an adapted process such that $V^\tau - V^t$ is a martingale, $V(\tau) = X(\tau)$, and $V(t) = M(t)$ on F for some martingale (or at least supermartingale) M with $M \geq X$ on F, then V coincides on $[\![t, \tau]\!] \cap (F \times [0, T])$ with the Snell envelope of Proposition 7.8 and τ is optimal on F for (7.5), i.e. it maximises $E(X(\tau)|\mathscr{F}_t)$ on F.*

Proof This follows along the same lines as its discrete counterpart Proposition 1.52. □

Remark 1.54 on the necessity of the conditions in the previous proposition remains valid in continuous time because Theorem 1.53 translates almost verbatim to continuous time as well:

Theorem 7.10

1. *Let τ be a stopping time with values in $[0, T]$. If $V \geq X$ is a supermartingale such that V^τ is a martingale and $V(\tau) = X(\tau)$, then τ is optimal for (7.4) and V coincides on $[\![0, \tau]\!]$ with the Snell envelope of X.*
2. *More generally, let τ be a stopping time with values in $[t, T]$. If $V \geq X$ is an adapted process such that $V - V^t$ is a supermartingale, $V^\tau - V^t$ is a martingale, and $V(\tau) = X(\tau)$, then V coincides on $[\![t, \tau]\!]$ with the Snell envelope of X and τ is optimal for (7.5), i.e. it maximises $E(X(\tau)|\mathscr{F}_t)$.*
3. *If τ is optimal for (7.5) and V denotes the Snell envelope of X, they have the properties in statement 2.*

Proof For the proof we refer to the discrete counterpart Theorem 1.53. □

The following result corresponds to Proposition 1.55. The first statement is the stopping version of Proposition 7.4.

Proposition 7.11

1. *The Snell envelope V is the smallest supermartingale dominating X.*
2. *The Snell envelope V is the unique supermartingale dominating X that satisfies $V(T) = X(T)$ and $1_{\{X_- \neq V_-\}} \bullet A^V = 0$, i.e. A^V decreases only when $X_- = V_-$.*
3. *The stopping times*

$$\underline{\tau}_t := \inf\{s \in [t, T] : V(s) = X(s)\}$$

and

$$\overline{\tau}_t := T \wedge \inf\{s \in [t, T] : A^V(s) < 0\}$$

are optimal in (7.5), i.e. they maximise $E(X(\tau)|\mathscr{F}_t)$.

Proof

1. By Proposition 7.8, V is a supermartingale dominating X. Let $\widetilde{V}$ denote another such supermartingale which dominates X. For any stopping time $\tau \geq t$ we have $E(X(\tau)|\mathscr{F}_t) \leq E(\widetilde{V}(\tau)|\mathscr{F}_t) \leq \widetilde{V}(t)$ by Proposition 2.7. Since $V(t) = \operatorname{ess\,sup}\{E(X(\tau)|\mathscr{F}_t) : \tau$ stopping time with values in $[t, T]\}$, we obtain $V(t) \leq \widetilde{V}(t)$.
2. *Step 1:* V is a supermartingale with $X - V \leq 0$ and $X(T) = V(T)$. In order for $1_{\{X_- \neq V_-\}} \bullet A^V = 0$ to hold, it suffices to show $A^V(\tau_t) - A^V(t) = 0$ a.s. for any $t \in [0, T]$ and $\tau_t := \inf\{s \geq t : V(s) = X(s)\}$. Indeed, $\{X_- \neq V_-\}$ differs from $\cup_{t\in[0,T]\cap\mathbb{Q}}[\![t, \tau_t[\![$ by a subset of the at most countably many jump times of $X - V$.
Step 2: Fix $t \in [0, T]$ and set $\tau := \inf\{s \geq t : V(s) = X(s)\}$. We want to show that $A^V(\tau) - A^V(t) = 0$ a.s.
Define $\tau^\varepsilon := \inf\{s \geq t : V(s) - X(s) \leq \varepsilon\}$ for $\varepsilon > 0$ and set

$$\tau^0 := \lim_{\varepsilon\to 0} \tau^\varepsilon = \inf\{s \geq t : V(s-) - X(s-) = 0\}.$$

Moreover, we define the predictable set $G := \{(V_- - X_-)^{\tau^0} = 0\} \cap (\Omega \times (t, \infty))$ and its *début* $\sigma^0(\omega) := \inf\{s \geq 0 : (\omega, s) \in G\}$. Since $[\![\sigma^0, \sigma^0]\!] \subset G$, the mapping σ^0 is a predictable stopping time by [154, Proposition I.2.13]. According to [154, Lemma I.2.27] we have $E(\Delta M^{V-X}(\sigma^0)|\mathscr{F}_{\sigma^0 -}) = 0$ on the set

$$F := \{(V - X)(\tau^0-) = 0\} = \{\sigma^0 < \infty\} \in \mathscr{F}_{\sigma^0-},$$

where $\mathscr{F}_{\sigma^0-} := \sigma(\mathscr{F}_0 \cup \{A \cap \{s < \tau^0\} : s \in \mathbb{R}_+, A \in \mathscr{F}_s\})$. Therefore

$$0 = E(\Delta M^{V-X}(\sigma^0)1_F) = E(\Delta M^{V-X}(\tau^0)1_F) = E(\Delta(V - X)(\tau^0)1_F),$$

where the last equality follows from $\Delta A^V(\tau^0) = 0 = \Delta A^V(\tau^0)$, cf. Proposition 7.8. Since $\Delta(V - X)(\tau^0) = (V - X)(\tau^0) \geq 0$ on F, we conclude that $(V - X)(\tau^0) = \Delta(V - X)(\tau^0) = 0$ on F. Together with $(V - X)(\tau^0) = 0$ on F^C we obtain $V(\tau^0) - X(\tau^0) = 0$. Thus $\tau^0 = \tau$ and consequently $\lim_{\varepsilon\to 0} \tau^\varepsilon \to \tau$. Therefore it suffices to show that $A^V(\tau^\varepsilon) - A^V(t) = 0$ a.s. for any $\varepsilon > 0$.
By Lemma A.4 we have $E(V(t)) = \sup\{E(X(\tau)) : \tau$ stopping time with values in $[t, T]\}$. Let $\tau_n, n \in \mathbb{N}$ denote a sequence of stopping times with values in $[t, T]$ and such that $E(X(\tau_n)) \to E(V(t))$. Then

$$E(X(\tau_n)) \leq E(V(\tau_n)) = E(V(t)) + E\big(A^V(\tau_n) - A^V(t)\big)$$

implies that

$$0 \leq \liminf_{n\to\infty} E\big(A^V(\tau_n) - A^V(t)\big) \leq \limsup_{n\to\infty} E\big(A^V(\tau_n) - A^V(t)\big) \leq 0.$$

Consequently, $A^V(\tau_n) - A^V(t)$ converges to 0 in L^1 and hence almost surely for a subsequence, again denoted as τ_n.
On the other hand,

$$\begin{aligned} E(V(\tau_n) - X(\tau_n)) &= E(V(\tau_n) - V(t)) + E(V(t) - X(\tau_n)) \\ &= E\big(A^V(\tau_n) - A^V(t)\big) + E(V(t)) - E(X(\tau_n)) \to 0 \end{aligned}$$

implies L^1-convergence of the nonnegative sequence $V(\tau_n) - X(\tau_n)$ to 0 and hence

$$V(\tau_n) - X(\tau_n) \to 0 \quad \text{a.s.} \tag{7.6}$$

for a subsequence, again denoted as τ_n. From (7.6) we conclude that $\tau^\varepsilon \leq \tau_n$ a.s. for fixed $\varepsilon > 0$ and n large enough. Consequently, $A^V(t) \geq A^V(\tau^\varepsilon) \geq A^V(\tau_n) \to A^V(t)$ as $n \to \infty$, which implies that $A^V(\tau^\varepsilon) = A^V(t)$ as desired.
Step 3: Let $\widetilde{V}$ denote another supermartingale as in statement 2 and set $D = V - \widetilde{V}$. Fix $t \in [0, T]$ and consider the event $N := \{D(t) < 0\}$. Set $\tau := \inf\{s \geq t : D(s) \geq 0\}$. Note that $X \leq V < \widetilde{V}$ on $[\![t, \tau[\![\cap (N \times [t, T])$, which implies that $A^D(s) = A^V(s) - A^{\widetilde{V}}(s)$ decreases on $[\![t, \tau[\![$. Therefore $(D(s \wedge \tau)1_N)_{s\in[t,T]}$ is a nonpositive supermartingale with nonnegative terminal value. We conclude $D(t)1_N = 0$ and hence $P(N) = 0$. The corresponding statement $P(D(t) < 0) = 0$ follows accordingly.

3. This follows from statement 2 and Theorem 7.10(2). □

As in discrete time we often have that the earliest optimal stopping time $\underline{\tau}_t$ for (7.5) and the latest optimal stopping time $\overline{\tau}_t$ coincide. Moreover, the solution to the original problem (7.4) is obtained for $t = 0$.

Remark 7.12 If X is nonnegative, problems with infinite time horizon $T = \infty$ can be treated as well by setting $X(\infty) := 0$. Theorem 7.10 and Proposition 7.11(1) still apply in this case if we make sure that the martingale resp. supermartingale property always refers to the time set $[0, T]$ including $T = \infty$.

A classical example allowing for a closed-form solution is the perpetual put option in the Black–Scholes model.

Example 7.13 (Perpetual Put) Let the price of a stock be modelled by geometric Brownian motion $S(t) = S(0)\exp(\mu t + \sigma W(t))$, where W denotes a Wiener process and $r > 0, \mu \leq 0, \sigma > 0$ parameters. We consider a perpetual put option on the stock with discounted payoff $X(t) = e^{-rt}(K - S(t))^+$ at time $t > 0$, where $K > 0$ denotes the *strike price*. The goal is to maximise the expected discounted payoff $E(X(\tau))$ over all stopping times τ. As in discrete time we make the ansatz that the Snell envelope is of the form $V(t) = e^{-rt}v(S(t))$ for some function v. It is natural to guess that it is optimal to stop when $S(t)$ falls below some threshold $s^\star \leq K$. This leads to $v(s) = (K - s)^+$ for $s \leq s^\star$. As long as $S(t) > s^\star$, the process $e^{-rt}v(S(t))$ should have vanishing drift, i.e.

$$\begin{aligned} 0 &= a^V(t) \\ &= e^{-rt}\left(-rv(S(t)) + \left(\mu + \frac{\sigma^2}{2}\right)S(t)v'(S(t)) + \frac{\sigma^2}{2}S(t)^2 v''(S(t))\right) \end{aligned} \tag{7.7}$$

for $S(t) > s^\star$. In order for the big parenthesis to vanish, we try a power ansatz for v, i.e.

$$v(s) = \begin{cases} K - s & \text{for } s \leq s^\star, \\ cs^{-a} & \text{for } s > s^\star \end{cases}$$

with some constants $s^\star, a, c > 0$. Indeed, (7.7) holds for

$$a = \frac{\sqrt{\mu^2 + 2r\sigma^2} - \mu^2}{\sigma^2}.$$

We choose $s^\star$ and c such that v is continuously differentiable in $s^\star$ because some kind of *smooth fit* condition is known to often lead to the solution. This holds for $s^\star = Ka/(1 + a)$ and $c = K^{1+a}a^a/(1 + a)^{1+a}$, which leads to the candidate Snell envelope

$$V(t) = e^{-rt}v(S(t)) = \begin{cases} e^{-rt}(K - S(t)) & \text{for } S(t) \leq \frac{Ka}{1+a}, \\ e^{-rt}\frac{K^{1+a}a^a}{(1+a)^{1+a}}S(t)^{-a} & \text{for } S(t) > \frac{Ka}{1+a}. \end{cases}$$

The candidate optimal stopping time is $\tau := \inf\{t \geq 0 : S(t) \leq Ka/(1 + a)\}$. The assumption $\mu \leq 0$ warrants that τ is finite.

We want to show optimality of τ using Proposition 7.9(1). For $S(0) \geq s^\star$ this follows from considering the martingale $M(t) := e^{-rt}cS(t)^{-a}$. For $S(0) < s^\star$ set instead $M(t) := e^{-rt}\widetilde{c}S(t)^{-\widetilde{a}}$, where $c > 0$ and $\widetilde{a} \in [0, a]$ are chosen such that the linear function $s \mapsto K - s$ and the concave function $s \mapsto \widetilde{c}s^{-\widetilde{a}}$ touch precisely at $s = S(0)$ and nowhere else. In this case, one needs to choose $\tau = 0$ and $V(t) = X(0)$, $t \in \mathbb{R}_+$ in Proposition 7.9(1).

As in discrete time the Snell envelope is the solution to some minimisation problem over a set of martingales.

Proposition 7.14 *The Snell envelope V of X satisfies*

$$V(0) = \min\left\{E\left(\sup_{t\in[0,T]}(X(t) - M(t))\right) : M \text{ martingale with } M(0) = 0\right\}$$

and, more generally,

$$V(t) = \min\left\{E\left(\sup_{s\in[t,T]}(X(s) - M(s))\middle|\mathscr{F}_t\right) : M \text{ martingale with } M(t) = 0\right\}$$

for any $t \in [0, T]$.

Proof This follows along the same lines as Proposition 1.59. □

7.3 The Markovian Situation

As in discrete time the value process can be expressed as a function of some state variables in a Markovian context. The role of the Bellman equation (1.77) for the value function is now played by the *Hamilton–Jacobi–Bellman* equation (7.15).

7.3.1 Stochastic Control

We work in the setup of Sect. 7.1 with a reward function as in Example 7.1. We assume that the controlled processes $X^{(\alpha)}$ are semimartingales with common initial value x_0 and local characteristics $b^{h,(\alpha)}(t), c^{(\alpha)}, K^{(\alpha)}$ of the form

$$b^{h,(\alpha)}(t) = \beta^h(\alpha(t), t, X^{(\alpha)}(t-)), \tag{7.8}$$

$$c^{(\alpha)}(t) = \gamma(\alpha(t), t, X^{(\alpha)}(t-)), \tag{7.9}$$

$$K^{(\alpha)}(t, dx) = \kappa(\alpha(t), t, X^{(\alpha)}(t-), dx) \tag{7.10}$$

for some deterministic functions β^h, γ, κ. In other words, the local characteristics are of Markovian type in the sense that they depend on ω only through $X^{(\alpha)}(t-)$ and $\alpha(t)$.

As in discrete time we suppose that—up to some measurability or integrability conditions—the set of admissible controls contains essentially all predictable processes with values in some set $A \subset \mathbb{R}^m$. These assumptions lead to the following natural ansatz for the value process

$$\mathscr{V}(t, \alpha) = \int_0^t f(s, X^{(\alpha)}(s-), \alpha(s))ds + v(t, X^{(\alpha)}(t)), \tag{7.11}$$

where $v : [0, T] \times \mathbb{R}^d \to \mathbb{R}$ denotes the **value function** of the control problem.

By Itō's formula the drift rate of $\mathscr{V}(\cdot, \alpha)$ amounts to

$$\begin{aligned} a^{\mathscr{V}(\cdot,\alpha)}(t) &= f(t, X^{(\alpha)}(t-), \alpha(t)) + D_1 v(t, X^{(\alpha)}(t-)) \\ &\quad + \sum_{i=1}^d D_{1+i} v(t, X^{(\alpha)}(t-))\beta_i^h(\alpha(t), t, X^{(\alpha)}(t-)) \\ &\quad + \frac{1}{2} \sum_{i,j=1}^d D_{1+i,1+j} v(t, X^{(\alpha)}(t-))\gamma_{ij}(\alpha(t), t, X^{(\alpha)}(t-)) \\ &\quad + \int \left(v(t, X^{(\alpha)}(t-) + y) - v(t, X^{(\alpha)}(t-)) - \sum_{i=1}^d D_{1+i} v(t, X^{(\alpha)}(t-))h_i(y) \right) \\ &\quad \times \kappa(\alpha(t), t, X^{(\alpha)}(t-), dy) \\ &= (f^{(\alpha(t))} + D_1 v + G^{(\alpha(t))} v)(t, X^{(\alpha)}(t-)), \end{aligned} \tag{7.12}$$

where we denote

$$\begin{aligned} (G^{(a)} v)(t, x) := &\sum_{i=1}^d D_{1+i} v(t, x)\beta_i^h(a, t, x) + \frac{1}{2} \sum_{i,j=1}^d D_{1+i,1+j} v(t, x)\gamma_{ij}(a, t, x) \\ &+ \int \left(v(t, x + y) - v(t, x) - \sum_{i=1}^d D_{1+i} v(t, x)h_i(y) \right) \kappa(a, t, x, dy) \end{aligned} \tag{7.13}$$

and

$$f^{(a)}(t, x) := f(t, x, a).$$

Since we want $\mathscr{V}(\cdot, \alpha)$ to be a supermartingale for any control and a martingale for some control, we naturally obtain the following counterpart of Theorem 1.60.

Theorem 7.15 *As suggested above we assume that the differential characteristics of $X^{(\alpha)}$ are of the form (7.8–7.10) with some deterministic measurable functions β^h, γ, κ. In addition, let $A \subset \mathbb{R}^m$ be such that any admissible control is A-valued. Suppose that*

1. *the function $v : [0, T]\times\mathbb{R}^d \to \mathbb{R}$ is twice continuously differentiable and satisfies*

$$v(T, x) = g(x) \tag{7.14}$$

as well as the **Hamilton–Jacobi–Bellman (HJB) equation**

$$\sup_{a\in A} \big(f^{(a)} + D_1 v + G^{(a)} v\big)(t, x) = 0, \quad t \in (0, T], \tag{7.15}$$

2. *for any $\alpha \in \mathscr{A}$ with $E(u(\alpha)^-) < \infty$ and any $t \in [0, T]$ there is some admissible control $\overline{\alpha}$ with $\overline{\alpha}^t = \alpha^t$ and*

$$\overline{\alpha}(t) = \arg\max_{a\in A} \big(f^{(a)} + D_1 v + G^{(a)} v\big)(t, X^{(\overline{\alpha})}(t-)), \quad t \in (0, T], \tag{7.16}$$

where we use the notation (7.13),

3. *the process $(\mathscr{V}(t, \alpha))_{t\in[0,T]}$ in (7.11) is of class (D) for any admissible control α.*

Then (7.11) coincides with the value process for any $\alpha \in \mathscr{A}$ with $E(u(\alpha)^-) < \infty$ and $\overline{\alpha}$ is optimal for the problem (7.1). In particular, $\alpha^\star := \overline{\alpha}$ is an optimal control if $t = 0$ is chosen in 2.

Proof Define $\mathscr{V}$ as in (7.11). By assumption on $X^{(\alpha)}$ we have $\mathscr{V}(t, \widetilde{\alpha}) = \mathscr{V}(t, \alpha)$ if $\widetilde{\alpha}^t = \alpha^t$. From (7.12, 7.15, 7.16) and Propositions 4.22, 4.23 we conclude that $\mathscr{V}(\cdot, \overline{\alpha}) - \mathscr{V}(\cdot, \overline{\alpha})^t$ is a martingale and $\mathscr{V}(\cdot, \alpha) - \mathscr{V}(\cdot, \alpha)^t$ is a supermartingale for any admissible control α. The statement follows now from Theorem 7.2(2) and the continuous-time version of Remark 1.45(5). □

One may wonder whether the converse of Theorem 7.15 also holds, i.e. whether the value process is always of the form (7.11) for some function v which solves the HJB equation. The answer is generally no because the value function may fail to be twice continuously differentiable. But if one allows for generalised solutions to the HJB equation in the so-called *viscosity* sense, the answer is typically yes. The corresponding theory is beyond the scope of this introductory text.

By Theorem 7.15 the natural approach to obtain the value function is to solve the PIDE

$$D_1 v(t, x) = -\sup_{a\in A} \big(f^{(a)} + G^{(a)} v\big)(t, x), \quad (t, x) \in (0, T] \times \mathbb{R}^d \tag{7.17}$$

with terminal condition $v(T, x) = g(x)$, $x \in \mathbb{R}^d$. It may happen that the value function is not twice continuously differentiable at some critical boundaries, in particular in optimal stopping problems. In this case contact conditions such as smooth fit may lead to the candidate value function. One can then try to verify the optimality directly, e.g. by using generalised versions of Itō's formula and Theorem 7.2.

Similarly to Proposition 7.4 and to Remark 1.62, the value function has some minimality property.

Proposition 7.16 *In the situation of Theorem 7.15, the value function is the smallest twice continuously differentiable function v such that (7.14) as well as*

$$(f^{(a)} + D_1 v + G^{(a)} v)(t,x) \leq 0, \quad (t,x) \in (0,T] \times \mathbb{R}^d, a \in A \tag{7.18}$$

hold and $v(t, X^{(\alpha^\star)})(t)$ is of class (D). More precisely, $v(t, X^{(\alpha^\star)})(t) \leq \widetilde{v}(t, X^{(\alpha^\star)})(t)$ holds almost surely for any $t \in [0,T]$ and any other such function $\widetilde{v}$.

Proof Let $\widetilde{v}$ denote another such function and $t \in [0,T]$. Define the event $N := \{\widetilde{v}(t, X^{(\alpha^\star)}(t)) < v(t, X^{(\alpha^\star)}(t))\}$ and set $D(s) := (v(s, X^{(\alpha^\star)})(s) - \widetilde{v}(s, X^{(\alpha^\star)})(s))1_N$ for $s \in [t,T]$. By (7.15, 7.16, 7.18) $(D(s))_{s \in [0,T]}$ is a submartingale with terminal value 0. This implies $0 \leq E(D(s)) \leq E(D(T)) = 0$ and hence $D(s) = 0$, i.e. $P(N) = 0$ as desired. □

Along the same lines as in Remark 1.63, we can define **sub-** and **supersolutions** to the Hamilton–Jacobi–Bellman equation. Based on the continuous-time version of Remark 1.45(2), they yield upper and lower bounds to the value process, provided that the class (D) condition in Theorem 7.15 holds.

The following variation of Example 7.5 is the continuous-time analogue of Example 1.64.

Example 7.17 (Power Utility of Terminal Wealth) We consider the counterpart of Example 7.5 for power utility $u(x) = x^{1-p}/(1-p)$ with some $p \in (0,\infty) \setminus \{1\}$. Parallel to Example 1.64 we work with the simplifying assumption that the stock price $S(t) = S(0)\mathscr{E}(X)(t)$ follows geometric Brownian motion, i.e. $X(t) = \mu t + \sigma W(t)$ for some Wiener process W and parameters $\mu \in \mathbb{R}, \sigma > 0$. Recall that the investor's goal is to maximise expected utility of terminal wealth $E(u(V_\varphi(T)))$ over all investment strategies φ. As in Example 1.64 we consider relative portfolios π instead, related to φ via $\pi(t) = \varphi(t)S(t)/V_\varphi(t)$. It is easy to see that the investor's wealth expressed in terms of π equals

$$V_\varphi(t) := v_0 + \varphi \bullet S(t) = v_0 \mathscr{E}(\pi \bullet X)(t),$$

cf. Example 7.5. Using slightly ambiguous notation, it is denoted by $V_\pi(t)$ in the sequel. As the set of admissible portfolio fractions we consider

$$\mathscr{A} := \left\{\pi \in L(X) : \pi(0) = 0 \text{ and } (u(V_\varphi(t)))_{t \in [0,T]} \text{ is of class (D)}\right\}.$$

The uniform integrability condition could be avoided at the expense of a more involved verification. Moreover, we let $A := \mathbb{R}$, which is needed only for the application of Theorem 7.15. Motivated by Example 1.64, we make the ansatz

$v(t,x) = e^{c(T-t)}u(x)$ for the value function, with some constant c which is yet to be determined. In view of Propositions 4.6 and 4.8, the generator of V_π in the sense of (7.13) equals

$$(G^{(a)}v)(t,x) = xa\mu D_2v(t,x) + x^2a^2\frac{\sigma^2}{2}D_{22}v(t,x).$$

The Hamilton–Jacobi–Bellman equation (7.17) then reads as

$$-cv(t,x) = -\sup_{a\in\mathbb{R}}\left(e^{c(T-t)}\left(a\mu xu'(x) + a^2\frac{\sigma^2}{2}x^2u''(x)\right)\right). \tag{7.19}$$

Since $u'(x)x = (1-p)u(x)$ and $u''(x)x^2 = -(1-p)pu(x)$, the supremum is attained for $a^\star := \mu/(p\sigma^2)$, leading to the constant candidate process

$$\pi^\star(t) = \frac{\mu}{p\sigma^2} \tag{7.20}$$

for the optimal control. Moreover, (7.19) yields $c = (1-p)\mu^2/(2p\sigma^2)$ and hence the candidate value function

$$v(t,x) = \exp\left(\frac{(1-p)\mu^2}{2p\sigma^2}(T-t)\right)u(x).$$

Optimality follows if these candidates satisfy the conditions of the verification theorem 7.15, i.e. if $\pi^\star$ is an admissible control. It is easy to see that $V_{\pi^\star}(t) = f(t)M(t)$ for some deterministic function f and the geometric Brownian motion $M := \mathscr{E}(a^\star\sigma W)$, which is a martingale by Theorem 2.22. From Proposition 4.22(5) it follows that M and hence $V_{\pi^\star}$ is of class (D).

We observe that it is optimal to invest a constant fraction of wealth in the risky asset. Quite consistent with common sense, this fraction is proportionate to the stock's drift and in inverse proportion to the stock's variance parameter σ^2, which can be interpreted as risk. The *risk aversion* parameter p appears in the denominator as well. As a side remark, the optimal portfolio for logarithmic utility in Example 7.5 corresponds to (7.20) with $p = 1$ in the present setup.

7.3.2 Optimal Stopping

We turn now to the counterpart of Theorem 1.66 in continuous time. The process Y is supposed to be a time-inhomogeneous Lévy–Itō diffusion. Denote by $\overline{G}$ the generator of the space-time process $\overline{Y}(t) := (t, Y(t))$. The reward process is assumed to be of the form $X(t) = g(Y(t))$ with $g : \mathbb{R}^d \to \mathbb{R}$. As in Sect. 7.2 we suppose that $E(\sup_{t\in[0,T]}|X(t)|) < \infty$.

Theorem 7.18 *If the function* $v : [0, T] \times \mathbb{R}^d \to \mathbb{R}$ *is in the domain* $\mathscr{D}_{\overline{G}}$ *of the extended generator and satisfies*

$$v(T, y) = g(y)$$

$$\max\{g(y) - v(t, y), \overline{G}v(t, y)\} = 0, \qquad t \in [0, T], \tag{7.21}$$

then $V(t) := v(t, Y(t))$, $t \in [0, T]$ *is the Snell envelope of* X, *provided that* V *is of class (D).*

Proof By definition of the extended generator, V is a local supermartingale. Proposition 4.23 yields that it is a supermartingale because it is bounded from below by X, which is a process of class (D). Fix $t \in [0, T]$ and define the stopping time $\tau := \inf\{s \geq t : V(s) = X(s)\}$. By (7.21), $1_{[\![t,T]\!]} \bullet V^\tau$ is a local martingale. Since it is of class (D), it is a martingale, cf. Proposition 4.22(6). The assertion now follows from Theorem 7.10. □

Similarly as in Sect. 7.3.1 one may wonder whether (7.21) generally holds for stopping problems in a Markovian context. A positive answer is to be found in [14, Theorem 11] if (7.21) is interpreted in the generalised sense of *viscosity solutions.*

Optimal stopping times are obtained from Proposition 7.11. $\underline{\tau}_t$ can be interpreted as the first entrance time of Y into the **stopping region**

$$S := \{(t, y) \in [0, T] \times \mathbb{R}^d : v(t, y) = g(y)\}.$$

The complement

$$C := S^c = \{(t, y) \in [0, T] \times \mathbb{R}^d : v(t, y) > g(y)\} \tag{7.22}$$

is called the **continuation region** because stopping is suboptimal as long as Y is inside C.

In view of (5.11), the extended generator of $\overline{Y}(t) = (t, Y(t))$ is typically of the form

$$\overline{G}v(t, y) := D_1 v(t, y) + \sum_{i=1}^d \beta_i^h(y) D_{1+i} v(t, y) + \frac{1}{2} \sum_{i,j=1}^d \gamma_{ij}(y) D_{1+i,1+j} v(t, y)$$

$$+ \int \Big(v(t, y+z) - v(t, y) - \sum_{i=1}^d h_i(z) D_{1+i} v(t, y)\Big) K(y, dz)$$

for sufficiently regular v. Therefore, the *linear complementarity problem* (7.21) is often solved by considering the associated free-boundary problem

$$\begin{aligned} -D_1 v(t,y) &= \sum_{i=1}^d \beta_i^h(y) D_{1+i} v(t,y) + \frac{1}{2} \sum_{i,j=1}^d \gamma_{ij}(y) D_{1+i,1+j} v(t,y) \\ &\quad + \int \Big(v(t,y+z) - v(t,y) - \sum_{i=1}^d h_i(z) D_{1+i} v(t,y) \Big) \kappa(y,dz) \quad \text{on } C, \\ v(t,y) &= g(y) \quad \text{on } C^c \end{aligned} \tag{7.23}$$

with some unknown domain $C \subset [0,T] \times \mathbb{R}^d$, which then yields the continuation region introduced in (7.22). However, the value function typically fails to be twice differentiable at the boundary. But for continuous processes, the solution is at least C^1 in most cases. Therefore a contact condition at the stopping boundary leads to the solution, cf. Example 7.13, as in the Stefan problem for phase transitions. Put differently, the C^1-solution of (7.23) is the natural candidate for the value function. Sometimes, generalised versions of Itō's formula are applied in order to verify that this candidate v indeed belongs to the domain of $\overline{G}$ so that Theorem 7.18 can be applied. But the C^1-property may also fail, as is illustrated in Example 7.21 below.

Similarly to Remark 1.68 and Proposition 7.16, the value function has some minimality property.

Proposition 7.19 *The function $v : [0,T] \times \mathbb{R}^d \to \mathbb{R}$ in Theorem 7.18 is the smallest one that belongs to the domain of the generator of $\overline{Y}(t) = (t, Y(t))$ and satisfies $v(t,\cdot) \geq g(\cdot)$, $t \in [0,T]$ as well as $\overline{G} v \leq 0$. More precisely, $v(t, Y(t)) \leq \widetilde{v}(t, Y(t))$ holds almost surely for any $t \in [0,T]$ and any other such function $\widetilde{v}$.*

Proof Let $\widetilde{v}$ be another such function and $t \in [0,T]$. Define the event $N := \{\widetilde{v}(t,Y(t)) < v(t,Y(t))\}$ and the stopping time $\tau := \inf\{s \geq t : v(s,Y(s)) = g(Y(s))\}$. By (7.21), $(v(s \wedge \tau, Y(s \wedge \tau)))_{s \in [t,T]}$ is a martingale and $(\widetilde{v}(s \wedge \tau, Y(s \wedge \tau)))_{s \in [t,T]}$ a local supermartingale which is bounded from below by a process of class (D), namely X. This implies that $U(s) := (v(s \wedge \tau, Y(s \wedge \tau)) - \widetilde{v}(s \wedge \tau, Y(s \wedge \tau)))1_N$, $s \in [t,T]$ defines a submartingale with terminal value $U(T) = (g(Y(\tau)) - \widetilde{v}(\tau, Y(\tau)))1_N \leq 0$. Since $0 \leq E(U(s)) \leq E(U(T)) \leq 0$, it follows that $U(s) = 0$ almost surely, which yields the claim. □

Remark 7.20 Proposition 7.19 indicates that $v(t,x)$ can be obtained by minimising $\widetilde{v}(t,x)$ over all sufficiently regular functions $\widetilde{v} \geq g$ which are *superharmonic* in the sense that $\overline{G}\widetilde{v} \leq 0$. If one minimised instead only over *harmonic* functions $\widetilde{v} \geq g$ in the sense that $\overline{G}\widetilde{v} = 0$, one would expect to typically obtain a strict upper bound of $v(t,x)$. In some cases, however, there exists a harmonic function $\widetilde{v} \geq g$ which coincides with the value function v in the continuation region or, more precisely, in a given connected component of the continuation region. Therefore, minimising over the smaller and more tractable set of harmonic functions may in fact yield the true value and not just an upper bound. This happens, for example, for the perpetual put in Example 7.13, but also in the following Example 7.21. It is not yet understood in which cases this phenomenon occurs, i.e. when minimisation over harmonic functions leads to the value function.

This can also be viewed in the context of the continuous-time version of Remark 1.54. As noted there, the value of the stopping problem is obtained by minimising over all martingales dominating X. Unfortunately, the set of martingales is rather large. The smaller set of *Markovian-type martingales* of the form $M(t) = m(t, Y(t))$ with some deterministic function m is more tractable because these processes are determined by their terminal payoff function $y \mapsto m(T, y)$. Since $(m(t, Y(t)))_{t\in[0,T]}$ is a martingale if and only if $\overline{G}m = 0$, these martingales correspond to the harmonic functions above. Again, it is not *a priori* clear why minimising over this smaller set should lead to the true value instead of an upper bound. Indeed, recall that the canonical candidate for M in Proposition 7.9 is the martingale part $V_0 + M^V$ of the Snell envelope V. This martingale is typically not of Markovian type in the above sense.

Example 7.21 (American Butterfly in the Bachelier Model) Using the notation of Theorem 7.18 we consider $Y = Y(0) + W$ with standard Brownian motion W and the payoff function $g(y) := (1 - |y|)^+$.

It is evident that one should stop whenever $Y(t)$ reaches 0 because the payoff attains its maximal value there. We show that the stopping time $\tau := \inf\{s \in [t, T] : Y(s) = 0\} \wedge T$ is in fact optimal for (7.5). This is done by finding a harmonic function v such that $V(t) := v(t, Y(t))$ equals the Snell envelope of $X = g(Y)$ on $\{Y \geq 0\}$. Since $v(t, Y(t))$ is a martingale, v is determined by $v(T, \cdot)$ via $v(t, y) = E(v(T, Y(T))|Y(t) = y)$. In order to achieve $v(\tau, Y(\tau)) = V(\tau) = g(Y(\tau))$ we need that $v(T, y) = 1 - y$ for any $y \geq 0$ and $v(t, 0) = 1$ for any $t \in [0, T]$. These properties are shared by the function

$$\begin{aligned} v(t, y) &= E\big(f(Y(T))\big|Y(t) = y\big) \\ &= \int f(z)\frac{1}{\sqrt{2\pi(T-t)}}\exp\left(-\frac{(z-y)^2}{2(T-t)}\right)dz \\ &= \begin{cases} \Phi\Big(\frac{-1-|y|}{\sqrt{T-t}}\Big)(1+|y|) + \Phi\Big(\frac{1-|y|}{\sqrt{T-t}}\Big)(1-|y|) & \\ \quad + \sqrt{T-t}\Big(\varphi\big(\frac{1-|y|}{\sqrt{T-t}}\big) - \varphi\big(\frac{-1-|y|}{\sqrt{T-t}}\big)\Big) & \text{for } t < T, \\ f(y) & \text{for } t = T, \end{cases} \end{aligned}$$

and Φ, φ denote the cumulative distribution function and the probability density function of the standard normal distribution.

We show now that v is the value function on $[0, T]\times\mathbb{R}_+$ in the sense that $V(t) := v(t, Y(t))$ is the Snell envelope of $X = g(Y)$ on $\{Y \geq 0\}$. Along the same lines one verifies that $(t, y) \mapsto v(t, -y)$ is the value function on $[0, T] \times \mathbb{R}_-$, i.e. $V(t) = v(t, -Y(t))$ equals the Snell envelope on $\{Y \leq 0\}$.

Fix $t \in [0, T)$ and consider the set $F := \{Y(t) \geq 0\}$. The statement for $\{Y(t) \leq 0\}$ follows by symmetry. We define a nonnegative martingale M via $M(t) := E(f(Y(T))|\mathscr{F}_t)$. A straightforward calculation shows that $M(t) = v(t, Y(t)) = V(t)$ on the set F. It is easy to see that $v(t, 0) = 1 = g(0)$ for $t < T$ and

$$v(t, y) > \int \widetilde{f}(z) \frac{1}{\sqrt{2\pi(T-t)}} \exp\Big(-\frac{(z - Y(t))^2}{(T-t)}\Big) dz = 1 - y$$

for $y > 0, t < T$ and

$$\widetilde{f}(z) = \begin{cases} 2 & \text{if } z \leq -1, \\ 1 - z & \text{if } -1 < z < 1 + 2y, \\ -2y & \text{if } z \geq 1 + 2y. \end{cases}$$

Since $v(t, y) \geq 0$, this yields $v(t, y) \geq g(y)$ for $y \geq 0$. Moreover, $v(t, y) \geq v(t, 0) = 1 \geq g(y)$ for $y \leq 0$ because f is decreasing. Set

$$\tau := \inf\big\{s \in [t, T] : Y(s) = 0\big\} \wedge T = \inf\big\{s \in [t, T] : V(s) = X(s)\big\}.$$

The statement follows now from Proposition 7.9(2).

7.4 The Stochastic Maximum Principle

As in discrete time, control problems can be tackled by some kind of dual approach as well. The setup is as in Sect. 7.1 with a reward function as in Example 7.1. We assume that the function g is concave and that the set of admissible controls contains only predictable processes with values in some convex set $A \subset \mathbb{R}^m$.

7.4.1 Sufficient Criterion

The continuous-time counterpart of Theorem 1.69 reads as follows:

Theorem 7.22 *Suppose that Y is an $\mathbb{R}^d$-valued semimartingale and $\alpha^\star$ an admissible control with the following properties, where we suppress as usual the argument $\omega \in \Omega$ in random functions.*

1. *$E(|u(\alpha^\star)|) < \infty$.*
2. *The processes Y_i, $X_i^{(\alpha)}$, $Y_i X_i^{(\alpha)}$ are special semimartingales with absolutely continuous compensator for $i = 1, \dots, d$ and any $\alpha \in \mathscr{A}$. We use the notation (2.9, 3.7).*
3. *g is differentiable in $X^{(\alpha^\star)}(T)$ with derivative denoted by $\nabla g(X^{(\alpha^\star)}(T))$.*

4. *For any* $\alpha \in \mathscr{A}$ *we have*

$$\sum_{i=1}^{d}\Big(Y_i(t-)a^{X_i^{(\alpha)}}(t) + \widetilde{c}^{Y_i,X_i^{(\alpha)}}(t)\Big) + f\big(t, X^{(\alpha)}(t-), \alpha(t)\big)$$

$$= H(t, X^{(\alpha)}(t-), \alpha(t)), \quad t \in [0, T] \tag{7.24}$$

for some function $H : \Omega \times [0, T] \times \mathbb{R}^d \times A \to \mathbb{R}$ *such that* $(x, a) \mapsto H(t, x, a)$ *is concave and* $x \mapsto H(t, x, \alpha^\star(t))$ *is differentiable in* $X^{(\alpha^\star)}(t-)$. *We denote the derivative by* $\nabla_x H(t, X^{(\alpha^\star)}(t-), \alpha^\star(t)) = (\partial_{x_i} H(t, X^{(\alpha^\star)}(t-), \alpha^\star(t)))_{i=1,\dots,d}$.

5. *The drift rate and the terminal value of* Y *are given by*

$$a^Y(t) = -\nabla_x H(t, X^{(\alpha^\star)}(t-), \alpha^\star(t)), \quad t \in [0, T] \tag{7.25}$$

and

$$Y(T) = \nabla g(X^{(\alpha^\star)}(T)).$$

6. $\alpha^\star(t)$ *maximises* $a \mapsto H(t, X^{(\alpha^\star)}(t-), a)$ *over* A *for any* $t \in [0, T]$.
7. *The processes* $\int_0^\cdot f(t, X^{(\alpha)}(t), \alpha(t))dt + Y^\top X^{(\alpha)}$ *are bounded from below by a process of class (D) for any* $\alpha \in \mathscr{A}$.
8. *The process* $\int_0^\cdot f(t, X^{(\alpha^\star)}(t-), \alpha^\star(t))dt + Y^\top X^{(\alpha^\star)}$ *is of class (D).*

Then $\alpha^\star$ *is an optimal control, i.e. it maximises*

$$E\bigg(\int_0^T f(t, X^{(\alpha)}(t-), \alpha(t))dt + g(X^{(\alpha)}(T))\bigg)$$

over all $\alpha \in \mathscr{A}$.

Proof Let $\alpha \in \mathscr{A}$ be a competing control. We show that

$$\begin{aligned}
&\int_0^T E\Big(f(s, X^{(\alpha)}(s-), \alpha(s)) - f(s, X^{(\alpha^\star)}(s-), \alpha^\star(s))\Big|\mathscr{F}_t\Big)\, ds \\
&\quad + E\big(g(X^{(\alpha)}(T)) - g(X^{(\alpha^\star)}(T))\big|\mathscr{F}_t\big) \\
&\le \int_0^t \Big(f(s, X^{(\alpha)}(s-), \alpha(s)) - f(s, X^{(\alpha^\star)}(s-), \alpha^\star(s))\Big)ds \\
&\quad + Y(t)^\top\Big(X^{(\alpha)}(t) - X^{(\alpha^\star)}(t)\Big) \qquad (7.26)\\
&=: U(t)
\end{aligned}$$

for $t \in [0, T]$. Since $\alpha(0) = \alpha^\star(0)$ implies $X^{(\alpha)}(0) = X^{(\alpha^\star)}(0)$, we obtain optimality of $\alpha^\star$ by inserting $t = 0$.

In order to verify (7.26) we introduce the notation

$$h(t,x) := \sup_{a\in A} H(t,x,a).$$

Concavity of g implies

$$g(X^{(\alpha)}(T)) - g(X^{(\alpha^\star)}(T)) \leq Y(T)^\top \big(X^{(\alpha)}(T) - X^{(\alpha^\star)}(T)\big),$$

which in turn yields (7.26) for $t = T$.

The drift rate of U equals

$$\begin{aligned}
a^U(t) &= f(t, X^{(\alpha)}(t-), \alpha(t)) - f(t, X^{(\alpha^\star)}(t-), \alpha^\star(t)) \\
&\quad + \sum_{i=1}^d \left(a^{Y_i X_i^{(\alpha)}}(t) - a^{Y_i X_i^{(\alpha^\star)}}(t)\right) \\
&= f(t, X^{(\alpha)}(t-), \alpha(t)) - f(t, X^{(\alpha^\star)}(t-), \alpha^\star(t)) \\
&\quad + \sum_{i=1}^d \Big(X_i^{(\alpha)}(t-)a^{Y_i}(t) + Y_i(t-)a^{X_i^{(\alpha)}}(t) + \tilde{c}^{Y_i, X_i^{(\alpha)}}(t) \\
&\quad - X_i^{(\alpha^\star)}(t-)a^{Y_i}(t) - Y_i(t-)a^{X_i^{(\alpha^\star)}}(t) - \tilde{c}^{Y_i, X_i^{(\alpha^\star)}}(t)\Big) \\
&= H(t, X^{(\alpha)}(t-), \alpha(t)) - h(t, X^{(\alpha)}(t-)) \qquad (7.27) \\
&\quad + \Big(\big(h(t, X^{(\alpha)}(t-)) + X^{(\alpha)}(t-)^\top a^Y(t)\big) \\
&\quad - \big(h(t, X^{(\alpha^\star)}(t-)) + X^{(\alpha^\star)}(t-)\big)^\top a^Y(t)\Big) \\
&\leq 0,
\end{aligned}$$

where the second equality follows from integration by parts, cf. Proposition 3.25. The inequality holds because (7.27) is nonpositive by definition and

$$\begin{aligned}
&h(t, X^{(\alpha)}(t-)) - h(t, X^{(\alpha^\star)}(t-)) + \big(X^{(\alpha)}(t-) - X^{(\alpha^\star)}(t-)\big)^\top a^Y(t) \\
&= h(t, X^{(\alpha)}(t-)) - h(t, X^{(\alpha^\star)}(t-)) - \big(X^{(\alpha)}(t-) - X^{(\alpha^\star)}(t-)\big)^\top \nabla_x h(t, X^{(\alpha^\star)}(t-)) \\
&\leq 0
\end{aligned}$$

follows from (7.25) and Proposition 1.42.

From $a^U \leq 0$ we conclude that U is a σ-supermartingale. By assumptions 7 and 8 it is bounded from below by a process of class (D) and hence a supermartingale, cf. Proposition 4.23. This implies that (7.26) holds for $t < T$ as well. □

Remark 1.70 holds accordingly in the present setup.

7.4.2 BSDE Formulation

The previous theorem plays a key role in the martingale characterisation of optimal investment problems in Part II. We refer the reader to Chap. 10 for Examples. Instead we provide a link to the literature, where the stochastic maximum principle is usually discussed in a Brownian setup and expressed in terms of a backward stochastic differential equation (BSDE) or rather *forward-backward stochastic differential equation (FBSDE)*. The natural extension to processes with jumps involves representable processes similar to those in Sect. 4.8.

To this end, let L be an $\mathbb{R}^n$-valued semimartingale with local characteristics (b, c, K). We assume that the controlled processes are all driven by L in the sense that

$$\begin{aligned} dX^{(\alpha)}(t) &= \beta(t, X^{(\alpha)}(t-), \alpha(t))dt + \gamma(t, X^{(\alpha)}(t-), \alpha(t))dL^c(t) \\ &\quad + \delta(t, X^{(\alpha)}(t-), \alpha(t), x) * (\mu^L - \nu^L)(dt) \end{aligned} \tag{7.28}$$

with deterministic or at least predictable functions β, γ, δ.

We look for a semimartingale Y as in Theorem 7.22. We expect it to be of similar form

$$dY(t) = a^Y(t)dt + \gamma^Y(t)dL^c(t) + \delta^Y(t, x) * (\mu^L - \nu^L)(dt)$$

with some predictable integrands a^Y, γ^Y, δ^Y. Since

$$\begin{aligned} \widetilde{c}^{Y,X^{(\alpha)}}(t) &= \gamma^Y(t)^\top c(t)\gamma(t, X^{(\alpha)}(t-), \alpha(t)) \\ &\quad + \int \delta^Y(t, x)\delta(t, X^{(\alpha)}(t-), \alpha(t), x)K(t, dx), \end{aligned}$$

(7.24) boils down to

$$\begin{aligned} H(t, X^{(\alpha)}(t-), \alpha(t)) &= Y(t-)^\top \beta(t, X^{(\alpha)}(t-), \alpha(t)) \\ &\quad + \gamma^Y(t)^\top c(t)\gamma(t, X^{(\alpha)}(t-), \alpha(t)) \\ &\quad + \int \delta^Y(t, x)\delta(t, X^{(\alpha)}(t-), \alpha(t), x)K(t, dx) \\ &\quad + f(t, X^{(\alpha)}(t-), \alpha(t)) \\ &= \widetilde{H}(t, X^{(\alpha)}(t-), \alpha(t), Y(t-), \gamma^Y(t), \delta^Y(t)) \end{aligned}$$

with a deterministic or at least predictable function

$$\widetilde{H}(t,x,a,y,\varphi,\xi) := y^\top \beta(t,x,a) + \varphi^\top c(t)\gamma(t,x,a) + \int \xi(\widetilde{x})\delta(t,x,a,\widetilde{x})K(t,d\widetilde{x}) + f(t,x,a), \qquad (7.29)$$

which is usually called **Hamiltonian** in this context. We suppose that $f(t,x,a)$, $g(x)$, $\beta(x,a)$, $\gamma(t,x,a)$, $\delta(t,x,a,\widetilde{x})$ or, more precisely, $\int \delta^Y(t,\widetilde{x})\delta(t,x,a,\widetilde{x}) K(t,d\widetilde{x})$ are differentiable in x. For given control α, the BSDE

$$\begin{aligned} dY(t) &= -\nabla_x \widetilde{H}(t, X^{(\alpha)}(t-), \alpha(t), Y(t-), \gamma^Y(t), \delta^Y(t,\cdot))dt \\ &\quad + \gamma^Y(t)dL^c(t) + \delta^Y(t,x) * (\mu^L - \nu^L)(dt), \\ Y(T) &= \nabla g(X^{(\alpha)}(T)) \end{aligned} \qquad (7.30)$$

is called the **adjoint equation**, where $\nabla_x \widetilde{H}$ denotes the derivative with respect to the second argument. Equation (7.30) is a *backward* stochastic differentiable equation because the terminal rather than the initial value of Y is given. But since the solution (Y, γ^Y, δ^Y) depends on the solution to the forward SDE (7.28), we are in the situation of a *forward-backward* stochastic differential equation.

Theorem 7.22 can now be rephrased as follows. If $\alpha^\star$ is an admissible control, Y an adapted $\mathbb{R}^d$-valued process, and γ^Y, δ^Y predictable such that for any $t \in [0,T]$

- (Y, γ^Y, δ^Y) solves BSDE (7.30) for $\alpha = \alpha^\star$,
- $(x,a) \mapsto \widetilde{H}(t,x,a,Y(t-),\gamma^Y(t),\delta^Y(t,\cdot))$ is concave,
- $\alpha^\star(t)$ maximises $a \mapsto \widetilde{H}(t, X^{(\alpha^\star)}(t-), a, Y(t-), \gamma^Y(t), \delta^Y(t,\cdot))$ over A,
- assumptions 7 and 8 in Theorem 7.22 hold,

then $\alpha^\star$ is an optimal control. Indeed, one easily verifies that all the assumptions in Theorem 7.22 hold.

As a concrete example we have a look at optimal investment in a market of several assets with continuous price processes.

Example 7.23 (Mutual Fund Theorem) We consider the generalisation of the stock market model in Example 7.17 to $d \geq 1$ risky assets. More specifically, we assume that

$$S_i(t) = S_i(0) + \mathscr{E}(\mu_i \bullet I + \sigma_{i\cdot} \bullet W)(t), \quad i = 1, \ldots, d$$

with $\mu \in \mathbb{R}^d$, invertible $\sigma \in \mathbb{R}^{d\times d}$ and a Wiener process W in $\mathbb{R}^d$. We write $c := \sigma\sigma^\top$. As in the earlier examples, the investor's wealth can be written as

$$V_\varphi(t) := v_0 + \varphi \bullet S(t),$$

where v_0 denotes her initial endowment and φ her dynamic portfolio. We assume that she strives to maximise her expected utility $E(u(V_\varphi(T)))$ for some continuously differentiable, increasing, strictly concave utility function $u : \mathbb{R}_+ \to [-\infty, \infty)$ with $\lim_{x\to 0} u'(x) = \infty$, $\lim_{x\to\infty} u'(x) = 0$. To be more precise, we specify the set

$$\mathscr{A} := \{\varphi \in L(S) : \varphi(0) = 0 \text{ and } V_\varphi \geq 0\}$$

of admissible controls in this context.

Motivated by the related optimisation problem in Example 1.71 and Theorem 1.72, we look for a process Y such that both Y and YS are martingales. It turns out that there is—up to some multiplicative constant y—only one natural candidate, namely $Y^\star := \mathscr{E}(-(\sigma^{-1}\mu)^\top W)$. Since it is a positive martingale, it is the density process of some probability measure $Q \sim P$ on $\mathscr{F}_T$. By Remark 3.72, $\widetilde{W}(t) := W(t) + \sigma^{-1}\mu t$ is an $\mathbb{R}^d$-valued Wiener process under Q. In terms of $\widetilde{W}$, the stock price process can be written as

$$S_i(t) = S_i(0)\mathscr{E}(\sigma_{i\cdot} \bullet \widetilde{W})(t), \quad i = 1, \ldots, d.$$

Observe that S_i are Q-martingales for $i = 1, \ldots, d$, which is why Q is called an *equivalent martingale measure* in Sect. 9.3. $Y^\star S$ is a martingale by Proposition 3.69.

Again motivated by Example 1.71 and Theorem 1.72, we look for a strategy $\varphi^\star$ and a constant y with $u'(V_{\varphi^\star}(T)) = Y(T) := yY^\star(T)$ or, equivalently,

$$(u')^{-1}(Y(T)) = v_0 + \varphi^\star \bullet S(T). \tag{7.31}$$

To this end, we consider

$$F := \frac{1}{Y^\star} = \mathscr{E}\big((\sigma^{-1}\mu)^\top \widetilde{W}\big) = 1 + \psi \bullet S$$

for

$$\psi_i(t) := \frac{F(t)(c^{-1}\mu)_i}{S_i(t)}, \quad i = 1, \ldots, d.$$

One can think of F as the value of a dynamic portfolio with initial value 1 which holds $\psi_i(t)$ shares of asset S_i at any time t. We call this composite asset F a *mutual fund*. Observe that $F(T) = \exp(\sqrt{\mu^\top c^{-1}\mu}U(T) - \frac{1}{2}\mu c^{-1}\mu T)$ for the univariate Q-Wiener process $U := (\mu^\top c^{-1}\mu)^{-1/2}(\sigma^{-1}\mu)^\top \widetilde{W}$. We assume that one can choose a positive constant y such that $E_Q((u')^{-1}(Y(T))) = v_0$, which is typically possible because $y \mapsto (u')^{-1}(Y(T))$ is continuous, increasing, and has limits 0 resp. ∞. By Theorem 3.61 there exists some U-integrable process ξ with

$$(u')^{-1}(Y(T)) = (u')^{-1}(y/F(T)) = v_0 + \xi \bullet U(T),$$

i.e.

$$(u')^{-1}(Y(T)) = v_0 + \chi \bullet F(T) \tag{7.32}$$

for $\chi(t) = \xi(t)/(F(t)\sqrt{\mu^\top c^{-1}\mu})$. In financial terms, (7.32) can be interpreted as the terminal value of an investment trading only in the riskless asset and the mutual fund, holding $\chi(t)$ shares of the latter at time t. In the language of Part II, χ corresponds to a *perfect hedge* or *replicating strategy* of the *contingent claim* with terminal payoff $(u')^{-1}(y/F(T))$. It can be determined numerically by methods which are discussed in Sects. 11.5 and 11.6.

By setting $\varphi^\star(t) := \chi(t)\psi(t)$ the above hedge can be interpreted as an investment in the original assets as well, i.e. (7.31) holds. This dynamic portfolio turns out to maximise the expected utility. Indeed, straightforward calculations yield that the control $\varphi^\star$ and the processes $L := S$, $Y := yY^\star$, $\widetilde{H} = 0$, and $\gamma_i^Y := -yY^\star S_i^{-1}(c^{-1}\mu)_i$, $i = 1, \dots, d$ meet the requirements stated before this example. For assumption 7 in Theorem 7.22 one must show that $YV_{\varphi^\star}$ is of class (D). Since it is a nonnegative local martingale, it is a supermartingale. Hence it suffices to verify that $E(Y(T)V_{\varphi^\star}(T)) = Y(0)V_{\varphi^\star}(0) = yv_0$. But this follows from $E(Y(T)V_{\varphi^\star}(T)) = yE_Q((u')^{-1}(Y(T))) = yv_0$ as desired.

Besides having determined the concrete solution, we observe that all utility maximisers in this market agree to invest only in the risky asset and the mutual fund F, regardless of their specific utility function and their time horizon.

7.4.3 The Markovian Situation

In discrete time we observed that the derivative of the value function evaluated at the optimal controlled processes is the natural candidate for the process Y in Theorem 7.22. We consider here the corresponding statement in the BSDE setup above, cf. [28] for an early reference. For ease of notation, we focus on the univariate case $d = 1 = n$. Our goal is to show that

$$\begin{aligned}
Y(t) &= v_x(t, X^{(\alpha^\star)}(t)),\\
\gamma^Y(t) &= \gamma(t, X^{(\alpha^\star)}(t-), \alpha^\star(t))v_{xx}(t, X^{(\alpha^\star)}(t-)),\\
\delta^Y(t,x) &= v_x\big(t, X^{(\alpha^\star)}(t-) + \delta(t, X^{(\alpha^\star)}(t-), \alpha^\star(t), x)\big) - v_x\big(t, X^{(\alpha^\star)}(t-)\big)
\end{aligned}$$

typically satisfy the conditions in Sect. 7.4.2, where v denotes the value function of Sect. 7.3.1 and v_x, v_{xx} derivatives with respect to the second argument. To this end, we make the following assumptions.

1. The objective function g is concave and the state space A of admissible controls is convex.

2. The controlled processes are of the form (7.28). The Hamiltonian $\widetilde{H}$ is defined as in (7.29) and the mapping $(x, a) \mapsto \widetilde{H}(t, x, a, Y(t), \gamma^Y(t), \delta^Y(t, \cdot))$ is assumed to be concave.
3. We are in the situation of Sect. 7.3.1 and in particular Theorem 7.15. $v : \Omega \times [0, T] \times \mathbb{R}^d \to \mathbb{R}$ denotes the value function of the problem. We assume that the supremum in (7.15) is attained at some $a^\star(t, x)$ such that the optimal control satisfies $\alpha^\star(t) = a^\star(t, X^{(\alpha^\star)}(t-))$ in line with (7.16).
4. The functions $(t, x, a) \mapsto f(t, x, a)$, $g(x)$, $\beta(t, x, a)$, $\gamma(t, x, a)$, $\delta(t, x, a, \widetilde{x})$, $v(t, x)$, $a^\star(t, x)$ are three times continuously differentiable with partial derivatives denoted as f_t, f_x, f_a etc. Moreover, we assume that

$$\begin{aligned}
0 = {} & f_a(t, x, a^\star(t, x)) + v_x(t, x)\beta_a(t, x, a^\star(t, x)) \\
& + v_{xx}(t, x)\gamma(t, x, a^\star(t, x))\gamma_a(t, x, a^\star(t, x))c(t) \\
& + \int \Big(v_x(t, x + \delta(t, x, a^\star(t, x), \widetilde{x})) - v_x(t, x)\Big)\delta_a(t, x, a^\star(t, x), \widetilde{x})K(t, d\widetilde{x}),
\end{aligned}$$

which follows from (7.16) if $a^\star(t, x)$ is in the interior of A.
5. Integrals are finite and derivatives and integrals are exchangeable whenever needed.

We must show that Y, γ^Y, δ^Y solve BSDE (7.30) and $\alpha^\star(t)$ maximises $a \mapsto \widetilde{H}(t, X^{(\alpha^\star)}(t-), a, Y(t), \gamma^Y(t), \delta^Y(t, \cdot))$ over A. The Hamilton–Jacobi–Bellman equation (7.15) in this setup reads as

$$\begin{aligned}
0 = {} & f(t, x, a^\star(t, x)) + v_t(t, x) \\
& + v_x(t, x)\beta(t, x, a^\star(t, x)) + \frac{1}{2}v_{xx}(t, x)\gamma^2(t, x, a^\star(t, x))c(t) \\
& + \int \Big(v(t, x + \delta(t, x, a^\star(t, x), \widetilde{x})) - v(t, x) - v_x(t, x)\delta(t, x, a^\star(t, x), \widetilde{x})\Big)K(t, d\widetilde{x}).
\end{aligned}$$

A key step is to differentiate this HJB equation with respect to x. Together with assumption 4 we obtain

$$\begin{aligned}
0 = {} & f_x(t, x, a^\star(t, x)) + v_{tx}(t, x) + v_x(t, x)\beta_x(t, x, a^\star(t, x)) \\
& + v_{xx}(t, x)\beta(t, x, a^\star(t, x)) \\
& + v_{xx}(t, x)\gamma(t, x, a^\star(t, x))\gamma_x(t, x, a^\star(t, x))c(t) \\
& + \frac{1}{2}v_{xxx}(t, x)\gamma^2(t, x, a^\star(t, x))c(t) \\
& + \int \Big(\big(v_x(t, x + \delta(t, x, a^\star(t, x), \widetilde{x})) - v_x(t, x)\big)\big(1 + \delta_x(t, x, a^\star(t, x), \widetilde{x})\big) \\
& - v_{xx}(t, x)\delta(t, x, a^\star(t, x), \widetilde{x})\Big)K(t, d\widetilde{x}).
\end{aligned}$$

Now, Itō's formula (3.53) yields

$$\begin{aligned}
dY(t) &= dv_x(t, X^{(\alpha^\star)}(t-)) \\
&= \Big(v_{xt}(t, X^{(\alpha^\star)}(t-)) + v_{xx}(t, X^{(\alpha^\star)}(t-))\beta(t, X^{(\alpha^\star)}(t-), \alpha^\star(t)) \\
&\quad + \frac{1}{2} v_{xxx}(t, X^{(\alpha^\star)}(t-))\gamma^2(t, X^{(\alpha^\star)}(t-), \alpha^\star(t))c(t) \\
&\quad + \int \Big(v_x\big(t, X^{(\alpha^\star)}(t-) + \delta(t, X^{(\alpha^\star)}(t-), \alpha^\star(t), \widetilde{x})\big) \\
&\quad - v_x(t, X^{(\alpha^\star)}(t-)) - v_{xx}(t, X^{(\alpha^\star)}(t-))\delta(t, X^{(\alpha^\star)}(t-), \alpha^\star(t), \widetilde{x}) \Big) K(t, d\widetilde{x}) \Big) dt \\
&\quad + v_{xx}(t, X^{(\alpha^\star)}(t-))\gamma(t, X^{(\alpha^\star)}(t-), \alpha^\star(t)) dL^c(t) \\
&\quad + \Big(v_x\big(t, X^{(\alpha^\star)}(t-) + \delta(t, X^{(\alpha^\star)}(t-), \alpha^\star(t), \widetilde{x})\big) \\
&\quad - v_x(t, X^{(\alpha^\star)}(t-)) \Big) * (\mu^L - \nu^L)(dt, d\widetilde{x}) \\
&= -\Big(f_x(t, X^{(\alpha^\star)}(t-), \alpha^\star(t)) - v_x(t, X^{(\alpha^\star)}(t-))\beta_x(t, X^{(\alpha^\star)}(t-), \alpha^\star(t)) \\
&\quad - v_{xx}(t, X^{(\alpha^\star)}(t-))\gamma(t, X^{(\alpha^\star)}(t-), \alpha^\star(t))\gamma_x(t, X^{(\alpha^\star)}(t-), \alpha^\star(t))c(t) \\
&\quad - \int \Big(v_x\big(t, X^{(\alpha^\star)}(t-) + \delta(t, X^{(\alpha^\star)}(t-), \alpha^\star(t), \widetilde{x})\big) - v_x(t, X^{(\alpha^\star)}(t-)) \Big) \\
&\quad \times \delta_x(t, X^{(\alpha^\star)}(t-), \alpha^\star(t), \widetilde{x}) K(t, d\widetilde{x}) \Big) dt \\
&\quad + \gamma^Y(t) dL^c(t) + \delta^Y(t) * (\mu^L - \nu^L)(dt, d\widetilde{x}) \\
&= -\widetilde{H}_x(t, X^{(\alpha^\star)}(t-), \alpha^\star(t), Y(t-), \gamma^Y(t), \delta^Y(t)) dt \\
&\quad + \gamma^Y(t) dL^c(t) + \delta^Y(t) * (\mu^L - \nu^L)(dt, d\widetilde{x})
\end{aligned}$$

as desired. Moreover, observe that

$$\begin{aligned}
&\widetilde{H}_a(t, X^{(\alpha^\star)}(t-), \alpha^\star(t), Y(t-), \gamma^Y(t), \delta^Y(t)) \\
&\quad = Y(t)\beta_a(X^{(\alpha^\star)}(t-), \alpha^\star(t)) + \gamma^Y(t)c(t)\gamma_a(t, X^{(\alpha^\star)}(t-), \alpha^\star(t)) \\
&\qquad + \int \delta^Y(t, \widetilde{x})\delta_x(t, X^{(\alpha^\star)}(t-), \alpha^\star(t), \widetilde{x}) K(t, d\widetilde{x}) + f_a(t, X^{(\alpha^\star)}(t-), \alpha^\star(t)) \\
&\quad = 0
\end{aligned}$$

by assumption 4 because $a^\star(t, X^{(\alpha^\star)}(t-)) = \alpha^\star(t)$. Since $a \mapsto \widetilde{H}(t, X^{(\alpha^\star)}(t-), a, Y(t-), \gamma^Y(t), \delta^Y(t))$ is concave, it follows that its maximum is attained at $\alpha^\star(t)$.

Appendix 1: Problems

The exercises correspond to the section with the same number.

7.1 In the situation of Example 7.5 suppose that the goal is to maximise the expected utility of consumption $E(\log(c) \bullet I(T))$ subject to the affordability constraint that the investor's aggregate consumption cannot exceed her cumulative profits, i.e. $0 \leq V_{\varphi,c}(t) := v_0 + \varphi \bullet S(t) - c \bullet I(t)$, cf. the discrete-time analogue in Example 1.49. The investor's set of admissible controls is $\mathscr{A} := \{(\varphi, c)$ predictable: $\varphi \in L(S)$, $c \in L(I)$, $(\varphi, c)(0) = 0$, $V_{\varphi,c} > 0$, and $\log V_{\varphi,c}$ as well as $\log(c) \bullet I$ are of class (D)$\}$. Determine the optimal control $(\varphi^\star, c^\star)$ using the ansatz of Examples 7.5 and 1.49.

7.2 Let the price of a stock be modelled by a geometric Brownian motion $S = S(0)\mathscr{E}(W)$ with a Wiener process W. We consider a perpetual American option on the stock with payoff $X(t) = e^{-\varrho t}(S(t) - 1)$ at time $t > 0$, where $\varrho > 0$ is a given discount factor. Determine the stopping time τ^* maximising the expected payoff $E(X(\tau))$ over all stopping times τ.

Hint: Consider the function

$$f(t, s) = e^{-\varrho t}(s_0 - 1)\left(\frac{s}{s_0}\right)^\gamma$$

with

$$\gamma = \frac{1}{2} + \sqrt{\frac{1}{4} + 2\varrho}, \quad s_0 = \frac{\gamma}{\gamma - 1},$$

the process $M(t) = f(t, S(t))$, and Proposition 7.9. You may use without proof—or verify yourself—that Proposition 7.9 is applicable in this infinite-horizon problem, where we set $X(\infty) = 0$.

7.3 Assume that an investor with initial capital $v_0 > 0$ trades in a market consisting of a constant bank account and a stock with price $S(t) = S(0)\exp(L(t))$ for some Lévy process L. The investor has to pay a holding fee with rate $\lambda\varphi(t)^2$, where $\varphi(t)$ denotes the number of shares of stock in the portfolio at time t and $\lambda > 0$ is a constant. Determine the trading strategy φ^* that maximises the expected terminal wealth

$$E\left(v_0 + \varphi \bullet S(T) - \lambda \int_0^T \varphi(t)^2 dt\right)$$

over all $\varphi \in L(S)$ with $\varphi(0) = 0$ and such $\varphi \bullet S$ and $\varphi^2 \bullet I$ are of class (D). Use dynamic programming.

7.4 Solve Problem 7.3 by applying the stochastic maximum principle.

Appendix 2: Notes and Comments

Textbook treatments of stochastic optimal control include [112, 226, 227, 234, 271, 287, 295]. We follow the approach of [96] of starting from a non-Markovian point of view. As an authoritative reference to optimal stopping one may consult [232].

The solution to the portfolio problems in Examples 7.5, 7.17 goes back to [217], cf. [126] for references on the general semimartingale version of Example 7.5. We refer to [233, 270] for an overview of the quadratic hedging literature, cf. also Chap. 12. The perpetual American put is treated in [219]. Proposition 7.14 is based on [135, 249]. The American butterfly in the Bachelier model is discussed in [206, Example 2.31]. The stochastic maximum principle is typically stated in terms of BSDEs, cf. [234, 271, 287, 295] for Itō processes and [227] for processes with jumps. For a thorough discussion of the mutual fund theorem we refer to [218, 260].

Part II
Mathematical Finance

Overview and Notation

This second part of this monograph concerns mathematical models for markets of exchange-traded assets. The goal is to address questions such as valuation of derivative securities, hedging or more generally portfolio selection, and assessment of risks.

Chapter 8 covers some families of equity models as examples of mathematically tractable models which take stylised features of real data into account. The subsequent chapter introduces basic concepts of financial mathematics for traded assets, in particular the fundamental notion of *arbitrage*. Its consequences on valuation and hedging of derivatives are discussed in Chap. 11, regardless of any specific model.

Advanced models involving processes with jumps are typically *incomplete*, i.e. perfect hedging strategies do not exist and superhedging strategies are forbiddingly expensive. As a way out, Chap. 12 provides an approach how to hedge as efficiently as possible in the sense that the variance of the portfolio is minimised. The alternative and at the same time related concept of utility-based hedging and pricing is discussed in Chap. 13. The latter relies on portfolio optimisation, which is discussed in Chap. 10 because it is interesting in its own right.

Chapter 14 turns to arbitrage-free models for fixed-income markets. It is explained how major approaches to interest rate markets, namely the short-rate, Heath–Jarrow–Morton, forward process and rational market models extend to discontinuous processes. As in Chap. 8 the focus is on tractable and at the same time flexible families.

We generally use the notation of Part I. In most chapters the notion of semimartingale characteristics is used. Given an $\mathbb{R}^d$-valued semimartingale X, we generally denote its local characteristics relative to the "truncation" function $h(x) = x$ by (a^X, c^X, K^X) and its modified second characteristics by $\widetilde{c}^X$, cf. Definition 4.3, (2.9), and (4.31). This implies secretly that we assume X to be a special semimartingale whose characteristics are absolutely continuous, cf. the end of Sect. 4.1 and note that $b^{X,\mathrm{id}} = a^X$.

Given an $\mathbb{R}^{d+n}$-valued semimartingale (X, Y) we write $c^{X,Y}$ and $\widetilde{c}^{X,Y}$ for the $\mathbb{R}^{d\times n}$-valued predictable processes satisfying

$$c_{ij}^{X,Y} \bullet I = \langle X_i^c, Y_j^c \rangle, \quad i = 1, \ldots, d, \quad j = 1, \ldots, n$$

and

$$\widetilde{c}_{ij}^{X,Y} \bullet I = \langle X_i, Y_j \rangle, \quad i = 1, \ldots, d, \quad j = 1, \ldots, n,$$

in line with e.g. (3.7). This implies that the local characteristics of (X, Y) are of the form

$$\left(a^{(X,Y)}, c^{(X,Y)}, K^{(X,Y)}\right) = \left(\begin{pmatrix} a^X \\ a^Y \end{pmatrix}, \begin{pmatrix} c^X & c^{X,Y} \\ c^{Y,X} & c^Y \end{pmatrix}, K^{(X,Y)}\right) \tag{II.1}$$

and its modified second characteristics can be written as

$$\widetilde{c}^{(X,Y)} = \begin{pmatrix} \widetilde{c}^X & \widetilde{c}^{X,Y} \\ \widetilde{c}^{Y,X} & \widetilde{c}^Y \end{pmatrix}. \tag{II.2}$$

Finally, we write c^{-1} for the *Moore–Penrose pseudoinverse* of a matrix or matrix-valued process c, which is a particular matrix satisfying $cc^{-1}c = c$. It allows us to use the formulas of Part II in cases where the corresponding matrices fail to be invertible. One can show that c^{-1} is non-negative and symmetric if this holds for c.

As a side remark, recall that $K^X = 0$ and $\widetilde{c}^X = c^X$ for continuous processes X. Moreover, $b^{X,h}$ does not depend on the truncation function h and coincides with the growth rate a^X in this case. Finally, both the covariation and the predictable covariation are of the form $[X_i, X_j] = \langle X_i, X_j \rangle = c_{ij}^X \bullet I$ for continuous X.

Chapter 8
Equity Models

The theory of Mathematical Finance in the following chapters rests on price processes of assets such as stocks, currencies, bonds, and commodities. They are often assumed to be given exogenously. Therefore we start by discussing what kinds of processes are suggested by statistical properties of real data. Specifically, we consider two families of stochastic models for equity data, namely geometric Lévy processes in Sect. 8.1 and the more general class of affine stochastic volatility models in Sect. 8.2. More involved objects such as interest rates are considered in Chap. 14 of this monograph.

8.1 Geometric Lévy Processes

In spite of its weaknesses, geometric Brownian motion still constitutes a benchmark for modelling financial data such as equity and currency prices. It is the only possible choice if we make two seemingly weak assumptions on the behaviour of prices, namely continuity of price paths and constant relative growth in a stochastic sense as explained in Sect. 3.7. Surprisingly enough, this leaves only a three parameter set of processes in the univariate case, namely $S(t) = S(0)\exp(\mu t + \sigma W(t))$ with a standard Brownian motion W and parameters $S(0), \mu, \sigma$. At first glance, this extremely simple model produces price paths that look similar to real data, cf. Fig. 8.1.

Closer examination reveals that the fit is not entirely convincing. If we compare daily log returns of real prices with increments of geometric Brownian motion (cf. Fig. 8.2), we observe typical differences, which have been termed *stylised facts*. Firstly, large absolute returns occur much more often than in a simulation of geometric Brownian motion with comparable mean and variance. Secondly, real data lacks the homogeneity of iid returns; specifically, volatility changes over time. Thirdly, equity—but not currency—returns tend to be skewed to the left and negatively correlated with changes in volatility.

© Springer Nature Switzerland AG 2019

E. Eberlein, J. Kallsen, *Mathematical Finance*, Springer Finance,
https://doi.org/10.1007/978-3-030-26106-1_8

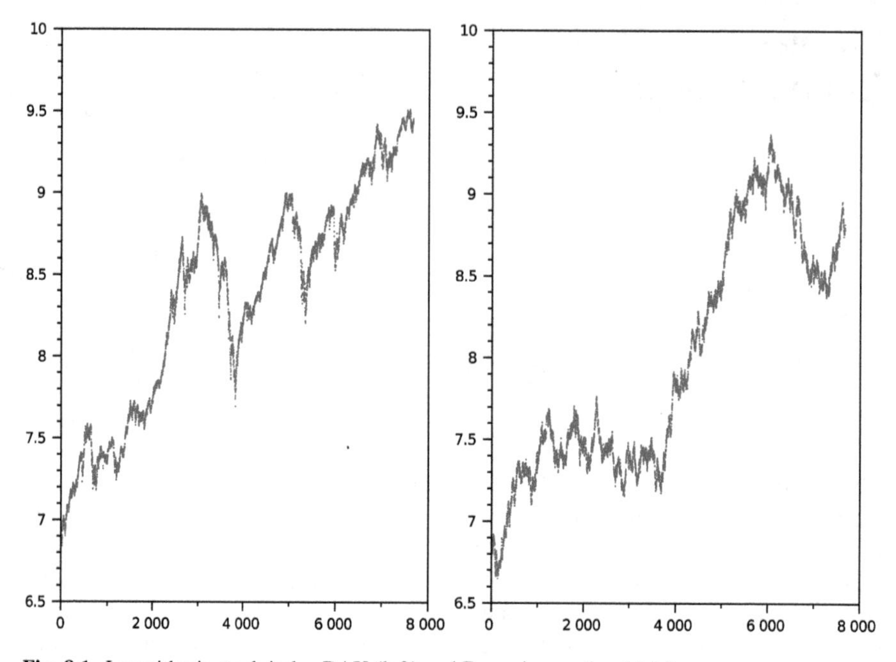

Fig. 8.1 Logarithmic stock index DAX (left) and Brownian motion (right)

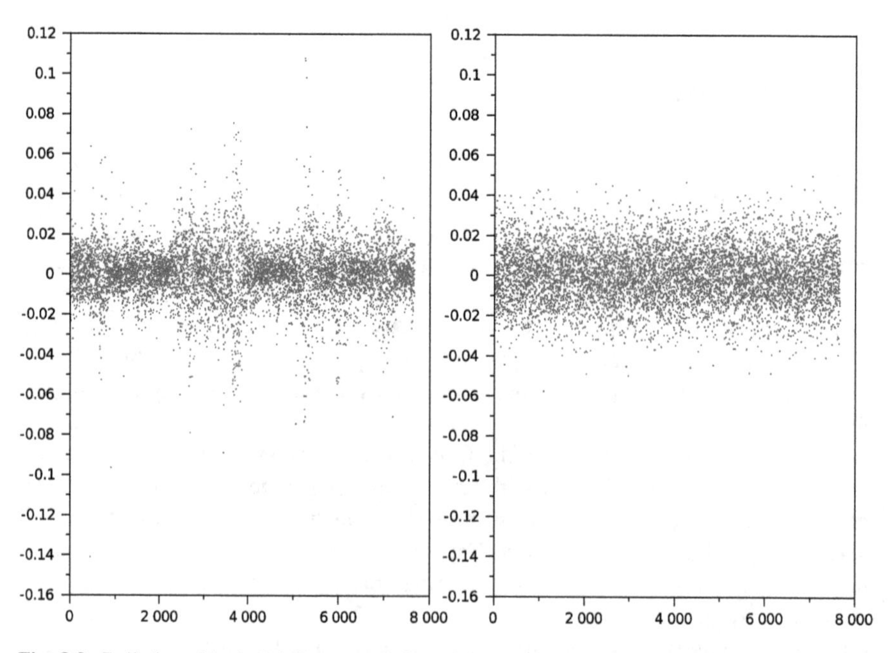

Fig. 8.2 Daily logarithmic DAX returns (left) and increments of Brownian motion (right)

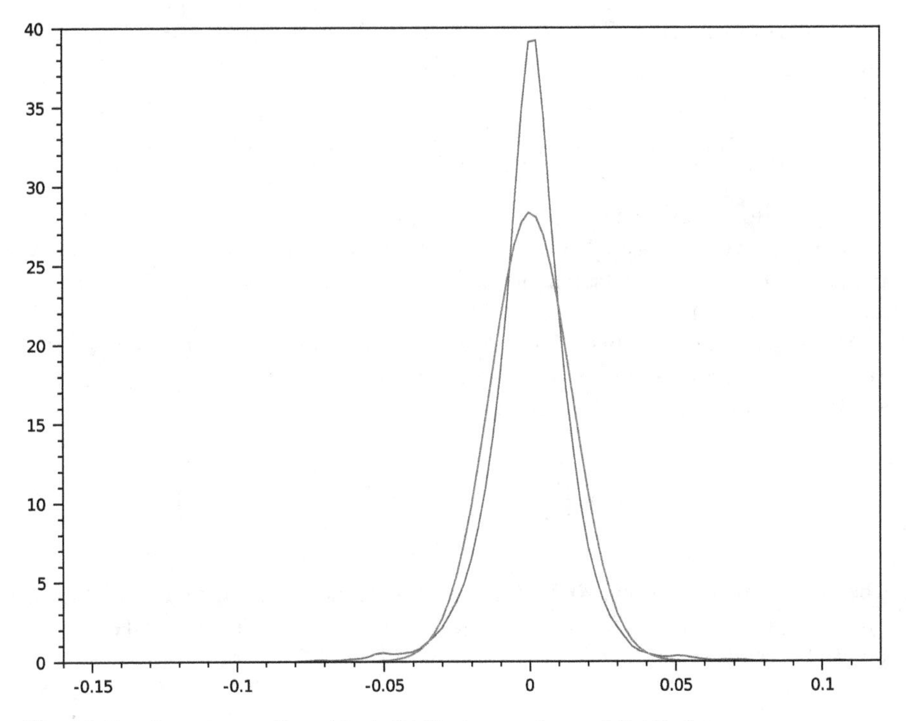

Fig. 8.3 Density estimate of logarithmic DAX returns and normal distribution

A density estimate for logarithmic stock index returns is shown in Fig. 8.3 together with the density of a normal distribution with the same mean and variance. The decay in the tails is much slower for the data than for the normal distribution. On the other hand, the empirical law is typically less fat tailed than stable distributions, which motivates the expression *heavy tails* in the context of financial data. A closer look reveals that the negative tail tends to be heavier than the positive for equity data.

Geometric Lévy processes provide a relatively simple and natural means to account for the heavy tails of financial data. For moderate time horizons they may serve as a reasonable approximation to real data even if they do not allow for time-varying volatility. The latter becomes increasingly pronounced and important on larger time scales, cf. Sect. 8.2.

8.1.1 Estimation

Various classes of Lévy processes are discussed in Sect. 2.4. An important question is how to estimate the involved parameters based on an observed time series of return data. Let us assume that the price process of a stock is of the form

$$S(t) = S(0)e^{X(t)} \tag{8.1}$$

with some Lévy process X. Moreover, suppose that prices $S(n\Delta)$, $n = 0, \dots, N$ on an equally spaced time grid are observed. The logarithmic increments

$$x_n := \log \frac{S(n\Delta)}{S((n-1)\Delta)}, \quad n = 1, \dots, N$$

correspond to a realisation of $X(n\Delta) - X((n-1)\Delta)$, $n = 1, \dots, N$. These are iid random variables with law $\mathscr{L}(X(\Delta))$ because X is assumed to be a Lévy process. For such iid data the maximum likelihood estimator has typically desirable asymptotic properties.

Suppose that the density of $\mathscr{L}(X(\Delta))$ is known up to some unknown parameter vector ϑ, say of the form $x \mapsto \varrho(x, \vartheta)$. By independence, the density of a vector of N observations $x_1, \dots, x_N$ is given by

$$(x_1, \dots, x_N) \mapsto \prod_{n=1}^{N} \varrho(x_j, \vartheta) = \exp\left(\sum_{n=1}^{N} \log \varrho(x_j, \vartheta)\right).$$

The **maximum likelihood (ML) estimator** $\widehat{\vartheta}$ is the parameter vector ϑ maximising this expression given the data $(x_1, \dots, x_N)$. Provided that sufficient regularity in ϑ holds, a necessary condition is

$$0 = \sum_{n=1}^{N} \left.\frac{d \log \varrho(x_j, \vartheta)}{d\vartheta}\right|_{\vartheta=\widehat{\vartheta}}.$$

For some families closed-form expressions of the density are not available. A simple way out is to consider moment estimators if the characteristic function is known explicitly, which is the case for most classes in Sect. 2.4. By the strong law of large numbers, *empirical moments* $\frac{1}{N}\sum_{n=1}^{N} f(x_n)$ converge to $E(f(X(\Delta)))$ as $n \to \infty$, regardless of the function f. **Moment estimators** lead to an estimate $\widehat{\vartheta}$ for the parameter vector ϑ by equating empirical and theoretical moments, i.e.

$$\frac{1}{N}\sum_{n=1}^{N} f(x_n) = E_{\widehat{\vartheta}}(f(X(\Delta))),$$

where the left-hand side depends on the data and the right on the parameter vector. Here, the notation $E_\vartheta(\cdot)$ stands for the expectation provided that ϑ is the true parameter vector. For parametric families such as VG, NIG with four parameters, the first four moments $f(x) = x^k$, $k = 1, 2, 3, 4$ are typically chosen for estimation, i.e. $\widehat{\vartheta}$ solves

$$\frac{1}{N}\sum_{n=1}^{N} x_n^k = E_{\widehat{\vartheta}}\left(X(\Delta)^k\right), \quad k = 1, 2, 3, 4.$$

In general, the moments should be chosen such that they correspond to the most important aspects. For example, for a symmetric three-parametric family, we should not consider skewness but rather concentrate on $k = 1, 2, 4$.

Although moment estimators are theoretically less efficient than maximum likelihood estimation, they may be very useful in practice even if the likelihood is known in closed form. Firstly, numerical routines for ML estimates tend to be stuck in suboptimal local maxima. As a way out, one may use moment estimates as reasonable starting values for algorithms maximising the likelihood. An advantage of moment estimators is their relative robustness under model misspecification. If the data is not generated from the model under consideration, they still provide a model whose moments are close to the data. This may not be the case for ML estimates.

8.1.2 Multivariate Models and Lévy Copulas

The general model (8.1) allows for a straightforward extension to multiple assets by working with multivariate geometric Lévy processes as in Sect. 3.7. Specifically, we consider price processes

$$S_i(t) = S_i(0)e^{X_i(t)}, \quad i = 1, \dots, d$$

for d stocks, where $X = (X_1, \dots, X_d)$ denotes an $\mathbb{R}^d$-valued Lévy process.

However, compared to the univariate case fewer parametric families have been considered in the literature. Even though some classes of Lévy processes in Sect. 2.4 allow for natural multivariate extensions, they do not offer the same kind of flexibility as in the univariate case. Since the vector-valued versions of VG, NIG and GH, for example, rest intrinsically on Brownian motion, they do not allow for the often observed *tail dependence* of asset return data. This means that joint extreme events of the components occur only with relatively small probability—even if the correlation between the marginal Lévy processes is large. Moreover, some of the parameters in the multivariate versions of VG, NIG, and GH must coincide for all components, i.e. two VG models for two single stocks are compatible with a joint VG model for both stocks only under restrictive assumptions on the coefficients.

Alternatively, one can choose the marginal laws and the dependence structure of a multivariate Lévy process separately. The key role in this context is played by *Lévy copulas*. This notion is inspired by copulas for random vectors, which also allow us to model the dependence structure independently of the marginal laws. The starting point is the Lévy–Khintchine triplet (a^X, c^X, K^X) of an $\mathbb{R}^d$-valued Lévy process $X = (X^1, \dots, X^d)$. Our aim is to describe this triplet and hence the law of X uniquely by the marginal Lévy–Khintchine triplets $(a^{X_i}, c^{X_i}, K^{X_i})$ of the components $X_i, i = 1, \dots, d$ and some additional information on the dependence structure. Here we obviously have $K^{X_i}(B) = K^X(\{x \in \mathbb{R}^d : x_i \in B \setminus \{0\})$.

The drift vector a^X is entirely determined by the drift coefficients a^{X_i} of the components X_i. The second characteristic c^X is just the covariance matrix of the Brownian motion part of X. It is uniquely specified by the variances $c^{X_i}, i = 1, \dots, d$ and the correlation matrix $\varrho = (\varrho_{ij})_{i,j=1,\dots,d}$, which is defined as

$$\varrho_{ij} := \begin{cases} \frac{c_{ij}^X}{\sqrt{c_{ii}^X c_{jj}^X}} & \text{if } c_{ii}^X, c_{jj}^X \neq 0, \\ 0 & \text{if } c_{ii}^X = 0 \text{ or } c_{jj}^X = 0. \end{cases}$$

ϱ is a symmetric nonnegative matrix with unit diagonal elements. Observe that any variances $c^{X_i} \geq 0, i = 1, \dots, d$ and any given symmetric nonnegative matrix ϱ with unit diagonal elements can be combined to yield a symmetric nonnegative matrix c^X via

$$c_{ij}^X := \sqrt{c_{ii}^X c_{jj}^X} \varrho_{ij}.$$

Consequently, we are left to specify the joint Lévy measure K^X given its marginal Lévy measures K_i^X. The dependence structure of probability measures on $\mathbb{R}^d$ can be described in terms of *copulas*, i.e. cumulative distribution functions for laws with uniform marginals on the unit square. A similar construction exists for Lévy measures as well. It is slightly complicated by the fact that Lévy measures are generally infinite with a singularity at the origin.

In order to define *Lévy copulas* we need some technical tools. Let us write $\overline{\mathbb{R}} := \mathbb{R} \cup \{\infty\}$ and

$$\operatorname{sgn}(x) := \begin{cases} 1 & \text{for } x \geq 0 \\ -1 & \text{for } x < 0 \end{cases}$$

in this section. Finite or more generally σ-finite measures μ on $\overline{\mathbb{R}}^d$ are uniquely determined by the volume $\mu((a, b])$ of intervals $(a, b] = (a_1, b_1] \times \cdots \times (a_d, b_d]$ with $a_i \leq b_i, i = 1, \dots, d$. Due to the σ-additivity of μ, this volume can in turn be written as a difference

$$\mu((a, b]) = F(b) - F(a) \tag{8.2}$$

with some function F which is unique up to some additive constant. For probability measures on the real line, F coincides with the cumulative distribution function. In order for a general function $F : \overline{\mathbb{R}}^d \to \mathbb{R}$ to correspond to a measure via (8.2), it must be increasing in some proper sense in order for $\mu((a, b])$ to be nonnegative.

Specifically, a function $F : \overline{\mathbb{R}}^d \to \overline{\mathbb{R}}$ is called d-**increasing** if $V_F((a,b]) \geq 0$ for all $a, b \in \overline{\mathbb{R}}^d$, where the F-**volume** of $(a,b]$ is defined as

$$V_F((a,b]) := \sum_{u \in \{a_1,b_1\} \times \cdots \times \{a_d,b_d\}} (-1)^{N(u)} F(u)$$

with $N(u) := |\{k \in \{1,\dots,d\} : u_k = a_k\}|$. A d-increasing function F corresponds to a finite measure on $\overline{\mathbb{R}}^d$ similarly as the cumulative distribution function corresponds to a probability measure on $\mathbb{R}$.

Given a measure μ on $\mathbb{R}^d$, its marginals μ^I on $\overline{\mathbb{R}}^{|I|}$ for subsets $I = \{i_1,\dots,i_n\} \subset \{1,\dots,d\}$ can be defined via

$$\mu^I(B_1 \times \cdots \times B_n) := \mu\Big(\big\{x \in \overline{\mathbb{R}}^d : x_{i_1} \in B_1, \dots, x_{i_n} \in B_n\big\}\Big)$$

for $B_1,\dots,B_n \in \mathscr{B}(\overline{\mathbb{R}})$, in particular

$$\mu^i(B) := \mu^{\{i\}}(B) := \mu\Big(\big\{x \in \overline{\mathbb{R}}^d : x_i \in B\big\}\Big)$$

for $B \in \mathscr{B}(\overline{\mathbb{R}})$. If μ corresponds to the d-increasing function F on $\overline{\mathbb{R}}^d$, what $|I|$-increasing function F^I corresponds to its marginal μ^I on $\mathbb{R}^I$?

To this end, let $F : \overline{\mathbb{R}}^d \to \overline{\mathbb{R}}$ be a d-increasing function which is **grounded**, i.e. $F(u_1,\dots,u_d) = 0$ if $u_i = 0$ for at least one $i \in \{1,\dots,d\}$. For any non-empty index set $I \subset \{1,\dots,d\}$, the I-**margin** $F^I : \overline{\mathbb{R}}^{|I|} \to \overline{\mathbb{R}}$ of F is defined by

$$F^I((u_i)_{i\in I}) := \lim_{a\to\infty} \sum_{(u_i)_{i\in I^C} \in \{-a,\infty\}^{|I^C|}} F(u_1,\dots,u_d) \prod_{i\in I^C} \operatorname{sgn}(u_i),$$

where $I^C := \{1,\dots,d\} \setminus I$.

It is easy to see that this F^I corresponds to μ^I if F belongs to μ via (8.2). Given these notions, we can define Lévy copulas, which are the d-increasing grounded functions corresponding via (8.2) to measures on $\overline{\mathbb{R}}^d$ with uniform marginals.

Specifically, a function $F : \overline{\mathbb{R}}^d \to \overline{\mathbb{R}}$ is called a **Lévy copula** if

1. $F(u_1,\dots,u_d) \neq \infty$ for $(u_1,\dots,u_d) \neq (\infty,\dots,\infty)$,
2. $F(u_1,\dots,u_d) = 0$ if $u_i = 0$ for at least one $i \in \{1,\dots,d\}$,
3. F is d-increasing,
4. $F^{\{i\}}(u) = u$ for any $i \in \{1,\dots,d\}$, $u \in \mathbb{R}$.

The next step is to relate Lévy measures and Lévy copulas. For Lévy measures K^X of an $\mathbb{R}^d$-valued Lévy process X the quantity $K^X((a,b])$ may be infinite for $0 \in \overline{(a,b]}$ because of a singularity at the origin. Consequently, defining a grounded d-increasing definition via $F(b) := K^X((0,b])$ does not make sense for general

Lévy measures. A way out is to consider the **tail integral** of X. This is the function $U : (\mathbb{R} \setminus \{0\})^d \to \mathbb{R}$ defined by

$$U(x_1, \ldots, x_d) := \prod_{i=1}^{d} \operatorname{sgn}(x_i) K^X\left(\prod_{j=1}^{d} \mathscr{I}(x_j)\right).$$

Here we use the notation

$$\mathscr{I}(x) := \begin{cases} (x, \infty) & \text{for } x \geq 0, \\ (-\infty, x] & \text{for } x < 0. \end{cases}$$

Moreover, for non-empty $I \subset \{1, \ldots, d\}$ the I**-marginal tail integral** U^I of X is the tail integral of the process $X^I := (X^i)_{i \in I}$. To simplify the notation, we denote one-dimensional margins by $U_i := U^{\{i\}}$.

There is a one-to-one correspondence between Lévy measures and their marginal tail integrals.

Proposition 8.1 *Let X be an $\mathbb{R}^d$-valued Lévy process. Its marginal tail integrals $\{U^I : I \subset \{1, \ldots, d\}$ non-empty$\}$ are uniquely determined by its Lévy measure K^X. Conversely, its Lévy measure is uniquely determined by the set of its marginal tail integrals.*

Idea $\Rightarrow$: By Propositions 4.6 and 4.13, the Lévy measure of X^I can be expressed in terms of K^X, which in turn implies that U^I is determined by K^X.

$\Leftarrow$: It suffices to show that $K^X((a, b])$ is completely specified by the tail integrals for any $a, b \in \mathbb{R}^d$ with $a \leq b$ and $0 \notin (a, b]$. We prove by induction on $k = 0, \ldots, d$ that $(K^X)^I(\prod_{i \in I}(a_i, b_i])$ is determined by the tail integrals for any $a, b \in \mathbb{R}^d$ such that $a \leq b$ and $a_i b_i \leq 0$ for at most k indices and any non-empty $I \subset \{1, \ldots, d\}$ with $0 \notin \prod_{i \in I}(a_i, b_i]$.

If $k = 0$, the above definitions yield that

$$(K^X)^I\left(\prod_{i \in I}(a_i, b_i]\right) = (-1)^{|I|} V_{U^I}\left(\prod_{i \in I}(a_i, b_i]\right).$$

Let $a, b \in \mathbb{R}^d$ such that $a_i b_i \leq 0$ for at most k indices. For ease of notation we suppose that $a_i b_i \leq 0$ for $i = 1, \ldots, k$. Let $I \subset \{1, \ldots, d\}$ be non-empty with $0 \notin \prod_{i \in I}(a_i, b_i]$. By induction hypothesis, $(K^X)^I(\prod_{i \in I}(a_i, b_i])$ is uniquely determined if $k \notin I$. Suppose that $k \in I$. If $a_k = 0$, then

$$(K^X)^I\left(\prod_{i \in I}(a_i, b_i]\right) = \lim_{\alpha \downarrow 0}(K^X)^I\left(\prod_{i \in I, i < k}(a_i, b_i] \times (\alpha, b_k] \times \prod_{i \in I, i > k}(a_i, b_i]\right)$$

and the right-hand side is uniquely specified by the induction hypothesis. If $a_k \neq 0$, then

$$
\begin{aligned}
(K^X)^I\left(\prod_{i\in I}(a_i,b_i]\right) &= (K^X)^{I\setminus\{k\}}\left(\prod_{i\in I\setminus\{k\}}(a_i,b_i]\right) \\
&\quad - \lim_{\beta\downarrow b_k;c\uparrow\infty}(K^X)^I\left(\prod_{i\in I,i<k}(a_i,b_i]\times(\beta,c]\times\prod_{i\in I,i>k}(a_i,b_i]\right) \\
&\quad - \lim_{c\downarrow-\infty}(K^X)^I\left(\prod_{i\in I,i<k}(a_i,b_i]\times(c,a_k]\times\prod_{i\in I,i>k}(a_i,b_i]\right),
\end{aligned}
$$

which is uniquely determined as well. □

We are now ready to state the main result in this context, namely *Sklar's theorem for Lévy copulas*. It states that Lévy measures on $\mathbb{R}^d$ are generally obtained by combining d arbitrary Lévy measures on $\mathbb{R}$ with an arbitrary Lévy copula on $\mathbb{R}^d$.

Theorem 8.2 (Sklar's Theorem for Lévy Copulas)

1. *Let $X = (X^1, \ldots, X^d)$ be an $\mathbb{R}^d$-valued Lévy process. Then there exists a Lévy copula F such that the tail integrals of X satisfy:*

$$U^I((x_i)_{i\in I}) = F^I((U_i(x_i))_{i\in I}) \tag{8.3}$$

   for any non-empty $I \subset \{1, \ldots, d\}$ and any $(x_i)_{i\in I} \in (\mathbb{R}\setminus\{0\})^{|I|}$. The Lévy copula F is unique on $\prod_{i=1}^d \overline{\text{Ran } U_i}$, where Ran U_i *denotes the range of U_i and the bar its closure.*
2. *Let F be a d-dimensional Lévy copula and $U_i, i = 1, \ldots, d$ be tail integrals of real-valued Lévy processes. Then there exists an $\mathbb{R}^d$-valued Lévy process X whose components have tail integrals $U_1, \ldots, U_d$ and whose marginal tail integrals satisfy equation (8.3) for any non-empty $I \subset \{1, \ldots, d\}$ and any $(x_i)_{i\in I} \in (\mathbb{R} \setminus \{0\})^{|I|}$. The Lévy measure K^X of X is uniquely determined by F and $U_i, i = 1, \ldots, d$.*

Idea

1. If the one-dimensional marginal Lévy measures are infinite and have no atoms, we have Ran $U_i = (-\infty, 0)\cup(0, \infty)$ for every i and one can construct F directly via $F((x_i)_{i=1}^d) = U((U_i^{-1}(x_i))_{i=1}^d)$. It is straightforward to verify that F has the desired properties in this case. For the general case see [183, Theorem 3.6].
2. A full proof is again to be found in [183, Theorem 3.6]. The idea is to define a measure μ on $\overline{\mathbb{R}}^d \setminus \{\infty, \ldots, \infty\}$ by $\mu((a, b]) := V_F((a, b])$ for any $a, b \in \overline{\mathbb{R}}^d \setminus \{\infty, \ldots, \infty\}$ with $a \le b$. Moreover, let

$$U_i^{-1}(u) := \begin{cases} \inf\{x > 0 : u \ge U_i(x)\} & \text{if } u \in [0, \infty], \\ \inf\{x < 0 : u \ge U_i(x)\} \wedge 0 & \text{if } u \in (-\infty, 0] \end{cases}$$

and denote by ν' the push-forward measure of μ under

$$f : (u_1, \dots, u_d) \mapsto (U_1^{-1}(u_1), \dots, U_d^{-1}(u_d)).$$

The final step is to show that the restriction ν' of ν to $\mathbb{R}^d \setminus \{0\}$ is a Lévy measure whose tail integrals U_ν^I are of the form

$$U_\nu^I((x_i)_{i\in I}) = F^I((U_i(x_i))_{i\in I}).$$

□

In principle, parametric families of multivariate Lévy processes can now be constructed by combining parametric families of univariate Lévy processes with a parametric family of Lévy copulas. The latter should be chosen parsimoniously to allow for efficient estimation without losing the ability to capture the properties of real data. The challenge to construct and determine such families of Lévy copulas still requires research in the area.

8.2 Stochastic Volatility Models

Processes with constant growth cannot account for the clustering of volatility that is present in real data. The latter is demonstrated by the behaviour of the sample autocorrelation function of squared log returns in Fig. 8.4. For geometric Brownian motion and in fact most geometric Lévy processes one would rather expect it to look as in Fig. 8.5. Models with time-varying volatility are more consistent with real data in this respect. Moreover, they lead to the heteroskedasticity (heavy tails) discussed above even if jumps are absent.

Coming from the geometric Lévy process (8.1), two different paths lead to models with time-varying volatility. A first approach is to choose the *return process*

$$X(t) := \log \frac{S(t)}{S(0)}$$

as a stochastic integral

$$X(t) = \sigma \bullet L(t) \tag{8.4}$$

of a Lévy process L. The volatility process $\sigma(t)$ leads to increments whose size changes over time. Alternatively, we consider time-changed Lévy processes

$$X(t) = L(Y(t)) \tag{8.5}$$

with an increasing process

$$Y(t) = \int_0^t y(s)ds. \tag{8.6}$$

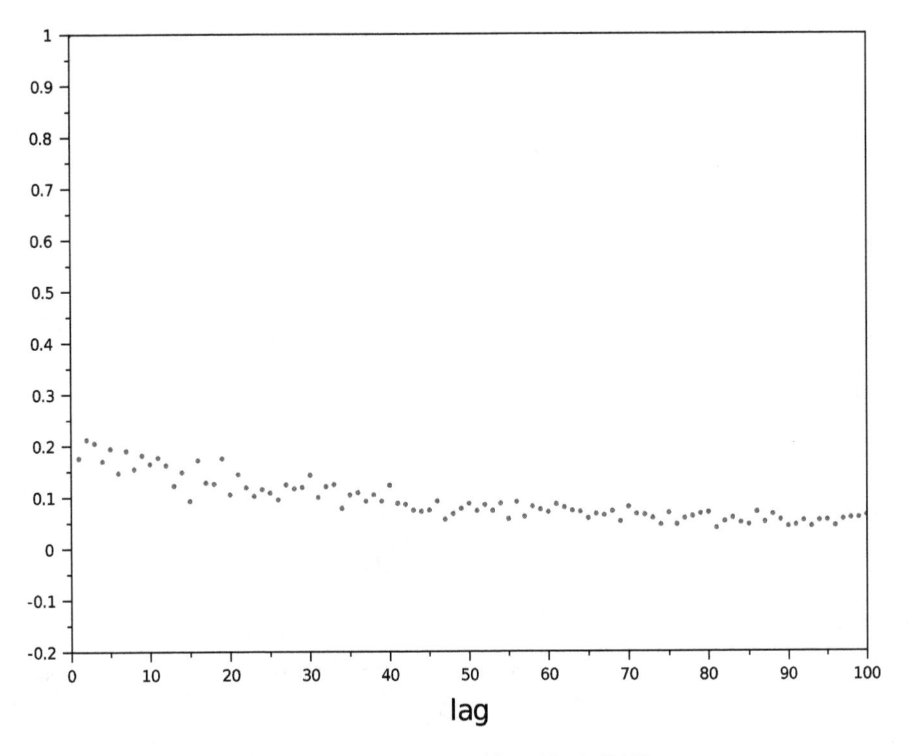

Fig. 8.4 Sample autocorrelation function of squared logarithmic DAX returns

The latter can be interpreted as a kind of operational time, which may, for example, correspond to the cumulative trading volume. In the context of (8.5), prices move homogeneously in operational time but not in calendar time due to randomly changing trading activity y. In contrast to (8.4), it is the speed rather than the size of price changes that varies over time.

For Brownian motion or more generally stable Lévy motions, both approaches lead to essentially the same class of models. This is due to the self-similarity of these Lévy processes, which implies that speeding up and rescaling the process have the same effect on its law. For general Lévy processes, however, the two approaches lead to different classes of models.

Below we discuss a number of models featuring an affine structure in the sense of Chap. 6. This leads to an explicit representation of the characteristic function which proves to be very useful for many purposes such as estimation, calibration, option pricing, and hedging. For ease of exposition we consider single assets but the models allow for an extension to multiple securities as in the case of Lévy processes. In all cases, we assume the filtration to be generated by the multivariate affine process under consideration. Moreover, we choose the "truncation" function $h(x) = x$ for all triplets in the following subsections, which is indicated by using the letter a instead of b^h and leads to less involved formulas. Using the conversion rules (2.21) resp. (4.13) one easily obtains the representation for arbitrary true truncation

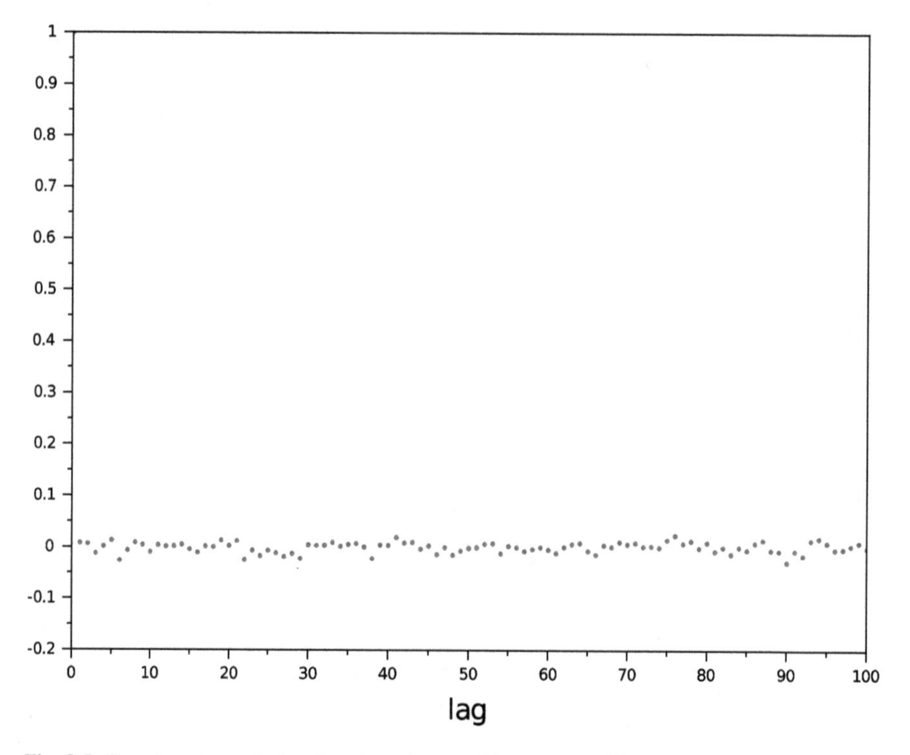

Fig. 8.5 Sample autocorrelation function of squared increments of Brownian motion

functions. The resulting formulas hold even if the expressions for $h = \mathrm{id}$ do not make sense, i.e. if the involved Lévy measures do not have first moments.

8.2.1 Inhomogeneous Lévy Processes via Integration

We start by considering a logarithmic asset price of the form (8.4) with deterministic "volatility" σ and a Lévy process L with Lévy–Khintchine triplet (a^L, c^L, K^L). Using Proposition 4.8 we obtain that the differential characteristics of X are of the form

$$a(t) = \sigma(t)a^L, \quad c(t) = \sigma(t)^2 c^L, \quad K(t, B) = K^L(\sigma(t)B)$$

for $B \in \mathscr{B}$ with $0 \notin B$. In particular, they are deterministic which by Proposition 4.7 means that such processes are instances of time-inhomogeneous Lévy processes in the sense of Sect. 3.8. Again by Proposition 4.7, the conditional characteristic function of X is of the form

$$E\Big(\exp\big(iuX(s+t)\big)\Big|\mathscr{F}_s\Big) = \exp\bigg(iuX(s) + \int_s^{s+t} \psi(u\sigma(r))dr\bigg), \tag{8.7}$$

where ψ denotes the characteristic exponent of L. A drawback of these models is that a time-varying but deterministic volatility process may seem somewhat unrealistic or at least hard to justify economically.

8.2.2 Inhomogeneous Lévy Processes via Time Change

Alternatively, we consider (8.5, 8.6) with deterministic activity y and a Lévy process L with Lévy–Khintchine triplet (a^L, c^L, K^L). By Proposition 4.14 the differential characteristics of X are of the form

$$a(t) = y(t)a^L, \quad c(t) = y(t)c^L, \quad K(t) = y(t)K^L.$$

These are again characteristics of a time-inhomogeneous Lévy processes. By Proposition 4.7 the conditional characteristic function of X is of the form

$$E\big(\exp(iuX(s+t))\big|\mathscr{F}_s\big) = \exp\left(iuX(s) + \psi(u)\int_s^{s+t} y(r)dr\right),$$

where ψ denotes the characteristic exponent of L. This expression is slightly less involved than (8.7) but the demur concerning deterministic activity applies to this kind of model as well.

8.2.3 Stein and Stein (1991)

References [280] and [261] suggest a volatility process of Ornstein–Uhlenbeck type. Their model is of the form

$$dX(t) = (\mu + \delta\sigma^2(t))dt + \sigma(t)dW(t), \tag{8.8}$$

$$d\sigma(t) = (\kappa - \lambda\sigma(t))dt + \alpha dZ(t) \tag{8.9}$$

with constants $\kappa \geq 0, \mu, \delta, \lambda, \alpha$ and Wiener processes W, Z having constant correlation ϱ. As can be seen from straightforward application of Propositions 4.6, 4.8, 4.13, neither (σ, X) nor (σ^2, X) have affine characteristics in the sense of (6.23–6.25) unless the parameters are chosen in a very specific way, e.g. $\kappa = 0$. However, since

$$d\sigma^2(t) = 2\big(\sigma(t)\kappa - \lambda\sigma^2(t)\big)dt + 2\alpha\sigma(t)dZ(t),$$

the $\mathbb{R}^3$-valued process (σ, σ^2, X) is "almost" the solution to an affine martingale problem related to (6.23–6.25), namely for $(a_{(j)}, c_{(j)}, K_{(j)})$, $j = 0, \ldots, 3$ given by

$$(a_{(0)}, c_{(0)}, K_{(0)}) = \left(\begin{pmatrix} \kappa \\ \alpha^2 \\ \mu \end{pmatrix}, \begin{pmatrix} \alpha^2 & 0 & 0 \\ 0 & 0 & 0 \\ 0 & 0 & 0 \end{pmatrix}, 0 \right),$$

$$(a_{(1)}, c_{(1)}, K_{(1)}) = \left(\begin{pmatrix} -\lambda \\ 2\kappa \\ 0 \end{pmatrix}, \begin{pmatrix} 0 & 2\alpha^2 & \alpha\varrho \\ 2\alpha^2 & 0 & 0 \\ \alpha\varrho & 0 & 0 \end{pmatrix}, 0 \right),$$

$$(a_{(2)}, c_{(2)}, K_{(2)}) = \left(\begin{pmatrix} 0 \\ -2\lambda \\ \delta \end{pmatrix}, \begin{pmatrix} 0 & 0 & 0 \\ 0 & 4\alpha^2 & 2\alpha\varrho \\ 0 & 2\alpha\varrho & 1 \end{pmatrix}, 0 \right),$$

$$(a_{(3)}, c_{(3)}, K_{(3)}) = (0, 0, 0)\,.$$

Since $c_{(1)}$ is not nonnegative definite, $(a_{(1)}, c_{(1)}, K_{(1)})$ is not a Lévy–Khintchine triplet in the usual sense and hence Theorem 6.6 cannot be applied. Nevertheless, the Riccati-type equations (6.30, 6.31) lead to the correct characteristic function in this case, cf. e.g. the derivation in [261]. It is given by

$$\begin{aligned} &E\Big(\exp\big(iu_1\sigma^2(s+t) + iu_2\sigma(s+t) + iu_3X(s+t)\big) \Big| \mathscr{F}_s \Big) \\ &= \exp\Big(\Psi_0(t,u) + \Psi_1(t,u)\sigma(s) + \Psi_2(t,u)\sigma^2(s) + \Psi_3(t,u)X(s) \Big). \end{aligned} \tag{8.10}$$

$\Psi_0, \ldots, \Psi_3$ are obtained from the ODEs (6.30, 6.31), which here read as

$$\begin{aligned} \dot{\Psi}_0 &= \kappa\Psi_1 + \frac{1}{2}\alpha^2\Psi_1^2 + \alpha^2\Psi_2 + \mu\Psi_3, \quad \Psi_0(0,u) = 0, \\ \dot{\Psi}_1 &= 2\kappa\Psi_2 + \Psi_1\big(2\alpha^2\Psi_2 + \alpha\varrho\Psi_3 - \lambda\big), \quad \Psi_1(0,u) = iu_1, \\ \dot{\Psi}_2 &= \frac{1}{2}\Psi_3^2 + \delta\Psi_3 + 2\Psi_2\,(\alpha\varrho\Psi_3 - \lambda) + 2\alpha^2\Psi_2^2, \quad \Psi_2(0,u) = iu_2, \\ \dot{\Psi}_3 &= 0, \quad \Psi_3(0,u) = iu_3. \end{aligned}$$

The dot refers to the derivative with respect to the time variable t. These ODEs are solved by

$$\Psi_0(t,u) = \alpha^2 y(t) + \mu tiu_3 + Et - \frac{F}{4D(D-w_0)}\left(\frac{e^{-Dt}}{v_2(t)} - \frac{1}{D}\right) + \frac{G}{2D}\left(\frac{1}{v_2(t)} - \frac{1}{D}\right),$$

$$\Psi_1(t,u) = \frac{\kappa}{C}\left(\frac{B}{D}v(t) - 2 + \frac{2z_0}{v_2(t)}\right),$$

$$\Psi_2(t,u) = w(t),$$

$$\Psi_3(t,u) = iu_3,$$

where

$$A := \delta i u_3 - \frac{1}{2}u_3^2,$$

$$B := 2(\alpha\varrho i u_3 - \lambda),$$

$$C := 2\alpha^2,$$

$$D := \sqrt{\frac{B^2}{4} - AC} = \sqrt{\lambda^2 + \alpha^2 u_3^2(1-\varrho^2) - 2iu_3(\lambda\alpha\varrho + \alpha^2\delta)},$$

$$w_0 := \frac{B}{2} + Ciu_2 = \alpha\varrho i u_3 - \lambda + 2\alpha^2 i u_2,$$

$$z_0 := 1 - \frac{B}{2D^2}w_0 + \frac{C}{2\kappa}iu_1,$$

$$v_1(t) := w_0\cosh(Dt) - D\sinh(Dt),$$

$$v_2(t) := D\cosh(Dt) - w_0\sinh(Dt),$$

$$v(t) := \frac{v_1(t)}{v_2(t)},$$

$$w(t) := -\frac{B}{2C} + \frac{D}{C}v(t),$$

$$y(t) := -\frac{B}{2C}t - \frac{1}{C}\log\frac{v_2(t)}{D},$$

$$E := \frac{\kappa^2}{C}\left(\frac{B^2}{4D^2} - 1\right),$$

$$F := \frac{\kappa^2}{C}\left(z_0^2 D^2 + B^2\left(\frac{w_0^2}{D^2} - 1\right)\right),$$

$$G := \frac{\kappa^2}{C}2Bz_0.$$

Apart from the fact that the results of Chap. 6 are not literally applicable, a drawback of this model is that σ is not adapted to the natural filtration of X, which is the only observable process in practice. Only σ^2 but not the sign of σ can be recovered from X. Hence the affine form of the characteristics and the validity of (8.10) is lost if one works with this more realistic filtration.

8.2.4 Heston (1993)

If κ is chosen to be 0 in the Ornstein–Uhlenbeck equation (8.9), the Stein and Stein model reduces to a special case of the model of [142]:

$$dX(t) = (\mu + \delta Y(t))dt + \sqrt{Y(t)}dW(t), \tag{8.11}$$

$$dY(t) = (\kappa - \lambda Y(t))dt + \sigma\sqrt{Y(t)}dZ(t). \tag{8.12}$$

Here $\kappa \geq 0$, $\mu, \delta, \lambda, \sigma$ denote constants and W, Z Wiener processes with constant correlation ϱ. Calculation of the characteristics yields that (Y, X) is an affine process as in Theorem 6.6 with triplets $(a_{(j)}, c_{(j)}, K_{(j)})$, $j = 0, 1, 2$ given by

$$\begin{aligned}
(a_{(0)}, c_{(0)}, K_{(0)}) &= \left(\begin{pmatrix} \kappa \\ \mu \end{pmatrix}, 0, 0\right), \\
(a_{(1)}, c_{(1)}, K_{(1)}) &= \left(\begin{pmatrix} -\lambda \\ \delta \end{pmatrix}, \begin{pmatrix} \sigma^2 & \sigma\varrho \\ \sigma\varrho & 1 \end{pmatrix}, 0\right), \\
(a_{(2)}, c_{(2)}, K_{(2)}) &= (0, 0, 0)\,.
\end{aligned}$$

As in Sect. 8.2.3 there is no need to specify a truncation function. Theorem 6.6 leads to the conditional characteristic function

$$\begin{aligned}
&E\big(\exp\big(iu_1Y(s+t) + iu_2X(s+t)\big)\big|\,\mathscr{F}_s\big) \\
&\qquad = \exp\Big(\Psi_0(t,u) + \Psi_1(t,u)Y(s) + \Psi_2(t,u)X(s)\Big).
\end{aligned}$$

Ψ_0, Ψ_1, Ψ_2 solve the ODEs (6.30, 6.31), namely

$$\begin{aligned}
\dot\Psi_0 &= \kappa\Psi_1 + \mu\Psi_2, \quad \Psi_0(0) = 0, \\
\dot\Psi_1 &= \frac{1}{2}\Psi_2^2 + \delta\Psi_2 + \Psi_1\left(\sigma\varrho\Psi_2 - \lambda\right) + \frac{1}{2}\sigma^2\Psi_1^2, \quad \Psi_1(0) = iu_1, \\
\dot\Psi_2 &= 0, \quad \Psi_2(0) = iu_2.
\end{aligned}$$

Similarly as in the previous model, we have

$$\begin{aligned}
\Psi_0(t,u) &= \kappa x(t) + \mu t i u_2, \\
\Psi_1(t,u) &= w(t), \\
\Psi_2(t,u) &= iu_2,
\end{aligned}$$

where

$$
\begin{aligned}
A &:= \delta i u_2 - \frac{1}{2}u_2^2, \\
B &:= \sigma \varrho i u_2 - \lambda, \\
C &:= \frac{1}{2}\sigma^2, \\
D &:= \sqrt{\frac{B^2}{4} - AC} = \frac{1}{2}\sqrt{\lambda^2 + \sigma^2 u_2^2(1-\varrho^2) - 2iu_2(\lambda\sigma\varrho + \sigma^2\delta)}, \\
w_0 &:= \frac{B}{2} + Ciu_1 = \frac{\sigma\varrho i u_2 - \lambda}{2} + \frac{1}{2}\sigma^2 i u_1, && (8.13) \\
v_1(t) &:= w_0 \cosh(Dt) - D\sinh(Dt), && (8.14) \\
v_2(t) &:= D\cosh(Dt) - w_0 \sinh(Dt), && (8.15) \\
v(t) &:= \frac{v_1(t)}{v_2(t)}, && (8.16) \\
w(t) &:= -\frac{B}{2C} + \frac{D}{C}v(t), && (8.17) \\
x(t) &:= -\frac{B}{2C}t - \frac{1}{C}\log\frac{v_2(t)}{D}. && (8.18)
\end{aligned}
$$

Unlike for (8.8, 8.9), the affine structure of the characteristics is preserved under the natural filtration of the observed process X because Y is adapted to this filtration. Moreover, a two-dimensional process suffices in order to obtain an affine structure and the results of Chap. 6 need not be extended to cover (8.11, 8.12). In this sense, the Heston model does not share the theoretical drawbacks of the Stein and Stein model.

8.2.5 Bates (1996)

Reference [15] combines the Heston model with jumps as in the Merton model from Sect. 2.4.5, which leads to a model of the form

$$
\begin{aligned}
dX(t) &= \delta Y(t)dt + \sqrt{Y(t)}dW(t) + L(t), \\
dY(t) &= (\kappa - \lambda Y(t))dt + \sigma\sqrt{Y(t)}dZ(t).
\end{aligned}
$$

As in the Heston model, $\kappa \geq 0$, $\mu, \delta, \lambda, \sigma$ are constants and W, Z Wiener processes with constant correlation ϱ. Moreover, L denotes an independent Lévy process, say with triplet (a^L, c^L, K^L) and characteristic exponent ψ^L. In [15] this Lévy process is of the form

$$L(t) = \mu t + \sum_{k=1}^{N(t)} U_k$$

with a Poisson process N with parameter η and a sequence of iid Gaussian random variables $U_1, U_2, \ldots$ with mean β and variance γ^2, which means that the characteristic exponent of L is

$$\psi^L(u) = i\mu u + \eta(e^{i\beta u - \frac{1}{2}\gamma^2 u^2} - 1), \quad u \in \mathbb{R}.$$

By applying Propositions 4.6 and 4.8, we have that the characteristics of (Y, X) are of the form (6.23–6.25) with triplets $(a_{(j)}, c_{(j)}, K_{(j)})$, $j = 0, 1, 2$ given by

$$a_{(0)} = \begin{pmatrix} \kappa \\ a^L \end{pmatrix}, \quad c_{(0)} = \begin{pmatrix} 0 & 0 \\ 0 & c^L \end{pmatrix}, \quad K_{(0)}(B) = \int 1_B(0, x) F^L(dx) \quad \forall B \in \mathscr{B}^2,$$

$$(a_{(1)}, c_{(1)}, K_{(1)}) = \left(\begin{pmatrix} -\lambda \\ \delta \end{pmatrix}, \begin{pmatrix} \sigma^2 & \sigma\varrho \\ \sigma\varrho & 1 \end{pmatrix}, 0 \right),$$

$$(a_{(2)}, c_{(2)}, K_{(2)}) = (0, 0, 0).$$

Theorem 6.6 yields the conditional characteristic function

$$E\big(\exp\big(iu_1 Y(s+t) + iu_2 X(s+t)\big) \big| \mathscr{F}_s\big) = \exp\Big(\Psi_0(t, u) + \Psi_1(t, u) Y(s) + \Psi_2(t, u) X(s) \Big).$$

The ODEs

$$\dot{\Psi}_0 = \kappa \Psi_1 + \psi^L(\Psi_2), \quad \Psi_0(0) = 0,$$

$$\dot{\Psi}_1 = \frac{1}{2}\Psi_2^2 + \delta\Psi_2 + \Psi_1(\sigma\varrho\Psi_2 - \lambda) + \frac{1}{2}\sigma^2\Psi_1^2, \quad \Psi_1(0) = iu_1,$$

$$\dot{\Psi}_2 = 0, \quad \Psi_2(0) = iu_2.$$

for Ψ_0, Ψ_1, Ψ_2 are solved by

$$\Psi_0(t, u) = \kappa x(t) + t\psi^L(u_2),$$

$$\Psi_1(t, u) = w(t),$$

$$\Psi_2(t, u) = iu_2,$$

where $w(t)$ and $x(t)$ are defined as in the Heston model.

This model allows for both jumps and correlation between volatility and asset returns. Hence it accounts for most of the stylised facts of return data. As a slight drawback one may note that stochastic volatility and hence also the dependence between volatility and asset return movements pertains only to the continuous part of X, but not to its jumps. From an economical point of view, however, one would rather expect big price impacts to come along with rising volatility.

8.2.6 *Barndorff-Nielsen and Shephard (2001), Henceforth BNS*

Reference [10] suggest a stochastic volatility model with jumps in the activity process. It is of the form

$$dX(t) = (\mu + \delta Y(t-))dt + \sqrt{Y(t-)}dW(t) + \varrho dZ(t), \tag{8.19}$$

$$dY(t) = -\lambda Y(t-)dt + dZ(t) \tag{8.20}$$

with constants $\mu, \delta, \varrho, \lambda$, a Wiener process W, and a subordinator (i.e. an increasing Lévy process) Z with Lévy–Khintchine triplet $(a^Z, 0, F^Z)$ and characteristic exponent ψ^Z. Compared to the Heston model, the square-root process (8.12) is replaced with a Lévy-driven Ornstein–Uhlenbeck (OU) process. Since W and Z are necessarily independent, leverage is introduced by the $\varrho dZ(t)$ term. Propositions 4.6 and 4.8 yield that (Y, X) is an affine process in the sense of Theorem 6.6 with triplets $(a_{(j)}, c_{(j)}, K_{(j)})$, $j = 0, 1, 2$ of the form

$$a_{(0)} = \begin{pmatrix} a^Z \\ \mu + \varrho a^Z \end{pmatrix}, \quad c_{(0)} = 0,$$

$$K_{(0)}(B) = \int 1_B(x, \varrho x) F^Z(dx) \quad \forall B \in \mathscr{B}^2,$$

$$(a_{(1)}, c_{(1)}, K_{(1)}) = \left(\begin{pmatrix} -\lambda \\ \delta \end{pmatrix}, \begin{pmatrix} 0 & 0 \\ 0 & 1 \end{pmatrix}, 0 \right),$$

$$(a_{(2)}, c_{(2)}, K_{(2)}) = (0, 0, 0)\,.$$

Solving the involved linear ODEs leads to the conditional characteristic function

$$E\big(\exp(iu_1 Y(s+t) + iu_2 X(s+t))\big| \mathscr{F}_s\big) \\ = \exp\Big(\Psi_0(t, u) + \Psi_1(t, u)Y(s) + \Psi_2(t, u)X(s)\Big).$$

Specifically, the ODEs (6.30, 6.31) here read as

$$\dot{\Psi}_0 = \psi^Z(\Psi_1 + \varrho\Psi_2) + \mu\Psi_2, \quad \Psi_0(0) = 0,$$

$$\dot{\Psi}_1 = \frac{1}{2}\Psi_2^2 + \delta\Psi_2 - \lambda\Psi_1, \quad \Psi_1(0) = iu_1,$$

$$\dot{\Psi}_2 = 0, \quad \Psi_2(0) = iu_2.$$

They are solved by

$$\Psi_0(t, u) = \int_0^t \psi^Z(-i\Psi_1(s, u) + \varrho u_2)ds + \mu t i u_2, \tag{8.21}$$

$$\Psi_1(t, u) = e^{-\lambda t} iu_1 + \frac{1 - e^{-\lambda t}}{\lambda}\psi^L(u_2), \tag{8.22}$$

$$\Psi_2(t, u) = iu_2, \tag{8.23}$$

where

$$\psi^L(u_2) := \delta i u_2 - \frac{1}{2}u_2^2.$$

We consider two particular examples of subordinators Z.

Example 8.3 (Gamma-OU Process) If we choose

$$\psi^Z(u) = \frac{iu\beta}{\alpha - iu} \tag{8.24}$$

for some $\alpha, \beta > 0$, we end up with the Gamma-OU process of Example 3.57. In this case, the integral in (8.21) can be computed in closed form:

$$\int_0^t \psi^Z(-i\Psi_1(s, u) + \varrho u_2)ds = \begin{cases} \frac{\beta}{\alpha - k_2}\left(\frac{\alpha}{\lambda}\log\varphi(t, u) + k_2 t\right) & \text{if } \alpha \neq k_2, \\ -\beta\left(\frac{\alpha}{k_1}\frac{e^{\lambda t} - 1}{\lambda} + t\right) & \text{if } \alpha = k_2, \end{cases}$$

where

$$k_1 := iu_1 - \frac{\psi^L(u_2)}{\lambda}, \tag{8.25}$$

$$k_2 := \frac{\psi^L(u_2)}{\lambda} + \varrho i u_2, \tag{8.26}$$

$$\varphi(t, u) := \frac{\alpha - \Psi_1(t, u) - \varrho i u_2}{\alpha - k_1 - k_2}$$

and the branch of the complex logarithm is to be chosen such that $t \mapsto \log \varphi(t, u)$ is continuous and equals 0 for $t = 0$.

Example 8.4 (Inverse-Gaussian OU Process) In this second example we consider

$$\psi^Z(u) = \frac{iu\beta}{\sqrt{\alpha^2 - 2iu}} \tag{8.27}$$

for some $\alpha, \beta > 0$, leading to the inverse-Gaussian OU process of Example 3.58. For this choice, the integral in (8.21) reads as

$$\int_0^t \psi^Z(-i\Psi_1(s,u) + \varrho u_2)ds$$
$$= \begin{cases} \frac{\beta}{\lambda}\left(\frac{k_2}{B}(\log\varphi(t,u) + \lambda t) + A(t) - A(0)\right) & \text{if } \alpha^2 \neq 2k_2 \text{ and } k_1 \neq 0, \\ \frac{\sqrt{2}\beta}{\sqrt{-k_1}}\left(k_1 \frac{1-e^{-\lambda t/2}}{\lambda} + k_2 \frac{e^{\lambda t/2}-1}{\lambda}\right) & \text{if } \alpha^2 = 2k_2, \\ \frac{\beta k_2 t}{B} & \text{if } k_1 = 0, \end{cases}$$

where k_1, k_2 are as in (8.25, 8.26),

$$A(t) := \sqrt{\alpha^2 - 2\Psi_1(t,u) - 2\varrho i u_2},$$
$$B = \sqrt{\alpha^2 - 2k_2},$$
$$\varphi(t,u) := \frac{(B(B - A(0)) - k_1)\left(B(B + A(t)) - e^{-\lambda t}k_1\right)}{k_1^2},$$

and the branch of the complex logarithm is chosen such that $t \mapsto \log \varphi(t, u)$ is continuous and equals 0 for $t = 0$.

Barndorff-Nielsen and Shephard also consider a slightly extended version of the above model. They argue that the autocorrelation pattern of volatility is not appropriately matched by a single OU process. As a way out they suggest a linear combination of independent OU processes, i.e. a model of the form

$$dX(t) = (\mu + \delta Y(t-))dt + \sqrt{Y(t-)}dW(t) + \sum_{k=1}^n \varrho_k dZ_k(t),$$

$$Y(t) = \sum_{k=1}^n \alpha_k Y_k(t),$$

$$dY_k(t) = -\lambda_k Y_k(t-)dt + dZ_k(t)$$

with constants $\alpha_1, \dots, \alpha_n \geq 0, \mu, \delta, \varrho_1, \dots, \varrho_n, \lambda_1, \dots, \lambda_n$, a Wiener processes W, and an $\mathbb{R}^n$-valued Lévy process Z whose components are subordinators. We denote the triplet of Z by $(a^Z, 0, F^Z)$ and its characteristic exponent by ψ^Z. The $\mathbb{R}^{n+1}$-valued process $(Y_1, \dots, Y_n, X)$ is affine in the sense of Theorem 6.6. Its triplets $(a_{(j)}, c_{(j)}, K_{(j)})$, $j = 0, \dots, n+1$ are of the form

$$a_{(0)} = \begin{pmatrix} a_1^Z \\ \vdots \\ a_n^Z \\ \mu + \sum_{k=1}^n \varrho_k a_k^Z \end{pmatrix}, \quad c_{(0)} = 0,$$

$$K_{(0)}(B) = \int 1_B\big(x_1, \dots, x_n, \textstyle\sum_{k=1}^n \varrho_k x_k\big) F^Z(dx) \quad \forall B \in \mathscr{B}^{n+1},$$

$$(a_{(k)}, c_{(k)}, K_{(k)}) = \left(\begin{pmatrix} 0 \\ \vdots \\ 0 \\ -\lambda_k \\ 0 \\ \vdots \\ 0 \\ \delta\alpha_k \end{pmatrix}, \begin{pmatrix} 0 & \dots & 0 & 0 \\ \vdots & \ddots & \vdots & \vdots \\ 0 & \cdots & 0 & 0 \\ 0 & \cdots & 0 & \alpha_k \end{pmatrix}, 0 \right), \quad k = 1, \dots, n,$$

$$(a_{(n+1)}, c_{(n+1)}, K_{(n+1)}) = (0, 0, 0).$$

The resulting conditional characteristic function is

$$E\left(\exp\Big(\sum_{k=1}^n i u_k Y_k(s+t) + i u_{n+1} X(s+t) \Big) \Big| \mathscr{F}_s \right)$$
$$= \exp\left(\Psi_0(t,u) + \sum_{k=1}^n \Psi_k(t,u) Y_k(s) + \Psi_{n+1}(t,u) X(s) \right)$$

with functions $\Psi_0, \dots, \Psi_{n+1}$ solving the ODEs

$$\dot{\Psi}_0 = \psi^Z(\Psi_1 + \varrho_1 \Psi_{n+1}, \dots, \Psi_n + \varrho_n \Psi_{n+1}) + \mu \Psi_{n+1}, \quad \Psi_0(0) = 0,$$

$$\dot{\Psi}_k = \frac{1}{2}\alpha_k \Psi_{n+1}^2 + \delta\alpha_k \Psi_{n+1} - \lambda_k \Psi_k, \quad \Psi_k(0,u) = i u_k, \quad k = 1, \dots, n$$

$$\dot{\Psi}_{n+1} = 0, \quad \Psi_{n+1}(0,u) = i u_{n+1}.$$

We have

$$\Psi_0(t,u) = \int_0^t \psi^Z\big(-i\Psi_1(s,u)+\varrho_1 u_{n+1},\dots,-i\Psi_n(s,u)+\varrho_n u_{n+1}\big)ds + \mu t i u_{n+1},$$

$$\Psi_k(t,u) = e^{-\lambda_k t} i u_k + \frac{1-e^{-\lambda_k t}}{\lambda_k}\psi^{L_k}(u_{n+1}), \quad k=1,\dots,n,$$

$$\Psi_{n+1}(t,u) = i u_{n+1},$$

where

$$\psi^{L_k}(u_{n+1}) := \alpha_k\left(\delta i u_{n+1} - \frac{1}{2}u_{n+1}^2\right).$$

If the subordinators $Z_1, .., Z_n$ are chosen to be independent, we obtain

$$\Psi_0(t,u) = \sum_{k=1}^n \int_0^t \psi^{Z_k}\big(-i\Psi_k(s,u)+\varrho_k u_{n+1}\big)ds + \mu t i u_{n+1}.$$

For exponents ψ^{Z_k}, $k=1,\dots,n$ of the form (8.24) resp. (8.27), the integrals can be expressed in closed form, cf. Examples 8.3 and 8.4. However, in order to preserve this affine structure, the subordinators $Z_1,\dots,Z_n$ need not be independent. If one considers the other extreme case $Z_1 = \dots = Z_n$, the activity process Y will be a Lévy-driven CARMA process as is briefly discussed at the end of Sect. 3.9.

Up to the leverage terms $\varrho dZ(t)$ resp. $\varrho_k dZ_k(t)$, the asset price in the BNS model does not jump. In its extension involving superpositions of OU processes, the BNS model allows for a more flexible autocorrelation structure than, say, Heston, while preserving its computational tractability.

8.2.7 Carr et al. (2003), Henceforth CGMY

Reference [52] generalise both the Heston and the BNS model by allowing for jumps in the asset price. In order to preserve the affine structure, one must consider time changes unless the driver of the asset price changes is a stable Lévy motion, cf. Sect. 8.2.8.

The analogue of the Heston model is

$$\begin{aligned} X(t) &= X(0) + \mu t + L(Y(t)) + \varrho(y(t)-y(0)),\\ dY(t) &= y(t)dt,\\ dy(t) &= (\kappa - \lambda y(t))dt + \sigma\sqrt{y(t)}dZ(t), \end{aligned} \tag{8.28}$$

where $\kappa \geq 0, \mu, \varrho, \lambda, \sigma$ denote constants, L a Lévy process with triplet (a^L, c^L, F^L) and exponent ψ^L, and Z an independent Wiener process. Again, (y, X) is an affine process whose triplets $(a_{(j)}, c_{(j)}, K_{(j)})$, $j = 0, 1, 2$ meet the equations

$$(a_{(0)}, c_{(0)}, K_{(0)}) = \left(\begin{pmatrix} \kappa \\ \mu + \varrho\kappa \end{pmatrix}, 0, 0 \right),$$

$$a_{(1)} = \begin{pmatrix} -\lambda \\ a^L - \varrho\lambda \end{pmatrix}, \quad c_{(1)} = \begin{pmatrix} \sigma^2 & \sigma^2\varrho \\ \sigma^2\varrho & \sigma^2\varrho^2 + c^L \end{pmatrix},$$

$$K_{(1)}(B) = \int 1_B(0, x) F^L(dx) \quad \forall B \in \mathscr{B}^2,$$

$$(a_{(2)}, c_{(2)}, K_{(2)}) = (0, 0, 0) .$$

Observe that we recover the characteristics of the Heston model—up to a rescaling of the volatility process y—if L is chosen to be Brownian motion with drift. Applying Theorem 6.6 yields the conditional characteristic function

$$E\big(\exp\big(iu_1 y(s+t) + iu_2 X(s+t)\big)\big|\, \mathscr{F}_s\big) = \exp\Big(\Psi_0(t, u) + \Psi_1(t, u) y(s) + \Psi_2(t, u) X(s) \Big). \tag{8.29}$$

The ODEs (6.30, 6.31)

$$\dot\Psi_0 = \kappa\Psi_1 + \mu + \varrho\kappa\Psi_2, \quad \Psi_0(0, u) = 0,$$

$$\dot\Psi_1 = \frac{1}{2}\left(\sigma^2\varrho^2 + c_L\right)\Psi_2^2 + (b_L - \varrho\lambda)\Psi_2 + \Psi_1\left(\sigma^2\varrho\Psi_2 - \lambda\right) + \frac{1}{2}\sigma^2\Psi_1^2,$$

$$\Psi_1(0, u) = iu_1,$$

$$\dot\Psi_2 = 0, \quad \Psi_2(0, u) = iu_2.$$

for Ψ_0, Ψ_1, Ψ_2 are solved by

$$\begin{aligned} \Psi_0(t, u) &= \kappa x(t) - (\mu + \varrho\kappa) t i u_2, \\ \Psi_1(t, u) &= w(t), \\ \Psi_2(t, u) &= iu_2, \end{aligned} \tag{8.30}$$

where

$$A := -\frac{1}{2}\sigma^2\varrho^2 u_2^2 - \varrho\lambda i u_2 + \psi^L(u_2),$$

$$B := \sigma\varrho i u_2 - \lambda,$$

$$C := \frac{1}{2}\sigma^2,$$
$$D := \frac{1}{2}\sqrt{\lambda^2 - 2\sigma^2\psi^L(u_2)}$$

and w_0, v_1, v_2, v, w, x are defined as in (8.13–8.18).

In order to generalise the BNS model, the square-root process (8.28) is replaced with a Lévy-driven OU process:

$$\begin{aligned} X(t) &= X(0) + \mu t + L(Y(t)) + \varrho Z(t), \\ dY(t) &= y(t-)dt, \\ dy(t) &= -\lambda y(t-)dt + dZ(t). \end{aligned} \tag{8.31}$$

Here, μ, ϱ, λ denote constants and L, Z independent Lévy processes with triplets (a^L, c^L, F^L), $(a^Z, 0, F^Z)$ and exponents ψ^Z, ψ^L, respectively. Z is assumed to be increasing. The triplets $(a_{(j)}, c_{(j)}, K_{(j)})$, $j = 0, 1, 2$, of the affine process (y, X) are given by

$$a_{(0)} = \begin{pmatrix} a^Z \\ \mu + \varrho a^Z \end{pmatrix}, \quad c_{(0)} = 0, \quad K_{(0)}(B) = \int 1_B(x, \varrho x)F^Z(dx) \quad \forall B \in \mathscr{B}^2,$$

$$a_{(1)} = \begin{pmatrix} -\lambda \\ a^L \end{pmatrix}, \quad c_{(1)} = \begin{pmatrix} 0 & 0 \\ 0 & c^L \end{pmatrix}, \quad K_{(1)}(B) = \int 1_B(0, x)F^L(dx) \quad \forall B \in \mathscr{B}^2,$$

$$(a_{(2)}, c_{(2)}, K_{(2)}) = (0, 0, 0),$$

where the integration variable x should not be confused with the deterministic function in (8.30). For a Brownian motion with drift L, we recover the dynamics of the BNS model (8.20). The relevant functions in (8.29) are now solutions to

$$\begin{aligned} \dot{\Psi}_0 &= \psi^Z(\Psi_1 + \varrho\Psi_2) + \mu\Psi_2, \quad \Psi_0(0, u) = 0, \\ \dot{\Psi}_1 &= \psi^L(\Psi_2) - \lambda\Psi_1, \quad \Psi_1(0, u) = iu_1, \\ \dot{\Psi}_2 &= 0, \quad \Psi_2(0, u) = iu_2. \end{aligned}$$

They are of the form (8.21–8.23), but with ψ^L denoting the exponent of L.

Remark 8.5 (Gamma-OU and Inverse-Gaussian OU Processes) As in the BNS case, we can express Ψ_0 more explicitly for particular subordinators Z. Indeed, Examples 8.3 and 8.4 apply verbatim in the present context.

The CGMY models allow for jumps in the asset price even if the leverage parameter ϱ vanishes. The OU version could be generalised as in Sect. 8.2.6 to a superposition of Ornstein–Uhlenbeck processes to obtain a more flexible autocorrelation structure.

8.2.8 Carr and Wu (2003)

Reference [49] consider a modification of the Heston model where the Wiener process W is replaced by an α-stable Lévy motion L with $\alpha \in (1, 2)$, Lévy–Khintchine triplet $(0, 0, F^L)$ (relative to $h(x) = x$ as usual), and Lévy exponent ψ^L:

$$dX(t) = \mu dt + Y(t)^{1/\alpha} dL(t),$$

$$dY(t) = (\kappa - \lambda Y(t))dt + \sigma\sqrt{Y(t)}dZ(t).$$

L and the Wiener process Z must be independent. The Lévy measure of L has the scaling property $\int 1_B(\gamma^{1/\alpha}x)F^L(dx) = \gamma F^L(B)$ for $\gamma > 0$, $B \in \mathscr{B}$, which is essential in order to obtain an affine structure.

An application of Propositions 4.6 and 4.8 shows that (Y, X) is an affine process with triplets $(a_{(j)}, c_{(j)}, K_{(j)})$, $j = 0, 1, 2$ of the form

$$(a_{(0)}, c_{(0)}, K_{(0)}) = \left(\begin{pmatrix} \kappa \\ \mu \end{pmatrix}, 0, 0\right),$$

$$a_{(1)} = \begin{pmatrix} -\lambda \\ 0 \end{pmatrix}, \quad c_{(1)} = \begin{pmatrix} \sigma^2 & 0 \\ 0 & 0 \end{pmatrix}, \quad K_{(1)}(B) = \int 1_B(0, x)F^L(dx) \quad \forall B \in \mathscr{B}^2,$$

$$a_{(2)}, c_{(2)}, K_{(2)}) = (0, 0, 0)\,.$$

With Theorem 6.6 we obtain the conditional characteristic function

$$E\big(\exp\big(iu_1Y(s+t) + iu_2X(s+t)\big)\big|\,\mathscr{F}_s\big)$$
$$= \exp\Big(\Psi_0(t, u) + \Psi_1(t, u)Y(s) + \Psi_2(t, u)X(s)\Big),$$

where Ψ_0, Ψ_1, Ψ_2 solve

$$\dot{\Psi}_0 = \kappa\Psi_1 + \mu\Psi_2, \quad \Psi_0(0, u) = 0,$$

$$\dot{\Psi}_1 = \psi^L(\Psi_2) - \lambda\Psi_1 + \frac{1}{2}\sigma^2\Psi_1^2, \quad \Psi_1(0, u) = iu_1,$$

$$\dot{\Psi}_2 = 0, \quad \Psi_2(0, u) = iu_2.$$

We have

$$\Psi_0(t, u) = \kappa x(t) + \mu t i u_2,$$

$$\Psi_1(t, u) = w(t),$$

$$\Psi_2(t, u) = iu_2,$$

with

$$\begin{aligned}
A &:= \psi^L(u_2), \\
B &:= -\lambda, \\
C &:= \frac{1}{2}\sigma^2, \\
D &:= \frac{1}{2}\sqrt{\lambda^2 - 2\sigma^2\psi^L(u_2)}, \\
\varrho &:= 0,
\end{aligned}$$

and w_0, v_1, v_2, v, w, x are defined as in (8.13–8.18). The asset price process $S(t) = e^{X(t)}$ has finite expectation only if L is totally skewed to the left ($\beta = -1$ in Theorem 2.31), which means that it has only negative jumps. In contrast to Sect. 8.2.4, the above model does not allow for correlation between changes in asset prices and volatility because L and Z are independent.

8.2.9 *Carr and Wu (2004) and Affine ARCH-Like Models*

Reference [50] consider a number of models, two of which could be written in the form

$$X(t) = X(0) + \mu t + L(Y(t)), \tag{8.32}$$

$$dY(t) = y(t-)dt, \tag{8.33}$$

$$y(t) = y(0) + \kappa t + Z(Y(t)), \tag{8.34}$$

where $\kappa \geq 0, \mu$ denote constants and (Z, L) a Lévy process in $\mathbb{R}^2$ with triplet $(a^{(Z,L)}, c^{(Z,L)}, F^{(Z,L)})$ and exponent $\psi^{(Z,L)}$. Moreover, Z is assumed to have only non-negative jumps and finite expected value $E(Z_1)$.

Note that the above time change equation $y(t) = y(0) + \kappa t + Z(\int_0^t y(s-)ds)$ is implicit. It may not be obvious whether a unique solution to this equation exists. But if it does, Propositions 4.6, 4.8, 4.14 yield the differential characteristics of (y, X), which are of the form (6.23–6.25) with triplets $(a_{(j)}, c_{(j)}, K_{(j)})$, $j = 0, 1, 2$ given by

$$\begin{aligned}
(a_{(0)}, c_{(0)}, K_{(0)}) &= \left(\begin{pmatrix}\kappa \\ \mu\end{pmatrix}, 0, 0\right), \\
(a_{(1)}, c_{(1)}, K_{(1)}) &= \left(a^{(Z,L)}, c^{(Z,L)}, F^{(Z,L)}\right), \\
(a_{(2)}, c_{(2)}, K_{(2)}) &= (0, 0, 0)\,.
\end{aligned}$$

By Theorem 6.6 a solution (y, X) to the corresponding affine martingale problem exists and its law is uniquely determined. Hence we may work with this process regardless of the question of whether it allows for a pathwise representation as in (8.32–8.34). The conditional characteristic function is of the form

$$E\big(\exp(iu_1 y(s+t) + iu_2 X(s+t))\big| \mathscr{F}_s\big)$$
$$= \exp\Big(\Psi_0(t,u) + \Psi_1(t,u)y(s) + \Psi_2(t,u)X(s)\Big)$$

with solutions Ψ_0, Ψ_1, Ψ_2 to the ODEs

$$\dot{\Psi}_0 = \kappa\Psi_1 + \mu\Psi_2, \quad \Psi_0(0,u) = 0,$$
$$\dot{\Psi}_1 = \psi^{(Z,L)}(\Psi_1, \Psi_2), \quad \Psi_1(0,u) = iu_1,$$
$$\dot{\Psi}_2 = 0, \quad \Psi_2(0,u) = iu_2,$$

namely

$$\Psi_0(t,u) = \kappa \int_0^t (\Psi_1(s,u) + \varrho iu_2)ds + \mu t iu_2,$$
$$\Psi_1(t,u) = iu_1 + \int_0^t \psi^{(Z,L)}(-i\Psi_1(s,u), u_2)ds, \tag{8.35}$$
$$\Psi_2(t,u) = iu_2.$$

Reference [50] discuss two particular cases of the above setup, namely a joint compound Poisson process (Z, L) with drift and, alternatively, the completely dependent case $Z(t) = -\lambda t - \sigma L(t)$ with constants λ, σ and some totally skewed α-stable Lévy motion L, where $\alpha \in (1, 2]$. The latter model has an ARCH-like structure in the sense that the same source of randomness L drives both the volatility and the asset price process. This extends to the more general situation where L is an arbitrary Lévy process and $\Delta Z(t) = f(\Delta L(t))$ for some deterministic function $f : \mathbb{R} \to \mathbb{R}_+$ such as $f(x) = x^2$. If L or f are asymmetric, such models allow for leverage.

A drawback of this very flexible setup (8.32–8.34) is that it is not obvious how to solve the generalised Riccati equation (8.35) explicitly. Therefore the characteristic function of (y, X) is only known up to the solution of some ODE.

Appendix 1: Problems

The exercises correspond to the section with the same number.

8.1 (Siegel's Paradox) Denote by $\$(t)$ the price at time t of one US-Dollar expressed in Euro and by $€(t)$ the price of one Euro in US-Dollars. Suppose that

$$\$(0) = 1, \quad d\$(t) = \$(t)(0.02dt + 0.2dW(t))$$

for some Wiener process W.

1. Compute $E(\$(t) - \$(0))$, i.e. the expected profit in Euro from an investment of 1 US-Dollar. Is the US-Dollar an attractive investment from this perspective? Compute also $E(€(t) - €(0))$, i.e. the expected profit of one Euro form the perspective of a US investor. Is the Euro an attractive investment for the latter?
2. Show that $t^{-1}\log(\$(t)/\$(0))$ converges as $t \to \infty$ towards a deterministic number, which is called *long-term growth rate of wealth*. Determine this rate.

8.2 Consider the Heston model (8.11, 8.12) or, alternatively, the BNS model (8.19, 8.20). Suppose that the goal is to compute the autocovariance function of squared daily returns, i.e.

$$\alpha(h) := \mathrm{Cov}\Big(\big(X(t\Delta) - X((t-1)\Delta)\big)^2, \big(X((t+h)\Delta) - X((t+h-1)\Delta)\big)^2\Big)$$

for $\Delta > 0$, $t, h \in \mathbb{N}$. It can be determined explicitly either using Propositions 6.8 and A.5(2) or with Theorem 6.15. Compare the effort needed to actually compute the autocovariance function.

Appendix 2: Notes and Comments

The use of Lévy processes in finance starts with [5], who invented Brownian motion for this purpose. The transition to geometric Brownian motion can be attributed to [228, 257]. A number of the models in this chapter are discussed in [263]. Others are reviewed from an affine perspective in [169]. The formulas in Examples 8.3, 8.4 are taken from [231]. Section 8.1.2 is based on [183] which generalises the Lévy copula concept of [60].

Chapter 9
Markets, Strategies, Arbitrage

We study several problems of Mathematical Finance in a common framework, which is laid down in this chapter. We start by introducing notions such as price processes, trading strategies, discounting, and dividends. The definitions and statements in Sect. 9.1 warrant a proper bookkeeping even in complex stochastic models.

Market prices do not move in an entirely arbitrary fashion. Reference [132] shows that if a securities market is *viable* in the sense of economic theory, discounted asset prices are martingales relative to some equivalent probability measure (called an *equivalent martingale measure*). Subsequently, viability has been replaced by the simpler concept of *absence of arbitrage*, which does not involve detailed assumptions on investors' behaviour, cf. [133]. This concept and its mathematical implications are studied in Sect. 9.3.

9.1 Mathematical Framework

This section provides the general abstract setup for all concrete problems in this book.

9.1.1 Price Processes and Trading Strategies

Prices of securities, currencies, commodities etc. constitute the key object of interest in Mathematical Finance. We model them in terms of stochastic processes $S_0(t), \dots, S_d(t)$, $t \in [0, T]$. The random variable $S_i(t)$ denotes the price of security i at time t, expressed in some common currency that is fixed in the first place. In order to avoid confusing technicalities, we work mostly with a fixed finite time

© Springer Nature Switzerland AG 2019

E. Eberlein, J. Kallsen, *Mathematical Finance*, Springer Finance,
https://doi.org/10.1007/978-3-030-26106-1_9

horizon $T < \infty$. Specifically, we assume that $S = (S_0(t), \dots, S_d(t))_{t\in[0,T]}$ is a vector-valued semimartingale on some filtered probability space $(\Omega, \mathscr{F}, \mathbf{F}, P)$. For ease of notation we generally assume $\mathscr{F}_0$ to be trivial in Part II, i.e. $\mathscr{F}_0$-measurable random variables are actually constant. The semimartingale assumption on S hardly limits the set of possible price evolutions. The smallest index 0 indicates that this security serves as a natural reference—typically because it is simple. But for the theory to work it could be any tradable asset.

Example 9.1 (Geometric Brownian Motion) In the standard model underlying the Black–Scholes option pricing theory, we have $S_0(t) = e^{rt}$ for a deterministic bank account (slightly incorrectly called a *bond*) and

$$S_1(t) = S_1(0)\exp(\mu t + \sigma W(t)) \tag{9.1}$$

$$= S_1(0)\mathscr{E}\left(\left(\mu + \frac{\sigma^2}{2}\right)I + \sigma W\right)(t) \tag{9.2}$$

for a stock. Here, r, μ, σ denote constants and W a standard Brownian motion.

This setup is reconsidered in Examples 9.16 (absence of arbitrage), 10.10 (optimal investment), 11.4 (completeness), 11.19 (delta hedge), 11.22 (Black–Scholes formula), 11.24 (Black–Scholes PDE), 11.32, 11.37 (pricing by integral transform), 11.46 (impossible arbitrage), 12.5 (mean-variance hedging), 13.3, 13.11 (basis risk), and Problems 10.5 (transaction costs), 12.1, 13.4 (basis risk). It also appeared in Examples 7.13 (perpetual put) and 7.17 (optimal investment). Its multivariate extension is used in the mutual fund theorem of Examples 7.23, 10.9, 10.27. In the context of optimal investment see also the references in Examples 9.2, 9.3. Occasionally, the parameter $\mu+\sigma^2/2$ in (9.2) is denoted as μ, namely in the mutual fund theorems and in Examples 7.17, 13.3, 13.11 as well as Problem 12.1.

As an alternative to (9.1) one may replace S_1 by the more advanced processes from Chap. 8, which allow for jumps and/or stochastic volatility. Models with stochastic interest rate r are discussed in Chap. 14. In order to illustrate the concepts, methods, and results of the following chapters, we apply them repeatedly to the following three generalisations of Example 9.1.

Example 9.2 (Geometric Lévy Model) Besides the riskless bond $S_0(t) = e^{rt}$ we consider a single risky asset as above. Its price process is assumed to be of the form

$$S_1(t) = S_1(0)\exp(X(t)) = S_1(0)\mathscr{E}(\widetilde{X})(t),$$

where X denotes a Lévy process with Lévy–Khintchine triplet (a^X, c^X, K^X) resp. $\widetilde{X}$ a Lévy process with Lévy–Khintchine triplet $(a^{\widetilde{X}}, c^{\widetilde{X}}, K^{\widetilde{X}})$, cf. Theorem 3.49. In the particular case $K^X = 0 = K^{\widetilde{X}}$, the Lévy processes $X, \widetilde{X}$ reduce to Brownian motion with drift. More specifically, we have $X(t) = a^X t + (c^X)^{1/2}W(t)$ and $\widetilde{X}(t) = a^{\widetilde{X}}t + (c^{\widetilde{X}})^{1/2}W(t)$ for some Wiener process W, whence we are in the setup of Example 9.1 for $\mu = a^X = a^{\widetilde{X}} - c^{\widetilde{X}}/2$ and $\sigma^2 = c^X = c^{\widetilde{X}}$. For ease of notation we assume $r = 0$ in the application of this example in Chaps. 10, 12, 13, i.e. we work effectively with discounted prices in the sense of Sect. 9.1.2 below. In Sects. 11.5, 11.6 and Chap. 12, on the other hand, the initial value $S_1(0)$ is put into X, i.e. we write $S_1(t) = \exp(X(t))$ with a *shifted* Lévy process X.

The present Lévy setup is considered in Examples 9.16 (absence of arbitrage), 10.13, 10.19, 10.22, 10.30, 10.41, 10.48 (optimal investment), 11.16 (price range), 11.22, 11.24, 11.32, 11.35, 11.37 (option pricing), 12.9, 12.13 (mean-variance hedging), 13.7, 13.13 (indifference pricing), 13.29 (option valuation and hedging).

For Itō processes some of the results in subsequent chapters simplify or they can be stated in terms of backward differential equations. Therefore we consider this case separately.

Example 9.3 (Itō Process Model) We assume that the filtration is generated by an $\mathbb{R}^n$-valued Wiener process W and that the risky asset S_1 follows an Itō process of the form

$$dS_1(t) = \mu(t)dt + \sigma(t)dW_1(t) \tag{9.3}$$

with predictable processes μ, σ. Moreover, we suppose that $S_0(t) = 1$, i.e. we work effectively with discounted prices in the sense of Sect. 9.1.2. The process in (9.1) is of the form (9.3) if we choose $\mu(t) := S_1(t)(\mu + \sigma^2/2)$, $\sigma(t) := S_1(t)\sigma$. The Heston model (8.11, 8.12) provides another example of (9.3).

For applications of the Itō process setup see Examples 9.17 (absence of arbitrage), 10.14, 10.20, 10.31, 10.42, 10.49, 10.53, 10.56 (optimal investment), 11.17 (price range), 11.20 (hedging), 13.10 (indifference pricing), and Problems 10.1, 11.4, 13.1, 13.2.

By adding some Markovian structure to the previous example, results can often be expressed in terms of partial differential equations.

Example 9.4 (Continuous Markovian Model) Suppose that X denotes an $\mathbb{R}^n$-valued continuous Markov process with extended generator $\widetilde{G}$. More specifically, we assume that there are functions $\beta : \mathbb{R}^n \to \mathbb{R}^n$ and $\gamma : \mathbb{R}^n \to \mathbb{R}^{n\times n}$ with

$$\widetilde{G}f(x) = \beta(x)^\top Df(x) + \frac{1}{2}\sum_{i,j=1}^{n} \gamma_{ij}(x)D_{ij}f(x)$$

for any smooth function $f : \mathbb{R}^n \to \mathbb{R}^n$ with compact support, cf. Theorem 5.19. As in Example 9.3, we consider a market with a constant bond $S_0(t) = 1$ and one risky asset S_1, here given by $S_1(t) := X_1(t)$.

This can in fact be viewed as a special case of the previous example. Indeed, the local characteristics of X are of the form (5.36–5.38) with $K = 0$ and hence $\delta = 0$. In Sect. 5.7.3 we noted that such processes can be written as solutions to (5.35), where the integrals relative to $\mu - \nu$ and μ vanish in this case.

For $n = 1$, $\beta(x) := x(\mu + \sigma^2/2)$, $\gamma(x) := x^2\sigma^2$ we recover the geometric Brownian motion of Example 9.1. The Heston model of Sect. 8.2.4 falls into the present framework as well.

For applications of the continuous Markovian model see Examples 10.15, 10.21, 10.32, 10.50 (optimal investment) and Sect. 11.4 (hedging).

At this stage we remain unspecific about the particular choice of S and assume it to be given exogenously. Two concrete approaches to modelling come into mind. One could derive a specific price process based on assumptions on the behaviour, endowment and particular preferences of all market participants. However, this economically appealing approach is hard to implement in practice. Even if the assumptions came sufficiently close to reality, it would probably be next to impossible to obtain the quantitative information on endowments, preferences etc. that are needed to come up with a concrete price process.

Alternatively, one may construct a model on purely statistical grounds based on the analysis of observed price data. This is more or less the origin of the models in Chap. 8. But an extreme form of this approach tends to ignore economic patterns behind the data, which may no longer be present in the fitted model. Ideally, one constructs models which rely on both statistical inference and on economic considerations limiting the conceivable class of processes.

These economic considerations and also most issues in Mathematical Finance involve dynamic trading of the securities in the market. This is modelled mathematically in terms of *trading strategies*. A **trading strategy** or **portfolio** is a predictable vector-valued stochastic process $\varphi = (\varphi_0(t), \dots, \varphi_d(t))_{t\in[0,T]}$ which represents the dynamic evolution of a trader's portfolio. The random variable $\varphi_i(t)$ stands for the number of shares of asset i which the investor holds at time t. It is crucial to allow for random trading strategies because the investor may want to base her portfolio on the past evolution of market prices, which are themselves random.

But why do we assume φ to be predictable? Adaptedness should naturally hold because the trader cannot take information and events into account that have not yet happened. Predictability must be imposed for almost the same reason. At a particular time t both the portfolio and prices may change. As indicated in Sect. 1.2, we must agree upon what happens first. Otherwise it is not clear whether assets are bought or sold at the old price $S(t-)$ or the new price $S(t)$. The theory implicitly assumes that changes of the portfolio take place prior to price changes. Consequently, the information on the price change $S(t-) \to S(t)$ *at* time t cannot be incorporated in $\varphi(t)$ because it has not yet happened. In other words, $\varphi(t)$ may only depend on the information up to but excluding t. This intuition is made precise by the notion of predictability, cf. Sect. 2.1.

The **value** or **wealth process** of a portfolio φ is naturally defined as

$$V_\varphi := \varphi^\top S = \sum_{i=0}^{d} \varphi_i S_i. \tag{9.4}$$

In this book we consider only *self-financing* strategies, which means that no money is added to or withdrawn from the portfolio after inception at time 0. Securities can only be bought if others are sold at the same time. This condition is made mathematically precise by considering the *gains process* $\varphi \bullet S(t)$ of a portfolio. From the very definition of the stochastic integral in Sects. 1.2 and 3.2 it follows that this quantity naturally describes the cumulative amount of money that the investor has earned due to price changes from time 0 up to and including t. If no money is added or withdrawn after 0, the value of the portfolio must be the sum of initial endowment and gains from trade. Put differently, we have

$$V_\varphi = V_\varphi(0) + \varphi \bullet S, \tag{9.5}$$

where it is implicitly assumed that φ is S-integrable. Consequently, it is natural to call trading strategies φ with this property **self-financing**.

9.1.2 Discounting

So far, we have expressed all prices in terms of a common currency. It turns out that concrete calculations simplify considerably if we replace this currency by some tradable asset. For ease of notation we suppose this **numeraire** asset to be the zeroth security S_0 in our price process $S = (S_0, \dots, S_d)$. From now on we assume the numeraire S_0 and also its left limit $(S_0)_-$ to be strictly positive. In this case we can define the **discounted price process** as

$$\widehat{S} := \frac{1}{S_0} S = \left(1, \frac{S_1}{S_0}, \dots, \frac{S_d}{S_0}\right).$$

Occasionally, we use the same letter $\widehat{S}$ to denote the $\mathbb{R}^d$-valued process $(\widehat{S}_1, \dots, \widehat{S}_d)$ because $\widehat{S}_0 = 1$ is trivial. Accordingly,

$$\widehat{V}_\varphi := \frac{1}{S_0} V_\varphi = \varphi^\top \widehat{S}$$

defines the **discounted value** or **discounted wealth process** of a trading strategy φ.

If S_0 is a bank account with fixed interest rate r as in Example 9.1, we have $\widehat{S}(t) = e^{-rt} S(t)$, which means that $\widehat{S}(t)$ coincides with the *present value* of $S(t)$. Put differently, *discounting* coincides with the usual operation in bookkeeping. But we also use this term for arbitrary securities or even dynamic portfolios as numeraires. Especially in interest rate theory an ingenious choice of the numeraire leads to much simpler expressions.

Whether a strategy is self-financing can be expressed in terms of discounted processes.

Rule 9.5 *A strategy φ is self-financing if and only if φ is $\widehat{S}$-integrable and*

$$\widehat{V}_\varphi = \widehat{V}_\varphi(0) + \varphi \bullet \widehat{S}. \tag{9.6}$$

Idea Suppose that (9.6) holds. Integration by parts of $\varphi^\top S = (\varphi^\top \widehat{S})S_0$ and Rule 3.14(3,4) yield

$$\begin{aligned}\varphi^\top S &= \varphi(0)^\top S(0) + (\varphi^\top \widehat{S})_- \bullet S_0 + (S_0)_- \bullet (\varphi \bullet \widehat{S}) + [\varphi \bullet \widehat{S}, S_0] \\ &= \varphi(0)^\top S(0) + (\varphi^\top \widehat{S})_- \bullet S_0 + (\varphi(S_0)_-) \bullet \widehat{S} + \varphi \bullet [\widehat{S}, S_0].\end{aligned}$$

Note that $\Delta(\varphi^\top \widehat{S}) = \Delta(\varphi \bullet \widehat{S}) = \varphi^\top \Delta \widehat{S}$ and hence $(\varphi^\top \widehat{S})_- = \varphi^\top \widehat{S}_-$. Again using integration by parts and Rule 3.14(3,4), we obtain

$$\begin{aligned}\varphi^\top S &= \varphi(0)^\top S(0) + \varphi \bullet \big(\widehat{S}_- \bullet S_0 + (S_0)_- \bullet \widehat{S} + [\widehat{S}, S_0]\big) \\ &= \varphi(0)^\top S(0) + \varphi \bullet \big(\widehat{S} S_0\big).\end{aligned}$$

The converse statement follows from replacing $S, 1/S_0$ with $\widehat{S}, S_0$. □

Observe that the integral $\varphi \bullet \widehat{S}$ does not depend on the numeraire part φ_0 because $\widehat{S}_0$ is constant.

Defining a self-financing strategy means choosing $1 + d$ processes such that a nontrivial constraint holds. This task can be reduced to choosing d processes without constraint. Indeed, the following rule shows that it suffices to fix the initial endowment $V_\varphi(0)$ and arbitrary investment strategies $\varphi_1, \dots, \varphi_d$ for the non-numeraire assets in order to define a unique self-financing strategy. This is intuitively obvious because any money that is not invested in the "risky" securities $S_1, \dots, S_d$ remains on the "money market account" S_0—or whatever the numeraire stands for.

Rule 9.6 *Suppose that $(\varphi_1, \dots, \varphi_d)$ is an $\widehat{S}$-integrable process and v_0 a real number. Then there exists a unique predictable process φ_0 such that $\varphi := (\varphi_0, \varphi_1, \dots, \varphi_d)$ is a self-financing strategy with initial value $V_\varphi(0) = v_0$.*

Idea By the previous rule φ is self-financing if and only if

$$\begin{aligned}&\varphi_0(t)\widehat{S}_0(t) + (\varphi_1, \dots, \varphi_d)(t)^\top \big(\widehat{S}_1, \dots, \widehat{S}_d\big)(t) \\ &\qquad = \frac{v}{S_0(0)} + (\varphi_1, \dots, \varphi_d) \bullet \big(\widehat{S}_1, \dots, \widehat{S}_d\big)(t),\end{aligned}$$

i.e. if and only if

$$\begin{aligned}\varphi_0(t) &= \frac{v}{S_0(0)} + (\varphi_1, \dots, \varphi_d) \bullet \big(\widehat{S}_1, \dots, \widehat{S}_d\big)(t) - (\varphi_1, \dots, \varphi_d)(t)^\top \big(\widehat{S}_1, \dots, \widehat{S}_d\big)(t) \\ &= \frac{v}{S_0(0)} + (\varphi_1, \dots, \varphi_d) \bullet \big(\widehat{S}_1, \dots, \widehat{S}_d\big)(t-) - (\varphi_1, \dots, \varphi_d)(t)^\top \big(\widehat{S}_1, \dots, \widehat{S}_d\big)(t-).\end{aligned}$$

This is a predictable process. □

The preceding rule indicates why discounting simplifies calculations considerably. In order to define φ, we only need to specify $\varphi_1, \ldots, \varphi_d$ and $V_\varphi(0)$. By (9.6) we can compute its discounted value $\widehat{V}_\varphi$ without knowing φ_0 because the integral $\varphi \bullet \widehat{S}$ does not depend on the numeraire component. Should φ_0 really be needed, it is readily obtained from

$$\varphi_0 = \widehat{V}_\varphi - \sum_{i=1}^{d} \varphi_i \widehat{S}_i.$$

From now on we often identify $\mathbb{R}^d$-valued processes $(\varphi_1, \ldots, \varphi_d)$ with their unique self-financing extension $(\varphi_0, \varphi_1, \ldots, \varphi_d)$ satisfying $V_\varphi(0) = 0$. Moreover, we use the letter φ for both processes.

9.1.3 Dividends

So far, we have implicitly assumed that the securities do not pay dividends. This restriction must be relaxed in particular for assets such as coupon bonds. We will see that futures contracts can be viewed as securities with dividend payments as well. Therefore we extend the notions from Sects. 9.1.1–9.1.2 to the dividend case. There are two ways to accomplish this. The first approach explicitly keeps track of dividends. Alternatively, one can replace dividend-paying assets by fictitious dividend-free securities that yield the same profits and losses.

9.1.3.1 Explicit Modelling

As before, prices are modelled by an $\mathbb{R}^{1+d}$-valued semimartingale S. Dividends are expressed in terms of a **dividend process** $D = (D_0(t), \ldots, D_d(t))_{t\in[0,T]}$. It is assumed to be a vector-valued semimartingale satisfying $D(0) = (0, \ldots, 0)$. The random variable $D_i(t)$ stands for the cumulative dividends that are distributed up to time t on one share of security i. The investor's dynamic portfolio is modelled as before by some $\mathbb{R}^{1+d}$-valued predictable process $\varphi = (\varphi_0(t), \ldots, \varphi_d(t))_{t\in[0,T]}$.

In order to get the bookkeeping right, we have to be very careful about the order of events. Three events may happen at time t: the portfolio is restructured by the investor, prices move and dividends are paid. We assume that they happen in precisely this order. But note that the portfolio changes again when dividends are paid because the latter are paid to the investor's account. We interpret $\varphi(t)$ as the investor's portfolio *before* dividends are paid at time t. On the other hand, the random variable $V_\varphi(t)$ is supposed to refer to the value of the portfolio *after* prices have changed and dividends are paid at time t. This leads to the definition

$$V_\varphi := \varphi^\top (S + \Delta D) \tag{9.7}$$

of the **value** or **wealth process** of φ. It differs from (9.4) because $\varphi(t)$ only incorporates the dividends paid before but not *at* time t.

Since the investor earns money due to price changes and dividend payments, the *gains* of a portfolio up to and including time t are naturally given by $\varphi \bullet (S + D)$. As in Sect. 9.1.1 we call a strategy **self-financing** if its value is the sum of initial endowment and gains, i.e. if φ is $(S + D)$-integrable and satisfies

$$V_\varphi = V_\varphi(0) + \varphi \bullet (S + D).$$

In order to define discounted prices we assume $D_0 = 0$, i.e. no dividends are paid on the numeraire asset. The concept of discounting turns out to be nontrivial in the presence of dividends. For a better understanding we look at discrete time first.

Example 9.7 (Discrete-Time Markets) Consider an asset price process $S = (S_0(t), \dots, S_d(t))_{t=0,1,2,\dots}$ and a dividend process $D=(D_0(t), \dots, D_d(t))_{t=0,1,2,\dots}$ in discrete time. Discounting means to express prices as multiples of the numeraire. Since the incremental dividend $\Delta D(t) = D(t) - D(t-1)$ for the period $t-1$ to t is paid at time t, it must be discounted by $S_0(t)$, which leads to the definition

$$\Delta \widehat{D}(t) := \frac{1}{S_0(t)} \Delta D(t), \quad t = 1, 2, \dots \tag{9.8}$$

or, equivalently,

$$\widehat{D}_i(t) := \frac{1}{S_0} \bullet \Delta D_i(t) \tag{9.9}$$

in the sense of (1.11) for $i = 0, \dots, d$ and $t = 0, 1, 2, \dots$ Since the integrand $1/S_0$ may not be predictable, the continuous-time counterpart of (9.9) is generally undefined. Therefore we rewrite (9.8) as

$$\Delta \widehat{D}_i(t) := \frac{1}{S_0(t-)} \Delta D_i(t) + \Delta \frac{1}{S_0}(t) \Delta D_i(t), \quad t = 1, 2, \dots$$

or, equivalently,

$$\widehat{D}_i(t) := \frac{1}{(S_0)_-} \bullet D_i(t) + \left[\frac{1}{S_0}, D_i\right](t) \tag{9.10}$$

for $i = 0, \dots, d$ and $t = 0, 1, 2, \dots$ Expression (9.10) makes sense in continuous time as well.

The previous example motivates us to define the **discounted dividend process** as

$$\widehat{D} := \frac{1}{(S_0)_-} \bullet D + \left[\frac{1}{S_0}, D\right], \tag{9.11}$$

where both the integral and the covariation are to be understood componentwise in $D = (D_0, \dots, D_d)$. If the dividend process D is of finite variation or the numeraire price S_0 is predictable and of finite variation, this reduces to

$$\widehat{D} = \frac{1}{S_0} \bullet D,$$

which could have been inspired by (9.9). Observe that the jumps of $\widehat{D}$ are in any case given by

$$\Delta\widehat{D}(t) = \frac{1}{S_0(t-)}\Delta D(t) + \Delta\frac{1}{S_0(t)}\Delta D(t) = \frac{\Delta D(t)}{S_0(t)},$$

which parallels (9.8).

The **discounted value** or **discounted wealth process** of φ is defined naturally as

$$\widehat{V}_\varphi := \frac{V_\varphi}{S_0} = \varphi^\top\big(\widehat{S} + \Delta\widehat{D}\big).$$

The counterpart of Rule 9.5 now reads as follows.

Rule 9.8 *A strategy φ is self-financing if and only if φ is $\big(\widehat{S}+\widehat{D}\big)$-integrable and*

$$\widehat{V}_\varphi = \widehat{V}_\varphi(0) + \varphi \bullet \big(\widehat{S}+\widehat{D}\big).$$

Idea The proof is similar to that of Rule 9.5. □

Moreover, Rule 9.6 holds almost literally in the dividend case as well.

Rule 9.9 *Suppose that $(\varphi_1, \dots, \varphi_d)$ is an $(\widehat{S}+\widehat{D})$-integrable process and v_0 a real number. Then there exists a unique predictable process φ_0 such that $\varphi := (\varphi_0, \varphi_1, \dots, \varphi_d)$ is a self-financing strategy with initial value $V_\varphi(0) = v_0$.*

Idea The proof is similar to that of Rule 9.6. □

9.1.3.2 Implicit Modelling

As we have seen above, dividend processes can be naturally embedded in the general mathematical framework. But many results from Mathematical Finance have not yet been stated for processes with dividends. Therefore the use of such statements would require extending them to the dividend case in the first place. We briefly discuss an alternative approach which avoids this potentially tedious work. The idea is to replace the true dividend-paying assets by fictitious dividend-free securities.

For an investor a dividend payment of 1€ differs from a rise of the stock price by 1€ only by the fact that the dividend is paid in currency whereas the stock price increase is not. We formalise this intuition by introducing dividend-free fictitious securities $\underline{S}_0, \dots, \underline{S}_d$ which yield the same profits and losses as the original assets $S_0, \dots, S_d$. To this end, assume that the price process S and the dividend process D are as in the previous section and that no dividends are paid for the numeraire asset, i.e. $D_0 = 0$.

The **fictitious dividend-adjusted price process** is defined as

$$\underline{S} := S + S_0 \widehat{D}$$

with $\widehat{D}$ as in (9.11). If φ denotes a strategy, we define the corresponding **dividend-adjusted trading strategy** $\underline{\varphi}$ for $\underline{S}$ by $\underline{\varphi}(0) := \varphi(0)$ and

$$\underline{\varphi}_i(t) := \begin{cases} \varphi_i(t) & \text{for } i = 1, \dots, d, \\ \varphi_0(t) - \sum_{i=0}^d \varphi_i(t) \frac{\underline{S}_i(t-) - S_i(t-)}{S_0(t-)} & \text{for } i = 0 \end{cases} \tag{9.12}$$

for $t > 0$. Its value or wealth process is defined as usual by

$$\underline{V}_{\underline{\varphi}} := \underline{\varphi}^\top \underline{S}.$$

The following rule shows that the above definitions make sense.

Rule 9.10

1. *For any strategy φ we have $\underline{V}(\underline{\varphi}) = V(\varphi)$.*
2. *For the discounted processes we have $\widehat{\underline{S}} = \widehat{S} + \widehat{D}$.*
3. *If $Q \sim P$ denotes another probability measure, then $\widehat{\underline{S}}$ is a Q-martingale if and only if $\widehat{S} + \widehat{D}$ is a Q-martingale.*
4. *A strategy φ is self-financing relative to S, D if and only if the corresponding dividend-adjusted strategy $\underline{\varphi}$ is self-financing for $\underline{S}$.*
5. *Any predictable $\mathbb{R}^{1+d}$-valued process ψ can be written as $\psi = \underline{\varphi}$ for some predictable $\mathbb{R}^{1+d}$-valued process φ.*

Idea

1. From $\Delta D = S_0 \Delta \widehat{D}$ it follows that

$$\begin{aligned} \underline{V}_{\underline{\varphi}}(t) &= \sum_{i=0}^d \underline{\varphi}_i(t) \underline{S}_i(t) \\ &= \sum_{i=0}^d \varphi_i(t) \left(\underline{S}_i(t) - \frac{\underline{S}_i(t-) - S_i(t-)}{S_0(t-)} S_0(t) \right) \\ &= \sum_{i=0}^d \varphi_i(t) \left(S_i(t) + S_0(t) \widehat{D}_i(t) - \widehat{D}_i(t-) S_0(t) \right) \end{aligned}$$

$$= \sum_{i=0}^{d} \varphi_i(t) \left(S_i(t) + \Delta D_i(t)\right)$$

$$= V_\varphi(t)$$

as desired.

2. This is evident because $\widehat{\underline{S}} = \underline{S}/S_0 = \widehat{S} + \widehat{D}$.
3. This follows immediately from statement 2.
4. By Rule 9.5 strategy $\underline{\varphi}$ is self-financing if and only if

$$\widehat{\underline{V}}_{\underline{\varphi}} - \widehat{\underline{V}}_{\underline{\varphi}}(0) = \underline{\varphi} \bullet \widehat{\underline{S}}. \tag{9.13}$$

Similarly, φ is self-financing if and only if

$$\widehat{V}_\varphi - \widehat{V}_\varphi(0) = \varphi \bullet \left(\widehat{S} + \widehat{D}\right), \tag{9.14}$$

cf. Rule 9.8. Since the left-hand sides of (9.13, 9.14) coincide by statement 1 and the right-hand sides by statement 2, the asserted equivalence follows.

5. This follows from solving (9.12) for φ. □

9.2 Trading with Consumption

In applications to optimal investment and valuation of derivative contracts we need to consider trading strategies allowing for consumption. We work with the general setup of an asset price process $S = (S_0, \ldots, S_d)$ without dividend payments as in Sect. 9.1.1. We call (φ, C) a **strategy/consumption pair** if the *trading strategy* φ denotes a predictable $\mathbb{R}^{1+d}$-valued process as before and C a real-valued semimartingale. $C(t)$ stands for the **cumulative consumption** of the investor, i.e. for the money she spends up to time t. As in the case of dividends, we must be careful about the order in which things happen at time t, We assume that the portfolio is rebalanced before asset prices change, which in turns precedes consumption. Similarly as in (9.7), the **value process** of (φ, C) is defined as

$$V_{\varphi,C} = \varphi^\top (S - \Delta C). \tag{9.15}$$

The last term is due to the fact that $\varphi(t)$ reflects consumption *before* but not *at* time t. Parallel to Sects. 9.1.1 or 9.1.3.1, we call the pair (φ, C) **self-financing** (or φ C**-financing**) if

$$V_{\varphi,C} = V_{\varphi,C}(0) + \varphi \bullet S - C,$$

i.e. if the current wealth equals the difference of initial endowment plus gains and consumption. The same motivation as in Sect. 9.1.3.1 leads to the definition

$$\widehat{C} := \frac{1}{(S_0)_-} \bullet C + \left[\frac{1}{S_0}, C \right] \tag{9.16}$$

of **discounted cumulative consumption**, which simplifies to

$$\widehat{C} = \frac{1}{S_0} \bullet C$$

if S_0 is predictable. Moreover, the **discounted value** or **discounted wealth process** of (φ, C) is defined as

$$\widehat{V}_{\varphi,C} := \frac{V_{\varphi,C}}{S_0} = \varphi^\top \big(\widehat{S} - \Delta \widehat{C}\big).$$

This leads to the following counterpart of Rules 9.5 and 9.8.

Rule 9.11 *A strategy/consumption pair (φ, C) is self-financing if and only if φ is $\widehat{S}$-integrable and*

$$\widehat{V}_{\varphi,C} = \widehat{V}_{\varphi,C}(0) + \varphi \bullet \widehat{S} + \widehat{C}.$$

Idea The proof is similar to that of Rule 9.5. □

Given the initial endowment v_0, the "risky" part $(\varphi_1, \dots, \varphi_d)$ of the portfolio, and the cumulative consumption process C, the missing component φ_0 of the portfolio can always be chosen such that φ is C-financing. This is to be expected from the similar Rules 9.6 and 9.9.

Rule 9.12 *Suppose that $(\varphi_1, \dots, \varphi_d)$ is an $\widehat{S}$-integrable process and v_0 a real number. Then there exists a unique predictable process φ_0 such that $\varphi := (\varphi_0, \varphi_1, \dots, \varphi_d)$ is a C-financing strategy with initial value $V_{\varphi,C}(0) = v_0$.*

Idea The proof is similar to that of Rule 9.6. □

As a side remark note that the definition of $V_{\varphi,C}$ differs from $V_{\varphi,c}$ in (1.51) in the sense that the latter excludes consumption at time t. Moreover, C stands for *cumulative* consumption as opposed to the consumption *rate* c.

9.3 Fundamental Theorems of Asset Pricing

Market prices result from the interplay of supply and demand, which may lead to complex and unpredictable random fluctuations. But even if we do not know the true mechanisms and processes underlying observed data, simple economic considerations yield relations which should hold in general. Consider, for example, two call options on the same stock with the same maturity but with a different strike price. It seems evident that the option with smaller strike should have a larger market price—regardless of any other circumstances. This natural postulate can be justified and formalised in terms of riskless gains. If the above relation did not hold, you could buy the option with low strike and sell the other one with higher strike at no cost. At maturity the revenues of the long position more than suffice to meet the obligations involved in the short option. You would have produced a perfectly riskless gain (*arbitrage*). Since it involves no initial capital, it could theoretically be scaled up to arbitrary amounts. Economic common sense considers such riskless gains as impossible: if such an arbitrage opportunity exists, somebody will probably detect and exploit it immediately. Therefore *absence of arbitrage* seems to be a reasonable and very modest assumption about real markets. Nevertheless, it has far-reaching implications in particular for derivative pricing.

Mathematically, an *arbitrage* is defined as a self-financing trading strategy φ with zero initial endowment and non-negative terminal wealth which is strictly positive with positive probability:

Definition 9.13 An **arbitrage opportunity** is a self-financing strategy with $V_\varphi(0) = 0$, $V_\varphi(T) \geq 0$ and $P(V_\varphi(T) > 0) > 0$.

Unfortunately, arbitrage in the proper sense of Definition 9.13 exists even in quite innocent and ubiquitous market models such as in the geometric Brownian motion setup of Example 9.1, cf. Example 11.46 in this context. Therefore the definition needs to be modified in order to be useful. Similarly, the rules in this section hold literally true only in discrete-time markets. Their rigorous statement in continuous time requires a careful specification of admissible trading strategies, a distinction of martingales and σ-martingales etc. While a thorough mathematical treatment of these issues is mathematically utmost desirable, we do not feel that it contributes much to the understanding of the dynamics of real markets. Therefore we stick to a somewhat informal engineering-style attitude. For rigorous statements we refer the reader to Sect. 11.7 and to the notes and comments.

As noted above, one generally assumes that real markets do not allow for riskless gains. This implies that assets must have the same price on different exchanges. Otherwise buying low in one place and selling high in the other produces such a *free lunch*. This idea can be transferred to a kind of identity in time rather than location.

Rule 9.14 (Law of One Price) *Suppose that the market does not allow for arbitrage. If φ, ψ are self-financing strategies with the same final value $V_\varphi(T) = V_\psi(T)$, then $V_\varphi(t) = V_\psi(t)$ for all $t \in [0, T]$. Similarly, $V_\varphi(T) \leq V_\psi(T)$ implies $V_\varphi \leq V_\psi$.*

Idea Otherwise there is a $t \in [0, T]$ with $P(V_\varphi(t) \neq V_\psi(t)) > 0$, e.g. $P(A) > 0$ for $A := \{V_\varphi(t) > V_\psi(t)\}$. Set

$$(\vartheta^1, \ldots, \vartheta^d)(s) := \begin{cases} 0 & \text{for } s \leq t, \\ \left((\psi^1, \ldots, \psi^d)(s) - (\varphi^1, \ldots, \varphi^d)(s)\right) 1_A & \text{for } s > t \end{cases}$$

and denote by ϑ the corresponding self-financing strategy from Rule 9.6 for $v = 0$. Then $\widehat{V}_\vartheta(0) = 0$ and

$$\begin{aligned} \widehat{V}_\vartheta(T) &= \left((\psi - \varphi)1_{A\times(t,T]}\right) \bullet \widehat{S}(T) \\ &= 1_{A\times(t,T]} \bullet ((\psi - \varphi) \bullet \widehat{S})(T) \\ &= \left((\psi - \varphi) \bullet \widehat{S}(T) - (\psi - \varphi) \bullet \widehat{S}(t)\right) 1_A \\ &= \left(\widehat{V}_\psi(T) - \widehat{V}_\varphi(T) - \widehat{V}_\psi(t) + \widehat{V}_\varphi(t)\right) 1_A \\ &\geq \left(\widehat{V}_\varphi(t) - \widehat{V}_\psi(t)\right) 1_A. \end{aligned}$$

Since $\widehat{V}_\vartheta(T)$ is almost surely nonnegative and positive on A, we have that ϑ is an arbitrage. □

Absence of arbitrage can be rephrased in purely mathematical terms. The corresponding theorem opens the door to the use of martingale methods in Mathematical Finance. It has far-reaching implications for applications such as option pricing.

Rule 9.15 (First Fundamental Theorem of Asset Pricing, FTAP) *The following statements are equivalent.*

1. *There are no arbitrage opportunities.*
2. *There is a* ***state price density process****, i.e. a semimartingale Z with $Z, Z_- > 0$ and such that ZS is a martingale.*
3. *There exists an* ***equivalent martingale measure (EMM)****, i.e. some probability measure $Q \sim P$ such that the discounted price process $\widehat{S}$ is a Q-martingale.*

Martingale in 2. and 3. has to be understood componentwise, i.e. $ZS_0, \ldots, ZS_d$ and $\widehat{S}_0, \ldots, \widehat{S}_d$ are martingales resp. Q-martingales.

Idea $1 \Rightarrow 3$: Denote by

$$K := \{\varphi \bullet \widehat{S}(T) : \varphi \text{ self-financing with } \varphi(0) = 0\} \tag{9.17}$$

the subspace of all discounted terminal wealths which can be attained with zero initial capital. By absence of arbitrage it does not intersect with the convex set $M := \{X \text{ nonnegative random variable: } E(X) = 1\}$. If a separation theorem similar to Theorem A.14 can be applied, there is some random variable Y with $E(YX) = 0$ for all $X \in K$ and $E(YX) > 0$ for all $X \in M$. The second property implies that $Y > 0$ almost surely. Hence $\frac{dQ}{dP} := Y/E(Y)$ defines a probability measure $Q \sim P$. The first property yields $E_Q(X) = 0$ for all $X \in K$ and in particular for $X = \varphi \bullet \widehat{S}(T)$, where $i \in \{1, \dots, d\}$, $t \in [0, T]$, $C \in \mathscr{F}_t$, and the strategy φ is defined as

$$\varphi_j(s)(\omega) := \begin{cases} 1_C(\omega)1_{(t,T]}(s) & \text{for } j = i, \\ 0 & \text{otherwise.} \end{cases}$$

Since $0 = E_Q(X) = E_Q(\widehat{S}_i(T)1_C) - E_Q(\widehat{S}_i(t)1_C)$, we have $\widehat{S}_i(t) = E_Q(\widehat{S}_i(T)|\mathscr{F}_t)$, which means that $\widehat{S}_i$ is a Q-martingale.

$3 \Rightarrow 1$: Denote by φ a self-financing strategy with $V_\varphi(0) = 0$ and $V_\varphi(T) \geq 0$, which implies $\varphi \bullet \widehat{S}(T) = \widehat{V}_\varphi(T) \geq 0$. Since $\varphi \bullet \widehat{S}$ is a Q-martingale by Rule 3.14(6), we have $E_Q(\varphi \bullet \widehat{S}(T)) = E_Q(\varphi \bullet \widehat{S}(0)) = 0$. This yields $\widehat{V}_\varphi(T) = \varphi \bullet \widehat{S}(T) = 0$ almost surely because this random variable is nonnegative.

$3 \Rightarrow 2$: Take $Z := Z_0/S_0$ where Z_0 denotes the density process of some EMM relative to numeraire S_0. The claim follows from Proposition 3.69.

$2 \Rightarrow 3$: Since $Z_0 := ZS_0$ is a positive martingale, it is the density process of some probability measure $Q \sim P$. This measure is an EMM because Z_0S is a martingale, cf. Proposition 3.69. □

Under the *real*, *statistical* or *physical* probability measure P one typically assumes discounted stock prices to have a positive drift or excess return in order to compensate for the risk. This is not true relative to the hypothetical probability measure Q in Rule 9.15, which is why Q is often called *risk neutral measure*.

As an application of the first fundamental theorem, we observe that models of exponential Lévy type are typically arbitrage free.

Example 9.16 (Geometric Lévy Model) Let us consider the bond and stock model of Example 9.2 with a Lévy process X such that $X(t) - rt$ is neither increasing nor decreasing. We show that this model does not allow for arbitrage. Indeed, by Rule 9.15 and Theorem 2.22 it suffices to find some probability measure $Q \sim P$ such that X is a Q-Lévy process whose Q-Lévy–Khintchine triplet $(\widetilde{a}^X, \widetilde{c}^X, \widetilde{K}^X)$ satisfies

$$0 = \widetilde{a}^X - r + \frac{1}{2}\widetilde{c}^X + \int (e^x - 1 - x)\widetilde{K}^X(dx). \tag{9.18}$$

This can be achieved by applying Theorem 4.20 twice. Indeed, in a first step one may choose κ in this theorem such that the new Lévy measure $\widetilde{K}^X$ satisfies $\int_{\{|x|>1\}} e^{px}\widetilde{K}^X(dx) < \infty$ for any $p \in \mathbb{R}$. In a second step we apply Theorem 4.20 once more in order for (9.18) to hold. If the diffusion coefficient c^X does not vanish, this is achieved by choosing β in Theorem 4.20 appropriately. Otherwise we can, for example, consider $\kappa(x) = e^{\vartheta x}$ with some parameter $\vartheta \in \mathbb{R}$. By the intermediate value theorem, β resp. ϑ can be chosen such that (9.18) holds relative to the new measure.

Absence of arbitrage can also be expected in Itō process models if the diffusion coefficient does not vanish.

Example 9.17 (Itō Process Model) We consider the situation of Example 9.3 with the additional assumption that σ^2 is strictly positive. Define a probability measure $Q \sim P$ with density process $Z := \mathscr{E}(-\frac{\mu}{\sigma} \bullet W)$. By Corollary 3.71 or Theorem 4.15 the drift rate of S_1 under Q vanishes, which means that Q is an EMM.

Observe that the martingale measure in the above theorems is generally not unique. Moreover, it depends on the choice of the numeraire. The following result indicates how it is affected by a numeraire change.

Rule 9.18 (Change of Numeraire) *Let $Q \sim P$ be a probability measure and $i \in \{1, \dots, d\}$ such that $S_i, (S_i)_- > 0$. Then Q is an EMM for numeraire S_0 if and only if Q_i is an EMM for numeraire S_i, where*

$$\frac{dQ_i}{dQ} := \frac{\widehat{S}_i(T)}{E(\widehat{S}_i(0))}.$$

In this case the density process of Q_i relative to Q is given by

$$Z := \frac{\widehat{S}_i}{E(\widehat{S}_i(0))}.$$

Idea This follows immediately from Proposition 3.69. □

In principle, the previous result can be applied to any tradable asset $\widetilde{S}$ in the sense that $\widetilde{S} = V_\varphi$ for some self-financing strategy φ.

9.3.1 Dividends

So far, we assumed the situation without dividend payments. We will need a dividend version of the FTAP in the following chapter for the valuation of futures contracts. It reads as follows.

Rule 9.19 (FTAP with Dividends) *Suppose that S and D are price and dividend processes as in Sect. 9.1.3. Then the following statements are equivalent.*

1. *There are no arbitrage opportunities.*
2. *There exists an equivalent martingale measure for $\widehat{S} + \widehat{D}$, i.e. some probability measure $Q \sim P$ such that $\widehat{S} + \widehat{D}$ is a Q-martingale.*

Idea By Rule 9.10 an arbitrage for the market with dividends corresponds to an arbitrage for the dividend-free asset $\underline{S}$. Likewise, Q is an EMM for $\widehat{S} + \widehat{D}$ if and only if it is an EMM for $\underline{S}$. The assertion follows now by applying Rule 9.15 to $\underset{\wedge}{S}$. □

9.3.2 Constraints

Let us return to the dividend-free case. In some instances one may want to restrict trading to some subset Γ of self-financing strategies, e.g. if short-selling of certain assets is prohibited. This leads to the following version of the FTAP. It turns out to be useful for the valuation of American options.

Rule 9.20 (FTAP Under Constraints) *Suppose that Γ is a convex cone of self-financing strategies. Then the following statements are equivalent.*

1. *There are no* Γ-arbitrage opportunities, *i.e. no arbitrage opportunities that belong to Γ.*
2. *There exists some probability measure $Q \sim P$ such that $E_Q(\varphi \bullet \widehat{S}(T)) \leq 0$ for all strategies $\varphi \in \Gamma$.*

Idea This follows similarly as in the proof of Rule 9.15 if we replace the subspace (9.17) with the convex cone $K := \{\varphi \bullet \widehat{S}(T) : \varphi \in \Gamma \text{ and } \varphi(0) = 0\}$. □

9.3.3 Bid-Ask Spreads

Finally we have a brief look at models that allow for transaction costs or, more precisely, bid-ask spreads. In this framework we are facing a **bid price process** $\underline{S} = (\underline{S}_0(t), \dots, \underline{S}_d(t))_{t\in[0,T]}$ and an **ask price process** $\overline{S} = (\overline{S}_0(t), \dots, \overline{S}_d(t))_{t\in[0,T]}$ rather then a universal price process S as before. We naturally assume $0 \leq \underline{S}_i \leq \overline{S}_i$ for $i = 0, \dots, d$. The meaning should be obvious: if one wants to buy security i, one must pay the higher price $\overline{S}_i(t)$ whereas one receives only $\underline{S}_i(t)$ for selling it. Moreover, we suppose that the numeraire asset is strictly positive and traded without spread, i.e. $S_0(t) := \underline{S}_0(t) = \overline{S}_0(t) > 0$.

Trading strategies are defined as before as predictable processes $\varphi = (\varphi_0(t), \dots, \varphi_d(t))_{t\in[0,T]}$. For simplicity we assume them to be of finite variation. Indeed, if the spread is strictly positive, strategies of infinite variation typically lead to infinite transaction costs, which makes them unfeasible in practice.

Condition (9.5) cannot be used any more to define self-financing strategies because S no longer exists. Instead we use an alternative equation based on the intuition that no funds are added or withdrawn. To this end, we write the components of any strategy φ as a difference $\varphi_i = \varphi_i^{\uparrow} - \varphi_i^{\downarrow}$ of two increasing processes $\varphi_i^{\uparrow}, \varphi_i^{\downarrow}$ which do not grow at the same time. In order to understand self-financability in the context of bid-ask spreads we look at the discrete-time situation first.

Example 9.21 (Discrete-Time Market) Consider bid and ask price processes $\underline{S} = (\underline{S}_0(t), \dots, \underline{S}_d(t))_{t\in\{0,1,\dots T\}}$ and $\overline{S} = (\overline{S}_0(t), \dots, \overline{S}_d(t))_{t\in\{0,1,\dots T\}}$ with $\underline{S}_i \leq \overline{S}_i$ for $i = 0, \dots, d$. The investor chooses a trading strategy, i.e. a predictable process $\varphi = (\varphi_0(t), \dots, \varphi_d(t))_{t\in\{0,1,\dots,T\}}$. As noted above, we write $\varphi_i = \varphi_i^{\uparrow} - \varphi_i^{\downarrow}$ with increasing processes $\varphi_i^{\uparrow}, \varphi_i^{\downarrow}$ which do not grow at the same time. Self-financability means that proceeds and expenses from rebalancing at any time t sum up to 0, i.e.

$$\sum_{i=0}^{d} \left(\overline{S}_i(t-1)\Delta\varphi_i^{\uparrow}(t) - \underline{S}_i(t-1)\Delta\varphi_i^{\downarrow}(t)\right) = 0$$

or, in integral terms,

$$\overline{S}_- \bullet \varphi^{\uparrow} - \underline{S}_- \bullet \varphi^{\downarrow} = 0. \tag{9.19}$$

However, the continuous-time counterpart of the terms in (9.19) may not be obvious because the stochastic integral in Sect. 3.2 is defined only for càdlàg processes. Therefore we rewrite the integrals in (9.19) as

$$\begin{aligned}
\overline{S}_- \bullet \varphi^{\uparrow}(t) &= \sum_{s=1}^{t} \overline{S}_i(s-1)\big(\varphi_i^{\uparrow}(s) - \varphi_i^{\uparrow}(s-1)\big) \\
&= \sum_{i=0}^{d}\sum_{s=1}^{t} \Big(\varphi_i^{\uparrow}(s)\overline{S}_i(s) - \varphi_i^{\uparrow}(s-1)\overline{S}_i(s-1) - \varphi_i^{\uparrow}(s)\left(\overline{S}_i(s) - \overline{S}_i(s-1)\right)\Big) \\
&= \sum_{i=0}^{d} \left(\varphi_i^{\uparrow}(t)\overline{S}_i(t) - \varphi_i^{\uparrow}(0)\overline{S}_i(0) - \varphi_i^{\uparrow} \bullet \overline{S}_i(t)\right) \\
&= \varphi^{\uparrow}(t)^{\top}\overline{S}(t) - \varphi^{\uparrow}(0)^{\top}\overline{S}(0) - \varphi^{\uparrow} \bullet \overline{S}(t)
\end{aligned}$$

and likewise for $\underline{S}_- \bullet \varphi^{\downarrow}$.

The discussion in the previous example motivates to call strategy φ **self-financing** if

$$\overline{S}_- \bullet \varphi^{\uparrow} - \underline{S}_- \bullet \varphi^{\downarrow} = 0, \tag{9.20}$$

where the integrals are defined as

$$\begin{aligned}
\overline{S}_- \bullet \varphi^{\uparrow} &:= (\varphi^{\uparrow})^{\top}\overline{S} - \varphi^{\uparrow}(0)^{\top}\overline{S}(0) - \varphi^{\uparrow} \bullet \overline{S}, \\
\underline{S}_- \bullet \varphi^{\downarrow} &:= (\varphi^{\downarrow})^{\top}\underline{S} - \varphi^{\downarrow}(0)^{\top}\underline{S}(0) - \varphi^{\downarrow} \bullet \underline{S}.
\end{aligned}$$

Consequently, Condition (9.20) for self-financing strategies is to be understood as a shorthand for

$$(\varphi^{\uparrow})^{\top}\overline{S} - (\varphi^{\downarrow})^{\top}\underline{S} - \varphi^{\uparrow}(0)^{\top}\overline{S}(0) + \varphi^{\downarrow}(0)\underline{S}(0) - \varphi^{\uparrow} \bullet \overline{S} + \varphi^{\downarrow} \bullet \underline{S} = 0.$$

Observe that this equation reduces to the usual self-financing condition (9.5) if $\underline{S} = \overline{S}$, i.e. if bid and ask prices coincide.

The value of a portfolio is not obvious either because the securities do not have a unique price. The **liquidation value (process)** V_φ refers to the proceeds if the portfolio were liquidated immediately, which amounts to

$$V_\varphi := \sum_{i=0}^{d} \left(\varphi_i^+ \underline{S}_i - \varphi_i^- \overline{S}_i\right) \tag{9.21}$$

with $\varphi_i^+(t) := \varphi_i(t) \vee 0$ and $\varphi_i^-(t) := (-\varphi_i(t)) \vee 0$. A self-financing strategy starting at $\varphi(0) = (0, \dots, 0)$ is called **arbitrage** if its liquidation value $V_\varphi(T)$ is nonnegative and positive with positive probability.

A version of the FTAP in this setup can be stated in terms of **consistent price systems**. This notion refers to some fictitious process $S = (S_0(t), \dots, S_d(t))_{t\in[0,T]}$ without transaction costs such that $\underline{S}_i \leq S_i \leq \overline{S}_i$ holds for $i = 0, \dots, d$. An investment in a consistent price system is at least as attractive as in the true market with transaction costs because the investor buys at lower and sells at higher prices.

Rule 9.22 (FTAP Under Bid-Ask Spreads) *The following statements are equivalent.*

1. *There are no arbitrage opportunities.*
2. *There exists a consistent price system S such that there are no arbitrage opportunities for this price process.*

Idea $1 \Rightarrow 2$: Define

$$K := \big\{(\varphi_0(T) - \xi_0, \dots, \varphi_d(T) - \xi_d) : \varphi \text{ self-financing strategy with}$$
$$\varphi(0) = (0, \dots, 0) \text{ and } \xi_0, \dots, \xi_d \text{ nonnegative random variables}\big\}. \tag{9.22}$$

This represents the *attainable cone* of terminal portfolio holdings which can be achieved by zero initial investment if it is allowed to remove assets at time T. Moreover, set

$$M := \Bigg\{(\xi_0, \dots, \xi_d) \text{ random variable :}$$
$$v_\xi := \sum_{i=0}^{d} \left(\xi_i^+ \underline{S}_i(T) + x_i^- \overline{S}_i(T)\right) \geq 0 \text{ and } E(v_\xi) = 1\Bigg\}.$$

Observe that v_ξ is the liquidation value of a time-T portfolio ξ. This *solvency cone* contains all portfolios which can be liquidated at time T without having to add money. Absence of arbitrage implies that $K \cap M = \varnothing$. If a strict separation theorem similar to Theorem A.14 applies in this setup, there exists a random vector $(Y_0, \dots, Y_d)$ such that $E(\sum_{i=0}^d Y_i X_i) \leq 0$ for any $X \in K$ and $E(\sum_{i=0}^d Y_i X_i) > 0$ for any $X \in M$. Denoting by $e_0 = (1, 0, \dots, 0) \in \mathbb{R}^{1+d}$ the zeroth unit vector, we have $e_0 1_C / E(S_0(T)1_C) \in M$ and hence $E(Y_0 1_C) \geq 0$ for any event C with $P(C) > 0$. This implies $Y_0 > 0$ almost surely. By normalising we may assume $E(Y_0) = 1$ without loss of generality.

The density $\frac{dQ}{dP} := Y_0$ defines a probability measure $Q \sim P$. Consider martingales $Z_i(t) := E(Y_i|\mathscr{F}_t)$, $t \in [0, T]$ for $i = 1, \dots, d$ and price processes $S_i(t) := S_0(t)Z_i(t)/Z_0(t)$, $t \in [0, T]$ for $i = 1, \dots, d$. The corresponding discounted price processes are denoted $\widehat{S}_i := S_i/S_0$ as usual. Since $Z_0\widehat{S}_i = Z_i$ is a martingale, $\widehat{S}_i$ is a Q-martingale. Put differently, Q is an EMM for $S = (S_0, \dots, S_d)$, which implies that the frictionless market with price process S does not allow for arbitrage.

In order to verify that S is a consistent price system, fix $i \in \{1, \dots, d\}$ and $t \in [0, T]$. For fixed $C \in \mathscr{F}_t$ consider the self-financing portfolio φ with zero initial position which buys at time t and afterwards holds one share of asset i if event C occurs. Its terminal state

$$(\varphi_0(T), \dots, \varphi_d(T)) = \big(-1_C\overline{S}_i(t)/S_0(t), 0, \dots, 0, 1_C, 0, \dots, 0\big)$$

lies in K, which yields $E((Z_i(T) - Z_0(T)\overline{S}_i(t)/S_0(t))1_C) \leq 0$. Since Z_i, Z_0 are martingales, we conclude $E((Z_i(t) - Z_0(t)\overline{S}_i(t)/S_0(t))1_C) \leq 0$, which in turn implies $Z_i(t) - Z_0(t)\overline{S}_i(t)/S_0(t) \leq 0$ almost surely because C was an arbitrary set in $\mathscr{F}_t$. The definition of Z_i and S_i yields $S_i(t) \leq \overline{S}_i(t)$. By considering a strategy that sells rather than buys asset i at time t, we obtain $S_i(t) \geq \underline{S}_i(t)$ along the same lines. This finishes the proof.

$2 \Rightarrow 1$: If φ denotes a self-financing strategy with $\varphi(0) = (0, \dots, 0)$ for $\underline{S}, \overline{S}$, consider the corresponding self-financing strategy $\widetilde{\varphi}$ in the market S with $\widetilde{\varphi}(0) = (0, \dots, 0)$ and $\widetilde{\varphi}_i = \varphi_i$ for $i = 1, \dots, d$. The difference of the numeraire parts $\widetilde{\varphi}_0(t) - \varphi_0(t)$ increases in time because whenever assets are bought and sold, $\widetilde{\varphi}$ trades at prices S rather than the possibly less attractive bid-ask prices $\overline{S}$ and $\underline{S}$ in the original market. Consequently, $\widetilde{V}_{\widetilde{\varphi}}(T) \geq V_\varphi(T)$, where $\widetilde{V}_{\widetilde{\varphi}}$ denotes the value process of $\widetilde{\varphi}$ relative to the frictionless price process S. Now suppose that $V_\varphi(T) \geq 0$ and hence $\widetilde{V}_{\widetilde{\varphi}}(T) \geq 0$. By absence of arbitrage, $\widetilde{\varphi}$ cannot be an arbitrage for S. This implies $\widetilde{V}_{\widetilde{\varphi}}(T) = V_\varphi(T) = 0$. □

Of course, the second condition could be rephrased in terms of EMMs as in Rule 9.15. Moreover, one could extend the setup to multi-currency models where bid-ask spreads are defined for any pair of assets. We refer to the notes for such issues.

Appendix 1: Problems

The exercises correspond to the section with the same number.

9.1 (Relative Portfolio) Let $S = (S_0, \dots, S_d)$ be a price process with $S_i, (S_i)_- > 0$, $i = 1, \dots, d$ and fix an initial endowment $v > 0$. For predictable $\pi = (\pi_0, \dots, \pi_d)$ we define $V_{(\pi)} := v\mathscr{E}(\pi \bullet (\mathscr{L}(S_0), \dots, \mathscr{L}(S_d)))$.

1. For self-financing φ with $V_\varphi, (V_\varphi)_- > 0$ set $\pi_i(t) := \varphi_i(t)S_i(t-)/V_\varphi(t-)$, $i = 0, \dots, d$. In other words, π_i stands for the fraction of wealth invested in asset i. Show that $\pi_0 + \dots + \pi_d = 1$ and $V_{(\pi)} = V_\varphi$.
2. Suppose that $\pi_0 + \dots + \pi_d = 1$ and set $\varphi_i(t) := \pi_i(t)V_{(\pi)}(t-)/S_i(t-)$, $i = 0, \dots, d$. Show that $\varphi = (\varphi_0, \dots, \varphi_d)$ is a self-financing strategy with $V_\varphi = V_{(\pi)}$.

9.2

1. Consider a market $S(t) = (S_0, S_1, S_2)(t)$, $t \in [0, T]$ with time horizon $T = 1$ and three assets of the form

$$\begin{aligned} S_0(t) &= 1, \\ dS_1(t) &= S_1(t)(0.02dt + 0.2dW(t)), \quad S_1(0) = 100, \\ dS_2(t) &= S_2(t)(0.03dt + 0.2dW(t)), \quad S_2(0) = 50 \end{aligned}$$

for some Wiener process W. Are there arbitrage opportunities in this market?
2. Consider the same market as in 1. but with transaction costs of 1% on the risky assets, i.e.

$$\underline{S}_1(t) = 0.99S_1(t), \quad \overline{S}_1(t) = 1.01S_1(t)$$

and likewise for S_2. Are there arbitrage opportunities in this market?

Appendix 2: Notes and Comments

Of many excellent textbooks in Mathematical Finance let us mention [29, 223] as classical and very accessible introductions, [60] as a title including jump processes, and [73], which covers the subtleties of the fundamental theorems in detail.

A full proof of Rule 9.5 can be found in [125, Proposition 2.1]. Rule 9.6 is mentioned in [133, Remark 3.27]. Dividends are covered in [29, Section 6.3]. For a discussion of the law of one price see [61] and Theorem 11.53. Rule 9.15 goes back to [133], cf. Sect. 11.7 for rigorous versions. Absence of arbitrage in Lévy models is derived under different conditions in [81, Proposition 1]. For Rule 9.18 see also [29, Proposition 24.4] and [294, Theorem 2.2]. A version of Rule 9.20 can be found in [185, Theorem 4.4]. Rule 9.22 goes back to [161], cf. [130] for a rigorous result in continuous time. Problem 9.1 is related to [29, Section 6.2].

Chapter 10
Optimal Investment

As an investor in a securities market you have to decide how to arrange your portfolio. In this chapter we consider the natural situation that you want to maximise your profits. It is not entirely obvious how to formalise this goal because the payoff of investments is typically random. A seemingly sensible approach is to maximise the expected profit $E(X)$ over all attainable payoffs, represented by random variables X. This criterion, however, does not reflect that we are typically willing to sacrifice some profit in order to avoid the risk of losses: most investors would prefer a sure 2% return to some high-risk investment with an average return of 2.2%.

A classical way out is to maximise expected *utility* $E(u(X))$ instead. Here u denotes some increasing, concave function which quantifies the degree of happiness derived from money. Monotonicity means that you prefer more to less. Concavity, on the other hand, implies that the same absolute change in wealth affects your well-being more when you are poor than when you are rich. Logarithmic utility $u(x) = \log x$ allows for a particularly intuitive interpretation. It implies that the additional happiness derived from becoming richer by a fixed percentage (e.g. 5%) is largely independent of your wealth today. In any case, the combination of monotonicity and concavity in the optimisation of $X \mapsto E(u(X))$ means that you try to maximise profits, but controlled by a certain aversion to risk.

Remark 10.1 (Von Neumann–Morgenstern Rationality) Are there "rational" alternatives to expected utility maximisation? In their pioneering work [290], von Neumann and Morgenstern show that the following seemingly innocent and natural axioms already imply that your preferences can be expressed in terms of expected utility.

1. *(Finitely many values attained with known probabilities)* You can choose between random payoffs X having values in some fixed finite set A. The law of any payoff X is assumed to be known.

© Springer Nature Switzerland AG 2019
E. Eberlein, J. Kallsen, *Mathematical Finance*, Springer Finance,
https://doi.org/10.1007/978-3-030-26106-1_10

2. *(Law invariance)* You are indifferent between any two payoffs X and Y with the same law.
3. *(Completeness)* For any two payoffs X, Y you either prefer X to Y, Y to X, or you are indifferent between the two.
4. *(Transitivity)* If you like Z at least as much as Y and Y at least as much as X, then you also like Z at least as much as X.
5. *(Archimedean Property)* Let X, Y, Z be payoffs such that you prefer Z to Y and Y to X. Consider a new compound payoff U which yields the attractive Z with probability $p \in (0, 1)$ and the unattractive X with probability $1 - p$, decided by a previous independent Bernoulli experiment. The axiom states that you prefer U to Y for some sufficiently large $p < 1$ and Y to U for some sufficiently small $p > 0$.
6. *(Independence)* Let X, Y, Z be payoffs such that Z is independent of both X and Y. Consider compound payoffs $\widetilde{X}, \widetilde{Y}$, which yield payoff X resp. Y with probability $p \in (0, 1)$ and payoff Z with probability $1 - p$, again decided by a previous independent Bernoulli experiment. The independence axiom assumes that you prefer $\widetilde{Y}$ to $\widetilde{X}$ if you prefer Y to X.

Under these assumptions there is some function $u : A \to \mathbb{R}$ such that you prefer any payoff Y to any other payoff X if and only if $E(u(Y)) > E(u(X))$. The converse is also true.

Empirical research, notably by Kahneman and Tversky, has shown that most people's decisions are not in line with the axioms in Remark 10.1, cf. [162]. This observation casts doubts on the significance of economic models relying on von Neumann–Morgenstern rationality of all market participants. Consequently, we avoid statements based on this assumption in this monograph.

On the other hand, the findings of behavioural finance do not imply that it is unreasonable or even irrational to base one's decisions on expected utility maximisation—even if many people do not. Quite the opposite, observed human preferences appear at times as somewhat inconsistent. We discuss different variations of the expected utility theme in the following sections. We mostly focus on logarithmic, power, and exponential utility functions, mainly because of their superior mathematical tractability.

10.1 Utility of Terminal Wealth

In this section we assume that you want to maximise your wealth or rather utility of wealth at some fixed finite time horizon. We consider the general setup in Chap. 9 of an arbitrage-free market with $1+d$ liquidly traded assets. For simplicity, we assume that prices are discounted relative to the zeroth asset, i.e. $S_0 = 1$. Your initial endowment is denoted by v_0. Your goal is to maximise the expected utility $E(u(V(T)))$ of terminal wealth at the future date T. Your preferences are modelled in terms of some increasing, strictly concave, differentiable **utility**

function $u : \mathbb{R} \to [-\infty, \infty)$. The most prominent examples comprise the logarithm $u(x) = \log x$, power functions $u(x) = x^{1-p}/(1-p)$ for risk aversion parameter $p \in (0, \infty) \setminus \{1\}$, and exponential utility $u(x) = -\exp(-px)$ for risk aversion parameter $p > 0$.

We call a self-financing strategy $\varphi^\star$ with initial value v_0 **optimal for expected utility of terminal wealth** if it maximises

$$\varphi \mapsto E(u(V_\varphi(T))) \tag{10.1}$$

over all such strategies φ.

Remark 10.2

1. If the numeraire asset S_0 is not constant or, put differently, the investor is interested in utility of *undiscounted* wealth, this can still be treated in the present setup as long as S_0 is deterministic. Indeed, since $u(V_\varphi(T)) = \widetilde{u}(V_\varphi(T)/S_0(T))$ for $\widetilde{u}(x) := u(xS_0(T))$, one just has to consider a slightly modified utility function. For logarithmic and power utility, the factor $S_0(T)$ does not affect the optimiser; for exponential utility it amounts to changing the risk aversion parameter.

   In view of $E(\log V_\varphi(T)) = E(\log \widehat{V}_\varphi(T)) + E(\log S_0(T))$, discounting with respect to an arbitrary random numeraire changes expected logarithmic utility only by a constant. Therefore the optimal portfolio for logarithmic utility does not depend on the chosen numeraire or currency at all.
2. Suppose that the maximal expected utility $\sup_\varphi E(u(V_\varphi(T)))$ is finite. Strict concavity of u implies that any two optimal strategies $\varphi, \widetilde{\varphi}$ have the same expected utility of terminal wealth. Indeed, otherwise the mean $(\varphi + \widetilde{\varphi})/2$ would have even higher expected utility. The law of one price in turn yields that the wealth processes V_φ and $V_{\widetilde{\varphi}}$ coincide, cf. Rule 9.14.

In Chap. 7 we discussed different methods of tackling such problems. The following characterisation of optimality can be viewed as a version of the stochastic maximum principle in Sect. 7.4.

Rule 10.3

1. *If $\varphi^\star$ is a self-financing trading strategy with initial value v_0 and if Y is a martingale with $Y(T) = u'(V_{\varphi^\star}(T))$ and such that YS is a martingale as well, then $\varphi^\star$ is optimal for expected utility of terminal wealth.*
2. *If $\varphi^\star$ is optimal for expected utility of terminal wealth and if Y is the martingale generated by the random variable $u'(V_{\varphi^\star}(T))$, then YS is a martingale as well.*
3. *If $\tilde{u}$ denotes the convex dual of u in the sense of (A.7, 1.94), then Y from above minimises*

$$Z \mapsto E(\tilde{u}(Z(T))) + v_0 Z(0) \tag{10.2}$$

among all nonnegative martingales Z such that ZS is a martingale.

4. *The optimal values of the primal maximisation problem* $\varphi \mapsto E(u(V_\varphi(T)))$ *and the dual minimisation problem (10.2) coincide in the sense that*

$$E(u(V_{\varphi^\star}(T))) = E(\tilde{u}(Y(T))) + v_0 Y(0).$$

Idea

1. The idea is to apply Theorem A.17 and Remark A.18(2), boldly assuming that they extend to the present infinite-dimensional setup. To this end let U and likewise V denote the set of real-valued random variables, where V is viewed as the dual space of U via $y(x) = E(xy)$ for $x \in U$, $y \in V$. Set $K := \{\varphi \bullet S(T) : \varphi$ predictable strategy$\}$, which is the set of terminal wealths attainable with zero initial investment. Its polar cone is $K^\circ := \{y \in V : E(xy) \leq 0$ for any $x \in K\}$. The problem at hand can be rephrased as the task to maximise $f(x) := E(u(v_0 + x))$ over K.

   For Y as in statement 1 set $y^\star := Y(T) = u'(v_0 + \varphi^\star \bullet S(T))$ and $x^\star := \varphi^\star \bullet S(T)$, which implies that the first-order condition (A.9) in Remark A.18(2) holds. For predictable φ integration by parts yields

$$\begin{aligned}\varphi \bullet S(T) y^\star &= \varphi \bullet S(T) Y(T)\\ &= \varphi \bullet (Y_- \bullet S(T) + [Y, S](T)) + (\varphi \bullet S)_- \bullet Y(T)\\ &= \varphi \bullet (YS - S_- \bullet Y)(T) + (\varphi \bullet S)_- \bullet Y(T). \end{aligned} \tag{10.3}$$

   Since both Y and YS are martingales, we conclude that $E(xy^\star) = 0$ for all $x \in K$. Put differently, $y^\star \in K^\circ$ and it solves the complementary slackness condition (A.8) in Remark A.18(2). In view of Theorem A.17, 3c)⇒3a) we have that $\varphi^\star$ is optimal for expected utility of terminal wealth.
2. Using the same notation as above, Theorem A.17, 3a)⇒3c) yields that $y^\star = u'(v_0 + x^\star) = u'(v_0 + \varphi^\star \bullet S(T))$ is optimal for the dual problem and that $E(xy^\star) \leq 0$ for any $x \in K$. Since K is a vector space, this even implies $E(xy^\star) = 0$ for any $x \in K$.

   Choose $s \leq t \leq T$ and $F \in \mathscr{F}_s$. For $\varphi := 1_{F \times (s,t]}$ and $x := \varphi \bullet S(T)$ we obtain

$$\begin{aligned} 0 &= E(xy^\star)\\ &= E(\varphi \bullet S(T) Y(T))\\ &= E(Y(T)S(t)1_F) - E(Y(T)S(s)1_F)\\ &= E(Y(t)S(t)1_F) - E(Y(s)S(s)1_F), \end{aligned}$$

   which implies that YS is a martingale.
3. Recall from the proof of statement 2 that $y^\star = Y(T)$ minimises the conjugate function over K°, which here reads as $\widetilde{f}(y) = E(\widetilde{u}(y)) - v_0 E(y)$, cf. Example A.19. If Z, ZS are martingales, we have $Z(T) \in K^\circ$ by the same

argument as in the proof of statement 1. The martingale property of Z yields

$$\begin{aligned}\widetilde{f}(Z(T)) &= E(\widetilde{u}(Z(T))) + v_0 E(Z(T)) \\ &= E(\widetilde{u}(Z(T))) + v_0 Z(0).\end{aligned} \tag{10.4}$$

4. According to Theorem A.17(1) and (10.4) we have

$$E(u(v_0 + \varphi^\star \bullet S(T))) = f(x^\star) = \widetilde{f}(y^\star) = E(\widetilde{u}(Y(T))) + v_0 Y(0). \qquad \square$$

The following corollary clarifies the link between optimal investment and equivalent martingale measures.

Rule 10.4 (Fundamental Theorem of Utility Maximisation) *For self-financing strategies $\varphi^\star$ with initial value $V_{\varphi^\star}(0) = v_0$ the following statements are equivalent:*

1. *$\varphi^\star$ is optimal for expected utility of terminal wealth.*
2. *The probability measure $Q^\star \sim P$ defined by its density*

$$\frac{dQ^\star}{dP} := \frac{u'(V_{\varphi^\star}(T))}{E(u'(V_{\varphi^\star}(T)))} \tag{10.5}$$

   is an equivalent martingale measure (EMM), i.e. S is a $Q^\star$-martingale.

Idea 1⇒2: If $\varphi^\star$ is optimal, YS is a martingale for Y as in Rule 10.3(2). Consequently, ZS is a martingale as well for the density process Z of $Q^\star$. Hence S is a $Q^\star$-martingale by Proposition 3.69.

2⇒1: If Z denotes the density process of $Q^\star$, the $Q^\star$-martingale property of S implies that ZS is a P-martingale, cf. Proposition 3.69. This yields that YS is a martingale for $Y = ZE(u'(V_{\varphi^\star}(T)))$. Thus $\varphi^\star$ is optimal by Rule 10.3(1). □

We have taken the liberty to call the above result a *fundamental theorem* because it bears some similarity with the fundamental theorems of asset pricing. In both cases an economical issue is characterised in terms of equivalent martingale measures. But similarly to the FTAP, Rule 10.4 is literally true only in finite probability spaces, the general statement being more involved, cf. the notes and comments.

Note that essentially any strategy φ defines a probability measure Q via (10.5). The optimiser is characterised only by the fact that this measure turns S into a martingale. From a very intuitive point of view, this condition can be compared to the condition that the maximiser x of a concave function $f : \mathbb{R} \to \mathbb{R}$ is characterised by $f'(x) = 0$. Indeed, the martingale property means vanishing drift of S, which corresponds at least roughly to the vanishing derivative in ordinary calculus. Unfortunately, Rule 10.4 does not point directly to the solution, in particular because the corresponding EMM $Q^\star$ is not known in advance. It rather helps to check whether some candidate is indeed optimal. In so-called complete markets, however, there exists only one EMM by Rule 11.3, which means that the left-hand side of (10.5) is known. It only remains to find the strategy $\varphi^\star$ of the form

$$\frac{dQ^\star}{dP} = cu'(V_{\varphi^\star}(T)) \tag{10.6}$$

or, equivalently,

$$\varphi^\star \bullet S(T) = (u')^{-1}\left(\frac{1}{c}\frac{dQ^\star}{dP}\right) - v_0$$

with some normalising constant c.

In incomplete markets the measure $Q^\star$ in (10.5, 10.6) or, more precisely, the process Y in Rule 10.3 solves a dual minimisation problem. However, the latter is not necessarily easier to solve than the original utility maximisation corresponding to (10.1). In order to obtain more concrete results we focus on standard utility functions of logarithmic, power and exponential type. For these choices, the minimisation problem (10.2) can actually be rephrased in terms of the corresponding EMM $Q^\star$ in Rule 10.4. Moreover, it turns out that the EMM $Q^\star$ in Rule 10.4 does not depend on the initial endowment v_0 for logarithmic, power, and exponential utility.

Rule 10.5 (Dual Minimisation Problems)

1. *If $u(x) = \log x$, the measure $Q^\star$ in (10.5) minimises the* reverse relative entropy

$$H(P|Q) := E\left(\log\frac{dP}{dQ}\right)$$

among all EMMs Q. The minimal value is related to the maximal utility U_0 in (10.1) via

$$U_0 = \log v_0 + H(P|Q^\star). \tag{10.7}$$

2. *If $u(x) = x^{1-p}/(1-p)$ with $p \in (0,\infty)\setminus\{1\}$, the measure $Q^\star$ in (10.5) minimises*

$$\operatorname{sgn}(1-p)E\left(\left(\frac{dQ}{dP}\right)^q\right)$$

among all EMMs Q, where $q := 1 - \frac{1}{p}$. The minimal value is related to the maximal utility in (10.1) via

$$U_0 = \frac{v_0^{1-p}}{1-p}\left(E\left(\left(\frac{dQ^\star}{dP}\right)^q\right)\right)^p \tag{10.8}$$

3. *If* $u(x) = -\exp(-px)$ *with* $p > 0$*, the measure* $Q^\star$ *in (10.5) minimises the* relative entropy *or* Kullback–Leibler distance

$$H(Q|P) := E\left(\frac{dQ}{dP}\log\frac{dQ}{dP}\right) = E_Q\left(\log\frac{dQ}{dP}\right)$$

among all EMMs Q*. Therefore it is called the* ***minimal entropy martingale measure (MEMM)*** *in the literature. The minimal value is related to the maximal utility in (10.1) via*

$$U_0 = -\exp\big(-pv_0 - H(Q^\star|P)\big). \tag{10.9}$$

Idea Straightforward calculations yield that the transforms $\widetilde{u}(y) := \sup_{x\in\mathbb{R}}(u(x) - yx)$ of u are given by

$$\widetilde{u}(y) = \begin{cases} -\log y - 1 & \text{for } u(x) = \log x, \\ -\frac{1}{q}y^q & \text{for } u(x) = \frac{x^{1-p}}{1-p}, \\ \frac{y}{p}(\log\frac{y}{p} - 1) & \text{for } u(x) = -e^{-px}. \end{cases}$$

By Rule 10.3 the martingale Y minimises $E(\widetilde{u}(Z(T)))+v_0 Z(0)$ over all nonnegative martingales Z such that ZS is a martingale. Put differently, $(Y(0), Q^\star)$ minimises $E(\widetilde{u}(c\frac{dQ}{dP})) + v_0 c$ over all absolutely continuous martingale measures Q and all $c \geq 0$, the optimal value being $U_0 = E(u(V_{\varphi^\star}(T)))$ by Rule 10.3(4). Since we have

$$\begin{aligned} &E\left(\widetilde{u}\left(c\frac{dQ}{dP}\right)\right) + v_0 c \\ &= \begin{cases} E(\log\frac{dP}{dQ}) - 1 - \log c + v_0 c & \text{for } u(x) = \log x, \\ -\frac{c}{q}E((\frac{dQ}{dP})^q) + v_0 c & \text{for } u(x) = \frac{x^{1-p}}{1-p}, \\ \frac{c}{p}(v_0 p + E_Q(\log\frac{dQ}{dP}) + \log\frac{c}{p} - 1) & \text{for } u(x) = -e^{-px}, \end{cases} \end{aligned} \tag{10.10}$$

it follows that $Q^\star$ minimises the respective expressions in statements 1–3. Optimising (10.10) over c yields the minimisers

$$c^\star = \begin{cases} 1/v_0 & \text{for } u(x) = \log x, \\ \left(v_0/E\big((\frac{dQ}{dP})^q\big)\right)^{-p} & \text{for } u(x) = \frac{x^{1-p}}{1-p}, \\ p\exp\big(-pv_0 - E_Q\big(\log\frac{dQ}{dP}\big)\big) & \text{for } u(x) = -e^{-px}, \end{cases}$$

which leads to the expressions (10.7, 10.8, 10.9) for U_0. □

10.1.1 Logarithmic Utility

According to the Weber–Fechner law, human beings react to many stimuli such as brightness, taste and weight on a logarithmic rather than a linear scale. If we extend this perception to satisfaction derived from wealth, it makes sense to consider logarithmic utility $u(x) = \log x$. The popularity of this utility function in the academic literature, however, may partly be due to the fact that it turns out to be mathematically particularly tractable. For instance, the optimal portfolio can be determined almost explicitly as we already observed in Example 7.5 in the case of a single risky asset. In order to state the multivariate case as well, suppose that $S_i = S_i(0)\mathscr{E}(X_i)$, $i = 1, \dots, d$, where X has local characteristics (a^X, c^X, K^X).

Rule 10.6 *Let the $\mathbb{R}^d$-valued process π be given by the equation*

$$0 = a^X(t) - c^X(t)\pi(t) - \int \frac{(\pi(t)^\top x)x}{1 + \pi(t)^\top x} K^X(t, dx) \tag{10.11}$$

and set

$$\varphi_i^\star(t) := \frac{v_0 \mathscr{E}(\pi \bullet X)(t-)}{S_i(t-)} \pi_i(t), \quad i = 1, \dots, d.$$

As usual, we identify $\varphi_1^\star, \dots, \varphi_d^\star$ with a self-financing strategy, here with initial value v_0. Then $\varphi^\star$ is an optimal portfolio for terminal wealth with wealth process $v_0 \mathscr{E}(\pi \bullet X)$.

Idea Firstly, note that

$$\begin{aligned}
V_\varphi^\star &= v_0 + \varphi^\star \bullet S \\
&= v_0 \left(1 + \sum_{i=1}^d \left(\frac{\mathscr{E}(\pi \bullet X)_-}{(S_i)_-} \pi_i\right) \bullet S_i\right) \\
&= v_0 \big(1 + \mathscr{E}(\pi \bullet X)_- \bullet (\pi \bullet X)\big) \\
&= v_0 \mathscr{E}(\pi \bullet X).
\end{aligned}$$

By Problem 3.6 we have

$$Y := \frac{1}{V_\varphi^\star} = \frac{1}{(V_\varphi^\star)_-} \bullet \left(-\pi \bullet X + [\pi \bullet X^c, \pi \bullet X^c] + \frac{(\pi^\top x)^2}{1 + \pi^\top x} * \mu^X\right),$$

which implies

$$a^Y = \frac{\pi^\top}{(V_\varphi^\star)_-} \left(-a^X + c^X \pi + \int \frac{x \pi^\top x}{1 + \pi^\top x} K^X(dx)\right) = 0$$

for the drift rate of Y, i.e. Y is a martingale. Moreover, integration by parts yields

$$\begin{aligned} YS_i &= (S_i)_- \bullet Y + Y_- \bullet S_i + [Y, S_i] \\ &= (S_i)_- \bullet Y + (YS_i)_- \bullet \left(X_i + \left[\frac{1}{Y_-} \bullet Y, X_i \right] \right), \end{aligned}$$

which equals

$$\begin{aligned} &(YS_i)_- \bullet \left(A^{X_i} + \left\langle \frac{1}{Y_-} \bullet Y, X_i \right\rangle \right) \\ &= (YS_i)_- \bullet \left(\left(a^{X_i} - (c^X \pi)_i + \int \left(-\pi^\top x x_i + \frac{(\pi^\top x)^2 x_i}{1 + \pi^\top x} \right) K^X(dx) \right) \bullet I \right) = 0 \end{aligned}$$

up to a martingale, cf. (4.29) and (10.11). We conclude that YS is a martingale as well. In view of $u'(V_\varphi^\star(T)) = 1/V_\varphi^\star(T) = Y(T)$, the assertion follows from Rule 10.3(1). □

Since (10.11) is, at any time t, a system of d equations with d unknowns, one can expect it to have a unique solution in the generic case. $\pi_i(t)$ can be interpreted as the fraction of the investor's wealth that is invested in asset i at time t. Note that it depends only on the local characteristics of X at time t. Put differently, the problem is *myopic* in the sense that the solution does not involve the past or long-term dynamics of the stock. Moreover, it is easy to see that the solution does not depend on the chosen numeraire security either. Finally and interestingly, the optimal portfolio for logarithmic utility is independent of the investor's time horizon T. As has been noted already by Samuelson [258], this is in sharp contrast to common wisdom which recommends to invest a larger proportion of wealth in the risky asset if the time horizon is large. These properties do not generally hold for arbitrary utility functions, but see Examples 10.13, 10.19 in this context.

Since the optimal portfolio does not depend on the time horizon, it makes sense to investigate its long-term properties. Surprisingly, it turns out that it outperforms any other portfolio with probability one in the long run. In order to make this vague statement precise, note that wealth typically grows exponentially in time, i.e. $\frac{1}{T} \log V_\varphi(T)$ can be expected to stay in the same order of magnitude. The quantity $\frac{1}{T} \log(V_\varphi(T)/v_0)$ can be interpreted as the average growth rate of wealth in the period $[0, T]$.

Rule 10.7 *Denote by $\varphi^\star$ the optimal portfolio for logarithmic utility of terminal wealth which, by Rule 10.6, does not depend on the time horizon T. Then we have*

$$\limsup_{T \to \infty} \frac{1}{T} \log V_\varphi(T) \leq \limsup_{T \to \infty} \frac{1}{T} \log V_{\varphi^\star}(T)$$

for any self-financing strategy φ, i.e. $\varphi^\star$ maximises the ***long-term growth rate of wealth*** *with probability one. For this reason we call $\varphi^\star$ the* ***growth-optimal portfolio****.*

Idea Set $Z := v_0/V_{\varphi^\star}$. The proof of Rule 10.6 yields that Z and ZS are martingales, which implies that Z is the density process of some martingale measure $Q^\star$, cf. Proposition 3.69. Hence $V_\varphi = V_\varphi(0) + \varphi \bullet S$ is a $Q^\star$-martingale, which in turn means that ZV_φ is a martingale. Doob's martingale inequality as in [98, Corollary 4.8 and Theorem 4.2] implies

$$e^{\delta n} P\Big(\sup_{t\geq n} Z(t)V_\varphi(t) > e^{\delta n}\Big) \leq E(Z(0)V_\varphi(0)) = v_0$$

for any $\delta \in (0,1)$, $n \in \mathbb{N}$ and hence

$$\sum_{n=1}^{\infty} P\Big(\sup_{t\geq n} \frac{1}{n} \log\big(Z(t)V_\varphi(t)\big) > \delta\Big) \leq v_0 \sum_{n=1}^{\infty} e^{-\delta n} < \infty.$$

From the Borel–Cantelli lemma [153, Theorem 10.5] it follows that P-almost surely there exists some random $n_0 \in \mathbb{N}$ such that $\sup_{t\geq n} \frac{1}{n} \log(Z(t)V_\varphi(t)) \leq \delta$ for any $n \geq n_0$. Since $Z = v_0/V_{\varphi^\star}$ we have that

$$\limsup_{t\to\infty} \frac{1}{t} \log V_\varphi(t) \leq \limsup_{t\to\infty} \frac{1}{t} \log V_{\varphi^\star}(t) + \delta$$

almost surely. □

Let us illustrate the benefit of diversification with a simple example.

Example 10.8 (Geometric Brownian Motion) Consider a single stock as in Example 9.1 with discounted price

$$\log S_1(t) = \log S_1(0) + \mu t + \sigma W(t)$$

or

$$S_1 = S_1(0)\mathscr{E}\Big(\Big(\mu + \frac{\sigma^2}{2}\Big)I + \sigma W\Big).$$

From $a^X = \mu + \sigma^2/2$ and $c^X = \sigma^2$ in Rule 10.6 we obtain $\pi = \mu/\sigma^2 + 1/2$ for the optimal fraction of wealth that is invested in the stock. Note that the discounted wealth process of the growth-optimal portfolio in Rules 10.6, 10.7 is a geometric Brownian motion as well, namely

$$\begin{aligned} V_\varphi &= v_0 \mathscr{E}\Big(\pi\Big(\Big(\mu + \frac{\sigma^2}{2}\Big)I + \sigma W\Big)\Big) \\ &= v_0 \mathscr{E}\Big(\Big(\frac{\mu + \sigma^2/2}{\sigma}\Big)^2 I + \frac{\mu + \sigma^2/2}{\sigma} W\Big) \end{aligned}$$

or

$$\log V_{\varphi^\star}(t) = \log v_0 + \frac{(\mu + \sigma^2/2)^2}{2\sigma^2} t + \frac{\mu + \sigma^2/2}{\sigma} W(t),$$

cf. (3.67). Its long-term growth rate of discounted wealth $(\mu + \sigma^2/2)^2/(2\sigma^2)$ exceeds both the values 0 for the riskless investment and μ for the stock. An interesting phenomenon happens if we assume, say, $r = -0.001$ for the riskless rate and $\mu = 0$, $\sigma = 0.2$. Then the long-term growth rate of wealth in undiscounted terms equals $r = -0.001 < 0$ for both the money market account and the stock, i.e. wealth tends to zero in the long run for these investments. The corresponding rate for the growth-optimal portfolio, however, amounts to $\sigma^2/8 - r = 0.004 > 0$, which implies that wealth tends to infinity in the long run! The effect that a portfolio grows faster than its components increases with the number of securities. This illustrates the benefit of diversification.

The growth-optimal portfolio is also called the **numeraire portfolio** because it is the unique self-financing portfolio with initial value 1 such that discounted prices $S/V_\varphi = (S_0/V_\varphi, \dots, S_d/V_\varphi)$ are martingales. Within a geometric Brownian motion setup it becomes surprisingly relevant for general utility functions:

Example 10.9 (Mutual Fund Theorem Revisited, cf. Example 7.23) Suppose that the market consists of a riskless bond $S_0(t) = e^{r_0 t}$ and d risky assets of the form $S_i(t) = S_i(0)\mathscr{E}(X_i(t))$, $i = 1, \dots, d$ with $X(t) = \mu t + \sigma W(t)$, where $r_0 \in \mathbb{R}$, $\mu \in \mathbb{R}^d$, $\sigma \in \mathbb{R}^{d\times d}$ is invertible, and W is a Wiener process in $\mathbb{R}^d$. This model is the multivariate extension of Example 9.1 but note that μ here corresponds to $\mu + \sigma^2/2$ in (9.2).

By ψ we denote the growth-optimal portfolio in this market, say for initial capital 1. Applying Rule 10.6 we have $V_\psi = 1 + \psi \bullet S = \mathscr{E}(\mu^\top(\sigma\sigma^\top)^{-1}X)$, cf. also Example 9.1. We consider now a general investor with time horizon T, initial endowment v_0, and increasing, strictly concave utility function $u : \mathbb{R} \to \mathbb{R} \cup \{-\infty\}$. The mutual fund theorem states that this investor's optimal portfolio is of the form

$$\varphi_i^\star(t) = \pi(t)\frac{V_{\varphi^\star}(t)}{V_\psi(t)}\psi_i(t), \quad i = 1, \dots, d \tag{10.12}$$

for some predictable real-valued process π. Moreover, we have

$$\varphi_0^\star(t) = \pi(t)\frac{V_{\varphi^\star}(t)}{V_\psi(t)}\psi_0(t) + (1 - \pi(t))V_{\varphi^\star}(t)$$

in order for $\varphi^\star$ to be self-financing. In other words, the investor invests a—possibly randomly changing—fraction $\pi(t)$ of her wealth in the growth-optimal portfolio and the remaining money in the bond. This has been shown in Example 7.23, but we repeat the main arguments here.

Considering discounted price processes we may assume $r_0 = 0$ without loss of generality. Using Girsanov's theorem 4.15 we observe that the probability measure $Q \sim P$ with density process $Z = \mathscr{E}(-(\sigma^{-1}\mu)^\top W)$ is an EMM for S. Define an auxiliary market (S_0, F) with risky asset price process

$$\begin{aligned} F(t) &= 1 + \psi \bullet S(t) \\ &= \mathscr{E}\Big(\mu^\top (\sigma\sigma^\top)^{-1} X\Big)(t) \\ &= \mathscr{E}\Big((\sigma^{-1}\mu)^\top (\sigma^{-1}\mu) I + (\sigma^{-1}\mu)^\top W\Big)(t) \end{aligned}$$

and filtration generated by F. Measure Q is an EMM with density process Z for this smaller market as well because F is a Q-martingale and $Z = 1/F$ is adapted to the filtration generated by F, cf. Problem 3.6. By the martingale representation theorem 3.61 and Theorem 4.15 we conclude that Q is the only EMM for F, cf. Example 11.4 in this context. Let $\widetilde{\varphi} = (\widetilde{\varphi}_0, \widetilde{\varphi}_1)$ denote the optimal strategy for expected terminal wealth in the market (S_0, F). By Rule 10.4 we have

$$\frac{dQ}{dP} = \frac{u'(V_{\widetilde{\varphi}}(T))}{E(u'(V_{\widetilde{\varphi}}(T)))}. \tag{10.13}$$

Since the mutual fund F is replicable in the market S by the self-financing strategy ψ, the portfolio $\widetilde{\varphi}$ naturally corresponds to a self-financing strategy in the market S with the same wealth process as $\widetilde{\varphi}$, namely $\varphi_i^\star(t) := \widetilde{\varphi}_1(t)\psi_i(t)$ for $i = 1, \dots, d$. Note that $\varphi^\star$ is of the form (10.12). By (10.13) and Rule 10.4, it is optimal for expected utility of terminal wealth for the market S.

The point of the mutual fund theorem is that the fund F or, put differently, the relative weights in the risky part of the portfolio are the same for all investors, i.e. they do not depend on the utility function or the time horizon. This becomes even more remarkable if we consider the following extension of the above setup to international stocks.

Suppose that $S_1, \dots, S_n$ above denote in fact bonds in foreign currencies $1, \dots, n$, with respective constant interest rates $r_1, \dots, r_n$. The remaining $d - n$ processes $S_{n+1}, \dots, S_d$ stand for domestic or international stocks, denoted in our given currency 0. Expressed in terms of currency $j \in \{0, \dots, n\}$, the prices of the $1 + d$ assets are $\widetilde{S} = (\widetilde{S}_0, \dots, \widetilde{S}_d)$ with $\widetilde{S}_i(t) = e^{r_j t} S_i(t)/S_j(t)$, $i = 0, \dots, d$. The factor $e^{r_j t}$ is due to the fact that S_j denotes the price of the currency-j bond rather than the exchange rate. Hence it is easy to see that the model has—up to changing parameters—the same structure from the perspective of any currency.

By Remark 10.2(1) the growth-optimal portfolio does not depend on the chosen currency. This does not hold for the optimal investment strategy relative to an arbitrary utility function, i.e. an investor maximising the expected utility $\varphi \mapsto E(u(\widetilde{V}_\varphi(T)))$ relative to the currency-j wealth $\widetilde{V}_\varphi(t) := e^{r_j t} V_\varphi(t)/S_j(t)$ typically obtains a different optimiser than $\varphi^\star$. But since the mutual fund F above corresponds to the growth-optimal portfolio ψ, we immediately conclude that it does *not* depend on the investor's reference currency. The currency plays a role only in so far as the investor melds the fund with a riskless investment in her own currency.

This may seem surprising. Indeed, since foreign stocks involve a foreign exchange rate risk on top of their intrinsic randomness, it may seem plausible that any investor's optimal portfolio should assign relatively higher weights to domestic stocks. But at least for expected utility maximisers in this standard setup this is not the case, as the above analysis shows.

Unfortunately, the application of Rules 10.6 and 10.7 faces some obstacles in practice. The optimal portfolio in Rule 10.6 depends sensitively on a^X, which is hard to estimate. Secondly, with realistic parameters it may take a long time to even outperform the investment in the riskless bond with, say, 90% probability:

Example 10.10 (Geometric Brownian Motion with Estimated Parameters) For simplicity we consider the model of Example 9.1 with only two assets, namely a bond with fixed interest rate r and a risky asset following geometric Brownian motion. Time t is measured in years. As a well-diversified and hence reasonable risky asset we consider the German stock index DAX along with the German government bond index REXP as an approximation of the riskless asset. Based on data covering the period from December 1987 to April 2018 for both the DAX and the REXP, standard estimators for the Gaussian law yield point estimates of $\mu - r = 3.2\%$ and $\sigma = 21.6\%$.

Based on these figures, the growth-optimal portfolio would invest a fraction of $(\mu - r + \sigma^2/2)/\sigma^2 = 119\%$ of wealth in the index. In order to outperform the pure bond investment with, say, 95% probability, the investment horizon should exceed at least 41 years.

What happens if we take the estimation error into account? Assuming for simplicity that σ has been observed without error, the standard Gaussian 90%-confidence interval for $\mu - r$ amounts to $[-3.2\%, 9.6\%]$, implying a 90%-confidence interval $[-20\%, 257\%]$ for the optimal fraction of wealth to be invested in the risky asset. In other words, even using more than 30 years of data and based on the assumption of constant parameters that are still relevant for the future to come, we cannot be relatively sure whether we should be long or short in the risky asset!

Of course, we could tighten the confidence interval by working with longer time series. For example, we obtain an interval of $[93\%, 252\%]$ for the optimal fraction in the risky asset if we repeat the above study for the S&P 500 index, based on data collected by [273] covering the 140-year period from 1871 to 2011. However, it may seem somewhat bold to assume that the excess growth rate $\mu - r$ is firstly constant over such a long time and secondly still relevant for the future. Indeed, the above period turned out to be quite exceptional to the United States, which rose to the leading economic and political power during that time. In this context we refer to the literature on the *equity premium puzzle*, which designates the observation that the average investor seems to hold a lower fraction of wealth in stock than suggested by utility theory.

As a side remark, note that the optimal fraction of the risky asset never exceeds 100% if, instead of Brownian motion in Example 9.1, we assume a Lévy process with unbounded negative jumps, cf. Example 9.2. Indeed, if wealth jumps to negative values with positive probability, we immediately end up with an expected logarithmic utility of $-\infty$.

10.1.2 Power Utility

In this section we consider utility functions of the form $u(x) := x^{1-p}/(1-p)$ with $p \in (0, \infty) \setminus \{1\}$. They are called of *constant relative risk aversion (CRRA)* because their *relative risk aversion coefficient* $-u''(x)x/u'(x)$ equals p for all x. Observe that the excluded relative risk aversion coefficient $p = 1$ is obtained for logarithmic utility. In order to study optimal investment for terminal wealth, we introduce the so-called *opportunity process*.

For power utility $u(x) = x^{1-p}/(1-p)$ the **opportunity process** L is defined as

$$L(t) := (1-p) \operatorname*{ess\,sup}_{\varphi} E\big(u(V_\varphi(T))\big|\mathscr{F}_t\big), \tag{10.14}$$

where maximisation extends over all self-financing strategies that start with endowment 1 at time t, i.e.

$$V_\varphi(T) = 1 + \varphi 1_{]\!]t,T]\!]} \bullet S(T) = 1 + \int_t^T \varphi(s) dS(s).$$

The opportunity process stands for the maximal expected utility attainable by trading between t and T if one starts with wealth 1. As such it is closely related to the value process in stochastic control theory and it plays a similar role for solving the optimal investment problem.

Rule 10.11 *Let L be a positive semimartingale with $L(T) = 1$ and $\varphi^\star$ a self-financing strategy with initial value $v_0 > 0$. The following are equivalent:*

1. *L is the opportunity process and $\varphi^\star$ is the optimal strategy in (10.1).*
2. *The value process for the control problem related to (10.1) is of the form $\mathscr{V}(t, \varphi) = L(t)u(V_\varphi(t))$ for any $t \in [0, T]$ and any self-financing strategy φ with initial value v_0. Moreover, $\varphi^\star$ is optimal for expected utility of terminal wealth.*
3. *The process Y in Rule 10.3 equals*

$$Y(t) := L(t)u'(V_{\varphi^\star}(t)), \quad t \in [0, T]. \tag{10.15}$$

In particular, the density process of the EMM $Q^\star$ in (10.5) amounts to

$$Z(t) := \frac{L(t)}{L(0)} u'(V_{\varphi^\star}(t)/v_0), \quad t \in [0, T].$$

4. *$L(t)u'(V_{\varphi^\star}(t))$ and $L(t)u'(V_{\varphi^\star}(t))S(t)$ are martingales.*
5.

$$\frac{a^L(t)}{L(t-)} = -\frac{1}{2}p(1-p)\gamma(t)^\top c^S(t)\gamma(t)$$
$$-\int\Big((1+\gamma(t)^\top x)^{-p}(1+p\gamma(t)^\top x)-1\Big)\Big(1+\frac{y}{L(t-)}\Big)K^{(S,L)}(d(x,y)) \tag{10.16}$$

and

$$0 = a^S(t) + \frac{c^{S,L}(t)}{L(t-)} - pc^S(t)\gamma(t)$$
$$+\int\Big((1+\gamma(t)^\top x)^{-p}\Big(1+\frac{y}{L(t-)}\Big)-1\Big)xK^{(S,L)}(d(x,y)), \tag{10.17}$$

where $\gamma_i(t) := \varphi_i^\star(t)/V_{\varphi^\star}(t-)$ for $i = 1,\dots,d$ denotes the number of shares invested in risky asset i per unit of wealth *and the notation as in (II.1) is used.*

Idea We focus on portfolios with positive wealth. Indeed if $V_\varphi(t) < 0$ occurs with positive probability, absence of arbitrage implies that $V_\varphi(T) < 0$ with positive probability, which leads to expected utility $-\infty$.

Fix $t \in [0,T]$ and a self-financing strategy φ with $V_\varphi > 0$. Note that there is a one-to-one correspondence between self-financing portfolios $\widetilde{\varphi}$ with $V_{\widetilde{\varphi}}(t) = V_\varphi(t)$ and self-financing portfolios ψ with $V_\psi(t) = 1$, namely via $\widetilde{\varphi} = \psi V_\varphi(t)$. The utilities of terminal wealth are naturally linked by $u(V_{\widetilde{\varphi}}(T)) = u(V_\psi(T))V_\varphi(t)^{1-p}$.

1⇒2: The optimality of $\varphi^\star$ implies that it maximises $\varphi \mapsto E(u(V_\varphi(T))|\mathscr{F}_t)$ over all self-financing portfolios φ starting with wealth $V_{\varphi^\star}(t)$ at time t, cf. Remark 1.45(5). By the above one-to-one correspondence we have that $\varphi^\star/V_{\varphi^\star}(t)$ is optimal for the maximisation in (10.14). Moreover, we conclude that $\mathscr{V}(t,\varphi) = L(t)u(V_{\varphi^\star}(t))$.

2⇒1: If $\psi^\star$ is optimal for the maximisation problem in (10.14), the above one-to-one correspondence yields that $\varphi^\star = \psi^\star V_\varphi(t)$ solves the conditional optimisation problem in the definition of $\mathscr{V}(t,\varphi)$. Moreover, the optimal values are related to each other via $L(t)u(V_{\varphi^\star}(t)) = \mathscr{V}(t,\varphi)$.

2⇒3: Since $L(T) = 1$, we have $Y(T) = L(T)u'(V_{\varphi^\star}(T)) = u'(V_{\varphi^\star}(T))$. Moreover,

$$\begin{aligned} Y(t)V_{\varphi^\star}(t) &= L(t)u'(V_{\varphi^\star}(t))V_{\varphi^\star}(t) \\ &= (1-p)L(t)u(V_{\varphi^\star}(t)) \\ &= (1-p)\mathscr{V}(t,\varphi^\star) \end{aligned}$$

is a martingale by Theorem 7.2(1). Consequently, Y is the process in Rule 10.3 because $YV_{\varphi^\star}$ is a martingale with the corresponding terminal value. Finally, we have $Y(0) = L(0)u'(v_0)$, which implies

$$Z(t) = \frac{Y(t)}{Y(0)} = \frac{L(t)}{L(0)}u'(V_{\varphi^\star}/v_0)$$

as desired.

3⇒2: Since $Y(T) = u'(V_{\varphi^\star}(T))$, Rule 10.3(1) yields that $\varphi^\star$ is optimal. By (10.3) we have that $YV_{\varphi^\star}$ and hence also

$$\begin{aligned} L(t)u(V_{\varphi^\star}(t)) &= L(t)u'(V_{\varphi^\star}(t))V_{\varphi^\star}(t)(1-p)^{-1} \\ &= Y(t)V_{\varphi^\star}(t)(1-p)^{-1} \end{aligned}$$

are martingales. Since $\mathscr{V}(t,\varphi^\star)$ is a martingale with the same terminal value, we conclude that $\mathscr{V}(t,\varphi^\star) = L(t)u(V_{\varphi^\star}(t))$ for all $t \in [0,T]$. The considerations in the beginning of this proof yield $\mathscr{V}(t,\varphi) = \mathscr{V}(t,\varphi^\star)u(V_\varphi(t))/u(V_{\varphi^\star}(t))$ for any self-financing strategy φ, which implies $\mathscr{V}(t,\varphi) = L(t)u(V_\varphi(t))$ as desired.

3⇒4: This follows immediately from the martingale property of Y and YS.

4⇒3: Since Y and YS are martingales with $Y(T) = u'(V_{\varphi^\star}(T))$ the assertion follows from Rule 10.3(1).

4⇒5: First note that $V_{\varphi^\star} = v_0 + \varphi^\star \bullet S = v_0 + (V_{\varphi^\star})_- \bullet(\gamma \bullet S)$, which implies that $V_{\varphi^\star} = v_0\mathscr{E}(\gamma \bullet S)$. Applying Ito's formula (3.43) to $u'(V_{\varphi^\star}) = V_{\varphi^\star}^{-p}$ yields $u'(V_{\varphi^\star}) = v_0^{-p} + (V_{\varphi^\star}^{-p})_- \bullet U$ and hence $u'(V_{\varphi^\star}) = v_0\mathscr{E}(U)$ for

$$U := -p\gamma \bullet S + \frac{p(p+1)}{2}(\gamma^\top c^S\gamma)\bullet I + \Big((1+\gamma^\top x)^{-p} - 1 + p\gamma^\top x\Big) * \mu^S.$$

Integration by parts leads to

$$\begin{aligned} Lu'(V_{\varphi^\star}) &= L_- \bullet u'(V_{\varphi^\star}) + (u'(V_{\varphi^\star}))_- \bullet L + [u'(V_{\varphi^\star}), L] \\ &= (Lu'(V_{\varphi^\star}))_- \bullet \left(U + \frac{1}{L_-}\bullet L + \left[U, \frac{1}{L_-}\bullet L\right]\right). \end{aligned} \tag{10.18}$$

The drift coefficient of this process equals

$$\begin{aligned} &(Lu'(V_{\varphi^\star}))_-\Big(-p\gamma^\top a^S + \frac{p(p+1)}{2}\gamma^\top c^S\gamma + \int\Big((1+\gamma^\top x)^{-p} - 1 + p\gamma^\top x\Big)K^S(dx) \\ &\quad + \frac{a^L}{L_-} - \frac{1}{L_-}p\gamma^\top c^{S,L} + \int\Big(\Big((1+\gamma^\top x)^{-p} - 1\Big)\frac{y}{L_-}\Big)K^{(S,L)}(d(x,y))\Big), \end{aligned} \tag{10.19}$$

which vanishes because $Lu'(V_{\varphi^\star})$ is a martingale. Once more applying integration by parts yields

$$Lu'(V_{\varphi^\star})S = (Lu'(V_{\varphi^\star}))_- \bullet S + S_- \bullet (Lu'(V_{\varphi^\star})) + [Lu'(V_{\varphi^\star}), S],$$

which equals $(Lu'(V_{\varphi^\star}))_- \bullet (S + [U + \frac{1}{L_-} \bullet L + [U, \frac{1}{L_-} \bullet L], S])$ up to a martingale, cf. (10.18). The drift coefficient of this process amounts to

$$(Lu'(V_{\varphi^\star}))_- \Big(a^S - p\gamma^\top c^S + \frac{1}{L_-} c^{L,S} + \int \Big(\big((1+\gamma^\top x)^{-p} - 1 \big) x \\ + \frac{xy}{L_-} + \big((1+\gamma^\top x)^{-p} - 1 \big) \frac{y}{L_-} x \Big) K^{(S,L)}(d(x,y)) \Big), \tag{10.20}$$

which vanishes because $Lu'(V_{\varphi^\star})S$ is a martingale. Since (10.19) and (10.20) vanish if and only if (10.16, 10.17) hold, the assertion follows.

5⇒4: Equations (10.16, 10.17) imply that (10.19, 10.20) vanish, which means that $Lu'(V_{\varphi^\star})$ and $Lu'(V_{\varphi^\star})S$ are martingales. □

Remark 10.12 Equations (10.16, 10.17) simplify slightly if we consider $K = \log L$ instead of L. In view of Proposition 4.13, the conditions then read as $K(T) = 0$,

$$a^K(t) = -\frac{1}{2} p(1-p)\gamma(t)^\top c^S(t)\gamma(t) - \frac{1}{2} c^K(t) \\ - \int \Big((1+\gamma(t)^\top x)^{-p}(1+p\gamma(t)^\top x)e^y - 1 - y \Big) K^{(S,K)}(d(x,y)) \tag{10.21}$$

and

$$0 = a^S(t) + c^{K,S}(t) - pc^S(t)\gamma(t) \\ + \int \big((1+\gamma(t)^\top x)^{-p} e^y - 1 \big) x K^{(S,K)}(d(x,y)), \tag{10.22}$$

where γ is defined as above.

Let us illustrate the use of opportunity processes in a geometric Lévy model.

Example 10.13 (Geometric Lévy Model) We consider the model of Example 9.2. Since X has independent increments, we guess that the opportunity process is deterministic. More specifically, we make an ansatz

$$L(t) = \exp(-\alpha(T-t))$$

with some constant $\alpha \geq 0$. Instead of γ in Rule 10.11(5) we consider the *fraction of wealth* $\pi := (S_1)_- \gamma$ invested in the risky asset. In terms of π, (10.17) or equivalently (10.22) reduce to

$$0 = a^{\widetilde{X}} - p\pi c^{\widetilde{X}} + \int ((1+\pi x)^{-p} - 1) x K^{\widetilde{X}}(dx) \tag{10.23}$$

or, using the triplet of X as computed in Theorem 3.49,

$$\begin{aligned} 0 &= a^X + \frac{1}{2}c^X + \int (e^x - 1 - x) K^X(dx) \\ &\quad - p\pi c^X + \int \big((1+\pi(e^x-1))^{-p} - 1\big)(e^x - 1) K^X(dx), \end{aligned}$$

which simplifies to

$$0 = a^X - \Big(p\pi - \frac{1}{2}\Big)c^X + \int \big((1+\pi(e^x-1))^{-p}(e^x-1) - x\big) K^X(dx). \tag{10.24}$$

Conditions (10.16) or (10.21) are satisfied if we choose $\alpha = a^L(t)/L(t) = a^K(t)$ as

$$\begin{aligned} \alpha &= -\frac{1}{2}p(1-p)c^{\widetilde{X}}\pi^2 - \int \Big((1+\pi x)^{-p}(1+p\pi x) - 1\Big) K^{\widetilde{X}}(dx) \qquad (10.25) \\ &= -\frac{1}{2}p(1-p)c^X\pi^2 - \int \Big((1+\pi(e^x-1))^{-p}(1+p\pi(e^x-1)) - 1\Big) K^X(dx). \end{aligned}$$

Consequently, in order to maximise expected utility you should invest the fraction π of wealth in the risky asset and the remaining money in the bond, where π denotes the root of (10.23) or equivalently (10.24), which we assume to exist. To see this, we first observe that the self-financing strategy $\varphi^\star = (\varphi_0^\star, \varphi_1^\star)$ corresponding to this investment is of the form $\varphi_1^\star(t) = \pi V(t-)/S_1(t-)$, where $V(t) = v_0 \mathscr{E}(\pi \widetilde{X})(t)$ denotes the wealth at time t. Indeed, since

$$dV(t) = V(t-)\pi d\widetilde{X}(t) = \pi \frac{V(t-)}{S_1(t-)} dS_1(t) = \varphi_1^\star(t) dS_1(t),$$

we see that V is really the wealth process corresponding to $\varphi^\star$ and hence π. Since the conditions in Rule 10.11(5) are satisfied, we conclude that $\varphi^\star$ is optimal.

The solution generalises the one in Example 7.17, where it was obtained using dynamic programming. Observe that the investor's fraction of wealth in the risky asset is not only constant over time but does not depend on the investor's time horizon either. As a side remark one may note that the investor's wealth process is a geometric Lévy process as well.

The extension to $d > 1$ risky assets is straightforward.

The ansatz to assume an opportunity process of exponential form also works for some of the affine stochastic volatility models in Sect. 8.2. It is typically of the form

$$L(t) = \exp(\alpha_1(t) + \alpha_2 Y(t))$$

with deterministic functions α_1, α_2, where (X, Y) denotes as in Sect. 8.2 the bivariate affine process driving the model.

For general Itō processes as in Example 9.3, there is typically no simple explicit solution to the investment problem. But we can rephrase (10.16, 10.17) in terms of a backward stochastic differential equation.

Example 10.14 (Itō Process Model) For a single continuous risky asset the optimal relative investment in (10.22) is obtained simply as

$$\gamma(t) = \frac{a^S(t) + c^{K,S}(t)}{pc^S(t)} \tag{10.26}$$

and the logarithmic opportunity process $K = \log L$ satisfies

$$a^K(t) = -\frac{1-p}{2p}\frac{(a^S(t) + c^{K,S}(t))^2}{c^S(t)} - \frac{1}{2}c^K(t). \tag{10.27}$$

In the context of Example 9.3 the filtration is generated by an $\mathbb{R}^n$-valued Wiener process W. The martingale representation theorem 3.61 suggests that L should be of the form

$$dK(t) = a^K(t)dt + \eta(t)dW(t)$$

with $K(T) = 0$ and some $\mathbb{R}^n$-valued predictable process η. Since (10.27) can be rewritten as

$$a^K(t) = -\frac{1-p}{2p}\left(\frac{\mu(t)}{\sigma(t)} + \eta_1(t)\right)^2 - \frac{1}{2}\sum_{i=1}^n \eta_i(t)^2,$$

we obtain a BSDE for K. By (10.26) the optimal portfolio is obtained from

$$\gamma(t) = \frac{1}{p}\left(\frac{\mu(t)}{\sigma(t)^2} + \frac{\eta_1(t)}{\sigma(t)}\right).$$

Recall from Example 10.13 that K is deterministic and hence $\eta = 0$ in the special case of geometric Brownian motion as in Example 9.1.

Alternatively, we can obtain the opportunity process as a solution to a PDE if the Itō process is of Markovian type as in Example 9.4, which is typically the case.

Example 10.15 (Continuous Markovian Model) In the setup of Example 9.4 it is natural to conjecture that the logarithmic opportunity process and the number of shares of the risky asset per unit of wealth are of the form $K(t) = \ell(T-t, X(t))$ and $\gamma(t) = \xi(T-t, X(t))$ for some functions $\ell, \xi : [0,T]\times\mathbb{R}^n \to \mathbb{R}$. Note that $\gamma(t)$ here refers to the process from Rule 10.11 whereas $\gamma(x)$ below denotes the diffusion coefficient matrix from Example 9.4. By Proposition 4.13, the local characteristics appearing in (10.27) can be expressed in terms of β, γ and the derivatives of ℓ. This leads to the following PDE for ℓ:

$$\begin{aligned}
\ell(0,x) &= 0,\\
D_1\ell(\vartheta,x) &= \sum_{i=1}^n D_{1+i}\ell(\vartheta,x)\beta_i(x)\\
&\quad + \frac{1}{2}\sum_{i,j=1}^n \big(D_{1+i,1+j}\ell(\vartheta,x) + D_{1+i}\ell(\vartheta,x)D_{1+j}\ell(\vartheta,x)\big)\gamma_{ij}(x)\\
&\quad + \frac{1-p}{2p}\frac{\big(\beta_1(x)+\sum_{i=1}^n D_{1+i}\ell(\vartheta,x)\gamma_{i1}(x)\big)^2}{\gamma_{11}(x)}.
\end{aligned}$$

Having solved this PDE, the function ξ is obtained from (10.26) as

$$\xi(\vartheta,x) = \frac{\beta_1(x)+\sum_{i=1}^n D_{1+i}\ell(\vartheta,x)\gamma_{i1}(x)}{p\gamma_{11}(x)}.$$

10.1.3 Exponential Utility

For the purposes of utility-based hedging, exponential utility $u(x) = -\exp(-px)$ has certain advantages, both conceptual and computational. This is discussed in Sect. 13.1 in greater detail. In view of this later application, we allow for a random terminal endowment, i.e. we assume that the investor tries to maximise

$$\varphi \mapsto E\big(u(V_\varphi(T)-H)\big) \tag{10.28}$$

over all self-financing portfolios φ with initial values $v_0 \in \mathbb{R}$, where H denotes an $\mathscr{F}_T$-measurable random variable. Observe that the optimality of a strategy φ does not depend on the initial endowment v_0, whereas the optimal utility (10.28) is proportional to $\exp(-pv_0)$. Moreover, if φ is optimal for $H=0$ and risk aversion p, then $p\varphi$ is optimal for risk aversion parameter 1. The optimal utility for $H=0$ does not depend on p.

We proceed similarly as in Sect. 10.1.2. For exponential utility $u(x) = -\exp(-px)$ the **opportunity process** L^H is defined as

$$L^H(t) := -\operatorname*{ess\,sup}_{\varphi} E\big(u(V_\varphi(T) - H)\big|\mathscr{F}_t\big),$$

where maximisation extends over all self-financing strategies that start with endowment 0 at time t, i.e.

$$V_\varphi(T) = \varphi 1_{]\!]t,T]\!]} \bullet S(T) = \int_t^T \varphi(t) dS(t).$$

Remark 10.16 Statements 1 and 2 in Rule 10.3 hold in the presence of a random endowment if we replace $u'(V_{\varphi^\star}(T))$ by $u'(V_{\varphi^\star}(T) - H)$. The same is true for Rule 10.4. In this case we denote the EMM in (10.5) by Q^H.

As for power utility, the opportunity process is linked in several ways to the terminal wealth problem.

Rule 10.17 *Let L^H be a positive semimartingale with $L^H(T) = e^{pH}$ and φ^H a self-financing strategy with initial value $v_0 > 0$. The following are equivalent:*

1. *L^H is the opportunity process and φ^H is the optimal strategy for (10.28).*
2. *The value process for the control problem (10.28) is of the form $\mathscr{V}(t, \varphi) = L^H(t)u(V_\varphi(t))$ for any $t \in [0, T]$ and any self-financing strategy φ with initial value v_0. Moreover, φ^H is optimal for (10.28).*
3. *The density process of the EMM Q^H in Remark 10.16 equals*

$$Z(t) := \frac{L^H(t)}{L^H(0)} \frac{1}{p} u'(V_{\varphi^H}(t) - v_0), \quad t \in [0, T].$$

4. *$L^H(t)u'(V_{\varphi^H}(t))$ and $L^H(t)u'(V_{\varphi^H}(t))S(t)$ are martingales.*
5.

$$\frac{a^{L^H}(t)}{L^H(t-)} = \frac{1}{2} p^2 \varphi^H(t)^\top c^S(t) \varphi^H(t)$$
$$- \int \Big(e^{-p\varphi^H(t)^\top x}\big(1 + p\varphi^H(t)^\top x\big) - 1\Big)\Big(1 + \frac{y}{L^H(t-)}\Big) K^{(S,L^H)}(d(x, y)) \tag{10.29}$$

and

$$\begin{aligned}0 &= a^S(t) + \frac{c^{L^H,S}(t)}{L^H(t-)} - pc^S(t)\varphi^H(t)\\&\quad + \int \left(e^{-p\varphi^H(t)^\top x}\left(1 + \frac{y}{L^H(t-)}\right) - 1\right) x K^{(S,L^H)}(d(x,y)),\end{aligned} \tag{10.30}$$

where we use the notation as in (II.1).

Idea Fix $t \in [0, T]$ and a self-financing strategy φ. Similarly as in the proof of Rule 10.11 there is a one-to-one correspondence between self-financing strategies $\widetilde{\varphi}$ with $V_{\widetilde{\varphi}}(t) = V_\varphi(t)$ and self-financing ψ with $V_\psi(t) = 0$, namely via $(\varphi_1, \dots, \varphi_d) = (\psi_1, \dots, \psi_d)$. The corresponding utilities of terminal wealth are related via $u(V_{\widetilde{\varphi}}(T)) = u(V_\psi(T)) \exp(-pV_\varphi(t))$. The assertions now follow along the same lines as Rule 10.11, where $Y(t) = L^H(t)u'(V_{\varphi^H}(t)) = -p\mathscr{V}(\varphi^H, t)$ in the present context. □

Remark 10.18 As in the power utility case, (10.29, 10.30) simplify by turning to $K^H := \log L^H$. In view of Proposition 4.13, the conditions in statement 5 then read as $K^H(T) = pH$,

$$\begin{aligned}a^{K^H}(t) &= \frac{1}{2}p^2\varphi^H(t)^\top c^S(t)\varphi^H(t) - \frac{1}{2}c^{K^H}(t)\\&\quad - \int \left(e^{-p\varphi^H(t)^\top x + y}(1 + p\varphi^H(t)^\top x) - 1 - y\right) K^{(S,K^H)}(d(x,y))\end{aligned} \tag{10.31}$$

and

$$\begin{aligned}0 &= a^S(t) + c^{K^H,S}(t) - pc^S(t)\varphi^H(t)\\&\quad + \int (e^{-p\varphi^H(t)^\top x + y} - 1) x K^{(S,K^H)}(d(x,y)).\end{aligned} \tag{10.32}$$

We reconsider Example 10.13 for exponential utility.

Example 10.19 (Geometric Lévy Model) We work in the setup of Example 9.2. As for power utility, we make the ansatz

$$L(t) = \exp(-\alpha(T - t)) \tag{10.33}$$

with some $\alpha \geq 0$ for the opportunity process L in the case $H = 0$. For exponential utility the bookkeeping simplifies if we work with $\psi = (S_1)_-\varphi_1^\star$, i.e. with the amount of money invested in the stock. In terms of ψ, (10.30) or (10.32) reads as

$$0 = a^{\widetilde{X}} - c^{\widetilde{X}} p\psi + \int (e^{-p\psi x} - 1) x K^{\widetilde{X}}(dx) \tag{10.34}$$

or, using the triplet of X, as

$$\begin{aligned}0 &= a^X + \frac{1}{2}c^X + \int (e^x - 1 - x)K^X(dx)\\ &\quad - c^X p\psi + \int \big(\exp(-p\psi(e^x-1)) - 1\big)(e^x - 1)K^X(dx),\end{aligned}$$

which is equivalent to

$$0 = a^X + \Big(\frac{1}{2} - p\psi\Big)c^X + \int \big(\exp(-p\psi(e^x-1))(e^x-1) - x\big)K^X(dx). \tag{10.35}$$

In order to satisfy condition (10.29) or (10.31) choose $\alpha = a^L(t)/L(t-) = a^K$ as

$$\begin{aligned}\alpha &= \frac{1}{2}p^2\psi^2 c^{\widetilde{X}} - \int \big(e^{-p\psi x}(1 + p\psi x) - 1\big)K^{\widetilde{X}}(dx)\\ &= \frac{1}{2}p^2\psi^2 c^X - \int \Big(\exp\big(-p\psi(e^x-1)\big)\big(1 + p\psi(e^x-1)\big) - 1\Big)K^X(dx).\end{aligned} \tag{10.36}$$

We conclude that for exponential utility it is optimal to invest the constant amount ψ of money in the risky asset, where ψ solves (10.34) or equivalently (10.35). To verify this, define the self-financing strategy $\varphi^\star = (\varphi_0^\star, \varphi_1^\star)$ with $\varphi_1^\star(t) = \psi/S_1(t-)$. Since it satisfies the conditions in Rule 10.17(5), it is optimal for exponential utility of terminal wealth. One may note that the risk aversion parameter p appears in the solution ψ just as a scaling factor, i.e. it suffices to solve (10.34) resp. (10.35) for $p = 1$. The initial endowment v_0 does not affect $\varphi_1^\star$ at all. Moreover, the optimal wealth process $V_{\varphi^\star}(t) = v_0 + \psi\widetilde{X}(t)$ follows a Lévy process, shifted by the initial endowment v_0. As a main difference to Example 10.13 we observe that now the *amount of money* invested in the risky asset rather than the *fraction of wealth* remains constant over time. But as in the previous case, the solution does not depend on the time horizon.

As for power utility, the example carries over easily to $d > 1$ assets.

As in the case of power utility (10.31, 10.32) can be rephrased in terms of a BSDE for Itō processes.

Example 10.20 (Itō Process Model) For a single continuous risky asset, (10.31, 10.32) reduce to

$$\varphi_1^H(t) = \frac{a^S(t) + c^{K^H,S}(t)}{pc^S(t)} \tag{10.37}$$

and

$$a^{K^H}(t) = \frac{1}{2}\frac{(a^S(t) + c^{K^H,S}(t))^2}{c^S(t)} - \frac{1}{2}c^{K^H}(t). \tag{10.38}$$

Along the same lines as in Example 10.14, we obtain the BSDE

$$K^H(T) = pH, \quad dK^H(t) = a^{K^H}(t)dt + \eta(t)dW(t)$$

for the logarithmic opportunity process $K^H := \log L^H$, where

$$a^{K^H}(t) = \frac{1}{2}\left(\frac{\mu(t)}{\sigma(t)} + \eta_1(t)\right)^2 - \frac{1}{2}\sum_{i=1}^n \eta_i(t)^2.$$

The optimal portfolio is obtained from (10.37) as

$$\varphi_1^H(t) = \frac{1}{p}\left(\frac{\mu(t)}{\sigma(t)^2} + \frac{\eta_1(t)}{\sigma(t)}\right).$$

From Example 10.19 we conclude that K^0 is deterministic and hence $\eta = 0$ in the special case of geometric Brownian motion S_1 without random endowment H.

In a diffusion-type framework, we obtain a PDE as in Example 10.15.

Example 10.21 (Continuous Markovian Model) If the random endowment is of the form $H = f(X(T))$ for some function $f : \mathbb{R}^n \to \mathbb{R}$, the ansatz $K^H(t) = \ell(T - t, X(t))$ for the logarithmic opportunity process leads to the PDE

$$\begin{aligned}
\ell(0,x) &= pf(x),\\
D_1\ell(\vartheta,x) &= \sum_{i=1}^n D_{1+i}\ell(\vartheta,x)\beta_i(x)\\
&\quad + \frac{1}{2}\sum_{i,j=1}^n \big(D_{1+i,1+j}\ell(\vartheta,x) + D_{1+i}\ell(\vartheta,x)D_{1+j}\ell(\vartheta,x)\big)\gamma_{ij}(x)\\
&\quad - \frac{1}{2}\frac{\big(\beta_1(x) + \sum_{i=1}^n D_{1+i}\ell(\vartheta,x)\gamma_{i1}(x)\big)^2}{\gamma_{11}(x)}
\end{aligned} \tag{10.39}$$

for $t \in [0,T], x \in \mathbb{R}^n$. Indeed, along the same lines as in Example 10.15 we apply Proposition 4.13 in order to rephrase (10.38) in terms of ℓ, β, γ. The optimal portfolio is also of the form $\varphi_1^H(t) = \xi(T - t, X(t))$ for some function ξ. It is obtained from (10.37) as

$$\xi(\vartheta, x) = \frac{\beta_1(x) + \sum_{i=1}^n D_{1+i}\ell(\vartheta, x)\gamma_{i1}(X(\vartheta))}{p\gamma_{11}(x)}. \tag{10.40}$$

In the presence of a claim H, we do not get an explicit solution any more for geometric Lévy processes. But at least we can reformulate the problem in terms of some partial integro-differential equation (PIDE).

Example 10.22 (Geometric Lévy Model with Claim) We reconsider Example 10.19 in the presence of a contingent claim of the form $H = f(S(T))$ for some sufficiently regular function $f : \mathbb{R} \to \mathbb{R}$. We want to determine the H-opportunity process L^H and the optimal portfolio $\varphi^H = (\varphi_0^H, \varphi_1^H)$ in Rule 10.17. Because of the Markovian structure of S it is reasonable to assume that $K^H := \log L^H$ and φ_1^H are functions of time and the current stock price, i.e. $K^H(t) = \ell(T-t, S(t))$ and $\varphi_1^H(t) = \xi(T-t, S(t-))$ with functions $\ell, \xi : [0,T] \times \mathbb{R}_+ \to \mathbb{R}$. Applying Proposition 4.13 we can express the characteristics of (S, K^H) in terms of ℓ and the Lévy–Khintchine triplet (a^X, c^X, K^X) of X. Equations (10.31, 10.32) turn into

$$\begin{aligned} D_1\ell(\vartheta, x) = {} & D_2\ell(\vartheta, x)x\left(a^X + \frac{1}{2}c^X + \int (e^y - 1 - y)K^X(dx)\right) \\ & + \frac{1}{2}(D_2\ell(\vartheta, x))^2x^2c^X - \frac{1}{2}p^2\xi(\vartheta, x)^2x^2c^X \\ & + \int \Big(e^{-p\xi(\vartheta,x)x(e^y-1)+\ell(\vartheta,xe^y)-\ell(\vartheta,x)}\big(1 + p\xi(\vartheta, x)x(e^y - 1)\big) \\ & - 1 - \ell(\vartheta, xe^y) + \ell(\vartheta, x)\Big)K^X(dy), \end{aligned} \tag{10.41}$$

where $\xi(\vartheta, x)$ is the implicit function defined by

$$\begin{aligned} 0 = {} & x\left(a^X + \frac{1}{2}c^X + \int (e^y - 1 - y)K^Y(dy)\right) + \left(D_2\ell(t, x)x - p\xi(t, x)x^2\right)c^X \\ & + \int e^{-p\xi(t,x)x(e^y-1)+\ell(t,xe^y)-\ell(t,x)}x(e^y - 1)K^X(dy). \end{aligned} \tag{10.42}$$

Together with (10.42), (10.41) is a PIDE for ℓ, which is complemented by the initial condition $\ell(0, x) = pf(x)$.

For exponential utility, optimal investment with random endowment can in fact be reduced to a pure investment problem under a different probability measure. To this end, observe that

$$E(u(V_\varphi(T) - H)) = E(e^{pH})E_{P^H}(u(V_\varphi(T))) \tag{10.43}$$

for the probability measure $P^H \sim P$ defined by

$$\frac{dP^H}{dP} = \frac{e^{pH}}{E(e^{pH})}. \tag{10.44}$$

In terms of distances, this leads to the following generalisation of Rule 10.5(3).

Rule 10.23 *If* $u(x) = -\exp(-px)$ *with* $p > 0$, *the measure* Q^H *in Remark 10.16 resp. Rule 10.17(3) minimises the relative entropy*

$$H(Q|P^H) = H(Q|P) - pE_Q(H) + \log E(e^{pH})$$

among all EMMs Q. *The minimal value is related to the maximal utility* U_H *in (10.28) via*

$$\begin{aligned} U_H &= -E(e^{pH}) \exp\Big(-pv_0 - H(Q^H|P^H)\Big) \\ &= -\exp\Big(-pv_0 - H(Q^H|P) + pE_{Q^H}(H)\Big). \end{aligned}$$

Idea Firstly, note that

$$\begin{aligned} H(Q|P^H) &= E_Q\left(\log \frac{dQ}{dP^H}\right) \\ &= E_Q\left(\log \frac{dQ}{dP}\right) - E_Q\left(\log \frac{dP^H}{dP}\right) \\ &= H(Q|P) - E_Q(pH) + \log E\big(e^{pH}\big). \end{aligned} \tag{10.45}$$

The first expression for U_H follows from (10.9) together with (10.43), the second from (10.45). □

10.2 Utility of Consumption

In real life it is not obvious why wealth should matter only at a fixed time T as in the previous section. Moreover, one could argue that utility is not derived from money but rather from spending it. This leads to the idea of considering expected utility of *consumption* instead, where the latter takes place continuously over time. In this framework, you not only optimise your investment decision but also how to spend your proceeds over time.

Parallel to Sect. 10.1 we consider a market with $1 + d$ assets as in Sect. 9.1.1 with $S_0 = 1$ and a finite time horizon T. Your initial endowment is denoted by v_0. On top of a portfolio φ you choose a **consumption rate**, i.e. a predictable process

$(c(t))_{t\in[0,T]}$. We call a pair (φ, c) of a trading strategy φ and a consumption rate process c **self-financing** (or we call φ c-**financing**) if

$$V_\varphi = v_0 + \varphi \bullet S - c \bullet I, \tag{10.46}$$

where $V_\varphi = \sum_{i=0}^d \varphi_i S_i$ denotes the wealth process as usual. Obviously, $c \bullet I(t) = \int_0^t c(s)ds$ represents the money that is consumed or spent up to time t.

Parallel to Rule 9.6, self-financability can always be achieved by choosing φ_0 appropriately: given an initial endowment v_0, positions $\varphi_1, \dots, \varphi_d$ in the risky assets, and a consumption rate process c, there is a unique predictable process φ_0 such that the strategy/consumption pair $((\varphi_0, \dots, \varphi_d), c)$ is self-financing with initial endowment v_0. Indeed, in order for (10.46) to hold, you choose

$$\varphi_0(t) := v_0 + \sum_{i=1}^d \varphi_i \bullet S_i(t-) - c \bullet I(t) - \sum_{i=1}^d \varphi_i S_i(t-).$$

Observe that a strategy/consumption pair (φ, c) as above corresponds to a pair $(\varphi, c \bullet I)$ in Sect. 9.2 because *cumulative* consumption is considered in the latter. We hope that this slight abuse of terminology does not lead to confusion. The wealth process V_φ above coincides with $V_{\varphi,C}$ in (9.15) because $\Delta C = 0$ in the present setup of absolutely continuous consumption $C = c \bullet I$.

Since utility is now assumed to be derived from consumption, we need a function $u : [0, T] \times \mathbb{R} \to [-\infty, \infty)$, where $u(t, c(t))$ represents the incremental utility that is derived from consuming with rate $c(t)$ at time t. Parallel to Sect. 10.1 we assume that $x \mapsto u(t, x)$ is an increasing, strictly concave function whose derivative we denote as $u'(t, x)$. Similarly as in the terminal wealth case, the most prominent cases are $u(t, x) = \log(x)e^{-\delta t}$ and $u(t, x) = x^{1-p}(1 - p)^{-1}e^{-\delta t}$ with some *impatience rate* $\delta \geq 0$ and some risk aversion parameter $p \in (0, \infty) \setminus \{1\}$. Your goal is to behave optimally in an appropriate sense.

Specifically, we call a self-financing strategy/consumption pair $(\varphi^\star, c^\star)$ with initial value v_0 and nonnegative terminal wealth $V_\varphi(T) \geq 0$ **optimal for expected utility of consumption** if it maximises

$$(\varphi, c) \mapsto E\left(\int_0^T u(t, c(t))dt\right) \tag{10.47}$$

over all such strategy/consumption pairs (φ, c).

Remark 10.24

1. As in Remark 10.2(1), the consumption problem with deterministic numeraire asset S_0 can be reduced to the present setup of constant $S_0 = 1$ by slightly modifying the utility function. And as in that case, arbitrary numeraires S_0 can be considered for logarithmic utility $u(t, x) = \log(x)e^{-\delta t}$, in which case the solution does not depend on S_0 at all.

2. Parallel to Remark 10.2(2), any two optimal pairs (φ, c), $(\widetilde{\varphi}, \widetilde{c})$ have the same wealth and the same consumption process, i.e. $V_\varphi = V_{\widetilde{\varphi}}$ and $c \bullet I = \widetilde{c} \bullet I$.

Rules 10.3 and 10.4 read as follows in the present context.

Rule 10.25

1. *If $(\varphi^\star, c^\star)$ is a self-financing strategy/consumption pair with initial value $V_{\varphi^\star}(0) = v_0$, terminal value $V_{\varphi^\star}(T) = 0$ and if the process $Y(t) = u'(t, c^\star(t))$, $t \in [0, T]$ is a martingale such that YS is a martingale as well, then $(\varphi^\star, c^\star)$ is optimal for expected utility of consumption.*
2. *If $(\varphi^\star, c^\star)$ is optimal for expected utility of consumption and if we set $Y(t) := u'(t, c^\star(t))$, $t \in [0, T]$, then both Y and YS are martingales.*
3. *If $x \mapsto \tilde{u}(t, x)$ denotes the convex dual of $x \mapsto u(t, x)$ in the sense of (A.7, 1.99), then Y from above minimises*

$$Z \mapsto E\left(\int_0^T \tilde{u}(t, Z(t))dt\right) + v_0 Z(0) \tag{10.48}$$

over all nonnegative martingales Z such that ZS is a martingale.

4. *The optimal values of the primal maximisation problem (10.47) and the dual minimisation problem (10.48) coincide in the sense that*

$$E\left(\int_0^T u(t, c^\star(t))dt\right) = E\left(\int_0^T \tilde{u}(t, Y(t))dt\right) + v_0 Y(0).$$

Idea Similarly as for Rule 10.3, the idea is to apply Remark A.18(2). Here U and V are the set of adapted real-valued processes, V being viewed as the dual space of U via $y(x) := E(\int_0^T x(t)y(t)dt)$ for $x \in U$, $y \in V$. Set $K := \{c \in U :$ there exists a predictable process φ with $\varphi \bullet S(T) - c \bullet I(T) \geq 0\}$, which is the set of consumption rates attainable with zero initial investment. Its polar cone is $K^\circ = \{y \in V : E(\int_0^T c(t)y(t)dt) \leq 0$ for any $c \in K\}$. The problem at hand can be rephrased as the task to maximise $f(x) = E(\int_0^T u(t, x(t) + v_0/T)dt)$ over K.

1. Set $y^\star := Y$ and $x^\star := c^\star(t) - v_0/T$ for Y, $(\varphi^\star, c^\star)$ as in statement 1, which implies that the first-order condition (A.9) in Remark A.18(2) holds. For any $c \in K$ and any corresponding predictable process φ we have

$$\begin{aligned} E\left(\int_0^T c(t)y^\star(t)dt\right) &= \int_0^T E(c(t)Y(t))dt \\ &= \int_0^T E(c(t)Y(T))dt \\ &= E\left(\int_0^T c(t)dtY(T)\right) \\ &\leq E(\varphi \bullet S(T)Y(T)) = 0, \end{aligned} \tag{10.49}$$

where the last equality is derived as in the proof of Rule 10.3. It follows that $y^\star \in K^\circ$. For $(\varphi, c) = (\varphi^\star, c^\star - v_0/T)$ we have $\varphi \bullet S(T) - c \bullet I(T) = V_{\varphi^\star}(T) - v_0 = 0$ and hence equality in (10.49). Therefore, $x^\star, y^\star$ solve the complementary slackness condition (A.8) in Remark A.18(2). In view of Theorem A.17, 3c)$\Rightarrow$3a) we have that $(\varphi^\star, c^\star) = Y$ is optimal for expected utility of consumption.

2. Using the same notation as above, Theorem A.17, 3a)$\Rightarrow$3c) yields that $y^\star = u'(t, x^\star(t) + v_0/T) = u'(t, c^\star(t))$ is optimal for the dual problem and that $E(\int_0^T c(t)y^\star(t)dt) \leq 0$ for any $c \in K$. For $0 \leq t_1 \leq t_2 < T$ and $F \in \mathscr{F}_{t_1}$ consider the consumption rate $c \in K$ with

$$c \bullet I = 1_{F \times (t_1, t_2]}, \tag{10.50}$$

which of course exists only in a limiting sense because of the jumps of the right-hand side of (10.50) at t_1 and t_2. Since

$$\begin{aligned} E\big((Y(t_2) - Y(t_1))1_F\big) &= E(Y \bullet (c \bullet I)(T)) \\ &= E((Yc) \bullet I(T)) \\ &= E\left(\int_0^T c(t)y^\star(t)dt\right) \leq 0, \end{aligned}$$

we have that Y is a supermartingale. Replacing c by $-c$, we conclude that Y is a submartingale as well and hence a martingale. For a predictable process φ consider $c \in K$ with $c \bullet I(T) = \varphi \bullet S(T)$. Since $c \in K$, we have

$$\begin{aligned} E(\varphi \bullet S(T)Y(T)) &= E(c \bullet I(T)Y(T)) \\ &= \int_0^T E(c(t)Y(T))dt \\ &= \int_0^T E(c(t)Y(t))dt \\ &= E\left(\int_0^T c(t)y^\star(t)dt\right) \leq 0 \end{aligned}$$

and likewise ≥ 0 by considering $-\varphi, -c$ instead of φ, c. As in the proof of Rule 10.3(2) we conclude that YS is a martingale.

3. Recall from the proof of statement 2 that $y^\star = Y$ minimises $y \mapsto \widetilde{f}(y) = \sup_{x \in U}(f(x) - y(x))$ over K°. If Z, ZS are martingales, we have $Z \in K^\circ$ by the same argument as in the proof of statement 1. From Example A.19 applied to

the product probability measure $\widetilde{P}(A) := \int \frac{1}{T} \int_0^T 1_A(\omega, t) dt P(d\omega)$ we obtain

$$\widetilde{f}(Z) = E\left(\int_0^T \widetilde{u}(t, Z(t)) dt\right) + \frac{v_0}{T} E\left(\int_0^T Z(t) dt\right)$$
$$= E\left(\int_0^T \widetilde{u}(t, Z(t)) dt\right) + v_0 Z(0), \qquad (10.51)$$

where the second equality follows from the fact that $E(Z(t)) = Z(0)$.

4. According to Theorem A.17, 3b) and (10.51) we have

$$E\left(\int_0^T u(t, c^\star(t)) dt\right) = f(x^\star) = \widetilde{f}(y^\star) = E\left(\int_0^T \widetilde{u}(t, Y(t)) dt\right) + v_0 Y(0). \qquad \square$$

The link between optimal consumption and equivalent martingale measures is provided by the following result.

Rule 10.26 (Fundamental Theorem for Optimal Utility of Consumption) *Suppose that $T < \infty$. For self-financing strategy/consumption pairs $(\varphi^\star, c^\star)$ with initial value $V_{\varphi^\star}(0) = v_0$ and terminal wealth $V_\varphi(T) = 0$, the following statements are equivalent:*

1. *$(\varphi^\star, c^\star)$ is optimal for expected utility of consumption.*
2. *For some normalising constant $k > 0$, the process*

$$Z(t) := k u'(t, c^\star(t)), \quad t \in [0, T] \qquad (10.52)$$

is the density process of some equivalent martingale measure $Q^\star \sim P$.

Idea 1⇒2: If $(\varphi^\star, c^\star)$ is optimal, Y and YS are martingales for Y as in Rule 10.25(1). Consequently the normalised martingale $Z = Y/Y(0)$ is the density process of some equivalent martingale measure $Q^\star$.

2⇒1: The $Q^\star$-martingale property of S implies that ZS is a martingale, cf. Proposition 3.69. This yields that the martingale $Y(t) = Z(t)/k = u'(t, c^\star(t))$ satisfies the requirements of Rule 10.25(1). □

Similarly to the terminal wealth case, the above fundamental theorem is not generally true in the present continuous-time setup. For details we refer to the notes and comments.

The mutual fund theorem continues to hold in the present setup.

Example 10.27 (Mutual Fund Theorem Revisited, cf. Example 10.9) We consider the same market with price process $S = (S_0, \dots, S_d)$ as in Example 10.9, but now in the context of maximising expected utility of consumption. To this end, fix an initial endowment v_0, a time horizon T, and a utility function u.

We argue essentially as in Example 10.9 and use the same notation as in that case. Denote by $(\widetilde{\varphi}, \widetilde{c})$ the optimal portfolio/consumption pair for the small market

(S_0, F) with filtration generated by F. By Rule 10.26 we have that the density process of the EMM Q is of the form

$$Z(t) = \frac{u'(t, \widetilde{c}(t))}{E(u'(t, \widetilde{c}(t)))}. \tag{10.53}$$

Since $\widetilde{\varphi}$ trades in the replicable fund F, the pair $(\widetilde{\varphi}, \widetilde{c})$ naturally corresponds to a self-financing portfolio/consumption pair $(\varphi^\star, c^\star)$ in the market S, namely with $\varphi_i^\star(t) := \widetilde{\varphi}_1(t)\psi_i(t)$ for $i = 1, \dots, d$ and $c^\star(t) := \widetilde{c}(t)$. By (10.53) and Rule 10.26, we have that $(\varphi^\star, c^\star)$ is optimal for expected utility of consumption in the market S. Since $\widetilde{\varphi}$ and hence $\varphi^\star$ trade only in F and the numeraire asset, the mutual fund theorem extends to the consumption case, using the same fund as in Example 10.9.

As in Sect. 10.1, more can be said for the standard utility functions, where counterparts exist for Rules 10.6, 10.11 and Examples 10.13, 10.14, 10.15.

10.2.1 Logarithmic Utility

For logarithmic utility $u(t, x) = \log(x)e^{-\delta t}$ we can solve the utility maximisation problem quite explicitly. As in Sect. 10.1.1, we write $S_i = S_i(0)\mathscr{E}(X_i)$ and denote the local characteristics of X by (a^X, c^X, K^X).

Rule 10.28 *Let the $\mathbb{R}^d$-valued process π be given by the equation*

$$0 = a^X(t) - c^X(t)\pi(t) - \int \frac{(\pi(t)^\top x)x}{1 + \pi(t)^\top x} K^X(t, dx) \tag{10.54}$$

and set

$$\kappa(t) := \begin{cases} \delta/(1 - e^{-\delta(T-t)}) & \text{if } \delta \neq 0, \\ 1/(T-t) & \text{if } \delta = 0, \end{cases}$$

$$V(t) := v_0\mathscr{E}(\pi \bullet X - \kappa \bullet I)(t)$$

$$= v_0\mathscr{E}(\pi \bullet X)(t) \exp\left(-\int_0^t \kappa(s)ds\right),$$

$$\varphi_i^\star(t) := \frac{V(t-)}{S_i(t-)}\pi_i(t), \quad i = 1, \dots, d,$$

$$c^\star(t) := V(t)\kappa(t).$$

As usual, we identify $\varphi_1^\star, \dots, \varphi_d^\star$ with a self-financing strategy, here with initial value v_0. Then $(\varphi^\star, c^\star)$ is optimal for consumption with wealth process $v_0\mathscr{E}(\pi \bullet X - c^\star \bullet I)$.

Idea Process V satisfies

$$V = v_0 + V_- \bullet \left(\sum_{i=1}^{d} \pi_i \bullet X_i \right) - V_- \bullet (\kappa \bullet I) = v_0 + \varphi \bullet S - c \bullet I,$$

which implies that it is the wealth process of (φ, c). Moreover, Problem 3.6 yields

$$\frac{1}{V} = \frac{1}{V(0)} \mathscr{E}\left(-\pi \bullet X + \kappa \bullet I + \langle \pi \bullet X^c, \pi \bullet X^c \rangle + \frac{(\pi^\top x)^2}{1 + \pi^\top x} * \mu^X \right) \tag{10.55}$$

and hence

$$a^{1/V} = \frac{1}{V_-} \left(-\pi^\top a^X + \pi^\top c^X \pi + \int \frac{(\pi^\top x)^2}{1 + \pi^\top x} K^X(dx) + \kappa \right) = \frac{\kappa}{V_-} \tag{10.56}$$

by (10.54).
Set $Y(t) := u'(t, c(t)) = e^{-\delta t}/c(t) = R(t)/V(t)$ for

$$R(t) := \frac{e^{-\delta t}}{\kappa(t)} = \begin{cases} \frac{e^{-\delta t} - e^{-\delta T}}{\delta} & \text{for } \delta \neq 0, \\ T - t & \text{for } \delta = 0. \end{cases}$$

Note that $dR(t) = -e^{-\delta t} dt$. Since

$$Y = \frac{R}{V} = R \bullet \frac{1}{V} + \frac{1}{V_-} \bullet R, \tag{10.57}$$

we obtain

$$a^Y(t) = \frac{R(t)\kappa(t)}{V(t-)} - \frac{1}{V(t-)} e^{-\delta t} = 0 \tag{10.58}$$

by (10.56). Another application of integration by parts yields

$$YS_i = Y(0)S_i(0) + (S_i)_- \bullet Y + (S_i)_- \bullet (Y_- \bullet X + [Y, X]).$$

By (10.58, 10.55, 10.57, 10.54, 4.29) its drift rate equals

$$a^{YS_i} = (YS_i)_- \left(a^X - \pi^\top c^X + \int \left(-(\pi^\top x)x + \frac{(\pi^\top x)^2 x}{1 + \pi^\top x} \right) K^X(dx) \right) = 0.$$

In view of Rule 10.25(1) we obtain the optimality of (φ, c). □

Note that the optimal fraction of wealth invested in the various assets coincides for logarithmic utility in the terminal wealth and the consumption case. Moreover, the optimal consumption rate does not depend on the asset price dynamics at all.

10.2.2 Power Utility

In this section we study utility functions of the form $u(t, x) = x^{1-p}(1-p)^{-1}e^{-\delta t}$ with $\delta \geq 0$ and $p \in (0, \infty) \setminus \{1\}$. As in Sect. 10.1.2 we can characterise the solution to the optimisation problem in terms of an *opportunity process*.

For power utility the **opportunity process** L is defined as

$$L(t) := (1-p) \operatorname*{ess\,sup}_{(\varphi,c)} E\left(\int_t^T u(s, c(s))ds \middle| \mathscr{F}_t \right),$$

where maximisation extends over all self-financing strategy/consumption pairs (φ, c) with value $V_\varphi(t) = 1$ at time t. Similarly to the terminal wealth case, the opportunity process stands for the maximal expected utility of consumption between t and T if one starts with wealth 1 at time t. The following characterisation parallels Rule 10.11.

Rule 10.29 *Let L be a positive semimartingale with $L(T) = 0$ and $(\varphi^\star, c^\star)$ a self-financing strategy/consumption pair with initial value $v_0 > 0$. The following are equivalent:*

1. *L is the opportunity process and $(\varphi^\star, c^\star)$ is the optimal strategy for (10.47).*
2. *The value process for the control problem related to (10.47) is of the form $\mathscr{V}(t, (\varphi, c)) = \int_0^t u(s, c(s))ds + L(t)u(t, V_\varphi(t))$ for any $t \in [0, T]$ and any self-financing strategy/consumption pair (φ, c) with initial value v_0. Moreover, $(\varphi^\star, c^\star)$ is optimal for expected utility of consumption.*
3. *The process Y in Rule 10.25 equals*

$$Y(t) := L(t)u'(t, V_{\varphi^\star}(t)), \quad t \in [0, T]$$

and we have

$$c^\star(t) = V_{\varphi^\star}(t)L(t)^{-1/p}, \quad t \in [0, T].$$

In particular, the density process of the EMM $Q^\star$ in (10.52) equals

$$Z(t) := \frac{L(t)}{L(0)}u'(t, V_{\varphi^\star}(t)/v_0), \quad t \in [0, T].$$

4. *$L(t)u'(t, V_{\varphi^\star}(t))$ and $L(t)u'(t, V_{\varphi^\star}(t))S(t)$ are martingales and*

$$c^\star(t) = V_{\varphi^\star}(t)L(t)^{-1/p}, \quad t \in [0, T].$$

5.

$$\frac{a^L(t)}{L(t-)} = \delta - pL(t)^{-1/p} - \frac{1}{2}p(1-p)\gamma(t)^\top c^S(t)\gamma(t)$$
$$- \int \Big((1+\gamma(t)^\top x)^{-p}(1+p\gamma(t)^\top x) - 1\Big)\Big(1+\frac{y}{L(t-)}\Big)K^{(S,L)}(d(x,y)) \tag{10.59}$$

as well as

$$0 = a^S(t) + \frac{c^{L,S}(t)}{L(t-)} - pc^S(t)\gamma(t)$$
$$+ \int \Big((1+\gamma(t)^\top x)^{-p}\Big(1+\frac{y}{L(t-)}\Big) - 1\Big)xK^{(S,L)}(d(x,y)) \tag{10.60}$$

and

$$c^\star(t) = V_{\varphi^\star}(t)L(t)^{-1/p}, \quad t \in [0,T], \tag{10.61}$$

where $\gamma_i(t) := \varphi_i^\star(t)/V_{\varphi^\star}(t-)$ *for* $i = 1, \ldots, d$ *denotes the number of shares invested in risky asset* i per unit of wealth *and the notation as in (II.1) is used.*

Idea 1⇔2: This follows along the same lines as in the proof of Rule 10.11.
2⇒3: Set $\widetilde{L} := (c^\star/V_{\varphi^\star})^{-p} = u'(I,c^\star)/u'(I,V_{\varphi^\star})$. Since $V_{\varphi^\star}(T) = 0$, we have $\widetilde{L}(T) = 0$. Moreover,

$$u(I,c^\star)\bullet I + \widetilde{L}u(I,V_{\varphi^\star}) = (1-p)^{-1}(u'(I,c^\star)c^\star)\bullet I + (1-p)^{-1}u'(I,c^\star)V_{\varphi^\star}$$
$$= (1-p)^{-1}\Big(u'(I,c^\star)(c^\star\bullet I) - (c^\star\bullet I)\bullet u'(I,c^\star) + u'(I,c^\star)V_{\varphi^\star}\Big). \tag{10.62}$$

Since $u'(I,c^\star)$ is a martingale by Rule 10.25(2), (10.62) equals

$$(1-p)^{-1}u'(I,c^\star)(V_{\varphi^\star} + c^\star\bullet I) = (1-p)^{-1}Y(v_0 + \varphi^\star\bullet S)$$
$$= (1-p)^{-1}\big((v_0+\varphi^\star\bullet S)_-\bullet Y + \varphi^\star\bullet(YS - S_-\bullet Y)\big) \tag{10.63}$$

up to a martingale, cf. (10.3). Rule 10.25(2) yields that (10.63) and hence (10.62) are martingales as well. Since both $\mathscr{V}(I,(\varphi^\star,c^\star)) = u(I,c^\star)\bullet I + Lu(I,V_{\varphi^\star})$ and $u(I,c^\star)\bullet I + \widetilde{L}u(I,V_{\varphi^\star})$ are martingales with the same terminal value, we conclude that $L = \widetilde{L}$. This directly yields $c^\star = V_{\varphi^\star}L^{-1/p}$ and $Y = u'(I,c^\star) = u'(I,V_{\varphi^\star})L$. Moreover, Z is the density process of Q because it starts at $u'(0,1) = 1^{-p}e^{-\delta 0} = 1$ and equals Y up to a multiplicative constant.

3⇒2: Since $Y = Lu'(I, V_{\varphi^\star}) = Lu'(I, c^\star)(L^{1/p})^{-p} = u'(I, c^\star)$, the optimality of $(\varphi^\star, c^\star)$ follows from Rule 10.25. As in the proof of 2⇒3 we conclude that L is the opportunity process.

3⇒4: This follows from the martingale property of Y and YS.

4⇒3: This follows from Rule 10.25(1).

4⇔5: This follows almost identically as Rule 10.11, 4⇔5. The counterpart of process U in (10.18) has the additional term $(pL^{-1/p} - \delta) \bullet I$. Therefore an additional drift rate $pL^{-1/p} - \delta$ appears in the analogue of (10.19) and hence in (10.59). □

We apply the previous rule in order to solve the optimisation problem in a geometric Lévy model.

Example 10.30 (Geometric Lévy Model) We consider the same setup as in Examples 9.2 and 10.13 but now the goal is to maximise expected power utility of consumption. As in Example 10.13 we make the assumption that L is deterministic. Since (10.60) coincides with (10.17), this implies that the fraction of wealth π solving (10.24) should also be optimal in the consumption case. (10.59) reduces to the ODE

$$\frac{L'(t)}{L(t)} = \delta - pL(t)^{-1/p} + \alpha$$

with α from (10.25). This ODE simplifies if we consider $L^{1/p}$ instead of L, namely to

$$(L^{1/p})'(t) = \frac{\delta + \alpha}{p}(L^{1/p})(t) - 1.$$

In view of the terminal condition $L(T) = 0$, this affine ODE is solved by

$$L(t)^{1/p} = \frac{1 - \exp(\frac{\delta+\alpha}{p}t)}{\frac{\delta+\alpha}{p}}.$$

If we write $\kappa(t) = c^\star(t)/V_{\varphi^\star}(t-)$ for the consumption rate *per unit of wealth*, (10.61) leads to the candidate

$$\kappa(t) = L(t)^{-1/p} = \frac{\frac{\delta+\alpha}{p}}{1 - \exp(\frac{\delta+\alpha}{p}t)}$$

for the optimal rate per unit of wealth. Reversing the arguments yields optimality of our candidates because they satisfy the conditions in Rule 10.29(5).

As in the terminal wealth case, the example allows for a straightforward extension to $d > 1$ risky assets.

For general Itō processes as in Example 9.3 we can rephrase (10.59, 10.60) in terms of a BSDE.

Example 10.31 (Itō Process Model) In the context of Example 9.3 we proceed similarly as in Example 10.14. The optimal relative investment $\gamma(t)$ is obtained from (10.60) as

$$\gamma(t) = \frac{a^S(t) + c^{L,S}(t)/L(t)}{pc^S(t)} \tag{10.64}$$

and the opportunity process L satisfies

$$a^L(t) = \left(\delta - pL(t)^{-1/p} - \frac{1-p}{2p}\frac{(a^S(t) + c^{L,S}(t)/L(t))^2}{c^S(t)}\right)L(t). \tag{10.65}$$

Since the filtration is generated by W, we expect L to be of the form

$$dL(t) = a^L(t)dt + \eta(t)dW(t)$$

with $L(T) = 1$ and some $\mathbb{R}^n$-valued predictable process η. Reformulating (10.65) as

$$a^L(t) = \left(\delta - pL(t)^{-1/p} - \frac{1-p}{2p}\left(\frac{\mu(t)}{\sigma(t)} + \frac{\eta_1(t)}{L(t)}\right)^2\right)L(t),$$

we obtain a BSDE for L, η. Equation (10.64) then reads as

$$\gamma(t) = \frac{1}{p}\left(\frac{\mu(t)}{\sigma(t)^2} + \frac{\eta_1(t)}{L(t)\sigma(t)}\right),$$

which yields the optimal portfolio. For geometric Brownian motion, L is deterministic and $\eta = 0$ by Example 10.30.

If the Itō process is of Markovian type, we obtain the opportunity process as a solution to a PDE as in the terminal wealth case.

Example 10.32 (Continuous Markovian Model) In the situation of Example 9.4, we argue parallel to Example 10.15. We write the opportunity process and the optimal number of shares of the risky asset per unit of wealth as $L(t) = \ell(T-t, X(t))$, $\gamma(t) = \xi(T-t, X(t))$ with some functions $\ell, \xi : [0,T] \times \mathbb{R}^n \to \mathbb{R}$. The reader should not be misled by our abuse of notation: $\gamma(t)$ here refers to the process from Rule 10.29 whereas $\gamma(x)$ below denotes the diffusion coefficient matrix from Example 9.4. Applying Proposition 4.13, equation (10.65) turns into the following

PDE for ℓ:

$$\ell(0, x) = 1,$$

$$D_1\ell(\vartheta, x) = \sum_{i=1}^{n} D_{1+i}\ell(\vartheta, x)\beta_i(x) + \frac{1}{2}\sum_{i,j=1}^{n} D_{1+i,1+j}\ell(\vartheta, x)\gamma_{ij}(x)$$
$$+ \ell(\vartheta, x)\Bigg(\delta - p\ell(\vartheta, x)^{-1/p}$$
$$- \frac{1-p}{2p}\frac{\left(\beta_1(x) + \sum_{i=1}^n D_{1+i}\ell(\vartheta, x)\gamma_{i1}(x)/\ell(\vartheta, x)\right)^2}{\gamma_{11}(x)}\Bigg).$$

The function ξ and the optimal consumption rate can be expressed in terms of ℓ as

$$\xi(\vartheta, x) = \frac{\beta_1(x) + \sum_{i=1}^n D_{1+i}\ell(\vartheta, x)\gamma_{i1}(x)/\ell(\vartheta, x)}{p\gamma_{11}(x)}$$

and

$$c^\star(t) = V_{\varphi^\star}(t)\ell(T-t, X(t))^{-1/p}.$$

Some of the results in this section such as Rule 10.28 and Example 10.30 can be extended to the limiting case of infinite time horizon, which is not discussed here.

10.3 Utility of P&L

Optimal control of consumption as in the previous section is probably not the main concern of, say, a trading desk. A distant time horizon as in Sect. 10.1, on the other hand, does not match reality well either because companies rather try to prosper on a continuous basis without a definitive end. The criterion below reflects this idea by focusing on the utility of day-to-day profits, integrated over the lifetime of the investor. This concept has hardly been considered in the academic literature, despite the fact that it probably comes closest to real investors' preferences. On top, it is often easier to compute optimal portfolios.

We consider the same setup as in Sect. 10.1 with a discounted price process S. As usual, we identify S with $(S_1, \dots, S_d)$ and self-financing strategies φ with the $\mathbb{R}^d$-valued process $(\varphi_1, \dots, \varphi_d)$. In contrast to Sects. 10.1 and 10.2, utility is now derived from period to period profits and losses. To this end, we assume that the discounted asset price process $(S_1, \dots, S_d)$ has local characteristics (a^S, c^S, K^S). We divide our investment horizon T into n subintervals of length $\Delta t := T/n$ and set $t_i := i\Delta t$, $i = 1, \dots, n$. Fix an increasing, strictly concave, smooth utility function $u : \mathbb{R} \to [-\infty, \infty)$ with $u(0) = 0$ and $u'(0) = 1$. Suppose that the investor's goal is to maximise

$$U^S(\varphi, n) := \sum_{i=1}^{n} E\big(u(V_\varphi(t_i) - V_\varphi(t_{i-1}))\big) \tag{10.66}$$

over all self-financing strategies φ, i.e. the cumulative expected utility of one-period profits resp. losses. For $n = 1$, this essentially boils down to utility maximisation of terminal wealth. What happens in the limit $n \to \infty$ of short time periods?

Rule 10.33 *For any self-financing strategy φ we have*

$$U^S(\varphi, n) \to U^S(\varphi) := \int_0^T E\big(\gamma^S(t, \varphi(t))\big)dt \tag{10.67}$$

as $n \to \infty$, where

$$\gamma^S(t, \psi) := \psi^\top a^S(t) + \frac{u''(0)}{2}\psi^\top c^S(t)\psi + \int \big(u(\psi^\top x) - \psi^\top x\big)K^S(t, dx) \tag{10.68}$$

for $t \in [0, T]$ and $\psi \in \mathbb{R}^d$.

Idea Set $Y_i(t) := \int_{t_{i-1}}^t \varphi(s)dS(s)$. Itō's formula (3.53) yields

$$\begin{aligned}
&u\big(V_\varphi(t_i) - V_\varphi(t_{i-1})\big) = u(Y_i(t_i))\\
&= \int_{t_{i-1}}^{t_i} u'(Y_i(t-))dY_i^c(t)\\
&\quad + \int_{[t_{i-1},t_i]\times\mathbb{R}^d} \Big(u(Y_i(t-) + \varphi(t)^\top x) - u(Y_i(t-))\Big)(\mu^S - \nu^S)(d(t, x))\\
&\quad + \int_{t_{i-1}}^{t_i} \Bigg(u'(Y_i(t-))\varphi(t)^\top a^S(t) + \frac{1}{2}u''(Y_i(t-))\varphi(t)^\top c^S(t)\varphi(t)\\
&\quad + \int \Big(u(Y_i(t-) + \varphi(t)^\top x) - u(Y_i(t-)) - u'(Y_i(t-))\varphi(t)^\top x\Big)K^S(t, dx)\Bigg)dt.
\end{aligned}$$

Its conditional expectation relative to $\mathscr{F}_{t_{i-1}}$ equals

$$\begin{aligned}
&\Bigg(u'(0)\varphi(t_{i-1})^\top a^S(t_{i-1}) + \frac{1}{2}u''(0)\varphi(t_{i-1})^\top c^S(t_{i-1})\varphi(t_{i-1})\\
&\qquad + \int \Big(u(\varphi(t_{i-1})^\top x) - u(0) - u'(0)\varphi(t_{i-1})^\top x\Big)K^S(t_{i-1}, dx)\Bigg)(t_i - t_{i-1})\\
&\qquad + o(t_i - t_{i-1})
\end{aligned}$$

$$= \left(\varphi(t_{i-1})^\top a^S(t_{i-1}) + \frac{1}{2} u''(0) \varphi(t_{i-1})^\top c^S(t_{i-1}) \varphi(t_{i-1}) \right.$$
$$\left. + \int \left(u(\varphi(t_{i-1})^\top x) - \varphi(t_{i-1})^\top x \right) K^S(t_{i-1}, dx) \right) \Delta t + o(\Delta t).$$

Summing these expressions up, we obtain the Riemann sum approximating the integral $\int_0^T \gamma^S(t, \varphi(t))dt$ up to terms of smaller order which vanish in the limit $\Delta t \to 0$. □

$E(\gamma^S(t, \psi))dt$ can be interpreted as expected utility of profits/losses in an infinitesimal time interval $(t - dt, t]$ if the portfolio vector in this period equals ψ. We call γ^S the **utility function of running profits and losses** or **utility of P&L** corresponding to the utility function u and price process S. Moreover, a strategy $\varphi^\star$ is **optimal for running profits and losses** or **optimal for P&L** if it maximises $U^S(\varphi)$ over all strategies φ. Optimality in a constrained set of strategies is defined accordingly.

Remark 10.34 If S is continuous, (10.68) reduces to

$$\gamma^S(t, \psi) = \psi^\top a^S(t) + \frac{u''(0)}{2} \psi^\top c^S(t) \psi, \tag{10.69}$$

which depends on u only through a single number $u''(0)$, namely the negative of the *risk aversion* of u in 0.

As in the terminal wealth case, optimal portfolios are essentially unique.

Rule 10.35 *Suppose that S does not allow for arbitrage. If both φ and $\widetilde{\varphi}$ are optimal portfolios for P&L, we have $\varphi \bullet S = \widetilde{\varphi} \bullet S$. The initial endowment does not play a role because it does not enter the utility of P&L.*

Idea $\psi \mapsto \gamma^S(t, \psi)$ is a sum of concave functions. If it is maximised by both ψ and $\widetilde{\psi}$, we have that $\lambda \to f(\lambda) := \gamma^S(t, \psi + \lambda(\widetilde{\psi} - \psi)) = f_1(\lambda) + f_2(\lambda) + f_3(\lambda)$ is constant on $[0, 1]$, where

$$f_1(\lambda) = (a^S)^\top (\lambda\psi + (1 - \lambda)\widetilde{\psi}),$$
$$f_2(\lambda) = \frac{u''(0)}{2} \left(\lambda\psi + (1 - \lambda)\widetilde{\psi}\right)^\top c^S \left(\lambda\psi + (1 - \lambda)\widetilde{\psi}\right),$$
$$f_3(\lambda) = \int \left(u\left(\left(\lambda\psi + (1 - \lambda)\widetilde{\psi}\right)^\top x\right) - \left(\lambda\psi + (1 - \lambda)\widetilde{\psi}\right)^\top x \right) K^S(dx).$$

Since f_1, f_2, f_3 are concave, $f_2 = f - f_1 - f_3$ is convex and hence affine. Thus

$$0 = f_2''(\lambda) = -u''(0)(\widetilde{\psi} - \psi)^\top c^S (\widetilde{\psi} - \psi) \tag{10.70}$$

and hence $(c^S)^{1/2}(\widetilde{\psi} - \psi) = 0$. This implies $f_2'(\lambda) = -u''(0)(\widetilde{\psi} - \psi)^\top c^S(\lambda\psi + (1-\lambda)\psi) = 0$, which means $f_2 = 0$. Thus

$$0 = f_2'(1) - f_2'(0) = -\int \Big(u'(\widetilde{\psi}^\top x) - u'(\psi^\top x)\Big)(\widetilde{\psi} - \psi)^\top x K^S(dx).$$

Since u' is strictly decreasing, the integrand is negative unless $(\widetilde{\psi} - \psi)^\top x = 0$. This implies that

$$\widetilde{\psi}^\top x - \psi^\top x = 0 \tag{10.71}$$

holds for K^S-almost all x and hence $f_3 = 0$. Therefore $f_1 = f - f_2 - f_3 = f$ is constant, which yields

$$(a^S)^\top(\widetilde{\psi} - \psi) = 0. \tag{10.72}$$

Altogether, it follows that

$$(\widetilde{\varphi} - \varphi) \bullet S = \big((\widetilde{\varphi} - \varphi)^\top a^S\big) \bullet I + (\widetilde{\varphi} - \varphi) \bullet S^c + (\widetilde{\varphi} - \varphi)^\top x * (\mu^S - \nu^S)$$

vanishes. Indeed, the first term is 0 by (10.72), the second because of (10.70) and hence

$$\big\langle(\widetilde{\varphi} - \varphi) \bullet S^c, (\widetilde{\varphi} - \varphi) \bullet S^c\big\rangle = \big((\widetilde{\varphi} - \varphi)^\top c^S(\widetilde{\varphi} - \varphi)\big) \bullet I = 0,$$

and the third by (10.71) and

$$\Big\langle(\widetilde{\varphi} - \varphi)^\top x * (\mu^S - \nu^S), (\widetilde{\varphi} - \varphi)^\top x * (\mu^S - \nu^S)\Big\rangle$$
$$= \int \big((\widetilde{\varphi} - \varphi)^\top x\big)^2 K^S(dx) \bullet I = 0. \qquad \square$$

Running P&L allows for a similar dual representation as the terminal wealth and the consumption cases.

Rule 10.36

1. A trading strategy $\varphi^\star$ is optimal for P&L if and only if YS is a martingale for

$$Y = \mathscr{E}\Big(u''(0)\varphi^\star \bullet S^c + \big(u'(\varphi^{\star\top} x) - 1\big) * (\mu^S - \nu^S)\Big). \tag{10.73}$$

2. *If $\tilde{u}$ denotes the convex dual of u in the sense of (A.7, 1.94), then Y from (10.73) minimises*

$$\begin{aligned} Z &\mapsto \widetilde{U}(Z) \\ &:= E\Big(-\frac{1}{2}\widetilde{u}''(1)\langle \mathscr{L}(Z)^c, \mathscr{L}(Z)^c\rangle(T) + \widetilde{u}(1+x) * \nu^{\mathscr{L}(Z)}(T)\Big) \end{aligned} \tag{10.74}$$

among all positive martingales Z such that ZS is a martingale.

3. *The optimal values of the primal maximisation problem $\varphi \mapsto E(U^S(\varphi))$ and the dual minimisation problem (10.74) coincide in the sense that*

$$E(U^S(\varphi^\star)) = E(\tilde{U}(Y)).$$

Idea

1. *Step 1:* By definition, $\varphi^\star$ is optimal if

$$E\Big(\int_0^T \gamma^S(t, \varphi^\star)dt\Big) \geq E\Big(\int_0^T \gamma^S(t, \varphi)dt\Big)$$

holds for all strategies φ, which is satisfied if $\gamma^S(t, \varphi^\star)(\omega) \geq \gamma^S(t, \varphi)(\omega)$ for all (ω, t) outside some $P \otimes \lambda$-null set. Since $\psi \mapsto \gamma^S(t, \psi)$ is concave, this is equivalent to the first-order condition

$$\begin{aligned} 0 &= (\gamma^S)'(t, \varphi^\star) \\ &= a^S(t) + u''(0)c^S(t)\varphi^\star(t) - \int \big(u'(\varphi^\star(t)^\top x) - 1\big)x K^S(t, dx), \end{aligned}$$

where $(\gamma^S)'$ denotes the derivative with respect to the second argument.

Step 2: Integration by parts yields

$$\begin{aligned} YS &= Y(0)S(0) + S_- \bullet Y + Y_- \bullet S + [Y, S] \\ &= Y(0)S(0) + S_- \bullet Y + Y_- \bullet (S + [\mathscr{L}(Y), S]). \end{aligned} \tag{10.75}$$

Since Y is a martingale and by (4.29), this equals

$$\Big(Y_-\Big(a^S + u''(0)c^S\varphi^\star + \int \big(u'((\varphi^\star)^\top x) - 1\big)x K^S(dx)\Big)\Big) \bullet I$$

up to a martingale. By step 1 this drift term vanishes if and only if $\varphi^\star$ is optimal. This yields the assertion.

2. As in the proofs of Rules 10.3 and 10.25, the idea is to apply Remark A.18(2). To this end, let both U and V denote the set of triplets (φ, ξ, η) of predictable $\mathbb{R}^d$-valued processes φ, ξ and a $\mathscr{P}\otimes\mathscr{B}^d$-measurable function $\eta : \Omega\times\mathbb{R}_+\times\mathbb{R}^d \to \mathbb{R}$. We view V as a dual space of U via

$$((\widetilde{\varphi}, \widetilde{\xi}, \widetilde{\eta}), (\varphi, \xi, \eta)) := E\left(\left(\sum_{i=1}^d \varphi_i a_i^S \widetilde{\varphi}_i + \xi c^S \widetilde{\xi} + \int \eta(x)\widetilde{\eta}(x)K^S(dx)\right)\bullet I(T)\right)$$

for $(\varphi, \xi, \eta) \in U$, $(\widetilde{\varphi}, \widetilde{\xi}, \widetilde{\eta}) \in V$. Set $K := \{(\varphi, \xi, \eta) \in U : \xi = \varphi, \eta(x) = \varphi^\top x$ for any $x \in \mathbb{R}^d\}$. Its polar cone $K^\circ := \{(\widetilde{\varphi}, \widetilde{\xi}, \widetilde{\eta}) \in V : ((\widetilde{\varphi}, \widetilde{\xi}, \widetilde{\eta}), (\varphi, \xi, \eta)) \leq 0$ for any $(\varphi, \xi, \eta) \in U\}$ contains all triplets $(\widetilde{\varphi}, \widetilde{\xi}, \widetilde{\eta})$ satisfying

$$E\left(\left(\sum_{i=1}^d \varphi_i\left(a_i^S \widetilde{\varphi}_i + c_{i\cdot}^S \widetilde{\xi} + \int x_i \widetilde{\eta}(x)K^S(dx)\right)\right)\bullet I(T)\right) \leq 0$$

for all predictable $\mathbb{R}^d$-valued φ or, equivalently,

$$a_i^S \widetilde{\varphi}_i + c_{i\cdot}^S \widetilde{\xi} + \int x_i \widetilde{\eta}(x)K^S(dx) = 0, \quad i = 1, \dots, d.$$

The portfolio optimisation problem amounts to maximising

$$\begin{aligned} &f(\varphi, \xi, \eta) \\ &:= E\left(\left(\varphi^\top a^S + \frac{u''(0)}{2}\xi^\top c^S \xi + \int \big(u(\eta(x)) - \eta(x)\big)K^S(dx)\right)\bullet I(T)\right) \end{aligned}$$

over K. A straightforward calculation shows that the conjugate function

$$\widetilde{f}(\widetilde{\varphi}, \widetilde{\xi}, \widetilde{\eta}) := \sup\left\{f(\varphi, \xi, \eta) - \big((\widetilde{\varphi}, \widetilde{\xi}, \widetilde{\eta}), (\varphi, \xi, \eta)\big) : (\varphi, \xi, \eta) \in U\right\}$$

equals

$$\begin{aligned} \widetilde{f}(\widetilde{\varphi}, \widetilde{\xi}, \widetilde{\eta}) = E\Bigg(\Bigg(&\infty \prod_{i=1}^d 1_{\{\widetilde{\varphi}_i = 1 \text{ or } a_i^S = 0\}^C} - \frac{1}{2u''(0)}\widetilde{\xi}^\top c^S \widetilde{\xi} \\ &+ \int \widetilde{u}(1 + \widetilde{\eta}(x))K^S(dx)\Bigg)\bullet I(T)\Bigg), \end{aligned}$$

where we set $\infty \cdot 0 := 0$ as usual. Suppose that Z is a positive martingale such that ZS is a martingale. By [154, Lemma III.4.24] we have that

$$\mathscr{L}(Z) = \widetilde{\xi}\bullet S^c + \widetilde{\eta} * (\mu^S - \nu^S) + N \tag{10.76}$$

for some integrands $\widetilde{\xi}, \widetilde{\eta}$ and some martingale N such that $\langle N^c, S^c\rangle = 0$ and $\langle N, W * (\mu^S - \nu^S)\rangle = 0$ for any $W \in G_{\mathrm{loc}}(\mu^S)$. The calculation in (10.75) and the martingale property of Z and ZS yield that

$$0 = a^S + \widetilde{c}^{\mathscr{L}(Z),S} = a^S + c^S\widetilde{\xi} + \int x\widetilde{\eta}(x)K^S(dx)$$

and hence $(\widetilde{\varphi}, \widetilde{\xi}, \widetilde{\eta}) \in K^\circ$ for $\widetilde{\varphi}_i = 1, i = 1, \dots, d$. We call $(\widetilde{\varphi}, \widetilde{\xi}, \widetilde{\eta}) \in K^\circ$ the triplet associated to Z. Moreover, $\widetilde{u}''(1) = 1/u''(0)$ and (10.76) yield

$$\begin{aligned}\widetilde{U}(Z) &= E\Big(\Big(-\frac{1}{2u''(0)}\widetilde{\xi}^\top c^S\widetilde{\xi} + \int \widetilde{u}(1+\widetilde{\eta}(x))K^S(dx)\Big)\bullet I(T)\Big)\\ &= \widetilde{f}(\widetilde{\varphi}, \widetilde{\xi}, \widetilde{\eta}).\end{aligned} \tag{10.77}$$

By Theorem A.17, 3a)⇒3c), the dual maximiser $(\widetilde{\varphi}, \widetilde{\xi}, \widetilde{\eta}) \in K^\circ$ satisfies the first-order conditions (A.9) in Remark A.18(2), which here read as

$$\begin{aligned} a_i^S &= a_i^S\widetilde{\varphi}_i, \quad i = 1, \dots, d,\\ u''(0)c^S\varphi^\star &= c^S\widetilde{\xi},\\ u'\big((\varphi^\star)^\top x\big) &= 1 + \widetilde{\eta}(x)\end{aligned}$$

or, equivalently,

$$\begin{aligned}\widetilde{\varphi}_i &= 1 \text{ or } a_i^S = 0, \quad i = 1, \dots, d,\\ \widetilde{\xi} &= u''(0)\varphi^\star,\\ \widetilde{\eta}(x) &= u'\big((\varphi^\star)^\top x\big) - 1, \quad x \in \mathbb{R}^d.\end{aligned}$$

A triple $(\widetilde{\varphi}, \widetilde{\xi}, \widetilde{\eta}) \in K^\circ$ maximises $\widetilde{f}$ on K°. By (10.77) we conclude that Y from (10.73) minimises $\widetilde{U}(Z)$ over all Z such that ZS is a martingale.

3. This follows from Theorem A.17, 3a)⇒3b), equation (10.77), and the fact that the minimum of $\widetilde{f}$ over K° is attained at the triplet $(\widetilde{\varphi}, \widetilde{\xi}, \widetilde{\eta})$ corresponding to Y. □

The related characterisation in terms of EMMs reads as follows.

Rule 10.37 (Fundamental Theorem of Utility Maximisation for P&L) *A strategy $\varphi^\star$ is optimal for P&L if and only if the martingale (10.73) is the density process of some equivalent martingale measure $Q^\star$.*

Idea This follows from Rule 10.36(1) and Proposition 3.69. □

Example 10.38 (Exponential Utility) As an example for the utility function u we consider

$$u(x) = \frac{1 - e^{-px}}{p}, \tag{10.78}$$

which leads to utility of P&L

$$\gamma^S(t, \psi) = \psi^\top a^S(t) - \frac{p}{2}\psi^\top c^S(t)\psi + \int \left(\frac{1 - e^{-p(\psi^\top x)}}{p} - \psi^\top x \right) K^S(t, dx). \tag{10.79}$$

The equivalent martingale measure $Q^\star$ from Rule 10.37 has density process

$$\begin{aligned} Y(t) &= \mathscr{E}\Big(-p\varphi^\star \bullet S^c + \big(\exp(-p(\varphi^{\star\top} x)) - 1\big) * (\mu^S - \nu^S)\Big)(t) \\ &= \exp\big(-p\varphi^\star \bullet S(t) - A(t)\big) \end{aligned} \tag{10.80}$$

in this case, where A is the unique predictable process of finite variation such that (10.80) is a martingale. Measure changes with density processes of the form $\exp(\varphi \bullet X(t) - A(t))$ for some given process X, some integrand φ, and some predictable process A of finite variation are sometimes called *Esscher transforms* in the literature, in particular if X is a Lévy process, φ constant and hence A a linear deterministic function.

Example 10.39 (Hyperbolic Utility) In order for the integral in (10.79) to exist, the jump measure of S needs to have exponential moments, which does not hold, for example, in many popular Lévy models discussed in the empirical literature. As an alternative we suggest a utility function which does not decrease faster than linearly for large negative values. Specifically, set

$$u(x) := \frac{1}{p} f(px)$$

for

$$f(x) := x + 1 - \sqrt{x^2 + 1}$$

and some risk aversion parameter $p > 0$. Since

$$u'(x) = 1 - \frac{x}{\sqrt{x^2 + p^{-2}}}$$

and $u''(0) = -p$, the utility of P&L equals

$$\gamma^S(t,\psi) = \psi^\top a^S(t) - \frac{p}{2}\psi^\top c^S(t)\psi + \int \left(p^{-1} - \sqrt{(\psi^\top x)^2 + p^{-2}}\right) K^S(t,dx)$$

and the equivalent martingale measure $Q^\star$ from Rule 10.37 has density process

$$Y = \mathscr{E}\left(-p\varphi^\star \bullet S^c - \frac{\varphi^{\star\top} x}{\sqrt{(\varphi^{\star\top} x)^2 + p^{-2}}} * (\mu^S - \nu^S)\right).$$

As a side remark observe that the hyperbolic utility function $u = u_p$ converges to

$$\lim_{p\to 0} u_p(x) = x$$

resp.

$$\lim_{p\to\infty} u_p(x) = \min\{2x, 0\}$$

for small resp. large risk aversion.

Similarly to logarithmic utility in the terminal wealth case, the P&L problem allows for a rather explicit solution.

Rule 10.40 *Strategy $\varphi^\star$ is optimal for P&L if and only if*

$$0 = a^S(t) + u''(0)c^S(t)\varphi^\star(t) + \int \left(u'(\varphi^\star(t)^\top x) - 1\right) x K^S(t,dx) \qquad (10.81)$$

for any $t \in [0,T]$.

More generally, $\varphi^\star \in C$ is optimal in a set of strategies of the form

$$C := \{\varphi^\star \in L(S) : \varphi(t) \in C(t) \textit{ for any } t \in [0,T]\}$$

if and only if $\varphi(t)$ maximises $\psi \mapsto \gamma^S(t,\psi)$ for almost any $t \in [0,T]$.

Idea This is shown in the first part of the proof of Rule 10.36. □

Example 10.41 (Geometric Lévy Model) We consider the geometric Lévy model from Example 9.2 for P&L optimisation with the exponential utility function from Example 10.38. In this case (10.81) reduces to (10.34) for $\psi = (S_1)_-\varphi_1$. This in turn implies that the optimal strategy coincides with the one for expected exponential utility of terminal wealth in Example 10.19.

As in Examples 10.13, 10.19, 10.30, the extension to $d > 1$ risky assets is straightforward. In particular, we obtain that the optimal portfolios for P&L coincide with those for utility of terminal wealth if exponential utility is chosen in both cases. This implies that the mutual fund theorem in Example 10.9 extends to utility of P&L

as well. Indeed, this is evident for exponential utility. But since the price process is continuous in Example 10.9, the utility function of P&L coincides with the one for exponential utility regardless of the choice of u, cf. Remark 10.34.

For hyperbolic utility from Example 10.39 we obtain very similar results as in the exponential case. Only (10.34) or the equivalent equation (10.35) for ψ have to be replaced by

$$0 = a^{\widetilde{X}} - c^{\widetilde{X}} p\psi - \int \frac{\psi x^2}{\sqrt{(\psi x)^2 + p^{-2}}} K^{\widetilde{X}}(dx)$$

resp.

$$0 = a^X + \left(\frac{1}{2} - p\psi\right)c^X - \int \left(\left(\frac{\psi(e^x - 1)}{\sqrt{(\psi(e^x - 1))^2 + p^{-2}}} - 1\right)(e^x - 1) + x\right) K^X(dx).$$

For Itō processes, the solution to (10.81) is explicit.

Example 10.42 (Itō Process Model) For a single continuous risky asset as in Example 9.3, (10.81) immediately yields

$$\varphi^\star(t) = \frac{a^S(t)}{-u''(0)c^S(t)} = \frac{\mu(t)}{-u''(0)\sigma(t)^2}$$

for the optimal strategy. The corresponding optimal utility of P&L equals

$$\gamma^S(t, \varphi^\star(t)) = \frac{a^S(t)^2}{-2u''(0)c^S(t)} = \frac{\mu(t)^2}{-2u''(0)\sigma(t)^2}.$$

10.4 Mean-Variance Efficient Portfolios

A classical approach to portfolio optimisation is to balance the expected return against the risk measured in terms of variance. This leads to the notion of *efficient* portfolios. We work in the same setup of $1 + d$ discounted assets as in Sect. 10.1.

Rule 10.43 *Suppose that S is not a martingale. Let $\varphi^\star$ be a self-financing strategy. The following statements are equivalent.*

1. *$\varphi^\star$ maximises $E(\varphi \bullet S(T))$ among all self-financing strategies φ satisfying*

$$\mathrm{Var}(\varphi \bullet S(T)) \leq \mathrm{Var}(\varphi^\star \bullet S(T)).$$

2. *$\varphi^\star$ minimises $\mathrm{Var}(\varphi \bullet S(T))$ among all self-financing strategies φ satisfying*

$$E(\varphi \bullet S(T)) \geq E(\varphi^\star \bullet S(T)).$$

3. *For some* $m \geq 0$*, strategy* $\varphi^\star$ *minimises* $E((m-\varphi \bullet S(T))^2)$ *among all self-financing strategies* φ*.*
4. *Unless* $\varphi^\star \bullet S(T)$ *vanishes identically,* $\varphi^\star$ *maximises the* ***Sharpe ratio***

$$\mathrm{SR}(\varphi) := \frac{E(\varphi \bullet S(T))}{\sqrt{\mathrm{Var}(\varphi \bullet S(T))}}$$

among all self-financing strategies φ*, where we set* $0/0 := 0$*.*

5. *Unless* $\varphi^\star \bullet S(T)$ *vanishes identically, we have* $\varphi^\star = m\widetilde{\varphi}$ *for some* $m > 0$*, where* $\widetilde{\varphi}$ *here denotes the minimiser of*

$$E\big((1-\varphi \bullet S(T))^2\big) \tag{10.82}$$

among all self-financing strategies φ*.*

Idea 1⇒4: Suppose that $\varphi^\star \bullet S(T) \neq 0$. Let φ be a competing strategy with $\varphi \bullet S(T) \neq 0$. W.l.o.g. we may assume $\mathrm{Var}(\varphi \bullet S(T)) = \mathrm{Var}(\varphi^\star \bullet S(T))$. Indeed, otherwise consider $\widetilde{\varphi} := \varphi\sqrt{\mathrm{Var}(\varphi^\star \bullet S(T))/\mathrm{Var}(\varphi \bullet S(T))}$, which has the same Sharpe ratio as φ. Statement 1 yields $E(\varphi \bullet S(T)) \leq E(\varphi^\star \bullet S(T))$, which in turn implies $\mathrm{SR}(\varphi) \leq \mathrm{SR}(\varphi^\star)$.

4⇒1: Suppose that $\varphi^\star \bullet S(T) \neq 0$ because otherwise statement 1 is obvious. Let φ be a competing strategy. W.l.o.g. we may assume $\mathrm{Var}(\varphi \bullet S(T)) = \mathrm{Var}(\varphi^\star \bullet S(T)) > 0$. Indeed, otherwise consider $\widetilde{\varphi} := \varphi\sqrt{\mathrm{Var}(\varphi^\star \bullet S(T))/\mathrm{Var}(\varphi \bullet S(T))}$, which satisfies $E(\widetilde{\varphi} \bullet S(T)) \geq E(\varphi \bullet S(T))$. Statement 4 states that $\mathrm{SR}(\varphi) \leq \mathrm{SR}(\varphi^\star)$, which by equality of the variances yields $E(\varphi \bullet S(T)) \leq E(\varphi^\star \bullet S(T))$.

2⇒4: Suppose that $\varphi^\star \bullet S(T) \neq 0$. Obviously $E(\varphi^\star \bullet S(T)) > 0$ because otherwise $\varphi^\star$ could not be optimal in 2. Let φ be a competing strategy. If $E(\varphi \bullet S(T)) \leq 0$, we immediately obtain $\mathrm{SR}(\varphi) \leq \mathrm{SR}(\varphi^\star)$. If $E(\varphi \bullet S(T)) > 0$, we may assume $E(\varphi \bullet S(T)) = E(\varphi^\star \bullet S(T))$ because $\widetilde{\varphi} := \varphi E(\varphi^\star \bullet S(T))/E(\varphi \bullet S(T))$ has the same Sharpe ratio as φ. Statement 2 yields $\mathrm{Var}(\varphi \bullet S(T)) \geq \mathrm{Var}(\varphi^\star \bullet S(T))$, which in turn implies $\mathrm{SR}(\varphi) \leq \mathrm{SR}(\varphi^\star)$.

4⇒2: We assume $\varphi^\star \bullet S(T) \neq 0$ because otherwise Statement 2 is obvious. Since $\varphi^\star$ has maximal Sharpe ratio, we have $E(\varphi^\star \bullet S(T)) > 0$. Let φ be a competing strategy for statement 2, i.e. with $E(\varphi \bullet S(T)) \geq E(\varphi^\star \bullet S(T))$. W.l.o.g. we may assume $E(\varphi \bullet S(T)) = E(\varphi^\star \bullet S(T))$. Indeed, otherwise consider $\widetilde{\varphi} := \varphi E(\varphi^\star \bullet S(T))/E(\varphi \bullet S(T))$, which satisfies $\mathrm{Var}(\widetilde{\varphi} \bullet S(T)) < \mathrm{Var}(\varphi \bullet S(T))$. Statement 4 states $\mathrm{SR}(\varphi) \leq \mathrm{SR}(\varphi^\star)$, which by equality of the expected values yields $\mathrm{Var}(\varphi \bullet S(T)) \geq \mathrm{Var}(\varphi^\star \bullet S(T))$.

2⇔3: If $\varphi^\star \bullet S(T)$ vanishes, both 2 and 3 hold. Suppose that $\varphi^\star \bullet S(T) \neq 0$ and set $\mu := E(\varphi^\star \bullet S(T))$. It is easy to see that $\mu \leq 0$ cannot hold if either $\varphi^\star$ minimises $\mathrm{Var}(\varphi \bullet S(T))$ in statement 2 or

$$E\big((m-\varphi \bullet S(T))^2\big) = \mathrm{Var}(\varphi \bullet S(T)) + \big(m - E(\varphi \bullet S(T))\big)^2$$

in statement 3. Indeed, the competing strategy $\varphi = 0$ would yield a better value. Therefore we may assume $\mu > 0$. We show the equivalence of the following four statements.

(a) $\varphi^\star$ minimises $\mathrm{Var}(\varphi \bullet S(T))$ over all φ with $E(\varphi \bullet S(T)) \geq \mu$.
(b) $\varphi^\star$ minimises $\mathrm{Var}(\varphi \bullet S(T))$ over all φ with $E(\varphi \bullet S(T)) = \mu$.
(c) $\varphi^\star$ minimises $E((\varphi \bullet S(T))^2)$ over all φ with $E(\varphi \bullet S(T)) = \mu$.
(d) For some $m \geq 0$, strategy $\varphi^\star$ minimises $E((m - \varphi \bullet S(T))^2)$ over all φ.

This immediately yields the equivalence of statements 2 and 3 in Rule 10.43. The implication (a)$\Rightarrow$(b) is obvious. For (b)$\Rightarrow$(a) note that for any φ with $E(\varphi \bullet S(T)) > \mu$ there is a strategy $\widetilde{\varphi}$ with $E(\widetilde{\varphi} \bullet S(T)) = \mu$ having smaller variance, namely $\widetilde{\varphi} := \varphi\mu / E(\varphi \bullet S(T))$. The equivalence (b)$\Leftrightarrow$(c) is obvious because $\mathrm{Var}(\varphi \bullet S(T)) = E((\varphi \bullet S(T))^2) - \mu^2$. For (d)$\Rightarrow$(c) observe that

$$E((m - \varphi \bullet S(T))^2) = E((\varphi \bullet S(T))^2) - 2m\mu + m^2$$

holds for all φ with $E(\varphi \bullet S(T)) = \mu$. In order to prove that (c) implies (d) set $m := E((\varphi^\star \bullet S(T))^2)/\mu$. Note first that the quadratic function $\mathbb{R} \to \mathbb{R}$, $c \mapsto E((m - c\varphi^\star \bullet S(T))^2)$ attains its minimal value at $c = 1$. Now consider an arbitrary strategy $\widetilde{\varphi}$. Since $E((\widetilde{\varphi}\mu / E(\widetilde{\varphi} \bullet S(T))) \bullet S(T)) = \mu$, we have $E((\varphi^\star \bullet S(T))^2) \leq E(((\widetilde{\varphi}\mu / E(\widetilde{\varphi} \bullet S(T))) \bullet S(T))^2)$ and hence $E((c\varphi^\star \bullet S(T))^2) \leq E((\widetilde{\varphi} \bullet S(T))^2)$ for $c := E(\widetilde{\varphi} \bullet S(T))/\mu$. Together we obtain

$$\begin{aligned}
E((m - \widetilde{\varphi} \bullet S(T))^2) &= E((\widetilde{\varphi} \bullet S(T))^2) - 2mE(\widetilde{\varphi} \bullet S(T)) + m^2 \\
&\geq E((c\varphi^\star \bullet S(T))^2) - 2mcE(\varphi^\star \bullet S(T)) + m^2 \\
&= E((m - c\varphi^\star \bullet S(T))^2) \\
&\geq E((m - \varphi^\star \bullet S(T))^2)
\end{aligned}$$

as desired.

3$\Leftrightarrow$5: This is obvious. □

We call portfolios as in Rule 10.43 **(mean-variance) efficient**. If S is a martingale, all strategies have the same expectation $E(\varphi \bullet S(T)) = 0$, which implies that they trivially satisfy statements 1 and 4 in Rule 10.43. In this case we call $\varphi^\star$ *(mean-variance) efficient* if $\varphi^\star \bullet S = 0$, which means that it satisfies all statements in Rule 10.43.

The characterisation in Rule 10.43(3) indicates that choosing efficient portfolios corresponds to utility maximisation for quadratic utility functions of the form $u(x) = -(x - v_0 - m)^2$ for some $m > 0$, where v_0 denotes the investor's initial endowment. This function is strictly increasing only on $(-\infty, v_0 + m]$, which occasionally leads to certain counterintuitive properties of efficient portfolios. Indeed, once the wealth exceeds the *bliss point m*, the investor is willing to take some risk in order to reduce her wealth. However, this phenomenon does not occur for continuous price processes because the optimal wealth process never surpasses m. The characterisation of mean-variance efficient portfolios via expected utility also shows that the mutual fund theorem in Example 10.9 extends to mean-variance optimisation as well.

In order to determine efficient portfolios, we introduce an opportunity process as in Sect. 10.1.2. The **opportunity process** L for quadratic utility is defined as

$$L(t) := \operatorname*{ess\,sup}_{\varphi} E\big((V_\varphi(T))^2 \big| \mathscr{F}_t\big), \tag{10.83}$$

where maximisation extends over all self-financing strategies φ that start with endowment 1 at time t, i.e.

$$V_\varphi(T) = 1 + \varphi 1_{]\!]t,T]\!]} \bullet S(T) = 1 + \int_t^T \varphi(s) dS(s).$$

The optimal portfolios in (10.83) for different starting times t can all be expressed in terms of a common process, called the *adjustment process* in the literature.

Rule 10.44 *There is an $\mathbb{R}^d$-valued predictable process a such that the optimiser $\varphi = (\varphi_0, \dots, \varphi_d)$ in (10.83) is of the form*

$$(\varphi_1, \dots, \varphi_d)(s) = -a(s) V_\varphi(s-), \quad s \in [t, T],$$

regardless of the initiation time t. The wealth process of φ above equals

$$V_\varphi = \mathscr{E}(-a 1_{]\!]t,T]\!]} \bullet S).$$

The process a is called the ***adjustment process****.*

Idea We prove the statement only in the case when $V_{\varphi^\star}$ does not attain the value 0, where $\varphi^\star$ denotes the optimiser in (10.82) with $V_{\varphi^\star}(0) = -1$. According to the continuous-time counterpart of Remark 1.45(5), $(\varphi^\star(s))_{s\in[t,T]}$ minimises

$$\widetilde{\varphi} \mapsto E\big((1 - \widetilde{\varphi} \bullet S(T))^2 \big| \mathscr{F}_t\big) = E\big((-1 + \widetilde{\varphi} \bullet S(T))^2 \big| \mathscr{F}_t\big)$$

over all self-financing strategies $(\widetilde{\varphi}(s))_{s\in[t,T]}$ with $V_{\widetilde{\varphi}}(t) = V_{\varphi^\star}(t)$. Observe that there is a natural one-to-one mapping between portfolios $(\widetilde{\varphi}(s))_{s\in[t,T]}$ with $V_{\widetilde{\varphi}}(t) = V_{\varphi^\star}(t)$ and portfolios $(\varphi(s))_{s\in[t,T]}$ with $V_\varphi(t) = 1$ namely via $\widetilde{\varphi} = V_{\varphi^\star}(t)\varphi$. The functional $\varphi \mapsto E(V_\varphi(T)^2|\mathscr{F}_t)$ scales with $V_{\varphi^\star}(t)^2$ by moving from φ to $\widetilde{\varphi}$. Now set

$$a := -\varphi^\star / (V_{\varphi^\star})_-, \tag{10.84}$$

which implies

$$V_{\varphi^\star} = -(1 - \varphi^\star \bullet S) = -\mathscr{E}(-a \bullet S). \tag{10.85}$$

The above considerations yield that

$$(\varphi^\star(s)/V_{\varphi^\star}(t))_{s\in[t,T]} = (-a(s)V_{\varphi^\star}(s-)/V_{\varphi^\star}(t))_{s\in[t,T]}$$

is the optimiser in (10.83), which in turn implies the claim. □

The optimal strategy and hence the adjustment process are unique up to irrelevant integrals in the sense that $(a - \widetilde{a}) \bullet S = 0$ for any two adjustments processes $a, \widetilde{a}$.

Using the notation (II.1), the opportunity and the adjustment process can be characterised as follows.

Rule 10.45 *Let L be a positive semimartingale with $L(T) = 1$ and a a predictable process. The following are equivalent:*

1. *L, a are the opportunity process and the adjustment process, respectively.*
2. *$L\mathscr{E}(-a \bullet S)$ and $L\mathscr{E}(-a \bullet S)S$ are martingales.*
3. *a) L is $(0, 1]$-valued,*
   b) the $\mathbb{R}^{d+1}$-valued stochastic process $(S, \widetilde{K}) := (S, \mathscr{L}(L))$ satisfies

   $$a^{\widetilde{K}} = \overline{a}^\top \overline{c}^{-1}\overline{a},$$

   where

   $$\overline{a} := a^S + c^{S,\widetilde{K}} + \int xyK^{(S,\widetilde{K})}(d(x, y))$$

   and

   $$\overline{c} := c^S + \int xx^\top(1 + y)K^{(S,\widetilde{K})}(d(x, y)),$$

   c) $a = \overline{c}^{-1}\overline{a}$.
4. *a) L is $(0, 1]$-valued,*
   b) the $\mathbb{R}^{d+1}$-valued stochastic process $(S, K) := (S, \log L)$ satisfies

   $$a^K = \overline{a}^\top \overline{c}^{-1}\overline{a} + \frac{1}{2}c^K(t) + \int (e^y - 1 - y)K^K(dy),$$

   where

   $$\overline{a} := a^S + c^{S,K} + \int x(e^y - 1)K^{(S,K)}(d(x, y))$$

and

$$\overline{c} := c^S + \int xx^\top e^y K^{(S,K)}(d(x,y)),$$

c) $a = \overline{c}^{-1}\overline{a}$.

Idea As in the proof of Rule 10.44 we suppose that $V_{\varphi^\star}$ does not vanish. In this case the statements can be derived by applying Rule 10.11 in the case $p = -1$. Strictly speaking, negative p are not permitted in Sect. 10.1.2 because the corresponding utility functions fail to be increasing. However, monotonicity is not needed for the reasoning below.

1⇒2: This follows as for Rule 10.11, 1⇒4.
2⇒1: This follows as for Rule 10.11, 4⇒1.
2⇔3: This can be viewed as a reformulation of Rule 10.11, 4⇔5. Indeed, since $a^{\widetilde{K}} = a^L/L_-$, $c^{S,\widetilde{K}} = c^{S,L}/L_-$, and

$$\int f(x,y)K^{(S,\widetilde{K})}(d(x,y)) = \int f(x,y/L_-)K^{(S,L)}(d(x,y))$$

for any $f : \mathbb{R}^2 \to \mathbb{R}$, (10.16) reads as

$$a^{\widetilde{K}} = \gamma^\top c^S \gamma + \int (\gamma^\top x)^2(1+y)K^{(S,\widetilde{K})}(d(x,y)) = \gamma^\top \overline{c}\gamma$$

and (10.17) as

$$0 = a^S + c^{S,\widetilde{K}} + c^S\gamma + \int \big((1+\gamma^\top x)(1+y) - 1\big)xK^{(S,\widetilde{K})}(d(x,y)) = \overline{a} + \overline{c}\gamma.$$

Together, these equations are equivalent to $a^{\widetilde{K}} = \overline{a}^\top(\overline{c})^{-1}\overline{a}$ and $\gamma = -(\overline{c})^{-1}\overline{a}$. With $-a = \gamma$ the assertion follows.
3⇔4: This equivalence follows from the calculus of semimartingale characteristics, cf. the rules in Sect. 4.2. □

One may view the control problem in (10.82) as a utility optimisation problem for power utility with initial endowment -1 and forbidden risk aversion coefficient $p = -1$ in Sect. 10.1.2. The following observations establish the links to Rules 10.11 and 10.45.

Remark 10.46

1. If L, a are the opportunity and the adjustment processes, the value process for the control problem in (10.82) is of the form $\mathscr{V}(t,\varphi) = L(t)(1 - \varphi \bullet S(t))^2$.

2. The martingale $Y := L\mathscr{E}(-a \bullet S)$ in Rule 10.45 corresponds to the one in (10.15). More specifically, YS is a martingale and Y minimises

$$Z \mapsto E\big(Z(T)^2\big) - 2Z(0) \tag{10.86}$$

over all martingales Z such that ZS is a martingale, similarly to Rule 10.3(3). And parallel to Rule 10.3(4), the optimal values of (10.82) and the dual problem (10.86) correspond to each other in the sense that

$$-\Big(E\big(Y(T)^2\big) - 2Y(0)\Big) = E\big((1 - \varphi^\star \bullet S(T))^2\big) = L(0). \tag{10.87}$$

Compared to Rule 10.3, the factor 2 in (10.86) and the minus sign in Rule (10.87) are due to the fact that we consider here minimisation for $u(x) = x^2$ rather than maximisation for $u(x) = -x^2/2$, which would correspond to Sect. 10.1.2.
3. Since Y is a martingale such that YS is a martingale as well, it should give rise to an equivalent martingale measure $Q^\star$, namely the one with density process $Y/Y(0)$. This measure is called the **variance-optimal martingale measure** because the density $\frac{dQ^\star}{dP}$ minimises the variance $\mathrm{Var}(\frac{dQ}{dP})$ among all EMMs Q, similarly to Rule 10.5(2). However, for processes with jumps, $\mathscr{E}(-a \bullet S)$ and hence $Y/Y(0)$ may turn negative, which means that $Q^\star$ is only a *signed measure* instead of a bona fide probability measure. The possible nonpositivity of $Q^\star$ can be viewed as the dual manifestation of the fact that quadratic utility is not increasing on its whole domain.

The link to efficient portfolios is provided by the following result.

Rule 10.47 *Let L and a denote the opportunity and the adjustment process, respectively. A self-financing strategy $\varphi^\star$ is mean-variance efficient if and only if $\varphi^\star = ma\mathscr{E}(-a \bullet S_-)$ for some $m \geq 0$. In this case*

$$E(\varphi^\star \bullet S_T) = m(1 - L(0)), \tag{10.88}$$

$$\mathrm{Var}(\varphi^\star \bullet S_T) = m^2 L(0)(1 - L(0)), \tag{10.89}$$

and

$$\mathrm{SR}(\varphi^\star \bullet \mathrm{S_T}) = \sqrt{\frac{1}{L(0)} - 1}. \tag{10.90}$$

Idea Recall from (10.84) and (10.85) that $a\mathscr{E}(-a \bullet S)_-$ is the minimiser of (10.82). The equivalence follows now from characterisation 5 of mean-variance efficient portfolios in Rule 10.43.

For $m = 1$ we have

$$L(0) = E\big((1 - \varphi^\star \bullet S(T))^2\big) \tag{10.91}$$

by definition. We show that the initial value of the opportunity process can also be represented as

$$L(0) = E(1 - \varphi^\star \bullet S(T)) \tag{10.92}$$

for $m = 1$ which, together with (10.91), immediately yields (10.88–10.90). In order to verify (10.92), observe that $c = 1$ minimises

$$\begin{aligned} c \mapsto\ & E\big((1 - c\varphi^\star \bullet S(T))^2\big) \\ &= c^2 E\big((1 - \varphi^\star \bullet S(T))^2\big) + 2c(1-c)E(1 - \varphi^\star \bullet S(T)) + (1-c)^2. \end{aligned}$$

The first-order condition is

$$0 = 2cE\big((1 - \varphi^\star \bullet S(T))^2\big) + (2 - 4c)E(1 - \varphi^\star \bullet S(T)) - 2(1-c),$$

which for $c = 1$ yields $E((1 - \varphi^\star \bullet S(T))^2) = E(1 - \varphi^\star \bullet S(T))$ as desired. □

In the following example the ingredients L and a of mean-variance efficient portfolios can be determined explicitly.

Example 10.48 (Geometric Lévy Model) We consider the geometric Lévy model from Example 9.2. Essentially the same arguments as in Example 10.13 show that the opportunity process L and the adjustment process a are of the form $L(t) = \exp(-\alpha(T - t))$ and $a(t) = \pi/S_1(t-)$ for

$$\pi = \frac{a^{\widetilde{X}}}{c^{\widetilde{X}} + \int x^2 K^{\widetilde{X}}(dx)} = \frac{E(\widetilde{X}(1))}{\mathrm{Var}(\widetilde{X}(1))}$$

and

$$\alpha := \left(c^{\widetilde{X}} + \int x^2 K^{\widetilde{X}}(dx)\right)\pi^2 = \frac{(a^{\widetilde{X}})^2}{c^{\widetilde{X}} + \int x^2 K^{\widetilde{X}}(dx)} = \frac{E(\widetilde{X}(1))^2}{\mathrm{Var}(\widetilde{X}(1))}.$$

Extending the characteristic exponent ψ of X naturally to complex numbers, (3.73–3.75) yield

$$\pi = \frac{\psi(-i)}{\psi(-2i) - 2\psi(-i)} \tag{10.93}$$

and

$$\alpha = \frac{\psi(-i)^2}{\psi(-2i) - 2\psi(-i)}.$$

The extension to $d > 1$ risky assets is straightforward as usual.

Parallel to the previous section, we state the BSDE and PDE versions of Rule 10.45 for Itō and continuous Markov process models.

Example 10.49 (Itō Process Model) By Rule 10.45(4) the adjustment process a and the logarithmic opportunity process $K = \log L$ satisfy

$$a(t) = \frac{a^S(t) + c^{S,K}(t)}{c^S(t)} \tag{10.94}$$

and

$$a^K(t) = \frac{(a^S(t) + c^{S,K}(t))^2}{c^S(t)} - \frac{1}{2}c^K(t) \tag{10.95}$$

for a single continuous risky asset. In the context of Example 9.3 the martingale representation theorem 3.61 warrants that the martingale part of K is an integral relative to W, i.e. we have $dK(t) = a^K(t)dt + \eta(t)dW(t)$ with some $\mathbb{R}^n$-valued predictable process η. Reformulating (10.94) and (10.95) in terms of μ, σ, η yields

$$a(t) = \frac{\mu(t)}{\sigma(t)^2} + \frac{\eta_1(t)}{\sigma(t)}$$

and the BSDE

$$dK(t) = \left(\left(\frac{\mu(t)}{\sigma(t)} + \eta_1(t)\right)^2 - \frac{1}{2}\sum_{i=1}^n \eta_i(t)^2\right)dt + \eta dW(t), \quad K(T) = 0$$

for K, η.

Example 10.48 shows that K is deterministic and hence $\eta = 0$ in the special case of geometric Brownian motion as in Example 9.1.

Example 10.50 (Continuous Markovian Model) In the Markovian setup of Example 9.4 we expect the adjustment process a and the logarithmic opportunity process $K = \log L$ to be of the form $a(t) = \xi(T - t, X(t))$ and $K(t) = \ell(T - t, X(t))$ for some functions $\xi, \ell : [0, T] \times \mathbb{R}^n \to \mathbb{R}$. By Proposition 4.13 the local characteristics appearing in (10.95) can be expressed in terms of β, γ and the derivatives of ℓ. This leads to the following PDE for ℓ:

$$\ell(0,x) = 0,$$

$$D_1\ell(\vartheta,x) = \sum_{i=1}^{n} D_{1+i}\ell(\vartheta,x)\beta_i(x) + \frac{1}{2}\sum_{i,j=1}^{n}\big(D_{1+i,1+j}\ell(\vartheta,x) + D_{1+i}\ell(\vartheta,x)D_{1+j}\ell(\vartheta,x)\big)\gamma_{ij}(x) - \frac{\big(\beta_1(x) + \sum_{i=1}^{n} D_{1+i}\ell(\vartheta,x)\gamma_{i1}(x)\big)^2}{\gamma_{11}(x)}.$$

Having solved this PDE, the function ξ is obtained from (10.94) as

$$\xi(\vartheta,x) = \frac{\beta_1(x) + \sum_{i=1}^{n} D_{1+i}\ell(\vartheta,x)\gamma_{i1}(x)}{\gamma_{11}(x)}.$$

As in the terminal wealth case, the example allows for a straightforward extension to $d > 1$ risky assets.

More in line with academic and industry practice, one could study the problems in Rule 10.43 on a sequence of short time horizons rather than T above. This essentially means considering the problem (10.66) with a quadratic utility function $u(x) = -\frac{1}{2m}((m-x)^2 - m^2)$ for some $m > 0$ even if this function is strictly increasing only on $(-\infty, m]$. As corresponding utility function of P&L in Sect. 10.3 we obtain

$$\begin{aligned}\gamma^S(t,\psi) &= \psi^\top a^S(t) - \frac{1}{2m}\left(\psi^\top c^S(t)\psi + \int(\psi^\top x)^2 K^S(t,dx)\right)\\ &= \psi^\top a^S(t) - \frac{1}{2m}\psi^\top \widetilde{c}^S(t)\psi\end{aligned}$$

for the modified second characteristic $\widetilde{c}^S$. This reduces to (10.69) for continuous price processes. We observe that for price processes without jumps, P&L optimisation for arbitrary utility functions u can be interpreted as choosing efficient portfolios repeatedly for very short time horizons. In the discontinuous case the choice of increasing utility functions u in Sect. 10.3 avoids the anomalies discussed in the beginning of this section.

10.5 Bid-Ask Spreads

In real markets traders face various kinds of frictions. Here we briefly discuss the effect of proportional transaction costs in the sense that bid and ask prices differ. The key concept in this context turns out to be the *shadow price*, which is a particular consistent price system in the sense of Sect. 9.3.3.

10.5.1 Utility of Terminal Wealth or Consumption

In principle, investment under transaction costs can be viewed as a particular kind of control problem, which can be tackled using the dynamic programming approach in Chap. 7. Rather than elaborating on this in detail, we confine ourselves to touching a duality result that bears some similarity to the FTAP in Rule 9.22.

We work in the setup of Sect. 9.3.3 with the additional assumption that the numeraire asset satisfies $\underline{S}_0 = \overline{S}_0 = 1$, i.e. prices are discounted with respect to the frictionless zeroth asset. The investor's goal is to maximise expected utility of terminal wealth as in Sect. 10.1 and more specifically (10.1) but now the wealth process V_φ is to be understood as the liquidation wealth (9.21).

Rule 10.51 *For a self-financing strategy φ the following statements are equivalent.*

1. *φ is optimal for expected utility of terminal wealth.*
2. *There exists a consistent price system $\widetilde{S}$ such that*
   a) *φ is self-financing for the frictionless market with price process $\widetilde{S}$ as well and*
   b) *φ is optimal for expected utility of terminal wealth in this frictionless market with price process $\widetilde{S}$.*

In this case we call $\widetilde{S}$ a ***shadow price process*** *for the problem.*

Idea 2⇒1: Let $\psi = (\psi_0, \dots, \psi_d)$ be any self-financing strategy. Denote by $\widetilde{\psi} = (\widetilde{\psi}_0, \dots, \widetilde{\psi}_d)$ the self-financing strategy in the frictionless market $\widetilde{S}$ with $\widetilde{\psi}_i = \psi_i, i = 1, \dots, d$. Its wealth process relative to $\widetilde{S}$ is denoted by

$$\widetilde{V}_{\widetilde{\psi}} := \widetilde{\psi}^\top \widetilde{S} = \widetilde{\psi}(0)^\top \widetilde{S}(0) + \widetilde{\psi} \bullet \widetilde{S}.$$

It is obvious that $\widetilde{V}_{\widetilde{\psi}} \geq V_\psi$ because, whenever a transaction is made, $\widetilde{\psi}$ buys/sells at prices $\widetilde{S}$, whereas ψ trades at the less advantageous prices $\overline{S}$, $\underline{S}$. By assumption 2a, however, φ is self-financing relative to $\widetilde{S}$ as well and hence $\widetilde{V}_\varphi(T) = V_\varphi(T)$. This yields

$$E\big(u(V_\psi(T))\big) \leq E\big(u(\widetilde{V}_{\widetilde{\psi}}(T))\big) \leq E\big(u(\widetilde{V}_\varphi(T))\big) \leq E\big(u(V_\varphi(T))\big)$$

by assumption 2b.

1⇒2: As for Rule 10.3, the idea is to apply Remark A.18(2). Here U and V denote the set of $\mathbb{R}^{1+d}$-valued random variables, where V is viewed as the dual space of U via $y(x) = E(\sum_{i=0}^d x_i y_i)$ for $x \in U, y \in V$. Define the attainable cone $K \subset U$ as in (9.22) and denote its polar cone by $K^\circ := \{y \in V : y(x) \leq 0\}$. The optimal investment problem under consideration amounts to maximising

$$f(x) := E(u(v_0 + V_x(T)))$$

over $x \in K$, where $V_x(T) := \sum_{i=0}^d (x_i^+ \underline{S}_i(T) - x_i^- \overline{S}_i(T))$ denotes the liquidation value of the portfolio x in line with (9.21).

Step 1: Denote by $x^\star = (\varphi_0^\star(T) - \xi_0^\star, \dots, \varphi_d^\star(T) - \xi_d^\star) \in K$ and $y^\star \in K^\star$ optimal solutions to the primal and dual problems in the sense of Theorem A.17 and Remark A.18(2). Since u is strictly increasing, we obviously have $\xi_i^\star = 0$, $i = 1, \dots, d$. Theorem A.17 resp. Remark A.18(2) yields the complementary slackness condition $E(\sum_{i=0}^d x_i^\star y_i^\star) = 0$. Moreover, the zeroth component of the first-order condition reads as

$$y_0^\star = u'(v_0 + V_{x^\star}(T)). \tag{10.96}$$

Since u is strictly increasing, the remaining d conditions imply $y_i^\star > 0$, $i = 1, \dots, d$. Denote by $Y_0, \dots, Y_d$ the martingales generated by the random variables $y_0^\star, \dots, y_d^\star$, i.e. $Y_i(t) = E(y_i^\star | \mathscr{F}_t)$, $i = 0, \dots, d$. Finally, set $\widetilde{S}_i := Y_i / Y_0$ for $i = 0, \dots, d$ and define a probability measure $Q \sim P$ by its density process $Y_0 / Y_0(0)$.

Step 2: We show that $(\widetilde{S}_0, \dots, \widetilde{S}_d)$ is a consistent price system. To this end, fix $j \in \{1, \dots, d\}$, $t \in [0, T]$, and $F \in \mathscr{F}_t$. Define a self-financing portfolio φ by $\varphi(s) := 0$ for $s \leq t$ and

$$\varphi_i(s) := \begin{cases} -\overline{S}_j(t) & \text{for } i = 0, \\ 1 & \text{for } i = j, \\ 0 & \text{otherwise} \end{cases}$$

for $s > t$. For $(\varphi_0(T), \dots, \varphi_d(T)) =: x \in K$ we obtain

$$\begin{aligned} E_Q\big((\widetilde{S}_j(t) - \overline{S}_j(t))1_F\big) &= Y_0(0) E\big(Y_0(t)(\widetilde{S}_j(t) - \overline{S}_j(t))1_F\big) \\ &= Y_0(0) E\big((Y_j(t) - Y_0(t)\overline{S}_j(t))1_F\big) \\ &= Y_0(0) E\big((Y_j(T) - Y_0(T)\overline{S}_j(t))1_F\big) \\ &= Y_0(0) E\big(y_j^\star \varphi_j(T) - y_0^\star \varphi_0(T)\big) \\ &= Y_0(0) y^\star(x) \leq 0 \end{aligned}$$

because $y^\star \in K^\circ$. By choosing $F := \{\widetilde{S}_j(t) > \overline{S}_j(t)\}$ we conclude $\widetilde{S}_j(t) \leq \overline{S}_j(t)$ almost surely. Along the same lines we derive the inequality $\widetilde{S}_j(t) \geq \underline{S}_j(t)$.

Step 3: Observe that Q is an EMM for $\widetilde{S}$. Indeed, this follows immediately from Proposition 3.69 because $\widetilde{S}_i Y_0/Y_0(0) = Y_i/Y_0(0)$ is a martingale.

Step 4: We show that $\varphi^\star$ is self-financing relative to the frictionless market $\widetilde{S}$. Since φ is self-financing in the market with bid-ask spread, we have

$$\varphi_0^\star(t) = -\sum_{i=1}^d \left((\overline{S}_i)_- \bullet(\varphi^\star)^\uparrow(t) - (\underline{S}_i)_- \bullet(\varphi^\star)^\downarrow(t)\right)$$

by (9.20). In order for φ to be self-financing relative to $\widetilde{S}$, it suffices to verify

$$\varphi_0^\star(t) = -\sum_{i=1}^d \left((\widetilde{S}_i)_- \bullet(\varphi^\star)^\uparrow(t) - (\widetilde{S}_i)_- \bullet(\varphi^\star)^\downarrow(t)\right) =: \widetilde{\varphi}_0^\star(t).$$

It is evident that $\widetilde{\varphi}_0^\star - \varphi_0^\star$ is nonnegative and increasing. Integration by parts yields

$$\begin{aligned}\sum_{i=1}^d \varphi_i^\star(T)\widetilde{S}_i(T) = \sum_{i=1}^d \varphi_i^\star(0)\widetilde{S}_i(0) &+ (\varphi_1^\star, \dots, \varphi_d^\star)\bullet(\widetilde{S}_1, \dots, \widetilde{S}_d)(T)\\ &+ (\widetilde{S}_1, \dots, \widetilde{S}_d)_- \bullet(\varphi_1^\star, \dots, \varphi_d^\star)(T).\end{aligned}$$

Since $\widetilde{S}$ is a Q-martingale and from the complementary slackness condition, we obtain

$$\begin{aligned}0 &= E_Q\big((\varphi_1^\star, \dots, \varphi_d^\star)\bullet(\widetilde{S}_1, \dots, \widetilde{S}_d)(T)\big)\\ &= Y_0(0)E\left(Y_0(T)\left(\sum_{i=1}^d \varphi_i^\star(T)\widetilde{S}_i(T) + \widetilde{\varphi}_0^\star(T)\right)\right)\\ &= Y_0(0)E\left(\sum_{i=1}^d \varphi_i^\star(T)Y_i(T) + \widetilde{\varphi}_0^\star(T)Y_0(T)\right)\\ &\geq Y_0(0)E\left(\sum_{i=1}^d \varphi_i^\star(T)Y_i(T) + \varphi_0^\star(T)Y_0(T)\right) \qquad (10.97)\\ &= Y_0(0)y^\star(x^\star) = 0.\end{aligned}$$

The equality in (10.97) yields $\widetilde{\varphi}_0^\star(T) = \varphi_0^\star(T)$, which a fortiori means $\widetilde{\varphi}_0^\star = \varphi_0^\star$ as desired.

Step 5: We show that $\varphi^\star$ is optimal for the frictionless market $\widetilde{S}$. To wit, recall that $\varphi^\star$ is self-financing relative to $\widetilde{S}$, that $Y_0\widetilde{S}$ is a martingale and that $Y_0(T) = u'(v_0 + V_{\varphi^\star}(T))$ by (10.96). The optimality now follows from Rule 10.3(1). □

Typically one would not expect a self-financing strategy in the market with bid-ask spread to be self-financing relative to a given consistent price system because it trades at more favourable prices in the frictionless case. Consequently, condition 2b means that φ only buys/sells assets when the shadow price equals the upper/lower price, i.e. it effectively trades at bid-ask prices.

In order to state the corresponding result for expected utility of consumption, we need to define self-financability in the present setup of bid-ask spreads. Generalising (9.20), we call a pair (φ, c) of a trading strategy φ and a consumption rate process c **self-financing** (or we call φ c-**financing**) if

$$\overline{S}_- \bullet \varphi^{\uparrow} - \underline{S}_- \bullet \varphi^{\downarrow} + c \bullet I = 0, \tag{10.98}$$

where $\varphi^{\uparrow}$ and $\varphi^{\downarrow}$ are defined as in Sect. 9.3.3.

The investor aims at maximising expected utility of consumption as in Sect. 10.2 but with self-financing pairs in the sense of (10.98) or, more specifically, (10.47). Not surprisingly we have the following counterpart of Rule 10.51.

Rule 10.52 *For a self-financing strategy/consumption pair (φ, c) the following statements are equivalent.*

1. *(φ, c) is optimal for expected utility of consumption.*
2. *There exists a consistent price system $\widetilde{S}$ such that*
   a) *(φ, c) is self-financing for the frictionless market with price process $\widetilde{S}$ as well and*
   b) *(φ, c) is optimal for expected utility of consumption in this frictionless market with price process $\widetilde{S}$.*

*As above we call $\widetilde{S}$ a **shadow price process** for the problem.*

Idea 2⇒1: This implication follows as for Rule 10.51.

1⇒2: As in Rules 10.25 and 10.51 we apply Theorem A.17. To this end, denote by U and V the sets of $\mathbb{R}^{1+d}$-valued adapted processes, V being viewed as the dual space of U via $y(x) = E(\int_0^T \sum_{i=0}^d x_i(t)y_i(t)dt)$. Set $K := \{x \in U :$ there exists a self-financing strategy $\psi = (\psi_0, \dots, \psi_d)$ with $V_{\psi - c \bullet I}(T) \geq 0\}$, where the liquidation wealth $V_{\psi - c \bullet I}$ is defined as in (9.21). The process x_i represents the rate at which asset i is sold in order to finance consumption. The polar of K is $K^\circ := \{y \in V : y(x) \leq 0$ for any $x \in U\}$. As the primal optimisation problem consider the maximisation of

$$f(x) := E\left(\int_0^T u\left(t, \frac{v_0}{T} + \sum_{i=0}^d \left(x_i(t)^+ \bullet \underline{S}_i(t) + x_i(t)^- \bullet \overline{S}_i(t)\right)\right)dt\right)$$

over $x \in K$.

Step 1: We may and do focus on $x \in K$ with $x_i = 0$ for $i \neq 0$. To wit, for an arbitrary $x \in K$ with corresponding self-financing strategy ψ set

$$\begin{aligned}
\widetilde{x}_0 &:= x_0 + \sum_{i=1}^d (x_i^+ \underline{S}_i - x_i^- \overline{S}_i), \\
\widetilde{x}_i &:= 0, \quad i = 1, \dots, d, \\
\widetilde{\psi}_0 &:= \psi_0 - \sum_{i=1}^d (x_i^+ \underline{S}_i - x_i^- \overline{S}_i) \bullet I, \\
\widetilde{\psi}_i &:= \psi_i - x_i \bullet I, \quad i = 1, \dots, d.
\end{aligned}$$

Note that $f(\widetilde{x}) = f(x)$. In order to show that $\widetilde{x} \in K$, suppose first that

$$(\psi_i + c_i \bullet I)^\uparrow = \varphi_i^\uparrow + (c_i \bullet I)^\uparrow, \quad i = 1, \dots, d \tag{10.99}$$

and hence also $(\psi_i + c_i \bullet I)^\downarrow = \varphi_i^\downarrow + (c_i \bullet I)^\downarrow$, $i = 1, \dots, d$. In this case one easily derives

$$(\overline{S}_i)_- \bullet \widetilde{\psi}^\uparrow - (\underline{S}_i)_- \bullet \widetilde{\psi}^\downarrow = (\overline{S}_i)_- \bullet \psi^\uparrow - (\underline{S}_i)_- \bullet \psi^\downarrow,$$

which means that $\widetilde{\psi}$ is self-financing as well. Moreover, we have $\widetilde{\psi} - \widetilde{x} \bullet I = \psi - x \bullet I$, which implies $\widetilde{x} \in K$.
If (10.99) does not hold, it is not hard to verify that

$$(\overline{S}_i)_- \bullet \widetilde{\psi}^\uparrow - (\underline{S}_i)_- \bullet \widetilde{\psi}^\downarrow - (\overline{S}_i)_- \bullet \psi^\uparrow - (\underline{S}_i)_- \bullet \psi^\downarrow$$

is an increasing process so that $\widetilde{\psi}$ can be made self-financing by increasing $\widetilde{\psi}_0$. Again we conclude that $\widetilde{x} \in K$.

Step 2: We show that the utility maximisation problem under consideration amounts to maximising $f(x)$ over $x \in K$ as in step 1. To this end, let (φ, c) be self-financing with $\varphi(0) = (v_0, 0, \dots, 0)$ and $V_\varphi(T) \geq 0$. Then

$$E\left(\int_0^T u(t, c(t))dt\right) = f(x) \tag{10.100}$$

for $x = (c(t) - v_0/T, 0, \dots, 0) \in K$ with corresponding self-financing strategy $\psi(t) = (\varphi_0(t) - v_0 + c \bullet I(t), \varphi_1(t), \dots, \varphi_d(t))$. Conversely, for $x \in K$ as in step 1 with corresponding self-financing ψ we have (10.100) for the self-financing pair (φ, c) defined as $\varphi(t) = (\psi_0(t) + v_0(T-t)/T - x_0(t), \psi_1(t), \dots, \psi_d(t))$ and $c(t) = x_0(t) + v_0/T$.

Step 3: Denote by $x^\star \in K$ corresponding to $\psi^\star$ the optimal solution to the primal problem and by $Y := y^\star \in K^\circ$ the optimal solution to the dual problem in the sense of Theorem A.17 and Remark A.18(2). The zeroth component of the first-order condition $\nabla f(x^\star) = y^\star$ in Theorem A.17 reads as

$$Y_0(t) = u'\Big(t, \frac{v_0}{T} + x_0^\star(t)\Big). \tag{10.101}$$

Since $u(t, \cdot)$ is strictly increasing, the derivatives of f with respect to the other components yield $Y_i > 0$, $i = 1, \dots, d$. Set $\widetilde{S}_i := Y_i / Y_0$ for $i = 0, \dots, d$ and define a probability measure $Q \sim P$ by its density process $Y_0/Y_0(0)$.

Step 4: We show that $Y_0, \dots, Y_d$ are martingales. Fix $j \in \{0, \dots, d\}$ and let $0 \leq t_1 \leq t_2 \leq T$, $F \in \mathscr{F}_{t_1}$. Consider the consumption rate $x \in K$ with $x_i = 0$ for $i \neq j$ and

$$x_j \bullet I(t) = -1_{F\times(t_1,t_2]}(t), \tag{10.102}$$

which of course exists only in a limiting sense because of the jumps of the right-hand side of (10.102) at t_1 and t_2. Since

$$\begin{aligned} E\big((Y_j(t_2) - Y_j(t_1))1_F\big) &= E\big(Y_j \bullet (x_j \bullet I)(T)\big) \\ &= E\big((Y_j x_j) \bullet I(T)\big) \\ &= y^\star(x) \leq 0, \end{aligned}$$

we conclude that Y_j is a martingale, as desired.

Step 5: We show that $(\widetilde{S}_0, \dots, \widetilde{S}_d)$ is a consistent price system. To this end, fix $j \in \{0, \dots, d\}$, $t \in [0, T]$, and $F \in \mathscr{F}_t$. Define the consumption rate $x \in K$ by $x_i = 0$ for $i \notin \{0, j\}$ and

$$\begin{aligned} x_0 \bullet I(s) &= -1_{F\times(t,T]}(s)\overline{S}_j(t), \\ x_j \bullet I(s) &= 1_{F\times\{T\}}(s), \end{aligned}$$

for $s \in [0, T]$, which, as in step 2, exists only in a limiting sense because of the jumps of aggregate consumption at t resp. T. This consumption is financed by the trading strategy $\varphi = (\varphi_0, \dots, \varphi_d)$ with $\varphi_i = 0$ for $i \neq j$ and $\varphi_j(s) = 1_{F\times(t,T]}(s)$, $s \in [0, T]$. By steps 1, 2 we obtain

$$\begin{aligned} E_Q\big((\widetilde{S}_j(t) - \overline{S}_j(t))1_F\big) &= Y_0(0)E\big(Y_0(t)(\widetilde{S}_j(t) - \overline{S}_j(t))1_F\big) \\ &= Y_0(0)E\big((Y_j(t) - Y_0(t)\overline{S}_j(t))1_F\big) \\ &= Y_0(0)E\big((Y_j(T) - Y_0(T)\overline{S}_j(t))1_F\big) \\ &= Y_0(0)E\big(x_j \bullet I(T)Y_j(T) + x_0 \bullet I(T)Y_0(T)\big) \\ &= Y_0(0)y^\star(x) \leq 0, \end{aligned}$$

which implies $\widetilde{S}_j(t) \leq \overline{S}_j(t)$ almost surely. The inequality $\widetilde{S}_j(t) \geq \underline{S}_j(t)$ is derived along the same lines.

Step 6: We show that Q is an EMM for $\widetilde{S}$. This follows immediately from Proposition 3.69 because $\widetilde{S}_i Y_0/Y_0(0) = Y_i/Y_0(0)$ is a martingale.

Step 7: Let $(\varphi^\star, c^\star)$ be the optimal self-financing pair that corresponds to $x^\star \in K$ in the sense of step 1. We show that it is self-financing with respect to the frictionless price process $\widetilde{S}$ as well. Since $\overline{S}_- \bullet (\varphi^\star)^\uparrow - \underline{S}_- \bullet (\varphi^\star)^\downarrow + c^\star \bullet I = 0$ and since $\overline{S}_- \bullet (\varphi^\star)^\uparrow - \underline{S}_- \bullet (\varphi^\star)^\downarrow - \widetilde{S}_- \bullet \varphi^\star$ is increasing, we have $\widetilde{c} \geq c^\star$ for the process $\widetilde{c}$ satisfying

$$\widetilde{S}_- \bullet \varphi^\star + \widetilde{c} \bullet I = 0. \tag{10.103}$$

Equation (10.103) means that $(\varphi^\star, \widetilde{c})$ is self-financing for the frictionless price process $\widetilde{S}$. If $\widetilde{V}_{\varphi^\star}$ denotes its wealth process relative to $\widetilde{S}$, integration by parts and (10.103) yield

$$\begin{aligned}
\widetilde{V}_{\varphi^\star}(T) &= v_0 + \varphi^\star \bullet \widetilde{S}(T) - \widetilde{c} \bullet I(T) \\
&= \varphi^\star(T)^\top \widetilde{S}(T) - \widetilde{S}_- \bullet \varphi^\star(T) - \widetilde{c} \bullet I(T) \\
&= \varphi^\star(T)^\top \widetilde{S}(T) \\
&\geq V_{\varphi^\star}(T) \geq 0,
\end{aligned}$$

where $V_{\varphi^\star}$ denotes the liquidation wealth of $(\varphi^\star, c^\star)$ as in (9.21). The complementary slackness condition $y^\star(x^\star) = 0$ and the fact that $\widetilde{S}$ is a Q-martingale yield

$$\begin{aligned}
0 &= Y_0(0) E_Q(\varphi^\star \bullet \widetilde{S}(T)) \\
&\geq Y_0(0) E_Q(\widetilde{c} \bullet I(T) - v_0) \\
&\geq Y_0(0) E_Q(c^\star \bullet I(T) - v_0) \\
&= Y_0(0)(E_Q(c^\star) \bullet I(T) - v_0) \\
&= E(Y c^\star) \bullet I(T) - v_0 \\
&= E\left(\int_0^T Y(t) x_0^\star(t) dt\right) \\
&= y^\star(x^\star) \geq 0.
\end{aligned} \tag{10.104}$$

Therefore equality holds in (10.104), i.e. $\widetilde{c} = c^\star$.

Step 8: We show that $(\varphi^\star, c^\star)$ is optimal for the frictionless market $\widetilde{S}$. Recall the first-order condition $Y_0(t) = u'(t, \frac{v_0}{T} + x_0^\star(t)) = u'(t, c^\star(t))$ from (10.101) and the fact that Y_0 and $Y_0\widetilde{S}$ are martingales from steps 4, 6. The assertion now follows from Theorem 10.25(1). □

In a certain sense, Rules 10.51 and 10.52 allow us to reduce the investment problem to the frictionless case with unique prices. However, it is not obvious how to determine the corresponding shadow price, which is part of the problem. There exists hardly any example with a simple explicit solution. The situation becomes easier if we content ourselves with determining the leading-order effect of transaction costs for small spreads. This is explained in more detail in the subsequent section for utility of P&L.

Example 10.53 (Itō Process Model with Small Bid-Ask Spread) We consider the market of Example 9.3 where transaction costs of size $\varepsilon S(t)$ have to be paid whenever the stock is bought or sold. Put differently, we are facing bid-ask bounds of the form $\underline{S}(t) = (1-\varepsilon)S(t)$ and $\overline{S}(t) = (1+\varepsilon)S(t)$ with constant $\varepsilon > 0$.

The goal is to determine a *nearly optimal* trading strategy $\varphi^{(\varepsilon)}$ for expected utility of terminal wealth if the transaction cost parameter ε is small. To this end fix the utility function $u(x) = -e^{-px}$ and the time horizon T. We denote the optimal strategy for the frictionless market with price process S by $\varphi^{(0)}$.

From related studies it is known that the optimal strategy for $[\underline{S}, \overline{S}]$ does not trade as long as the portfolio holdings $\varphi(t)$ are in a corridor around the frictionless optimal strategy $\varphi^{(0)}(t)$. If $\varphi(t)$ hits the boundary of this no-trade region, the investor buys resp. sells infinitesimal amounts of the asset, just enough to stay in the desired corridor. From the literature we may also expect that the candidate corridor

$$[\underline{\varphi}(t), \overline{\varphi}(t)] := [\varphi^{(0)}(t) - \varepsilon^{1/3}\delta(t), \varphi^{(0)}(t) + \varepsilon^{1/3}\delta(t)] \tag{10.105}$$

yields the maximal expected utility to the leading order if one chooses the corridor width process $\delta(t)$ appropriately. More specifically, denote by $\varphi^{(\varepsilon)}$ the continuous strategy obtained from trading at the boundaries of (10.105). The goal is to determine $\delta(t)$ such that

$$E\big(u(V_{\varphi^{(\varepsilon)}}(T))\big) = \sup\big\{E\big(u(V_\varphi(T))\big) : \varphi \text{ strategy}\big\} + o(\varepsilon^{2/3}). \tag{10.106}$$

Since the difference of the maximal expected utilities with and without transaction costs is typically of the order $O(\varepsilon^{2/3})$, the suboptimality of the order $o(\varepsilon^{2/3})$ can be considered as asymptotically negligible.

Below we sketch that (10.106) can be achieved by choosing the corridor width

$$\delta(t) = \left(\frac{3}{2p}\frac{c^{\varphi^{(0)}}(t)}{\sigma(t)^2}S(t)\right)^{1/3}. \tag{10.107}$$

Here $c^{\varphi^{(0)}}(t) := d[\varphi^{(0)}, \varphi^{(0)}](t)/dt$ denotes the second characteristic of the frictionless optimal strategy $\varphi^{(0)}$, which is assumed to be an Itō process as well. If the volatility of the frictionless optimal portfolio is large, the investor chooses a wide corridor in order to avoid large transaction costs. A high stock volatility or large risk aversion, however, leads to a narrow corridor width in order to avoid *displacement loss* resulting from being far away from the frictionless target portfolio $\varphi^{(0)}$.

Step 1: The idea to tackle the problem is to look for processes $\delta(t)$, $S^{(\varepsilon)}$, $Y^{(\varepsilon)}$ such that

1. $S^{(\varepsilon)}$ is a consistent price system;
2. $\varphi^{(\varepsilon)}$ constructed from δ trades only when $S^{(\varepsilon)}$ equals $\underline{S}$ resp. $\overline{S}$, i.e. the self-financing strategy corresponding to $\varphi^{(\varepsilon)}$ is self-financing for the frictionless price process $S^{(\varepsilon)}$ as well;
3. the drift rate $a^{Y^{(\varepsilon)}}$ of $Y^{(\varepsilon)}$ is of the order $O(\varepsilon^{2/3})$;
4. the drift rate $a^{Y^{(\varepsilon)}S^{(\varepsilon)}}$ of $Y^{(\varepsilon)}S^{(\varepsilon)}$ is of the order $O(\varepsilon^{2/3})$;
5. $Y^{(\varepsilon)}(T) = u'(x + \varphi^{(\varepsilon)} \bullet S^{(\varepsilon)}(T)) + O(\varepsilon^{2/3})$.

This yields $E\big(u(V_{\psi^{(\varepsilon)}}(T))\big) \leq E\big(u(V_{\varphi^{(\varepsilon)}}(T))\big) + o(\varepsilon^{2/3})$, at least for sequences of competing strategies $\psi^{(\varepsilon)}$ with $\psi^{(\varepsilon)} - \varphi^{(0)} = o(1)$.
Indeed, first observe that $V_{\psi^{(\varepsilon)}} \leq v_0 + \psi^{(\varepsilon)} \bullet S^{(\varepsilon)}$ because transactions are made at better prices when using $S^{(\varepsilon)}$ instead of bid/ask prices $\underline{S}$, $\overline{S}$. In addition, condition 2 implies that $V_{\varphi^{(\varepsilon)}}(T) = v_0 + \varphi^{(\varepsilon)} \bullet S^{(\varepsilon)}(T) + o(\varepsilon)$, where the error term stems from liquidating the portfolio at time T. Together with condition 5 and the concavity of u we obtain

$$\begin{aligned}
E\big(u(V_{\psi^{(\varepsilon)}}(T)\big) &\leq E\big(u(v_0 + \psi^{(\varepsilon)} \bullet S^{(\varepsilon)}(T))\big) \\
&\leq E\big(u(v_0 + \varphi^{(\varepsilon)} \bullet S^{(\varepsilon)}(T))\big) \\
&\quad + E\Big(u'(v_0 + \varphi^{(\varepsilon)} \bullet S^{(\varepsilon)}(T))\big((\psi^{(\varepsilon)} - \varphi^{(\varepsilon)}) \bullet S^{(\varepsilon)}(T)\big)\Big) \\
&= E\left(u(V_{\varphi^{(\varepsilon)}}(T))\right) + E\Big((\psi^{(\varepsilon)} - \varphi^{(\varepsilon)}) \bullet S^{(\varepsilon)}(T) Y^{(\varepsilon)}(T)\Big) + o(\varepsilon^{2/3}).
\end{aligned}$$

Applying integration by parts twice we obtain

$$\begin{aligned}
&(\psi^{(\varepsilon)} - \varphi^{(\varepsilon)}) \bullet S^{(\varepsilon)} Y^{(\varepsilon)} \\
&\quad = \big((\psi^{(\varepsilon)} - \varphi^{(\varepsilon)}) \bullet S^{(\varepsilon)}\big) \bullet Y^{(\varepsilon)} + \big(Y^{(\varepsilon)}(\psi^{(\varepsilon)} - \varphi^{(\varepsilon)})\big) \bullet S^{(\varepsilon)} \\
&\qquad + (\psi^{(\varepsilon)} - \varphi^{(\varepsilon)}) \bullet [S^{(\varepsilon)}, Y^{(\varepsilon)}] \\
&\quad = \Big((\psi^{(\varepsilon)} - \varphi^{(\varepsilon)}) \bullet S^{(\varepsilon)} - (\psi^{(\varepsilon)} - \varphi^{(\varepsilon)}) S^{(\varepsilon)}\Big) \bullet Y^{(\varepsilon)} \\
&\qquad + (\psi^{(\varepsilon)} - \varphi^{(\varepsilon)}) \bullet (S^{(\varepsilon)} Y^{(\varepsilon)})
\end{aligned}$$

and therefore

$$\begin{aligned}
E\big(u(V_{\psi^{(\varepsilon)}}(T)\big) \leq E\left(u(V_{\varphi^{(\varepsilon)}}(T))\right)& \\
+ E\bigg(\Big((\psi^{(\varepsilon)} - \varphi^{(\varepsilon)}) \bullet S^{(\varepsilon)} &- (\psi^{(\varepsilon)} - \varphi^{(\varepsilon)}) S^{(\varepsilon)}\Big) \bullet Y^{(\varepsilon)}(T) \\
+ (\psi^{(\varepsilon)} - \varphi^{(\varepsilon)}) &\bullet (S^{(\varepsilon)} Y^{(\varepsilon)})(T)\bigg) + o(\varepsilon^{2/3}).
\end{aligned}$$

Conditions 3 and 4 together with $\psi^{(\varepsilon)} - \varphi^{(0)} = o(1)$ yield that the second expectation on the right is of the order $o(\varepsilon^{2/3})$. Consequently, we obtain

$$E\big(u(V_{\psi^{(\varepsilon)}}(T))\big) \leq E\big(u(V_{\varphi^{(\varepsilon)}}(T))\big) + o(\varepsilon^{2/3})$$

as desired.

Step 2: Write $\varphi^{(\varepsilon)} = \varphi^{(0)} + \Delta\varphi$ and $S^{(\varepsilon)} = S + \Delta S$ for processes $\varphi^{(\varepsilon)}, S^{(\varepsilon)}$ as in step 1 that are yet to be found. Note that the notation Δ does not refer to jumps in this example where only continuous processes occur. From results in the literature we expect

$$\Delta\varphi = O(\varepsilon^{1/3}) \tag{10.108}$$

for the difference from the frictionless optimal strategy. The condition $\Delta S = O(\varepsilon)$ follows naturally from the fact that $(1-\varepsilon)S \leq S^{(\varepsilon)} \leq (1+\varepsilon)S$. Finally, we suppose that ΔS is an Itō process whose characteristics satisfy

$$a^{\Delta S} = O(\varepsilon^{1/3}), \quad c^{\Delta S} = O(\varepsilon^{4/3}). \tag{10.109}$$

For a justification of these assumptions see step 4 below. From (10.108, 10.109) we obtain $\Delta\varphi \bullet \Delta S = O(\varepsilon^{2/3})$. Integration by parts together with $\Delta S = O(\varepsilon)$ and (10.109) yields $\varphi^{(0)} \bullet \Delta S = O(\varepsilon^{2/3})$. Summing up we conclude that

$$\varphi^{(\varepsilon)} \bullet S^{(\varepsilon)} = \varphi^{(0)} \bullet S + \Delta\varphi \bullet S + O(\varepsilon^{2/3}),$$

which implies

$$\begin{aligned} u'(v_0 + \varphi^{(\varepsilon)} \bullet S^{(\varepsilon)}(T)) &= pe^{-p(v_0 + \varphi^{(\varepsilon)} \bullet S^{(\varepsilon)}(T))} \\ &= pe^{-p(x+\varphi^{(0)} \bullet S(T))}(1 - p\Delta\varphi \bullet S(T)) + O(\varepsilon^{2/3}). \end{aligned}$$

Note that $p\exp(-p(v_0 + \varphi^{(0)} \bullet S(T))) = u'(v_0 + \varphi^{(0)} \bullet S(T)) = Y(T)$ holds for the martingale Y in Rule 10.3 corresponding to the frictionless optimisation problem relative to the price process S. Recall from Rule 10.4 that $Y/Y(0)$ is the density process of the MEMM Q for S. Since S is a Q-martingale,

$$Y^{(\varepsilon)} := Y(1 - p\Delta\varphi \bullet S) \tag{10.110}$$

is a martingale under the original measure P which meets conditions 3 and 5 in step 1. In order for condition 4 to hold as well, we need to choose δ and ΔS appropriately.

Step 3: Integration by parts yields

$$Y^{(\varepsilon)}S^{(\varepsilon)} - Y^{(\varepsilon)}(0)S^{(\varepsilon)}(0) = S^{(\varepsilon)} \bullet Y^{(\varepsilon)} + Y^{(\varepsilon)} \bullet S^{(\varepsilon)} + [Y^{(\varepsilon)}, S^{(\varepsilon)}].$$

Since $Y^{(\varepsilon)}$ is a martingale, the drift rate of $Y^{(\varepsilon)}S^{(\varepsilon)}$ is given by

$$a^{Y^{(\varepsilon)}S^{(\varepsilon)}} = Y^{(\varepsilon)}(a^S + a^{\Delta S}) + c^{Y^{(\varepsilon)}, S+\Delta S}. \tag{10.111}$$

From (10.109) and (10.110) it follows that

$$Y^{(\varepsilon)}(a^S + a^{\Delta S}) = Y\left(a^S - p(\Delta\varphi \bullet S)a^S + a^{\Delta S}\right) + O(\varepsilon^{2/3}). \tag{10.112}$$

Moreover, (10.109) and integration by parts yield that

$$\begin{aligned}
[Y^{(\varepsilon)}, S + \Delta S] &= [Y(1 - p\Delta\varphi \bullet S), S] + O(\varepsilon^{2/3}) \\
&= Y \bullet ([\mathscr{L}(Y), S] - p(\Delta\varphi \bullet S) \bullet [\mathscr{L}(Y), S] - p\Delta\varphi \bullet [S, S]) + O(\varepsilon^{2/3}),
\end{aligned}$$

so that

$$c^{Y^{(\varepsilon)}, S+\Delta S} = Y\left(c^{\mathscr{L}(Y),S} - p(\Delta\varphi \bullet S)c^{\mathscr{L}(Y),S} - p\Delta\varphi c^S\right) + O(\varepsilon^{2/3}). \tag{10.113}$$

Inserting (10.112) and (10.113) into (10.111) and noting that $a^S + c^{\mathscr{L}(Y),S} = 0$ by Proposition 4.27 yields

$$\begin{aligned}
a^{Y^{(\varepsilon)}S^{(\varepsilon)}} &= Y\left(a^S + c^{\mathscr{L}(Y),S} - p(\Delta\varphi \bullet S)(a^S + c^{\mathscr{L}(Y),S}) + a^{\Delta S} - p\Delta\varphi c^S\right) + O(\varepsilon^{2/3}) \\
&= Y(a^{\Delta S} - p\Delta\varphi c^S) + O(\varepsilon^{2/3}).
\end{aligned}$$

Recall that we want the drift of $Y^{(\varepsilon)}S^{(\varepsilon)}$ to vanish up to terms of order $O(\varepsilon^{2/3})$, which means that

$$a^{\Delta S} = p\Delta\varphi c^S + O(\varepsilon^{2/3}). \tag{10.114}$$

This suggests an ansatz of the form $\Delta S(t) = f(t, \Delta\varphi)$ for some function f. In order for $\varphi^{(\varepsilon)}$ only to trade at bid/ask prices $\underline{S}, \overline{S}$ as required in condition 2 of step 1, the function f ought to satisfy the boundary conditions

$$f(t, -\varepsilon^{1/3}\delta(t)) = \varepsilon S(t), \quad f(t, \varepsilon^{1/3}\delta(t)) = -\varepsilon S(t). \tag{10.115}$$

The process $\Delta\varphi$ is reflected at the boundaries of the trading corridor, which leads to two local time terms in its dynamics. They resemble local time of Brownian motion, cf. the end of Sect. 3.2.3. However, such singular terms should not be present in the dynamics of $S^{(\varepsilon)} = S + \Delta S$ because they would lead to arbitrage possibilities relative to this frictionless price process. By Itō's formula this implies that the derivative $f'(t,x)$ of $x \mapsto f(t,x)$ should vanish at the boundaries:

$$f'(t, -\varepsilon^{1/3}\delta(t)) = 0, \quad f'(t, \varepsilon^{1/3}\delta(t)) = 0. \tag{10.116}$$

A simple family of functions matching these boundary conditions is of the form $f(t,x) = \alpha(t)x^3 - \gamma(t)x$. Equations (10.116) and (10.115) then boil down to

$$\varepsilon^{1/3}\delta(t) = \sqrt{\frac{\gamma(t)}{3\alpha(t)}} \tag{10.117}$$

and

$$\gamma(t) = \left(\frac{1}{2}\varepsilon S(t)\right)^{2/3} 3\alpha(t)^{1/3}. \tag{10.118}$$

Itō's formula applied to $\Delta S = f(t, \Delta\varphi)$ yields

$$\Delta S - \Delta S(0) = (3\alpha\Delta\varphi^2 - \gamma) \bullet \Delta\varphi + 3\alpha\Delta\varphi \bullet [\Delta\varphi, \Delta\varphi] \bullet \dot{f}(I, \Delta\varphi) \bullet I, \tag{10.119}$$

where $\dot{f}(t,x)$ denotes the derivative of $t \mapsto f(t,x)$. We assume

$$\dot{f}(t, \Delta\varphi) = O(\varepsilon^{2/3}). \tag{10.120}$$

Since $\varphi^{(\varepsilon)} = \varphi^{(0)} + \Delta\varphi$ is continuous and of finite variation, we have $[\Delta\varphi, \Delta\varphi] = [\varphi^{(0)}, \varphi^{(0)}]$. Moreover, $d(\Delta\varphi)(t) = -d\varphi^{(0)}(t)$ except at the boundaries because $\varphi^{(\varepsilon)}$ is constant inside the trading corridor. Since $(3\alpha\Delta\varphi^2 - \gamma)$ vanishes at these boundaries, (10.119) reads as

$$\Delta S - \Delta S(0) = -(3\alpha\Delta\varphi^2 - \gamma) \bullet \varphi^{(0)} + 3\alpha\Delta\varphi \bullet [\varphi^{(0)}, \varphi^{(0)}] + O(\varepsilon^{2/3}).$$

Thus $a^{\Delta S} = 3\alpha\Delta\varphi c^{\varphi^{(0)}} + O(\varepsilon^{2/3})$ because $3\alpha\Delta\varphi^2 - \gamma = O(\varepsilon^{2/3})$. In order to satisfy (10.114) we need $\alpha = pc^S/(3c^{\varphi^{(0)}})$. Equations (10.118) and (10.117) in turn yield

$$\gamma = \left(\frac{3p^{1/2}}{2}\sqrt{\frac{c^S}{c^{\varphi^{(0)}}}}S\right)^{2/3} \varepsilon^{2/3}$$

and

$$\delta = \sqrt{\frac{\gamma}{3\alpha}} \varepsilon^{-1/3} = \left(\frac{3}{2p} \frac{c^{\varphi^{(0)}}}{c^S} S \right)^{1/3},$$

in line with (10.107).

Step 4: In order to complete the proof, it remains to reverse the arguments for showing that the candidates meet assumptions (10.108, 10.109, 10.120) and the requirements 1–5 in step 1. We refer to [176, 177] for details where equations for the asymptotic loss of utility compared to the frictionless case can be found, too.

10.5.2 Utility of P&L

We assume the same setup as in the previous section, namely that of Sect. 9.3.3 with the additional assumption that the numeraire asset satisfies $\underline{S}_0 = \overline{S}_0 = 1$. Moreover, we fix a utility function u and a time horizon T as in Sect. 10.3 as well as the investor's initial portfolio holdings $\psi = (\psi_1, \dots, \psi_d) \in \mathbb{R}^d$ in assets $1, \dots, d$.

In the context of P&L, bid-ask spreads and illiquidity have more severe consequences than in the preceding section. The point is that unique asset prices at any time are needed in order to compare the performance of strategies. Indeed, the expected utility in (10.66) and (10.68) depends explicitly on the whole price process $(S(t))_{t\in[0,T]}$ even if trading takes place only at times 0 and T. In order to turn utility of P&L into a well-defined object, we need some *book value* for illiquid assets. Of course, such a value should lie within the bid/ask bounds $[\underline{S}, \overline{S}]$.

To begin with, we call $(\widetilde{S}, \varphi)$ a **consistent pair** if $\widetilde{S}$ is a consistent price system and φ a strategy with $\varphi(0) = \psi$ that buys/sells assets only when the consistent price coincides with the corresponding ask/bid price of the asset, i.e. when

$$\overline{S}_- \bullet \varphi^\uparrow - \underline{S}_- \bullet \varphi^\downarrow = \widetilde{S}_- \bullet \varphi,$$

where $\varphi^\uparrow, \varphi^\downarrow$ are defined as in Sect. 9.3.3. Among the many consistent pairs we single out a particular one.

Definition 10.54 Suppose that $(S^\star, \varphi^\star)$ is a consistent pair. We call $S^\star$ a **book value** or **shadow price process** and $\varphi^\star$ an **optimal strategy for expected utility of P&L** if $\varphi^\star$ maximises $\varphi \mapsto U^{S^\star}(\varphi)$ over all strategies φ, i.e. if $\varphi^\star$ is optimal for the frictionless market with price process $S^\star$.

Let us motivate why we consider this pair as natural from an economic point of view. In principle, a market with bid-ask spread $[\underline{S}, \overline{S}]$ offers less attractive investment opportunities than a frictionless market with price process $S^\star$ inside these bounds. But an investor trading optimally in $S^\star$ would not actually be bothered by the nonvanishing spread because she would buy effectively only when

$S^\star(t-) = \overline{S}(t-)$ and sell only when $S^\star(t-) = \underline{S}(t-)$. Therefore, $\varphi^\star$ should still be optimal with the same expected utility in the market $[\underline{S}, \overline{S}]$. This is precisely what Definition 10.54 requires.

Note that the book value process generally depends on the initial portfolio and on preferences, i.e. different investors in the same market may and should work with differing book values. But it is typically unique for a particular investor:

Rule 10.55 *Suppose that $\underline{S}(T) = \overline{S}(T)$. Then there exists a unique book value process $S^\star$. The corresponding optimal strategy is unique in the sense that the wealth process $\varphi^\star \bullet S^\star$ coincides for any two such optimisers.*

Idea Let $(S^\star, \varphi^\star)$ be as in Definition 10.54. Since $\varphi^\star$ trades at bid-ask prices, it is of the form

$$\varphi_i^\star = \varphi_i^\star(0) + 1_{\{(S_i^\star)_- = (\overline{S}_i)_-\}} \bullet \varphi^{\star,\uparrow} - 1_{\{(S_i^\star)_- = (\underline{S}_i)_-\}} \bullet \varphi^{\star,\downarrow}, \tag{10.121}$$

which can be viewed as a stochastic differential equation with reflection at two boundaries that depend on $S_i^\star$. Rule 10.40, on the other hand, requires that

$$\begin{aligned} &\left(a_i^{S^\star} + u''(0)c_{i\cdot}^{S^\star}\varphi^\star + \int \left(u'(\varphi^{\star\top}x) - 1\right)x_i K^{S^\star}(dx)\right)\left((\overline{S}_i)_- - (S_i^\star)_-\right) \geq 0, \\ &\left(a_i^{S^\star} + u''(0)c_{i\cdot}^{S^\star}\varphi^\star + \int \left(u'(\varphi^{\star\top}x) - 1\right)x_i K^{S^\star}(dx)\right)\left((\underline{S}_i)_- - (S_i^\star)_-\right) \geq 0 \end{aligned} \tag{10.122}$$

for $i = 1, \ldots, d$. Inequalities (10.122) yield $a_i^{S^\star}(t)$ in terms of $c^{S^\star}(t)$, $K^{S^\star}(t, \cdot)$, $\varphi^\star(t)$ unless $S_i^\star(t-) = \overline{S}_i(t-)$ or $S_i^\star(t-) = \underline{S}_i(t-)$. Therefore the system (10.121, 10.122) can be viewed as a forward-backward stochastic differential equation (FBSDE) with reflection. Given that this FBSDE has a unique solution, the assertion follows. □

Unfortunately, the optimal investment problem typically does not allow for a simple explicit solution in the presence of bid-ask spreads. Therefore, we consider the limit for small spreads in the following example. This allows us to understand the structure of the solution and to quantify it to the leading order.

Example 10.56 (Itō Process Model with Small Bid-Ask Spread) We consider the same setup as in Example 10.53 but now the goal is to determine a nearly optimal trading strategy $\varphi^{(\varepsilon)}$ for the expected utility of P&L if the transaction cost parameter ε is small. To this end fix the time horizon T and the risk aversion $p := -u''(0)$ of the utility function (10.78). We denote the optimal strategy for the frictionless market with price process S by $\varphi^{(0)}$. From Rule 10.40 we obtain that it is given by

$$\varphi^{(0)}(t) = \frac{\mu(t)}{p\sigma(t)^2}.$$

We suppose that $\varphi^{(0)}$ is an Itō process as well and denote its local characteristics by a^φ and c^φ.

From Example 10.53 for the terminal wealth case we expect that the optimal strategy for $[\underline{S}, \overline{S}]$ does not trade as long as the portfolio holdings $\varphi(t)$ are in a corridor around the frictionless optimal strategy $\varphi^{(0)}(t)$. If $\varphi(t)$ hits the boundary of this no-trade region, the investor buys resp. sells infinitesimal amounts of the asset, just enough to stay in the desired corridor. From the previous example we may also expect that the candidate corridor

$$[\underline{\varphi}(t), \overline{\varphi}(t)] := [\varphi^{(0)}(t) - \varepsilon^{1/3}\delta(t), \varphi^{(0)}(t) + \varepsilon^{1/3}\delta(t)] \tag{10.123}$$

yields the maximal expected utility to the leading order if one chooses the corridor width process $\delta(t)$ appropriately. We show below that there is a consistent pair $(S^{(\varepsilon)}, \varphi^{(\varepsilon)})$ such that

$$U^{S^{(\varepsilon)}}(\varphi^{(\varepsilon)}) = \sup\left\{U^{S^{(\varepsilon)}}(\varphi) : \varphi \text{ strategy}\right\} + o(\varepsilon^{2/3}). \tag{10.124}$$

Here, $\varphi^{(\varepsilon)}$ denotes the continuous strategy obtained from trading at the boundaries of (10.123), and the process $\delta(t)$ is given by

$$\delta(t) = \left(\frac{3}{2p}\frac{c^{\varphi^{(0)}}(t)}{\sigma(t)^2}S(t)\right)^{1/3},$$

which is identical to the one in (10.107). Put differently, $S^{(\varepsilon)}$ is **nearly** a book value process with corresponding optimiser $\varphi^{(\varepsilon)}$ in the sense that $(S^{(\varepsilon)}, \varphi^{(\varepsilon)})$ is consistent but $\varphi^{(\varepsilon)}$ reaches the maximal expected utility for $S^{(\varepsilon)}$ only to the leading order, i.e. up to some small error $o(\varepsilon^{2/3})$. Since the difference $U^{S^{(\varepsilon)}}(\varphi^{(\varepsilon)}) - U^S(\varphi^0)$ is typically of the order $O(\varepsilon^{2/3})$, this small error can be considered as negligible.

In order to derive (10.124), define $S^{(\varepsilon)}(t) := S(t) + \varepsilon S(t) f(\Delta\varphi(t)/(\varepsilon^{1/3}\delta(t)))$ with $\Delta\varphi(t) := \varphi^{(\varepsilon)}(t) - \varphi^{(0)}(t)$ and $f(x) := \frac{1}{2}x^3 - \frac{3}{2}x$. As in Example 10.53 the notation $\Delta\varphi$ does not refer to jumps here.

Step 1: We show that $S^{(\varepsilon)}(t) \in [\underline{S}(t), \overline{S}(t)]$. To wit, $\Delta\varphi(t) \in [-\delta(t)\varepsilon^{1/3}, \delta(t)\varepsilon^{1/3}]$ implies that $f(\Delta\varphi(t)/(\varepsilon^{1/3}\delta(t))) \in [f(1), f(-1)] = [-1, 1]$. This in turn yields the claim.

Step 2: $\varphi^{(\varepsilon)}$ trades only when $S^{(\varepsilon)} = \underline{S}$ or $S^{(\varepsilon)} = \overline{S}$ because $\varphi^{(\varepsilon)}(t) = \underline{\varphi}(t)$ happens if and only if $S^{(\varepsilon)}(t) = S(t) + f(-1)\varepsilon S(t) = \overline{S}(t)$ and likewise for $\varphi^{(\varepsilon)}(t) = \overline{\varphi}(t)$.

Step 3: Itō's formula and slightly tedious calculations yield

$$\begin{aligned} S^{(\varepsilon)}(t) = S^{(\varepsilon)}(0) + 1 \bullet S(t) + \frac{\varepsilon S f'(\Delta\varphi\varepsilon^{-1/3}\delta^{-1})}{\varepsilon^{1/3}\delta} \bullet \Delta\varphi \\ + \frac{\varepsilon S f''(\Delta\varphi\varepsilon^{-1/3}\delta^{-1})}{2\varepsilon^{2/3}\delta^2} \bullet [\Delta\varphi, \Delta\varphi] + O(\varepsilon). \end{aligned} \tag{10.125}$$

Since $\varphi^{(\varepsilon)} = \varphi^{(0)} + \Delta\varphi$ is continuous and of finite variation, we have $[\Delta\varphi, \Delta\varphi] = [\varphi^{(0)}, \varphi^{(0)}]$. Moreover, $\varphi^{(\varepsilon)}$ is constant outside $\{\varphi^{(\varepsilon)} = \underline{\varphi}$ or $\varphi^{(\varepsilon)} = \overline{\varphi}\}$ and hence $\xi \bullet \varphi^{(0)} + \xi \bullet \Delta\varphi = \xi \bullet \varphi^{(\varepsilon)} = 0$ for any integrand ξ that vanishes outside this set. Thus (10.125) reads as

$$\begin{aligned} S^{(\varepsilon)}(t) = S^{(\varepsilon)}(0) + 1 \bullet S(t) - \varepsilon^{2/3} \frac{Sf'(\Delta\varphi\varepsilon^{-1/3}\delta^{-1})}{\delta} \bullet \varphi^{(0)} \\ + \varepsilon^{1/3} \frac{Sf''(\Delta\varphi\varepsilon^{-1/3}\delta^{-1})}{2\delta^2} \bullet [\varphi^{(0)}, \varphi^{(0)}] + o(\varepsilon^{2/3}). \end{aligned}$$

We conclude that

$$\begin{aligned} a^{S^{(\varepsilon)}} &= a^S + \varepsilon^{1/3} \frac{Sf''(\Delta\varphi\varepsilon^{-1/3}\delta^{-1})}{2\delta^2} c^{\varphi^{(0)}} + o(\varepsilon^{1/3}) \\ &= a^S + pc^S \Delta\varphi + o(\varepsilon^{1/3}), \\ c^{S^{(\varepsilon)}} &= c^S + o(\varepsilon^{1/3}), \end{aligned}$$

and therefore

$$\begin{aligned} \gamma^{S^{(\varepsilon)}}(t, \varphi) &= a^{S^{(\varepsilon)}} \varphi - \frac{p}{2} c^{S^{(\varepsilon)}} \varphi^2 \\ &= -\frac{p}{2} c^{S^{(\varepsilon)}} \left(\varphi - \frac{a^{S^{(\varepsilon)}}}{pc^{S^{(\varepsilon)}}} \right)^2 + \frac{(a^{S^{(\varepsilon)}})^2}{2pc^{S^{(\varepsilon)}}} \\ &= -\frac{p}{2} c^{S^{(\varepsilon)}} \left(\varphi - \frac{a^S + pc^S \Delta\varphi}{pc^{S^{(\varepsilon)}}} + o(\varepsilon^{1/3}) \right)^2 + \frac{(a^{S^{(\varepsilon)}})^2}{2pc^{S^{(\varepsilon)}}} \\ &= -\frac{p}{2} c^{S^{(\varepsilon)}} \left(\varphi - (\varphi^{(0)} + \Delta\varphi) + o(\varepsilon^{1/3}) \right)^2 + \frac{(a^{S^{(\varepsilon)}})^2}{2pc^{S^{(\varepsilon)}}}. \end{aligned}$$

Up to an error of order $o(\varepsilon^{2/3})$ the maximal value is obtained at $\varphi = \varphi^{(0)} + \Delta\varphi = \varphi^{(\varepsilon)}$. Together with steps 1 and 2 we conclude that $S^{(\varepsilon)}$ is nearly a book value process with corresponding strategy $\varphi^{(\varepsilon)}$.

Remark 10.57 In Sect. 13.2 we consider markets where some assets with index $0, \ldots, d$ are liquid with no bid/ask spread and another one is entirely illiquid with uniquely determined price $S_{d+1}(t)$ only at time T. This can be embedded in the present framework by letting

$$[\underline{S}_{d+1}(t), \overline{S}_{d+1}(t)] := [-\infty, \infty], \quad t \in [0, T).$$

In this context a **consistent pair** $(\widetilde{S}, \varphi)$ consists of

1. a *consistent price system* $\widetilde{S}$, i.e. $\widetilde{S}_i = S_i$ for $i = 0, \dots, d$ and $\widetilde{S}_{d+1}(T) = S_{d+1}(T)$, and
2. a strategy φ such that $\widetilde{\varphi}_{d+1}(t) = \psi_{d+1}$ for all $t \in [0, T)$ because buying and selling the illiquid asset is impossible before T.

The definitions and statements of this section now extend naturally to this setup. In particular, $S^\star$ is the unique book value process and $\varphi^\star$ an optimal strategy for expected utility of P&L if $(S^\star, \varphi^\star)$ is a consistent pair and $\varphi^\star$ is optimal for the frictionless market with price process $S^\star$, cf. Definition 10.54.

Appendix 1: Problems

The exercises correspond to the section with the same number.

10.1 Consider an Itō process model as in Example 9.3 where μ, σ are adapted to the filtration of $W_2, \dots, W_n$ and hence independent of W_1. Determine the optimal portfolio for the expected utility of terminal wealth relative to power utility $u(x) = x^{1-p}/(1-p)$.

Hint: Show that $\eta_1 = 0$ in Example 10.14.

10.2 Generalise Example 10.30 to the case when the asset price process $S = (S_1, \dots, S_d)$ is a multivariate geometric Lévy process in the sense that

$$S_i(t) = S_i(0) \exp(X_i(t)) = S_i(0)\mathscr{E}(\widetilde{X}_i)(t), \quad i = 1, \dots, d$$

for multivariate Lévy processes $X, \widetilde{X}$ with triplets (a^X, c^X, K^X) resp. $(a^{\widetilde{X}}, c^{\widetilde{X}}, K^{\widetilde{X}})$.

10.3 Suppose that the discounted price process is a geometric Lévy process as in Example 9.2. Show that the martingale measure in Example 10.38 coincides with the minimal entropy martingale measure (MEMM).

10.4 (Time Consistency of Efficient Portfolios) Let $\varphi^\star$ be mean-variance efficient as in Sect. 10.4 and suppose that $\varphi^\star \bullet S \leq m$ for m in Rule 10.43. This holds, for example, if S is continuous. Show that $\varphi^\star$ is also **conditionally mean-variance efficient** in the sense that the following statements hold for any $t \in [0, T]$.

1. $\varphi^\star$ maximises $E(\varphi \bullet S(T)|\mathscr{F}_t)$ among all self-financing strategies φ satisfying

$$\mathrm{Var}(\varphi \bullet S(T)|\mathscr{F}_t) \leq \mathrm{Var}(\varphi^\star \bullet S(T)|\mathscr{F}_t).$$

2. $\varphi^\star$ minimises $\mathrm{Var}(\varphi \bullet S(T)|\mathscr{F}_t)$ among all self-financing strategies φ satisfying

$$E(\varphi \bullet S(T)|\mathscr{F}_t) \geq E(\varphi^\star \bullet S(T)|\mathscr{F}_t).$$

10.5 Consider the market in Example 9.1 with proportional transaction costs of size $\varepsilon S(t)$, i.e. $\underline{S}(t) = (1-\varepsilon)S(t)$ and $\overline{S}(t) = (1+\varepsilon)S(t)$ with constant $\varepsilon > 0$. The goal is to determine the optimal strategy/consumption pair for the utility function $u(t,x) = \log(x)e^{-\delta t}$ and infinite time horizon T. To this end, let the function $f : [\log(1-\varepsilon), \log(1+\varepsilon)] \to \mathbb{R}$ denote the solution to the ordinary differential equation

$$
\begin{aligned}
f''(x) &= \tfrac{2\delta}{\sigma^2}\left(1+e^{f(x)}\right) + \left(\tfrac{2\mu}{\sigma^2} - \tfrac{4\delta}{\sigma^2}\left(1+e^{f(x)}\right)\right) f'(x) \\
&\quad + \left(-\tfrac{4\mu}{\sigma^2} + \tfrac{2\delta}{\sigma^2}\left(1+e^{f(x)}\right) + \tfrac{1-e^{-f(x)}}{1+e^{-f(x)}}\right)(f'(x))^2 \\
&\quad + \left(\tfrac{2\mu}{\sigma^2} - \tfrac{1-e^{-f(x)}}{1+e^{-f(x)}}\right)(f'(x))^3
\end{aligned}
$$

with $f'(\log(1-\varepsilon)) = -\infty = f'(\log(1+\varepsilon))$, define

$$
\begin{aligned}
\widetilde{\mu} &:= -\mu + \frac{\sigma^2}{2}\left(\frac{f'}{f'-1}\right)^2 \frac{1-e^{-f}}{1+e^{-f}}, \\
\widetilde{\sigma} &:= \frac{\sigma}{f'-1},
\end{aligned}
$$

C as the solution to the SDE

$$
dC(t) = \widetilde{\mu}(C(t))dt + \widetilde{\sigma}(C(t))dW(t)
$$

for appropriate some $C(0)$,

$$
\begin{aligned}
\widetilde{S}(t) &:= S(t)\exp(C(t)), \\
\widetilde{\pi}(t) &:= \frac{\mu + \widetilde{\mu}(C(t))}{(\sigma + \widetilde{\sigma}(C(t)))^2} + \frac{1}{2}, \\
\widetilde{V}(t) &:= \widetilde{V}(0)\mathscr{E}\big(\widetilde{\pi} \bullet \mathscr{L}(\widetilde{S}) - \delta I\big)(t)
\end{aligned}
$$

for some appropriate $\widetilde{V}(0)$, and

$$
\varphi(t) := \frac{\widetilde{V}(t)\widetilde{\pi}(t)}{\widetilde{S}(t)}, \qquad c(t) := \delta\widetilde{V}(t).
$$

Show that (φ, c) is optimal for the expected utility of consumption and $\widetilde{S}$ is a corresponding shadow price as in Rule 10.52.

Hints:

1. Apply Rule 10.28 in order to show that (φ, c) is optimal for the expected utility of consumption in the frictionless market with price process $\widetilde{S}$.

2. Use Itō's formula in order to obtain that

$$1_{\{\widetilde{S}(t)\neq S(t)\}}d\varphi(t) = 0dt + 0dW(t).$$

Put differently, the otherwise constant stock holdings φ are adjusted only when $\widetilde{S}$ or, equivalently, $\widetilde{\pi}$ or φ reaches some no-trade bounds. What are these bounds?

Appendix 2: Notes and Comments

Monographs and textbooks on or covering optimal investment include [29, 187, 226, 227, 234, 236, 250, 287].

Classical references to expected utility maximisation in a dynamic context include [217, 218, 222, 258]. The dual approach goes back to [27, 136, 137, 189]. We refer to [24, 197] for versions of Rule 10.3 in a general semimartingale setup and to [188] in the consumption case. Rules 10.4 and 10.26 are literally true only in simple finite models, cf. [167]. For the connection to distances or divergences between martingale measures in Rule 10.5, we refer to [21, 127]. Rules 10.6, 10.28 are taken from [125, 126]. Rule 10.7 can be found in [187, Theorem 3.10.1] and [126, Lemma 5.3]. For mutual fund theorems we refer as in Chap. 7 to [218, 260]. Rule 10.7 is taken from [187, Theorem 3.10.1] and [126, Lemma 5.3]. The numeraire portfolio plays a key role in the benchmark approach to Mathematical Finance of [236]. Strictly speaking, S/V_φ is only a local martingale (for locally bounded processes) or even a supermartingale in the general case, cf. [197] for details.

For disillusions along the lines of Example 10.10 see [248]. The concept of an opportunity process is introduced in [55] for quadratic utility and extended to power utility in [225]. Its characterisation in Rules 10.11, 10.17, 10.29 is inspired by the parallel statement in Rule 10.45. Examples 10.13, 10.19, 10.30 are based on [165], whereas [243] considers the case with claim in Example 10.22. For BSDE characterisations similar to Example 10.14, 10.20, 10.31 we refer to [146, 253]. The PDEs in Examples 10.15, 10.21, 10.32 are related but not identical to the classical Hamilton–Jacobi–Bellman PDEs for the value function in a Markovian setup, cf. [218]. The measure P^H of (10.44) is considered, for example, in [180].

Utility maximisation of P&L has been introduced in [163, 164]. As an optimisation concept over in some sense infinitesimal time horizons it is related to *local risk minimisation* in the sense of [266, 267].

Portfolio choice based on mean and variance goes back to [213]. It is mostly studied in a one-period context. Its dynamic counterpart starts to be considered in [6, 244, 268]. We follow here the reasoning of [55], where the notion of an opportunity process as in (10.83) and the subsequent rules can be found. The terms *adjustment process* and *variance-optimal martingale measure* are taken from [269]. The adjustment process of Example 10.48 is determined in [148]. PDE characterisations of the opportunity process similar to Example 10.50 can be found in [205, 235].

References showing the duality of Rules 10.51, 10.52 include [22, 65, 67, 68, 161, 175, 208]. Example 10.53 goes back to [176, 177].

The discussion of book values for P&L in the sense of Definition 10.54 is new. A rigorous uniqueness proof remains to be given. Moreover, it is theoretically conceivable that the maximal utility of P&L in a model is increased by replacing the "true" price process with a positive bid-ask spread, i.e. by artificially considering a liquid market as illiquid. It is an open question to what extent this undesirable phenomenon really occurs in reasonable market models such as those in Chap. 8.

Utility maximisation for P&L is considered in [214, 215] in the presence of small proportional transaction costs. However, they consider profits and losses relative to the mid-price process instead of some book value. Transaction costs are deducted separately, essentially by adjusting the drift part of the wealth process. In spite of this slightly different setup, they obtain the same no-trade region as in Example 10.56.

Problem 10.1 is based on [174]. Problems 10.2 and 10.3 are solved in [165].

It is often claimed that dynamic mean-variance portfolio selection is time inconsistent in the sense that one later regrets the decision made at earlier times, cf. e.g. [13, 35, 66]. Problem 10.4 indicates that this may be primarily due to how the problem is phrased.

Problem 10.5 is treated in [173].

Chapter 11
Arbitrage-Based Valuation and Hedging of Derivatives

The valuation of derivative securities constitutes one of the main issues in modern Mathematical Finance. Economic theory has considered the genesis of asset prices for a long time. But here the focus is on *relative* pricing, i.e. valuation of *contingent claims* or *secondary assets* in relation to some *underlyings* or *primary assets* whose prices are assumed to be given exogenously. Several classes of derivatives are introduced in the following section. In many cases it is obvious whether to consider an asset as primary or secondary. In others, such as in interest rate theory, there may exist more than one reasonable choice.

But what does *valuation* mean? We distinguish two possible interpretations. Depending on the application one may prefer one or the other. In Sect. 11.2 we consider contingent claims as *traded securities*. If we exclude the possibility of riskless gains, the fundamental theorem of asset pricing restricts the set of possible derivative price processes, in some instances dramatically. But even in the others it helps to come up with reasonable models.

Instead of considering market prices one may also adopt the *option writer's point of view*, which is done in Sect. 11.3. It is particularly suited to derivatives that are traded over the counter (OTC), so that a liquid market does not exist. Suppose you are in the position of a bank that is approached by some customer who wants to buy a certain contingent claim. At this point you must decide at which minimum price you are willing to make the deal. Moreover, you may wonder how to invest the premium you receive in exchange for the option. The answer to these questions generally depends on your own endowment, beliefs, and attitude towards risk. But even without taking these into account, arbitrage theory provides at least partial information on the range of possible premia. In some lucky cases such as in the Black–Scholes model it even tells you exactly what to do; in other situations it does not help much. Alternatives in such cases are discussed later in Chaps. 12 and 13.

© Springer Nature Switzerland AG 2019
E. Eberlein, J. Kallsen, *Mathematical Finance*, Springer Finance,
https://doi.org/10.1007/978-3-030-26106-1_11

Surprisingly or not, the market and the individual approach to option pricing lead to similar results. But in any case, expressions such as conditional expectations must be evaluated in order to come up with concrete numbers. We discuss several possible paths in Sect. 11.5.

We generally work in the probabilistic framework of Chap. 9. But for ease of notation we assume the initial σ-field $\mathscr{F}_0$ to be trivial, i.e. $\mathscr{F}_0$-measurable random variables are actually deterministic in this chapter.

11.1 Derivative Securities

As in Chap. 9 we denote by $S_0, S_1, \dots$ price processes of traded securities. We consider the following kinds of derivative contracts written on S_1 as underlying.

European options pay a random amount at time T, for example $(S_1(T) - K)^+$ (or $(K - S_1(T))^+$) for a European call (resp. put) on S_1 with *strike price* K. Mathematically, such claims are modelled by some random variable H which represents their *payoff* at T, i.e. $H = (S_1(T) - K)^+$ for a call. Observe that this general framework allows for path-dependent options such as $H = \sup_{t\in[0,T]}(S_1(t) - K)^+$ and even for contracts that depend on untradable processes such as temperature. The *discounted payoff* $\widehat{H} := H/S_0(T)$ refers as usual to the value expressed in terms of the numeraire security. For ease of notation, we have assumed that the option expires at T. But the theory extends accordingly to arbitrary maturities $t \leq T$, in which case the *discounted payoff* must be defined in terms of the current numeraire value, i.e. $\widehat{H} := H/S_0(t)$.

A **forward** contract can be viewed as a special case of a European option with payoff $H = S_1(T) - K$. The **forward price** K, however, is determined in a very particular way. It is chosen so that the counterparties agree to enter the contract at no cost. It is denoted by $F(t)$ because it depends on the time t when the contract is signed. Obviously we have $F(T) = S_1(T)$ for a forward that expires immediately. Note that *forward price* does not refer to the price of an asset but rather to a reference value similar to the strike of a call.

American options give the owner the right to cancel the contract prematurely. The holder of an American call (or put) receives $(S_1(t)-K)^+$ (resp. $(K-S_1(t))^+$) if she decides to exercise the option at time t. Such contracts are modelled in terms of an adapted càdlàg **exercise process** $H = (H(t))_{t\in[0,T]}$. The random variable $H(t)$ stands for the payoff in case of exercise at time t, e.g. $H(t) = (K - S_1(t))^+$ for an American put. Since $H(t)$ is paid at time t, the corresponding *discounted exercise process* $\widehat{H}$ is defined as $\widehat{H}(t) := H(t)/S_0(t)$.

Finally, we turn to **futures** contracts, which are sometimes considered as liquidly traded versions of forwards. They are entered and terminated at no cost. As in the case of a forward, there is a **futures price process** $F = (F(t))_{t\in[0,T]}$ satisfying $F(T) = S_1(T)$ for a futures on S_1 with maturity T. Once more, this price is not the price of an asset but rather a reference value similar to the strike of a call. Here the similarity between forward and futures ends. With a forward contract, all payments are made at maturity, whereas a futures contract involves daily or—in theory—continuous resettlement. In our continuous-time framework we assume that the futures price increment $F(t) - F(t - dt)$ in period $(t - dt, t]$ is paid to the investor at time t. The total amount received by the investor between t and T equals $S_1(T) - F(t)$ as in the case of a forward contract entered at time t. But because of interest payments, this does not imply that forward and futures prices behave identically.

11.2 Liquidly Traded Derivatives

We generally distinguish two kinds of assets, called *underlyings* and *derivatives*. In this and the following section we consider all of them to be liquidly traded securities, which makes the distinction somewhat arbitrary. For ease of notation we study a market with three securities: the numeraire S_0, a further underlying S_1 and a derivative security S_2. In principle, however, the results do not depend on the number of underlyings and derivatives. The price processes S_0, S_1 are assumed to be given exogenously. Moreover, we know that the terminal payoff of the derivative equals the random variable H in the case of a European option. Its price process S_2 is yet to be derived.

11.2.1 Arbitrage-Based Pricing

As usual we assume that absence of arbitrage holds for the whole market $S = (S_0, S_1, S_2)$. What are the implications of the fundamental theorem of asset pricing for option prices? Suppose that S_2 denotes the price process of a European-style contingent claim with terminal payoff $S_2(T) = H$. If the market is arbitrage-free, then the discounted processes $\widehat{S}_1, \widehat{S}_2$ are Q-martingales relative to some $Q \sim P$, which implies

$$\widehat{S}_2(t) = E_Q\big(\widehat{H}\big|\mathscr{F}_t\big) \tag{11.1}$$

for the option price process and in particular

$$\widehat{S}_2(0) = E_Q\big(\widehat{H}\big) \tag{11.2}$$

for its initial value. We summarise this as a rule.

Rule 11.1 (European Options) *Suppose that the market (S_0, S_1) without derivative does not allow for arbitrage. A price process S_2 with $S_2(T) = H$ leads to an arbitrage-free market $S = (S_0, S_1, S_2)$ if and only if there exists an equivalent* *martingale measure for* (S_0, S_1) *such that*

$$\widehat{S}_2(t) = E_Q(\widehat{H}|\mathscr{F}_t), \quad t \in [0, T].$$

Idea $\Rightarrow$: This is derived above.

$\Leftarrow$: If Q is an EMM for (S_0, S_1) satisfying (11.1), it is a fortiori an EMM for (S_0, S_1, S_2). Hence this market is arbitrage free by the fundamental theorem of asset pricing 9.15. □

Rule 11.1 holds for any number of underlyings and derivatives. But note that the same EMM must be used for all contingent claims.

Recall that the law of the underlying processes S_0, S_1 relative to P is assumed to be given exogenously, e.g. based on statistical inference. The problem is that we do not know the market's EMM or *pricing measure* Q needed to compute option prices $S_2(t)$ via (11.1). If we are lucky, the conditional expectation in (11.1) does not depend on the choice of Q. Payoffs with this property can be characterised in terms of *replicating strategies*:

Rule 11.2 *There is only one arbitrage-free price process* S_2 *as in Rule 11.1 if and only if there exists some self-financing strategy* φ *which trades only in* S_0, S_1 *and satisfies* $V_\varphi(T) = H$. *In this case we call the claim* H **attainable** *and* φ *a* **replicating strategy** *or* **perfect hedge** *for* H.

Idea $\Leftarrow$: For any EMM Q we have that $\widehat{V}_\varphi = \widehat{V}_\varphi(0) + \varphi \bullet \widehat{S}$ is a Q-martingale and hence

$$E_Q(\widehat{H}|\mathscr{F}_t) = E_Q(\widehat{V}_\varphi(T)|\mathscr{F}_t) = \widehat{V}_\varphi(t),$$

which does not depend on Q.

$\Rightarrow$: Theorem 3.64 on optional decomposition states that $C := \widehat{S}_2(0)+\varphi \bullet \widehat{S}-\widehat{S}_2$ is increasing for some predictable $\mathbb{R}^d$-valued process φ. Taking expectations under some EMM Q yields $E_Q(C(T)) = \widehat{S}_2(0) + E_Q(\varphi \bullet \widehat{S}(T)) - E_Q(\widehat{S}_2(T)) = 0$. Since $C(T) \geq C(0) = 0$, we conclude $C(T) = 0$ and thus $C = 0$. Consequently $\widehat{S}_2 = \widehat{S}_2(0) + \varphi \bullet \widehat{S}$ as desired. □

Uniqueness of the option price holds in particular if there exists only one probability measure $Q \sim P$ such that $\widehat{S}_1$ is a Q-martingale. By Rule 11.2 this means that any contingent claim is attainable. Markets with this property are called **complete**.

Rule 11.3 (Second Fundamental Theorem of Asset Pricing) *An arbitrage-free market is complete if and only if there exists only one EMM.*

Idea $\Rightarrow$: For $F \in \mathscr{F}_t$ consider the claim with discounted payoff $\widehat{H} := 1_F$. Market completeness and Rule 11.2 yield that the $E_Q(\widehat{H})$ coincide for all EMMs Q.

$\Leftarrow$: This follows immediately from Rule 11.2. □

The prime example of a complete market is the Black–Scholes model which we encountered already in Example 9.1.

Example 11.4 (Black–Scholes Model) The Black–Scholes model from Example 9.1 is complete. Indeed, by Examples 9.16 or 9.17 there is an EMM Q such that the discounted price process is of the form $\widehat{S}_1(t) = \widehat{S}_1(0)\exp(-\frac{1}{2}\sigma^2 t + \sigma W^Q(t))$ for some Q-Wiener process W^Q. Put differently, we have

$$\widehat{S}_1 = \widehat{S}_1(0)\mathscr{E}(\sigma W^Q).$$

For an arbitrary contingent claim H define the Q-martingale $\widehat{S}_2$ as in (11.1). The martingale representation theorem 3.61 yields that it can be written as $\widehat{S}_2 = \widehat{S}_2(0) + \xi \bullet W^Q$ for some predictable process ξ. If we set $\varphi := \xi/(\sigma\widehat{S}_1)$, we obtain

$$\widehat{S}_2(T) = \widehat{S}_2(0) + (\varphi\sigma\widehat{S}_1) \bullet W^Q(T) = \widehat{S}_2(0) + \varphi \bullet \widehat{S}_1(T),$$

which means that H is attainable, as claimed. In Examples 11.19 and 11.22 we will see how to compute the perfect hedge φ in concrete cases.

If there exist more than one Q such that $\widehat{S}_1$ is a Q-martingale, (11.2) yields only an interval of possible initial values. Consider, for example, a discrete version of the Black–Scholes model, namely a discrete-time process $S_1(t), t = 0, 1, \ldots, T$ such that the log-returns $\log(S_1(t)/S_1(t-1))$ are i.i.d. Gaussian random variables. For simplicity we assume a constant numeraire $S_0(t) = 1, t = 0, 1, \ldots, T$. One can show that the interval of possible initial prices for a European call with strike K is given by $((S_1(0) - K)^+, S_1(0))$. This means that for an out-of-the-money option (i.e. $K < S_1(0)$) any positive option price below the stock price is consistent with the absence of arbitrage. In other words, arbitrage theory hardly provides any useful information on option prices in this case. The same holds for many continuous-time stock price models of the form $S_1(t) = S_1(0)\exp(X(t))$ with a jump-type Lévy process X rather than Brownian motion and also for stochastic volatility models, cf. Examples 11.16 and 11.17.

Remark 11.5 (Rule of Thumb) Absence of arbitrage is more likely to hold if there are only a few assets in the market because this reduces the set of portfolios. The converse holds for completeness. According to a vague rule of thumb we can typically expect absence of arbitrage if d, i.e. the number of assets besides the numeraire, does not exceed the number of "sources of randomness" in the filtered probability space. By contrast, for an arbitrage-free market to be complete, we need that this number of sources of randomness does not exceed d. In this context, any univariate Wiener process or any possible jump height of a Lévy process counts as one source of randomness.

The Black–Scholes model, for example, is expected to be complete according to this rule because $d = 1$ corresponds to one driving Wiener process and hence one source of randomness. The Heston model, however, is driven by two Wiener processes, which suggests its incompleteness. Even worse, Lévy-driven stock price models as in Example 9.2 contain infinitely many sources of randomness for a diffuse Lévy measure because any jump height can occur in this case.

The rule of thumb can be understood in terms of the martingale representation theorem 3.62, applied to the density process of a candidate EMM Q. For any (ω, t) the parameters $\varphi(\omega, t)$ and $\psi(\omega, t)$ can be freely chosen. On the other hand, d drift conditions in, say, (4.17) must be satisfied in order for the Q-drift rate of $\widehat{S}^1, \dots, \widehat{S}^d$ to vanish. So the above rule of thumb means to expect the existence of at least (resp. at most) one EMM if the number of conditions, namely d, equals at most (resp. at least) the number of degrees of freedom. Of course, absence of arbitrage may still hold if d exceeds the degrees of freedom. But in this case, the P-drift rates of the assets must be strongly linked in order to allow for a solution of the d Q-martingale constraints.

Sometimes it helps to consider a few derivatives as additional underlyings. Suppose that we model not only the stock but also a number of liquidly traded call options $S_2, \dots, S_k$ on the stock exogenously by statistical methods. This reduces the set of possible pricing measures Q in the FTAP because now $\widehat{S}_i$, $i = 1, \dots, k$ must be Q-martingales rather than only $\widehat{S}_1$. However, working with large numbers of primary assets leads to new difficulties. From Rule 11.5 we know that assets are strongly linked by arbitrage constraints if the number of assets exceeds the number of "sources of randomness." In other words, their price processes cannot be chosen arbitrarily and the set of P-dynamics that does not lead to arbitrage may be quite cumbersome to determine and work with. Even if we have found a feasible class of P-dynamics, we may still face the problem of a non-unique pricing measure Q. These problems also occur in the Heath–Jarrow–Morton approach to term structure modelling, which is why we come back to this issue in Chap. 14, in particular Sect. 14.1.3. An alternative to the possibly tedious approach of treating options as primary assets is discussed in Sect. 11.2.3 below.

11.2.2 American Options and Futures Contracts

The above arguments do not extend immediately to American options. The FTAP requires that assets can be held in arbitrary positive and negative amounts. There is nothing wrong with long positions in an American option. Shorting the option, however, is less obvious because the holder may exercise it, in which case the option and hence the short position vanishes immediately. It is reasonable to assume that this does not happen as long as the market price exceeds the exercise price. Indeed, in this case the holder will rather sell the option on the market than exercise it because she receives a larger profit. Moreover, the option price process will never fall below the exercise price. Otherwise one could buy the option and exercise it immediately, which would imply a trivial form of arbitrage. Hence, trading American options is constrained to nonnegative positions only in instances when the market price equals the exercise price. This is formalised by the set

$$\Gamma := \{\varphi \text{ self-financing strategy} : \varphi_2(t) \geq 0 \text{ if } S_2(t-) = H(t-)\},$$

where H denotes the exercise price process of the option and S_2 its market price. The constrained version of the FTAP leads to the following characterisation of arbitrage-free American option price processes.

Rule 11.6 (American Option) *Suppose that the market* (S_0, S_1) *without derivative does not allow for arbitrage. A price process* $S_2 \geq H$ *with* $S_2(T) = H(T)$ *leads to a* Γ*-arbitrage free market* $S = (S_0, S_1, S_2)$ *if and only if there exists an equivalent martingale measure for* (S_0, S_1) *such that* $\widehat{S}_2$ *is the* Q*-Snell envelope of* $\widehat{H}$. *In particular,*

$$\widehat{S}_2(t) = \sup_{\tau \in \mathscr{T}_{[t,T]}} E_Q(\widehat{H}(\tau)|\mathscr{F}_t), \quad t \in [0, T], \tag{11.3}$$

where $\mathscr{T}_{[t,T]}$ *denotes the set of stopping times with values in* $[t, T]$.

Idea $\Rightarrow$: If absence of arbitrage holds, Rule 9.20 yields the existence of some probability measure $Q \sim P$ such that $E_Q(\varphi \bullet \widehat{S}(T)) \leq 0$ for any $\varphi \in \Gamma$. Since $(\pm 1_{F \times (s,T]}, 0) \in \Gamma$ for any $s \in [0, T]$ and $F \in \mathscr{F}_s$, we have

$$\pm \int_F \widehat{S}_1(t) dQ \mp \int_F \widehat{S}_1(s) dQ \leq 0$$

and hence $\widehat{S}_1(s) = E_Q(\widehat{S}_1(t)|\mathscr{F}_s)$ for $s \leq t$. Put differently, $\widehat{S}_1$ is a Q-martingale. Similarly, considering $(0, 1_{F \times (s,T]}) \in \Gamma$ yields that $\widehat{S}_2$ is a Q-supermartingale. Finally, $(0, \pm 1_{(F \times (s,t]) \cap \{(S_2)_- \neq H_-\}}) \in \Gamma$ and

$$\begin{aligned} 1_{(F \times (s,t]) \cap \{(S_2)_- \neq H_-\}} \bullet \widehat{S}_2(T) &= 1_{F \times (s,T]} \bullet (1_{\{(S_2)_- \neq H_-\}} \bullet \widehat{S}_2)(T) \\ &= \left(1_{\{(S_2)_- \neq H_-\}} \bullet \widehat{S}_2(t) - 1_{\{(S_2)_- \neq H_-\}} \bullet \widehat{S}_2(s)\right) 1_F \end{aligned}$$

imply that $1_{\{(S_2)_- \neq H_-\}} \bullet \widehat{S}_2$ is a Q-martingale. By Proposition 7.11, $\widehat{S}_2$ is the Q-Snell envelope of H.

$\Leftarrow$: Let Q be an EMM such that $\widehat{S}_2$ is the Q-Snell envelope of $\widehat{H}$. By Proposition 7.11(2) we have that $1_{\{(S_2)_- \neq H_-\}} \bullet A^{\widehat{S}_2} = 0$, where $\widehat{S}_2 = \widehat{S}_2(0) + M^{\widehat{S}_2} + A^{\widehat{S}_2}$ denotes the Q-canonical decomposition of $\widehat{S}_2$. For $\varphi \in \Gamma$ we conclude that

$$\begin{aligned} \varphi \bullet \widehat{S} &= \varphi_1 \bullet \widehat{S}_1 + \varphi_2 \bullet \widehat{S}_2 \\ &= \varphi_1 \bullet \widehat{S}_1 + \varphi_2 \bullet M^{\widehat{S}_2} + \varphi_2 \bullet A^{\widehat{S}_2} \\ &= \varphi_1 \bullet \widehat{S}_1 + \varphi_2 \bullet M^{\widehat{S}_2} + (\varphi_2 1_{\{(\widehat{S}_2)_- = \widehat{H}_-\}}) \bullet A^{\widehat{S}_2} \\ &= \varphi_1 \bullet \widehat{S}_1 + \varphi_2 \bullet M^{\widehat{S}_2} + (\varphi_2 1_{\{\varphi_2 \geq 0\}}) \bullet A^{\widehat{S}_2} \end{aligned}$$

is a Q-supermartingale and hence $E_Q(\varphi \bullet \widehat{S}(T)) \leq 0$. This yields absence of Γ-arbitrage for the market $S = (S_0, S_1, S_2)$. □

The counterpart to Rule 11.2 reads as follows:

Rule 11.7 *There is only one arbitrage-free price process S_2 as in Rule 11.6 if and only if there exists some self-financing strategy/consumption pair (φ, C) such that*

1. *φ trades only in S_0, S_1,*
2. *$V_{\varphi,C} \geq H$,*
3. *$C(0) = 0$,*
4. *C is increasing,*
5. *for any $t \leq T$ we have $V_{\varphi,C}(t) + (1_{]\!]t,T]\!]}\varphi) \bullet (S_0, S_1)(\tau) = H(\tau)$ for some stopping time $\tau \geq t$.*

In this case we have $V_{\varphi,C} = S_2$. We call φ a ***replicating strategy*** *or* ***perfect hedge*** *for H.*

Idea $\Leftarrow$: Let Q be an EMM. Since $\widehat{V}_{\varphi,C} = \widehat{V}_{\varphi,C}(0) + \varphi \bullet \widehat{S}_1 - \widehat{C}$ is a Q-supermartingale that satisfies the conditions in Theorem 7.10(2), it is the Q-Snell envelope of $\widehat{H}$.

$\Rightarrow$: Since it is the Q-Snell envelope, $\widehat{S}_2$ is a Q-supermartingale for any EMM Q. By Theorem 3.64 there exists a predictable process φ_1 such that $\widehat{C} := \widehat{S}_2(0)+\varphi_1 \bullet \widehat{S}_1 - \widehat{S}_2$ is increasing. Set $C = (S_0)_- \bullet \widehat{C} + [S_0, \widehat{C}]$, which implies that C and $\widehat{C}$ are related as in (9.16). Choose φ_0 such that $\varphi = (\varphi_0, \varphi_1)$ is C-financing with initial value $S_2(0)$, cf. Rule 9.12. Since

$$\widehat{V}_{\varphi,C} = \widehat{S}_2(0) + \varphi_1 \bullet \widehat{S}_1 - \widehat{C} = \widehat{S}_2$$

is the Q-Snell envelope of $\widehat{H}$, the conditions 2 and 5 above are satisfied by Theorem 7.10(2). □

In the case of positive interest rates, the American call reduces to the corresponding European call because the price process of the latter never falls below the exercise price process:

Remark 11.8 (American vs. European Call) Consider a market $S = (S_0, S_1, S_2, S_3)$ where S_2, S_3 denote the price processes of an American and a European call option on S_1 with strike price K and maturity T, i.e. $H(t) = (S_1(t) - K)^+$ denotes the exercise process of S_2 and $H(T) = (S_1(T) - K)^+$ is the terminal payoff of S_3. Suppose that the numeraire S_0 is increasing and the market does not allow for Γ-arbitrage with

$$\Gamma := \{\varphi \text{ self-financing strategy} : \varphi_2(t) \geq 0 \text{ if } S_2(t-) = H(t-)\}.$$

Then $S_2 = S_3$, i.e. the price processes of the American and the European call coincide.

Indeed, by the counterpart of Rule 11.6 for the present setup of four traded assets there is some $Q \sim P$ such that $\widehat{S}_3$ is a Q-martingale and $\widehat{S}_2$ is the Q-Snell envelope of $\widehat{H}$. Since $f(x) = (x-K)^+$ is a convex function, Jensen's inequality for conditional expectations yields

$$\begin{aligned}
\widehat{S}_3(t) &= E_Q\big(f(S_1(T))/S_0(T)\big|\mathscr{F}_t\big)\\
&= E_Q\big(f(S_1(T))S_0(t)/S_0(T)\big|\mathscr{F}_t\big)/S_0(t)\\
&\geq E_Q\big(f(S_0(t)\widehat{S}_1(T))\big|\mathscr{F}_t\big)/S_0(t)\\
&\geq f\left(S_0(t)E_Q(\widehat{S}_1(T)|\mathscr{F}_t)\right)/S_0(t)\\
&= f(S_1(t))/S_0(t)\\
&= \widehat{H}(t).
\end{aligned}$$

Together with the Q-martingale property of $\widehat{S}_3$ we have

$$\begin{aligned}
\widehat{S}_2(t) &= \max_{\tau\in\mathscr{T}_{[t,T]}} E_Q(\widehat{H}(\tau)|\mathscr{F}_t)\\
&\leq \max_{\tau\in\mathscr{T}_{[t,T]}} E_Q(\widehat{S}_3(\tau)|\mathscr{F}_t)\\
&= \widehat{S}_3(t).
\end{aligned}$$

In view of

$$\widehat{S}_2(t) = \max_{\tau\in\mathscr{T}_{[t,T]}} E_Q(\widehat{H}(\tau)|\mathscr{F}_t) \geq E_Q(\widehat{H}(T)|\mathscr{F}_t) = \widehat{S}_3(t),$$

the assertion follows.

Futures contracts involve intermediate payments. This suggests to view them as securities with dividends. Since the incremental payments coincide with the increments of the futures price process F, the dividend process in the sense of Sect. 9.1.3 coincides with the futures price process. More precisely, we set $D_2 = F - F(0)$ because D is supposed to start at 0. The price of the futures contract itself is zero because entering and terminating the contract does not involve any payments. Consequently, we can view a futures as a dividend-paying asset with price process $S_2 = 0$ and dividend process $D_2 = F - F(0)$. The dividend version of the FTAP now yields the following result.

Rule 11.9 (Futures) *Suppose that the market (S_0, S_1) does not allow for arbitrage. Moreover, we assume that the numeraire S_0 is predictable. A futures price process F with $F(T) = S_1(T)$ leads to an arbitrage-free market if and only if there exists an equivalent martingale measure Q for (S_0, S_1) such that*

$$F(t) = E_Q(S_1(T)|\mathscr{F}_t), \quad t\in[0,T]. \tag{11.4}$$

Idea $\Rightarrow$: Under absence of arbitrage, the FTAP with dividends Rule 9.19 yields the existence of a probability measure $Q \sim P$ such that $\widehat{S}_1$ and

$$\widehat{S}_2 + \widehat{D}_2 = 0 + \frac{1}{S_0} \bullet D_2 = \frac{1}{S_0} \bullet F$$

are Q-martingales. Consequently, $F - F(0) = S_0 \bullet (\frac{1}{S_0} \bullet F)$ is a Q-martingale as well, which implies (11.4).

$\Leftarrow$: (11.4) means that F and hence also $\frac{1}{S_0} \bullet F = \widehat{S}_2 + \widehat{D}_2$ are Q-martingales. By Rule 9.19 we obtain absence of arbitrage for the market consisting of S_0, S_1 and the futures contract. □

Observe that neither S_1 nor F are discounted in (11.4).

If interest rates are deterministic, the price of a futures on S_1 is uniquely determined by absence of arbitrage.

Example 11.10 (Deterministic Interest Rate) Suppose that S_0 is deterministic. Then the futures price of a contract on S_1 with maturity T is given by

$$F(t) = \frac{S_1(t)}{S_0(t)} S_0(T). \tag{11.5}$$

The cashflow of the futures can be replicated by a self-financing strategy holding

$$\varphi_1(t) = \frac{S_0(T)}{S_0(t)} \tag{11.6}$$

shares of S_1. Indeed, $E_Q(S_1(T)|\mathscr{F}_t) = E_Q(\widehat{S}_1(T)|\mathscr{F}_t)S_0(T) = \widehat{S}_1(t)S_0(T)$ yields (11.5). Moreover, (11.6) follows from

$$\varphi_1 \bullet \widehat{S}_1 = \frac{\varphi_1}{S_0(T)} \bullet F = \frac{1}{S_0} \bullet F = \widehat{S}_2 + \widehat{D}_2.$$

As in Rules 11.2 and 11.7 unique arbitrage-free prices correspond to the existence of perfect hedging strategies.

Rule 11.11 *There is only one arbitrage-free futures price process F as in Rule 11.9 if and only if there exists a self-financing strategy/consumption pair (φ, C) such that*

1. *φ trades only in (S_0, S_1),*
2. *$V_{\varphi,C} = 0$,*
3. *$C(T) = S_1(T)$.*

*In this case we have $F = C$ for the futures price process. Moreover, we call the futures **replicable** and φ its **replicating strategy** or **perfect hedge**.*

Idea $\Leftarrow$: By Rule 9.11 we have $\widehat{V}_{\varphi,C}(0) + \varphi_1 \bullet \widehat{S}_1 - \widehat{C} = \widehat{V}_{\varphi,C} = 0$, which implies that $\widehat{C} = \widehat{V}_{\varphi,C}(0) + \varphi_1 \bullet \widehat{S}_1$ is a Q-martingale for any EMM Q for $\widehat{S}_1$. Since stochastic integration preserves the martingale property, we conclude that

$$C - C(0) = \frac{S_0}{S_0} \bullet C = S_0 \bullet \widehat{C} = (S_0\varphi_1) \bullet \widehat{S}_1$$

is a Q-martingale as well. Hence $F(t) = E_Q(S_1(T)|\mathscr{F}_t) = C(t)$ as claimed.

$\Rightarrow$: By Rule 11.2 there is a self-financing strategy ψ in the market (S_0, S_1) which replicates the discounted terminal payoff $S_1(T)$, i.e. $\widehat{V}_\psi(0) + \psi_1 \bullet \widehat{S}_1(T) = S_1(T)$. Taking conditional expectations relative to some EMM Q, we conclude $\widehat{V}_\psi(0) + \psi_1 \bullet \widehat{S}_1 = F$ and hence

$$\frac{\psi_1}{S_0} \bullet \widehat{S}_1 = \frac{1}{S_0} \bullet F =: \widehat{F}.$$

Denote by φ the F-financing trading strategy in the market (S_0, S_1) that has initial value 0 and risky part $\varphi_1 = \psi_1/S_0$, cf. Rule 9.12. Then

$$\widehat{V}_{\varphi,F} = \widehat{V}_{\varphi,F}(0) + \varphi_1 \bullet \widehat{S}_1 - \frac{1}{S_0} \bullet F = \frac{\psi_1}{S_0} \bullet \widehat{S}_1 - \widehat{F} = 0,$$

which implies $V_{\varphi,F} = 0$, as desired. □

Observe that the proof also reveals how to obtain the perfect hedge in terms of the replicating strategy of the option with discounted payoff $S_1(T)$.

11.2.3 Modelling the Market's Risk Neutral Measure

As can be seen from Example 11.16, arbitrage theory yields only limited information in many models. We discuss here a way out that enjoys popularity in practice and which is based on methods from statistics. The idea is as follows: if the theory does not tell us much about the true pricing measure Q, we ask the market, i.e. we make an inference on Q by observing option prices in the real world.

We proceed as follows. Suppose that S_1 is a stock and $S_2, \dots, S_d$ are liquidly traded derivatives on the stock. The FTAP tells us that there exists some probability measure Q such that all discounted assets—the underlying S_1 as well as the derivatives $S_2, \dots, S_d$—are Q-martingales. Unfortunately, we cannot make inference on Q with statistical methods because prices move according to the objective probability measure P.

But as in statistics we can start by postulating a particular parametric or non-parametric family of models. More precisely, let us *assume* an explicit parametric expression for the dynamics of S_1 under Q. Note that the parameter vector ϑ_Q must be chosen such that $\widehat{S}_1$ is a Q-martingale. By contrast, the dynamics of the options $S_2, \dots, S_d$ need *not* be specified explicitly. Since we want Q to be an EMM for $S_1, S_2, \dots, S_d$, the evolution of $S_2, \dots, S_d$ and in particular the initial prices $S_2(0), \dots, S_d(0)$ follow as in (11.1) and (11.2). The parameter vector ϑ_Q must be chosen such that the theoretical prices computed as in (11.2) match the observed market prices. This corresponds to a moment estimator for ϑ_Q. The "estimation" of ϑ_Q by equating theoretical and observed option prices is commonly called *calibration*. If the dimension of ϑ_Q is small compared to the number of observed prices, some approximation such as a least-squares fit will be necessary. In the non-parametric case, on the other hand, one may wish to rely on methods from non-parametric statistics in order to avoid very non-smooth or irregular solutions.

Strictly speaking, we should distinguish two situations. If one only wants to obtain information on the functional dependence of option prices on the underlying, it suffices to consider the martingale measure Q. There is basically no need to model the objective measure P as well. We call this approach **martingale modelling** because the martingale measure Q is chosen right from the start and not within the equivalence class of some given real-world measure P. If, on the other hand, one also wants to make statements on "real" probabilities, quantiles and expectations, one must model the underlying under both P and Q. Typically, one chooses the same parametric class for the dynamics of S_1 under both measures, e.g. one of those discussed in Chap. 8. The parameter set ϑ_P is obtained by statistical inference from past data (*estimation part*) whereas the corresponding parameters ϑ_Q are determined from option prices as explained above (*calibration part*).

Note that some parameters must coincide under P and Q in order to warrant the equivalence $Q \sim P$ stated in the FTAP. Corresponding conditions are discussed in Theorems 4.20 and Sect. 6.4 for Lévy and affine processes, respectively. Hence statistical inference does in fact yield at least partial information on Q-parameters. In concrete situations one may proceed in two different orders. One could, for example, start by estimating *all* P-parameters using statistical means. In a second step the *remaining* Q-parameters are determined by calibration. Alternatively, *all* Q-parameters may be fitted by calibration, which means that estimation concerns only those P-parameters that are not yet determined by the Q-model. The results of these two approaches will generally not be the same.

Can we get any evidence as to whether our postulated class of models for Q is appropriate to describe the market's "true" pricing measure? At first sight the answer seems to be no because statistical methods yield information on real-world probabilities but not on the theoretical object Q. However, some evidence may be obtained again from option prices. If no choice of the parameter vector ϑ_Q yields theoretical option prices that are reasonably close to observed market prices, the chosen class is obviously inappropriate. Moreover, if the right-hand side of (11.1) deviates more and more from observed option prices as time moves on, this indicates that Q does not describe the market appropriately. In practice, this problem is typically met by *recalibration* of parameters, i.e. by changing ϑ_Q through time. However, the need to recalibrate means that the original model is inconsistent with market data. Indeed, the parameters are not supposed to change over time in our probabilistic framework.

How can the model be applied to options that are not yet traded? This is obvious from a mathematical point of view: if we assume Q to be the pricing measure of the FTAP for the whole market—including the new claim that is yet to be priced—we can determine its initial value as Q-expectation of the discounted payoff as in (11.2).

However, in practice we should be aware of two problems. Application of the calibrated model to new payoffs means extrapolation and is consequently to be taken with care. References [59, 264] compare models that produce vastly different prices for exotic options even though they are calibrated successfully to the same large set of plain vanilla options in the first place.

Secondly, it is not clear what these extrapolated prices mean to the single investor. The fact that they are consistent with absence of arbitrage does not imply that the new options can be hedged well or involve only small risk. A more individual view on option pricing is discussed in the following section and in Chaps. 12 and 13.

11.3 Over-the-Counter Traded Derivatives

In this section we consider derivatives that are contracted between two counterparties and may not be exchange traded. In particular, it is not obvious whether and at what price the contract can be sold between inception and maturity. Hence we cannot base our theory on a derivative price *process* because this term usually refers to a market price at which the claim can be bought and sold at any time.

We work with the same mathematical setup as in Sect. 11.1, i.e. we assume price processes $S_0, \dots, S_d$ to be given, along with some random variable H which represents the random payoff at time T of a European option. As before, $\widehat{H} := H/S_0(T)$ denotes the corresponding discounted payoff.

We consider an asymmetric situation where the potential buyer is interested in the claim for unknown reasons. The seller (hereafter called *you*) acts as a pure service provider who is not interested in the derivative on its own account. You are facing two basic questions which are considered in the sequel.

1. What price do you need to charge from the potential buyer?
2. How can you hedge against the risk of losses that are involved in the unknown random payment which is due at maturity?

Generally, the answers to these questions depend on many factors, in particular on your attitude towards risk. Your minimal acceptable price cannot be determined unless you make up your mind in this respect. We will come back to this issue in the subsequent chapters. At this point, we only determine reasonable upper and lower bounds which are based on general considerations.

To this end we call

$$\begin{aligned}\pi_{\text{high}}(H) := \inf\big\{\pi \in \mathbb{R} : &\text{ There is a self-financing strategy } \varphi \\ &\text{satisfying } V_\varphi(0) = \pi \text{ and } V_\varphi(T) \geq H\big\}\end{aligned} \tag{11.7}$$

the **upper price** and

$$\pi_{\text{low}}(H) := \sup \{\pi \in \mathbb{R} : \text{There is a self-financing strategy } \varphi \\ \text{satisfying } V_\varphi(0) = \pi \text{ and } V_\varphi(T) \leq H\} \tag{11.8}$$

the **lower price** of any contingent claim H. Before justifying why these bounds make economic sense, we characterise them in terms of equivalent martingale measures.

Rule 11.12 (Superreplication Theorem) *Let H denote a contingent claim.*

1. *For the upper price we have*

$$\pi_{\text{high}}(H) = \min \{\pi \in \mathbb{R} : \textit{There exists some self-financing strategy } \varphi \\ \textit{with } V_\varphi(0) = \pi \textit{ and } V_\varphi(T) \geq H\} \tag{11.9}$$

$$= \sup \{S_0(0)E_Q(\widehat{H}) : Q \textit{ EMM}\}. \tag{11.10}$$

2. *Accordingly, the lower price satisfies*

$$\pi_{\text{low}}(H) = \max \{\pi \in \mathbb{R} : \textit{There exists some self-financing strategy } \varphi \\ \textit{with } V_\varphi(0) = \pi \textit{ and } V_\varphi(T) \leq H\} \\ = \inf \{S_0(0)E_Q(\widehat{H}) : Q \textit{ EMM}\}.$$

3. *If H is not attainable, then*

$$\{S_0(0)E_Q(\widehat{H}) : Q \textit{ EMM}\} \tag{11.11}$$

is an open interval (and a singleton otherwise).

Idea

1. *Step 1:* For any self-financing φ with $V_\varphi(T) \geq H$ and any EMM Q we have

$$V_\varphi(0) = S_0(0)\widehat{V}_\varphi(0) = S_0(0)E_Q(\widehat{V}_\varphi(T)) \geq S_0(0)E_Q(\widehat{H}).$$

This implies $\pi_{\text{high}}(H) \geq S_0(0)E_Q(\widehat{H})$ for any EMM Q and hence $\pi_{\text{high}}(H) \geq \sup\{S_0(0)E_Q(\widehat{H}) : Q \text{ EMM}\}$.

Step 2: Set $\pi := \sup\{S_0(0)E_Q(\widehat{H}) : Q \text{ EMM}\}$ and

$$\widehat{S}_{d+1}(t) := \operatorname{ess\,sup} \{E_Q(\widehat{H}|\mathscr{F}_t) : Q \text{ EMM}\}.$$

By Proposition 3.65 the process $\widehat{S}_{d+1}$ is a Q-supermartingale for any EMM Q. Theorem 3.64 on optional decomposition yields the existence of a predictable process φ such that $C := \widehat{S}_{d+1}(0) + \varphi \bullet \widehat{S} - \widehat{S}_{d+1}$ is increasing. We obtain

$$\widehat{H} = \widehat{S}_{d+1}(T) \leq \widehat{S}_{d+1}(0) + \varphi \bullet \widehat{S}(T) = \widehat{V}_\varphi(T)$$

if we identify φ with a self-financing strategy with discounted initial value $\widehat{S}_{d+1}(0) = \pi$, cf. Rule 9.6. This implies $\pi \geq \pi_{\text{high}}(H)$ and hence $\pi = \pi_{\text{high}}(H)$ by step 1. Moreover, it shows that the infimum in the definition of $\pi_{\text{high}}(H)$ is actually attained, namely by the strategy φ above.

2. This follows from statement 1 applied to $-H$ instead of H.
3. If H is attainable, the set (11.11) is a singleton by Rule 11.2. Convexity of the set of EMMs implies that the set (11.11) is an interval. Suppose that $\pi_{\text{high}}(H) = S_0(0)E_Q(\widehat{H})$ for some EMM Q. For φ as in (11.9) there is some random variable $C \geq 0$ such that

$$\widehat{H} = \widehat{V}_\varphi(T) - C = \pi_{\text{high}}(H)/S_0(0) + \varphi \bullet \widehat{S}(T) - C$$

and hence $E_Q(\widehat{H}) = E_Q(\widehat{H}) - E_Q(C)$. Consequently $C = 0$ almost surely, which implies that H is attainable. Along the same lines it follows that the infimum of the set (11.11) is only attained by some EMM Q if H is attainable. □

Corollary 11.13 $\pi_{\text{high}}(H) = \pi_{\text{low}}(H)$ *holds if and only if H is replicable with this initial endowment. In this case we call* $\pi(H) := \pi_{\text{high}}(H) = \pi_{\text{low}}(H)$ *the* ***unique fair price*** *of the claim.*

Idea This follows immediately from Rule 11.12(3). □

In what sense do $\pi_{\text{low}}(H)$, $\pi_{\text{high}}(H)$ represent rational bounds for the price that you can or will actually charge for the contingent claim? If you receive any premium $\pi \geq \pi_{\text{high}}(H)$ for the option, you can buy a self-financing portfolio whose terminal value $V_\varphi(T) \geq H$ suffices to meet your obligations at maturity. You do not face the risk of losing any money. So except for administrative costs, there is no reason not to accept if someone bids $\pi \geq \pi_{\text{high}}(H)$ for the option.

On the other hand, there is no reason to be satisfied with less than the lower price. Instead of selling the option at a premium $\pi < \pi_{\text{low}}(H)$, you should rather go short in a trading strategy with initial price $\pi_{\text{low}}(H)$ and terminal value $V_\varphi(T) \leq H$ (i.e. you must pay at most $V_\varphi(T) \leq H$ at time T). This deal on the stock market yields a higher profit than the suggested option trade with premium $\pi < \pi_{\text{low}}(H)$, which means that you should not accept the latter. Altogether it follows that any reasonably charged or offered price for the option should lie in the interval $[\pi_{\text{low}}(H), \pi_{\text{high}}(H)]$.

In Rule 11.12 it is stated that the infimum in the definition of the upper price is actually attained. Put differently, one needs exactly the upper price in order to afford such a *superhedge*, which provides perfect protection against the risk of losses from selling the claim. Therefore, a self-financing strategy φ with $V_\varphi(0) = \pi_{\text{high}}(H)$ and $V_\varphi(T) \geq H$ is called the **cheapest superhedge** or **cheapest superreplicating strategy** for a contingent claim H.

If the interval (11.11) reduces to a singleton, the option can be perfectly hedged by Corollary 11.13. In this case, the questions from the beginning of this section have a clear answer. You will charge the unique fair price—plus administrative costs—and you are perfectly protected against losses by investing in the replicating portfolio. In the general case, you could be tempted to request the upper price and invest this premium in the cheapest superhedge. However, Examples 11.16 and 11.17 below illustrate that the upper price is often far too high to be obtained in real markets. Therefore it seems misleading to call the difference $\pi_{\text{high}}(H) - \pi_{\text{low}}(H)$ the bid-ask spread as is sometimes done in textbooks. It rather corresponds to an extreme upper bound for the real spread.

Forward contracts allow for a unique fair price under very weak conditions:

Example 11.14 (Forward) Suppose that the market (S_0, S_1) does not allow for arbitrage. Moreover, we assume that the terminal value $S_0(T)$ of the numeraire is deterministic. We consider a forward on S_1 with maturity T. Its only reasonable forward price at time 0 is

$$F(0) := \widehat{S}_1(0)S_0(T) = S_1(0)\frac{S_0(T)}{S_0(0)}.$$

Indeed, the payoff $S_1(T) - F(0)$ can be replicated by the constant and hence self-financing strategy $\varphi := (-F(0)/S_0(T), 1)$ whose initial value equals

$$-\frac{F(0)}{S_0(T)}S_0(0) + S_1(0) = 0$$

as desired.

Hence a forward contract can be hedged perfectly whenever there exists a bond with deterministic payoff at time T. No assumptions must be made about the dynamics of the underlying security S_1. Moreover, the hedge is static in the sense that it does not require frequent rebalancing. In practice, however, forward contracts may involve a risk that does not appear in our mathematical model, namely the *counterparty risk* that the other party does not meet its obligations.

Observe that the forward price coincides with the futures price of Example 11.10. But interestingly, their replicating strategies differ unless the numeraire asset is constant.

The size of the interval (11.11) obviously depends on the model. However, there are some universal bounds which hold regardless of the underlying price process.

Example 11.15 (Call and Put Options) Suppose that $\widehat{H} = f(\widehat{S}_1(T))$ for $f(x) = (x - K)^+$ with some strike $K > 0$ or, more generally, for some convex function $f : \mathbb{R} \to \mathbb{R}_+$ with $f(x) < x$, $x \in \mathbb{R}$. Then

$$E_Q(\widehat{H}) \in [f(\widehat{S}_1(0)), \widehat{S}_1(0)) \tag{11.12}$$

for any EMM Q. To wit, for any such Q we have

$$E_Q(\widehat{H}) = E_Q(f(\widehat{S}_1(T))) \geq f(E_Q(\widehat{S}_1(T))) = f(\widehat{S}_1(0))$$

by Jensen's inequality, whereas the upper bound follows from

$$E_Q(\widehat{H}) = E_Q(f(\widehat{S}_1(T))) < E_Q(\widehat{S}_1(T)) = \widehat{S}_1(0).$$

We call (11.12) the *trivial no-arbitrage interval* because it does not depend on the specific model. Note that it is typically very large. Indeed, for example, for an out-of-the-money call (i.e. with $\widehat{S}(0) < K$) it only tells you that the option price could be anything between 0 and the present stock price.

Call and put options are related to each other by the **put-call parity**

$$(x - K)^+ - (K - x)^+ = x - K. \tag{11.13}$$

If $\widehat{J} = (K - \widehat{S}_1(T))^+$ represents a put option and $\widehat{H} = (\widehat{S}_1(T) - K)^+$ the corresponding call, equation (11.13) for the payoffs yields

$$E_Q(\widehat{H}) - E_Q(\widehat{J}) = E_Q(\widehat{S}_1(T)) - K = \widehat{S}_1(0) - K$$

after taking expectations under any EMM Q. Equation (11.12) now yields the trivial no-arbitrage interval

$$E_Q(\widehat{J}) \in [(K - \widehat{S}_1(0))^+, K)$$

for the put.

Perhaps surprisingly, the model-specific interval (11.11) essentially coincides with the trivial no-arbitrage interval for many price processes of practical relevance.

Example 11.16 (Geometric Lévy Model) We consider the geometric Lévy model of Example 9.2 with a Lévy process X such that the Lévy measure K^X has a strictly positive density in a neighbourhood of 0 and $c^X = 0$, i.e. X has no Brownian motion part. This covers, for example, the VG, NIG and hyperbolic Lévy processes of Sects. 2.4. In this case the interval of possible prices (11.11) for call and put options coincides with the trivial no-arbitrage interval up to its left limit. For ease of notation we suppose that $r = 0$, i.e. $S_1 = S_1(0)\exp(X)$ here represents the discounted asset price.

Indeed, by put-call parity it suffices to consider put options with arbitrary strike $K > 0$. First note that we may assume $\int_{\{|x|>1\}} e^{\pm x} K^X(dx) < \infty$ and $\psi(-i) = 0$ by Example 9.16, where ψ denotes the characteristic exponent of X. Since

$$\psi(u) = iua^X + \int (e^{iux} - 1 - iux)K^X(dx)$$

and $\psi(-i) = a^X + \int (e^x - 1 - x) K^X(dx)$, this means that

$$\psi(u) = \int \big(e^{iux} - 1 - iu(e^x - 1)\big) K^X(dx), \quad u \in \mathbb{R}.$$

Step 1: Suppose we find EMMs $Q_n \sim P$ such that X is a Lévy process under Q_n with characteristic exponent ψ_n satisfying $\lim_{n\to\infty} \psi_n(u) = 0, u \in \mathbb{R}$. By Lévy's continuity theorem A.5(8) this means that the law $Q_n^{X(T)}$ converges weakly to the Dirac measure ε_0 as $n \to \infty$. Since $f(x) = (K - \widehat{S}_1(0)e^x)^+$ is a bounded continuous function, we have that

$$E_{Q_n}\big((K - \widehat{S}_1(T))^+\big) = E_{Q_n}\big(f(X(T))\big) \to (K - \widehat{S}_1(0))^+$$

as $n \to \infty$. This yields the statement on the lower bound.

Step 2: Similarly, suppose we can find EMMs Q_n such that X is a Lévy process under Q_n whose characteristic exponent ψ_n satisfies $\lim_{n\to\infty} \psi_n(u/n) = iua$, $u \in \mathbb{R}$ for some $a < 0$. Put differently, $Q_n^{X(T)/n}$ converges weakly to the Dirac measure in aT as $n \to \infty$. This implies $X(T)/n \to aT$ in Q_n-measure and hence

$$E_{Q_n}\big((K - \widehat{S}_1(T))^+\big) = E_{Q_n}\Big(\big(K - \widehat{S}_1(0)\exp(X(T)/n)^n\big)^+\Big) \to E_{Q_n}(K) = K$$

by dominated convergence. So we obtain the statement for the upper bound.

Step 3: Suppose first that $\int |x| K^X(dx) = \infty$. We assume $\int x^+ K^X(dx) < \infty$, the case $\int x^- K^X(dx) = \infty$ being treated along the same lines. Fix $n \in \mathbb{N}$. Since K^X is continuous and $\int_{[0,1]}(e^x - 1)K^X(dx) = \infty$, we can find $\alpha_n \in (0, \frac{1}{n}]$ with

$$\left| \int_{\mathbb{R}\setminus[-\frac{1}{n},\frac{1}{n}]} (e^x - 1) K^X(dx) \right| = \frac{1}{2} \int_{[\alpha_n, \frac{1}{n}]} (e^x - 1) K^X(dx). \tag{11.14}$$

Let $Q_n \sim P$ be chosen as in Theorem 4.20 with

$$\kappa_n(x) := \begin{cases} 1 & \text{for } x \in [-\frac{1}{n}, \alpha_n), \\ 1 \mp (\frac{1}{n} - 1)/2 & \text{for } x \in [\alpha_n, \frac{1}{n}], \\ \frac{1}{n} & \text{for } x \in \mathbb{R} \setminus [-\frac{1}{n}, \frac{1}{n}]. \end{cases}$$

The sign $\mp$ depends on whether the integral on the left of (11.14) is positive or negative. In view of Theorem 4.20, the characteristic exponent ψ_n of X under Q_n is of the form

$$\begin{aligned} \psi_n(u) &= iu \left(a^X + \int x(\kappa_n(x) - 1) K^X(dx) \right) \\ &\quad + \int \big(e^{iux} - 1 - iux\big) \kappa_n(x) K^X(dx) \end{aligned}$$

$$= \int \Big((e^{iux} - 1)\kappa_n(x) - iu(e^x - 1) \Big) K^X(dx)$$

$$= \int \big(e^{iux} - 1 - iu(e^x - 1) \big) \kappa_n(x) K^X(dx),$$

where the last equality follows from the choice of α_n. Dominated convergence yields that $\psi_n(u) \to 0$ as $n \to \infty$. Moreover, $\psi_n(-i) = 0$. By Theorem 2.22 the process $\widehat{S}_1 = \widehat{S}_1(0)e^X$ is a Q_n-martingale and we are in the situation of step 1.

Step 4: Suppose now that $\int |x| K^X(dx) < \infty$. In this case we set

$$\kappa_n(x) := \begin{cases} 1 & \text{for } |x| \leq 1/n, \\ 1/n & \text{for } |x| \geq 2/n, \\ \alpha_n & \text{for } x \in (1/n, 2/n), \\ 1/\alpha_n & \text{for } x \in (-2/n, -1/n), \end{cases}$$

where α_n is chosen such that

$$a^X + \int \big((e^x - 1)\kappa_n(x) - x \big) K^X(dx) = 0. \tag{11.15}$$

Since $e^x - 1 > 0$ for $x > 0$ and $e^x - 1 < 0$ for $x < 0$, such an α_n can always be found. Let $Q_n \sim P$ be the probability measure corresponding to κ_n in the sense of Theorem 4.20. Relative to Q_n the process is still a Lévy process, but with characteristic exponent

$$\psi_n(u) = iua^X + \int \big((e^{iux} - 1)\kappa_n(x) - iux \big) K^X(dx).$$

Since $\psi_n(-i) = 0$, we have that e^X is a martingale and hence Q is an EMM. Suppose that $\alpha_n \geq 1$, the case $\alpha_n < 1$ being treated along the same lines. Equation (11.15) and $|\kappa_n(x)| \leq 1$ for $x \notin (1/n, 2/n)$ imply that

$$\alpha_n \int_{(\frac{1}{n}, \frac{2}{n})} (e^x - 1) K^X(dx)$$
$$= -a^X - \int_{\mathbb{R} \setminus (\frac{1}{n}, \frac{2}{n})} \big((e^x - 1)\kappa_n(x) - x \big) K^X(dx) + \int_{(\frac{1}{n}, \frac{2}{n})} x K^X(dx)$$

is bounded by some constant $c_1 < \infty$ which does not depend on n. Fix $u \in \mathbb{R}$. There is some $c_2 < \infty$ such that $|e^{iux} - 1 - iux| \leq c_2 x^2$ and $|e^x - 1 - x| \leq c_2 x^2$

holds for all $x \in [-1, 1]$. Since $x \leq e^x - 1$ for $x \geq 0$ and hence

$$\alpha_n \int_{(\frac{1}{n},\frac{2}{n})} x^2 K^X(dx) \leq \alpha_n \int_{(\frac{1}{n},\frac{2}{n})} x(e^x - 1)K^X(dx)$$
$$\leq 2\alpha_n \int_{(\frac{1}{n},\frac{2}{n})} (e^x - 1)K^X(dx) \leq 2c_1,$$

dominated convergence yields that $\psi_n(u) = \int (e^{iux} - 1 - iu(e^x - 1))\kappa_n(x)K^X(dx)$ converges to 0 as $n \to \infty$. Consequently we are again in the situation of step 1.

Step 5: Suppose that $K^X((-\infty, -1)) > 0$ and $K^X((1, \infty)) > 0$. Otherwise replace 1 below by some smaller threshold. Set

$$\alpha := \frac{\int_{\{x>1\}} (e^x - 1)K^X(dx)}{-\int_{\{x<-1\}} (e^x - 1)K^X(dx)},$$

which implies that

$$\alpha \int_{\{x<-1\}} (e^x - 1)K^X(dx) + \int_{\{x>1\}} (e^x - 1)K^X(dx) = 0.$$

Let $Q_n \sim P$ be defined as in Theorem 4.20 with

$$\kappa_n(x) := \begin{cases} 1 & \text{for } |x| \leq 1, \\ 1 + n & \text{for } x > 1, \\ 1 + \alpha n & \text{for } x < -1. \end{cases}$$

As above, X is still a Lévy process under Q_n whose characteristic exponent ψ_n is of the form

$$\psi_n(u) = iua^X + \int \big((e^{iux} - 1)\kappa_n(x) - iux\big)K^X(dx) = \psi(u) + n\widetilde{\psi}(u)$$

with

$$\widetilde{\psi}(u) := \alpha \int_{\{x<-1\}} (e^{iux} - 1)K^X(dx) + \int_{\{x>1\}} (e^{iux} - 1)K^X(dx).$$

Since $\psi_n(-i) = \psi(-i) + n\widetilde{\psi}(-i) = 0$, we conclude that e^X is a Q_n-martingale and hence Q_n is an EMM.

Fix $u \in \mathbb{R}$. Since $\psi(0) = 0$, we have $\psi(u/n) \to 0$ as $n \to \infty$. Moreover

$$n\widetilde{\psi}(u/n) = u\frac{\widetilde{\psi}(u/n) - \widetilde{\psi}(0)}{u/n} \to u\widetilde{\psi}'(0) = ui\widetilde{a}^X$$

as $n \to \infty$, where $(\widetilde{a}^X, 0, \widetilde{K}^X)$ is the Lévy–Khintchine triplet corresponding to exponent $\widetilde{\psi}$. Since

$$0 = \widetilde{\psi}(-i) = \widetilde{a}^X + \int (e^x - 1 - x)\widetilde{K}^X(dx) > \widetilde{a}^X,$$

we have $\widetilde{a}^X < 0$. Together, it follows that

$$\psi_n(u/n) = \psi(u/n) + n\widetilde{\psi}(u/n) \to iu\widetilde{a}^X$$

as $n \to \infty$, which means that we are in the situation of step 2.

Example 11.17 (Heston Model) We consider the Heston model or more generally any Itō process model as in Example 9.3 where

$$d\widehat{S}_1(t) = \widehat{S}_1(t)\left(\mu(Y(t))dt + \sqrt{Y(t)}dW_1(t)\right),$$
$$dY(t) = \alpha(Y(t))dt + \sigma(Y(t))dW_2(t). \tag{11.16}$$

Here (W_1, W_2) denotes a two-dimensional Brownian motion with drift vector $a = (0, 0)$ and diffusion matrix

$$c = \begin{pmatrix} 1 & \varrho \\ \varrho & 1 \end{pmatrix}$$

for some $\varrho < 1$. Moreover, $\mu : (0, \infty) \to \mathbb{R}$, $\alpha : (0, \infty) \to \mathbb{R}$, $\sigma : (0, \infty) \to \mathbb{R}$ are supposed to be continuously differentiable coefficients such that the SDE (11.16) has a strictly positive solution. As in the previous example, the interval of possible call and put prices (11.11) coincides with the trivial no-arbitrage interval from Example 11.15 up to its left limit.

Indeed, this is shown in [120, Theorem 3.4]. At this point we give an idea why this is plausible. As in Example 11.16 we focus on put options. Consider a probability measure $Q \sim P$ with density process

$$Z = \mathscr{E}\left(\left(-\frac{\mu(Y)}{\sqrt{Y}} - \varrho\eta\right) \bullet W_1, \eta \bullet W_2\right)$$

for some predictable process η. By Theorem 4.15

$$(W_1^Q, W_2^Q) := \left(W_1 + \frac{\mu(Y)}{\sqrt{Y}} \bullet I, W_2 + \left(\varrho\frac{\mu(Y)}{\sqrt{Y}} - (1 - \varrho^2)\eta\right) \bullet I\right)$$

is a two-dimensional Brownian motion with drift vector $a = (0, 0)$ and diffusion matrix c relative to Q. Since

$$d\widehat{S}_1(t) = \widehat{S}_1(t)\sqrt{Y(t)}dW_1^Q(t),$$
$$dY(t) = \alpha_Q(t)dt + \sigma(Y(t))dW_2^Q(t)$$

with $\alpha_Q = \alpha(Y) - \sigma(Y)\varrho\frac{\mu(Y)}{\sqrt{Y}} + \sigma(Y)(1-\varrho^2)\eta$, the probability measure Q is an EMM. If η is chosen such that $\alpha_Q(t) = -\lambda Y(t)$ with very large λ, this large mean reversion will move the process Y very close to 0. This implies $\widehat{S}_1(T) \approx \widehat{S}_1(0)$ with large Q-probability and hence

$$E_Q\big((K - \widehat{S}_1(T))^+\big) \approx E_Q\big((K - \widehat{S}_1(0))^+\big) = (K - \widehat{S}_1(0))^+,$$

which is the trivial lower bound. If η is chosen such that $\alpha_Q(t) = \lambda Y(t)$ with very large λ, the process Y and in particular $Y \bullet I(T)$ will be very large. Consequently, $\sqrt{Y} \bullet W_1^Q(T)$ will be small compared to $\frac{1}{2}Y \bullet I(T)$ and hence we have $\exp(\sqrt{Y} \bullet W_1^Q(T) - \frac{1}{2}Y \bullet I(T)) \approx 0$ with large Q-probability. Therefore

$$E_Q\big((K - \widehat{S}_1(T))^+\big) = E_Q\left(\left(K - \widehat{S}_1(0)\exp\left(\sqrt{Y} \bullet W_1^Q(T) - \frac{1}{2}Y \bullet I(T)\right)\right)^+\right)$$
$$\approx E_Q\big((K - 0)^+\big) = K,$$

which is the trivial upper bound.

Above we argued that your premium should belong to the interval

$$[\pi_{\text{low}}(H), \pi_{\text{high}}(H)]$$
$$= \left[\inf\left\{S_0(0)E_Q(\widehat{H}) : Q \text{ EMM}\right\}, \sup\left\{S_0(0)E_Q(\widehat{H}) : Q \text{ EMM}\right\}\right].$$

Rule 11.12(3) shows that this interval coincides except for the boundary with the set (11.11). The latter represents the set of initial values of arbitrage-free liquid derivative price processes in the sense of the previous section. Consequently, the two seemingly different situations and approaches lead ultimately to similar pricing formulas.

Finally, let us turn to lower and upper price bounds for American options. If H denotes the exercise process of some American option, we call

$$\pi_{\text{high}}(H) := \inf\big\{\pi \in \mathbb{R} : \text{There is a self-financing strategy } \varphi$$
$$\text{satisfying } V_\varphi(0) = \pi \text{ and } V_\varphi \geq H\big\}$$

the **upper price** and

$$\pi_{\text{low}}(H) := \sup\big\{\pi \in \mathbb{R} : \text{There exists some stopping time } \tau \text{ such that } \pi \text{ is the lower price of the European option with discounted payoff } \widehat{H}(\tau)\big\} \tag{11.17}$$

the **lower price** of the option. The upper price is justified by the same argument as in the European case. In order to motivate the lower bound we fix a stopping time τ and compare the American option with exercise process H to the European claim with discounted payoff $\widehat{H}(\tau)$. From the point of view of the holder, the first option is at least as attractive as the second. Indeed, by choosing τ as exercise time, she can use the American option to generate the same discounted payoff as the European claim. Claiming that more attractive options should not be sold for less, we argue for $\pi_{\text{low}}(H)$ in (11.17) as a reasonable lower price bound.

The counterpart of Rule 11.12 reads as follows.

Rule 11.18 (American Options) *Let H denote the exercise process of some American option. Then*

$$\begin{aligned}\pi_{\text{high}}(H) &= \min\big\{\pi \in \mathbb{R} : \textit{There exists some self-financing strategy } \varphi \\ &\qquad \textit{with } V_\varphi(0) = \pi \textit{ and } V_\varphi \geq H\big\} \\ &= \sup\left\{\sup_{\tau\in\mathscr{T}_{[0,T]}} S_0(0)E_Q(\widehat{H}(\tau)) : Q \textit{ EMM for } S\right\}\end{aligned} \tag{11.18}$$

and accordingly

$$\begin{aligned}\pi_{\text{low}}(H) &= \max\big\{\pi \in \mathbb{R} : \textit{There exists some stopping time } \tau \\ &\qquad \textit{such that } \pi \textit{ is the discounted lower price of the} \\ &\qquad \textit{European option with discounted payoff } \widehat{H}(\tau)\big\} \\ &= \inf\left\{\sup_{\tau\in\mathscr{T}_{[0,T]}} S_0(0)E_Q(\widehat{H}(\tau)) : Q \textit{ EMM for } S\right\},\end{aligned} \tag{11.19}$$

where $\mathscr{T}_{[s,t]}$ denotes the set of stopping times with values in $[s,t]$.

Idea We start with the upper price.

Step 1: For any self-financing φ with $V_\varphi(T) \geq H$, any EMM Q, and any stopping time τ we have

$$V_\varphi(0) = S_0(0)\widehat{V}_\varphi(0) = S_0(0)E_Q\big(\widehat{V}_\varphi(\tau)\big) \geq S_0(0)E_Q\big(\widehat{H}(\tau)\big).$$

This implies that $\pi_{\text{high}}(H)$ dominates (11.18).

Step 2: Set $\pi = \sup\{\sup_{\tau \in \mathscr{T}_{[0,T]}} S_0(0)E_Q(\widehat{H}(\tau)) : Q \text{ EMM for } S\}$ and

$$\widehat{S}_{d+1}(t) = \operatorname{ess\,sup}\left\{\operatorname*{ess\,sup}_{\tau \in \mathscr{T}_{[t,T]}} E_Q(\widehat{H}(\tau)|\mathscr{F}_t) : Q \text{ EMM for } S\right\}.$$

Similarly as in the proof of Proposition 3.65 it follows that $\widehat{S}_{d+1}$ is a Q-supermartingale for any EMM Q. Theorem 3.64 on optional decomposition yields the existence of a predictable process φ such that $C := \widehat{S}_{d+1}(0) + \varphi \bullet \widehat{S} - \widehat{S}_{d+1}$ is increasing. We obtain $\widehat{H}(t) \leq \widehat{S}_{d+1}(t) \leq \widehat{S}_{d+1}(0) + \varphi \bullet S(t) = \widehat{V}_\varphi(t)$ if we identify φ with a self-financing strategy with discounted value $\widehat{S}_{d+1}(0) = \pi$, cf. Rule 9.6. This implies $\pi \geq \pi_{\text{high}}(H)$ and hence $\pi = \pi_{\text{high}}(H)$ by step 1. Moreover, it shows that the infimum in the definition of $\pi_{\text{high}}(H)$ is actually attained, namely by the strategy φ above.

Step 3: We now turn to the lower price. If, for a given EMM Q, the supremum in $\pi := \sup_{\tau \in \mathscr{T}_{[0,T]}} E_Q(\widehat{H}(\tau))$ is attained in $\tau^\star \in \mathscr{T}_{[0,T]}$, we have

$$\pi = E_Q(\widehat{H}(\tau^\star)) \geq \frac{\pi_{\text{low}}(H)}{S_0(0)}.$$

This implies that $\pi_{\text{low}}(H)$ is dominated by (11.19).

Step 4: By Rule 11.12 we can rewrite the lower price as

$$\pi_{\text{low}}(H) = \sup_{\tau \in \mathscr{T}_{[0,T]}} \inf_{Q \text{ EMM}} S_0(0)E_Q(\widehat{H}(\tau)), \tag{11.20}$$

whereas (11.19) equals $\inf_{Q \text{ EMM}} \sup_{\tau \in \mathscr{T}_{[0,T]}} S_0(0)E_Q(\widehat{H}(\tau))$. Note that

$$\widehat{H}(\tau) = \widehat{H}(0) + 1_{[0,\tau]} \bullet \widehat{H}(T) \tag{11.21}$$

for any $\tau \in \mathscr{T}_{[0,T]}$. For any decreasing predictable process φ with $\varphi(0) = 1$ and $\varphi(1) = 0$, one can view $\widehat{H}(0) + \varphi \bullet \widehat{H}(T)$ as an average of random variables as in (11.21). To wit, if φ takes its values in the discrete set $\{0, \frac{1}{n}, \ldots, \frac{n-1}{n}, 1\}$, we have $\varphi = \frac{1}{n}\sum_{k=1}^n 1_{[0,\tau_k]}$ for $\tau_k := \inf\{t \in [0,1] : \varphi(t) < k/n\}$ and hence $\widehat{H}(0) + \varphi \bullet \widehat{H}(T) = \frac{1}{n}\sum_{k=1}^n \widehat{H}(\tau_k)$. For processes φ with infinitely many values one can argue by approximation.

Since $\varphi \mapsto S_0(0)E_Q(\widehat{H}(0) + \varphi \bullet \widehat{H}(T))$ is affine in φ, we conclude that the maximal value on $A := \{\varphi \text{ predictable: } \varphi \text{ decreasing}, \varphi(0) = 1, \varphi(1) = 0\}$ is attained by some φ of the form $\varphi = 1_{[\![0,\tau]\!]}$ with $\tau \in \mathscr{T}_{[0,T]}$. Consequently, we can rewrite (11.20) and (11.19) as

$$\pi_{\text{low}}(H) = \sup_{\varphi \in A} \inf_{Q \text{ EMM}} f(\varphi, Q)$$

and $\inf_{Q \text{ EMM}} \sup_{\varphi \in A} f(\varphi, Q)$ with $f(\varphi, Q) := S_0(0)E_Q(\widehat{H}(\tau))$. Observe that f is affine in both arguments and that both A and the set of equivalent martingale measures are convex. If we can apply Theorem A.20 for appropriate topologies, (A.10) here reads as

$$\max_{\varphi \in A} \inf_{Q \text{ EMM}} f(\varphi, Q) = \inf_{Q \text{ EMM}} \max_{\varphi \in A} f(\varphi, Q).$$

Since the maximum in φ is attained in $1_{[\![0,\tau]\!]}$ for some stopping time $\tau \in \mathscr{T}_{[0,T]}$, we obtain

$$\begin{aligned}\pi_{\text{low}}(H) &= \max_{\tau \in \mathscr{T}_{[0,T]}} \inf_{Q \text{ EMM}} S_0(0)E_Q(\widehat{H}(\tau)) \\ &= \inf_{Q \text{ EMM}} \max_{\tau \in \mathscr{T}_{[0,T]}} S_0(0)E_Q(\widehat{H}(\tau)),\end{aligned}$$

which yields the second assertion. □

As in the European case we observe that the price bounds coincide with those for liquidly traded options, cf. Rule 11.6.

11.4 Hedging Based on Sensitivities

In this section we discuss the computation of perfect hedging strategies in specific multivariate Markov models. To this end, consider a market with d liquidly traded assets $S_1, \dots, S_d$ besides the numeraire S_0. Denote by Q some martingale measure, i.e. a probability measure such that the discounted price processes $\widehat{S}_i(t)$, $i = 1, \dots, d$ are martingales. We assume that these processes $\widehat{S}_i(t)$ are of the form $g_i(t, X(t))$ for some $\mathbb{R}^d$-valued continuous Markov process X and deterministic functions $g_i : \mathbb{R}^{1+d} \to \mathbb{R}$. This holds, for example, if the $\widehat{S}_i$ are European-style options with discounted payoff $\widehat{f}_i(X(T_i))$ at time T_i. Indeed, in this case we have

$$\begin{aligned}\widehat{S}_i(t) &= E_Q\big(\widehat{f}_i(X(T_i))\big|\mathscr{F}_t\big) \\ &= \int \widehat{f}_i(x)P(T_i - t, X(t), dx),\end{aligned}$$

where P here denotes the transition function of the Markov process X under Q, cf. Sect. 5.1. Now consider an option with discounted payoff $\widehat{f}(X(T))$ at time $T \geq 0$. The Markov property of X implies that the martingale $\widehat{V}$ generated by the payoff can be written as

$$\widehat{V}(t) = E_Q(\widehat{f}(X(T)|\mathscr{F}_t) = v(t, X(t)),$$

where

$$v(t,x) := \int \widehat{f}(\xi) P(T-t,x,d\xi)$$

denotes the pricing function of the claim. Itō's formula yields that

$$\begin{aligned} d\widehat{V}(t) &= dv(t,X(t)) \\ &= \sum_{i=1}^{d} D_{1+i} v(t, X_1(t), \dots, X_d(t)) dX_i(t) + B \\ &= \sum_{i=1}^{d} D_{1+i} v(t, X_1(t), \dots, X_d(t)) dM_i^X(t) + \tilde{B} \end{aligned}$$

for some continuous processes of finite variation B and $\tilde{B}$, where M^X denotes the martingale part of X, cf. Theorem 2.9. Since $\widehat{V}$ is a Q-martingale, $\tilde{B}$ vanishes. The same argument applies to the liquid assets, i.e. we have

$$d(\widehat{S}_1, \dots, \widehat{S}_d)(t) = \Sigma(t) d(M_1^X, \dots, M_d^X)(t)$$

with

$$\Sigma_{ji}(t) = D_{1+i} g_j(t, X_1(t), \dots, X_d(t)), \quad i,j = 1, \dots, d.$$

We assume that the matrix $\Sigma(t)$ is regular with inverse $\Sigma^{-1}(t)$ for any t, which yields

$$d(M_1^X, \dots, M_d^X)(t) = \Sigma^{-1}(t) d(\widehat{S}_1, \dots, \widehat{S}_d)(t)$$

and hence

$$d\widehat{V}(t) = \sum_{i,j=1}^{d} D_{1+i} v(t, X_1(t), \dots, X_d(t)) \Sigma_{ij}^{-1}(t) d\widehat{S}_j(t). \tag{11.22}$$

In other words, the option can be perfectly hedged by trading in $S_0, \dots, S_d$, using the self-financing hedging strategy $\varphi = (\varphi_0, \dots, \varphi_d)$ given by $\widehat{V}_\varphi(0) = \widehat{V}(0)$ and

$$\varphi_j(t) = \sum_{i=1}^{d} D_{1+i} v(t, X_1(t), \dots, X_d(t)) \Sigma_{ij}^{-1}(t), \quad j = 1, \dots, d. \tag{11.23}$$

Note that the argument did not really require the option to be of European style. We only need that its discounted price process relative to Q is of the form $v(A(t), X(t))$

for some deterministic function v and some uni- or multivariate continuous process A of finite variation. Altogether, (11.22, 11.23) show that perfect hedging strategies in continuous Markovian setups can be based on derivatives of the option price with respect to *risk factors* $X_1, \dots, X_d$, i.e. on so-called *sensitivities*.

Example 11.19 (Black–Scholes Model) We consider the Black–Scholes model of Example 9.1 with a contingent claim whose discounted payoff is of the form $f(S_1(T))$. The above derivation is applicable with $n = 1$, $X = S_1$, $g_1(t, x) = xe^{-rt}$, $\Sigma(t) = e^{-rt}$ and hence

$$\varphi_1(t) = D_2 v(t, S_1(t))e^{rt}. \tag{11.24}$$

If we write the undiscounted fair price of the claim as

$$V(t) = \widehat{V}(t)e^{rt} = v(t, S_1(t))e^{rt} = w(t, S_1(t))$$

with $w(t, x) = v(t, x)e^{rt}$, (11.24) turns into

$$\varphi_1(t) = D_2 w(t, S_1(t)). \tag{11.25}$$

Put differently, the perfect hedge is obtained as the derivative of the option pricing function relative to the stock price. This quantity is commonly called the *delta* of the option. Some approaches to compute the function w are discussed in Sects. 11.5 and 11.6.

In the case of jumps, the above argument breaks down. The option is generally no longer replicable by trading in the $1 + d$ liquid instruments $S_0, \dots, S_d$. Note, however, that the computation above still holds if $g_1, \dots, g_d, v$ are linear. Therefore, one may expect that (11.22) still holds approximately if most of the jumps are small in the sense that the linear approximations

$$v(t, X(t) + \Delta X(t)) \approx v(t, X(t)) + \sum_{i=1}^{d} D_{1+i} v(t, X(t))\Delta X_i(t)$$

and likewise for $g_1, \dots, g_d$ hold reasonably well for typical jump sizes $\Delta X(t)$. In other words, this sensitivity-based hedge may also make sense in the case of incompleteness due to jumps. One should note, however, that such a hedge only provides protection against small risk factor movements, whereas the risk caused by big jumps remains. In Chaps. 12 and 13 we discuss alternatives in the general incomplete situation.

Example 11.20 (Heston Model) We consider a market where a stock and a European option on the stock are liquidly traded. We suppose that the stock has Heston-type dynamics under some equivalent martingale measures for both the stock and the call, cf. Sect. 8.2.4. More specifically, the bank account S_0, the stock

S_1 and the option S_2 are given by

$$S_0(t) = e^{rt},$$
$$S_1(t) = S_0(t)\widehat{S}_1(t) = S_0(t)e^{X(t)},$$
$$S_2(t) = S_0(t)\widehat{S}_2(t) = S_0(t)E_Q\left(\frac{f_2(X(T_2))}{S_0(T_2)}\middle|\mathscr{F}_t\right),$$

where $f_2(X(T_2))$ is the payoff of the option at time T_2 and r denotes the interest rate. Here X is assumed to satisfy the equations

$$dX(t) = -\frac{1}{2}Y(t)dt + \sqrt{Y(t)}dW(t),$$
$$dY(t) = (\kappa - \lambda Y(t))dt + \sigma\sqrt{Y(t)}dZ(t)$$

with constants $\kappa > 0, \lambda, \sigma$ and Q-Wiener processes W, Z with constant correlation ϱ. Since (X, Y) is a Markov process, we can write

$$\widehat{S}_2(t) = g_2(t, X(t), Y(t))$$

for some function g_2. Moreover, we have $\widehat{S}_1(t) = g_1(t, X(t), Y(t))$ with

$$g_1(t, x_1, x_2) = e^{x_1}.$$

Now we consider a contingent claim with payoff $f(X(T))$ at time $T \leq T_2$, which is to be hedged. Similar to $\widehat{S}_2$, the martingale $\widehat{V}$ generated by the discounted payoff is of the form

$$\widehat{V}(t) = S_0(t)E_Q\left(\frac{f(X(T))}{S_0(T)}\middle|\mathscr{F}_t\right) = v(t, X(t), Y(t))$$

for some function v. This option can be replicated by trading in the bank account, the stock and the liquid option. Equation (11.23) yields that the replicating portfolio $\varphi = (\varphi_0, \varphi_1, \varphi_2)$ satisfies

$$\begin{pmatrix} \varphi_1(t) \\ \varphi_2(t) \end{pmatrix} = \begin{pmatrix} e^{X(t)} & D_2 g_2(t, X(t), Y(t)) \\ 0 & D_3 g_2(t, X(t), Y(t)) \end{pmatrix}^{-1} \begin{pmatrix} D_2 v(t, X(t), Y(t)) \\ D_3 v(t, X(t), Y(t)) \end{pmatrix}$$
$$= \begin{pmatrix} e^{-X(t)}\left(D_2 v(t, X(t), Y(t)) - \frac{D_2 g_2(t,X(t),Y(t))}{D_3 g_2(t,X(t),Y(t))} D_3 v(t, X(t), Y(t))\right) \\ \frac{D_3 v(t,X(t),Y(t))}{D_3 g_2(t,X(t),Y(t))} \end{pmatrix}.$$

In Sect. 11.6, we discuss methods how to compute g_2, v and its derivatives efficiently.

11.5 Computation of Option Prices

We have seen that European option prices can be expressed as conditional expectations of their payoff. In this and the following section we discuss different ways of computing such expectations, namely integration, partial integro-differential equations, and integral transform approaches.

11.5.1 Integration

By (11.1) the discounted price of a European option is of the form $\widehat{S}_2(t) = E_Q(\widehat{H}|\mathscr{F}_t)$. This expectation can be computed naturally by integration. To this end, suppose that $\widehat{H} = f(\widehat{S}_1(T))$ for some function $f : \mathbb{R} \to \mathbb{R}$. Moreover, we assume that the conditional law of $\log \widehat{S}_1(T)$ given $\mathscr{F}_t$ has a probability density function (pdf) which is known in closed form. This immediately yields

Rule 11.21 *Denote by $\varrho_{t,T}$ the pdf of the conditional law of $\log \widehat{S}_1(T)$ given $\mathscr{F}_t$ and relative to the pricing measure Q. Then the time-t price of a European option with discounted payoff $\widehat{H} = f(\widehat{S}_1(T))$ is of the form*

$$S_2(t) = S_0(t) \int f(e^x)\varrho_{t,T}(x)dx. \tag{11.26}$$

Equation (11.26) can be evaluated numerically or, in lucky instances, analytically.

Example 11.22 (Black–Scholes and Geometric Lévy Models) We consider a geometric Lévy model as in Example 9.16. More specifically, we assume that $X = \log S_1$ in this setup is a shifted Q-Lévy process. Denote by ϱ_t the pdf of $X(t)$ relative to Q, which is known explicitly, for example, for VG and NIG Lévy processes, cf. Sect. 2.4. Then (11.26) can be rephrased as

$$S_2(t) = e^{-r(T-t)} \int_{\log(K/S_1(t))}^{\infty} (S_1(t)e^x - K)\varrho_{T-t}(x)dx$$

for a European call with payoff $H = (S_1(T) - K)^+$.

This formula can be simplified by a change of numeraire technique. Note that the payoff is of the form

$$H = S_1(T)1_{\{S_1(T)\geq K\}} - K1_{\{S_1(T)\geq K\}}. \tag{11.27}$$

The price of the second term at time t is given by

$$e^{rt}E_Q\Big(e^{-rT}K1_{\{S_1(T)\geq K\}}\Big|\mathscr{F}_t\Big)=e^{-r(T-t)}KQ\Big(S_1(t)e^{X(T)-X(t)}\geq K\Big|\mathscr{F}_t\Big)$$
$$=e^{-r(T-t)}K\left(1-\Phi\Big(\log\frac{K}{S_1(t)}\Big)\right),$$

where Φ denotes the cumulative distribution function (cdf) of $X(T-t)-X(0)$ under Q. In order to price the first term in (11.27), it turns out to be useful to use S_1 as a numeraire. By Rule 9.18 we must replace Q by $\widetilde{Q}$ with density

$$\frac{d\widetilde{Q}}{dQ}=\frac{\widehat{S}_1(T)}{\widehat{S}_1(0)}=e^{X(T)-X(0)-rT}$$

and density process

$$Z(t)=\frac{\widehat{S}_1(t)}{\widehat{S}_1(0)}=e^{X(t)-X(0)-rt}.$$

Note that this change of measure corresponds to an Esscher transform with $\beta=1$ in Proposition 2.26. The price of the first term at time t is hence given by

$$e^{rt}E_Q\big(\widehat{S}_1(T)1_{\{S_1(T)\geq K\}}\big|\mathscr{F}_t\big)=e^{rt}E_{\widetilde{Q}}\big(\widehat{S}_1(t)1_{\{S_1(T)\geq K\}}\big|\mathscr{F}_t\big)$$
$$=S_1(t)\widetilde{Q}\Big(S_1(t)e^{X(T)-X(t)}\geq K\Big|\mathscr{F}_t\Big)$$
$$=S_1(t)\left(1-\widetilde{\Phi}\Big(\log\frac{K}{S_1(t)}\Big)\right),$$

where $\widetilde{\Phi}$ denotes the cdf of $X(T)-X(t)$ under $\widetilde{Q}$. Consequently we have

$$S_2(t)=S_1(t)\left(1-\widetilde{\Phi}\Big(\log\frac{K}{S_1(t)}\Big)\right)-e^{-r(T-t)}K\left(1-\Phi\Big(\log\frac{K}{S_1(t)}\Big)\right).$$

Let us consider three particular models. We start with Example 9.1. In this case $X(t)-X(0)=\mu t+\sigma W(t)$ is Brownian motion under Q, where W denotes a Q-Wiener process and $\mu\in\mathbb{R},\sigma>0$ parameters. $\widehat{S}_1$ is a Q-martingale if $\mu=r-\sigma^2/2$. Proposition 2.26 yields that $X-X(0)$ is a Brownian motion with drift under $\widetilde{Q}$ with parameters $\widetilde{\mu}=r+\sigma^2/2$, $\widetilde{\sigma}:=\sigma$. Hence, Φ and $\widetilde{\Phi}$ can be expressed in terms of the standard normal distribution function, namely as

$$\Phi(x)=N\left(\frac{x-\mu(T-t)}{\sqrt{\sigma^2(T-t)}}\right)=1-N\left(\frac{-x+\mu(T-t)}{\sqrt{\sigma^2(T-t)}}\right)$$

and similarly for $\widetilde{\Phi}$. Here N denotes the cumulative distribution function of the standard normal law. This leads to

$$\begin{aligned}
S_2(t) &= S_1(t)\left(1-\widetilde{\Phi}\Big(\log\frac{K}{S_1(t)}\Big)\right)-e^{-r(T-t)}K\left(1-\Phi\Big(\log\frac{K}{S_1(t)}\Big)\right)\\
&= S_1(t)N\left(\frac{\log\frac{S_1(t)}{K}+(r+\frac{\sigma^2}{2})(T-t)}{\sqrt{\sigma^2(T-t)}}\right)\\
&\quad - e^{-r(T-t)}KN\left(\frac{\log\frac{S_1(t)}{K}+(r-\frac{\sigma^2}{2})(T-t)}{\sqrt{\sigma^2(T-t)}}\right).
\end{aligned} \tag{11.28}$$

We have thus derived the **Black–Scholes formula** for European call options. In view of Example 11.19 we can determine the perfect hedge $\varphi=(\varphi_1,\varphi_2)$ as well. Differentiating (11.28) as a function of $S_1(t)$ yields

$$\varphi_1(t)=N\left(\frac{\log\frac{S_1(t)}{K}+(r+\frac{\sigma^2}{2})(T-t)}{\sqrt{\sigma^2(T-t)}}\right).$$

Since $S_2(t)=\varphi_0(t)S_0(t)+\varphi_1(t)S_1(t)$, we obtain the numeraire part of the replicating portfolio as

$$\varphi_0(t)=-e^{-rT}KN\left(\frac{\log\frac{S_1(t)}{K}+(r-\frac{\sigma^2}{2})(T-t)}{\sqrt{\sigma^2(T-t)}}\right).$$

For the European put with payoff $H=(K-S_1(T))^+$ we obtain similarly

$$\begin{aligned}
S_2(t) &= e^{-r(T-t)}KN\left(\frac{\log\frac{K}{S_1(t)}-(r-\frac{\sigma^2}{2})(T-t)}{\sqrt{\sigma^2(T-t)}}\right)\\
&\quad - S_1(t)N\left(\frac{\log\frac{K}{S_1(t)}-(r+\frac{\sigma^2}{2})(T-t)}{\sqrt{\sigma^2(T-t)}}\right),\\
\varphi_1(t) &= -N\left(\frac{\log\frac{K}{S_1(t)}-(r+\frac{\sigma^2}{2})(T-t)}{\sqrt{\sigma^2(T-t)}}\right),\\
\varphi_0(t) &= e^{-rT}KN\left(\frac{\log\frac{K}{S_1(t)}-(r-\frac{\sigma^2}{2})(T-t)}{\sqrt{\sigma^2(T-t)}}\right).
\end{aligned}$$

This can alternatively be derived from the results for the call by applying the put-call parity.

Next, we consider a VG Lévy process $X - X(0)$ as in Sect. 2.4.7 rather than Brownian motion, say with parameters $\alpha, \lambda, \beta, \mu$ under Q. For $\widehat{S}_1$ to be a Q-martingale, we need that its drift vanishes, which means that

$$\mu = r + \lambda \log \frac{\alpha^2 - (\beta+1)^2}{\alpha^2 - \beta^2}.$$

Proposition 2.26 yields that $X - X(0)$ is a VG Lévy process under $\widetilde{Q}$ as well, specifically with parameters α, λ, $\beta + 1$, μ.

An expression for the pdf is also available for the NIG Lévy process as in Sect. 2.4.8 with parameters $\alpha, \delta, \beta, \mu$ under Q. In this case the martingale condition reads as

$$\mu = r + \delta \left(\sqrt{\alpha^2 - \beta^2} - \sqrt{\alpha^2 - (\beta+1)^2} \right).$$

Again Proposition 2.26 yields that $X - X(0)$ is an NIG Lévy process under $\widetilde{Q}$ as well, now with parameters α, δ, $\beta+1$, μ. By Example 11.16 and Rule 11.2 we know that a perfect hedge does not exist for calls and puts if $X - X(0)$ is a VG or NIG Lévy process.

11.5.2 Partial Integro-Differential Equations

The above approach requires the knowledge of probability density functions which are often unknown. Moreover, it does not work for American-style options. An alternative is provided by results from Markov process theory. The backward equation yields a PIDE for European option prices.

Rule 11.23 (European Options) *Suppose that X is an $\mathbb{R}^d$-valued Markov process with generator G relative to the pricing measure Q. Consider a European option with discounted payoff of the form $\widehat{H} = f(X(T))$ for some $f : \mathbb{R}^n \to \mathbb{R}$. The discounted price process (11.1) of the option satisfies*

$$\widehat{S}_2(t) = v(T - t, X(t)),$$

where v solves the partial integro-differential equation

$$\frac{\partial}{\partial t} v = Gv \tag{11.29}$$

subject to $v(0, x) = f(x)$.

Idea See Theorem 5.24. □

Example 11.24 (Black–Scholes and Geometric Lévy Models) As in Example 11.22 we consider a European call in a geometric Lévy model. There are at least two natural ways to derive a PIDE for the option price. The discounted payoff can be written as $(\widehat{S}_1(T) - Ke^{-rT})^+$, where we consider the geometric Lévy process $\widehat{S}_1$ as a Markov process. For the Black–Scholes model this leads to the PDE

$$\frac{\partial}{\partial t}v(t,x) = \frac{\sigma^2}{2}x^2\frac{\partial^2}{\partial x^2}v(t,x), \quad v(0,x) = (x - Ke^{-rT})^+ \tag{11.30}$$

for $\widehat{S}_2(t) = v(T-t, \widehat{S}_1(t))$.

In undiscounted terms, this yields the **Black–Scholes PDE**

$$\frac{\partial}{\partial t}v(t,x) = -rv(t,x) + rx\frac{\partial}{\partial x}v(t,x) + \frac{\sigma^2}{2}x^2\frac{\partial^2}{\partial x^2}v(t,x), \quad v(0,x) = (x-K)^+ \tag{11.31}$$

for $S_2(t) = v(T-t, S_1(t))$.

Alternatively, we write the discounted payoff as $(e^{\log S_1(T)} - K)^+e^{-rT}$, in which case we apply Rule 11.23 to $f(x) := (e^x - K)^+e^{-rT}$ and $X := \log S_1$. In the Black–Scholes model we obtain the PDE

$$\frac{\partial}{\partial t}v(t,x) = \left(r - \frac{\sigma^2}{2}\right)\frac{\partial}{\partial x}v(t,x) + \frac{\sigma^2}{2}\frac{\partial^2}{\partial x^2}v(t,x), \quad v(0,x) = (e^x - K)^+e^{-rT} \tag{11.32}$$

for $\widehat{S}_2(t) = v(T-t, \log S_1(t))$. This corresponds to (11.30, 11.31) after a change of variables. Note that (11.30–11.32) describe three different functions v.

For American options we must use Theorem 7.18 rather than the backward equation. Stated as an informal rule we obtain the following

Rule 11.25 (American Options) *Suppose that X is a $\mathbb{R}^n$-valued Markov process with generator G relative to the pricing measure Q. The discounted price process (11.3) of an American option with discounted exercise process $\widehat{H}(t) = f(t, X(t))$ satisfies*

$$\widehat{S}_2(t) = v(T-t, X(t)),$$

where v solves the linear complementarity problem

$$\min\left(-\frac{\partial}{\partial t}v - Gv, v - f\right) = 0$$

subject to $f(T,x) = f(x)$.

Idea See Theorem 7.18. □

11.6 Pricing via Laplace Transform

Closed-form expressions of marginal densities are not available for some classes of Lévy processes and for more complex models involving, say, stochastic volatility. In this case representation (11.26) does not help much. Efficient numerical methods for solving PIDEs related to pricing problems as (11.29) become more and more available but their implementation requires considerable effort. Direct Monte Carlo simulation of the expected values, on the other hand, suffers from a comparatively slow rate of convergence.

An alternative approach is based on the characteristic function of the underlying, which is known for many processes including most equity models in Chap. 8. It can be used to express the price of European options as an integral even if densities are not available in closed form. The method does not extend to American options and path-dependent claims such as Barrier and Asian options. However, it is fast, flexible and comparatively simple if it can be applied.

We consider a European-style contingent claim whose discounted payoff equals $\widehat{H} = f(X(T))$ for some $f : \mathbb{R} \to \mathbb{R}$ and some stochastic process X whose characteristic function under the pricing measure Q is known. Typically, this process is chosen to be the logarithmic underlying price $X := \log S_1$ but this is not essential.

For most payoff functions f the expectation $E_Q(f(X(T)))$ cannot be evaluated in closed form. An exception is exponential payoffs $\widehat{H} = \exp(zX(T))$. Indeed, the price $E_Q(\exp(zX(T))$ of such claims coincides with the value of the generalised characteristic function of $X(T)$ at $-iz$, which is often known explicitly. The attribute *generalised* refers to the fact that $-iz$ is typically not a real number.

The idea now is to view arbitrary payoffs as linear combinations or, more precisely, integrals of such exponential payoffs, i.e. we write

$$f(x) = \int e^{zx} \varrho(z) dz \tag{11.33}$$

with some integration kernel ϱ. Since pricing is linear in the claim, the discounted value of the option is obtained as the corresponding integral of the prices of the involved exponential claims, i.e.

$$E_Q(f(X(T))) = \int E_Q(e^{zX(T)}) \varrho(z) dz.$$

If the generalised characteristic function of $X(T)$ and the integral representation (11.33) of the payoff are known in closed form, the computation of option prices boils down to a univariate numerical integration as in Sect. 11.5.1.

11.6.1 Integral Representation of Payoffs

The above approach requires that we can express the payoff function in integral form (11.33). Specifically, we use the representation

$$f(x) = \int_{R-i\infty}^{R+i\infty} e^{zx}\varrho(z)dz := \int_{-\infty}^{\infty} e^{(R+iv)x}\varrho(R+iv)idv \tag{11.34}$$

for some function $\varrho : R + i\mathbb{R} \to \mathbb{C}$. The real number R is chosen such that $E_Q(\exp(RX(T))) < \infty$.

The integration kernel ϱ can be determined by Laplace transform. If the bilateral Laplace transform $\widetilde{f}$ of f exists for $R \in \mathbb{R}$ and if $v \mapsto \widetilde{f}(R+iv)$ is integrable, then

$$f(x) = \frac{1}{2\pi i}\int_{R-i\infty}^{R+i\infty} e^{zx}\widetilde{f}(z)dz \tag{11.35}$$

by Proposition A.9. Consequently, $\varrho(z) := \widetilde{f}(z)/(2\pi i)$ is the function we are looking for. It can be determined explicitly for practically all relevant payoff functions. We state a number of examples which are obtained by straightforward integration. Recall that the argument x refers typically to the logarithmic asset price.

Example 11.26 (Plain Vanilla Call and Put) Fix a strike price $K > 0$. The Laplace transform of the call payoff function $f(x) = (e^x - K)^+$ is given by

$$\widetilde{f}(z) = \frac{K^{1-z}}{z(z-1)} \quad \text{for } R = \text{Re}(z) > 1. \tag{11.36}$$

Interestingly, the put payoff function $f(x) = (K - e^x)^+$ has the same Laplace transform, but with negative $R = \text{Re}(z)$:

$$\widetilde{f}(z) = \frac{K^{1-z}}{z(z-1)} \quad \text{for } R = \text{Re}(z) < 0.$$

As a side remark, the remaining values of $R = \text{Re}(z)$ belong to the payoff function $f(x) = (e^x - K)^+ - e^x$ with Laplace transform

$$\widetilde{f}(z) = \frac{K^{1-z}}{z(z-1)} \quad \text{for } 0 < \text{Re}(z) < 1,$$

which leads to another representation of the call.

Example 11.27 (Power and Self-Quanto Call) Fix a strike $K > 0$ and an integer $n \geq 1$. The corresponding power call $f(x) = ((e^x - K)^+)^n$ has Laplace transform

$$\widetilde{f}(z) = \frac{n!K^{n-z}}{z(z-1)\cdots(z-n)} \quad \text{for } R = \text{Re}(z) > n.$$

The payoff function $f(x) = (e^x - K)^+ e^x$ belongs to the self-quanto call with strike K. Its Laplace transform equals

$$\widetilde{f}(z) = \frac{K^{2-z}}{(z-1)(z-2)} \quad \text{for } R = \mathrm{Re}(z) > 2.$$

Example 11.28 (Digital Option) The digital or binary option with strike $K > 0$ corresponds to the payoff function $f(x) = 1_{[K,\infty)}(e^x)$. Its Laplace transform equals

$$\widetilde{f}(z) = \frac{K^{-z}}{z} \quad \text{for } R = \mathrm{Re}(z) > 0.$$

Since $\widetilde{f}$ is not integrable, we do not exactly have representation (11.35) in this case. However, statement 2 in Proposition A.9 yields the similar equation

$$\frac{1}{2} 1_{\{K\}}(e^x) + 1_{(K,\infty)}(e^x) = \lim_{c\to\infty} \frac{1}{2\pi i} \int_{R-ic}^{R+ic} e^{zx} \frac{K^{-z}}{z} dz \quad \text{for } R = \mathrm{Re}(z) > 0. \tag{11.37}$$

Unless $X(T)$ attains the value $\log K$ with strictly positive probability, the option with payoff function (11.37) has the same value as the original digital call.

Example 11.29 (Call Options on Realised Variance) If $x > 0$ stands for realised variance, a call on the latter has payoff function $f(x) = (x-K)^+$ with some $K \geq 0$. Its Laplace transform equals

$$\widetilde{f}(z) = \frac{e^{-zK}}{z^2} \quad \text{for } R = \mathrm{Re}(z) > 0.$$

Example 11.30 (Volatility Swaps) If $x > 0$ denotes once more realised variance, a volatility swap has payoff function $f(x) = \sqrt{x} - K$ with some $K \geq 0$. It can be written in integral form as

$$\sqrt{x} - K = \frac{1}{2\sqrt{\pi}} \int_0^\infty \frac{1 - e^{-zx}}{z^{3/2}} dz - K, \tag{11.38}$$

cf. e.g. [265, (1.2.3)]. This does not correspond literally to a representation as in (11.35), but it can be used similarly. In our later applications in Chaps. 11–14, expressions of the kind

$$\frac{1}{2\pi i} \int_{R-i\infty}^{R+i\infty} g(z) \widetilde{f}(z) dz$$

have to be replaced by

$$\frac{1}{2\sqrt{\pi}}\int_0^\infty \frac{g(0)-g(-z)}{z^{3/2}}dz - Kg(0) \tag{11.39}$$

for this payoff function.

11.6.2 Pricing Formulas

As a second important ingredient on the way to option prices we need the generalised characteristic function of $X(T)$ on the line $\mathbb{R}-iR$ in the complex plane. More specifically, we write

$$\varphi_{t,T}(z) := E_Q\big(e^{izX(T)}\big|\mathscr{F}_t\big), \quad z \in \mathbb{R}-iR \tag{11.40}$$

for the generalised conditional characteristic function of $X(T)$ given the information up to time t. For $R = 0$ this function reduces to the conditional characteristic function, which is known explicitly for most Lévy processes or, more generally, components of affine processes, cf. Sects. 2.2, 2.4, 6.1, 8.2. For example, we have

$$\varphi_{t,T}(z) = \exp(izX(t) + \psi(z)(T-t)), \quad z \in \mathbb{R} \tag{11.41}$$

if X is a Q-Lévy process with characteristic exponent ψ. By Proposition A.5(4) the values of $\varphi_{t,T}$ for arbitrary z in some strip in the complex plane are typically obtained by inserting complex values in the analytic expression of the characteristic function. Consequently, we may assume (11.40) to be known for many models of interest.

The knowledge of $\varphi_{t,T}$ and $\widetilde{f}$ above now leads to the desired pricing formula.

Rule 11.31 *Fix $0 \leq t \leq T$. Suppose that the discounted payoff of an option is of the form $\widehat{H} := f(X(T))$, where f has Laplace transform $\widetilde{f}$ on $R+i\mathbb{R}$ with $R \in \mathbb{R}$ such that $E(\exp(RX(T))|\mathscr{F}_t) < \infty$. We denote the generalised conditional characteristic function of the process X under the pricing measure Q as in (11.40). Then the option price at time t is given by*

$$\begin{aligned} S_2(t) &:= S_0(t)E_Q\big(\widehat{H}\big|\mathscr{F}_t\big) \\ &= \frac{S_0(t)}{2\pi i}\int_{R-i\infty}^{R+i\infty} \varphi_{t,T}(-iz)\widetilde{f}(z)dz \\ &= \frac{S_0(t)}{\pi}\int_0^\infty \mathrm{Re}\big(\varphi_{t,T}(u-iR)\widetilde{f}(R+iu)\big)du. \end{aligned} \tag{11.42}$$

Idea The first integral representation follows from the linearity of the expected value. Since f and X are real, we have $\widetilde{f}(\overline{z}) = \overline{\widetilde{f}(z)}$ and $\varphi_{t,T}(-i\overline{z}) = \overline{\varphi_{t,T}(-iz)}$ for $z \in \mathbb{C}$. This yields

$$\begin{aligned}
\widehat{S}_2(t) &= \frac{1}{2\pi i}\int_{R-i\infty}^{R+i\infty} \varphi_{t,T}(-iz)\widetilde{f}(z)dz \\
&= \frac{1}{2\pi}\int_{-\infty}^{\infty} \varphi_{t,T}(u-iR)\widetilde{f}(R+iu)du \\
&= \frac{1}{2\pi}\int_{0}^{\infty} \Big(\varphi_{t,T}(u-iR)\widetilde{f}(R+iu) + \varphi_{t,T}(-u-iR)\widetilde{f}(R-iu)\Big)du \\
&= \frac{1}{2\pi}\int_{0}^{\infty} \Big(\varphi_{t,T}(u-iR)\widetilde{f}(R+iu) + \overline{\varphi_{t,T}(u-iR)\widetilde{f}(R+iu)}\Big)du \\
&= \frac{1}{\pi}\int_{0}^{\infty} \operatorname{Re}\big(\varphi_{t,T}(u-iR)\widetilde{f}(R+iu)\big)du
\end{aligned}$$

as claimed. □

Example 11.32 (Black–Scholes and Geometric Lévy Models) For the European call $H = (S_1(T) - K)^+$ in the Black–Scholes model (cf. Examples 9.1, 11.4) we obtain the option price

$$S_2(t) = \frac{1}{2\pi i}\int_{R-i\infty}^{R+i\infty} \exp\left(\frac{\sigma^2}{2}z(z-1)(T-t)\right)\frac{S_1(t)^z(Ke^{-r(T-t)})^{1-z}}{z(z-1)}dz \tag{11.43}$$

$$= \frac{1}{\pi}\int_0^{\infty} \operatorname{Re}\left(\exp\left(\frac{\sigma^2}{2}(R+iu)(R-1+iu)(T-t)\right)\right.$$
$$\left.\times\frac{S_1(t)^{R+iu}(Ke^{-r(T-t)})^{1-R-iu}}{(R+iu)(R-1+iu)}\right)du \tag{11.44}$$

with an arbitrary $R > 1$, cf. Example 11.26. Since $1/(z(z-1)) = 1/(z-1) - 1/z$, (11.43) can be rewritten as

$$\begin{aligned}
S_2(t) = &-Ke^{-r\tau}\frac{1}{2\pi i}\int_{R-i\infty}^{R+i\infty}\frac{\exp\left(\frac{\sigma^2\tau}{2}z^2 + \left(-\frac{\sigma^2\tau}{2} + \log\frac{S_1(t)}{K} + r\tau\right)z\right)}{z}dz \\
&+ S_1(t)\frac{1}{2\pi i}\int_{R-1-i\infty}^{R-1+i\infty}\frac{\exp\left(\frac{\sigma^2\tau}{2}z^2 + \left(\frac{\sigma^2\tau}{2} + \log\frac{S_1(t)}{K} + r\tau\right)z\right)}{z}dz
\end{aligned}$$

with $\tau := T - t$. Applying Lemma A.10 we recover the Black–Scholes formula (11.28).

For other geometric Lévy processes $S_1(t) = e^{X(t)}$ one just has to replace the factor $\frac{\sigma^2}{2}z(z-1)$ for $z = R + iu$ in the exponentials of (11.43, 11.44) with $\psi(-iz) - zr$, where ψ denotes the characteristic exponent of $X - X(0)$. This exponent equals (2.62) for the VG Lévy process and (2.63) for the NIG Lévy process from Example 11.22.

In rare cases one can proceed more directly without using Rule 11.31:

Example 11.33 (The Log Contract) The log contract with payoff $H := \log S_1(T)$ can be priced without numerical integration if $S_0(T)$ is deterministic and conditional first moments of $X := \log S_1$ are known. Indeed, we have

$$S_0(t)E_Q(\widehat{H}|\mathscr{F}_t) = \frac{S_0(t)}{S_0(T)}E_Q(X(T)|\mathscr{F}_t). \tag{11.45}$$

In view of Proposition A.5(2) these moments may be obtained by differentiating the characteristic function.

In order to apply the results of Chap. 12 to the log contract, we need an integral representation (11.35) of $f(x) = x$. However, the Laplace transform $\widetilde{f}$ does not exist as a function because $|f(x)|$ diverges for $|x| \to \infty$. But in the generalised sense of distributions, $z \mapsto \widetilde{f}(z)$ corresponds to a multiple of the derivative at 0, more specifically we have

$$\frac{1}{2\pi i}\int_{R-i\infty}^{R+i\infty} g(z)\widetilde{f}(z)dz = g'(0)$$

for any test function g. Indeed, for $g(z) = e^{zx}$, this interpretation of the right-hand side of (11.35) yields the correct result x for any $x \in \mathbb{R}$. As a sanity check note that (11.42) for $f(x) = x/S_0(T)$ turns into $-i\varphi'_{t,T}(0)S_0(t)/S_0(T) = E_Q(X(T)|\mathscr{F}_t)S_0(t)/S_0(T)$ in line with (11.45).

Since the process X in Rule 11.31 need not be a tradable asset nor its logarithm, we can use the approach for the computation of options on volatility as well.

Example 11.34 (Derivatives Involving Realised Volatility and Variance) Options on realised volatility or variance have a payoff that depends on

$$\sigma_R^2 := \sum_{n=1}^{N}(X_{n\Delta t} - X_{(n-1)\Delta t})^2,$$

where $X := \log S_1$ denotes the logarithmic underlying price and Δt a fixed short time interval, say a day. For $\Delta t \to 0$ and $N\Delta t \to T$, this expression converges to the quadratic variation $[X, X](T)$, cf. Definition 3.1. For simplicity we neglect the difference between σ_R^2 and $[X, X](T)$ and consider instead options on $[X, X](T)$ or $\sqrt{[X, X](T)}$ as an approximation to reality. In this sense, a variance swap has payoff $[X, X](T) - K$ for some appropriately chosen constant $K > 0$. Similarly, a call option on realised variance pays $([X, X](T) - K)^+$ and a volatility swap $\sqrt{[X, X](T)} - K$.

Let us consider the valuation of these instruments in an affine stochastic volatility model, namely the CGMY model from Sect. 8.2.7. Specifically, we assume that the bank account is of the simple form $S_0(t) = e^{rt}$ and the logarithmic stock price $X := \log S_1$ satisfies (8.31) with $\varrho = 0$ under the pricing measure Q. In order for $\widehat{S}_1$ to be a Q-martingale, we need $\mu = r$ and

$$\psi^L(-i) := a^L + \frac{c^L}{2} + \int (e^x - 1 - x) F^L(dx) = 0.$$

Proposition 6.18 yields that $[X, X]$ is a component of an affine process as well. Its generalised conditional characteristic function equals

$$\begin{aligned}\varphi_{t,T}(z) &= E\big(\exp(iz[X, X](T)) \big| \mathscr{F}_t\big) \\ &= \exp\Big(\Psi_0(T-t, z) + \Psi_1(T-t, z)Y(t) + \Psi_2(T-t, z)[X, X](t)\Big)\end{aligned}$$

with Ψ_0, Ψ_1, Ψ_2 of the form (8.21–8.23), but with ψ^L replaced by the Lévy exponent

$$\psi^{[L,L]}(u) := iuc^L + \int (e^{iux^2} - 1) K^L(dx)$$

in these formulas. Rule 11.31 and Example 11.29 now yield the price of call options $H := ([X, X](T) - K)^+$ on variance, namely

$$\begin{aligned}e^{rt} E_Q(\widehat{H}|\mathscr{F}_t) &= \frac{e^{-r(T-t)}}{2\pi i} \int_{R-i\infty}^{R+i\infty} \varphi_{t,T}(-iz) \frac{\exp(-zK)}{z^2} dz \\ &= \frac{e^{-r(T-t)}}{\pi} \int_0^\infty \mathrm{Re}\left(\varphi_{t,T}(u - iR) \frac{\exp(-(R+iu)K)}{(R+iu)^2}\right) du,\end{aligned} \tag{11.46}$$

where $R > 0$ is chosen sufficiently small.

The variance swap $H := [X, X](T) - K$ could be obtained by setting $K = 0$ in (11.46) and subtracting the price of a bond which pays K at T. This yields

$$\begin{aligned}e^{rt} E_Q(\widehat{H}|\mathscr{F}_t) &= e^{-r(T-t)} \left(\frac{1}{2\pi i} \int_{R-i\infty}^{R+i\infty} \varphi_{t,T}(-iz) z^{-2} dz - K\right) \\ &= e^{-r(T-t)} \left(\frac{1}{\pi} \int_0^\infty \mathrm{Re}\left(\frac{\varphi_{t,T}(u - iR)}{(R+iu)^2}\right) du - K\right)\end{aligned}$$

for sufficiently small $R > 0$. However, a more efficient approach is to compute $E_Q([X,X](T)|\mathscr{F}_t)$ directly, as it is the first moment of an affine process. Applying Theorem 6.15 yields

$$E_Q([X,X](T)|\mathscr{F}_t) - [X,X](t)$$
$$= \left(c^L + \int x^2 F^L(dx)\right)\left(\frac{a^Z}{\lambda}(T-t) - \left(\frac{a^Z}{\lambda} - Y(t)\right)\frac{1-e^{-\lambda(T-t)}}{\lambda}\right).$$

In view of

$$e^{rt}E_Q(\widehat{H}|\mathscr{F}_t) = e^{-r(T-t)}\left(E_Q([X,X](T)|\mathscr{F}_t) - K\right)$$

we obtain a simpler formula for the option price.

Finally, we turn to the volatility swap with payoff $H := \sqrt{[X,X](T)} - K$. Rule 11.31 is not literally applicable but its heuristic generalisation leads to the correct formula. The right-hand side of (11.38) can be viewed as a linear combination of payoffs 1 and $e^{-z[X,X](T)}$ at time T. The bond 1 has price

$$S_0(t)E_Q(1/S_0(T)|\mathscr{F}_t) = e^{-r(T-t)}$$

at time t, whereas the value of the exponential claims equals

$$S_0(t)E_Q(e^{-z[X,X](T)}/S_0(T)|\mathscr{F}_t) = e^{-r(T-t)}\varphi_{t,T}(iz).$$

Altogether, we obtain

$$e^{rt}E_Q(\widehat{H}|\mathscr{F}_t) = e^{-r(T-t)}\left(\frac{1}{2\sqrt{\pi}}\int_0^\infty \frac{1-\varphi_{t,T}(iz)}{z^{3/2}}dz - K\right)$$

for the price of H at time t, which is in line with (11.39). The swap rate at time t refers to the value K such that this expression equals 0.

Forward-start options depend on the stock price at two different times. If the return process of the underlying is a component of an affine process, they can nevertheless be treated with the help of Rule 11.31 or Rule 11.44 below. For ease of exposition we consider here a geometric Lévy model.

Example 11.35 (Forward-Start Call Options in Geometric Lévy Models) Let us consider a geometric Lévy model $S_1(t) = \exp(X(t))$ as in Example 9.16. A forward-start call has payoff $H := (S_1(T_2) - S_1(T_1))^+$ at time $T_2 > T_1$. We want to determine its price at time $t < T_2$. If $t \geq T_1$, it corresponds to an ordinary call because its strike $S_1(T_1)$ is already known. If ψ denotes the characteristic exponent of the Lévy process $X - X(0)$ under the pricing measure Q, the value of the forward-start call at time T_1 equals

$$S_2(T_1) = \frac{S_1(T_1)}{2\pi i} \int_{R-i\infty}^{R+i\infty} \frac{\exp((\psi(-iz) - zr)(T_2 - T_1))}{z(z-1)} dz$$

for sufficiently small $R > 1$ according to Example 11.32. This corresponds to the value of a deterministic number of stocks. Hence we obtain

$$\begin{aligned} S_2(t) &= S_0(t) E_Q(\widehat{H}|\mathscr{F}_t) \\ &= S_0(t) E_Q\big(E_Q(\widehat{H}|\mathscr{F}_{T_1})\big|\mathscr{F}_t\big) \\ &= S_0(t) E_Q(\widehat{S}_2(T_1)|\mathscr{F}_t) \\ &= \frac{S_1(t)}{2\pi i} \int_{R-i\infty}^{R+i\infty} \frac{\exp((\psi(-iz) - zr)(T_2 - T_1))}{z(z-1)} dz \\ &= \frac{S_1(t)}{\pi} \int_0^\infty \mathrm{Re}\left(\frac{\exp((\psi(u-iR) - (R+iu)r)(T_2 - T_1))}{(R+iu)(R-1+iu)}\right) du \end{aligned}$$

for the price of the forward-start call at time $t \leq T_1$. Comparing with Example 11.32 we observe that this equals the value of an ordinary European call option with strike $S_1(t)$ and maturity $T_2 - T_1$.

11.6.3 Vanilla Options as Fourier Integrals

The integral in (11.42) can be written in terms of the Fourier transform of some function. Consequently, a quadrature based on the discrete *fast Fourier transform (FFT)* may be used for its evaluation. However, there is no advantage to doing so as long as one focuses on a single option price. But if a plain vanilla call or put prices must be computed for many strikes, a Fourier integral representation turns out to be useful because FFT allows us to evaluate a number of integrals simultaneously at little extra cost.

Rule 11.36 *Fix Q, t, T, X as in Rule 11.31. Let $R > 1$ such that $E_Q(\exp RX(T)|\mathscr{F}_t) < \infty$. Denote by*

$$\pi(t, K) := S_0(t) E_Q\big((e^{X(T)} - K)^+\big|\mathscr{F}_t\big)$$

the price at time t of an option with discounted payoff $(e^{X(T)} - K)^+$, i.e. of a European call on e^X with discounted strike K. Then we have

$$\pi(t, K) = S_0(t) \frac{K^{1-R}}{2\pi} \widehat{g}(-\log K), \tag{11.47}$$

where

$$g(x) := \frac{-\varphi_{t,T}(x - iR)}{(x - iR)(x + i(1 - R))}$$

and $\widehat{g}(u) = \int_{-\infty}^{\infty} e^{iux} g(x)dx$ *denotes its Fourier transform.*

For a put with discounted payoff $(K - e^{X(T)})^+$ *we obtain the same formula if* $R < 0$ *is chosen such that* $E_Q(\exp RX(T)|\mathscr{F}_t) < \infty$.

Idea The right-hand side of (11.47) equals

$$S_0(t)\frac{K^{1-R}}{2\pi}\int_{-\infty}^{\infty} K^{-ix}\frac{\varphi_{t,T}(-i(R+ix))}{(R+ix)(R+ix-1)}dx.$$

Substituting $z := R + ix$, we obtain

$$\frac{S_0(t)}{2\pi i}\int_{R-i\infty}^{R+i\infty}\frac{K^{1-z}}{z(z-1)}\varphi_{t,T}(-iz)dz,$$

which is the price of the call by (11.42, 11.36). The put is treated along the same lines. □

Example 11.37 (Black–Scholes and Geometric Lévy Models) In the situation of Examples 11.22, 11.32 we have

$$\pi(t, K) = \frac{e^{rt}}{2\pi}K^{1-R}\widehat{g}(-\log K)$$

with

$$g(x) = -\frac{\exp\big(\psi(x - iR)(T - t)\big)S_1(t)^{R+ix}}{(x - iR)(x + i(1 - R))}.$$

11.6.4 Options on Multiple Assets

The approach in Sects. 11.6.1–11.6.2 can be extended to options on multiple underlyings. Similarly as above, we consider a contingent claim with discounted payoff $\widehat{H} = f(X(T))$ for some $f : \mathbb{R}^d \to \mathbb{R}$, where X denotes an $\mathbb{R}^d$-valued stochastic process whose characteristic function is known. Typically one chooses $X_i(t) = \log \widehat{S}_i(t)$, where $S_1(t), \ldots, S_d(t)$ denote the price processes of the underlyings. Parallel to (11.33) suppose that the payoff function is of the form

$$f(x) = \int_C e^{z^\top x}\varrho(z)dz, \tag{11.48}$$

with some function ϱ that is to be integrated on a subset C of

$$\{z \in \mathbb{C}^d : \mathrm{Re}(z_j) = R_j \text{ for } j = 1, \ldots, d\} \tag{11.49}$$

or of a union of such rectangles. The vector $R = (R_1, \ldots, R_d) \in \mathbb{R}^d$ is chosen such that $E(e^{R^\top X(T)}) < \infty$. As in Sect. 11.6.1 such a representation is obtained through the multivariate bilateral Laplace transform $\widetilde{f}$ of f. If this transform exists for $R \in \mathbb{R}^d$ and if the mapping f is integrable, then

$$f(x) = \frac{1}{(2\pi i)^d} \int_{R_1 - i\infty}^{R_1 + i\infty} \cdots \int_{R_d - i\infty}^{R_d + i\infty} e^{z_1 x_1 + \cdots + z_d x_d} \widetilde{f}(z_1, \ldots, z_d) dz_d \ldots dz_1 \tag{11.50}$$

by Proposition A.12. We turn to particular examples which are obtained by a more or less straightforward application of this inversion formula.

Example 11.38 (Spread Option) A call with strike price $K \geq 0$ on the difference of two stocks is known as a *spread option*. Its payoff function can be represented as

$$\begin{aligned}&(e^{x_1} - e^{x_2} - K)^+ \\ &\quad = -\frac{1}{4\pi^2} \int_{R_1 - i\infty}^{R_1 + i\infty} \int_{R_2 - i\infty}^{R_2 + i\infty} e^{x_1 z_1 + x_2 z_2} \frac{B(z_1 + z_2 - 1, -z_2)}{K^{z_1 + z_2 - 1} z_1 (z_1 - 1)} dz_2 dz_1\end{aligned}$$

for $R_1 + R_2 > 1$ and $R_2 < 0$, where

$$B(x, y) = \frac{\Gamma(x)\Gamma(y)}{\Gamma(x + y)}$$

denotes the Euler beta function.

Example 11.39 (Margrabe Option) For $K = 0$ the spread option reduces to the *Margrabe option* with integral representation

$$(e^{x_1} - e^{x_2})^+ = \frac{1}{2\pi i} \int_{R - i\infty}^{R + i\infty} e^{x_1 z + x_2 (1 - z)} \frac{1}{z(z - 1)} dz \quad \text{for } R > 1.$$

Example 11.40 (Quanto Options) A quanto call with payoff $S_1(T)(S_2(T) - K)^+$ corresponds to the payoff function

$$e^{x_1}(e^{x_2} - K)^+ = \frac{1}{2\pi i} \int_{R - i\infty}^{R + i\infty} e^{x_1 + z x_2} \frac{K^{1 - z}}{z(z - 1)} dz \quad \text{for } R > 1.$$

For the put $e^{x_1}(K - e^{x_2})^+$ we get the same formula with $R < 0$.

Example 11.41 (Asset-or-Nothing Option) An asset-or-nothing option paying the amount $S_1(T)1_{\{S_2(T)>K\}}$ can be identified with the payoff function

$$e^{x_1}\left(\frac{1}{2}1_{\{K\}}(e^{x_2})+1_{(K,\infty)}(e^{x_2})\right)$$
$$=\lim_{c\to\infty}\frac{1}{2\pi i}\int_{R-ic}^{R+ic}e^{x_1+zx_2}\frac{K^{-z}}{z}dz \quad \text{for } R>0$$

unless the value K is attained with strictly positive probability.

Example 11.42 (Basket Option) A put on the sum of several assets can be represented as

$$(K-e^{x_1}-\cdots-e^{x_d})^+$$
$$=\frac{1}{(2\pi i)^d}\int_{R_1-i\infty}^{R_1+i\infty}\cdots\int_{R_d-i\infty}^{R_d+i\infty}e^{z_1x_1+\cdots+z_dx_d}\frac{B_{d+1}(-z_1,\ldots,-z_d,2)}{K^{1-z_1-\cdots-z_d}}dz_d\cdots dz_1$$

for $R_1,\ldots,R_d<0$ and

$$B_d(x_1,\ldots,x_d)=\frac{\Gamma(x_1)\cdots\Gamma(x_d)}{\Gamma(x_1+\cdots+x_d)}.$$

The corresponding call is easily obtained from the put-call parity

$$(e^{x_1}+\cdots+e^{x_d}-K)^+=(K-e^{x_1}-\cdots-e^{x_d})^+-(K-e^{x_1}-\cdots-e^{x_d}).$$

Example 11.43 (Options on the Minimum or Maximum of Several Assets) A call with strike price $K>0$ on the minimum of a number of stocks can be represented as

$$(\min(e^{x_1},\ldots,e^{x_d})-K)^+=\frac{1}{(2\pi i)^d}\int_{R_1-i\infty}^{R_1+i\infty}\cdots\int_{R_d-i\infty}^{R_d+i\infty}e^{z_1x_1+\cdots+z_dx_d}$$
$$\times\frac{K^{1-z_1-\cdots-z_d}}{z_1\cdots z_d(z_1+\cdots+z_d-1)}dz_d\cdots dz_1$$

for $R_1,\ldots,R_d>0$ with $R_1+\cdots+R_d>1$.

In order to derive an expression for the put, we start with

$$\min(e^{x_1},\ldots,e^{x_d},K)=\frac{1}{(2\pi i)^d}\int_{R_1-i\infty}^{R_1+i\infty}\cdots\int_{R_d-i\infty}^{R_d+i\infty}e^{z_1x_1+\cdots+z_dx_d}$$
$$\times\frac{K^{1-z_1-\cdots-z_d}}{z_1\cdots z_d(z_1+\cdots+z_d-1)}dz_d\cdots dz_1$$

for $R_1, \dots, R_d > 0$ with $R_1 + \dots + R_d < 1$. In a second step, we decompose the put as

$$(K - \min(e^{x_1}, \dots, e^{x_d}))^+ = K - \min(e^{x_1}, \dots, e^{x_d}, K).$$

In the case of the maximum of several assets it is easier to consider the put

$$(K - \max(e^{x_1}, \dots, e^{x_d}))^+ = \left(\frac{-1}{2\pi i}\right)^d \int_{R_1-i\infty}^{R_1+i\infty} \cdots \int_{R_d-i\infty}^{R_d+i\infty} e^{z_1x_1+\dots+z_dx_d} \times \frac{K^{1-z_1-\dots-z_d}}{z_1\cdots z_d(1-z_1-\dots-z_d)} dz_d \cdots dz_1$$

for $R_1, \dots, R_d < 0$ with $R_1 + \dots + R_d > -1$.

The integral representation of the call on the maximum can be derived from the put-call parity

$$(\max(e^{x_1}, \dots, e^{x_d}) - K)^+ = (K - \max(e^{x_1}, \dots, e^{x_d}))^+ - (K - \max(e^{x_1}, \dots, e^{x_d}))$$

if one uses the following formula for the maximum

$$\begin{aligned}
&\max(e^{x_1}, \dots, e^{x_d}) \\
&= \frac{1}{(2\pi i)^{d-1}} \sum_{k=1}^d \int_{R_1-i\infty}^{R_1+i\infty} \cdots \int_{R_{k-1}-i\infty}^{R_{k-1}+i\infty} \int_{R_{k+1}-i\infty}^{R_{k+1}+i\infty} \cdots \int_{R_d-i\infty}^{R_d+i\infty} \\
&\quad e^{z_1x_1} \dots e^{z_{k-1}x_{k-1}} e^{(1-\sum_{j\neq k} z_j)x_k} e^{z_{k+1}x_{k+1}} \dots e^{z_dx_d} \\
&\quad \times \prod_{j\neq k} z_j^{-1} dz_d \cdots dz_{k+1} dz_{k-1} \cdots dz_1
\end{aligned}$$

for $R_1, \dots, R_d < 0$.

As in Sect. 11.6.2 we also need the generalised conditional characteristic function of $X(T)$ on the set (11.49). We denote it by

$$\varphi_{t,T}(z) := E_Q\big(e^{iz^\top X(T)} \big| \mathscr{F}_t\big) \tag{11.51}$$

for $t \leq T$ and z in (11.49). Parallel to (11.41), we have

$$\varphi_{t,T}(z) = \exp\Big(iz^\top X(t) + \psi(z)(T-t)\Big)$$

if $X - X(0)$ is an $\mathbb{R}^d$-valued Q-Lévy process with characteristic exponent ψ. The counterpart of Rule 11.31 reads as

Rule 11.44 *Fix $0 \leq t \leq T$. Suppose that the discounted payoff of an option is of the form $\widehat{H} := f(X(T))$, where f has a representation (11.48). We denote the generalised conditional characteristic function of the process X under the pricing measure Q as in (11.51). Then the option price at time t is given by*

$$S_0(t)E_Q(\widehat{H}|\mathscr{F}_t) = S_0(t) \int_C \varphi_{t,T}(-iz)\varrho(z)dz, \tag{11.52}$$

where the domain C of the integral equals the domain for the representation (11.48).

Idea This follows from linearity of the expectation. □

For payoffs as in (11.50), (11.52) refers to

$$S_0(t)E_Q(\widehat{H}|\mathscr{F}_t) = \frac{S_0(t)}{(2\pi i)^d} \int_{R_1-i\infty}^{R_1+i\infty} \cdots \int_{R_d-i\infty}^{R_d+i\infty} \varphi_{t,T}(-iz_1, \ldots, -iz_d)\widetilde{f}(z_1, \ldots, z_d)dz_d \ldots dz_1.$$

Example 11.45 (Margrabe Option) In the case of a Margrabe option with payoff $H = (S_1(T) - S_2(T))^+$ we obtain

$$S_0(t)E_Q(\widehat{H}|\mathscr{F}_t) = \frac{S_0(t)}{2\pi i} \int_{R-i\infty}^{R+i\infty} \varphi_{t,T}(-iz, -i(1-z))\frac{1}{z(z-1)}dz$$

for $R > 1$, where the generalised conditional characteristic function (11.51) refers to the bivariate process $(X_1, X_2) := (\log S_1, \log S_2)$.

11.7 Arbitrage Theory More Carefully

The law of one price 9.14, the first and second fundamental theorems of asset pricing 9.15, 11.3, and the superreplication theorem 11.12 constitute arguably the most important results of arbitrage theory. Unfortunately they are literally true only in discrete-time markets with a finite set of times $t = 0, 1, \ldots, T$. In continuous time, unexpected phenomena occur even in seemingly innocent standard models.

Example 11.46 (Geometric Brownian Motion) We consider the bond and stock model of Example 9.1 with $r = 0$ and $\mu + \sigma^2/2 = 0$, i.e. the discounted stock $\widehat{S}_1 = S_1$ is actually a martingale. We show there exists a self-financing strategy $\varphi = (\varphi_0, \varphi_1)$ with $V_\varphi(0) = 0$ and $V_\varphi(1) = 1$. Put differently, a riskless profit of 1€ can be obtained from zero initial investment.

Step 1: Recall that standard Brownian motion W is a martingale which reaches the level 1 in finite time, i.e. the stopping time $\tau := \inf\{t \geq 0 : W(t) = 1\}$ is finite with probability 1, cf. e.g. [204, Proposition 3.3.6]. This observation is crucial for the construction below.

Step 2: In order to reach the level 1 before $T = 1$, we consider $\xi \bullet W$ for the deterministic integrand $\xi(t) = 1/(1-t)$ and $t < 1$. This process can be viewed as a time-changed Wiener process which exists only up to time 1. Specifically, the law of $(\xi \bullet W(t))_{t\in[0,1)}$ coincides with the law of $(W(t/(1-t)))_{t\in[0,1)}$. By step 1, $\xi \bullet W$ reaches the level 1 with probability one, i.e. $\tau := \inf\{t \in [0,1) : \xi \bullet W(t) = 1\}$ is less than one. We set $\widetilde{\xi} := \xi 1_{[\![0,\tau]\!]}$.

Step 3: Recall that our market is given by the assets $S_0 = 1$ and $S_1 = \mathscr{E}(W) = 1 + S_1 \bullet W$. Let $\varphi = (\varphi_0, \varphi_1)$ be the self-financing strategy of Rule 9.6 with $V_\varphi(0) = 0$ and risky position $\varphi_1 := \widetilde{\xi}/S_1$. Since φ is self-financing and $S_0 = 1$, we conclude that

$$V_\varphi(1) = V_\varphi(0) + \varphi \bullet S(1) = \varphi_1 \bullet S_1(1) = \widetilde{\xi} \bullet W(1) = 1$$

as desired.

Note that the arbitrage strategy φ can be turned into a *suicide strategy* by considering $\widetilde{\varphi} = (\widetilde{\varphi}_0, \widetilde{\varphi}_1) := (1 - \varphi_0, -\varphi_1)$. We have that $\widetilde{\varphi}$ is self-financing with $V_{\widetilde{\varphi}} \geq 0$, $V_{\widetilde{\varphi}}(0) = 1$, and $V_{\widetilde{\varphi}}(1) = 1 - V_\varphi(0) = 0$. In particular, it produces a certain loss of 1€.

The previous example contradicts the law of one price and the first FTAP as stated in Sect. 9.3. Of course, continuous-time trading exists only as a limit in practice. A real trader would face a certain risk of losing money if she were trying to follow the strategy in Example 11.46. In order to rule out arbitrage in theory as well, we must limit the choice to, in some sense, *admissible* strategies. One could, for example, consider only piecewise constant strategies, which are the only ones that are feasible in real life. Such a reasonable choice, however, results in an unsatisfactory theory. Indeed, the replicating strategy of a European call option in the Black–Scholes model would no longer be admissible. The ultimate goal is to choose a notion of admissibility that excludes arbitrage opportunities as in Example 11.46 but allows for a satisfactory mathematical theory. Unfortunately, no set of trading strategies seems to be appropriate for all applications of interest such as option pricing, hedging, and portfolio optimisation. Therefore, one works with several notions of admissibility, depending on the particular problem. A widely accepted notion is given in Definition 11.48 below.

We work in the setup of Sect. 9.1, i.e. we consider a finite time horizon $T < \infty$ and a semimartingale $S = (S_0(t), \dots, S_d(t))_{t\in[0,T]}$ on some filtered space with trivial initial σ-field $\mathscr{F}_0$. We suppose that S_0 and $(S_0)_-$ are positive so that the discounted asset price process $\widehat{S} := S/S_0 = (1, S_1/S_0, \dots, S_d/S_0)$ makes sense. Recall that a trading strategy is a predictable process $\varphi = (\varphi_0(t), \dots, \varphi_d(t))_{t\in[0,T]}$ whose wealth process is defined in (9.4). In line with Sect. 9.1 we call φ

self-financing if φ is S-integrable and (9.5) holds. Recall from Rule 9.5 that self-financability is stable under discounting.

Definition 11.47 A self-financing trading strategy φ is **admissible** in the sense of [70] if the discounted wealth $\widehat{V}_\varphi$ is bounded from below by a constant.

Note that this notion of admissibility depends on the choice of the numeraire.

The finite credit line in Definition 11.47 suffices to rule out the phenomenon in Example 11.46. It serves particularly well for so-called *fundamental theorems of asset pricing (FTAPs)*. For optimisation problems one considers different sets, e.g. all portfolios with nonnegative wealth V_φ or all strategies which can be approximated in some sense by piecewise constant strategies.

Unfortunately, the absence of admissible arbitrage strategies does not suffice for the existence of an equivalent martingale measure. In order to obtain a rigorous version of the first FTAP, we need to relax the notion of arbitrage slightly.

Definition 11.48 A random variable $H \geq 0$ with $P(H > 0) > 0$ is called a *free lunch with vanishing risk* if, for any $\varepsilon > 0$, there exists some admissible strategy φ with $\widehat{V}_\varphi(0) \leq \varepsilon$ and $\widehat{V}_\varphi(T) \geq H$. The market is said to satisfy the condition **no free lunch with vanishing risk (NFLVR)** if no such random variable H exists.

The random variable H in the previous definition stands for a discounted payoff that is not really "free" but can be acquired arbitrarily cheaply. We are now ready to state a rigorous version of the first FTAP.

Theorem 11.49 (FTAP, Admissible Version) *The following statements are equivalent.*

1. *Condition NFLVR holds, i.e. there exists no free lunch with vanishing risk.*
2. *There exists an* ***equivalent*** σ***-martingale measure*** Q*, i.e. some probability measure* $Q \sim P$ *such that the discounted price process* $\widehat{S}$ *is a* σ*-martingale relative to* Q*.*

Proof This is shown in [72, Theorem 1.1].

This version of the FTAP may be the most prominent and satisfactory one but it is by no means the only possibility. If the notion of admissibility or arbitrage is relaxed, the dual set of probability measures in the FTAP narrows. One may, for example, consider riskless gains in the sense that some positive random variable can be approximated in L^2 by the payoff of piecewise constant strategies. From Théorème 2 in [281] it follows that the absence of such L^2*-free lunches* is equivalent to the existence of some true EMM Q with square-integrable density dQ/dP. Formally, an L^2*-free lunch* refers to a random variable $H \geq 0$ with $P(H > 0) > 0$ such that for any $\varepsilon > 0$ there exist some nonnegative random variable Y and some bounded self-financing strategy φ which is piecewise constant between finitely many stopping times and satisfies $\widehat{V}_\varphi(0) = 0$ as well as

$$E\left((\widehat{V}_\varphi(T) - Y - H)^2\right) \leq \varepsilon.$$

This means that the riskless gain H can, after possibly throwing away some money, be approximated arbitrarily well in the mean square sense by the payoff of a simple trading strategy.

Theorem 11.50 (FTAP, L^2-Version) *Let $E(|S(t)|^2) < \infty$, $t \in [0, T]$. The following statements are equivalent.*

1. *There are no L^2-free lunches.*
2. *There exists an equivalent martingale measure Q with square-integrable density dQ/dP.*

Proof This is shown in [281, Théorème 2]. □

Unfortunately, the concept of admissible strategies in Definition 11.47 does not allow for a law of one price. Indeed, the suicide strategy in Example 11.46 is an admissible strategy in a market satisfying NFLVR. Since the law of one price cannot be dispensed with if one wants to get anywhere in applications, we use two slight modifications of the notion of admissibility such that the four main rules 9.14, 9.15, 11.3, 11.12 can be stated rigorously. To this end, we suppose that the price processes $S_1, \ldots, S_d$ are nonnegative as well.

Definition 11.51

1. A self-financing strategy φ is called **allowable** or **market-admissible** if it is admissible relative to the numeraire $S_0 + \cdots + S_d$, i.e. if $V_\varphi(t) \geq -c(S_0(t) + \cdots + S_d(t))$, $t \in [0, T]$ holds for some $c < \infty$.
2. We call an allowable strategy φ **maximal allowable, maximal market-admissible** or **regular** if its terminal value is not dominated by any other allowable strategy ψ with the same initial value. That is, for any allowable ψ with $V_\psi(0) = V_\varphi(0)$ and $V_\psi(T) \geq V_\varphi(T)$ we have $V_\psi(T) = V_\varphi(T)$ almost surely.
3. We say that the market satisfies **NFLVR'** if there exists no free lunch as in Definition 11.48, but defined in terms of allowable rather than admissible strategies.

The counterpart of Theorem 9.15 now reads as

Theorem 11.52 (FTAP, Allowable Version) *The following statements are equivalent.*

1. *Condition NFLVR' holds, i.e. there are no allowable free lunches with vanishing risk.*
2. *There exists an equivalent martingale measure Q, i.e. some $Q \sim P$ such that $\widehat{S}$ is a Q-martingale.*

Proof We denote by $\widetilde{S} := S/(S_0 + \cdots + S_d)$ the discounted price process relative to the numeraire $S_0 + \cdots + S_d$ instead of S_0. The corresponding wealth process of a strategy φ is denoted by $\widetilde{V}_\varphi := V_\varphi/(S_0 + \cdots + S_d)$.

Step 1: Condition NFLVR' holds if and only if it holds relative to the numeraire $S_0 + \cdots + S_d$ instead of S_0. This follows immediately from $\widetilde{V}_\varphi = \widehat{V}_\varphi S_0/(S_0 + \cdots + S_d)$.

Step 2: There exists an EMM Q for $\widehat{S}$ if and only if there is an EMM $\widetilde{Q}$ for $\widetilde{S}$. Indeed, this follows from the construction in Rule 9.18 applied to $S_0 + \cdots + S_d$ instead of S_0.

Step 3: By steps 1 and 2, we may choose $S_0 + \cdots + S_d$ instead of S_0 as the numeraire asset. The admissible version Theorem 11.49 of the FTAP states that NFLVR' holds if and only if $\widetilde{S}$ is a $\widetilde{Q}$-σ-martingale relative to some probability measure $\widetilde{Q} \sim P$. But since $\widetilde{S}$ is bounded, this is actually equivalent to $\widetilde{S}$ being a $\widetilde{Q}$-martingale, as desired. □

One cannot replace allowable by maximal allowable in Definition 11.51(3) if one wants Theorem 11.52 to hold. Indeed, maximal allowable strategies do not even exist in some markets that allow for a free lunch.

If NFLVR' holds, a given allowable strategy φ is maximal allowable if and only if $\widehat{V}_\varphi$ is a Q-martingale relative to some EMM Q. This follows from [71, Theorem 2.5] after switching to the numeraire $S_0 + \cdots + S_d$.

We are now ready to state a rigorous version of Theorem 9.14.

Theorem 11.53 (Law of One Price, Maximal Allowable Version) *Suppose that the market satisfies NFLVR'. If φ, ψ are maximal allowable strategies with the same final value $V_\varphi(T) = V_\psi(T)$, then $V_\varphi(t) = V_\psi(t)$ for all $t \in [0, T]$. Accordingly, we have $V_\varphi \leq V_\psi$ if we only assume $V_\varphi(T) \leq V_\psi(T)$ for the final values.*

Proof As in the previous proof, we consider the numeraire $S_0 + \cdots + S_d$ and discounted values $\widetilde{V}_\varphi = V_\varphi/(S_0 + \cdots + S_d)$, $\widetilde{V}_\psi = V_\psi/(S_0 + \cdots + S_d)$ relative to this numeraire. According to [71, Corollary 2.19] there is an EMM $\widetilde{Q}$ for $\widetilde{S}$ such that both $\widetilde{V}_\varphi$ and $\widetilde{V}_\psi$ are $\widetilde{Q}$-martingales. Since they have the same final value, the first assertion follows. In the second case we have

$$\widetilde{V}_\varphi(t) = E_{\widetilde{Q}}\big(\widetilde{V}_\varphi(T)\big|\mathscr{F}_t\big) \leq E_{\widetilde{Q}}\big(\widetilde{V}_\psi(T)\big|\mathscr{F}_t\big) = \widetilde{V}_\psi(t), \quad t \in [0, T]$$

as desired. □

In order to state a version of the second FTAP, we call a random variable H an **allowable** or **market-admissible claim** if $|H| \leq c(S_0(T) + \cdots + S_d(T))$ for some $c < \infty$. This holds, for example, if the discounted payoff $\widehat{H} := H/S_0(T)$ is bounded. Moreover, we say that the market is **complete** if any allowable claim H is **attainable** by some maximal allowable strategy φ, i.e. $H = V_\varphi(T)$. We obtain the

Theorem 11.54 (Second FTAP, Maximal Allowable Version) *Suppose that the market satisfies NFLVR'. It is complete if and only if there exists only one EMM.*

Proof This follows immediately from Theorem 11.56 below. □

Finally, we state rigorous versions of Rules 11.12 and 11.2.

Theorem 11.55 (Superreplication Theorem, Maximal Allowable Version) *Suppose that the market satisfies NFLVR' and let H denote the payoff of some allowable claim. Then*

$$\begin{aligned}\pi_{\text{high}}(H) := & \inf\big\{\pi \in \mathbb{R} : \textit{There exists some maximal allowable strategy } \varphi \\ & \textit{with } V_\varphi(0) = \pi \textit{ and } V_\varphi(T) \geq H\big\} \\ = & \min\big\{\pi \in \mathbb{R} : \textit{There exists some maximal allowable strategy } \varphi \\ & \textit{with } V_\varphi(0) = \pi \textit{ and } V_\varphi(T) \geq H\big\} \\ = & \sup\big\{S_0(0)E_Q(\widehat{H}) : Q \textit{ EMM}\big\}\end{aligned}$$

and accordingly

$$\begin{aligned}\pi_{\text{low}}(H) := & \sup\big\{\pi \in \mathbb{R} : \textit{There exists some maximal allowable strategy } \varphi \\ & \textit{with } V_\varphi(0) = -\pi \textit{ and } V_\varphi(T) \geq -H\big\} \\ = & \max\big\{\pi \in \mathbb{R} : \textit{There exists some maximal allowable strategy } \varphi \\ & \textit{with } V_\varphi(0) = -\pi \textit{ and } V_\varphi(T) \geq -H\big\} \\ = & \inf\big\{S_0(0)E_Q(\widehat{H}) : Q \textit{ EMM}\big\}.\end{aligned}$$

As in (11.7, 11.8) we call $\pi_{\text{high}}(H)$ and $\pi_{\text{low}}(H)$ ***upper*** *and* ***lower price*** *of H.*

Proof The proof is essentially the same as indicated for Rule 11.12. We use the notation of the previous proofs.

Step 1: For any allowable φ with $V_\varphi(T) \geq H$ and any EMM $\widetilde{Q}$ for $\widetilde{S} = S/(S_0 + \cdots + S_d)$, the discounted wealth $\widetilde{V}_\varphi = \widetilde{V}_\varphi(0) + \varphi \bullet \widetilde{S}$ is a $\widetilde{Q}$-σ-martingale which is bounded from below. Consequently it is a Q-supermartingale by Proposition 3.27. This implies

$$\begin{aligned}V_\varphi(0) = (S_0 + \cdots + S_d)(0)\widetilde{V}_\varphi(0) &\geq (S_0 + \cdots + S_d)(0)E_{\widetilde{Q}}(\widetilde{V}_\varphi(T)) \\ &\geq (S_0 + \cdots + S_d)(0)E_{\widetilde{Q}}(\widetilde{H}).\end{aligned}$$

In view of Rule 9.18, we have

$$\begin{aligned}&\big\{S_0(0)E_Q(\widehat{H}) : Q \text{ EMM}\big\} \\ &\quad = \Big\{(S_0 + \cdots + S_d)(0)E_{\widetilde{Q}}(\widetilde{H}) : \widetilde{Q} \text{ EMM for numeraire } S_0 + \cdots + S_d\Big\}\end{aligned}$$

for $\widetilde{H} = H/(S_0 + \cdots + S_d)(T)$. This yields $\pi_{\text{high}}(H) \geq \sup\{S_0(0)E_Q(\widehat{H}) : Q \text{ EMM}\}$.

Step 2: As in step 1, we consider the numeraire $S_0 + \cdots + S_d$ instead of S_0. Set

$$\pi := \sup\left\{S_0(0)E_Q(\widehat{H}) : Q \text{ EMM}\right\}$$
$$= \sup\left\{(S_0 + \cdots + S_d)(0)E_{\widetilde{Q}}(\widetilde{H}) : \widetilde{Q} \text{ EMM for numeraire } S_0 + \cdots + S_d\right\}$$

and

$$\widetilde{S}_{d+1}(t) := \operatorname{ess\,sup}\left\{E_{\widetilde{Q}}(\widetilde{H}|\mathscr{F}_t) : \widetilde{Q} \text{ EMM for numeraire } S_0 + \cdots + S_d\right\}.$$

By Proposition 3.65 the process $\widetilde{S}_{d+1}$ is a $\widetilde{Q}$-supermartingale relative to any EMM $\widetilde{Q}$ for $\widetilde{S}$. Theorem 3.64 on optional decomposition yields the existence of an $\widetilde{S}$-integrable process φ such that $C := \widetilde{S}_{d+1}(0) + \varphi \bullet \widetilde{S} - \widetilde{S}_{d+1}$ is increasing. As usual we can identify φ with the self-financing strategy with discounted initial value $\widetilde{S}_{d+1}(0) = \pi/(S_0 + \cdots + S_d)(0)$, cf. Rule 9.6. Since $\widetilde{S}_{d+1}$ is bounded, $\widetilde{V}_\varphi$ is bounded from below and hence φ is allowable.
Moreover, $\widetilde{H} = \widetilde{S}_{d+1}(T) \leq \widetilde{S}_{d+1}(0) + \varphi \bullet \widetilde{S}(T) = \widetilde{V}_\varphi(T)$. By [71, Theorem 2.3] there is some maximal allowable ψ with $V_\psi(0) = V_\varphi(0)$ and $V_\psi(T) \geq V_\varphi(T)$. Altogether, this yields $\pi \geq \pi_{\text{high}}(H)$ and hence $\pi = \pi_{\text{high}}(H)$ by step 1. Moreover, it shows that the infimum in the definition of $\pi_{\text{high}}(H)$ is actually attained, namely by the strategy φ above.
Step 3: The second statement follows by considering $-H$ instead of H. □

Theorem 11.56 *Suppose that the market satisfies NFLVR' and let H denote the payoff of some allowable claim. We have $\pi_{\text{high}}(H) = \pi_{\text{low}}(H)$ in Theorem 11.55 if and only if H is attainable by some maximal allowable strategy. In this case we call $\pi := \pi_{\text{high}} = \pi_{\text{low}}$ the* ***unique fair price*** *of the claim.*

Proof ⇒ By Theorem 11.55 there exist maximal allowable strategies $\overline{\varphi}, \underline{\varphi}$ with $\pi_{\text{high}}(H) = V_{\overline{\varphi}}(0)$, $-\pi_{\text{low}}(H) = V_{\underline{\varphi}}(0)$, $V_{\overline{\varphi}}(T) \geq H$, $V_{\underline{\varphi}}(T) \geq -H$. According to [71, Corollary 2.13] there is an EMM $\widetilde{Q}$ for $\widetilde{S} = S/(S_0 + \cdots + S_d)$ such that both $\widetilde{V}_{\overline{\varphi}}$ and $\widetilde{V}_{\underline{\varphi}}$ are $\widetilde{Q}$-martingales. This implies

$$\begin{aligned}\frac{\pi_{\text{low}}(H)}{(S_0 + \cdots + S_d)(0)} &= -\widetilde{V}_{\underline{\varphi}}(0) = -E_{\widetilde{Q}}\big(\widetilde{V}_{\underline{\varphi}}(T)\big)\\ &\leq E_{\widetilde{Q}}(\widetilde{H})\\ &\leq E_{\widetilde{Q}}\big(\widetilde{V}_{\overline{\varphi}}(T)\big) = \widetilde{V}_{\overline{\varphi}}(0) = \frac{\pi_{\text{high}}(H)}{(S_0 + \cdots + S_d)(0)}\end{aligned}$$

for $\widetilde{H} := H/(S_0 + \cdots + S_d)(T)$. If $\pi_{\text{low}}(H) = \pi_{\text{high}}(H)$, we conclude that $E_{\widetilde{Q}}(\widetilde{V}_{\overline{\varphi}}(T)) - E_{\widetilde{Q}}(\widetilde{H}) = 0$, which implies $\widetilde{V}_{\overline{\varphi}}(T) = \widetilde{H}$ because $\widetilde{V}_{\overline{\varphi}}(T) \geq \widetilde{H}$. Hence $\overline{\varphi}$ replicates H.
⇐ *Step 1:* Suppose that $V_\varphi(T) = H$ for some maximal allowable φ. Since $\widetilde{V}_\varphi$ is a $\widetilde{Q}$-martingale for some EMM $\widetilde{Q}$ for $\widetilde{S}$ and since $\widetilde{H} = H/(S_0 + \cdots + S_d)(T)$

is bounded, $\widetilde{V}_\varphi$ is bounded as well. Consequently, $-\varphi$ is allowable. Since $\widetilde{V}_{-\varphi} = -\widetilde{V}_\varphi$ is a $\widetilde{Q}$-martingale, $-\varphi$ is maximal allowable due to the characterisation in [71, Theorem 2.5]. We conclude that $-H = V_{-\varphi}(T)$ is attainable by some maximal allowable strategy as well.

Step 2: Recall from step 1 that $\widetilde{H} = \widetilde{V}_\varphi(T)$ and $-\widetilde{H} = \widetilde{V}_{-\varphi}(T)$, where φ and $-\varphi$ are maximal allowable. Moreover, Theorem 11.55 yields that there exist maximal allowable strategies $\overline{\varphi}, \underline{\varphi}$ with $V_{\overline{\varphi}}(0) = \pi_{\text{high}}(H)$, $V_{\underline{\varphi}}(0) = -\pi_{\text{low}}(H)$ and $V_{\overline{\varphi}}(T) \geq H$, $V_{\underline{\varphi}}(T) \geq -H$. According to [71, Corollary 2.13] there exists an EMM $\widetilde{Q}$ for $\widetilde{S} = S/(S_0 + \cdots + S_d)$ such that $\widetilde{V}_\varphi$, $\widetilde{V}_{\overline{\varphi}}$, $\widetilde{V}_{\underline{\varphi}}$ are $\widetilde{Q}$-martingales. This implies

$$\begin{aligned}\frac{\pi_{\text{high}}(H)}{(S_0 + \cdots + S_d)(0)} &= \widetilde{V}_{\overline{\varphi}}(0) = E_{\widetilde{Q}}\big(\widetilde{V}_{\overline{\varphi}}(T)\big)\\ &\geq E_{\widetilde{Q}}(\widetilde{H}) = E_{\widetilde{Q}}\big(\widetilde{V}_\varphi(T)\big) = \widetilde{V}_\varphi(0).\end{aligned}$$

Since $V_\varphi(0)$ allows us to superreplicate H, we have $V_\varphi(0) \geq \pi_{\text{high}}(H)$ and hence $\pi_{\text{high}}(H)/(S_0 + \cdots + S_d)(0) = \widetilde{V}_\varphi(0)$. Similarly, we obtain

$$-\frac{\pi_{\text{low}}(H)}{(S_0 + \cdots + S_d)(0)} \geq E_{\widetilde{Q}}(-\widetilde{H}) = \widetilde{V}_{-\varphi}(0).$$

Since $V_{-\varphi}(0)$ allows us to superreplicate $-H$, we have $-V_{-\varphi}(0) \leq \pi_{\text{low}}(H)$ and hence $\pi_{\text{low}}(H) = -V_{-\varphi}(0) = V_\varphi(0)$. Altogether, we conclude that $\pi_{\text{high}}(H) = \pi_{\text{low}}(H)$, as desired. □

Appendix 1: Problems

The exercises correspond to the section with the same number.

11.1 (Static Replication) Suppose that call options with payoff $(S_1(T) - K)^+$ for any $K \geq 0$ are available on the market in arbitrary, even fractional and negative amounts. How can a payoff $H = f(S_1(T))$ with differentiable $f : \mathbb{R}_+ \to \mathbb{R}$ be obtained as a portfolio of such calls and the bond with payoff 1? Express the value of this portfolio in terms of the market prices $\pi(K)$ of the calls with strike $K \geq 0$ and of the market price of a bond with payoff 1.

11.2 (Futures on a Dividend-Paying Asset) Suppose that we are in the situation of Example 11.10, but dividends $D_1 = S_1 \bullet A$ are paid on S_1 by some deterministic process A, i.e. dividends are paid in proportion to the present asset price. Derive the futures price and the replicating strategy for a futures contract on S_1 with maturity T.

Hint: Show that $\widehat{S}_1\mathscr{E}(A)$ is a Q-martingale relative to some EMM Q for $\widehat{S}_1 + \widehat{D}_1$.

11.3 (Variance Swap) Suppose that $S = (S_0, S_1, S_2)$ is an arbitrage-free market such that S_0, S_1 are continuous, S_0 is deterministic, and S_2 is the price process of an option with payoff $\log S_1(T)$ at time T. Show that the variance swap with time-T-payoff

$$H = \frac{1}{T}[\log S_1, \log S_1](T) - K$$

for $K > 0$ is attainable in this market. How is the *variance strike* to be chosen in order for the fair price of H at time 0 to vanish? Does replicability hold for a discontinuous process S_1, too?

Hint: Apply Itō's formula to $\log \widehat{S}_1$.

11.4 (Model Risk) In a Black–Scholes model as in Example 9.1 the fair price of a claim with payoff $f(S_1(T))$ is of the form $w(t, S_1(t))$ for some function w, cf. Examples 11.19 and 11.24. Moreover, $\varphi_1(t) = D_2 w(t, S_1(t))$ provides a perfect hedge for the option, i.e.

$$w(0, S_1(0)) + \varphi_1 \bullet \widehat{S}_1(t) - e^{-rt} w(t, S_1(t)) \tag{11.53}$$

is constant over time, where the first two terms represent the discounted value of the hedge portfolio and the last is the discounted fair option price. Now suppose that the trader wrongly believes to be exposed to the above market, but the stock price S_1 really follows Heston dynamics

$$d(\log S_1)(t) = -\frac{1}{2} Y(t)dt + \sqrt{Y(t)} dW(t),$$

$$dY(t) = (\kappa - \lambda Y(t))dt + \widetilde{\sigma}\sqrt{Y(t)} dZ(t)$$

as in Sect. 8.2.4. She will realise her error by the fact that the perceived overall value (11.53) of the portfolio is not constant. How does it change over time, i.e. what is its Itō process representation? What do you observe?

11.5 Suppose that the discounted prices of European calls $E_Q((\widehat{S}_1(T) - K)^+)$ are known for all strikes $K > 0$, e.g. because they are observed as liquid market prices. Show that this determines the law of $S_1(T)$ under the pricing measure Q uniquely.

11.6 (Barrier Option) Determine an integral representation (11.50) for $f(x_1, x_2) = (e^{x_1} - K)^+ 1_{\{x_2 \leq B\}}$ with $B > K > 0$. Moreover, suppose that the generalised characteristic function

$$\varphi_{0,T}(z_1, z_2) = E_Q(e^{iz_1 X(T) + iz_2 M(T)})$$

is known for some process (X, M). Use this to obtain an integral representation for $E_Q(H)$ with $H := (e^{X(T)} - K)^+ 1_{\{M(T) \leq B\}}$ and deterministic $S_0(T)$. If $S_1(T) =$

$e^{X(T)}$ and $M(T) = \sup_{t \leq T} X(t)$, the random variable H is the discounted payoff of an *up-and-out call* with knock-out-barrier e^B.

11.7 Consider the market $S = (S_0, S_1)$ with $S_0 = 1$ and $S_1 = 1 - (\widetilde{\xi}/2) \bullet W$, where $\widetilde{\xi}$ is defined as in Example 11.46 relative to same Wiener process W. Show that this market allows for free lunch with vanishing risk. Alternatively, let $S_0 = 1 - (\widetilde{\xi}/2) \bullet W$ and $S_1 = 1$, i.e. the same market as above but with $1 - (\widetilde{\xi}/2) \bullet W$ considered as the numeraire asset. Show that this market satisfies condition NFLVR.

Appendix 2: Notes and Comments

The concepts of Sect. 11.1 are introduced in most textbooks on Mathematical Finance, cf. e.g. [29]. Rule 11.1 is a corollary of the first fundamental theorem of asset pricing, at least if the latter is stated in terms of true martingale measures. We refer to the discussion in Sect. 11.7 for details. Rule 11.2 goes back to [150, Theorem 3.1] and is reconsidered in Theorem 11.56. Rule 11.3 is due to [134], see also Theorem 11.54 for an exact statement. Example 11.4 is considered in [133]. The content of Remark 11.5 is stated for Itō processes in [29, Meta Theorem 8.3.1], for an early discussion in discrete time see [285]. For Rule 11.6 we refer to [171, Theorem 2.9]. Remark 11.8 is observed already in [219]. Versions of Rule 11.9 and Example 11.10 can be found in [29, Proposition 26.6]. Rule 11.12 is the subject of [196], see also Theorem 11.55 for a precise statement. Corollary 11.13 is reconsidered in Theorem 11.56. For a longer discussion of Example 11.15 see, for example, [279, Section 4.5.6]. Example 11.16 is shown under more restrictive assumptions on the Lévy process in [81, Theorem 2]. For the essence of Rule 11.18 we refer once more to [196]. Example 11.19 is discussed in any textbook on Mathematical Finance. For a discussion of Example 11.20 see also [69]. A classical reference to change of numeraire techniques as in Example 11.22 is [122]. The monograph [143] is devoted to solving the PIDEs in Sect. 11.5.2 by finite element methods. For numerical aspects of option pricing in general see [272] and in particular [123] for Monte Carlo methods. Integral transform methods for computing option prices have been introduced by [48, 240]. The integral representations of the payoffs in Sect. 11.6 are taken from [47, 147, 148, 240], see also [92]. Forward-start options as in Example 11.35 are discussed in [60, Section 11.2]. For a related approach to using integral transforms in the valuation of option prices see [101].

For an in-depth discussion of Sect. 11.7 we recommend [73], complemented by [293]. Definition 11.48 originates from [70]. The concept of allowable strategies is due to [294]. Regular strategies are introduced in [293] as allowable strategies whose discounted wealth process is a martingale under at least one EMM. Our equivalent definition, however, is inspired by the related notion of *maximal admissible* strategies from [71]. The equivalence in Theorem 11.52 is considered in [294, Theorem 3.2]. For an alternative short proof of Theorem 11.53 see [293, Lemma 6.1]. Theorem 11.54 can also be derived from the original statement in

[134]. Superreplication dualities as in Theorem 11.55 served as the motivation for studying optional decompositions in [196]. As observed in [293, Theorems 6.2, 6.3], Theorem 11.56 follows almost immediately from the original version in [150, Theorem 3.1].

Problems 11.1 and 11.5 correspond to an observation of [40]. A thorough discussion of Problem 11.3 can be found in [53]. The interesting phenomenon of Problem 11.4 has been observed in [97]. For background on Problem 11.6 we refer to [92, 93].

Chapter 12
Mean-Variance Hedging

In Sect. 11.3 we discussed pricing and hedging of OTC derivatives from the option writer's point of view. The approach was based on arbitrage arguments. However, the superreplication price π_{high} and the corresponding strategy are often unrealistic for practical purposes, as we noted after Corollary 11.13. Put differently, arbitrage arguments alone do not provide satisfactory answers to how one should proceed as an option writer.

In the following two chapters we discuss alternative concepts which take individual preferences into account. First we consider quadratic hedging, which may be viewed as an intuitive, easy to understand approach which is mathematically tractable. Its symmetric treatment of profits and losses is sometimes criticised from an economical point of view. As an alternative, we discuss the conceptually more satisfactory approach of utility-based valuation and hedging in Chap. 13. This mathematically more involved problem simplifies if one considers a first-order approximation, which reduces to a quadratic hedging problem.

We place ourselves into the setup of Sect. 11.3. More specifically, we consider liquidly traded securities $S_0, \dots, S_d$ and a contingent claim with payoff H at time T, which will typically be a deterministic function of the underlyings such as a call $H = (S_1(T) - K)^+$ on S_1. At this point we only assume H to be an $\mathscr{F}_T$-measurable random variable. For ease of exposition we assume $S_0, \dots, S_d$ to refer to discounted quantities, i.e. they are expressed relative to the zeroth asset as numeraire. Put differently, we have $S_0 = 1$ and it suffices to work with $S = (S_1, \dots, S_d)$ as explained in Sect. 9.1.2.

As an option writer you face the problem of what premium to charge if a potential buyer arrives at time 0 and how to hedge the risk involved in selling the claim. Quadratic hedging starts from the second question. If the claim were attainable, the risk could be eliminated entirely by investing in the replicating portfolio. In general such a perfect hedge does not exist. A natural way out is to try and hedge the option as efficiently as possible and to require some compensation for the remaining unhedgable risk. For reasons of mathematical tractability we measure

© Springer Nature Switzerland AG 2019

E. Eberlein, J. Kallsen, *Mathematical Finance*, Springer Finance,
https://doi.org/10.1007/978-3-030-26106-1_12

risk in terms of the mean squared error. In order to hedge the option, you choose the **variance-optimal hedging strategy**, i.e. the self-financing strategy φ^* minimising the **quadratic risk**

$$\varepsilon^2(\varphi) := E\big((V_\varphi(T) - H)^2\big). \tag{12.1}$$

As in Remark 10.2(2) it follows that the optimiser is essentially unique. More specifically, we have $V_{\varphi^\star} = V_{\widetilde{\varphi}}$ if $\varphi^\star, \widetilde{\varphi}$ are both variance-optimal hedging strategies.

The minimal quadratic risk is 0 for attainable claims because the option can be replicated by some φ^*. This in turn holds if and only if the interval $[\pi_{\text{low}}(H), \pi_{\text{high}}(H)]$ of Sect. 11.3 reduces to a singleton, namely $V_{\varphi^*}(0)$. In general, however, the minimal quadratic risk $\varepsilon^2(\varphi^*)$ does not vanish. A reasonable suggestion may be to charge

$$\pi := V_{\varphi^*}(0) + \frac{p}{2}\varepsilon^2(\varphi^*) \tag{12.2}$$

(or maybe $\pi := V_{\varphi^*}(0) + p\varepsilon(\varphi^*)$) as the option premium, where $\frac{p}{2}\varepsilon^2(\varphi^*)$ (resp. $p\varepsilon(\varphi^*)$) compensates for the unhedgable risk and $p \geq 0$ denotes a parameter chosen according to your personal risk aversion. The somewhat odd factor $1/2$ in (12.2) is introduced for consistency with related pricing formulas in Sect. 13.1.3 and at the end of Sect. 13.2. We call $V_{\varphi^*}(0)$ the **optimal initial endowment** for the hedging problem (12.1).

The question of how to determine the variance-optimal hedging strategy φ^*, the optimal initial endowment $V_{\varphi^*}(0)$, and the minimal quadratic risk $\varepsilon^2(\varphi^*)$ turns out to be considerably easier if S is a martingale under the objective measure P. We discuss this case in the following section. The more involved general setup is treated in Sect. 12.2.

12.1 The Martingale Case

12.1.1 General Structure

If S happens to be a martingale, the solution to the quadratic hedging problem can be expressed in terms of the Galtchouk–Kunita–Watanabe (GKW) decomposition of H relative to S. More precisely, denote by

$$V(t) := E(H|\mathscr{F}_t) \tag{12.3}$$

the martingale generated by the option payoff H, which we call the **mean value process** of the claim. But since the option is not assumed to be traded between times 0 and T, this does not necessarily refer to a value in any precise economical sense. The initial value $V(0)$ coincides with the optimal initial endowment of the quadratic hedging problem. The variance-optimal hedging strategy $\varphi^\star$ is given by the integrand φ in the GKW decomposition

$$V = V(0) + \varphi \bullet S + N$$

of V relative to S, cf. Theorem 3.63. The minimal quadratic risk equals $\varepsilon^2(\varphi^\star) = E(\langle N, N\rangle(T))$. More specifically, we have the following

Rule 12.1 *Define the value process V of the option as in (12.3). Denote by $(\varphi_1^\star, \ldots, \varphi_d^\star)$ the integrand in the GKW decomposition of V relative to S, i.e. a predictable process satisfying*

$$\langle V, S\rangle = (\varphi_1^\star, \ldots, \varphi_d^\star) \bullet \langle S, S\rangle, \tag{12.4}$$

cf. Theorem 3.63. Moreover, let $\varphi^\star = (\varphi_0^\star, \ldots, \varphi_d^\star)$ be the associated self-financing trading strategy in the sense of Rule 9.6 for initial wealth $V(0)$. Then $\varphi^\star$ minimises (12.1) and the minimal quadratic risk equals

$$\varepsilon^2(\varphi^\star) = E\big(\langle V, V\rangle(T) - \langle \varphi^* \bullet S, \varphi^* \bullet S\rangle(T)\big).$$

Idea For any competing self-financing strategy φ we have

$$\begin{aligned} E\big(V_\varphi(T)(V(T) - V_{\varphi^\star}(T))\big) &= E\big(V_\varphi(T)N(T)\big) \\ &= V_\varphi(0)N(0) + E\big(\langle V_\varphi, N\rangle(T)\big) \\ &= 0 + E\big(\varphi \bullet \langle S, N\rangle(T)\big) = 0, \end{aligned}$$

which means that $V_{\varphi^\star}(T)$ is the orthogonal projection of $V(T) = H$ on the set of all attainable final wealths $V_\varphi(T)$. Since the orthogonal projection minimises the L^2-distance and hence the quadratic risk, $\varphi^\star$ is variance-optimal. The expression for the minimal quadratic risk follows from

$$\begin{aligned} \varepsilon^2(\varphi^\star) &= E\big((V_{\varphi^\star}(T) - V(T))^2\big) \\ &= E\big([V - V_{\varphi^\star}, V - V_{\varphi^\star}](T)\big) \\ &= E\big(\langle V - V_{\varphi^\star}, V - V_{\varphi^\star}\rangle(T)\big) \\ &= E\big(\langle V, V\rangle(T) - \langle V, V_{\varphi^\star}\rangle(T)\big) \\ &= E\big(\langle V, V\rangle(T) - \langle V_{\varphi^\star}, V_{\varphi^\star}\rangle(T)\big) \\ &= E\big(\langle V, V\rangle(T) - \langle \varphi^\star \bullet S, \varphi^\star \bullet S\rangle(T)\big), \end{aligned}$$

where we used the orthogonality $\langle V_{\varphi^\star}, V - V_{\varphi^\star}\rangle = 0$ twice. □

Recall that we generally use the notation (II.2) for the modified second characteristic and c^{-1} for the Moore–Penrose pseudoinverse of a matrix c. This allows us to rephrase the formulas in Rule 12.1 as

$$(\varphi_1^\star, \dots, \varphi_d^\star)(t) = \widetilde{c}^S(t)^{-1}\widetilde{c}^{S,V}(t) \tag{12.5}$$

and

$$\varepsilon^2(\varphi^\star) = E\left(\int_0^T \left(\widetilde{c}^V(t) - \widetilde{c}^{S,V}(t)^\top \widetilde{c}^S(t)^{-1}\widetilde{c}^{S,V}(t)\right) dt\right). \tag{12.6}$$

12.1.2 PIDEs for Markov Processes

Although Rule 12.1 provides a satisfactory characterisation from a structural point of view, it may not be obvious how to compute the relevant quantities in concrete models. Similarly as for the valuation problem in Chap. 11 we discuss two different approaches, namely partial integro-differential equations in this and integral transform techniques in the following section.

We assume that the market is driven by some $\mathbb{R}^n$-valued Markov process X with generator G and carré du champ operator Γ, cf. Sect. 5.3. We assume the underlying price process to be of the form $S(t) = f(X(t))$ and consider a claim $H = g(X(T))$ where $f : \mathbb{R}^n \to \mathbb{R}^d$ and $g : \mathbb{R}^n \to \mathbb{R}$ denote deterministic functions. The solution to the quadratic hedging problem involves three objects, namely the initial value of the mean value process, the variance-optimal hedge and the minimal quadratic risk. Since the value process is the martingale generated by the payoff H, it can be determined as in Sect. 11.5.2 by solving a PIDE.

Rule 12.2 (Value Process) *The mean value process of the option satisfies*

$$V(t) = v(T - t, X(t)),$$

where $v : \mathbb{R}^{1+n} \to \mathbb{R}$ *solves the partial integro-differential equation*

$$\frac{\partial}{\partial t} v = Gv$$

subject to $v(0, x) = g(x)$.

Idea This follows immediately from Rule 11.23. □

The hedging strategy is obtained from the value process by a "local" operation as can be seen from (12.5). In the present Markovian context it can be expressed as follows.

Rule 12.3 (Hedging Strategy) *The variance-optimal hedging strategy* $\varphi^\star$ *is given by*

$$(\varphi_1^\star, \dots, \varphi_d^\star)(t) = \big(\Gamma(f, f)(X(t-))\big)^{-1}\Gamma(f, v(T-t, \cdot))(X(t-)) \qquad (12.7)$$

and $V_{\varphi^\star}(0) = v(T, X(0))$, *where* v *denotes the function from Rule 12.2.*

Idea Since $S(t) = f(X(t))$ and $V(t) = v(T-t, X(t))$, we have $\widetilde{c}^S(t) = \Gamma(f, f)(X(t-))$ and $\widetilde{c}^{S,V}(t) = \Gamma(f, v(T-t, \cdot))(X(t-))$ by Proposition 5.21. Hence (12.7) follows from (12.5). The second equation holds because $V_{\varphi^\star}(0) = V(0)$, cf. Rule 12.1. □

The minimal quadratic risk (12.6) corresponds once more to an expectation. Similarly as the mean value process, it can be computed by solving a PIDE. This equation itself involves the solution to the PIDE from Rule 12.2.

Rule 12.4 (Minimal Quadratic Risk) *The minimal quadratic risk equals* $\varepsilon^2(\varphi^\star) = h(T, X(0))$, *where* $h : \mathbb{R}^{1+n} \to \mathbb{R}$ *solves the partial integro-differential equation*

$$\frac{\partial}{\partial t}h - Gh = \Gamma(v, v) - \Gamma(f, v)^\top \Gamma(f, f)^{-1}\Gamma(f, v) \qquad (12.8)$$

subject to $h(0, x) = 0$. *Here,* v *denotes the function from Rule 12.2.*

Idea Using the notation $N(t) = V(t) - V(0) - \varphi^\star \bullet S$ for the orthogonal martingale in the GKW decomposition of V relative to S and

$$U(t) := E\left(\left(H - V(t) - \int_t^T \varphi^\star(s)dS(s)\right)^2 \middle| \mathscr{F}_t\right)$$

for the conditional quadratic risk between t and T, we have

$$\begin{aligned}
U(t) &= E\big((H - V(0) - \varphi^\star \bullet S(T))^2 \big| \mathscr{F}_t\big) \\
&\quad - 2E\big((H - V(0) - \varphi^\star \bullet S(T))\big| \mathscr{F}_t\big)\big(V(t) - V(0) - \varphi^\star \bullet S(t)\big) \\
&\quad + \big(V(t) - V(0) - \varphi^\star \bullet S(t)\big)^2 \\
&= E\big((H - V(0) - \varphi^\star \bullet S(T))^2 \big| \mathscr{F}_t\big) - \big(V(t) - V(0) - \varphi^\star \bullet S(t)\big)^2 \\
&= E\big((H - V(0) - \varphi^\star \bullet S(T))^2 \big| \mathscr{F}_t\big) \\
&\quad - (V(t) - V(0))^2 + 2\varphi^\star \bullet S(t)(V(t) - V(0)) - (\varphi^\star \bullet S(t))^2.
\end{aligned}$$

This equals

$$\begin{aligned}
-\langle V, V\rangle(t) + 2\varphi^\star \bullet \langle S, V\rangle(t) - \sum_{i,j=1}^{d} (\varphi_i^\star \varphi_j^\star) \bullet \langle S_i, S_j\rangle(t) \\
= -\left(\widetilde{c}^V - 2(\varphi^\star)^\top \widetilde{c}^{S,V} + (\varphi^\star)^\top \widetilde{c}^S \varphi^\star\right) \bullet I(t) \\
= -\left(\widetilde{c}^V - (\widetilde{c}^{S,V})^\top (\widetilde{c}^S)^{-1} \widetilde{c}^{S,V}\right) \bullet I(t)
\end{aligned}$$

up to a martingale. Recall from the proof of Rule 12.3 that $\widetilde{c}^S(t) = \Gamma(f, f)(X(t-))$, $\widetilde{c}^{S,V}(t) = \Gamma(f, v(T-t, \cdot))(X(t-))$ and similarly $\widetilde{c}^V(t) = \Gamma(v(T-t, \cdot), v(T-t, \cdot))(X(t-))$. Therefore the growth rate of U equals

$$\begin{aligned}
a^U(t) &= -\Gamma(v(T-t, \cdot), v(T-t, \cdot))(X(t-)) \\
&\quad + \Gamma(f, v(T-t, \cdot))(X(t-))^\top (\Gamma(f, f)(X(t-)))^{-1} \Gamma(f, v(T-t, \cdot))(X(t-)).
\end{aligned} \tag{12.9}$$

On the other hand, the Markovian structure of the model implies that $U(t) = h(T-t, X(t))$ for some function $h : [0, T] \times \mathbb{R}^d \to \mathbb{R}$. Proposition 4.31 yields that the growth rate of U is of the form

$$a^U(t) = a(t, X(t-)) \tag{12.10}$$

with $a(t, \cdot) = (\frac{d}{dt} + G)h(T-t, \cdot)$. Equating (12.9) and (12.10) yields (12.8). □

We consider the Black–Scholes model for illustration. Since perfect hedging clearly minimises the quadratic risk, we know that the variance-optimal hedge must be of the form as in Example 11.24 with vanishing quadratic risk.

Example 12.5 (Black–Scholes Model) Consider a European call in the Black–Scholes model as in Example 11.4. We write $S_1(t) = \exp(X(t))$, where

$$X(t) := \log S_1(t) = \log S_1(0) + \sigma W(t) - \frac{\sigma^2}{2} t$$

is the Markov process under consideration. Its generator G and carré du champ operator Γ are given by

$$G\xi(x) = \frac{\sigma^2}{2}\left(\xi''(x) - \xi'(x)\right)$$

and

$$\Gamma(\xi, \eta)(x) = \sigma^2 \xi'(x) \eta'(x). \tag{12.11}$$

The value process of a European call is of the form $V(t) = v(T-t, \log S_1(t))$, where v satisfies the PDE (11.32) as we know already from Example 11.24. The variance-optimal number of shares of stocks is

$$\varphi_1^\star(t) = \frac{\Gamma(\exp, v(T-t,\cdot))(X(t))}{\Gamma(\exp,\exp)(X(t))} = \frac{v'(T-t, \log S_1(t))}{S_1(t)},$$

where $v'(t,x)$ denotes the derivative of $x \mapsto v(t,x)$. Observe that this agrees with the perfect hedge φ_1 in (11.25). In order to compute the minimal quadratic risk, we must integrate an expression of the form

$$\Gamma(u,u) - \frac{\Gamma(\exp,u)^2}{\Gamma(\exp,\exp)}$$

for some function u, which vanishes due to (12.11).

12.1.3 The Laplace Transform Approach for Affine Processes

In the context of option pricing, solving PIDEs can often be avoided by using integral transform techniques, cf. Sect. 11.6. The same holds for quadratic hedging. Again, the idea is to solve the problem in closed form for exponential claims and to obtain the general solution by integration. This approach works because quadratic hedging is linear in the claim. For ease of notation we focus on the univariate case $d = 1$ although generalisation is possible along the same lines as in Sect. 11.6.4. In any case, we need that the traded security is of the form

$$S_1(t) = \exp(X(t)), \tag{12.12}$$

where X denotes a component of a multivariate affine semimartingale as in Theorem 6.6. For ease of notation we focus here on bivariate affine processes (Y, X) as in most of the models in Sect. 8.2, which satisfy $m = 1$ in Theorem 6.6. This implies that its generalised conditional characteristic function is of the form

$$E\big(e^{iu_1Y(T)+iu_2X(T)}\big|\mathscr{F}_t\big) = \exp\big(\Psi_0(T-t,u_1,u_2) + \Psi_1(T-t,u_1,u_2)Y(t) + iu_2X(t)\big) \tag{12.13}$$

for some functions Ψ_0, Ψ_1 which are known in concrete models. The function Ψ_2 in Theorem 6.6 is of the form $\Psi_1(T-t,u_1,u_2) = iu_2$ because e^X is a martingale, which implies that $\psi_2 = 0$. We set

$$\psi_0(u_1,u_2) := \left.\frac{\partial\Psi_0(t,u_1,u_2)}{\partial t}\right|_{t=0}, \quad \psi_1(u_1,u_2) := \left.\frac{\partial\Psi_1(t,u_1,u_2)}{\partial t}\right|_{t=0}.$$

These are the exponents corresponding to the triplets defining the affine process (Y, X), cf. Sects. 6.1 and 8.2.

The payoff is assumed to be of the form $H = f(X(T))$ for some function f with an integral representation as in (11.34, 11.35). The latter is explicitly stated in Sect. 11.6.1 for a number of payoffs. As in the previous section our task is to determine the optimal initial endowment, the variance-optimal hedge and the minimal quadratic risk. The mean value process has been determined in Rule 11.31. In the present notation, we obtain the following

Rule 12.6 (Value Process) *The mean value process of the option satisfies*

$$V(t) = \frac{1}{2\pi i} \int_{R-i\infty}^{R+i\infty} V(t, z)\widetilde{f}(z)dz$$

with

$$V(t, z) := S_1(t)^z \exp\big(\Psi_0(T-t, 0, -iz) + \Psi_1(T-t, 0, -iz)Y(t)\big). \tag{12.14}$$

Idea This follows from the linearity of the expected value. □

Applying (12.5) leads us to an integral representation of the optimal hedge.

Rule 12.7 (Hedging Strategy) *The variance-optimal hedging strategy $\varphi^\star$ is given by $V_{\varphi^\star}(0) = V(0)$ and*

$$\varphi_1^\star(t) = \frac{1}{2\pi i} \int_{R-i\infty}^{R+i\infty} \frac{V(t-, z)}{S_1(t-)} \frac{\kappa_0(t, z) + \kappa_1(t, z)Y(t-)}{\delta_0 + \delta_1 Y(t-)} \widetilde{f}(z)dz,$$

with $V(t, z)$ from (12.14) and

$$\kappa_j(t, z) := \psi_j(-i\Psi_1(T-t, 0, -iz), -i(z+1)) - \psi_j(-i\Psi_1(T-t, 0, -iz), -iz), \tag{12.15}$$

$$\delta_j := \psi_j(0, -2i), \quad j = 0, 1. \tag{12.16}$$

Idea Step 1: The generator G of (Y, X) acts on functions of the form $h(\xi_1, \xi_2) = \exp(iu_1\xi_1 + iu_2\xi_2)$ as $Gh(\xi_1, \xi_2) = h(\xi_1, \xi_2)(\psi_0(u_1, u_2) + \psi_1(u_1, u_2)\xi_1)$, cf. (6.28) and Sect. 5.7.1. Since $S_1 = e^X$ is a martingale, we have $\psi_0(0, -i) = 0$ and $\psi_1(0, -i) = 0$. Moreover, recall that $\Gamma(f, g) = G(fg) - fGg - gGf$.

Step 2: We first consider a "payoff" of the form $e^{zX(T)}$ whose "value process" is given by (12.13) with $u_1 = 0, u_2 = -iz$. In view of Rule 12.3 we need to apply the carré du champ operator to functions $f(\xi_1, \xi_2) = e^{\xi_2}$ and

$$v(t, \xi_1, \xi_2) = \exp\big(\Psi_0(t, 0, -iz) + \Psi_1(t, 0, -iz)\xi_1 + z\xi_2\big).$$

Using step 1, we obtain after a straightforward but slightly tedious calculation that

$$\frac{\Gamma(f, v(T-t,\cdot))(\xi_1,\xi_2)}{\Gamma(f,f)(\xi_1,\xi_2)}$$
$$= \exp\left(\Psi_0(T-t,0,-iz)+\Psi_1(T-t,0,-iz)\xi_1+(z-1)\xi_2\right)\frac{\kappa_0(t,z)+\kappa_1(t,z)\xi_1}{\delta_0+\delta_1\xi_1}.$$

Step 3: The assertion follows now from (12.13), Rule 12.3, and linearity of the variance-optimal hedging strategy in the claim. □

The minimal quadratic risk (12.6) is bilinear in the value process. Moreover, it involves an integration over time. Therefore we end up with a triple integral for $\varepsilon^2(\varphi^\star)$.

Rule 12.8 (Minimal Quadratic Risk) *The minimal quadratic risk equals*

$$\varepsilon^2(\varphi^\star) = \begin{cases} -\dfrac{1}{4\pi^2}\displaystyle\int_{R-i\infty}^{R+i\infty}\int_{R-i\infty}^{R+i\infty}\int_0^T J_1(t,z_1,z_2)dt\widetilde{f}(z_1)dz_1\widetilde{f}(z_2)dz_2 \\ \qquad\qquad\qquad\qquad\qquad\qquad \textit{if } \delta_0\neq 0, \delta_1\neq 0, \\ -\dfrac{1}{4\pi^2}\displaystyle\int_{R-i\infty}^{R+i\infty}\int_{R-i\infty}^{R+i\infty}\int_0^T J_2(t,z_1,z_2)dt\widetilde{f}(z_1)dz_1\widetilde{f}(z_2)dz_2 \\ \qquad\qquad\qquad\qquad\qquad\qquad \textit{if } \delta_0=0, \\ -\dfrac{1}{4\pi^2}\displaystyle\int_{R-i\infty}^{R+i\infty}\int_{R-i\infty}^{R+i\infty}\int_0^T J_3(t,z_1,z_2)dt\widetilde{f}(z_1)dz_1\widetilde{f}(z_2)dz_2 \\ \qquad\qquad\qquad\qquad\qquad\qquad \textit{if } \delta_1=0, \end{cases} \tag{12.17}$$

with

$$J_1(t,z_1,z_2)$$
$$:= S(0)^{z_1+z_2}e^{\xi_0}\Bigg(\exp\left(\Psi_0(t,-i\xi_1,-i(z_1+z_2))+\Psi_1(t,-i\xi_1,-i(z_1+z_2))Y(0)\right)$$
$$\times\left(\frac{\eta_2}{\delta_1}\left(D_2\Psi_0(t,-i\xi_1,-i(z_1+z_2))+D_2\Psi_1(t,-i\xi_1,-i(z_1+z_2))Y(0)\right)+\frac{\eta_1\delta_1-\eta_2\delta_0}{\delta_1^2}\right)$$
$$+\frac{\eta_0\delta_1^2-\eta_1\delta_0\delta_1+\eta_2\delta_0^2}{\delta_1^3}e^{-\frac{\delta_0}{\delta_1}\xi_1}\int_0^1\left(\frac{\delta_1}{\delta_0}+\xi_1 s\right)$$
$$\times e^{\frac{\delta_0}{\delta_1}\xi_1 s+\Psi_0\left(t,-i(\frac{\delta_1}{\delta_0}\log s+\xi_1 s),-i(z_1+z_2)\right)+\Psi_1\left(t,-i(\frac{\delta_1}{\delta_0}\log s+\xi_1 s),-i(z_1+z_2)\right)Y(0)}ds\Bigg),$$

$$J_2(t, z_1, z_2)$$
$$:= \frac{S(0)^{z_1+z_2}e^{\xi_0}}{\delta_1}\exp\Big(\Psi_0(t, -i\xi_1, -i(z_1+z_2)) + \Psi_1(t, -i\xi_1, -i(z_1+z_2))Y(0)\Big)$$
$$\times\Big(\eta_1 + \big(D_2\Psi_0(t, -i\xi_1, -i(z_1+z_2)) + D_2\Psi_1(t, -i\xi_1, -i(z_1+z_2))Y(0)\big)\eta_2\Big),$$

$$J_3(t, z_1, z_2)$$
$$:= \frac{S(0)^{z_1+z_2}\eta_0}{\delta_0}\exp\Big((\psi_0(0, -iz_1) + \psi_0(0, -iz_2))(T-t) + \psi_0(0, -i(z_1+z_2))t\Big).$$

The constants δ_0, δ_1 and functions κ_0, κ_1 are defined as in (12.16, 12.15). The remaining variables are specified as follows:

$$\begin{aligned}
\alpha_j &= \alpha_j(t, z_1, z_2)\\
&:= \psi_j(-i\xi_1(t, z_1, z_2), -i(z_1+z_2)) - \psi_j(-i\Psi_1(T-t, 0, -iz_1), -iz_1)\\
&\quad - \psi_j(-i\Psi_1(T-t, 0, -iz_2), -iz_2),\\
\eta_0 &= \eta_0(t, z_1, z_2) := \delta_0\alpha_0(t, z_1, z_2) - \kappa_0(t, z_1)\kappa_0(t, z_2),\\
\eta_1 &= \eta_1(t, z_1, z_2)\\
&:= \delta_0\alpha_1(t, z_1, z_2) + \delta_1\alpha_0(t, z_1, z_2) - \kappa_1(t, z_1)\kappa_0(t, z_2) - \kappa_1(t, z_2)\kappa_0(t, z_1),\\
\eta_2 &= \eta_2(t, z_1, z_2) := \delta_1\alpha_1(t, z_1, z_2) - \kappa_1(t, z_1)\kappa_1(t, z_2),\\
\xi_j &= \xi_j(t, z_1, z_2) := \Psi_j(T-t, 0, -iz_1) + \Psi_j(T-t, 0, -iz_2), \quad j = 0, 1.
\end{aligned}$$

Idea Step 1: By (12.6) and the representation of $\widetilde{c}^V, \widetilde{c}^{S,V}, \widetilde{c}^S$ in terms of the carré du champ operator Γ we have that

$$\varepsilon^2(\varphi^\star) = E\Bigg(\int_0^T \Bigg(\Gamma(v(T-t, \cdot), v(T-t, \cdot))((Y, X)(t-)) - \frac{\big(\Gamma(f, v(T-t, \cdot))((Y, X)(t-))\big)^2}{\Gamma(f, f)((Y, X)(t-))}\Bigg)dt\Bigg)$$

with v from Rule 12.2. Since Γ is linear in its two arguments and $v(t, \xi) = \frac{1}{2\pi i}\int_{R-i\infty}^{R+i\infty} g(z, t, \xi)\widetilde{f}(z)dz$ with

$$g(z, t, \xi_1, \xi_2) = \exp\big(\Psi_0(T-t, 0, -iz) + \Psi_1(T-t, 0, -iz)\xi_1 + z\xi_2\big)$$

by Rule 12.6, this can be written as

$$\varepsilon^2(\varphi^\star) = -\frac{1}{4\pi^2}\int_{R-i\infty}^{R+i\infty}\int_{R-i\infty}^{R+i\infty}\int_0^T E\Big(\Gamma(g(z_1,T-t,\cdot),g(z_2,T-t,\cdot))((Y,X)(t-)) - \frac{\big(\Gamma(f,g(z_1,T-t,\cdot))((Y,X)(t-))\big)^2}{\Gamma(f,f)((Y,X)(t-))}\Big)dt\,\widetilde{f}(z_1)dz_1\,\widetilde{f}(z_2)dz_2. \tag{12.18}$$

Therefore it suffices to determine the expectation inside the triple integral.

Step 2: Fix z_1, z_2. Straightforward but tedious calculations yield that the expressions $\Gamma(g(z_1,T-t,\cdot),g(z_2,T-t,\cdot))(\xi_1,\xi_2)$, $\Gamma(f,g(z_1,T-t,\cdot))(\xi_1,\xi_2)$ etc. are all of the form

$$\exp(\det+\det\xi_1+\mathrm{cst}\,\xi_2)(\det+\det\xi_1),$$

where "det" stands for various deterministic functions of time and "cst" for a constant which depends only on z_1, z_2. In order to compute the expected value in (12.18), we need to evaluate expressions of the form

$$E\big(\exp(\det+\det Y(t)+\mathrm{cst}\,X(t))(\det+\det Y(t))\big) \tag{12.19}$$

and

$$\begin{aligned} &E\Big(\exp(\det+\det Y(t)+\mathrm{cst}\,X(t))\frac{\det+\det Y(t)+\det Y(t)^2}{\det+\det Y(t)}\Big)\\ &= E\Big(\exp(\det+\det Y(t)+\mathrm{cst}\,X(t))\Big(\det+\det Y(t)+\frac{\det}{\det+\det Y(t)}\Big)\Big). \end{aligned} \tag{12.20}$$

Since (Y, X) is an affine process, we have an explicit representation of $\alpha(y,z) := E(\exp(yY(t)+zX(t)))$. Its derivative with respect to y is of the form

$$\frac{\partial\alpha(y,z)}{\partial y} = E\big(\exp(yY(t)+zX(t))Y(t)\big).$$

Consequently, (12.19) can be expressed in terms of α and its derivative with respect to its first argument. The first two summands in (12.20) also lead to an expression as in (12.19). For the third consider an affine ordinary differential equation of the form

$$a\frac{\partial}{\partial y}\beta(y,z)+b\beta(y,z)-\alpha(y,z)=0 \tag{12.21}$$

for $y\mapsto\beta(y,z)$. It is easy to see that it is solved by

$$\beta(y,z) := E\Big(\frac{\exp(yY(t)+zX(t))}{b+aY(t)}\Big).$$

Since the solution to (12.21) can be calculated in terms of α by variation of parameters, we obtain an integral representation for the third summand in (12.20). Tedious but straightforward calculations yield that the resulting expressions for the expected value in (12.18) are of the form J_1, J_2, J_3 stated in Rule 12.8. □

We illustrate the approach in the case of a geometric Lévy process, cf. Example 9.2.

Example 12.9 (Geometric Lévy Model) If we assume $X - X(0)$ in (12.12) to be a Lévy process, we have $\Psi_1 = 0$ and w.l.o.g. $Y = 0$ in (12.13). Moreover, $\Psi_0(u_1, u_2, 0) = \psi(u_2)t$, where ψ denotes the characteristic exponent of $X - X(0)$. Inserting this into the previous rules yields

$$V(t) = \frac{1}{2\pi i}\int_{R-i\infty}^{R+i\infty} S_1(t)^z e^{\psi(-iz)(T-t)}\widetilde{f}(z)dz,$$

and

$$\varphi_1^\star(t) = \frac{1}{2\pi i}\int_{R-i\infty}^{R+i\infty} S_1(t-)^{z-1}e^{\psi(-iz)(T-t)}\frac{\psi(-i(z+1)) - \psi(-iz)}{\psi(-2i)}\widetilde{f}(z)dz \tag{12.22}$$

for the mean value process and the variance-optimal hedging strategy, respectively. The time integral in (12.17) can be evaluated explicitly, which leads to the expression

$$\varepsilon^2(\varphi^\star) = -\frac{1}{4\pi^2}\int_{R-i\infty}^{R+i\infty}\int_{R-i\infty}^{R+i\infty} J_0(z_1, z_2)\widetilde{f}(z_1)dz_1\widetilde{f}(z_2)dz_2 \tag{12.23}$$

with

$$J_0(z_1, z_2) := \begin{cases} S(0)^{z_1+z_2}\beta(z_1, z_2)\dfrac{e^{(\psi(-iz_1)+\psi(-iz_2))T} - e^{\psi(-i(z_1+z_2))T}}{\psi(-iz_1) + \psi(-iz_2) - \psi(-i(z_1+z_2))} \\ \qquad\qquad \text{if } \psi(-iz_1) + \psi(-iz_2) \neq \psi(-i(z_1+z_2)), \\ S(0)^{z_1+z_2}\beta(z_1, z_2)Te^{\psi(-i(z_1+z_2))T} \\ \qquad\qquad \text{if } \psi(-iz_1) + \psi(-iz_2) = \psi(-i(z_1+z_2)) \end{cases}$$

and

$$\beta(z_1, z_2) := \psi(-i(z_1+z_2)) - \psi(-iz_1) - \psi(-iz_2) - \frac{(\psi(-i(z_1+1)) - \psi(-iz_1))(\psi(-i(z_2+1)) - \psi(-iz_2))}{\psi(-2i)}$$

for the minimal quadratic risk.

12.2 The Semimartingale Case

We turn now to the more general situation where S is not necessarily a martingale. It has proven useful to consider the slightly more general problem to minimise (12.1) where the initial endowment $V_\varphi(0)$ may or may not be fixed, i.e. we possibly maximise only over the smaller set of self-financing strategies with given initial value $V_\varphi(0) = v_0$.

Remark 12.10 If $V_\varphi(0) = v_0$ is fixed in the preceding section, we end up with the same optimal hedge up to a difference $v_0 - V_0(\varphi^\star)$ invested in the numeraire. The corresponding quadratic risk equals

$$\left(v_0 - V_0(\varphi^\star)\right)^2 + \varepsilon^2(\varphi^\star)$$

in the martingale case. Indeed, this follows from

$$\begin{aligned} E\left((V_\varphi(T) - H)^2\right) &= E\left((E(H) + \varphi \bullet S(T) - H)^2\right) + \left(V_\varphi(0) - E(H)\right)^2 \\ &\quad + 2E\left(E(H) + \varphi \bullet S(T) - H)(V_\varphi(0) - E(H)\right) \\ &= E\left((E(H) + \varphi \bullet S(T) - H)^2\right) + \left(V_\varphi(0) - E(H)\right)^2. \end{aligned}$$

In order to understand the structure of the solution to the hedging problem, we need the opportunity process L and the adjustment process a from Sect. 10.4. Moreover, we write $\widetilde{K} := \mathscr{L}(L)$ as in Rule 10.45. The process

$$Z^{Q^\star} := \frac{L\mathscr{E}(-a \bullet S)}{L(0)}$$

is a martingale starting at 1. If it is positive, it is the density process of an EMM $Q^\star$ called the **variance-optimal martingale measure**, cf. Remark 10.46(3). In general, however, $Z^{Q^\star}$ may attain negative values. But in any case, it can be written as $Z^{Q^\star} = \mathscr{E}(N)$ where the stochastic logarithm N is of the form

$$N := \widetilde{K} - a \bullet S - [\widetilde{K}, a \bullet S],$$

cf. Proposition 3.44. We call the process N the **variance-optimal logarithm process**.

The opportunity process leads to the definition of a so-called opportunity-neutral measure $P^\star \sim P$, which is needed below for representing the optimal hedging strategy. It is typically not a martingale measure. Specifically, this **opportunity-neutral measure** $P^\star$ is defined in terms of its density process

$$Z^{P^\star} := \mathscr{E}\left(M^{\widetilde{K}}\right) = \frac{L}{L(0)\mathscr{E}\left(A^{\widetilde{K}}\right)}, \tag{12.24}$$

where $\widetilde{K} = M^{\widetilde{K}} + A^{\widetilde{K}}$ denotes the canonical decomposition of $\widetilde{K}$ as in (2.6). The second equality in (12.24) holds because we assume our semimartingales to have local characteristics. Indeed, this implies $[M^{\widetilde{K}}, A^{\widetilde{K}}] = 0$ by Proposition 3.4(5) and hence

$$\mathscr{E}(M^{\widetilde{K}})\mathscr{E}(A^{\widetilde{K}}) = \mathscr{E}\Big(M^{\widetilde{K}} + A^{\widetilde{K}} + [M^{\widetilde{K}}, A^{\widetilde{K}}]\Big) = \mathscr{E}(\widetilde{K}) = \frac{L}{L(0)}.$$

Obviously we have $P^\star = P$ if L and hence $\widetilde{K}$ are deterministic. This happens, for example, in geometric Lévy models, cf. Example 12.13. If S is a martingale, we even have $Q^\star = P^\star = P$ because $L = 1$ and $a = 0$.

The role of the mean value process in Sect. 12.1 is played in general by the conditional expectation of H under the variance-optimal martingale measure $Q^\star$, i.e.

$$V(t) := E_{Q^\star}(H|\mathscr{F}_t).$$

But since $Q^\star$ may not be defined as an ordinary probability measure, we define the **mean value process** of the option with payoff H more generally as

$$V(t) := E\big(H\mathscr{E}(N - N^t)(T)\big|\mathscr{F}_t\big).$$

The mean value process can also be expressed as

$$V(t) = \frac{E(HZ^{Q^\star}(T)|\mathscr{F}_t)}{Z^{Q^\star}(t)}$$

if it makes sense, i.e. unless the denominator attains the value 0.

The optimal hedging strategy consists of two parts, a *pure hedge* and an investment term in feedback form. We write

$$\widetilde{c}^{(S,V)\star} = \begin{pmatrix} \widetilde{c}^{S\star} & \widetilde{c}^{S,V\star} \\ (\widetilde{c}^{S,V\star})^\top & \widetilde{c}^{V\star} \end{pmatrix} \tag{12.25}$$

for the modified second characteristic of the $\mathbb{R}^{d+1}$-valued process (S, V) relative to $P^\star$, cf. (II.2). We call the $\mathbb{R}^d$-valued predictable process

$$\xi := (\widetilde{c}^{S\star})^{-1}\widetilde{c}^{S,V\star} \tag{12.26}$$

the **pure hedge coefficient** of the option. It satisfies

$$\xi \bullet \langle S, S\rangle^{P^\star} = \langle S, V\rangle^{P^\star},$$

which generalises (12.4). The superscript indicates that the predictable covariation processes refer to the measure $P^\star$.

We are now ready to express the optimal hedging strategy in feedback form.

Rule 12.11 (Hedging Strategy) *The variance-optimal hedging strategy is the unique self-financing portfolio $\varphi^\star$ with initial value $V(0)$ that satisfies the feedback equation*

$$(\varphi_1^\star, \dots, \varphi_d^\star)(t) = \xi(t) - \big(V_{\varphi^\star}(t-) - V(t-)\big)\, a(t). \tag{12.27}$$

If we consider the problem (12.1) with fixed initial endowment v_0, the statement holds accordingly with v_0 instead of $V(0)$.

The wealth process of the optimal hedge satisfies the SDE

$$dV_{\varphi^\star}(t) = (\xi(t) + V(t-)a(t))\, dS(t) - V_{\varphi^\star}(t-)a(t)dS(t). \tag{12.28}$$

Hence it is a generalised stochastic exponential in the sense of Definition 3.47.

Idea Step 1: Let $\varphi^\star$ be the optimal strategy and set $Z(t) := E(V_{\varphi^\star}(T) - H|\mathscr{F}_t)$. Recall from Remark 10.16 that Rule 10.3(1,2) holds in the context of random endowment as well. They even apply to $u(x) = -x^2$ because the proof does not require u to be increasing. Consequently, we have that ZS and hence also ZV_ψ is a martingale for any self-financing strategy ψ. Define the process V as the solution to

$$Z = (V_{\varphi^\star} - V)L. \tag{12.29}$$

Step 2: We show that $V(t) = E(H\mathscr{E}(N)(T)|\mathscr{F}_t)/E(N)(t)$, which implies that $V\mathscr{E}(N)$ is a martingale.
To wit, since $Z\mathscr{E}(-a \bullet S)$ is a martingale, $V\mathscr{E}(N) = VL\mathscr{E}(-a \bullet S)/L(0)$ equals $V_{\varphi^\star}L\mathscr{E}(-a \bullet S)/L(0) = V_{\varphi^\star}Z^{Q^\star}$ and hence 0 up to a martingale. Moreover, $V(T)\mathscr{E}(N)(T) = H\mathscr{E}(N)(T)$, which yields the assertion.
Step 3: We show that N is a martingale. Indeed, by Rule 10.45 we have that

$$\begin{aligned} a^N &= a^{\widetilde{K}} - aa^S - a\widetilde{c}^{\widetilde{K},S} \\ &= \overline{a}^\top \overline{c}^{-1}\overline{a} - \overline{a}^\top \overline{c}^{-1}(a^S + \widetilde{c}^{\widetilde{K},S}) \\ &= \overline{a}^\top \overline{c}^{-1}\overline{a} - \overline{a}^\top \overline{c}^{-1}\overline{a} = 0, \end{aligned}$$

which yields the claim.
Step 4: We show that $S + [S, N]$ is a martingale. Recall from Remark 10.46(2) that

$$L(0)\mathscr{E}(N)_- \bullet \big(S + N + [N, S]\big) = L(0)\mathscr{E}(N)S = L\mathscr{E}(-a \bullet S)S$$

is a martingale. By step 2 this process equals $L(0)\mathscr{E}(N)_- \bullet (S + [S, N])$ up to a martingale. Integrating with $(L(0)\mathscr{E}(N)_-)^{-1}$ yields the assertion.

Step 5: We show that $V+[V,N]$ is a martingale. To this end, recall from step 2 that $\mathscr{E}(N)_-\bullet(V+V_-\bullet N+[V,N])=V\mathscr{E}(N)$ is a martingale. By step 3 this process equals $\mathscr{E}(N)_-\bullet(V+[V,N])$ up to a martingale. Integrating with $1/\mathscr{E}(N)_-$ yields the claim.

Step 6: Theorem 4.15 and $Z^{P^\star}=\mathscr{E}(M^{\widetilde{K}})$ yield

$$\widetilde{c}^{S,V\star}=c^{SV}+\int xz(1+y)K^{(S,\widetilde{K},V)}(d(x,y,z)),$$

$$\widetilde{c}^{V\star}=c^{V}+\int z^2(1+y)K^{(\widetilde{K},V)}(d(y,z)),$$

$$\widetilde{c}^{S\star}=c^{S}+\int xx^\top(1+y)K^{(S,\widetilde{K})}(d(x,y)).$$

Step 7: Define $\xi=\varphi+(V_{\varphi^\star}-V)_-a$. For any self-financing strategy ψ we have $ZV_\psi=L(V_{\varphi^\star}-V)V_\psi$. Since $L=L(0)\mathscr{E}(\widetilde{K})$, this equals

$$\begin{aligned}
&L_-\bullet\Big((V_{\varphi^\star}-V)V_\psi+\big((V_{\varphi^\star}-V)V_\psi\big)_-\bullet\widetilde{K}+\big[(V_{\varphi^\star}-V)V_\psi,\widetilde{K}\big]\Big)\\
&=L_-\bullet\Big((V_{\varphi^\star}-V)_-\bullet V_\psi+(V_\psi)_-\bullet(V_{\varphi^\star}-V)+[V_{\varphi^\star}-V,V_\psi]\\
&\quad+\big((V_{\varphi^\star}-V)V_\psi\big)_-\bullet\widetilde{K}+(V_{\varphi^\star}-V)_-\bullet[V_\psi,\widetilde{K}]\\
&\quad+(V_\psi)_-\bullet[V_{\varphi^\star}-V,\widetilde{K}]+\big[[V_{\varphi^\star}-V,V_\psi],\widetilde{K}\big]\Big)\\
&=\big(L(V_{\varphi^\star}-V)\big)_-\bullet\big((V_\psi)_-\bullet N+\psi\bullet(S+[S,N])\big)\\
&\quad+(L_-\xi)\bullet\Big((V_\psi)_-\bullet(S+[S,\widetilde{K}])+\big[V_\psi,S+[S,\widetilde{K}]\big]\Big)\\
&\quad-(LV_\psi)_-\bullet(V+[V,\widetilde{K}])-\big[L_-\bullet V_\psi,V+[V,\widetilde{K}]\big].
\end{aligned}$$

Steps 3 and 4 yield that the first term is a martingale. The remaining two terms equal

$$\begin{aligned}
&(LV_\psi)_-\bullet\Big(\xi\bullet(S+[S,N])-(V+[V,N])\Big)\\
&\quad+L_-\bullet\Big(\big(a(V_\psi)_-+\psi\big)\bullet\big(\xi\bullet\big[S+[S,\widetilde{K}],S\big]-\big[S+[S,\widetilde{K}],V\big]\big)\Big).
\end{aligned}$$

The first term is a martingale by steps 4 and 5. Note that

$$\big[S+[S,\widetilde{K}],S\big](t)=[S^c,S^c](t)+\int_{[0,t]\times\mathbb{R}^{d+1}}xx^\top(1+y)\mu^{(S,\widetilde{K})}(d(s,(x,y)))$$

equals

$$\left(c^S + \int xx^\top (1+y) K^{(S,\widetilde{K})}(d(x,y))\right) \bullet I = \widetilde{c}^{S\star} \bullet I$$

up to a martingale, cf. step 6. Similarly,

$$\left[S+[S,\widetilde{K}],V\right](t) = [S^c,V^c](t) + \int_{[0,t]\times\mathbb{R}^{d+1}} xz(1+y)\mu^{(S,\widetilde{K},V)}(d(s,(x,y,z)))$$

equals

$$\left(c^{S,V} + \int xz(1+y) K^{(S,\widetilde{K},V)}(d(x,y,z))\right) \bullet I = \widetilde{c}^{S,V\star} \bullet I$$

up to a martingale. Finally,

$$\left[V,V+[V,\widetilde{K}]\right](t) = [V^c,V^c](t) + \int_{[0,t]\times\mathbb{R}^{d+1}} z^2(1+y)\mu^{(\widetilde{K},V)}(d(s,(y,z)))$$

equals

$$\left(c^{V,V} + \int z^2(1+y) K^{(\widetilde{K},V)}(d(y,z))\right) \bullet I = \widetilde{c}^{V\star} \bullet I$$

up to a martingale, which is needed only in the proof of Rule 12.12. Since ZV_ψ is a martingale by step 1, we conclude that

$$\left(\left(a(V_\psi)_- + \psi\right)\left(\xi\widetilde{c}^{S\star} - \widetilde{c}^{S,V\star}\right)\right) \bullet I$$

is a martingale and hence zero. Since this holds for any strategy ψ, we conclude that $\xi = (\widetilde{c}^{S\star})^{-1}\widetilde{c}^{S,V\star}$, as desired.

Step 8: The SDE (12.28) follows directly from (12.27). □

Finally we state a formula for the hedging error.

Rule 12.12 (Quadratic Risk) *The minimal quadratic risk for fixed initial endowment v_0 equals*

$$E\left((V_{\varphi^\star}(T) - H)^2\right)$$
$$= E\left((v - V(0))^2 L(0) + \int_0^T \left(\widetilde{c}^{V\star}(t) - (\widetilde{c}^{S,V\star}(t))^\top (\widetilde{c}^{S\star}(t))^{-1}\widetilde{c}^{S,V\star}(t)\right) L(t)dt\right) \tag{12.30}$$
$$= E\left((v_0 - V(0))^2 L(0) + L \bullet \left(\langle V,V\rangle^{P^\star} - \langle \xi \bullet S, \xi \bullet S\rangle^{P^\star}\right)(T)\right), \tag{12.31}$$

where $\varphi^\star$ denotes the strategy of Rule 12.1. The minimal quadratic risk for optimally chosen initial endowment is obtained for $v_0 = V(0)$.

Idea Denote by A the compensator of $(V_{\varphi^\star} - V)^2 L$. Since $V(T) = H$ and $L(T) = 1$, we have

$$\begin{aligned}E\big((V_{\varphi^\star}(T) - H)^2\big) &= E\big((V_{\varphi^\star}(T) - V(T))^2 L(T)\big)\\ &= E\big((V_{\varphi^\star}(0) - V(0))^2 L(0)\big) + E(A(T)).\end{aligned}$$

Equation (12.29) implies $(V_{\varphi^\star} - V)^2 L = (V_{\varphi^\star} - V)Z$. Since $ZV_{\varphi^\star}$ and Z are martingales, $(V - V_{\varphi^\star})Z$ equals VZ and, using integration by parts,

$$\begin{aligned}Z_- \bullet V + [V, Z] &= Z_- \bullet V + \big[V, (V_{\varphi^\star} - V)L\big]\\ &= ((V_{\varphi^\star} - V)L)_- \bullet V\\ &\quad + \big[V, (V_{\varphi^\star} - V)_- \bullet L + L_- \bullet (V_{\varphi^\star} - V) + [V_{\varphi^\star} - V, L]\big]\end{aligned} \tag{12.32}$$

up to a martingale. By (12.27) we have $V_{\varphi^\star} - V = v_0 + \xi \bullet S - \big((V_{\varphi^\star} - V)_- a\big) \bullet S - V$. Hence (12.32) equals

$$\big((V_{\varphi^\star} - V)L\big)_- \bullet \Big(V + \big[V, \widetilde{K} - a \bullet S - [a \bullet S, \widetilde{K}]\big]\Big) + L_- \bullet \big[V + [V, \widetilde{K}], \xi \bullet S - V\big].$$

Since $V + [V, \widetilde{K} - a \bullet S - [a \bullet S, \widetilde{K}]] = V + [V, N]$ is a martingale by step 5 in the proof of Rule 12.11, we obtain altogether that $(V_{\varphi^\star} - V)^2 L$ differs from a martingale by

$$\begin{aligned}L_- \bullet \big[V + [V, \widetilde{K}], V - \xi \bullet S\big] &= L_- \bullet \big(\big[V + [V, \widetilde{K}], V\big] - \xi \bullet \big[V + [V, \widetilde{K}], S\big]\big)\\ &= L_- \bullet \big(\big[V + [V, \widetilde{K}], V\big] - \xi \bullet \big[V, S + [S, \widetilde{K}]\big]\big).\end{aligned}$$

In view of step 7 in the proof of Rule 12.11, this equals $(L_-(\widetilde{c}^{V\star} - \xi^\top \widetilde{c}^{S,V\star})) \bullet I$ up to a martingale. We conclude that $A = (L_-(\widetilde{c}^{V\star} - \xi^\top \widetilde{c}^{S,V\star})) \bullet I$, which yields (12.30).

The second representation (12.31) follows from $\langle V, V\rangle^{P^\star} = \widetilde{c}^{V\star} \bullet I$ and

$$\langle \xi \bullet S, \xi \bullet S\rangle^{P^\star} = \sum_{i,j}^{d} (\xi_i \xi_j \widetilde{c}_{ij}^{S\star}) \bullet I. \qquad \square$$

We illustrate the result in the geometric Lévy model of Example 9.2, which extends Example 12.9 to the non-martingale case.

Example 12.13 (Geometric Lévy Model) Recall that $X = \log S_1$. We denote the characteristic exponent of the Lévy process $X - X(0)$ by ψ. In order to characterise the variance-optimal hedging strategy $\varphi^\star$ in Rule 12.11 explicitly, we need the adjustment process a, the mean value process V, and the pure hedge ξ. They are given by

$$a(t) = \frac{\psi(-i)}{\psi(-2i) - 2\psi(-i)} \Big/ S_1(t-), \tag{12.33}$$

$$V(t) = \frac{1}{2\pi i} \int_{R-i\infty}^{R+i\infty} S_1(t)^z e^{\eta(z)(T-t)} \widetilde{f}(z) dz, \tag{12.34}$$

$$\xi_1(t) = \frac{1}{2\pi i} \int_{R-i\infty}^{R+i\infty} S_1(t-)^{z-1} e^{\eta(z)(T-t)} \frac{\psi(-i(z+1)) - \psi(-iz) - \psi(-i)}{\psi(-2i) - 2\psi(-i)} \widetilde{f}(z) dz$$

with

$$\eta(z) := \psi(-iz) - \psi(-i) \frac{\psi(-i(z+1)) - \psi(-iz) - \psi(-i)}{\psi(-2i) - 2\psi(-i)}.$$

The corresponding minimal quadratic risk equals

$$\varepsilon^2(\varphi^\star) = \frac{1}{2\pi i} \int_{R-i\infty}^{R+i\infty} \int_{R-i\infty}^{R+i\infty} J_0(z_1, z_2) \widetilde{f}(z_1) dz_1 \widetilde{f}(z_2) dz_2,$$

where

$$J_0(z_1, z_2) := \begin{cases} S(0)^{z_1+z_2} \beta(z_1, z_2) \dfrac{e^{\alpha(z_1,z_2)T} - e^{\psi(-i(z_1+z_2))T}}{\alpha(z_1, z_2) - \psi(-i(z_1+z_2))} \\ \qquad\qquad\qquad\qquad \text{if } \alpha(z_1, z_2) \neq \psi(-i(z_1+z_2)), \\ S(0)^{z_1+z_2} \beta(z_1, z_2) T e^{\psi(-i(z_1+z_2))T} \quad \text{if } \alpha(z_1, z_2) = \psi(-i(z_1+z_2)), \end{cases}$$

$$\alpha(z_1, z_2) := \eta(z_1) + \eta(z_2) - \frac{\psi(-i)^2}{\psi(-2i) - 2\psi(-i)},$$

and

$$\beta(z_1, z_2) := \psi(-i(z_1+z_2)) - \psi(-iz_1) - \psi(-iz_2) \\ - \frac{(\psi(-i(z_1+1)) - \psi(-iz_1) - \psi(-i))(\psi(-i(z_2+1)) - \psi(-iz_2) - \psi(-i))}{\psi(-2i) - 2\psi(-i)}.$$

Indeed, recall from Example 10.48 that $L(t) = e^{-\alpha(T-t)}$ is deterministic, a solves (12.33), and $a \bullet S = \pi \widetilde{X}$ with π from (10.93). In particular, $N = \widetilde{K} - a \bullet S - [\widetilde{K}, a \bullet S] = \pi(\widetilde{X} - A^{\widetilde{X}})$. A straightforward, but slightly tedious calculation yields that $V_z + [V_z, N]$ and hence $V_z \mathscr{E}(N) = \mathscr{E}(N)_- \bullet (N + V_z + [V_z, N])$ are martingales for $V_z(t) := \exp(zX(t) + \eta(z)(T-t))$. Therefore

$$V_z(t) = E\big(e^{zX(T)}\mathscr{E}(N-N^t)(T)\big|\mathscr{F}_t\big)$$

is the mean value process for the "payoff" $e^{zX(T)}$. By linearity, we obtain (12.34) for the mean value process of the payoff

$$H = \frac{1}{2\pi i}\int_{R-i\infty}^{R+i\infty} S_1(T)^z \widetilde{f}(z)dz = \frac{1}{2\pi i}\int_{R-i\infty}^{R+i\infty} V_z(T)\widetilde{f}(z)dz.$$

Note that $P^\star = P$ because L and hence $\widetilde{K}$ are deterministic. The expressions for ξ and $\varepsilon^2(\varphi^\star)$ follow again, after straightforward tedious calculations, from (12.26) and (12.30).

Appendix 1: Problems

The exercises correspond to the section with the same number.

12.1 (Basis Risk) Suppose that a claim is written on an asset $\widetilde{S}$ which cannot be traded. Instead, only a closely correlated security S_1 is liquid enough to be used for hedging purposes. Specifically, we consider discounted price processes

$$dS_1(t) = S_1(t)(\mu dt + \sigma dW(t)),$$
$$d\widetilde{S}(t) = \widetilde{S}(t)(\widetilde{\mu}dt + \widetilde{\sigma}d\widetilde{W}(t)),$$

where $\mu, \widetilde{\mu}, \sigma, \widetilde{\sigma}$ are constants and $(W, \widetilde{W})$ is a two-dimensional Brownian motion with $E(W(1)) = 0 = E(\widetilde{W}(1))$, $\mathrm{Var}(W(1)) = 1 = \mathrm{Var}(\widetilde{W}(1))$, and $\mathrm{Cov}(W(1), \widetilde{W}(1)) = \varrho$. Only S_1 is assumed to be tradable. For simplicity we assume S_1 to be a martingale, i.e. $\mu = 0$.

The claim under consideration is of the form $H = f(\widetilde{S}(T))$ for some function $f : \mathbb{R}_+ \to \mathbb{R}$. Derive expressions for the variance-optimal hedging strategy $\varphi^\star$ of H and for the minimal quadratic risk $\varepsilon^2(\varphi^\star)$. How do these quantities depend on ϱ?

12.2 (Reduction to the Martingale Case) We consider the setup and the notation of Sect. 12.2. Suppose that $\mathscr{E}(-a \bullet S)$ is positive. We denote by

$$\widetilde{S} := (\widetilde{S}_0, \dots, \widetilde{S}_d) := \frac{(S_0, \dots, S_d)}{\mathscr{E}(-a \bullet S)}$$

the original price process, but discounted relative to the value process of the self-financing strategy $\widetilde{\varphi}$ with $\widetilde{\varphi}_i := a_i\mathscr{E}(-a \bullet S)_-$, $i = 1, \dots, d$ and initial value 1. Accordingly, we write $\widetilde{H} := H/\mathscr{E}(-a \bullet S)(T)$ for the payoff H, discounted relative to the same numeraire. Finally, we need the probability measure $Q \sim P$ with density

$$\frac{dQ}{dP} := \frac{\mathscr{E}(-a \bullet S)(T)^2}{E(\mathscr{E}(-a \bullet S)(T)^2)}.$$

Show that

1. $\widetilde{S}$ is a Q-martingale, i.e. Q is an EMM for S relative to the tradable numeraire $\mathscr{E}(-a \bullet S)$;
2. the variance-optimal hedging strategy of $\widetilde{H}$ relative to $\widetilde{S}$ and measure Q coincides with the variance-optimal hedging strategy $\varphi^\star$ of H relative to S and the original measure P;
3. the minimal quadratic risks are related via

$$\widetilde{\varepsilon}^2(\varphi^\star)L(0) = \varepsilon^2(\varphi^\star)$$

where we write $\widetilde{V}_\varphi = \varphi^\top \widetilde{S}$ for the discounted value of any self-financing strategy φ and $\widetilde{\varepsilon}^2(\varphi) := E_Q((\widetilde{V}_\varphi(T) - \widetilde{H})^2)$ for its quadratic risk.

Note that the hedging problem for $\widetilde{H}$ relative to $\widetilde{S}$ and Q is of the simpler structure of Sect. 12.1.

Appendix 2: Notes and Comments

A hands-on introduction to regression in finance can be found in the textbook [54]. Quadratic hedging started with [118] in the martingale case and [79] for securities with drift. We refer to [233, 270] for an overview over the extensive literature in the field and to [55] for more recent publications.

As far as rigorous statements are concerned, Rule 12.1 is a special case of the general setup in Sect. 12.2. The operator formulation of the solution can be found in [37]. In particular, we refer to [37, Theorems 3 and 5] for rigorous versions of Rules 12.3, 12.4. Section 12.1.3 is based on the results of [178, 179, 231], but the idea to apply Laplace transforms to quadratic hedging goes back to [148]. The latter covers Example 12.9 as a special case. Section 12.2 closely follows the reasoning in [55], where rigorous statements and proofs can be found. Example 12.13 can be found in [148]. For extensions to the affine case see [56, 184]. Problem 12.1 is taken from [147]. The numeraire change approach of Problem 12.2 has been put forward by [128] in the continuous case and extended by [3] in the presence of jumps.

Chapter 13
Utility-Based Valuation and Hedging of Derivatives

Valuation based on quadratic hedging is easy to understand and mathematically tractable compared to other approaches. Its economic justification is less obvious. The minimisation criterion (12.1) penalises gains and losses alike. This is unreasonable from an economic point of view and occasionally leads to counterintuitive results. In this chapter we discuss an economically better founded alternative. It rests on the suggestion that the option writer should base her decisions on utility maximisation. In the context of over-the-counter trades this leads to the concept of utility indifference pricing and hedging which is introduced in Sects. 13.1 and 13.2 for utility of terminal wealth and utility of P&L, respectively. Viewed from a different angle, utility indifference pricing can be interpreted in terms of convex risk measures. This link is discussed in Sect. 13.3. Finally, we compare various approaches to pricing and hedging in the last section of this chapter.

13.1 Exponential Utility of Terminal Wealth

We consider the general setup of Sect. 11.3 and Chap. 12 with $1 + d$ liquidly traded securities and an option H that is to be sold. As in Chap. 12 we assume that prices are discounted relative to the zeroth asset, i.e. $S_0 = 1$. Your initial endowment is denoted by v_0.

We assume that the option writer (hereafter called *you*) aims at maximising expected utility of terminal wealth for $u(x) = -\exp(-px)$ with $p > 0$. Maximisation here refers to all investment opportunities that are feasible with given initial endowment v_0. These include dynamic trading of all liquid assets and a possible over-the-counter trade of the option. If a customer approaches you at time 0 and offers a premium π in exchange for the contingent claim H, you have two

© Springer Nature Switzerland AG 2019

E. Eberlein, J. Kallsen, *Mathematical Finance*, Springer Finance,
https://doi.org/10.1007/978-3-030-26106-1_13

choices: either to accept the offer or to reject it. If you decline, you are facing the pure investment problem and your expected utility of terminal wealth amounts to

$$U_0 := \sup_{\varphi} E(u(V_\varphi(T))), \tag{13.1}$$

where the supremum is taken over all self-financing strategies φ with initial value $V_\varphi(0) = v_0$. If, on the other hand, you accept the deal to sell the claim for π and additionally invest in a self-financing portfolio φ with initial value $V_\varphi(0) = v_0$, your terminal wealth equals instead $V_\varphi(T) + \pi - H$. This leads to the expected utility

$$U_H(\pi) := \sup_{\varphi} E\big(u(V_\varphi(T) + \pi - H)\big), \tag{13.2}$$

the supremum referring to the same set of strategies as in (13.1). Of course you will only accept the deal if it leads to higher utility, i.e. if $U_H(\pi) \geq U_0$. This is the case if and only if the premium π paid by the customer is large enough. The threshold π satisfying $U_H(\pi) = U_0$ is called the **utility indifference price** of the claim. Below this price the trade will not be made, above the threshold it becomes increasingly attractive.

The utility indifference approach also tells you how to invest your money if you decide to sell the option. You trade according to the optimal strategy φ^H in (13.2). Note that the optimal trading strategies φ^H in (13.2) and φ^0 in (13.1) typically differ. The difference

$$\vartheta := \varphi^H - \varphi^0 \tag{13.3}$$

can be interpreted as a **utility-based hedging strategy**. It corresponds to the adjustment of your portfolio made in order to account for the option.

13.1.1 Exact Solution

The utility maximisation problems (13.1) and (13.2) are studied in Sect. 10.1.3. This leads to the following results for indifference pricing and hedging.

Rule 13.1 *Let $u(x) = -\exp(-px)$ with $p > 0$. Define φ^H, Q^H, L^H as in Rules 10.17, 10.23 and accordingly φ^0, Q^0, L^0 for random endowment 0 instead of H. Moreover, define P^H as in (10.44). Then the utility indifference price of H equals*

$$\pi = \frac{1}{p}\Big(\log E(e^{pH}) + H(Q^0|P) - H(Q^H|P^H)\Big) \tag{13.4}$$

$$= \frac{1}{p}\Big(H(Q^0|P) - H(Q^H|P)\Big) + E_{Q^H}(H) \tag{13.5}$$

$$= -\frac{1}{p}\left(\inf_{Q \text{ EMM}} \big(H(Q|P) - pE_Q(H)\big) - \inf_{Q \text{ EMM}} H(Q|P)\right) \tag{13.6}$$

$$= \frac{1}{p} \log E_{Q^0}\big(\exp(-p(\vartheta \bullet S(T) - H))\big) \tag{13.7}$$

$$= \frac{1}{p} \log \frac{L^H(0)}{L^0(0)}, \tag{13.8}$$

where the infima are taken over the set of all equivalent martingale measures and ϑ denotes the utility-based hedging strategy, which is given by (13.3).

Idea By Rules 10.5(3) and 10.23 we have $U_0 = -\exp(-pv_0 - H(Q^0|P))$ and

$$\begin{aligned} U_H(\pi) &= -E(e^{pH}) \exp\big(-p(v_0+\pi) - H(Q^H|P^H)\big) \\ &= -\exp\big(-p(v_0+\pi) - H(Q^H|P) + pE_{Q^H}(H)\big). \end{aligned}$$

Equating these expected utilities and solving for π yields (13.4) and (13.5), respectively. Equation (13.6) follows from Rule 10.23. By optimality of φ^0 and $\varphi^H = \varphi^0 + \vartheta$, we have

$$U_0 = E\big(u(V_{\varphi^0}(T))\big) = -\frac{1}{p}E\big(u'(V_{\varphi^0}(T))\big) \tag{13.9}$$

and

$$\begin{aligned} U_H(\pi) &= E\big(u(V_{\varphi^H}(T) + \pi - H)\big) \\ &= E\big(-\exp(-p(V_{\varphi^0}(T) + \vartheta \bullet S(T) + \pi - H))\big) \\ &= -\frac{1}{p}E\big(u'(V_{\varphi^0}(T))\exp(-p(\vartheta \bullet S(T) + \pi - H))\big) \\ &= -\frac{1}{p}E_{Q^0}\big(\exp(-p(\vartheta \bullet S(T) + \pi - H))\big)E\big(u'(V_{\varphi^0}(T))\big). \end{aligned} \tag{13.10}$$

Equating (13.9, 13.10) and solving for π yields (13.7). By definition of L^0 and L^H we have $U_0 = L^0(0)u(v_0)$ and $U_H(\pi) = L^H(0)u(v_0 + \pi)$. Equating the two and solving for π yields (13.8). □

Let us study utility-based pricing and hedging in a few specific situations. We start with the trivial case of a replicable claim.

Rule 13.2 *If the claim is attainable, the utility indifference price equals its unique fair price in the sense of Rule 11.2. Moreover, the utility-based hedge coincides with the replicating strategy.*

Idea Suppose that H is attainable, i.e. $H = V_\psi(T)$ for some self-financing strategy ψ. Then

$$\begin{aligned} U_H(\pi) &= \sup_\varphi E\big(u(V_\varphi(T) + \pi - H)\big) \\ &= \sup_\varphi E\big(u(v_0 + (\varphi - \psi) \bullet S(T) + \pi - V_\psi(0))\big) \qquad (13.11) \\ &= \exp\big(-p(\pi - V_\psi(0))\big) \sup_\varphi E\big(u(V_\varphi(T))\big) \qquad (13.12) \\ &= \exp\big(-p(\pi - V_\psi(0))\big) U_0, \end{aligned}$$

where the supremum refers to all self-financing strategies φ with initial value $V_\varphi(0) = v_0$. This equals U_0 if and only if $\pi = V_\psi(0)$, i.e. if the utility indifference price π is the unique fair price. Moreover, the optimiser φ^H in (13.11) differs from the optimiser φ^0 in (13.12) by ψ. Hence ψ is the utility-based hedging strategy. □

In the general case the main difficulty is to solve the utility maximisation problems (13.1) and in particular (13.2). The latter rarely allows for an explicit or at least simple numerical solution, cf. Sect. 10.1.3. As an exception we consider the following example of hedging with basis risk.

Example 13.3 (Basis Risk) We suppose that a claim is written on an asset $\widetilde{S}$ which cannot be traded. Instead, only a closely correlated security S_1 is liquid enough to be used for hedging purposes. Specifically, we consider discounted price processes

$$\begin{aligned} dS_1(t) &= S_1(t)(\mu dt + \sigma dW(t)), \\ d\widetilde{S}(t) &= \widetilde{S}(t)(\widetilde{\mu} dt + \widetilde{\sigma} d\widetilde{W}(t)), \end{aligned}$$

where $\mu, \widetilde{\mu}, \sigma, \widetilde{\sigma}$ are constants and $(W, \widetilde{W})$ is a two-dimensional Brownian motion with $E(W(1)) = 0 = E(\widetilde{W}(1))$, $\mathrm{Var}(W(1)) = 1 = \mathrm{Var}(\widetilde{W}(1))$, and $\mathrm{Cov}(W(1), \widetilde{W}(1)) = \varrho$. Only S_1 is supposed to be tradable. The claim under consideration is of the form $H = f(\widetilde{S}(T))$ for some function $f : \mathbb{R}_+ \to \mathbb{R}$. We want to determine the utility-based price and hedge for exponential utility $u(x) = -\exp(-px)$. This setup can be viewed as a particular case of Example 9.4 with $(X_1, X_2) = (S_1, \widetilde{S})$,

$$\beta(x) = \begin{pmatrix} x_1\mu \\ x_2\widetilde{\mu} \end{pmatrix}, \quad \gamma(x) = \begin{pmatrix} x_1^2\sigma^2 & x_1x_2\varrho\sigma\widetilde{\sigma} \\ x_1x_2\varrho\sigma\widetilde{\sigma} & x_2^2\widetilde{\sigma}^2 \end{pmatrix}.$$

The opportunity process and the optimal portfolio in the presence of H are obtained by solving the PDE in Example 10.21. In the present setup an ansatz

$K^H(t) = \ell(T-t, X_2(t))$ is successful, i.e. the function in Example 10.21 does not depend on the first component of X. PDE (10.39) then reads as

$$\ell(0, x_2) = pf(x_2),$$
$$D_1\ell(\vartheta, x_2) = D_2\ell(\vartheta, x_2)x_2\widetilde{\mu} + \left(D_{22}\ell(\vartheta, x_2) + (D_2\ell(\vartheta, x_2))^2\right)\frac{(x_2\widetilde{\sigma})^2}{2}$$
$$- \frac{\left(\mu + D_2\ell(\vartheta, x_2)x_2\varrho\sigma\widetilde{\sigma}\right)^2}{2\sigma^2}.$$

The quadratic terms can be removed by switching to

$$g(\vartheta, x_2) = \exp\left(\beta(\ell(\vartheta, x_2) - \alpha\vartheta)\right) \tag{13.13}$$

with $\alpha = \mu^2/(2\sigma^2)$ and $\beta = 1 - \varrho^2$. Indeed, a straightforward calculation yields that g solves

$$g(0, x_2) = \exp\left((1-\varrho^2)pf(x_2)\right),$$
$$D_1g(\vartheta, x_2) = D_2g(\vartheta, x_2)x_2\left(\widetilde{\mu} - \frac{\mu\varrho\widetilde{\sigma}}{\sigma}\right) + \frac{1}{2}D_{22}g(\vartheta, x_2)x_2^2\widetilde{\sigma}^2. \tag{13.14}$$

We want to write the solution to this PDE in terms of an expected value. To this end, define a probability measure $Q \sim P$ via its density process

$$Z(t) = \mathscr{E}(-(\mu/\sigma) \bullet W)(t).$$

According to Theorem 4.15, the drift coefficient $S_1(t)\mu$ vanishes under Q whereas $\widetilde{S}(t)\widetilde{\mu}$ changes to $\widetilde{S}(t)(\widetilde{\mu} - \mu\varrho\widetilde{\sigma}/\sigma)$. Put differently, the process (X_1, X_2) is a Markov process relative to Q as well, but with coefficients

$$\beta_Q(x) = \begin{pmatrix} x_1\mu \\ x_2(\widetilde{\mu} - \frac{\mu\varrho\widetilde{\sigma}}{\sigma}) \end{pmatrix}, \quad \gamma_Q(x) = \gamma(x).$$

Moreover, $X_2 = \widetilde{S}$ is a Markov process on its own. Set

$$g(\vartheta, x_2) := E_Q\left(\exp\left((1-\varrho^2)pf(\widetilde{S}(T))\right)\middle|\widetilde{S}(T-\vartheta) = x_2\right).$$

Applying Theorem 5.24 to X_2 and g yields that g solves the PDE (13.14). We obtain

$$K^H(0) = \ell(T, \widetilde{S}(0)) = \frac{\log g(T, \widetilde{S}(0))}{1-\varrho^2} - \alpha T$$

by (13.13) and

$$\varphi^H(t) = \frac{\mu}{p\sigma^2 S_1(t)} + \frac{D_2(\log g(t, \widetilde{S}(t)))}{(1-\varrho^2)p} \frac{\varrho\widetilde{\sigma}\widetilde{S}(t)}{\sigma S_1(t)}$$

for the optimal portfolio by (10.40). The opportunity process and the optimal portfolio without claim are derived naturally by replacing H with 0. Rule 13.1 now yields the indifference price for H, which equals

$$\begin{aligned}\pi &= \frac{1}{p}\big(K^H(0) - K^0(0)\big)\\ &= \frac{1}{p}\big(\ell(T, \widetilde{S}(0)) + \alpha T\big)\\ &= \frac{1}{(1-\varrho^2)p} \log E_Q\Big(\exp\big((1-\varrho^2)pf(\widetilde{S}(T))\big)\Big).\end{aligned}$$

Similarly, we obtain

$$\vartheta = \varphi_1^H(t) - \varphi_1^0(t) = \frac{D_2(\log g(t, \widetilde{S}(t)))}{(1-\varrho^2)p} \frac{\varrho\widetilde{\sigma}\widetilde{S}(t)}{\sigma S_1(t)}$$

for the utility-based hedge.

13.1.2 Large Number of Claims

One should note that utility indifference prices are typically non-linear in the claim: if you sell two shares of an option rather than one you require more than twice the premium because of your non-linear attitude towards risk. In this and the following section we study the extreme cases of a very large resp. small number of contracts.

We consider the general setup of Sect. 13.1. We denote by $\pi(\lambda)$ the utility indifference price per unit if the customer wants to buy $\lambda > 0$ options. Equivalently, one may view $\pi(\lambda)$ as the indifference price for one claim as before but with risk aversion parameter λp instead of p. The utility-based hedge per claim is denoted by $\vartheta(\lambda)$, i.e. $\vartheta(\lambda, t) := \frac{1}{\lambda}(\varphi^{\lambda H}(t) - \varphi^0(t))$.

The dependence of $\pi(\lambda)$ on λ is not obvious, but we can easily show that it increases with λ.

Lemma 13.4 *$\pi(\lambda)$ is increasing in λ.*

Idea Rules 13.1 and 10.23 yield

$$\pi(\lambda) = E_{Q^H}(H) + \frac{1}{\lambda p}\big(H(Q^0|P) - H(Q^{\lambda H}|P)\big) \tag{13.15}$$

$$= \sup_{Q \text{ EMM}} \left(E_Q(H) + \frac{1}{\lambda p}\big(H(Q^0|P) - H(Q|P)\big) \right).$$

Since Q_0 is the minimal entropy martingale measure, we have

$$H(Q^0|P) - H(Q|P) \leq 0 \tag{13.16}$$

for any EMM Q. Consequently, the above expression is increasing in λ. □

In the limit of large numbers of claims or, equivalently, large risk aversion, we end up with the upper price of Sect. 11.3 and the cheapest superreplicating strategy.

Rule 13.5 *We have $\pi(\lambda) \to \pi_{\text{high}}(H)$ as $\lambda \to \infty$, where $\pi_{\text{high}}(H)$ denotes the upper price in Sect. 11.3. Moreover, $\vartheta(\lambda)$ is an asymptotic superreplicating strategy in the sense that*

$$\liminf_{\lambda\to\infty} V_{\vartheta(\lambda)}(T) \geq H, \tag{13.17}$$

where $V_{\vartheta(\lambda)}(T) := \pi(\lambda) + \vartheta(\lambda) \bullet S(T)$ denotes the terminal value of the utility-based hedge with initial endowment $\pi(\lambda)$.

Idea Equations (13.15, 13.16, 11.10) yield $\pi(\lambda) \leq E_{Q^H}(H) \leq \pi_{\text{high}}(H)$ for any $\lambda > 0$. Fix $\varepsilon > 0$. For $\lambda > 0$ consider the event $F_\lambda := \{H - V_{\vartheta(\lambda)}(T) \geq \varepsilon\}$. By definition of the utility indifference price and of Q^0 we obtain

$$\begin{aligned} U_0 &= U_{\lambda H}(\pi(\lambda)) \\ &= E\big(u(V_{\varphi^0}(T) + \lambda\vartheta(\lambda) \bullet S(T) + \lambda\pi(\lambda) - \lambda H)\big) \\ &= E_{Q_0}\big(\exp(-p\lambda(V_{\vartheta(\lambda)}(T) - H))\big)E\big(u(V_{\varphi^0}(T))\big) \\ &\leq Q^0(F_\lambda)e^{p\lambda\varepsilon}E\big(u(V_{\varphi^0}(T))\big) \end{aligned}$$

for any $\lambda > 0$, which implies that $Q^0(F_\lambda) \to 0$ as $\lambda \to \infty$. Put differently, we have $\lim_{\lambda\to\infty}(H - V_{\vartheta(\lambda)}(T))^+ = 0$ in Q^0-probability and hence in Q-probability for any $Q \sim P$. In particular, (13.17) holds.

By (11.10) there is an EMM Q such that $\pi_{\text{high}}(H) \leq E_Q(H) + \varepsilon$. Provided that enough uniform integrability holds, we have $E_Q((H - V_{\vartheta(\lambda)}(T))^+) \to 0$ as $\lambda \to \infty$ and hence

$$\begin{aligned} \pi_{\text{high}}(H) - \pi(\lambda) &\leq E_Q(H) + \varepsilon - E_Q\big(V_{\vartheta(\lambda)}(T)\big) \\ &\leq E_Q\big((H - V_{\vartheta(\lambda)}(T))^+\big) + \varepsilon, \end{aligned}$$

which converges to ε as $\lambda \to \infty$. This yields $\pi_{\text{high}}(H) \leq \lim_{\lambda\to\infty}\pi(\lambda)$, as desired. □

13.1.3 Small Number of Claims

We turn now to the limiting case as $\lambda \to 0$. We conjecture that the indifference price allows for a smooth dependence on λ, i.e.

$$\pi(\lambda) = \pi(0) + \lambda\delta + o(\lambda)$$

with constants $\pi(0), \delta$. In this case we call $\pi(0) + \lambda\delta$ the **approximate utility indifference price**. The quantity $\pi(0)$ represents a zeroth-order approximation or limiting price for a very small number of claims. δ can be viewed as a derivative of the option price relative to the number λ that are to be sold.

Similarly, we expect the maximal utility in the corresponding optimisation problem

$$U_{\lambda H}(\lambda\pi(\lambda)) = \sup_{\varphi} E\big(u(V_\varphi(T) + \lambda\pi(\lambda) - \lambda H)\big) \tag{13.18}$$

to be approximately attained by a strategy of the form $\varphi^0 + \lambda\xi$, i.e. we conjecture that

$$U_{\lambda H}(\lambda\pi(\lambda)) = E\big(u(V_\varphi(T) + \lambda\pi(\lambda) - \lambda H)\big) + o(\lambda^2)$$

holds for $\varphi = \varphi^0 + \lambda\xi$ with some predictable process ξ. Here φ^0 denotes the optimal strategy in the pure investment problem (13.1). We call $\varphi^0 + \lambda\xi$ a **nearly optimal strategy** for the utility maximisation problem. The process ξ deserves the name **approximate utility-based hedge** per unit of H because $\lambda\xi$ represents the adjustment of your portfolio due to the presence of λ claims with payoff H.

The goal now is to derive formulas for $\varphi^0, \pi(0), \delta$, and ξ. Since φ^0 is the optimal portfolio in the pure investment problem, we can refer to Sect. 10.1.3 in this respect. It turns out that the remaining quantities are obtained by solving the quadratic hedging problem of Sect. 12.1 relative to the minimal entropy martingale measure.

Rule 13.6 *Let Q^0 denote the minimal entropy martingale measure of Rules 10.5(3) and 10.17. If we choose*

1. *$\pi(0) := E_{Q^0}(H)$, i.e. the optimal initial endowment from Sect. 12.1 for the claim H and the hedging problem relative to Q^0,*
2. *ξ as the variance-optimal hedging strategy for the same problem, and*
3. *$\delta := \frac{p}{2} E_{Q^0}((V_\xi(T) - H)^2)$ as the corresponding minimal quadratic risk,*

then $\pi(0) + \lambda\delta$ is an approximate utility indifference price, $\varphi^0 + \lambda\xi$ is a nearly optimal strategy for (13.18) and hence ξ an approximate utility-based hedge per claim.

Idea We must verify that

$$E\big(u(V_\varphi(T) + \lambda(\pi(0) + \lambda\delta) - \lambda H)\big) \le E\big(u(V_{\varphi^0(T)})\big) + o(\lambda^2) \tag{13.19}$$

for any self-financing strategy φ with $V_\varphi(0) = v_0$ and that equality holds in (13.19) for $\varphi = \varphi^0 + \lambda\xi$. We consider only strategies of the form $\varphi = \varphi^0 + \lambda\psi + o(\lambda)$ for some predictable process ψ, loosely arguing that the optimiser φ in (13.18) should be close to φ^0 and allowing for a first-order expansion. Since

$$\frac{dQ^0}{dP} = \frac{u'(V_{\varphi^0}(T))}{E\big(u'(V_{\varphi^0}(T))\big)} = \frac{u(V_{\varphi^0}(T))}{E\big(u(V_{\varphi^0}(T))\big)},$$

we have

$$\begin{aligned} &u\big(V_\varphi(T) + \lambda(\pi(0) + \lambda\delta) - \lambda H\big) \\ &\qquad = \frac{dQ^0}{dP}\exp\big(-p\lambda(\psi \bullet S(T) + \pi(0) + \lambda\delta - H)\big)E\big(u(V_{\varphi^0}(T))\big) \end{aligned}$$

and hence

$$\begin{aligned} &E\big(u(V_\varphi(T) + \lambda(\pi(0) + \lambda\delta) - \lambda H\big) \\ &= E\big(u(V_{\varphi^0}(T))\big)E_{Q^0}\big(\exp(-p(\lambda(\psi \bullet S(T) + \pi(0) - H) + \lambda^2\delta))\big). \end{aligned} \tag{13.20}$$

A Taylor expansion $e^x = 1 + x + x^2/2 + o(x^2)$ yields that (13.20) equals

$$\begin{aligned} E\big(u(V_{\varphi^0}(T))\big)E_{Q^0}\Big(1 &- p\big(\lambda(\psi \bullet S(T) + \pi(0) - H) + \lambda^2\delta\big) \\ &+ \frac{p^2}{2}\lambda^2(\psi \bullet S(T) + \pi(0) - H)^2\Big) + o(\lambda^2). \end{aligned}$$

Since S is a Q^0-martingale and $\pi(0) = E_{Q^0}(H)$, this reduces to

$$E\big(u(V_{\varphi^0}(T))\big)\Big(1 - p\lambda^2\delta + \frac{p^2}{2}\lambda^2 E_{Q^0}\big((E_Q(H) + \psi \bullet S(T) - H)^2\big)\Big) + o(\lambda^2).$$

This is maximised in ψ by the variance-optimal hedging strategy ξ of H relative to Q^0. Hence (13.20) is dominated by

$$E\big(u(V_{\varphi^0}(T))\big)\Big(1 + p\lambda^2\Big(\delta - \frac{p}{2}E_{Q^0}\big((V_\xi(T) - H)^2\big)\Big)\Big) + o(\lambda^2).$$

In view of $\delta = \frac{p}{2}E_{Q^0}((V_\xi(T) - H)^2)$, this equals $E(u(V_{\varphi^0}(T))) + o(\lambda^2)$, as desired. □

Let us illustrate the approach in a geometric Lévy model. A number of affine stochastic volatility models can be treated similarly.

Example 13.7 (Geometric Lévy Model) We consider the setup of Example 9.2. The pure investment problem is solved in Example 10.19 for exponential utility. In particular, we have $V_{\varphi^\star}(t) = V_{\varphi^\star}(0) + \psi \widetilde{X}(t)$ for the optimal investment strategy, where the constant ψ solves (10.35). In order to apply Rule 13.6, we must know the dynamics of S under the MEMM Q^0. To this end, observe that the density process of Q^0 equals

$$Z(t) = \exp\left(\alpha t - p\psi \widetilde{X}(t)\right) \tag{13.21}$$

by Rule 10.17(3). Since $\widetilde{X}^c = X^c$ and $\Delta\widetilde{X}(t) = e^{\Delta X(t)} - 1$, Proposition 4.16 yields that the differential characteristics of X relative to Q^0 are of the form

$$\begin{aligned}
a^{X,Q^0}(t) &= a^X - p\psi c^X + \int x\left(\exp(-p\psi(e^x - 1)) - 1\right)K^X(dx),\\
c^{X,Q^0}(t) &= c^X,\\
K^{X,Q^0}(t,B) &= \int 1_B(x)\exp(-p\psi(e^x - 1))K^X(dx), \quad B \in \mathscr{B},
\end{aligned}$$

for any $t \in [0,T]$. By Proposition 4.6 this means that X is a Lévy process under Q^0 as well. The solution to the quadratic hedging problem can now be obtained as in Example 12.9.

The results in Sects. 13.1.2 and 13.1.3 establish a link between the different approaches to valuing and hedging OTC-derivatives. For exponential utility one obtains a continuum of prices and hedging strategies. Superreplication is located at one end of this spectrum, often requiring—as discussed in Sect. 11.3—an exceedingly high option premium. At the other end we have obtained expressions which are closely linked to variance-optimal hedging as presented in Chap. 12. So we observe in hindsight that this approach is of interest even if one rejects the symmetric treatment of profits and losses inherent in quadratic loss functions.

13.2 Utility of P&L

If OTC contracts with different maturities are traded, it is not obvious how to choose the time horizon T in Sect. 13.1. As an alternative we suggest to work instead with utility of P&L as introduced in Sect. 10.3. Contrary to the case of expected utility of terminal wealth, Rule 10.40 shows that the time horizon T in this setup does not affect the solution. Therefore it can be chosen in the distant future.

We work with the same setup of $1 + d$ traded assets and a claim with discounted payoff H as in Sect. 13.1. Moreover, we fix a utility function u as in Sect. 10.3. At first glance, utility of P&L seems to be incompatible with OTC contracts because they do not have a well-defined market price before maturity. In order to define profits and losses on a continuous basis, they must be assigned an appropriate book value. In Sect. 10.5.2 and Remark 10.57 we argued that the shadow price is the right concept for that purpose. Utility indifference prices and hedging strategies can then be defined similarly as in Sect. 13.1.

We interpret the OTC contract as an extremely illiquid asset which has a well-defined price only at maturity, namely $S_{d+1}(T) = H$. As a financial service provider, you have two choices, either to enter the option trade or not. In both cases we are in the setup of Remark 10.57, but with a different position ψ_{d+1} in the option, namely $\psi_{d+1} = -1$ or $\psi_{d+1} = 0$. The initial positions $\psi_1, \dots, \psi_d$ in the liquid assets $S_1, \dots, S_d$ do not matter because they can be converted into each other without transaction costs. We assume them to be 0.

If $\psi_{d+1} = 0$, we are effectively trading only in the original market with assets $S_1, \dots, S_d$, leading to an optimal portfolio φ^0 with expected utility of P&L $U^S(\varphi^0)$. For the alternative case $\psi_{d+1} = -1$ we consider the shadow price $S^H = (S_1, \dots, S_d, S^H_{d+1})$ and the optimal portfolio $\varphi^H = (\varphi^H_1, \dots, \varphi^H_d, -1)$ in the sense of Sect. 10.5.2. Denote by $U^{S^H}(\varphi^H)$ the corresponding expected utility of P&L.

The utilities $U^S(\varphi^0)$ and $U^{S^H}(\varphi^H)$ typically differ unless the claim H is replicable. We need to translate the gap into a utility indifference price. To this end, observe that expected utility of P&L can be viewed as a *certainty equivalent*, i.e. it is denoted in currency units. Indeed, owning one share of an asset with deterministic price process $\int_0^t \mu(s)ds$ effectively amounts to an aggregate profit of $\pi := \int_0^T \mu(t)dt$. At the same time, this asset contributes to precisely π in the expected utility of P&L on the right-hand side of (10.67), regardless of how the amount π is actually spread over time.

In our context, this means that the **utility indifference price** should be defined as

$$\pi = S^H_{d+1}(0) - \big(U^{S^H}(\varphi^H) - U^S(\varphi^0)\big). \tag{13.22}$$

To wit, $-S^H_{d+1}(0)$ is the book value of the shorted OTC contract in your portfolio. It is charged to the client. Without further adjustment, this would lead to an additional expected utility $U^{S^H}(\varphi^H) - U^S(\varphi^0)$ if you enter the trade rather than declining it. As argued above, this excess utility amounts to a certainty equivalent of the same size, which means that the premium π in (13.22) leaves you indifferent between accepting and refusing the option deal. The **utility-based hedge** is naturally defined as the difference

$$\vartheta := (\varphi^H_1, \dots, \varphi^H_d) - \varphi^0, \tag{13.23}$$

cf. (13.3).

Let us study this approach to pricing and hedging in a few specific situations. We start with the trivial case of a replicable claim.

Rule 13.8 *If the claim is attainable, the utility indifference price equals its unique fair price in the sense of Rule 11.2. Moreover, the utility-based hedge coincides with the replicating strategy.*

Idea Denote by S^H the unique fair price process of H and by ϑ the perfect hedging strategy. For any predictable $\mathbb{R}^{d+1}$-valued process ψ, the self-financing strategies corresponding to ψ and to $\psi + (\vartheta_1, \dots, \vartheta_d, -1)$ have the same wealth process and the same profits and losses. Consequently, the maximal utilities $U^S(\varphi^0)$ and $U^{S^H}(\varphi^H)$ in (13.22) coincide, which yields the assertion. □

In the general case the hardest step in order to compute utility-based prices and hedges is to obtain the solutions S^H, φ^H to the illiquid utility maximisation problem above. They can be characterised as follows.

Rule 13.9 *Let $\varphi^H = (\varphi_1^H, \dots, \varphi_d^H, -1)$ be a predictable process and S_{d+1}^H a semimartingale with $S_{d+1}^H(T) = H$. If $S^H := (S_1, \dots, S_d, S_{d+1}^H)$ has local characteristics $(a^{S^H}, c^{S^H}, K^{S^H})$ such that*

$$0 = a^{S^H}(t) + u''(0)c^{S^H}(t)\varphi^H(t) + \int \big(u'(\varphi^H(t)^\top x) - 1\big)x K^{S^H}(t, dx), \quad t \in [0, T], \tag{13.24}$$

φ^H is an optimal strategy for expected utility of P&L and S^H is a corresponding shadow price for the setup of Remark 10.57.

Idea This follows from Rule 10.40. □

Equation (13.24) resembles those for the opportunity processes in Rules 10.11, 10.17, 10.29, 10.45 and they reduce to a BSDE in an Itō process setup.

Example 13.10 (Itō Process Model) We consider utility-based pricing and hedging for P&L in the context of Example 9.3. By the martingale representation theorem 3.61, the shadow price process of H is of the form

$$S_2^H(T) = H, \quad dS_2^H(t) = a_2(t)dt + \sum_{i=1}^n \eta_i(t)dW_i(t)$$

for some predictable processes $a_2, \eta = (\eta_1, \dots, \eta_n)$. Equation (13.24) here reads as

$$\begin{aligned} 0 &= \mu(t) + u''(0)\left(\sigma(t)^2\varphi_1^H(t) - \sigma(t)\eta_1(t)\right), \\ 0 &= a_2(t) + u''(0)\left(\sigma(t)\eta_1(t)\varphi_1^H(t) - \sum_{i=1}^n \eta_i(t)^2\right) \end{aligned} \tag{13.25}$$

in this case. This yields

$$\varphi_1^H(t) = \varphi_1^0(t) + \frac{\eta_1(t)}{\sigma(t)}, \tag{13.26}$$

where

$$\varphi_1^0(t) = -\frac{\mu(t)}{u''(0)\sigma(t)^2}$$

corresponds to the investor's optimal number of shares of S_1 in absence of the claim, cf. Example 10.42. Inserting (13.26) in (13.25) yields

$$a_2(t) = \frac{\mu(t)\eta_1(t)}{\sigma(t)} + u''(0)\sum_{i=2}^{n} \eta_i(t)^2.$$

We have thus obtained a BSDE for S_2^H. It can be simplified by defining a probability measure $Q \sim P$ through its density process $Z := \mathscr{E}(-(\mu/\sigma) \bullet W_1)$. By Remark 3.72, $W^Q(t) = W(t) + (\mu/\sigma, 0, \dots, 0)^\top t$ is a Q-standard Wiener process in $\mathbb{R}^n$. Consequently, S_2^H, η solve the BSDE

$$S_2^H(T) = H, \quad dS_2^H(t) = u''(0)\sum_{i=2}^{n} \eta_i(t)^2 dt + \sum_{i=1}^{n} \eta_i(t) dW_i^Q(t).$$

In particular, we have

$$\begin{aligned} \pi_{u''(0)} &:= S_2^H(0) \\ &= E_Q(H) - u''(0) E_Q\left(\int_0^T \sum_{i=2}^{n} \eta_i(t)^2 dt\right). \end{aligned}$$

The second term can be interpreted as a risk premium for the claim. It vanishes in the limit of small risk aversion $u''(0) \to 0$, i.e. we obtain

$$\pi_{u''(0)} = S_2^H(0) \to E_Q(H) =: \pi_0$$

for $u''(0) \to 0$. From (10.69) we derive the optimal local utility $\gamma^H(t, \varphi^H(t))$ and $\gamma^0(t, \varphi^0(t))$ in the markets with and without claim. They are given by

$$\gamma^0(t, \varphi^0(t)) = \varphi^0(t)\mu(t) + \frac{u''(0)}{2}(\varphi^0(t))^2 \sigma(t)^2$$

and

$$
\begin{aligned}
\gamma^H(t, \varphi^H(t)) &= \varphi^H(t)\mu(t) - a_2(t) \\
&\quad + \frac{u''(0)}{2}\left((\varphi^H(t))^2\sigma(t)^2 - 2\varphi_1^H(t)\sigma(t)\eta_1(t) + \sum_{i=1}^n \eta_i(t)^2\right) \\
&= \gamma^0(t, \varphi^0(t)) + \frac{\eta_1(t)}{\sigma(t)}\mu(t) - a_2(t) \\
&\quad + \frac{u''(0)}{2}\left(\left(\frac{\eta_1(t)}{\sigma(t)}\right)^2\sigma(t)^2 - 2\frac{\eta_1(t)}{\sigma(t)}\sigma(t)\eta_1(t) + \sum_{i=1}^n \eta_i(t)^2\right) \\
&= \gamma^0(t, \varphi^0(t)) + \frac{u''(0)}{2}\sum_{i=2}^n \eta_i(t)^2.
\end{aligned}
$$

According to (13.22), the utility indifference price of H equals

$$
\begin{aligned}
\pi &= S_2^H(0) - E\left(\int_0^T \left(\gamma^H(t, \varphi^H(t)) - \gamma^0(t, \varphi^0(t))\right)dt\right) \\
&= E_Q(H) - \frac{u''(0)}{2}E_Q\left(\int_0^T \sum_{i=2}^n \eta_i(t)^2 dt\right) \\
&= \frac{\pi_0 + \pi_{u''(0)}}{2},
\end{aligned}
\tag{13.27}
$$

i.e. it is the average of the initial shadow prices for risk aversions $u''(0)$ and 0. The utility-based hedge satisfies

$$
\vartheta(t) = \varphi_1^H(t) - \varphi_1^0(t) = \frac{\eta_1(t)}{\sigma(t)},
$$

cf. (13.23).

One could express the utility-based price and hedge in continuous Markovian resp. Lévy models in terms of solutions to nonlinear PDEs resp. PIDEs. This parallels the terminal wealth case, where the equations in Examples 10.21 resp. 10.22 need to be solved in order to apply Rule 13.1. We leave their derivation to the interested reader. In rare cases, one can derive an explicit solution to the indifference problem. We reconsider the situation of Example 13.3 in the present context.

Example 13.11 (Basis Risk) We assume the setup of Example 13.3—but now relative to utility of P&L. We can view it as a particular case of Example 13.10

with $\mu(t) := S_1(t)\mu$, $\sigma(t) := S_1(t)\sigma$, $W_1 := W$, $W_2 := (\widetilde{W} - \varrho W)/\sqrt{1-\varrho^2}$. The BSDE for S_2^H, (η_1, η_2) reads as

$$
\begin{aligned}
S_2^H(t) &= H, \\
dS_2^H(t) &= \left(\frac{\mu\eta_1(t)}{\sigma} + u''(0)\eta_2(t)^2\right)dt + \eta_1(t)dW_1(t) + \eta_2(t)dW_2(t).
\end{aligned}
\tag{13.28}
$$

If we make the natural ansatz $S_2^H(t) = v(T-t, \widetilde{S}(t))$ for some function $v : [0, T] \times (0, \infty) \to \mathbb{R}$, we obtain

$$
\begin{aligned}
dS_2^H(t) = \Big(&- D_1 v(T-t, \widetilde{S}(t)) + D_2 v(T-t, \widetilde{S}(t))\widetilde{S}(t)\widetilde{\mu} \\
&+ \frac{1}{2} D_{22} v(T-t, \widetilde{S}(t))\widetilde{S}(t)^2\widetilde{\sigma}^2\Big)dt + D_2 v(T-t, \widetilde{S}(t))\widetilde{S}(t)\widetilde{\sigma}d\widetilde{W}(t)
\end{aligned}
$$

and hence

$$
\begin{aligned}
dS_2^H(t) = \Big(&- D_1 v(T-t, \widetilde{S}(t)) + D_2 v(T-t, \widetilde{S}(t))\widetilde{S}(t)\widetilde{\mu} \\
&+ \frac{1}{2} D_{22} v(T-t, \widetilde{S}(t))\widetilde{S}(t)^2\widetilde{\sigma}^2\Big)dt \\
&+ D_2 v(T-t, \widetilde{S}(t))\widetilde{S}(t)\widetilde{\sigma}\varrho dW_1(t) \\
&+ D_2 v(T-t, \widetilde{S}(t))\widetilde{S}(t)\widetilde{\sigma}\sqrt{1-\varrho^2}dW_2(t).
\end{aligned}
\tag{13.29}
$$

Equating the drift and diffusion coefficients of (13.28) and (13.29) yields

$$
\begin{aligned}
\eta_1(t) &= D_2 v(T-t, \widetilde{S}(t))\widetilde{S}(t)\widetilde{\sigma}\varrho, \\
\eta_2(t) &= D_2 v(T-t, \widetilde{S}(t))\widetilde{S}(t)\widetilde{\sigma}\sqrt{1-\varrho^2}
\end{aligned}
$$

and the PDE

$$
\begin{aligned}
v(0, x) &= f(x), \\
D_1 v(\vartheta, x) &= D_2 v(\vartheta, x)x\widetilde{\mu} + \frac{1}{2} D_{22} v(T-t, x(t))x^2\widetilde{\sigma}^2 \\
&\quad - D_2 v(\vartheta, x)x\frac{\mu\varrho\widetilde{\sigma}}{\sigma} - u''(0)\left(D_2 v(\vartheta, x)x\widetilde{\sigma}\right)^2(1-\varrho^2)
\end{aligned}
\tag{13.30}
$$

for v. The nonlinear term of this PDE vanishes if we change to $g(\vartheta, x) := \exp(\beta v(\vartheta, x))$ with $\beta = -2u''(0)(1-\varrho^2)$. Indeed, a straightforward calculation shows that (13.30) turns into the PDE

$$g(0,x) = \exp\big(-2u''(0)(1-\varrho^2)f(x)\big),$$
$$D_1 g(\vartheta, x) = D_2 g(\vartheta, x)x\left(\widetilde{\mu} - \frac{\mu\varrho\widetilde{\sigma}}{\sigma}\right) + \frac{1}{2}D_{22}g(\vartheta, x(t))x^2\widetilde{\sigma}^2.$$

As in Example 13.3, Kolmogorov's backward equation (5.17) yields that we have

$$g(\vartheta, x) = E_Q\Big(\exp\big(-2u''(0)(1-\varrho^2)f(\widetilde{S}(T))\big)\Big|\widetilde{S}(T-\vartheta) = x\Big)$$

with the same EMM Q as in Example 13.3. In particular, we observe

$$\begin{aligned} S_2^H(t) &= v(T-t, \widetilde{S}(t)) \\ &= \frac{\log g(T-t, \widetilde{S}(t))}{-2u''(0)(1-\varrho^2)} \\ &= \frac{\log E_Q\Big(\exp\big(-2u''(0)(1-\varrho^2)f(\widetilde{S}(T))\big)\Big|\mathscr{F}_t\Big)}{-2u''(0)(1-\varrho^2)}. \end{aligned}$$

By (13.27), we obtain

$$\begin{aligned} \pi &= \frac{E_Q(H) + S_2^H(0)}{2} \\ &= \frac{1}{2}\left(E_Q(H) + \frac{\log E_Q\big(\exp(-2u''(0)(1-\varrho^2)f(\widetilde{S}(T)))\big)}{-2u''(0)(1-\varrho^2)}\right) \end{aligned}$$

for the utility indifference price of H. The utility-based hedge is of the form

$$\vartheta(t) = \frac{\eta_1(t)}{S_1(t)\sigma} = D_2 v(T-t, \widetilde{S}(t))\frac{\varrho\widetilde{S}(t)\widetilde{\sigma}}{S_1(t)\sigma}.$$

It may be interesting to compare the utility indifference price to its counterpart for exponential utility of terminal wealth with risk aversion $p = -u''(0)$. Using the notation $\pi_{u''(0)} := v(T, \widetilde{S}_2(0))$ and $\pi_0 := E_Q(H)$ as in Example 13.10, the above indifference price equals $(\pi_0 + \pi_{u''(0)})/2$, whereas we obtained $\pi_{u''(0)/2}$ in Example 13.3.

13.2.1 Small Number of Claims

Due to their myopic character, investment problems for P&L are typically easier to solve than those for expected utility of terminal wealth and consumption. For utility-based valuation and hedging, however, we need to solve the BSDE-type equation (13.24), which resembles the problem of solving (10.29, 10.30) in the terminal wealth case.

This motivates us to study the first-order approximation for small numbers of claims, in parallel to Sect. 13.1.3 for expected utility of terminal wealth. We expect that for any $\lambda > 0$ there is a predictable process ξ and a semimartingale S^H_{d+1} such that $\varphi^{\lambda H} := (\varphi^0_1 + \lambda\xi_1, \dots, \varphi^0_d + \lambda\xi_d, -\lambda)$ and $S^H := (S_1, \dots, S_d, S^H_{d+1})$ *approximately* solve the optimisation problem in Remark 10.57 for $\psi_{d+1} = -\lambda$ in the sense that

1. $(\varphi^{\lambda H}, S^H)$ is a consistent pair,
2. $U^{S^H}(\varphi^{\lambda H})$ differs from the optimal value $\sup_\varphi U^{S^H}(\varphi)$ only by $o(\lambda^2)$.

In this case we call $\varphi^{\lambda H}$ a **nearly optimal strategy** for Remark 10.57 and ξ an **approximate utility-based hedge** per claim. The corresponding right-hand side of (13.22) then reads as

$$\lambda S^H_{d+1}(0) - \big(U^{S^H}(\varphi\lambda H) - U^S(\varphi^0)\big). \tag{13.31}$$

It stands for the price of λ options. As in Sect. 13.1.3, we denote the indifference price per claim for λ claims by $\pi(\lambda)$, now referring to utility of P&L. As before we conjecture that it is of the form

$$\pi(\lambda) = \pi(0) + \lambda\delta + o(\lambda)$$

for some constants $\pi(0), \delta$. In view of (13.31), we call $\pi(0) + \lambda\delta$ an **approximate indifference price** for utility of P&L if

$$\pi(0) + \lambda\delta = S^H_{d+1}(0) - \frac{1}{\lambda}\big(U^{S^H}(\varphi^{\lambda H}) - U^S(\varphi^0)\big) + o(\lambda).$$

For the following result we define a probability measure $P^\star \sim P$ with density process

$$Z = \mathscr{E}\left(u''(0)\varphi^0 \bullet S^c + \left(\frac{u''((\varphi^0)^\top x)}{u''(0)} - 1\right) * (\mu^S - \nu^S)\right)$$

and denote by $\widetilde{c}^{S^H\star}$ the modified second characteristic of S^H relative to $P^\star$. By Theorem 4.15 it is of the form

$$\widetilde{c}^{S^H\star}_{ij}(t) = c^{S^H}_{ij}(t) + \int \frac{u''((\varphi^0)^\top x)}{u''(0)} x_i x_j K^{S^H}(t, dx).$$

Let Q^0 denote the equivalent martingale measure of Rule 10.36. Define the semimartingale $S^H_{d+1}(t) := E_{Q^0}(H|\mathscr{F}_t)$ and set $S^H := (S_1, \dots, S_d, S^H_{d+1})$ as well as $\pi(0) = S^H_{d+1}(0) = E_{Q^0}(H)$. Similarly as in (12.25) we write

$$\widetilde{c}^{S^H\star} = \begin{pmatrix} \widetilde{c}^{S\star} & \widetilde{c}^{S,S^H_{d+1}\star} \\ (\widetilde{c}^{S,S^H_{d+1}\star})^\top & \widetilde{c}^{S^H_{d+1}\star} \end{pmatrix}$$

and call the $\mathbb{R}^d$-valued predictable process $\xi := (\widetilde{c}^{S\star})^{-1}\widetilde{c}^{S,S^H_{d+1}\star}$ the *hedge coefficient* of the claim H. It satisfies

$$\xi \bullet \langle S, S\rangle^{P^\star} = \langle S, S^H_{d+1}\rangle^{P^\star}.$$

Observe that

$$\begin{aligned}\delta &:= -\frac{u''(0)}{2} E\Big(\langle S^H_{d+1}, S^H_{d+1}\rangle^{P^\star}(T) - \langle \xi \bullet S, \xi \bullet S\rangle^{P^\star}(T)\Big) \\ &= -\frac{u''(0)}{2} \int_0^T E\Big(\widetilde{c}^{S^H_{d+1}\star}(t) - (\widetilde{c}^{S,S^H_{d+1}\star}(t))^\top (\widetilde{c}^{S\star}(t))^{-1}\widetilde{c}^{S,S^H_{d+1}\star}(t)\Big)\,dt.\end{aligned}$$

The counterpart of Rule 13.6 now reads as follows.

Rule 13.12 $\pi(0) + \lambda\delta$ *is an approximate utility indifference price and* ξ *an approximate utility-based hedge per claim.*

Idea As in the proof of Rule 13.6 we consider self-financing strategies of the form $\varphi = \varphi^0 + \lambda\psi + o(\lambda)$ with $\psi = (\psi_1, \dots, \psi_d, -1)$. A second-order Taylor expansion of γ^{S^H} yields

$$\begin{aligned}U^{S^H}(\varphi) &= U^S(\varphi^0) + \lambda E\big(\nabla\gamma^{S^H}(\varphi^0)^\top\psi\big) \bullet I(T) \\ &\quad + \frac{1}{2}\lambda^2 E\Big(\sum_{i,j=1}^{d+1} \psi_i\psi_j D_{ij}\gamma^{S^H}(\varphi^0)\Big) \bullet I(T) + o(\lambda^2) \\ &= U^S(\varphi^0) + \lambda E\Bigg(\bigg(a^S + u''(0)(\varphi^0)^\top c^S \\ &\quad + \int (u'((\varphi^0)^\top x) - 1)xK^S(dx)\bigg)^\top_{i=1,\dots,d} (\xi_1, \dots, \xi_d)\Bigg) \bullet I(T) \\ &\quad - \lambda E\Big(a^{S^H_{d+1}} + u''(0)\sum_{i=1}^d \varphi^0_i c^{S^H}_{i,d+1} + \int (u'((\varphi^0)^\top x) - 1)x_{d+1}K^{S^H}(dx)\Big) \bullet I(T)\end{aligned} \tag{13.32}$$

$$+\frac{\lambda^2}{2}u''(0)E\bigg(\sum_{i,j=1}^{d+1}\psi_i\bigg(c_{ij}^{S^H}+\int\frac{u''((\varphi^0)^\top x)}{u''(0)}x_ix_jK^{S^H}(dx)\bigg)\psi_j\bigg)\bullet I(T)$$

$$+o(\lambda^2).$$

The first linear term in λ vanishes by optimality of φ^0 and Rule 10.40. By Theorem 4.15 the parenthesis in (13.32) equals the Q^0-drift rate a^{S^H,Q^0} of S^H. But $a_{d+1}^{S^H,Q^0}=0$ because S_{d+1}^H is a Q^0-martingale. Consequently, we have

$$U^{S^H}(\varphi)=U^S(\varphi^0)+\frac{\lambda^2}{2}u''(0)\bigg(\sum_{i,j=1}^{d+1}\psi_i\widetilde{c}_{ij}^{S^H\star}\psi_j\bigg)\bullet I+o(\lambda^2).$$

The first-order condition yields that the quadratic function in $\psi=(\psi_1,\dots,\psi_d,-1)$ is maximised by $(\xi_1,\dots,\xi_d,-1)$, the maximal value being

$$\sup_\varphi U^{S^H}(\varphi)=U^S(\varphi^0)+\lambda^2\delta+o(\lambda^2). \tag{13.33}$$

This yields that $\varphi^{\lambda H}=\varphi^0+\lambda\xi$ and S^H approximately solve the optimisation problem mentioned in the beginning of this section. Moreover, (13.33) implies that $\pi(0)+\lambda\delta$ is an approximate utility indifference price. □

If S is continuous or $u(X)=-e^{-px}$, we have $P^\star=Q^0$. In this case ξ in Rule 13.12 coincides with the variance-optimal hedging strategy for the claim H relative to Q^0, cf. Sect. 12.1.

As in the case of Sect. 13.1.3, utility-based hedging is related to quadratic hedging under the dual martingale measure for the pure investment problem. In general, however, the EMMs in Rules 13.6 and 13.12 are not the same. Moreover, the formulas for the approximate indifference prices differ slightly. We consider the geometric Lévy model as an illustration.

Example 13.13 (Geometric Lévy Model) In the particular case of exponential utility $u(x)=-\exp(-px)$ in Sect. 10.3, the optimal strategy for the pure investment problem is the same as for exponential utility of terminal wealth, cf. Example 10.41. By comparing (10.73 ,13.21) and using (3.71), we have that the above EMM Q^0 and hence also $P^\star$ coincides with the minimal entropy martingale measure. The approximate utility-based prices, however, do not match precisely because

$$\delta=\frac{p}{2}E_{Q^0}\Big(\langle S_{d+1}^H,S_{d+1}^H\rangle^{Q_0}(T)-\langle\xi\bullet S,\xi\bullet S\rangle^{Q_0}(T)\Big)$$

in Example 13.7 whereas

$$\delta = \frac{p}{2} E\Big(\langle S_{d+1}^H, S_{d+1}^H \rangle^{P^\star}(T) - \langle \xi \bullet S, \xi \bullet S \rangle^{P^\star}(T)\Big)$$

in the present context, i.e. the angle brackets and the expected value are taken relative to $P = P^\star$ rather than Q^0.

13.3 Convex Risk Measures

If you enter a futures contract at an exchange, you must deposit a certain amount of money on a margin account as a collateral. A similar protection against losses is required for many risky investments. Starting with the seminal work of [4], formal criteria have been discussed how to set up a system of capital requirements in a consistent and economically reasonable way.

Similarly as before, we work with a fixed time horizon T where the outcome of risky investments is determined. At time 0 a certain amount of capital is required as a collateral to make a given investment acceptable. As in the previous sections we fix a reference asset such as a money market account and work with discounted values relative to this numeraire. Real-valued random variables H stand for the discounted obligation or loss at time T. We denote the set of such $\mathscr{F}_T$-measurable random variables by $\mathscr{X}$. Since they can assume negative values, these variables may actually mean a profit. In this section we consider mappings $H \mapsto \varrho(H)$ that assign to any obligation H the extra capital required—in whatsoever concrete sense—to make the investment acceptable to some responsible regulator. It may be negative, in which case money can even be withdrawn. One requires the following natural properties.

Definition 13.14 We call a mapping $\varrho : \mathscr{X} \to \mathbb{R}$ a **convex risk measure** if it satisfies the following axioms.

1. **(Convexity)** $\varrho(\lambda H + (1-\lambda)K) \leq \lambda\varrho(H) + (1-\lambda)\varrho(K)$ for any $H, K \in \mathscr{X}$ and any $\lambda \in [0, 1]$.
2. **(Translation invariance)** $\varrho(H + x) = \varrho(H) + x$ for any $H \in \mathscr{X}, x \in \mathbb{R}$.
3. **(Monotonicity)** $\varrho(H) \leq \varrho(K)$ for any $H, K \in \mathscr{X}$ with $H \leq K$.

If we require in addition

4. **(Positive homogeneity)** $\varrho(\lambda H) = \lambda\varrho(H)$ for any $H \in \mathscr{X}, \lambda \in \mathbb{R}_+$,

ϱ is called a **coherent risk measure**.

These properties should mostly be obvious. A larger loss requires more collateral (monotonicity) and subtracting a fixed amount x from the loss lowers the capital requirement by exactly this sum (translation invariance). Convexity is introduced because diversified portfolios are commonly perceived as less risky. Positive homogeneity stands for a certain scaling invariance, which may or may not be considered as a natural requirement.

Let us consider a number of examples.

Example 13.15 (Value at Risk) Fix a level $\alpha \in (0, 1)$, e.g. $\alpha = 0.99$. The mapping

$$\mathrm{VaR}_\alpha(H) := F_H^{\leftarrow}(\alpha) := \inf\{x \in \mathbb{R} : P(H \leq x) \geq \alpha\}$$

is called the **value at risk** at level α. As a function of α it coincides with the *quantile function* of H, i.e. the *generalised inverse* $F_H^{\leftarrow}$ of the cumulative distribution function F_H of H. This risk measure plays an important role in practice. However, it is not covered by Definition 13.14 because it fails to satisfy the convexity axiom.

Example 13.16 (Average Value at Risk) A related risk measure is the **average value at risk**, **expected shortfall**, or **conditional value at risk** at level α

$$\mathrm{AVaR}_\alpha(H) := \frac{1}{1-\alpha}\int_\alpha^1 \mathrm{VaR}_p(H)dp.$$

It turns out to be coherent in the sense of Definition 13.14.

We justify this assertion for the subset of random variables H with continuous law. To this end, consider the set of probability measures $\mathscr{Q} := \{Q \ll P : \frac{dQ}{dP} \leq \frac{1}{1-\alpha}\}$. We show that

$$\mathrm{AVaR}_\alpha(H) = \sup\left\{E_Q(H) : Q \in \mathscr{Q}\right\}. \tag{13.34}$$

By Rule 13.22 below this implies that AVaR_α is a coherent risk measure.

In order to verify (13.34) note that $P(H \geq \mathrm{VaR}_\alpha(H)) = 1 - \alpha$ implies that $Q_H \in \mathscr{Q}$ for

$$Q_H(F) = P(F|H \geq \mathrm{VaR}_\alpha(H)) = \int_F \frac{1_{\{H \geq \mathrm{VaR}_\alpha(H)\}}}{1-\alpha}dP.$$

Denoting by U a uniform random variable on $[0, 1]$, we conclude that

$$\begin{aligned}
(1-\alpha)\mathrm{AVaR}_\alpha(H) &= \int_\alpha^1 \mathrm{VaR}_p(H)dp \\
&= \int_0^1 F_H^{\leftarrow}(p)1_{[\alpha,1]}(F_H(F_H^{\leftarrow}(p)))dp \\
&= E\big(F_H^{\leftarrow}(U)1_{[\alpha,1]}(F_H(F_H^{\leftarrow}(U))\big) \\
&= E\big(H1_{[\alpha,1]}(F_H(H))\big) \\
&= E\big(H1_{\{H \geq \mathrm{VaR}_\alpha(H)\}}\big) \\
&= (1-\alpha)E_{Q_H}(H),
\end{aligned} \tag{13.35}$$

where we used that $F_H^{\leftarrow}(U)$ has the same law as H.

Now consider any probability measure $Q \in \mathcal{Q}$. Note that

$$(H - \mathrm{VaR}_\alpha(H))\left(1_{\{H \geq \mathrm{VaR}_\alpha(H)\}} - (1-\alpha)\frac{dQ}{dP}\right) \geq 0.$$

This yields

$$\begin{aligned}
&E\big(H 1_{\{H \geq \mathrm{VaR}_\alpha(H)\}}\big) - E\left(H(1-\alpha)\frac{dQ}{dP}\right) \\
&= E\left((H - \mathrm{VaR}_\alpha(H))\left(1_{\{H \geq \mathrm{VaR}_\alpha(H)\}} - (1-\alpha)\frac{dQ}{dP}\right)\right) \\
&\quad + E\big(\mathrm{VaR}_\alpha(H) 1_{\{H \geq \mathrm{VaR}_\alpha(H)\}}\big) - E\left(\mathrm{VaR}_\alpha(H)(1-\alpha)\frac{dQ}{dP}\right) \\
&\geq \mathrm{VaR}_\alpha(H)\big(P(H \geq \mathrm{VaR}_\alpha(H)) - (1-\alpha)\big) = 0.
\end{aligned}$$

In view of (13.35) we obtain

$$\begin{aligned}
(1-\alpha)\mathrm{AVaR}_\alpha(H) &= E\big(H 1_{\{H \geq \mathrm{VaR}_\alpha(H)\}}\big) \\
&\geq (1-\alpha) \sup\big\{E_Q(H) : Q \in \mathcal{Q}\big\}
\end{aligned}$$

as desired.

The next risk measure is related to utility indifference pricing, as we will see later.

Example 13.17 (Certainty Equivalent for Exponential Utility) Given an increasing, strictly concave utility function $u : \mathbb{R} \to \mathbb{R}$, the **certainty equivalent** of a random payoff H is defined as the fixed amount having the same expected utility. Since we consider losses H rather than payoffs in this section, this leads to the definition

$$\varrho(H) := -u^{-1}(E(u(-H))).$$

The certainty equivalent fails to be translation invariant—with one notable exception: for exponential utility $u(x) = -\exp(-px)$ with $p > 0$ the mapping ϱ is a convex risk measure in the sense of Definition 13.14.

To wit, note that

$$\begin{aligned}
\varrho(H) &= \inf\big\{x \in \mathbb{R} : -u^{-1}(E(u(-H))) \leq x\big\} \\
&= \inf\big\{x \in \mathbb{R} : E(e^{pH}) \leq e^{px}\big\} \\
&= \inf\big\{x \in \mathbb{R} : E(u(v - H + x)) \geq u(v)\big\}
\end{aligned}$$

for any fixed $v \in \mathbb{R}$. Consequently, ϱ is of the form (13.37) for $\mathscr{A}$ as in Example 13.19 below. By Rule 13.20(2) this implies that it is a convex risk measure.

Another approach to assessing risks is by specifying the set of investments that are acceptable with no extra deposit. This leads to the concept of *acceptance sets*.

Definition 13.18 We call a nonempty set $\mathscr{A} \subset \mathscr{X}$ an **acceptance set** if it satisfies the following conditions.

1. There exists some $x \in \mathbb{R}$ with $x \notin \mathscr{A}$.
2. For $H \in \mathscr{A}$ and $K \in \mathscr{X}$ with $K \leq H$ we have $K \in \mathscr{A}$.
3. $\mathscr{A}$ is convex.

If we also have that

4. $\mathscr{A}$ is a cone, i.e. $\lambda\mathscr{A} \subset \mathscr{A}$ for any $\lambda \in \mathbb{R}_+$,

we refer to $\mathscr{A}$ as a **coherent acceptance set**.

As above these axioms have a natural interpretation: large losses are unacceptable (1), reducing acceptable losses preserves acceptability (2), diversification is desirable (3), and, if applicable, acceptability is invariant under scaling (4).

As an example we consider the principle underlying utility indifference pricing, namely that a trade is only acceptable if it leads to higher expected utility than doing nothing.

Example 13.19 (Utility-Based Acceptance) Fix an increasing, strictly concave utility function $u : \mathbb{R} \to \mathbb{R}$ and an initial endowment $v_0 \in \mathbb{R}$. Then

$$\mathscr{A} := \{H \in \mathscr{X} : E(u(v_0 - H)) \geq u(v_0)\}$$

defines an acceptance set in the sense of Definition 13.18. This follows easily from verifying the axioms.

Since $\varrho(H)$ stands for the extra capital required to make H acceptable,

$$\mathscr{A}_\varrho := \{H \in \mathscr{X} : \varrho(H) \leq 0\} \tag{13.36}$$

naturally defines a set of obligations that can be entered at no cost. Conversely, an acceptance set $\mathscr{A}$ leads to a risk measure by considering the minimal amount that is to be put aside to make a given obligation H acceptable:

$$\varrho_{\mathscr{A}}(H) := \inf\{x \in \mathbb{R} : H - x \in \mathscr{A}\}. \tag{13.37}$$

Indeed, the following result states that Definitions (13.36) and (13.37) lead essentially to a one-to-one correspondence between convex risk measures and acceptance sets.

Rule 13.20

1. *Equation (13.36) defines an acceptance set for any convex risk measure* ϱ*. It is coherent if* ϱ *is coherent.*
2. *Equation (13.37) defines a convex risk measure for any acceptance set* $\mathscr{A}$*. It is coherent if* $\mathscr{A}$ *is coherent.*
3. *We have* $\varrho_{(\mathscr{A}_\varrho)} = \varrho$ *for convex risk measures* ϱ*.*
4. *If an acceptance set* $\mathscr{A}$ *is such that*

$$H - x \in \mathscr{A} \text{ for all } x > 0 \text{ implies } H \in \mathscr{A},$$

then $\mathscr{A}_{(\varrho_\mathscr{A})} = \mathscr{A}$.

Idea

1. The proof is straightforward.
2. This also follows by simply verifying the axioms.
3. Let $H \in \mathscr{X}$, $x \in \mathbb{R}$ with $\varrho_{(\mathscr{A}_\varrho)}(H) < x$. Then $H - x \in \mathscr{A}_\varrho$, which implies $\varrho(H - x) \leq 0$ and hence $\varrho(H) \leq x$. This shows $\varrho \leq \varrho_{(\mathscr{A}_\varrho)}$.

   Conversely, let $H \in \mathscr{X}$, $x \in \mathbb{R}$ with $\varrho(H) \leq x$. Then $\varrho(H - x) \leq 0$ and hence $H - x \in \mathscr{A}_\varrho$. This implies $\varrho_{(\mathscr{A}_\varrho)}(H - x) \leq 0$ and therefore $\varrho_{(\mathscr{A}_\varrho)}(H) \leq x$. So $\varrho_{(\mathscr{A}_\varrho)} \leq \varrho$ holds as well.
4. Let $H \in \mathscr{A}_{(\varrho_\mathscr{A})}$. It is easy to see that $H - x \in \mathscr{A}$ for any $x > 0$, which implies $H \in \mathscr{A}$ by assumption. On the other hand, for $H \in \mathscr{A}$ we have $\varrho_\mathscr{A}(H) \leq 0$ and hence $H \in \mathscr{A}_{(\varrho_\mathscr{A})}$. □

Example 13.21 (Utility Indifference Pricing) Consider the utility-based acceptance set from Example 13.19. The corresponding convex risk measure $\varrho_\mathscr{A}$ can be interpreted in terms of utility indifference pricing as in Sect. 13.1. Indeed, if an option writer sells a claim with random payoff H, then $\varrho_\mathscr{A}(H)$ is the minimal premium required to make the trade profitable from a utility point of view. At this point trading in the underlying securities is not yet taken into account. For exponential utility the above risk measure coincides with the certainty equivalent from Example 13.17, i.e. $\varrho = \varrho_\mathscr{A}$.

Convex risk measures can be characterised in terms of penalty functions on the set of probability measures.

Rule 13.22 (Dual Representation) *Denote by* $\mathscr{P}$ *the set of probability measures on* Ω*. For a mapping* $\varrho : \mathscr{X} \to \mathbb{R}$ *the following are equivalent:*

1. ϱ *is a convex risk measure.*
2. *There exists a* ***penalty function*** $\alpha : \mathscr{P} \to (-\infty, \infty]$ *such that*

$$\varrho(H) = \sup\{E_Q(H) - \alpha(Q) : Q \in \mathscr{P}\}, \quad H \in \mathscr{X}. \tag{13.38}$$

ϱ is a coherent risk measure if and only if the function α can be chosen $\{0, \infty\}$-valued, i.e.

$$\varrho(H) = \sup\{E_Q(H) : Q \in \mathscr{Q}\}$$

for some $\mathscr{Q} \subset \mathscr{P}$.

The function α in (13.38) can be chosen convex and lower semicontinuous. In this case it is uniquely determined by ϱ and given by

$$\alpha(Q) = \sup_{H \in \mathscr{A}_\varrho} E_Q(H). \tag{13.39}$$

Idea 2⇒1: Translation invariance, monotonicity and, in the second case, homogeneity are obvious. It remains to show convexity, which follows from

$$\begin{aligned}
\varrho\big(\lambda H + (1-\lambda)K\big) &= \sup\Big\{E_Q\big(\lambda H + (1-\lambda)K\big) - \alpha(Q) : Q \in \mathscr{P}\Big\} \\
&= \sup\Big\{\lambda\big(E_Q(H) - \alpha(Q)\big) + (1-\lambda)\big(E_Q(K) - \alpha(Q)\big) : Q \in \mathscr{P}\Big\} \\
&\leq \sup\Big\{\lambda\big(E_Q(H) - \alpha(Q)\big) : Q \in \mathscr{P}\Big\} \\
&\quad + \sup\Big\{(1-\lambda)\big(E_Q(K) - \alpha(Q)\big) : Q \in \mathscr{P}\Big\} \\
&= \lambda\varrho(H) + (1-\lambda)\varrho(K).
\end{aligned}$$

1⇒2: We consider the set $\mathscr{M}$ of signed measures on Ω as the dual space of $\mathscr{X}$ via $Q(H) := \int H dQ$ for $H \in \mathscr{X}$, $Q \in \mathscr{M}$. The convex conjugate of $\varrho : \mathscr{X} \to \mathbb{R}$ then reads as $\varrho^*(Q) = \sup_{H \in \mathscr{X}}(\int H dQ - \varrho(H))$ for $Q \in \mathscr{M}$. If we can apply Theorem A.15 in this general setup, the convexity of ϱ yields

$$\varrho(H) = \varrho^{**}(H) = \sup_{Q \in \mathscr{M}} \left(\int H dQ - \varrho^*(Q)\right). \tag{13.40}$$

Since ϱ is monotone, it is easy to see that $\varrho^\star(Q) = \infty$ unless Q is nonnegative. Therefore it suffices to take the supremum over all non-negative Q. Similarly, the translation invariance of ϱ yields that $\varrho^\star(Q) = \infty$ unless $Q(\Omega) = 1$. Hence we should have $Q(\Omega) = 1$ for all relevant Q in the supremum. Put differently, $\mathscr{M}$ in (13.40) can be replaced with the set of probability measures:

$$\varrho(H) = \varrho^{**}(H) = \sup_{Q \in \mathscr{P}} \big(E_Q(H) - \varrho^*(Q)\big). \tag{13.41}$$

Likewise, one easily sees that positive homogeneity in the coherent case yields $\varrho^*(Q) \in \{0, \infty\}$ for all $Q \in \mathscr{P}$. Considering the convex, lower semicontinuous penalty function $\alpha := \varrho^*|_{\mathscr{P}}$, the first part of the assertion follows.
Denoting by $\widetilde{\alpha} : \mathscr{P} \to (-\infty, \infty]$ an arbitrary convex, lower semicontinuous penalty function satisfying (13.38), we have

$$\varrho(H) = \widetilde{\alpha}^*(H) := \sup_{Q \in \mathscr{M}} \left(\int H dQ - \widetilde{\alpha}^*(H) \right)$$

by (13.38) if we identify $\widetilde{\alpha}$ with the mapping

$$Q \mapsto \begin{cases} \widetilde{\alpha}(Q) \text{ if } Q \in \mathscr{P}, \\ \infty \quad \text{otherwise} \end{cases}$$

on $\mathscr{M}$. Another application of Theorem A.15 yields $\widetilde{\alpha} = \widetilde{\alpha}^{**} = \varrho^*$ and hence the uniqueness statement in the assertion.
It remains to show that ϱ^* coincides with α in (13.39). To this end, note that

$$\begin{aligned} \varrho^*(Q) &= \sup_{H \in \mathscr{X}} \big(E_Q(H) - \varrho(H) \big) \\ &\geq \sup_{H \in \mathscr{A}_Q} \big(E_Q(H) - \varrho(H) \big) \\ &\geq \sup_{H \in \mathscr{A}_Q} E_Q(H). \end{aligned}$$

On the other hand, we have $H - \varrho(H) \in \mathscr{A}_Q$ for any $H \in \mathscr{X}$. This yields

$$\alpha(Q) \geq E_Q(H - \varrho(H)) = E_Q(H) - \varrho(H)$$

and hence $\varrho(H) \geq E_Q(H) - \alpha(Q)$ for any $H \in \mathscr{X}$, $Q \in \mathscr{P}$. In view of $\varrho^* \geq \alpha$ and (13.41), we conclude that

$$\begin{aligned} \varrho(H) &\geq \sup_{Q \in \mathscr{P}} \big(E_Q(H) - \alpha(Q) \big) \\ &\geq \sup_{Q \in \mathscr{P}} \big(E_Q(H) - \varrho^*(Q) \big) = \varrho(H). \end{aligned}$$

Therefore equality holds, which means that α can be chosen as penalty function in (13.38). Since it is convex and lower semicontinuous, it must actually coincide with $\varrho^*|_{\mathscr{P}}$. □

We consider two examples.

Example 13.23 (Average Value at Risk) The set $\mathscr{Q}$ of probability measures in Rule 13.22 for the coherent risk measure AVaR_α from Example 13.16 is given by

$$\mathscr{Q} := \left\{ Q \ll P : \frac{dQ}{dP} \leq \frac{1}{1-\alpha} \right\}.$$

Indeed, this has already been shown in Example 13.16.

Example 13.24 (Certainty Equivalent for Exponential Utility) The penalty function of the convex risk measure ϱ in Example 13.17 equals

$$\alpha(Q) = \begin{cases} \frac{1}{p} H(Q|P) & \text{if } Q \ll P, \\ \infty & \text{otherwise,} \end{cases}$$

where $H(Q|P) = E(\frac{dQ}{dP} \log \frac{dQ}{dP})$ denotes the entropy of Q relative to P.

Indeed, by Example 13.19, $\varrho(H)$ coincides with the utility indifference price π of H in the absence of tradable assets. Using the notation of Rule 13.1, we have $\pi = \frac{1}{p}(H(Q^0|P) - H(Q^H|P)) + E_{Q^H}(H)$ by (13.5), where Q^0 is the MEMM and Q^H minimises $Q \mapsto H(Q|P) - pE_Q(H)$ among all EMMs Q, cf. Rule 10.23. In the absence of tradable assets, any probability measure $Q \sim P$ is an EMM. In particular, we conclude that $Q^0 = P$ and consequently

$$\begin{aligned} \pi &= E_{Q^H}(H) - \frac{1}{p} H(Q^H|P) \\ &= \max\left\{ E_Q(H) - \frac{1}{p} H(Q|P) : Q \in \mathscr{P} \text{ with } Q \sim P \right\} \\ &= \max\left\{ E_Q(H) - \frac{1}{p} H(Q|P) : Q \in \mathscr{P} \text{ with } Q \ll P \right\}. \end{aligned} \tag{13.42}$$

The last equality in (13.42) follows because any $Q \ll P$ can be approximated by equivalent measures. Since α is convex and lower semicontinuous, it is the penalty function in (13.39).

So far we have not yet taken hedging aspects into account. The risk from selling a claim or from other investments may be reduced significantly by dynamic trading in liquid securities. To this end, we consider a market with $1 + d$ tradable assets as in Sect. 13.1. As before, we work in discounted terms (i.e. $S_0 = 1$) and with a fixed time horizon T.

Definition 13.25 Let ϱ denote a convex risk measure. The corresponding **convex risk measure with hedging** $\widetilde{\varrho} : \mathscr{X} \to \mathbb{R}$ is defined as

$$\begin{aligned} \widetilde{\varrho}(H) := \inf &\left\{ \varrho(H - V_\varphi(T)) : \varphi \text{ self-financing strategy with initial value } 0 \right\} \\ &- \inf \left\{ \varrho(-V_\varphi(T)) : \varphi \text{ self-financing strategy with initial value } 0 \right\}. \end{aligned}$$

It is straightforward to check that $\widetilde{\varrho}$ indeed satisfies Conditions 1–4 of Definition 13.14.

Denote by φ^H and φ^0 the optimisers in the first and second infimum, respectively. Similar to Sect. 13.1 the difference $\vartheta := \varphi^H - \varphi^0$ may be considered as a **risk-minimising hedging strategy** because it stands for the adjustment of the optimal portfolio due to the presence of the obligation H.

Example 13.26 (Utility Indifference Pricing) Consider the convex risk measure ϱ from Example 13.21 corresponding to utility indifference pricing, i.e.

$$\varrho(H) = \inf\{x \in \mathbb{R} : E(u(v - H + x)) \geq u(v)\}, \quad H \in \mathscr{X}$$

with $v = u^{-1}(\sup_\varphi E(u(V_\varphi(T))))$, where φ varies over the self-financing trading strategies with given initial endowment v_0. The associated risk measure with hedging $\widetilde{\varrho}$ corresponds directly to utility indifference pricing in the sense of Sect. 13.1. More precisely, $\pi(H) := \widetilde{\varrho}(H)$ is the minimal premium charged by the writer if the claim with payoff H is to be sold at time 0. Accordingly, the risk-minimising hedging strategy $\vartheta := \varphi^H - \varphi^0$ coincides with the utility-based hedging strategy from (13.3). In that sense Definition 13.25 generalises the concept of utility-based pricing and hedging to more general risk measures.

How can convex risk measures with hedging be represented in terms of penalty functions?

Rule 13.27 (Dual Representation) *Let ϱ be a convex risk measure with penalty function α as in (13.39). Then the corresponding risk measure with hedging $\widetilde{\varrho}$ is of the form*

$$\begin{aligned}\widetilde{\varrho}(H) = &\sup\{E_Q(H) - \alpha(Q) : Q \in \mathscr{P} \text{ such that } S \text{ is a } Q\text{-martingale}\} \\ &+ \inf\{\alpha(Q) : Q \in \mathscr{P} \text{ such that } S \text{ is a } Q\text{-martingale}\}, \quad H \in \mathscr{X}.\end{aligned}$$

Put differently,

$$\widetilde{\alpha}(Q) := \begin{cases} \alpha(Q) - \alpha(Q_0) & \text{if } S \text{ is a } Q\text{-martingale}, \\ \infty & \text{otherwise} \end{cases}$$

can be chosen as penalty function in the sense of Rule 13.22 for $\widetilde{\varrho}$, where Q_0 is the probability measure that minimises $\alpha(Q)$ among all $Q \in \mathscr{P}$ such that S is a Q-martingale.

Idea Obviously it suffices to show that

$$\begin{aligned}\varrho_1(H) &:= \inf\{\varrho(H - V_\varphi(T)) : \varphi \text{ self-financing strategy with initial value } 0\} \\ &= \sup\{E_Q(H) - \alpha(Q) : Q \in \mathscr{P} \text{ such that } S \text{ is a } Q\text{-martingale}\} =: \varrho_2(H)\end{aligned}$$

for any $H \in \mathscr{X}$. Since ϱ_1 differs from $\widetilde{\varrho}$ only by a constant, it is a convex risk measure. We have

$$\begin{aligned}\varrho_1(H) &= \inf_\varphi \varrho(H - V_\varphi(T)) \\ &= \inf_\varphi \sup_{Q\in\mathscr{P}} \big(E_Q(H - V_\varphi(T)) - \alpha(Q)\big) \\ &\geq \sup_{Q\in\mathscr{P}} \inf_\varphi \big(E_Q(H - V_\varphi(T)) - \alpha(Q)\big),\end{aligned}$$

where the infimum refers to all self-financing strategies φ with initial value 0. If S is not a Q-martingale, there exists some φ such that $E_Q(V_\varphi(T)) > 0$ and by rescaling we obtain $\inf_\varphi E_Q(H - V_\varphi(T)) = -\infty$. Consequently,

$$\begin{aligned}\varrho_1(H) &\geq \sup\Big\{\inf_\varphi E_Q(H - V_\varphi(T)) - \alpha(Q) : Q \in \mathscr{P} \text{ s.t. } S \text{ is a } Q\text{-martingale}\Big\} \\ &= \sup\big\{E_Q(H) - \alpha(Q) : Q \in \mathscr{P} \text{ such that } S \text{ is a } Q\text{-martingale}\big\} = \varrho_2(H).\end{aligned}$$

Conversely, suppose that $\varrho_1(H) < \pi \in \mathbb{R}$. Then there exists some φ such that $\varrho_2(H - V_\varphi(T)) < \pi$. This yields

$$\begin{aligned}\varrho_2(H) &= \sup\big\{E_Q(H) - \alpha(Q) : Q \in \mathscr{P} \text{ such that } S \text{ is a } Q\text{-martingale}\big\} \\ &= \sup\big\{E_Q(H - V_\varphi(T)) - \alpha(Q) : Q \in \mathscr{P} \text{ such that } S \text{ is a } Q\text{-martingale}\big\} \\ &\leq \sup\big\{E_Q(H - V_\varphi(T)) - \alpha(Q) : Q \in \mathscr{P}\big\} \\ &= \varrho(H - V_\varphi(T)) < \pi\end{aligned}$$

and hence $\varrho_2 \leq \varrho_1$. Altogether we obtain $\varrho_1 = \varrho_2$ as desired. □

Similarly to Rules 9.15, 11.3, 10.4, 10.26, Rules 13.22 and 13.27 express an economical concept in terms of probability resp. martingale measures. In that sense they could be called *fundamental theorems* as well, this time corresponding to risk measures.

We consider once more the certainty equivalent for exponential utility.

Example 13.28 (Utility Indifference Pricing for Exponential Utility) Let $u(x) = -\exp(-px)$ with $p > 0$. We consider the certainty equivalent from Example 13.17, which coincides with the risk measure ϱ from Example 13.26. As also noted in Example 13.26, the associated risk measure with hedging $\widetilde{\varrho}$ leads to utility indifference prices for exponential utility if we set $\pi(H) := \varrho(H)$. According to Rule 13.27, it can be represented by the penalty function

$$\widetilde{\alpha}(Q) := \begin{cases} \frac{1}{p}(H(Q|P) - H(Q_0|P)) & \text{if } Q \ll P \text{ is such that } S \text{ is a } Q\text{-martingale}, \\ \infty & \text{otherwise}, \end{cases}$$

where Q_0 minimises the entropy $H(Q|P)$ among all martingale measures $Q \ll P$. It can in fact be shown that it suffices to consider equivalent measures in both cases, i.e.

$$\widetilde{\alpha}(Q) = \begin{cases} \frac{1}{p}(H(Q|P) - H(Q_0|P)) & \text{if } Q \text{ is an EMM,} \\ \infty & \text{otherwise,} \end{cases}$$

where Q_0 denotes the minimal entropy martingale measure as in Rule 10.5(3). Consequently,

$$\widetilde{\varrho}(H) = \sup\left\{E_Q(H) - \frac{1}{p}(H(Q|P) - H(Q_0|P)) : Q \text{ is an EMM}\right\}.$$

By Rule 10.23 this is an equivalent way of writing (13.5).

13.4 Comparison of Valuation and Hedging Approaches

In order to compare the valuation and hedging approaches in the previous chapters, we distinguish two general situations. If the derivative contract is liquidly traded at an observable market price, there is no need for a model in order to determine its value. In this case we may instead be interested in the joint dynamics of the derivative assets and their underlyings, e.g. for the purpose of risk management of a portfolio containing underlyings and options. For OTC contracts, however, the situation is different. Here we are primarily interested in a reasonable price when the contract is settled, along with a hedge over the lifetime of the claim.

13.4.1 Liquidly Traded Assets

We discuss the pros and cons for the three approaches that are discussed in Chaps. 11–13.

- *Fair price of an attainable claim*
  Since it relies on the weak and economically very appealing assumption of absence of arbitrage, one may want to use this approach whenever possible. However, most conceivable models are incomplete so that the attainability of a claim constitutes an exception rather than the rule. Limiting one's attention to complete models may mean working with stochastic processes whose dynamics do not match those of real financial data very well. This may also be reflected by substantial differences between observed and theoretical option prices.

- *Ad-hoc choice of a pricing measure*
  Absence of arbitrage means that discounted derivative prices must be martingales relative to some EMM, even if the market is not complete. This consequence of the FTAP explains the common approach to pick a specific EMM for modelling derivatives, often one of a particularly simple structure. The advantage of this approach is that it leads to tractable models where contingent claims can be computed relatively easily. On the negative side, however, these specific choices often lack economic justification, and since theoretical derivative prices differ extensively over the set of all EMMs, the value derived from the ad-hoc choice may seem next to meaningless. In order to assess whether it makes sense, one should check whether computed derivative prices match observed market values reasonably well.
- *Calibrated martingale models*
  If derivative prices are used as model input as in Sect. 11.2.3, two birds are killed with one stone. Both a mismatch between model and observed prices as well as the ad-hoc choice of a martingale measure are avoided. But as discussed in Sect. 11.2.3, the common practice of frequent recalibration is somewhat at odds with the concept of a model and with arbitrage theory. Since recalibration implies that one does not believe in the originally calibrated model, one must be careful about what the latter is used for.

13.4.2 OTC Contracts

In the context of pricing and hedging OTC contracts, we have discussed a larger variety of competing approaches.

- *Fair price and replicating strategy of an attainable claim*
  Since it involves no risk for the option writer, the fair price and the corresponding perfect hedge seem to be the method of choice that should be used whenever possible. But since market completeness is a very strong requirement, we rarely encounter this pleasant situation. Focussing on complete market models, however, may be very dangerous. By definition, they are blind to potential losses in real markets that cannot be removed entirely by some smart choice of a hedge.
- *Superreplication and upper price*
  Superreplication is very appealing from a theoretical point of view because it allows us to extend the idea of perfect insurance to arbitrary models that need not be complete. In concrete cases, however, this approach tends to produce quite extreme results, as illustrated in Examples 11.16 and 11.17. As a consequence, it is unlikely that an option writer will be able to charge anything close to the superreplication price to the client.
- *Ad-hoc choice of a martingale measure*
  One could pick a mathematically tractable EMM Q as a pricing measure in order to come up with a premium for the option writer. In order to obtain a

hedging strategy as well, the approaches of Sect. 11.4 based on sensitivities or of Sect. 12.1 relying on the quadratic hedging error relative to Q may be used. As a drawback, the ad-hoc choice of Q may lack economic foundation, in particular in the common situation that an extreme range of values can be obtained from picking specific EMMs, cf. e.g. Examples 11.16 and 11.17. On top, the expected value relative to Q does not account for unhedgable risks threatening the option writer. If the approach of Sect. 12.1 is used, one can account for the latter by adding a multiple of the minimal quadratic risk to the option price.

- *Calibrated models*
  If derivatives are liquidly traded in the market, we may use a calibrated martingale measure as in Sect. 11.2.3 rather than an ad-hoc EMM. As before, it can be combined with the hedging strategies from Sects. 11.4 or 12.1 and a premium for the remaining risk in the latter case. All in all, this approach shares the pros and cons of the previous suggestion. Compared to the ad-hoc choice, prices may seem less arbitrary. On the other hand, their computation is typically more involved because the dynamics of the underlyings tend to be more complex.
- *Mean-variance hedging*
  The hedging concept of Chap. 12 is widely applicable, does not lead to extreme solutions as the superreplication approach, and seems quite natural from the option writer's perspective. By adding a multiple of the minimal quadratic risk to the optimal initial endowment, the option writer can account for the unhedgable risk. The punishment of profits, however, may be viewed as a conceptual drawback of this approach.
- *Indifference pricing and hedging for expected utility of terminal wealth*
  The indifference concept avoids the anomalies connected with the penalisation of profits in the mean-variance approach. Moreover, the utility-based price naturally incorporates a compensation for unhedgable risks. As a major drawback one may observe that, compared to the previous approaches, the solution is harder to compute in concrete models.
- *Valuation and hedging based on convex risk measures*
  The suggestion of Sect. 13.3 shares the advantages and disadvantages of utility-based valuation and hedging. Indeed, for the entropic risk measure resp. exponential utility, the two approaches coincide, cf. Example 13.28.
- *First-order approximation to utility-based pricing and hedging for P&L*
  The concept and properties of Sect. 13.2 resemble those of the indifference approach for utility of terminal wealth. If, however, the first-order approximation is used, we end up with a quadratic hedging problem in the martingale case. This can be solved more easily than in the terminal wealth case because the BSDE-type equations for the opportunity process need not be solved. Nevertheless, it avoids some anomalies of the variance-optimal hedge in Sect. 12.2 because it is based on increasing utility.

As a case study we consider an OTC derivative where prices and hedges can be computed quite explicitly.

Example 13.29 (Log Contract in a Variance-Gamma Lévy Model) We consider a market composed of a riskless bond $S_0(t) = e^{rt}$ with constant interest rate r and a stock with discounted price $\widehat{S}_1(t) = \widehat{S}_1(0)\exp(X(t))$, where X denotes a variance-gamma LÃľvy process as in Sect. 2.4.7 with parameters $\alpha = 270$, $\beta = -0.041$, $\lambda = 500$, $\mu = -0.0063$ and corresponding exponent (2.62). The first three parameters were chosen in accordance with values estimated in [211] from S&P 500 data in the period from January 1992 to September 1994. The remaining μ, however, is taken so that the discounted stock follows a martingale, i.e.

$$0 = \psi(-i) = \mu - \lambda \log\left(1 - \frac{1+2\beta}{\alpha^2 - \beta^2}\right)$$

or $\mu = \lambda \log(1 - \frac{1-2\beta}{\alpha^2-\beta^2})$. Indeed, the estimated discounted drift rate of the stock does not deviate from zero significantly enough to make this convenient simplifying assumption unreasonable. We consider a bank selling a contingent claim with discounted payoff $\widehat{H}_1 = \log \widehat{S}_1(T)$ or alternatively $\widehat{H}_2 = -\widehat{H}_1$ maturing in $T = 1$, i.e. in one year from now. The bank can hedge against the risk by trading in the stock and the bond. Let us compare the various approaches considered above.

1. *Fair price and replicating strategy of an attainable claim*

   Since the model is incomplete or, more precisely, the claims H_1, H_2 are unattainable, a perfect hedge and consequently a unique fair price for the options does not exist. Hence we cannot apply this approach here.

   As a way out, we may approximate the discounted price process by a geometric Brownian motion $\widehat{S}_1(t) = \widehat{S}_1(0)\exp(\mu_{\mathrm{BS}}t - \sigma W(t))$. We choose

$$\sigma^2 = \mathrm{Var}(X(1)) = \left(1 + \frac{2\beta^2}{\alpha^2 - \beta^2}\right)\frac{2\lambda}{\alpha^2 - \beta^2} = 0.0137$$

   so that the variance of the returns coincides with the one in the "true" model. The mean is taken by assuming once more that $\widehat{S}_1$ is a martingale, i.e. $\mu_{\mathrm{BS}} = -\sigma^2/2 = -0.00687$. This modified model is complete and we obtain

$$\begin{aligned}
\pi_{\mathrm{BS}} &= E(\log \widehat{S}_1(T)) \\
&= \log \widehat{S}_1(0) + E(\mu_{\mathrm{BS}}T + \sigma W(T)) \\
&= \log \widehat{S}_1(0) - \frac{\sigma^2}{2}T \\
&= \log \widehat{S}_1(0) - \left(1 + \frac{2\beta^2}{\alpha^2 - \beta^2}\right)\frac{\lambda}{\alpha^2 - \beta^2}T \\
&= \log \widehat{S}_1(0) - 0.00687
\end{aligned}$$

as a fair price in this Black–Scholes model. The discounted value at time t is obtained accordingly by replacing $\widehat{S}_1(0)$ with $\widehat{S}_1(t)$ and T with $T-t$. The perfect hedge in the Black–Scholes model coincides with the derivative of the pricing function relative to the stock price (cf. Example 11.19), which here leads to $\varphi_{1,\mathrm{BS}}(t) = 1/\widehat{S}_1(t-)$.

2. *Superreplication and upper price*

Recall from Example 11.16 that there is a sequence of equivalent martingale measures Q_n such that the law of $X(T)/n$ under Q_n converges weakly to the Dirac measure in aT for some $a < 0$. Since moreover $\sup_n E_{Q_n}((X(T)/n)^2) < \infty$, we deduce that $\lim_{n\to\infty} E_{Q_n}(X(T)/n) = aT < 0$ and hence

$$E_{Q_n}(\log \widehat{S}_1(T)) = \log \widehat{S}_1(0) + E_{Q_n}(X(T)) \to -\infty$$

as $n \to \infty$. Thus we obtain $\pi_L(H_1) = -\infty$ for the lower price of H_1 and accordingly $\pi_U(H_2) = -\pi_L(H_1) = \infty$ for the upper price of H_2. In particular, H_2 cannot be superreplicated starting with any finite initial endowment.

In order to obtain the upper price of H_1, recall once more from Example 11.16 that there is another sequence of EMMs Q_n such that the law of $X(T)$ under Q_n converges weakly to the Dirac measure in 0. Since $\sup_n E_{Q_n}(X(T)^2) < \infty$ for this sequence, we deduce $\lim_{n\to\infty} E_{Q_n}(X(T)) = 0$ and hence

$$E_{Q_n}\big(\log \widehat{S}_1(T)\big) = \log \widehat{S}_1(0) + E_{Q_n}(X(T)) \to \log \widehat{S}_1(0)$$

as $n \to \infty$. This implies $\pi_U(H_1) \leq \log \widehat{S}_1(0)$ for the upper price of H_1.

We show that in fact equality $\pi_U(H_1) = \log \widehat{S}_1(0)$ holds by constructing a superhedge $\varphi = (\varphi_0, \varphi_1)$ with initial value $V_\varphi(0) = \log \widehat{S}_1(0)$. Indeed, consider the constant portfolio $\varphi_0 := \log \widehat{S}_1(0) - 1$, $\varphi_1 := 1/\widehat{S}_1(0)$. Since $V_\varphi(0) = \widehat{V}_\varphi(0) = \varphi_0 + \varphi_1 \widehat{S}_1(0) = \log \widehat{S}_1(0)$ and

$$\begin{aligned}
\widehat{V}_\varphi(T) &= \varphi_0 + \varphi_1 \widehat{S}_1(T) \\
&= \log \widehat{S}_1(0) - 1 + \frac{\widehat{S}_1(T)}{\widehat{S}_1(0)} \\
&= \log \widehat{S}_1(0) + e^{X(T)} - 1 \\
&\geq \log \widehat{S}_1(0) + X(T) \\
&= \log \widehat{S}_1(T) = \widehat{H}_1,
\end{aligned}$$

strategy φ does indeed have the desired properties. By symmetry, we obtain $\pi_L(H_2) = -\log \widehat{S}_1(0)$ for the lower price of H_2.

3. *Ad-hoc choice of a martingale measure*

The most natural ad-hoc choice of an EMM in our setup is arguably the physical measure P because we assumed $\widehat{S}_1$ to be a P-martingale in the first place. This leads to

$$\begin{aligned}\pi_P &:= E(\widehat{H}_1)\\ &= E(\log \widehat{S}_1(T))\\ &= \log \widehat{S}_1(0) + E(X(T))\\ &= \log \widehat{S}_1(0) + \left(\mu + \frac{2\beta\lambda}{\alpha^2-\beta^2}\right)T\\ &= \log \widehat{S}_1(0) - 0.00687 \end{aligned} \tag{13.43}$$

as a suggested initial option value or, more generally,

$$E(\widehat{H}_1|\mathscr{F}_t) = f(t, \widehat{S}_1(t)) \tag{13.44}$$

with $f(t,x) = \log x + (\mu + \frac{2\beta\lambda}{\alpha^2-\beta^2})(T-t)$ for the discounted option price at time $t \in [0,T]$. Mimicking the perfect hedge in a Black–Scholes setup, one could use $\varphi_{1,\Delta}(t) := D_2 f(t, \widehat{S}_1(t-)) = 1/\widehat{S}_1(t-)$ as a dynamic hedging strategy for H_1, which happens to coincide with the perfect hedge if we wrongly assume the stock to follow geometric Brownian motion.

As an alternative to delta hedging we consider the variance-optimal hedging approach of Sect. 12.1, which may be more convincing in the presence of jumps. Note that $\widehat{H}_1 = f(\log \widehat{S}_1(T))$ with $f(x) = x$. In order to apply the results of Sect. 12.1, we follow the comment of using the Laplace transform in Example 11.33. The variance-optimal hedge is obtained from (12.22) as

$$\begin{aligned}\varphi_1^\star(t) &= \frac{1}{\widehat{S}_1(t-)}\frac{\psi'(-i)-\psi'(0)}{\psi(-2i)i}\\ &= \frac{1}{\widehat{S}_1(t-)}\frac{2\lambda\left(\frac{1+\beta}{\alpha^2-(1+\beta)^2}-\frac{\beta}{\alpha^2-\beta^2}\right)}{2\mu-\lambda\log\frac{\alpha^2-(2+\beta)^2}{\alpha^2-\beta^2}}\\ &= \frac{1}{\widehat{S}_1(t-)}\frac{\alpha^2+\beta^2+\beta}{(\alpha^2-\beta^2)(\alpha^2-(1+\beta)^2)\big(\mu/\lambda-\frac{1}{2}\log\big(1-\frac{4(1+\beta)}{\alpha^2-\beta^2}\big)\big)}\\ &= \frac{0.99997}{\widehat{S}_1(t-)}, \end{aligned} \tag{13.45}$$

which may be compared to the above delta hedge $\varphi_{1,\Delta}(t) = 1/\widehat{S}_1(t-)$. The minimal quadratic risk is derived from (12.23) as

$$\begin{aligned}
\varepsilon^2(\varphi^\star) &= \left(-\psi''(0) + \frac{(\psi'(-i) - \psi'(0))^2}{\psi(-2i)}\right) T \\
&= \left(2\lambda \frac{\alpha^2 + \beta^2}{(\alpha^2 - \beta^2)^2} - \frac{4\lambda^2 \left(\frac{1+\beta}{\alpha^2-(1+\beta)^2} - \frac{\beta}{\alpha^2-\beta^2}\right)^2}{2\mu - \lambda \log \frac{\alpha^2-(2+\beta)^2}{\alpha^2-\beta^2}}\right) T \\
&= \frac{2\lambda}{(\alpha^2 - \beta^2)^2}\left(\alpha^2 + \beta^2 - \frac{(\alpha^2 + \beta^2 + \beta)^2}{(\alpha^2 - (1+\beta)^2)^2(\mu/\lambda - \frac{1}{2}\log\left(1 - \frac{4(1+\beta)}{\alpha^2-\beta^2}\right))}\right) T \\
&= 0.0000003 \qquad (13.46)
\end{aligned}$$

in our case. It turns out to be linear in time to maturity T and independent of the initial stock price. The value of H_2 and the corresponding hedges are simply obtained as the negatives of those for H_1.

4. *Calibrated models*

For the calibration approach we need observed option prices in the first place. We follow [211] in this regard, who calibrated the VG model to S&P 500 options in the above period. Assuming that $\widehat{S}_1$ follows a variance-gamma Lévy process under the pricing measure Q chosen by the market, [211] obtains the parameter values $\alpha_Q = 30.0$, $\beta_Q = -9.76$, $\lambda_Q = 5.93$, $\mu_Q = 0.135$. The corresponding option price for H_1 is given by

$$\begin{aligned}
\pi_Q &:= E_Q(\widehat{H}_1) \\
&= \log \widehat{S}_1(0) + E_Q(X(T)) \\
&= \log \widehat{S}_1(0) + \left(\mu_Q + \frac{2\beta_Q\lambda_Q}{\alpha_Q^2 - \beta_Q^2}\right) T \\
&= \log \widehat{S}_1(0) - 0.00890,
\end{aligned}$$

which may be compared to $E(\widehat{H}_1) = \log \widehat{S}_1(0) - 0.00687$ from (13.43) above. The delta hedge and the variance-optimal hedge relative to Q are computed as before, but relative to Q-parameters of the Lévy process X. We obtain $\varphi_{1,Q}(t) = 1/\widehat{S}_1(t-)$ for the delta hedge,

$$\begin{aligned}
\varphi^\star_{1,Q}(t) &= \frac{1}{\widehat{S}_1(t-)} \frac{\alpha_Q^2 + \beta_Q^2 + \beta_Q}{(\alpha_Q^2 - \beta_Q^2)} \\
&\quad \times \frac{1}{(\alpha_Q^2 - (1+\beta_Q)^2)(\mu_Q/\lambda_Q - \frac{1}{2}\log\left(1 - \frac{4\beta_Q(1+\beta_Q)}{\alpha_Q^2-\beta_Q^2}\right))} \\
&= \frac{1.0312}{\widehat{S}_1(t-)},
\end{aligned}$$

for the variance-optimal hedge, and

$$\varepsilon_Q^2(\varphi_Q^\star) = \frac{2\lambda_Q}{(\alpha_Q^2 - \beta_Q^2)^2}\Bigg(\alpha_Q^2 + \beta_Q^2 - \frac{(\alpha_Q^2 + \beta_Q^2 + \beta_Q)^2}{(\alpha_Q^2 - (1+\beta_Q)^2)^2\big(\mu_Q/\lambda_Q - \frac{1}{2}\log\big(1 - \frac{4\beta_Q(1+\beta_Q)}{\alpha_Q^2-\beta_Q^2}\big)\big)}\Bigg)T$$

$$= 0.000034$$

for the minimal quadratic risk relative to Q. The latter appears to be larger than the minimal risk relative to P. This can be explained by the fact that observed returns are fitted well by a Lévy process which, in spite of its jumps, resembles Brownian motion quite closely. Since the latter would imply vanishing hedging errors, the minimal risk for the *estimated* model should naturally be very small. The *calibrated* Lévy process, however, has heavier tails and differs more severely from Brownian motion, which explains the larger hedging error.

However, one may wonder whether the calibrated pricing measure is relevant at all for the hedging problem at hand. Indeed, if the liquid options are not traded by the bank, they should not play a role in the bank's assessment of the risk. The calibrated model makes more sense if the options are actively used by the bank as hedging instruments.

5. *Mean-variance hedging*

Since the discounted stock $\widehat{S}_1$ follows a martingale in our setup, the mean-variance hedging approach can be applied in the simpler situation of Sect. 12.1. The relevant objects have already been determined above, namely the value process in (13.44), the variance-optimal hedging strategy in (13.45), and the minimal quadratic risk in (13.46). If we follow the suggestion (12.2) with risk aversion parameter $p = 100$, we end with

$$\pi_{\mathrm{MV}} := E(\widehat{H}_1) + \frac{p}{2}\varepsilon^2(\varphi^\star) = \log\widehat{S}_1(0) - 0.00687 + 0.0000003\frac{p}{2}$$

as a reasonable option premium that is to be charged to the buyer. The results for H_2 instead of H_1 are obtained accordingly.

6. *Indifference pricing and hedging for expected utility of terminal wealth*

In the setup of Sect. 13.1 we assume risk aversion $p = 100$ for exponential utility and start with $H := H_1$. Alternatively, we would obtain the same price and hedge per claim if the bank with risk aversion $p = 1$ sold 100 shares of the option.

First, we solve the optimal investment problem relative to P and P^H defined as in (10.44). This is almost trivial for P because it is a martingale measure. From Rule 10.3(1) or Example 10.19 we immediately conclude that the empty portfolio $\varphi_1^0 = 0$ is optimal for expected utility of terminal wealth and that the corresponding opportunity process equals $L_0 = 1$. This changes if we consider the measure P^H whose density equals

$$\frac{dP^H}{dP} = \frac{\exp(p\widehat{H}_1)}{E(\exp(p\widehat{H}_1))} = \frac{e^{pX(T)}}{E(e^{pX(T)})}.$$

Since $Z(t) = \exp(pX(t) - \psi(-ip)t)$ is a martingale by Theorem 2.22, it coincides with the density process of P^H. Moreover, it corresponds to a measure change as in Proposition 2.26. By Sect. 2.4.7 the Lévy–Khintchine triplet of X equals

$$\left(a^X, c^X, K^X(dx)\right) = \left(\mu + \frac{2\beta\lambda}{\alpha^2 - \beta^2}, 0, \frac{\lambda}{|x|}e^{\beta x - \alpha|x|}dx\right).$$

Relative to P^H, the process X is a variance-gamma Lévy process as well, but with exponent $\psi_H(u) = \psi(u - ip) - \psi(-ip)$, parameters $\alpha_H = \alpha$, $\beta_H = p + \beta$, $\lambda_H = \lambda$, $\mu_H = \mu$, and triplet

$$\left(a^{X,H}, c^{X,H}, K^{X,H}(dx)\right) = \left(\mu + \frac{2(p+\beta)\lambda}{\alpha^2 - (p+\beta)^2}, 0, e^{px}K^X(dx)\right),$$

cf. Sect. 2.4.7. From (10.35) in Example 10.19 we obtain the optimal strategy $\varphi_1^H = \chi/\widehat{S}_1(t-)$, where the constant χ solves

$$0 = a^{X,H} + \int \left(\exp(-p\chi(e^x - 1))(e^x - 1) - x\right)K^{X,H}(dx). \tag{13.47}$$

This leads to $\chi = 0.99997$ in our specific setup. By (10.43, 10.33, 10.36) the initial value of the opportunity process equals $L^H(0) = e^{-\ell T}E(e^{p\widehat{H}_1})$ with

$$\begin{aligned}\ell &= -\int \left(\exp\left(-p\chi(e^x - 1)\right)\left(1 + p\chi(e^x - 1)\right) - 1\right)K^{X,H}(dx) \\ &= 73.87\end{aligned}$$

and $E(e^{p\widehat{H}}) = \exp(p\log\widehat{S}_1(0) + \psi(-ip)T)$. Applying Rule 13.1, we obtain

$$\begin{aligned}\pi_{\mathrm{UI}} &= \frac{1}{p}\log\frac{L^H(0)}{L^0(0)} \\ &= \log\widehat{S}_1(0) + (\psi(-ip) - \ell)T/p \\ &= \log\widehat{S}_1(0) - 0.00685\end{aligned} \tag{13.48}$$

for the utility indifference price of H_1. The corresponding utility-based hedge equals

$$\vartheta_{\mathrm{UI}}(t) = \varphi_1^H(t) - \varphi_1^0(t) = \chi/\widehat{S}_1(t-) = 0.99997/\widehat{S}_1(t-), \tag{13.49}$$

which may be contrasted with the delta hedge $1/\widehat{S}_1(t-)$ and the variance-optimal hedge $0.99997/\widehat{S}_1(t-)$ from above.

Let us compare the utility-based price and hedge to its leading-order approximation from Sect. 13.1.3. Since the minimal-entropy martingale measure equals the physical measure P in our case, we simply have to determine the variance-optimal hedge and the minimal quadratic risk under P, which we already did above. Specifically, the approximate utility-based hedge coincides with the variance-optimal hedging strategy (13.45), which is close to the exact value (13.49). Likewise, the approximate utility indifference price

$$\pi_{\mathrm{AUI}} := E(\widehat{H}_1) + \frac{p}{2}\varepsilon^2(\varphi^\star) = \log \widehat{S}_1(0) - 0.00685$$

does not deviate from the exact value (13.48) in the significant three digits.

Let us have a look at H_2 as well. In this case the analogue of equation (13.47) does not have a solution because the discounted stock price process $\widehat{S}_1$ is lacking positive exponential moments. In other words, shortselling arbitrarily small amounts of the stock leads to expected utility $-\infty$. The optimal strategy is the empty portfolio, leading to discounted terminal wealth $v_0 + \log \widehat{S}_1(0) + X(T)$. Consequently, the condition for the utility indifference price is $e^0 = E(\exp(-p(\pi_{\mathrm{UI}} + \log \widehat{S}_1(0) + X(T))))$ or

$$\begin{aligned}\pi_{\mathrm{UI}} &= -\log \widehat{S}_1(0) + \frac{1}{p}\log E(e^{-pX(T)})\\ &= -\log \widehat{S}_1(0) + \frac{1}{p}\psi(ip)T\\ &= -\log \widehat{S}_1(0) + 0.746.\end{aligned} \tag{13.50}$$

The lack of exponential moments causes Rule 13.6 not to hold. Instead, the indifference price in the limit $p \to 0$ equals

$$\begin{aligned}\pi_{\mathrm{AUI}} &= -\log \widehat{S}_1(0) + \left(-a^X + \frac{1}{p}\int (e^{-px} - 1 + px)K^X(dx)\right)T\\ &= -\log \widehat{S}_1(0) + \left(-a^X + \frac{p}{2}\int x^2 K^X(dx) + o(p)\right)T\\ &= -\log \widehat{S}_1(0) + \left(-E(X(1)) + \frac{p}{2}\mathrm{Var}(X(1))\right)T + o(p)\end{aligned}$$

$$= -\log \widehat{S}_1(0) + \left(-\mu - \frac{2\beta\lambda}{\alpha^2 - \beta^2} + p\left(1 + \frac{2\beta^2}{\alpha^2 - \beta^2}\right)\frac{\lambda}{\alpha^2 - \beta^2}\right)T + o(p)$$

$$= -\log \widehat{S}_1(0) + 0.694 + o(p),$$

which is similar to (13.50). However, it differs substantially from $E(\widehat{H}_2) + \frac{p}{2}\varepsilon^2(\varphi^\star) = -\log \widehat{S}_1(0) + 0.00688$, which would have been expected from Rule 13.6. The utility-based hedge equals 0 for all p and hence in the limit $p \to 0$ as well. By contrast, one would have expected the negative of (13.45) from Rule 13.6.

7. *Valuation and hedging based on convex risk measures*

   If we consider the entropic risk measure from Example 13.17, Example 13.28 yields that the value obtained from the convex risk measure with hedging in Definition 13.25 equals the utility indifference price and the risk-minimising hedging strategy coincides with the utility-based hedging strategy of Sect. 13.1. Both have been determined above for H_1, H_2.

8. *Utility-based pricing and hedging for P&L*

   We consider the hyperbolic utility function $u(x) = (1 + px - \sqrt{1 + (px)^2})/p$ from Example 10.39 with $p = 100$. As above, the results can also be interpreted as price and hedge per share if the bank has risk aversion $p = 1$ and sells 100 claims. Contrary to exponential utility, the choice of the hyperbolic utility function avoids problems related to the non-existence of exponential moments.

   Let us first determine the indifference price for utility of P&L without the approximation for small numbers of claims. We start with the claim $H := H_1$. Firstly, note that $\varphi^0 = 0$ is an optimal strategy for P&L in the market without claim because $\widehat{S}_1$ is a martingale. This optimality can be derived, for example, by applying Rule 10.40 to $\varphi^\star = 0$. Now, denote by $\chi \in \mathbb{R}$ the solution to

$$\int \frac{(e^x - 1)(\chi(e^x - 1) - x)}{\sqrt{(\chi(e^x - 1) - x)^2 + p^{-2}}} K^X(dx) = 0, \tag{13.51}$$

which equals

$$\chi = 0.99997. \tag{13.52}$$

Since

$$Y := \mathscr{E}\left(\frac{\chi(e^x - 1) - x}{\sqrt{(\chi(e^x - 1) - x)^2 + p^{-2}}} * (\mu^X - \nu^X) \right)$$

is a positive martingale, it is the density process of some probability measure $Q \sim P$. Set $\widehat{S}_2(t) := E_Q(\widehat{H}_1 | \mathscr{F}_t)$. We claim that

$$\widehat{S}_2^H(t) = \log \widehat{S}_1(t) + \gamma(T - t) \tag{13.53}$$

for some $\gamma \in \mathbb{R}$. Indeed, Theorem 4.15 yields that X has constant deterministic characteristics under Q as well, which means that it is a Q-Lévy process. Consequently,

$$\begin{aligned}E_Q(X(T)|\mathscr{F}_t) &= X(t) + E_Q(X(T) - X(t)|\mathscr{F}_t)\\&= X(t) + E_Q(X(T-t))\\&= X(t) + \gamma(T-t).\end{aligned}$$

Since $\widehat{S}_2^H(t) = \log \widehat{S}_1(0) + E_Q(X(T)|\mathscr{F}_t)$ and $\log \widehat{S}_1(t) = \log \widehat{S}_1(0) + X(t)$, the claim (13.53) follows. Thus Y can be written as

$$Y = \mathscr{E}\left(\frac{\chi(\widehat{S}_1)_-^{-1}x_1 - x_2}{\sqrt{\left(\chi(\widehat{S}_1)_-^{-1}x_1 - x_2\right)^2 + p^{-2}}} * \left(\mu^{(\widehat{S}_1,\widehat{S}_2^H)} - \nu^{(\widehat{S}_1,\widehat{S}_2^H)}\right)\right).$$

An application of Theorem 4.15 yields that the Q-drift of $\widehat{S}_1$ vanishes, which implies that Q is an EMM for $(\widehat{S}_1, \widehat{S}_2^H)$. Since its density process Y is of the form (10.73) for the strategy $\varphi^H = (\chi/\widehat{S}_1(t-), -1)$, we conclude that φ^H is optimal for P&L in the market (S_0, S_1, S_2^H), which means that $\widehat{S}^H = (\widehat{S}_1, \widehat{S}_2^H)$ is a discounted shadow price with corresponding optimal strategy φ^H, cf. Sect. 10.5.2. In particular,

$$\vartheta_{\mathrm{PL}}(t) = \varphi_1^H(t) - \varphi_1^0(t) = \chi/\widehat{S}_1(t-) = 0.99997/\widehat{S}_1(t-) \tag{13.54}$$

is the utility-based hedge for P&L. In order to determine the utility indifference price, we must compute $U^S(\varphi^0)$ and $U^{S^H}(\varphi^H)$ in (13.22). The expected utility $U^S(\varphi^0)$ in the absence of the claim vanishes because $\varphi^0 = 0$. By (10.67) the other expected utility equals

$$U^{S^H}(\varphi^H) = \left(-(a^X - \gamma) + \int\left(p^{-1} - \sqrt{p^{-2} + (\chi(e^x - 1) - x)^2}\right)K^X(dx)\right)T.$$

Moreover, we have $\widehat{S}_2^H(0) = \log \widehat{S}_1(0) + \gamma T$, which yields

$$\begin{aligned}\pi_{\mathrm{PL}} &= \widehat{S}_2^H(0) - \left(U^{S^H}(\varphi^H) - U^S(\varphi^0)\right)\\&= \log \widehat{S}_1(0) + \left(\mu + \frac{2\beta\lambda}{\alpha^2 - \beta^2}\right.\\&\quad\left. - \int\left(p^{-1} - \sqrt{p^{-2} + (\chi(e^x - 1) - x)^2}\right)K^X(dx)\right)T\\&= \log \widehat{S}_1(0) - 0.00685\end{aligned} \tag{13.55}$$

by (13.22).

What happens in the approximation for small numbers of claims? Since P itself is a martingale measure, the same argument as in Example 13.13 yields that both the approximate utility-based hedge and the approximate utility indifference price for P&L coincide with the corresponding objects for exponential utility of terminal wealth, i.e. we obtain the price and the hedge from (13.48, 13.49).

The reasoning for $H = H_2$ instead of H_1 is essentially the same, leading to

$$\int \frac{(e^x - 1)(\chi(e^x - 1) + x)}{\sqrt{(\chi(e^x - 1) + x)^2 + p^{-2}}} K^X(dx) = 0,$$

$\chi = -0.99997$,

$$\vartheta_{\mathrm{PL}}(t) = \chi / \widehat{S}_1(t-) = -0.99997/\widehat{S}_1(t-),$$

and

$$\begin{aligned}\pi_{\mathrm{PL}} = & -\log \widehat{S}_1(0) - \left(\mu + \frac{2\beta\lambda}{\alpha^2 - \beta^2}\right. \\ & \left. + \int \left(p^{-1} - \sqrt{p^{-2} + (\chi(e^x - 1) + x)^2}\right) K^X(dx)\right) T \\ = & -\log \widehat{S}_1(0) + 0.00688\end{aligned}$$

instead of (13.51, 13.52, 13.54, 13.55). The approximate utility-based price and hedge for P&L are given by

$$\pi_{\mathrm{APL}} = E(\widehat{H}_2) + \frac{p}{2}\varepsilon^2(\varphi^\star) = -\log \widehat{S}_1(0) + 0.00688$$

and

$$\vartheta_{\mathrm{APL}}(t) = -\varphi_1^\star(t) = -0.99997/\widehat{S}_1(t-)$$

in this case. As noted above, we opted for the hyperbolic utility function in order to avoid problems linked to insufficient integrability, which occur in particular for H_2.

Comparing the figures in the columns of Table 13.1, we observe a large extent of similarity. This stems from the fact that the estimated model does not deviate greatly from geometric Brownian motion, in which case all approaches yield more or less the same result. Superreplication, however, is known to be quite sensitive to even small incidences of incompleteness, which is reflected by the extreme upper price for H_2. A related smaller effect is visible for exponential utility if sufficient moments fail to exist. The gap between the numbers in 3. and 4. can be traced back to the difference between the estimated physical and calibrated risk-neutral parameters. In contrast to 1., 3., 4., the approaches in 5.–8. include a risk premium for the

Table 13.1 Comparison of option prices and hedges for a log contract in a variance-gamma Lévy model, cf. Example 13.29 for details

	$\widehat{H}_1 = \log \widehat{S}_1(T)$		$\widehat{H}_2 = -\log \widehat{S}_1(T)$	
	price π:	hedge $\varphi_1(t)$:	price π:	hedge $\varphi_1(t)$:
Valuation and hedging approach	$\log \widehat{S}_1(0) + \dots$	$\frac{1}{\widehat{S}_1(t-)} \times \dots$	$-\log \widehat{S}_1(0) + \dots$	$\frac{-1}{\widehat{S}_1(t-)} \times \dots$
1. Black–Scholes approximation	−0.00687	1	0.00687	1
2. Upper price and superhedge	0	$\exp(X(t-))$	∞	—
3a. Ad-hoc EMM & delta hedging	−0.00687	1	0.00687	1
3b. Ad-hoc EMM & quadratic hedging	−0.00687	0.99997	0.00687	0.99997
4a. Calibrated MM & delta hedging	−0.00890	1	0.00890	1
4b. Calibrated MM & quadratic hedging	−0.00890	1.0312	0.00890	1.0312
5. Mean-variance hedging	−0.00685	0.99997	0.00688	0.99997
6a. Exponential utility of terminal wealth	−0.00685	0.99997	0.746	0
6b. Approx. utility of terminal wealth	−0.00685	0.99997	0.680	0
7. Entropic risk measure	−0.00685	0.99997	0.746	0
8a. Hyperbolic utility of P&L	−0.00685	0.99997	0.00688	0.99997
8b. Approx. hyperbolic utility of P&L	−0.00685	0.99997	0.00688	0.99997

unhedgable part, which leads to a nonlinearity of prices. Numerically, however, this effect remains mostly small in the above example.

Appendix 1: Problems

The exercises correspond to the section with the same number.

13.1 Determine the approximate price and hedge of Sect. 13.1.3 for a European call in the Heston model of Sect. 8.2.4. For simplicity assume that the discounted stock is a martingale or, more specifically, $S_0 = 1$ and $\mu = 0$, $\delta = -1/2$ in (8.11).

Hint: Apply the results of Sect. 12.1.3.

13.2 Determine the approximate price and hedge in the sense of Sect. 13.2 for a European call in the Heston model of Sect. 8.2.4. As in the previous exercise assume that the discounted stock is a martingale. Compare the results of Problems 13.1 and 13.2.

13.3 (Worst and Tail Conditional Expectation) Suppose that the loss H has a continuous law, i.e. a continuous cumulative distribution function. Show that in this case $\mathrm{AVaR}_\alpha(H) = \mathrm{WCE}_\alpha(H) = \mathrm{TCE}_\alpha(H)$, where the **worst conditional expectation** at level α is defined as

$$\mathrm{WCE}_\alpha(H) := \sup\left\{E(H|F) : F \text{ event with } P(F) \geq 1-\alpha\right\}$$

and the **tail conditional expectation** at level α as

$$\mathrm{TCE}_\alpha(H) := E(H|H \geq \mathrm{VaR}_\alpha(H)).$$

13.4 (Basis Risk) Repeat the comparative study of Example 13.29 for the basis risk example of Problem 12.1 and Examples 13.3, 13.11. Specifically, consider a market composed of a riskless bond $S_0(t) = e^{rt}$ with constant interest rate r and two assets with discounted price processes

$$dS_1(t) = S_1(t)(\mu dt + \sigma dW(t)),$$
$$d\widetilde{S}(t) = \widetilde{S}(t)(\widetilde{\mu} dt + \widetilde{\sigma} d\widetilde{W}(t)),$$

where $\mu, \widetilde{\mu}, \sigma, \widetilde{\sigma}$ are constants and $(W, \widetilde{W})$ is a two-dimensional Brownian motion with $E(W(1)) = 0 = E(\widetilde{W}(1))$, $\mathrm{Var}(W(1)) = 1 = \mathrm{Var}(\widetilde{W}(1))$, and $\mathrm{Cov}(W(1), \widetilde{W}(1)) = \varrho$. Only S_0, S_1 are supposed to be tradable. One may think of an index $\widetilde{S}$ and an exchange-traded fund (ETF) S_1 on this index. For simplicity, we assume $\mu = \widetilde{\mu} = 0$, i.e. $S_1, \widetilde{S}$ are martingales. The parameters $\sigma = 0.104$, $\widetilde{\sigma} = 0.103$, $\varrho = 0.993$ have been estimated from S&P 500 data and a corresponding ETF in the period January 1, 2016 till December 31, 2017. The initial values are $S_1(0) = 266.86$ and $\widetilde{S}(0) = 2673.61$.

We consider two contingent claims, namely an at-the-money put with discounted payoff $\widehat{H}_1 := (\widetilde{S}(0) - \widetilde{S}(T))^+$ and an at-the-money call with discounted payoff $\widehat{H}_2 := (\widetilde{S}(T) - \widetilde{S}(0))^+$. The goal is to determine the various option prices and hedging strategies parallel to Example 13.29. For the naive Black–Scholes approximation simply suppose that $S_1/S_1(0) \approx \widetilde{S}/\widetilde{S}(0)$, i.e. $\sigma \approx \widetilde{\sigma} = 0.103$, $\varrho \approx 1$. The calibration approach should be left out because it is not obvious how to obtain, say, the correlation parameter ϱ from liquid option prices. How do the resulting figures change if correlation amounts to $\varrho = 0.70$ instead of the value from above?

Appendix 2: Notes and Comments

Utility-based derivative pricing is discussed in [45, 117, 242]. For applied aspects of risk measures see also [216]. The utility indifference approach in finance goes back at least to [145]. Rigorous statements are to be found in [19, 25, 74, 253] and others. The dual characterisation (13.6) in Rule 13.1 is stated in [74, equation (5.6)]. For Rule 13.2 see [19, (3.8)]. Example 13.3 is due to [141]. For versions of Rule 13.5 we refer to [19, 74, 253]. The other limiting case of Rule 13.6 is treated in [20, 180, 212] in various levels of generality and in [198, 199] for general utility functions. The approach of running P&L is pursued in [164]. For pricing small numbers of claims see also [166] which, however, takes an equilibrium point of view which differs from the approach taken here.

The theory of risk measures goes back to [4] in the coherent case and to [116, 121, 139] in the convex extension. For Sect. 13.3 we suggest [117] as a prime reference where rigorous statements on risk measures can be found. For the convenience of the reader we point at corresponding statements: for Definition 13.14 see [117, Section 4.1], for Example 13.15 see [117, Definition 4.45], for Example 13.16 see [117, Definition 4.48], for Example 13.17 see [117, Example 4.13], for Definition 13.18 see [117, Section 4.1], for Example 13.19 see [117, Example 4.10], for Rule 13.20 see [117, Propositions 4.6, 4.7 and Exercise 4.1.4], for Example 13.21 see [117, Definition 4.112], for Rule 13.22 see [117, Proposition 4.15, Theorems 4.16, 4.22, 4.33, and Remark 4.17], for Example 13.23 see [117, Theorem 4.52], for Example 13.24 see [117, Example 4.34], for Definition 13.25 see [117, Section 4.8], and for Problem 13.3 see [117, Example 4.41 and Corollaries 4.54, 4.68]. Risk measures in the context of hedging is discussed in detail in [11, 12, 286]. Specifically, we refer to [12, Corollary 3.7] and [286, Theorem 3.2] for Rule 13.27, and to [12, Proposition 5.3] for Example 13.28.

Chapter 14
Interest Rate Models

From an abstract point of view interest rate markets fall into the reach of the general theory which is discussed in Chaps. 9–13. They differ from equity markets just in the kind of assets, namely bonds, swaps, caps and other interest rate products. On closer observation, however, one faces new phenomena. For example, the natural choice of a numeraire is not obvious in markets where interest rates change randomly over time. Moreover, one typically deals with a large number of bonds even in simple one-factor models. As soon as instantaneous forward rates come into play, one actually works implicitly or explicitly with an infinite number of securities. This may involve aspects of an infinite-dimensional stochastic calculus which, however, is beyond the scope of this monograph. All these aspects turn interest rate theory into a subject on its own.

14.1 Basics

In this section we introduce notions and discuss aspects of interest rate theory which do not depend on the particular model. This provides a common basis for the conceptually different approaches that are discussed in Sects. 14.2–14.6. As a general rule we assume or aim at market models that do not allow for arbitrage.

14.1.1 Interest Rates

In the examples in the previous chapters we typically assumed that a constant, continuously compounded interest rate r is paid to the investor on a money market account, leading to an exponentially growing amount $S_0(t) = e^{rt}$ if 1€ is initially invested at time 0. Such a **money market account** also plays a key role in several

© Springer Nature Switzerland AG 2019

E. Eberlein, J. Kallsen, *Mathematical Finance*, Springer Finance,
https://doi.org/10.1007/978-3-030-26106-1_14

approaches to interest rate theory but now the **short rate** $r(t)$ is assumed to change randomly over time. Then the money market earns interest according to the equation

$$dS_0(t) = S_0(t)r(t)dt. \tag{14.1}$$

Expressed differently, an initial endowment of 1€ grows to

$$S_0(t) = \mathscr{E}(r \bullet I)(t) = \exp\left(\int_0^t r(s)ds\right) \tag{14.2}$$

at time t, where $I(t) = t$ denotes the identity process as usual. One may argue that the money market account does not exist literally in real markets. It can only be approximated by a sequence of short term investments. Nevertheless, this idealised asset turns out to be very useful in interest rate theory. In some approaches such as in Sect. 14.4, however, it is avoided altogether.

Bonds constitute further basic objects in interest rate theory. A **zero coupon bond** with maturity T (also called a T-**bond**) pays 1€ at time T, i.e. the price process $B(\cdot, T) = (B(t, T))_{t\in[0,T]}$ satisfies

$$B(T, T) = 1.$$

There are no interest payments before maturity. After time T the bond ceases to exist. Bonds with intermediate coupon payments are considered in Sect. 14.1.2 below.

For the following we need the notion of simple interest rates in contrast to the continuously compounded rate r above. If a **simple rate** L is paid for a fixed period $[S, T]$, this means that x€ invested at time S runs up to

$$y = x(1 + L(T - S)) \tag{14.3}$$

euro at time $T > S$. Put differently, the simple rate corresponding to this investment satisfies

$$L = \frac{y - x}{x(T - S)}. \tag{14.4}$$

Recall from (14.1, 14.2) that a **continuously compounded rate** r paid on the initial endowment x€ at S leads to a payoff of

$$y = xe^{r(T-S)}$$

euro at time $T > S$. Hence the simple rate L from (14.3) corresponds to a continuously compounded rate

$$r = \frac{1}{T - S}\log(1 + L(T - S)), \tag{14.5}$$

which is equivalent to

$$L = \frac{e^{r(T-S)} - 1}{T - S}.$$

Note that the difference between these two kinds of rates tends to 0 for vanishing time horizon $T - S$.

Through buying and selling zero bonds one can fix an interest rate for an investment in some period in the future. Let us assume that zero bonds with maturities $S < T$ are liquidly traded, with initial prices $B(0, S)$ and $B(0, T)$, respectively. Suppose that you sell one S-bond and buy $B(0, S)/B(0, T)$ T-bonds at time 0. The net cash flow of this investment at time 0 equals

$$B(0, S) - \frac{B(0, S)}{B(0, T)} B(0, T) = 0.$$

At time S you must pay 1€ to the holder of the shorted S-bond. In exchange you receive $B(0, S)/B(0, T)$€ at time T for your investment in T-bonds. According to (14.4) this corresponds to a simple rate

$$L(0, S, T) := \frac{B(0, S) - B(0, T)}{B(0, T)(T - S)}.$$

Of course, this carries over to the case that the portfolio is set up at time $t > 0$ as long as $t \leq S < T$. In this case

$$L(t, S, T) := \frac{B(t, S) - B(t, T)}{B(t, T)(T - S)}$$

is the **simple forward rate** for the future period $[S, T]$ as implied by the bond prices at time t. The law of one price yields that it is the only riskless interest rate that is consistent with absence of arbitrage.

The yield from the above investment in S- resp. T-bonds can also be expressed in terms of continuously compounded rates. By (14.5) the simple rate $L(\cdot, S, T)$ corresponds to a **continuously compounded forward rate (cc forward rate)**

$$f(t, S, T) := \frac{1}{T - S} \log(1 + L(t, S, T)(T - S)) = \frac{1}{T - S} \log \frac{B(t, S)}{B(t, T)}.$$

It is determined by the ratio

$$B(t, S, T) := \frac{B(t, S)}{B(t, T)} = 1 + L(t, S, T)(T - S) = e^{f(t,S,T)(T-S)}$$

of bond prices, which we call **forward process** for the future period $[S, T]$. These notions will be of particular interest in Sect. 14.4 on forward process models. As noted above, the difference between $L(t, S, T)$ and $f(t, S, T)$ is likely to be small for short time horizons $T - S$.

If—at least in a mathematical idealisation—bonds are available for any maturity $T > 0$, we can consider the limit for $S \uparrow T$ or $T \downarrow S$. To this end we assume that bond prices $B(t, T)$ are differentiable as a function of maturity T. The limit

$$f(t, T) := \lim_{\varepsilon \downarrow 0} L(t, T - \varepsilon, T) = -\frac{d}{dT} \log B(t, T) = \lim_{\varepsilon \downarrow 0} f(t, T - \varepsilon, T) \qquad (14.6)$$

or, more or less equivalently,

$$f(t, T) := \lim_{\varepsilon \downarrow 0} L(t, T, T + \varepsilon) = -\frac{d}{dT} \log B(t, T) = \lim_{\varepsilon \downarrow 0} f(t, T, T + \varepsilon) \qquad (14.7)$$

is called the **(instantaneous) forward rate**. It corresponds to the interest rate paid for an investment in an infinitesimal time period around time T in the future as implied by bond prices at time $t < T$. Note that the difference between simple and continuously compounded rates vanishes here because of the short—in fact even "infinitesimal"—time horizon. Equation (14.6) and $B(T, T) = 1$ imply

$$B(t, T) = \exp\left(-\int_t^T f(t, s) ds\right), \quad t < T.$$

The argument T in (14.6, 14.7) stands for maturity. An alternative useful parametrisation is time *to* maturity $\tau := T - t$. In this so-called **Musiela parametrisation** instantaneous forward rates are defined as

$$r(t, \tau) := f(t, t + \tau)$$

for arbitrary $\tau \geq 0$. For $\tau = 0$ this interest rate corresponds to a short time investment that is to start immediately. This coincides with the interpretation of the short rate r above which underlies the money market account. From that perspective

$$f(t, t) = r(t, 0) = r(t) \qquad (14.8)$$

should hold in any reasonable market model. Indeed, heuristically this follows from arbitrage arguments. By starting with 1€ and investing successively in just maturing bonds one can create a portfolio whose value $S(t)$ grows according to

$$dS(t) = S(t) r(t, 0) dt.$$

By Proposition 3.43 this **roll-over portfolio** satisfies

$$S(t) = \mathscr{E}(r(\cdot, 0) \bullet I)(t) = \exp\left(\int_0^t r(s, 0) ds\right).$$

Alternatively, we may choose the more profitable strategy which invests momentarily either in the money market account or in the just maturing bond, depending on which interest rate is higher. This leads to the cashflow

$$dV(t) = V(t)\max(r(t), r(t,0))dt$$

and hence

$$V(t) = \exp\biggl(\int_0^t \max(r(s), r(s,0))ds\biggr).$$

Fix a time horizon T. Observe that $V(0) - S_0(0) = 0$, $V(T) - S_0(T) \geq 0$ and $P(V(T) - S_0(T) > 0) > 0$ unless $r(t,0) \leq r(t)$ on $[0,T]$. This implies that buying V and selling S_0 constitutes an arbitrage unless the instantaneous forward rate is dominated by the short rate. A similar comparison of V and the roll-over portfolio S yields that $r(t) \leq r(t,0)$ must hold as well. Hence absence of arbitrage implies (14.8). The above argument involves simultaneous trading with uncountable many securities, which exceeds the setup of Chap. 9. A rigorous justification requires a more general mathematical framework which is beyond the scope of this monograph, cf. [33, 34] in this context.

From a different point of view (14.8) can be used to *define* a short rate $r(t)$ if only the bond or forward rate processes are given in the first place. This means that the money market account is interpreted as the roll-over portfolio investing in just maturing bonds.

If the money market account and bonds of arbitrary time horizons do not exist, S_0 from (14.1) or (14.2) does not make sense. As an alternative we may repeatedly invest in the bonds with the shortest available maturity in the market. To this end assume that T_n-bonds are traded for $n = 1, \dots, N$ with $0 \leq T_1 < \dots < T_N$. We start with 1€ at $t = 0$ and keep our total wealth in T_n-bonds during period $(T_{n-1}, T_n]$, where we set $T_0 := 0$. This leads to a time-t wealth of

$$\begin{aligned}
\widetilde{S}_0(t) &:= \frac{B(t,T_n)}{\prod_{m=1}^n B(T_{m-1},T_m)} \\
&= \exp\biggl(\sum_{m=1}^n f(T_{m-1},T_{m-1},T_m)(T_m - T_{m-1}) - f(t,t,T_n)(T_n - t)\biggr) \\
&= \frac{\prod_{m=1}^n (1 + L(T_{m-1},T_{m-1},T_m)(T_m - T_{m-1}))}{1 + L(t,t,T_n)(T_n - t)}
\end{aligned} \tag{14.9}$$

for $t \in (T_{n-1}, T_n]$. We call the replicable asset with price process $\widetilde{S}_0$ a **discrete money market account** because it corresponds, like S_0, to a repeated investment in the shortest available time horizon. For vanishing mesh size $\sup_m(T_m - T_{m-1})$ one expects $f(T_{m-1}, T_{m-1}, T_m)$ to be a good approximation of the short rate $r(T_{m-1})$. In this sense (14.9) tends to (14.2).

14.1.2 Interest Rate Products

A variety of interest rate products is traded in real markets. We consider a few in this section. Some of them—e.g. coupon bonds and others—reduce to zero coupon bonds by simple arbitrage arguments—regardless of any distributional assumptions. Others such as bond options, caps, floors, and swaptions cannot be replicated by static portfolios. They require more sophisticated models which are introduced in Sects. 14.2–14.6.

14.1.2.1 Fixed Coupon Bonds

We consider a fixed coupon bond with coupon payments $c_1, \ldots, c_N$ to be made at dates $T_1, \ldots, T_N$, respectively. In addition, the nominal of 1€ is paid at maturity T_N. For ease of notation we assume that

$$T_n = T_0 + n\delta, \quad n = 1, \ldots, N$$

with $T_0 \geq 0$ and $\delta > 0$. The amount c_n stands for the interest to be paid on the nominal for the period from T_{n-1} to T_n, which has length δ. Again for simplicity we assume

$$c_n = \ell\delta, \quad n = 1, \ldots, N$$

i.e. the **coupon rate** ℓ is the same for all periods.

We assume zero bonds with maturities $T_1, \ldots, T_N$ to be liquidly traded on the market. In this case the cash flow of the coupon bond can be replicated by investing in a portfolio consisting of c_n T_n-bonds for $n = 1, \ldots, N$ and an additional T_N-bond. Indeed, the maturing T_n-bonds yield a payoff of c_n at $T_1, \ldots, T_{N-1}$ and $1+c_N$ at time T_N, which coincides with the payments of the coupon bond. The value of the replicating portfolio at time $t < T_1$ equals

$$V(t) = \sum_{n=1}^{N} \ell\delta B(t, T_n) + B(t, T_N).$$

The law of one price implies that this is the unique price of the fixed coupon bond which is consistent with absence of arbitrage—no matter how interest rates evolve. The price of the bond for $t \geq T_1$ is obtained accordingly by summing over the bonds that have not yet expired.

14.1.2.2 Floating Coupon Bonds

We turn now to a floating rate bond, which differs from the above fixed coupon bond only in the coupon payments. These are now assumed to be of the form

$$c_n = L(T_{n-1}, T_{n-1}, T_n)\delta,$$

i.e. the coupon rate is determined as the simple forward rate at the beginning of the corresponding period. Observe that it is known at time T_{n-1} but paid at time T_n. The simple rate $L(T_{n-1}, T_{n-1}, T_n)$ is typically called the **(spot) LIBOR rate**. This name refers to the *London interbank offered rate (LIBOR)* that is used for transactions between banks.

Again we assume that zero bonds with maturities $T_0, \dots, T_N$ are available on the market. The payoff stream of the floating rate bond can then be replicated by a dynamic portfolio composed of such bonds.

At time $t < T(0)$ one buys a T_0-bond which serves as the basis of the replicating portfolio.

At time T_0 the payoff of the T_0-bond is invested in $1/B(T_0, T_1)$ T_1-bonds. This transaction involves no net cash flow.

At time T_1 part of the payoff $1/B(T_0, T_1)$ is invested in $1/B(T_1, T_2)$ T_2-bonds. It remains a surplus of

$$\frac{1}{B(T_0, T_1)} - \frac{1}{B(T_1, T_2)} B(T_1, T_2) = L(T_0, T_0, T_1)\delta,$$

which coincides with the coupon payment c_1 due at T_1.

At times T_n for $n = 2, \dots, N-1$ one proceeds as at T_1.

At maturity T_N the $1/B(T_{N-1}, T_N)$ T_N-bonds are due. Their payoff equals

$$\frac{1}{B(T_{N-1}, T_N)} = 1 + L(T_{N-1}, T_{N-1}, T_N)\delta = 1 + c_N.$$

Consequently, the cashflow of the portfolio coincides with the payment stream of the floating rate bond. The unique no arbitrage price of the latter is therefore given by

$$V(t) = \begin{cases} B(t, T_0) & \text{if } t \le T_0, \\ \frac{B(t,T_n)}{B(T_{n-1},T_n)} & \text{if } t \in (T_{n-1}, T_n], \quad n = 1, \dots, N, \\ 0 & \text{if } t > T_N. \end{cases}$$

Surprisingly, it coincides with the value of a plain zero bond for $t \le T_0$.

14.1.2.3 Bond Options

Bond options are, for example, calls and puts on bonds whose maturity is yet to come. The payoff at T_0 of a call with strike K on a T_1-bond equals $(B(T_0, T_1) - K)^+$ and similarly $(K - B(T_0, T_1))^+$ for a put. Whether or not it can be replicated by a portfolio of bonds depends on the particular model. We discuss the valuation of such options in Sects. 14.2–14.6.

14.1.2.4 Swaps

Swaps allow us to trade a fixed interest rate for a floating one. We discuss here the *forward swap settled in arrears*. We consider dates

$$T_n = T_0 + n\delta, \quad n = 1, \dots, N$$

as in Sect. 14.1.2.1. At any of these time points the swap holder pays the fixed interest $\ell\delta$ (*fixed leg*) and receives the variable interest $L(T_{n-1}, T_{n-1}, T_n)\delta$ (*floating leg*) in exchange. The **swap rate** ℓ is chosen so that the contract value is 0 at initiation, i.e. no initial payment must be made.

The swap can easily be replicated by a combination of a fixed and a floating rate bond, which themselves can be expressed in terms of zero bonds. Indeed, consider a portfolio which is long one floating rate bond and short one fixed coupon bond as in Sects. 14.1.2.2 and 14.1.2.1. Up to the payment of the nominal 1 at time T_N, the coupons of these bonds coincide precisely with the floating resp. fixed leg of the swap. Suppose that the portfolio is set up at time $t < T_0$. Its value

$$\begin{aligned} V(t) &= B(t, T_0) - \left(\sum_{n=1}^{N} B(t, T_n)\ell\delta + B(t, T_N) \right) \\ &= \sum_{n=1}^{N} B(t, T_n)\left(L(t, T_{n-1}, T_n) - \ell\right)\delta \end{aligned} \tag{14.10}$$

vanishes if and only if

$$\ell = \frac{B(t, T_0) - B(t, T_N)}{\sum_{n=1}^{N} B(t, T_n)\delta}. \tag{14.11}$$

Therefore this is the unique swap rate at time t that is consistent with absence of arbitrage.

14.1.2.5 Caps

Caps are sequences of call options written on the spot LIBOR rate. They provide an insurance against high variable interest rates. As before we consider equidistant dates

$$T_n = T_0 + n\delta, \quad n = 1, \ldots, N.$$

At time T_n, $n = 1, \ldots, N$ the owner of the cap receives a payment

$$(L(T_{n-1}, T_{n-1}, T_n) - \ell)^+ \delta,$$

where ℓ denotes the **cap rate** and we have assumed a notional amount of 1. These single payments are called **caplets**.

Caplets and hence also the payment stream of caps can be replicated by a dynamic portfolio of zero bonds and put options on bonds. More specifically, we assume that T_n-bonds for $n = 1, \ldots, N$ and put options on these bonds with strike $1/(1 + \ell\delta)$ and maturity T_{n-1} are liquidly traded. The price processes of these put options are denoted as $S_1, \ldots, S_N$, respectively. The replicating strategy is set up as follows.

At time $t < T_0$ one buys $1 + \ell\delta$ put options $S_1, \ldots, S_N$. This portfolio costs

$$(1 + \ell\delta) \sum_{n=1}^{N} S_n(t).$$

At time T_0 the options S_1 mature. Their payoff

$$(1 + \ell\delta) \left(\frac{1}{1 + \ell\delta} - B(T_0, T_1) \right)^+$$

is invested in

$$\frac{1 + \ell\delta}{B(T_0, T_1)} \left(\frac{1}{1 + \ell\delta} - B(T_0, T_1) \right)^+$$

T_1-bonds so that the net cashflow is 0.

At time T_1 the options S_2 and the T_1-bonds are due. The holder receives

$$(1 + \ell\delta) \left(\frac{1}{1 + \ell\delta} - B(T_1, T_2) \right)^+ + \frac{1 + \ell\delta}{B(T_0, T_1)} \left(\frac{1}{1 + \ell\delta} - B(T_0, T_1) \right)^+ .$$

Part of this amount is invested in

$$\frac{1 + \ell\delta}{B(T_1, T_2)} \left(\frac{1}{1 + \ell\delta} - B(T_1, T_2) \right)^+$$

T_2-bonds. The remainder

$$\frac{1+\ell\delta}{B(T_0,T_1)}\left(\frac{1}{1+\ell\delta}-B(T_0,T_1)\right)^+ = \left(\frac{1}{B(T_0,T_1)}-(1+\ell\delta)\right)^+$$
$$= \left(\frac{1-B(T_0,T_1)}{B(T_0,T_1)\delta}-\ell\right)^+\delta$$
$$= (L(T_0,T_0,T_1)-\ell)^+\,\delta$$

coincides with the T_1-caplet of the cap.

At times T_n for $n = 2, \ldots, N-1$ one proceeds as at T_1. At time T_N the T_N-bonds are due. As above one shows that their payoff coincides with that of the last caplet.

Consequently, the price of the cap is uniquely given by

$$V(t) = \begin{cases} (1+\ell\delta)\sum_{m=1}^N S_m(t) & \text{if } t \leq T_0, \\ (1+\ell\delta)\sum_{m=n+1}^N S_m(t) + \left(\frac{1}{B(T_{n-1},T_n)}-1-\ell\delta\right)^+ B(t,T_n) & \\ \qquad\qquad \text{if } t \in (T_{n-1},T_n], & n = 1,\ldots,N, \\ 0 & \text{if } t > T_N. \end{cases}$$

provided that put options are available in addition to the zero bonds themselves.

14.1.2.6 Floors

Floors differ from caps only in the sense that **floorlets**

$$(\ell - L(T_{n-1},T_{n-1},T_n))^+\,\delta$$

are substituted for caplets, i.e. put rather than call options on the **floor rate** ℓ. The results of the previous section carry over directly to this case if the put options $S_1, \ldots, S_N$ are replaced with the corresponding calls.

14.1.2.7 Swaptions

The owner of a swaption has the right to enter at time T_0 a swap contract as in Sect. 14.1.2.4 with given swap rate ℓ. Since the value of this swap at $t \leq T_0$ equals

$$V(t) = \sum_{n=1}^N B(t,T_n)\,(L(t,T_{n-1},T_n)-\ell)\,\delta$$
$$= B(t,T_0) - B(t,T_N) - \sum_{n=1}^N B(t,T_n)\ell\delta$$

by (14.10), the swaption has a value

$$\left(1 - B(T_0, T_N) - \sum_{n=1}^{N} B(T_0, T_n)\ell\delta\right)^+ = \sum_{n=1}^{N} B(T_0, T_n)(L(T_0) - \ell)^+\delta$$

at maturity T_0, where

$$L(T_0) = \frac{1 - B(T_0, T_N)}{\sum_{n=1}^{N} B(T_0, T_n)\delta}$$

denotes the fair swap rate at T_0, cf. (14.11). As for bond options it depends on the particular model whether or not swaptions can be replicated by a portfolio of bonds.

However, in certain one-factor setups swaptions can be replicated by a static portfolio of bond options. To this end, consider a general claim with time T_0-payoff

$$H = \left(K - \sum_{n=1}^{N} c_n B(T_0, T_n)\right)^+ \tag{14.12}$$

for some $T_0 \leq \ldots \leq T_N$, $c_1, \ldots, c_N, K \geq 0$. The above case of a swaption is recovered for $K = 1$, $c_1, \ldots, c_{N-1} = \ell d$, $c_N = 1 + \ell d$. Now assume that there exists a univariate process X and strictly decreasing continuous functions $\pi(T_0, T_n, \cdot)$, $n = 1, \ldots, N$ satisfying

$$B(T_0, T_n) = \pi\big(T_0, T_n, X(T_0)\big).$$

Denote by x_0 the unique value of x such that

$$K - \sum_{n=1}^{N} c_n \pi(T_0, T_n, x) = 0$$

and set $K_n = \pi(T_0, T_n, x_0)$ for $n = 1, \ldots, N$. Then we can write the payoff in **Jamshidian's representation**

$$\begin{aligned} H &= \left(K - \sum_{n=1}^{N} c_n \pi\big(T_0, T_n, X(T_0)\big)\right)^+ \\ &= \left(\sum_{n=1}^{N} c_n \Big(\pi\big(T_0, T_n, x_0\big) - \pi\big(T_0, T_n, X(T_0)\big)\Big)\right)^+ \\ &= \sum_{n=1}^{N} c_n \big(K_n - B(T_0, T_n)\big)^+, \end{aligned} \tag{14.13}$$

where the last equality follows from the fact that $\pi(T_0, T_n, x_0) \geq \pi(T_0, T_n, X(T_0))$ holds if and only if $x_0 \leq X(T_0)$. In view of (14.13), the swaption or, more generally, any payoff as in (14.12) corresponds to a portfolio of put options on the bond.

14.1.3 Term Structure Modelling

Modelling typically means specifying the dynamics of relevant assets in a reasonable way. To fix ideas suppose that we want to come up with a model for the money market account S_0, a number of zero bonds with different maturities, and possibly some caps on these bonds.

14.1.3.1 Modelling Real-World and Risk-Neutral Measures

As we know from the general theory in Chaps. 9 and 11, two probability measures play an important role, namely the real world measure P and the market's risk-neutral measure Q relative to some numeraire, say S_0. While P is responsible for the "real" probability of events, the equivalent martingale measure Q is just a theoretical object which we know to exist from the fundamental theorem of asset pricing, cf. Rule 9.15. It may not be unique but absence of arbitrage warrants that at least one such measure Q exists that turns the discounted price process of all traded assets—underlyings as well as derivatives—into martingales. Concrete modelling consists of two steps. First one specifies—typically parametric—families of measures P, Q or, equivalently, stochastic processes. Secondly, the as yet unknown parameters are determined by estimation and/or calibration.

For the first step we need the dynamics of some underlying asset under both P and Q. Here two alternative procedures may be applied. We have discussed the obvious natural approach in Sect. 11.2.3. One starts with a parametric family for P which describes the dynamics of some primary assets. One then chooses a similar family for the pricing measure Q, where the set of parameters is constrained by equivalence to P and the martingale property of discounted prices.

However, this procedure faces difficulties if the number of primary assets is large, as we mentioned already at the end of Sect. 11.2. This happens in particular in the Heath–Jarrow–Morton approach to interest rate theory, where bonds of all maturities and hence uncountably many securities are considered as underlyings. In such a situation it may be difficult to check whether a given P-dynamics is consistent with absence of arbitrage.

An elegant alternative approach is to start with a parametric family for the pricing measure Q, without reference to any real-world measure P that has been chosen in the first place. As in Sect. 11.2.3 we use the term **martingale modelling** in this context. The relevant martingale constraints for the Q-dynamics are typically much easier to specify than those granting absence of arbitrage under the real-world measure P. In a second step a parametric family of real-world measures P is chosen if true probabilities are needed for risk management etc. The only constraint to be kept in mind here is the equivalence $P \sim Q$, which holds immediately if P is specified in terms of its density process relative to Q.

Having set up parametric families for the dynamics under both P and Q, we can move on to determine the actual parameters based on data. As noted in Sect. 11.2.3, parameters concerning real-world probabilities are estimated from past observations, whereas Q-parameters can only be inferred from derivative prices. Since some parameters belong to both the P- and Q-dynamics, we must proceed in a well-defined order. Either one starts by estimating all P-parameters and determines the remaining Q-parameters by calibration or the other way around. The order can be chosen regardless of whether the parametric families themselves were obtained by martingale modelling or by starting from real-world dynamics.

14.1.3.2 Change of Numeraire

In fixed income markets the natural choice of a numeraire is less obvious than in setups with a deterministic money market account. But in view of Rule 9.18 one can easily determine the new EMM if one changes from one numeraire to another.

We introduce special terms for the most frequently used risk-neutral measures. The market's pricing measure relative to the money market account is denoted by Q_0 and called the **spot martingale measure**. If we choose a T-bond as numeraire, we end up with the **T-forward (martingale) measure** Q_T, whose density process relative to Q_0 equals

$$\frac{dQ_T|_{\mathscr{F}_t}}{dQ_0|_{\mathscr{F}_t}} = \frac{B(t,T)}{B(0,T)S_0(t)}, \quad t \leq T \tag{14.14}$$

by Rule 9.18. Note that Q_T is only defined on $\mathscr{F}_T$ because T-bonds cease to exist after T.

In a discrete-tenor market where T_n-bonds exist only for finitely many maturities $T_n, n = 1, \ldots, N$, we consider the discrete money market account $\widetilde{S}_0$ as well. The risk-neutral measure $\widetilde{Q}_0$ relative to this numeraire is called the **discrete spot martingale measure**. It is related to Q_{T_n} via

$$\frac{d\widetilde{Q}_0|_{\mathscr{F}_t}}{dQ_{T_n}|_{\mathscr{F}_t}} = \frac{B(0,T_n)\widetilde{S}_0(t)}{B(t,T_n)}, \quad t \leq T_n.$$

14.1.3.3 Further Aspects of Modelling

We discuss five different general approaches to term structure modelling in the following sections. They differ mainly in their general perspective and less—if at all—in numerical results. Many concrete processes such as the Vasiček model can in fact be interpreted in more than one of these frameworks. In order to clarify the differences, we summarise the main features of the five approaches in a *profile* of the following kind:

Primary processes Some processes related to interest rates or prices are modelled right from the start, with given initial value. Others are derived later, for instance, by calculating expectations under a pricing measure. The approaches differ greatly in the choice of primary objects, which, for example, consists of only the short rate process $(r(t))_{t\geq 0}$ in short rate models and of the whole forward rate curve $(f(t,T))_{t\leq T, T\geq 0}$ in the framework of Heath, Jarrow, and Morton.

Necessary input Ultimately one would like to have a model for the whole interest rate market including, say, bonds, caps, swaptions. Since the prices of these products are in general not determined by the primary processes, further input such as the density process of the market's pricing measure must be specified.

Discrete vs. continuous tenor While some approaches rely on the idealised assumption that bonds of all maturities are traded (*continuous tenor* structure), others consider only finitely many (*discrete tenor* structure). Yet others get along with both continuous and discrete tenor structures.

Risk-neutral modelling The approaches differ in some aspects concerning the pricing measure Q. What is the natural numeraire for its definition? What drift conditions must be satisfied if one wants to put up a model directly under Q? Do we start with the real-world measure P and choose Q later or the other way around?

Stationarity Approaches may be distinguished as to whether they lead to stationary models for the short rate, bond prices etc. By *stationarity* we somewhat vaguely refer to the property that the present time does not appear explicitly in the dynamics, pricing formulas etc.—or at most in terms of time to maturity.

Tractability Estimation resp. calibration of the model requires that prices etc. can be computed relatively quickly. Which interest rate products allow for explicit or semiexplicit formulas? Which prices, on the other hand, can only be computed by Monte Carlo or sophisticated numerical methods?

Further aspects that play a role in the choice of a particular interest rate model include its number of "sources of randomness" or "factors", its ability to account for the stylised properties of real data, the choice of continuous versus discontinuous processes, or the question whether negative interest rates can occur in the model. These issues, however, depend on the concrete model rather than on the chosen general approach from Sects. 14.2–14.6.

14.2 Short Rate or Factor Model Approach

Short rate or *factor models* take the money market account

$$S_0(t) = \exp\left(\int_0^t r(s)ds\right)$$

and hence the instantaneous short rate r as a starting point. Typically, r is given as the solution to some SDE, as a Markov process, or at least as a component of a multivariate Markov process. We focus on models of a certain affine structure in order to obtain more or less explicit formulas for bond prices, derivatives etc.

As discussed in Sect. 14.1.3 we must decide whether we want to set up our model under the physical probability measure P or relative to the spot martingale measure Q_0. We start by modelling under Q_0 and turn to the other alternative in Sect. 14.2.2.

14.2.1 *Affine Short Rate Models*

We model the short rate as one component of some $\mathbb{R}^d$-valued semimartingale X, e.g. $r = X_d$. The process X is supposed to be affine or at least time-inhomogeneous affine under Q_0. We refer to a **one-factor model** if $d = 1$, i.e. if the driving process $X = r$ has no component except the short rate itself. A prime example is the Vasiček model, where r follows a shifted Gaussian Ornstein–Uhlenbeck process. We will consider this and other examples in Sect. 14.2.4.

The money market account is of the form $S_0(t) = \exp(J(t))$ with

$$J(t) := \int_0^t r(s)ds.$$

In principle, the interest rate market is now completely specified. Indeed, the price at time t of a financial instrument with payoff H at time T is given by

$$\pi(t) = E_{Q_0}\left(\frac{H}{S_0(T)}\middle|\mathscr{F}_t\right) S_0(t) = E_{Q_0}\big(H\exp(-J(T) + J(t))\big|\mathscr{F}_t\big).$$

In particular we have

$$\begin{aligned} B(t,T) &= E_{Q_0}\big(\exp(-J(T) + J(t))\big|\mathscr{F}_t\big) \\ &= E_{Q_0}\left(\exp\left(-\int_t^T r(s)ds\right)\middle|\mathscr{F}_t\right) \end{aligned}$$

for the price of a T-bond at $t \leq T$. Our goal is to obtain more explicit representations for bond prices and other objects of interest.

From Proposition 6.21 and its extension to the time-inhomogeneous case (cf. Sect. 6.5) it follows that the $\mathbb{R}^{d+1}$-valued process (X, J) is a time-inhomogeneous affine semimartingale under Q_0. Its generalised characteristic function can be computed using Theorems 6.6 resp. 6.23, see also Proposition A.5(4) and Theorem 6.10.

More specifically,

$$E_{Q_0}\Big(\exp(z^\top X(T) - J(T))\Big|\mathscr{F}_t\Big) = \exp\Big(\Psi_0(t, T, -iz) + \Psi_X(t, T, -iz)^\top X(t) - J(t)\Big) \tag{14.15}$$

holds for $z \in U \subset \mathbb{C}^d$, where $\Psi = (\Psi_0, \Psi_X) : \mathbb{R}_+ \times \mathbb{R}_+ \times U \to \mathbb{C}^{1+d}$ denotes the solution to some system of ODEs and the size of the set U depends on the process X. This function Ψ is calculated in Sect. 14.2.4 for a number of examples. If X is time-homogeneous, $\Psi_0(t, T, \cdot)$, $\Psi_X(t, T, \cdot)$ depend only on $T - t$, cf. Remark 6.24.

We are now ready to state the first main result.

Rule 14.1 *The price of a T-bond is given by*

$$B(t, T) = \exp\Big(\Psi_0(t, T, 0) + \Psi_X(t, T, 0)^\top X(t)\Big). \tag{14.16}$$

Consequently,

$$f(t, T) = r(t, T - t) = -D_2\Psi_0(t, T, 0) - D_2\Psi_X(t, T, 0)^\top X(t) \tag{14.17}$$

holds for forward rates, where $D_k\Psi$ stands for the derivative with respect to the kth argument. The dynamics of the preceding processes can be represented as

$$d\log B(t, T) = \Psi_X(t, T, 0)dX(t) + \big(D_1\Psi_0(t, T, 0) + D_1\Psi_X(t, T, 0)^\top X(t)\big)dt, \tag{14.18}$$

$$df(t, T) = -D_2\Psi_X(t, T, 0)dX(t) - \big(D_{12}\Psi_0(t, T, 0) + D_{12}\Psi_X(t, T, 0)^\top X(t)\big)dt, \tag{14.19}$$

and

$$\begin{aligned} dr(t, \tau) = &-\Big((D_{12} + D_{22})\Psi_0(t, t + \tau, 0) + (D_{12} + D_{22})\Psi_X(t, t + \tau, 0)^\top X(t)\Big)dt \\ &- D_2\Psi_X(t, t + \tau, 0)dX(t). \end{aligned} \tag{14.20}$$

Idea Equation (14.16) follows from inserting $z = 0$ in (14.15). The expression for $f(t, T)$ is evident from (14.6). Applying integration by parts to (14.16) and (14.17) yields (14.18) and (14.19).

In order to derive (14.20) note that

$$df(t, T) = a(t, T)dX(t) + (b(t, T) + c(t, T)^\top X(t))dt \tag{14.21}$$

for

$$a(t,T) := -D_2\Psi_X(t,T,0), \tag{14.22}$$

$$b(t,T) := -D_{12}\Psi_0(t,T,0), \tag{14.23}$$

$$c(t,T) := -D_{12}\Psi_X(t,T,0). \tag{14.24}$$

Hence

$$\begin{aligned}
r(t,\tau) &= f(t,t+\tau) \\
&= f(0,t+\tau) + \int_0^t a(s,t+\tau)dX(s) \\
&\quad + \int_0^t \big(b(s,t+\tau) + c(s,t+\tau)^\top X(s)\big)ds \\
&= f(0,\tau) + \int_0^t D_2 f(0,s+\tau)ds \\
&\quad + \int_0^t \left(a(s,s+\tau) + \int_s^t D_2 a(s,u+\tau)du\right)dX(s) \\
&\quad + \int_0^t \Bigg(b(s,s+\tau) + c(s,s+\tau)^\top X(s) \\
&\quad + \int_s^t \big(D_2 b(s,u+\tau) + D_2 c(s,u+\tau)^\top X(s)\big)du\Bigg)ds.
\end{aligned}$$

Applying Fubini's theorem we obtain

$$\begin{aligned}
r(t,\tau) = f(0,\tau) &+ \int_0^t D_2 f(0,u+\tau)du \\
&+ \int_0^t a(u,u+\tau)dX(u) + \int_0^t \big(b(u,u+\tau) + c(u,u+\tau)^\top X(u)\big)du \\
&+ \int_0^t \Bigg(\int_0^u D_2 a(s,u+\tau)dX(s) \\
&+ \int_0^u \big(D_2 b(s,u+\tau) + D_2 c(s,u+\tau)^\top X(s)\big)ds\Bigg)du.
\end{aligned} \tag{14.25}$$

Differentiating (14.21) with respect to the second argument yields

$$\begin{aligned}
D_2 f(u,u+\tau) = D_2 f(0,u+\tau) &+ \int_0^u D_2 a(s,u+\tau)dX(s) \\
&+ \int_0^u \big(D_2 b(s,u+\tau) + D_2 c(s,u+\tau)^\top X(s)\big)ds.
\end{aligned}$$

Together with (14.25) we obtain

$$r(t,\tau) = f(0,\tau) + \int_0^t \big(D_2 f(u,u+\tau) + b(u,u+\tau) + c(u,u+\tau)^\top X(u)\big)du$$
$$+ \int_0^t a(u,u+\tau)dX(u).$$

Inserting the expressions for $D_2 f$ from (14.17) and for a, b, c from (14.22–14.24), we obtain (14.20). □

If necessary, the transition from Q_0 to the real-world measure P can be specified in terms of its density process $\widetilde{Z}_0 = \exp(-Y(t))$. In order to obtain an affine structure under P as well, one may want to choose Y such that (X, Y) is an $\mathbb{R}^{d+1}$-valued time-inhomogeneous affine process under Q_0, cf. Proposition 6.20.

14.2.2 *Starting from the Physical Measure*

Instead of martingale modelling we may also specify the dynamics of the short rate under the real-world measure P. Similarly as before, we assume r in (14.1) to be one component of an $\mathbb{R}^d$-valued time-inhomogeneous affine semimartingale X, but now time-inhomogeneous affine relative to P.

Modelling the short rate does not yet say much about bond prices. So far, the money market account is the only traded asset. Since $S_0/S_0 = 1$ is a martingale under any law, we conclude that *any* probability measure $Q \sim P$ is an equivalent martingale measure. In this sense the market is extremely incomplete. Additional input is needed if we want to obtain the dynamics of bonds and other interest rate products.

According to the FTAP, the existence of a spot martingale measure Q_0 follows from absence of arbitrage. We denote its density process by $Z_0(t) = \exp(Y(t))$. In order to preserve the affine structure under Q_0, we assume that Y is one component of X, say $Y = X_{d-1}$. By Proposition 6.20 this implies that X and hence also (X, J) are time-inhomogeneous affine processes under Q_0 as was the case in Sect. 14.2.1.

The characteristic exponent of (X, J) under Q_0 can be computed from the characteristic exponent of (X, J) under P. Applying Proposition 6.20 resp. its extension to the time-inhomogeneous case yields that

$$E\Big(\exp(z^\top X(T) - J(T) + Y(T))\Big|\mathscr{F}_t\Big)$$
$$= \exp\Big(\widetilde{\Psi}_0(t,T,-iz) + \widetilde{\Psi}_X(t,T,-iz)^\top X(t) - J(t) + Y(t)\Big)$$

for $z \in \widetilde{U} \subset \mathbb{C}^d$, where $\widetilde{\Psi} = (\widetilde{\Psi}_0, \widetilde{\Psi}_X) : \mathbb{R}_+ \times \mathbb{R}_+ \times \widetilde{U} \to \mathbb{C}^{1+d}$ denotes the solution to some system of ODEs, cf. the discussion leading to (14.15). Since

$$\begin{aligned} E_{Q_0}\Big(\exp(z^\top X(T) - J(T))\Big|\mathscr{F}_t\Big) &= E\Big(\exp(z^\top X(T) - J(T))Z_0(T)\Big|\mathscr{F}_t\Big)/Z_0(t) \\ &= E\Big(\exp(z^\top X(T) - J(T) + Y(T))\Big|\mathscr{F}_t\Big)\exp(-Y(t)), \end{aligned}$$

we can represent the Q_0-coefficients $\Psi = (\Psi_0, \Psi_X)$ in (14.15) as

$$\Psi(t, T, z) = \widetilde{\Psi}(t, T, z).$$

Rule 14.1 applies now as in the previous section.

14.2.3 Derivatives

We turn now to the valuation of interest rate derivatives such as bond options, caps, floors, and swaptions. Bond options have a payoff $H = g(\log B(T_0, T_1))$ for some function $g : \mathbb{R} \to \mathbb{R}$, where $T_0 \leq T_1$ denote the payoff times of the option and the bond, respectively. As in Sect. 11.6.1 we assume that the Laplace transform $\widetilde{g}(z)$ exists for $z = R \in \mathbb{R}$ and hence on the line $R + i\mathbb{R}$. Since

$$g(x) = \frac{1}{2\pi i}\int_{R-i\infty}^{R+i\infty} e^{zx}\widetilde{g}(z)dz \qquad (14.26)$$

by Proposition A.9(1) and since pricing is linear, the time-t price of the claim is given by

$$\begin{aligned} \pi(t) &= E_{Q_0}\big(H\exp(-J(T_0) + J(t))\big|\mathscr{F}_t\big) \\ &= \frac{1}{2\pi i}\int_{R-i\infty}^{R+i\infty} \pi_{T_0,T_1}(t, z)\widetilde{g}(z)dz \end{aligned}$$

with

$$\pi_{T_0,T_1}(t, z) := E_{Q_0}\big(\exp(z\log B(T_0, T_1) - J(T_0))\big|\mathscr{F}_t\big)e^{J(t)}.$$

Hence it remains to compute the latter quantity more explicitly in our setup.

Rule 14.2 *Using the notation from Sect. 14.2.1 we have*

$$\pi_{T_0,T_1}(t,z) = \exp\Big(z\Psi_0(T_0,T_1,0) + \Psi_0\big(t,T_0,-iz\Psi_X(T_0,T_1,0)\big) + \Psi_X\big(t,T_0,-iz\Psi_X(T_0,T_1,0)\big)^\top X(t)\Big).$$

Idea This follows from (14.15, 14.16). □

As an example we consider vanilla options on bonds.

Example 14.3 (Call and Put on a Bond) For a call $H = (B(T_0,T_1)-K)^+$ on a T_1-bond with strike $K \geq 0$ and maturity $T_0 \leq T_1$, the Laplace transform of $g(x) = (e^x - K)^+$ is to be found in Example 11.26. This leads to the pricing formula

$$\begin{aligned}\pi(t) &= \frac{1}{2\pi i}\int_{R-i\infty}^{R+i\infty} \pi_{T_0,T_1}(t,z)\frac{K^{1-z}}{z(z-1)}dz \\ &= \frac{1}{\pi}\int_0^\infty \mathrm{Re}\left(\pi_{T_0,T_1}(t,R+iu)\frac{K^{1-R-iu}}{(R+iu)(R-1+iu)}\right)du\end{aligned} \tag{14.27}$$

for $R > 1$, cf. Rule 11.31. For puts we get the same formula with $R < 0$.

If the process (X, J) is Gaussian, the exponent on the right-hand side of (14.15) is a quadratic function of z. This implies

$$\pi_{T_0,T_1}(t,z) = \exp\left(\alpha z + \beta(z-1) + \frac{\gamma^2}{2}z(z-1)\right)$$

for some numbers $\alpha, \beta \in \mathbb{R}$, $\gamma > 0$ which depend on $t, T_0, T_1, X(t)$. The Q_0-martingale property of $\widehat{B}(\cdot, T_1)$ implies $e^\alpha = \pi_{T_0,T_1}(t,1) = B(t,T_1)$. Together with

$$e^{-\beta} = \pi_{T_0,T_1}(t,0) = E_{Q_0}\left(\frac{1}{S_0(T_0)}\middle|\mathscr{F}_t\right)S_0(t) = B(t,T_0)$$

we obtain

$$\pi_{T_0,T_1}(t,z) = \frac{B(t,T_1)^z}{B(t,T_0)^{z-1}}\exp\left(\frac{\gamma^2}{2}z(z-1)\right). \tag{14.28}$$

In view of $\frac{1}{z(z-1)} = \frac{1}{z-1} - \frac{1}{z}$ and hence

$$\begin{aligned}\pi_{T_0,T_1}(t,z)\frac{K^{1-z}}{z(z-1)} &= \exp\Big(\frac{\gamma^2}{2}z(z-1)\Big)\frac{B(t,T_1)^z K^{1-z}}{B(t,T_0)^{z-1}z(z-1)}\\ &= \frac{B(t,T_1)}{z-1}\exp\bigg(\Big(\log\frac{B(t,T_1)}{KB(t,T_0)}+\frac{\gamma^2}{2}\Big)(z-1)+\frac{\gamma^2}{2}(z-1)^2\bigg)\\ &\quad - \frac{KB(t,T_0)}{z}\exp\bigg(\Big(\log\frac{B(t,T_1)}{KB(t,T_0)}-\frac{\gamma^2}{2}\Big)z+\frac{\gamma^2}{2}z^2\bigg),\end{aligned}$$

an application of Lemma A.10 yields

$$\pi(t) = B(t,T_1)\Phi\bigg(\frac{1}{\gamma}\Big(\log\frac{B(t,T_1)}{KB(t,T_0)}+\frac{\gamma^2}{2}\Big)\bigg) - KB(t,T_0)\Phi\bigg(\frac{1}{\gamma}\Big(\log\frac{B(t,T_1)}{KB(t,T_0)}-\frac{\gamma^2}{2}\Big)\bigg),$$

where Φ denotes the cumulative distribution function of the standard normal distribution. For a put $H = (K - B(T_0,T_1))^+$ we have accordingly

$$\pi(t) = KB(t,T_0)\Phi\bigg(\frac{1}{\gamma}\Big(\log\frac{KB(t,T_0)}{B(t,T_1)}+\frac{\gamma^2}{2}\Big)\bigg) - B(t,T_1)\Phi\bigg(\frac{1}{\gamma}\Big(\log\frac{KB(t,T_0)}{B(t,T_1)}-\frac{\gamma^2}{2}\Big)\bigg).$$

Hence we obtain Black–Scholes type formulae in a Gaussian context.

As noted in Sect. 14.1.2, caps and floors can be reduced to put resp. call options on bonds. Swaptions do not generally allow for a simple representation because they depend on a whole portfolio of bonds. However, using Jamshidian's representation (14.13) they can be reduced to portfolios of put options in one-factor models. To this end suppose that $X = r$, i.e. the short rate is the only factor. By (14.16) we have $B(T_0,T_1) = \pi(T_0,T_1,r(T_0))$ for any $T_1 \geq T_0$, where

$$\pi(T_0,T_1,x) = \exp\big(\Psi_0(T_0,T_1,0) + \Psi_X(T_0,T_1,0)x\big).$$

Hence we are in the situation of Sect. 14.1.2.7 and can write options on portfolios of bonds and in particular swaptions as in (14.13).

14.2.4 Examples

In order to obtain concrete numbers, we must specify the short rate model in detail. Typically one assumes the short rate to have some mean-reverting behaviour, which suggests Ornstein–Uhlenbeck and related processes. We consider here a number of examples which allow for quite explicit formulas. We start with a rather general one-factor Lévy-driven model.

Example 14.4 (Lévy-Driven OU Model) In this one-factor model the short rate is assumed to follow a Lévy-driven Ornstein–Uhlenbeck process. More specifically, let

$$dr(t) = -\lambda r(t)dt + dL(t) \tag{14.29}$$

with $\lambda > 0$ and some time-inhomogeneous Lévy process L having Q_0-characteristic exponent ψ^L. Moreover, we set $X := r$. Observe that we obtain a positive short rate process if and only if $r(0) > 0$ and L is increasing.

Note that $r(t) = e^{-\lambda t}r(0) + \int_0^t e^{-\lambda(t-s)}dL(s)$ and

$$J(t) = \int_0^t r(s)ds = \frac{1 - e^{-\lambda t}}{\lambda} r(0) + \int_0^t \frac{1 - e^{-\lambda(t-s)}}{\lambda} dL(s)$$

by Proposition 3.53 and (3.95). An application of Proposition 3.54 (or Proposition 4.8 in the time-inhomogeneous case) yields that

$$\begin{aligned} &E\big(\exp(iur(T) - J(T))\big|\mathscr{F}_t\big) \\ &= \exp\left(f(T, u)r(0) + \int_0^t f(T - s, u)dL(s) + \int_t^T \psi^L(s, -if(T - s, u))ds\right) \\ &= \exp\left(f(T - t, u)r(t) - J(t) + \int_t^T \psi^L(s, -if(T - s, u))ds\right) \end{aligned}$$

with

$$f(s, u) := iue^{-\lambda s} - \Sigma_\lambda(s) \tag{14.30}$$

and

$$\Sigma_\lambda(s) := \frac{1 - e^{-\lambda s}}{\lambda}. \tag{14.31}$$

Hence the function $\Psi = (\Psi_0, \Psi_X)$ in (14.15) is given by

$$\Psi_0(t, T, u) = \int_t^T \psi^L(s, -if(T - s, u))ds, \tag{14.32}$$

$$\Psi_X(t, T, u) = f(T - t, u). \tag{14.33}$$

If the Lévy process is chosen to be Brownian motion, we obtain the Vasiček and Ho-Lee models.

Example 14.5 (Vasiček and Ho-Lee Model) In the Vasiček model the short rate is of the form

$$dr(t) = (\kappa - \lambda r(t))dt + \sigma dW(t) \tag{14.34}$$

with constants $\kappa \in \mathbb{R}$, $\lambda, \sigma > 0$, and a Q_0-Wiener process W. The Ho-Lee model corresponds to the limiting case $\lambda = 0$ where the short rate does not have a mean-reverting component. The process (14.34) is of the form (14.29) with $L(t) = \kappa t + \sigma W(t)$. The Q_0-characteristic exponent of L equals $\psi^L(u) = iu\kappa - \frac{\sigma^2}{2}u^2$. The integral in (14.32) can be evaluated in closed form, leading to

$$\begin{aligned}\Psi_0(t,T,u) = &\left(-\frac{\kappa}{\lambda} + \frac{\sigma^2}{2\lambda^2}\right)(T-t) + \left(\frac{\kappa}{\lambda} - \frac{\sigma^2}{\lambda^2}\right)(iu\lambda + 1)\Sigma_\lambda(T-t) \\ &+ \frac{\sigma^2}{2\lambda^2}(iu\lambda + 1)^2\Sigma_{2\lambda}(T-t)\end{aligned} \tag{14.35}$$

for $\lambda \neq 0$,

$$\Psi_0(t,T,u) = \kappa\left(iu(T-t) - \frac{1}{2}(T-t)^2\right) + \frac{\sigma^2}{2}\left(-u^2(T-t) - iu(T-t)^2 + \frac{1}{3}(T-t)^3\right)$$

for $\lambda = 0$, with

$$\Psi_X(t,T,u) = \begin{cases} iue^{-\lambda(T-t)} - \Sigma_\lambda(T-t) & \text{for } \lambda \neq 0, \\ iu + T - t & \text{for } \lambda = 0 \end{cases}$$

and Σ_λ as in Example 14.4. Both the Vasiček and the Ho-Lee model are Gaussian, which here means that the short rate r, the forward rates $f(\cdot, T), r(\cdot, \tau)$ and the logarithmic bond prices $\log B(\cdot, T)$ are all Gaussian processes. In view of Example 14.3 we obtain Black–Scholes type formulae for calls, puts, caplets and floorlets. The parameter γ in Example 14.3 equals

$$\gamma = \sigma\Sigma_\lambda(T-t)\sqrt{\Sigma_{2\lambda}(T-t)}$$

in the Vasiček and

$$\gamma = \sigma(T-t)^{3/2}$$

in the Ho-Lee case.

The following alternative model leads to positive interest rates but it still allows for a closed-form expression in (14.32).

Example 14.6 (Gamma and Inverse-Gaussian OU Model) As the driving Lévy process L in (14.29) we choose the sum of a linear drift κt and a compound Poisson process with exponential jump distribution as in Example 3.57. Then $r(t) - \frac{\kappa}{\lambda}$ follows the Gamma-OU process from Example 3.57. To be more explicit, let us assume that the characteristic exponent of L under the spot martingale measure equals

$$\psi(u) = iu\kappa + \frac{iu\beta}{\alpha - iu}. \tag{14.36}$$

In this case the integral in (14.32) can be evaluated in closed form. Since $J(t) = \int_0^t r(s)ds$, Proposition 3.59 and (3.97) yield

$$E\big(\exp(iur(T) - J(T))\big|\mathscr{F}_t\big) = \exp\big(\Upsilon(T-t,u)r(0) - f(T-t,u)r(t) - J(t)\big)$$

with f, Σ_λ from (14.30, 14.31) and

$$\Upsilon(t,u) = \frac{\beta}{\lambda\alpha - 1}\left(\alpha \log \frac{\alpha - f(t,u)}{\alpha - iu} + t\right) + \frac{\kappa}{\lambda}\big(t - (iu\lambda + 1)\Sigma_\lambda(t)\big)$$

for $\lambda \neq 0$. Hence

$$\begin{aligned}
\Psi_0(t,T,u) = {} & \frac{\beta}{\lambda\alpha - 1}\left(\alpha \log \frac{\lambda\alpha - 1 + (1 - iu\lambda)e^{-\lambda(T-t)}}{\lambda(\alpha - iu)} + T - t\right) \\
& - \frac{\kappa}{\lambda}\big(T - t - (iu\lambda + 1)\Sigma_\lambda(T-t)\big), \qquad (14.37) \\
\Psi_X(t,T,u) = {} & f(T-t,u).
\end{aligned}$$

For the inverse-Gaussian OU process from Example 3.58 instead of the Gamma-OU above one can also compute $\Psi_0(t,T,u)$ explicitly. Parallel to the Gamma-OU model, we consider the exponent

$$\psi(u) = iu\kappa + \frac{iu\beta}{\sqrt{\alpha^2 - 2iu}}. \tag{14.38}$$

In this case we use Proposition 3.59 and (3.98) in order to obtain

$$\begin{aligned}
\Psi_0(t,T,u) = {} & \frac{\beta}{\lambda}\left(\sqrt{\alpha^2 - 2f(T-t,u)} - \sqrt{\alpha^2 - 2iu}\right) \\
& + \frac{2\beta}{\lambda^2\sqrt{\alpha^2 - 2/\lambda}}\left(\operatorname{arctanh}\sqrt{\frac{\alpha^2 - 2f(T-t,u)}{\alpha^2 - 2iu}} - \operatorname{arctanh}\sqrt{\frac{\alpha^2 - 2iu}{\alpha^2 - 2iu}}\right) \\
& - \frac{\kappa}{\lambda}\big(T - t - (iu\lambda + 1)\Sigma_\lambda(T-t)\big), \qquad (14.39) \\
\Psi_X(t,T,u) = {} & f(T-t,u).
\end{aligned}$$

Hull and White suggested a time-inhomogeneous extension of Example 14.5.

Example 14.7 (Hull–White Model) In the Hull–White extension of the Vasiček resp. Ho-Lee model, the constant κ in (14.34) is allowed to be a deterministic function of time, i.e. we have

$$dr(t) = (\kappa(t) - \lambda r(t))dt + \sigma dW(t) \tag{14.40}$$

with $\kappa : \mathbb{R}_+ \to \mathbb{R}$. The relevant function $\Psi = (\Psi_0, \Psi_X)$ in (14.15) is obtained from (14.32, 14.33) with $\psi^L(t,u) = iu\kappa(t) - \frac{\sigma^2}{2}u^2$. As in the previous example, $r, f(\cdot,T), r(\cdot,\tau), \log B(\cdot,T)$ are Gaussian processes. Consequently, we again obtain Black–Scholes type formulae for calls, puts, caplets and floorlets. Interestingly, the parameter γ in Example 14.3 coincides with the parameter of the Vasček model because it does not depend on κ.

In the short rate approach, bonds play the role of derivatives rather than underlyings. Their initial prices cannot be specified directly as market input. In order to make model values coincide with observed bond prices $B(0,T)$, $T \geq 0$, the model parameters must be chosen accordingly, e.g. the function $\kappa(t)$ in the Hull–White approach. This procedure is called **inverting the yield curve**.

In the following popular one-factor model the short rate is a positive continuous affine process. It is not of Ornstein–Uhlenbeck type.

Example 14.8 (Cox–Ingersoll–Ross Model) The model suggested by Cox, Ingersoll, and Ross is another one-factor model driven by a Wiener process. The short rate follows the SDE

$$dr(t) = (\kappa - \lambda r(t))dt + \sigma\sqrt{r(t)}dW(t) \tag{14.41}$$

with positive constants κ, λ, σ and a Q_0-standard Wiener process W. This process is affine by Example 6.9. Applying Theorem 6.6 yields that the function $\Psi = (\Psi_0, \Psi_X)$ in (14.15) is given by

$$\begin{aligned}
\Psi_0(t,T,u) &= \frac{\kappa}{\sigma^2}\left(\lambda(T-t) - 2\log\frac{D\cosh(D(T-t)) - w_0\sinh(D(T-t))}{D}\right),\\
\Psi_X(t,T,u) &= \frac{\lambda}{\sigma^2} + \frac{2D}{\sigma^2}\frac{w_0\cosh(D(T-t)) - D\sinh(D(T-t))}{D\cosh(D(T-t)) - w_0\sinh(D(T-t))},\\
w_0 &:= -\frac{\lambda}{2} + \frac{\sigma^2}{2}iu,\\
D &:= \frac{\sqrt{\lambda^2 + 2\sigma^2}}{2}
\end{aligned} \tag{14.42}$$

because these expressions solve the corresponding ODEs.

In the above one-factor models the whole forward rate curve is driven by a single process. As a consequence it belongs to a one-parametric family, which amounts to a rather rough approximation of real term structures. The following two examples allows for a second factor. In such more flexible models the long and short ends of the forward rate curve may move in opposite directions.

Example 14.9 (Lévy-Driven Two-Factor Model) In a simple extension to Example 14.4 the short rate is assumed to follow the sum of two independent or even dependent Lévy-driven OU processes. More specifically, we consider an $\mathbb{R}^3$-valued affine process $X = (X_1, X_2, r)$ of the form

$$\begin{aligned} dX_1(t) &= -\lambda_1 X_1(t)dt + dL_1(t), \\ dX_2(t) &= -\lambda_2 X_2(t)dt + dL_2(t), \\ dr(t) &= dX_1(t) + dX_2(t) \\ &= -\big(\lambda_1 X_1(t) + \lambda_2 X_2(t)\big)dt + dL_1(t) + dL_2(t), \end{aligned}$$

where λ_1, λ_2 denote positive constants and $L = (L_1, L_2)$ an $\mathbb{R}^2$-valued Lévy process under Q_0. We denote the characteristic exponent of L relative to Q_0 by ψ^L. The processes X_1 and X_2 can be interpreted as two additive components of the short rate which move at different speeds.

Applying Theorem 6.6 yields that $\Psi = (\Psi_0, \Psi_X) = (\Psi_0, \Psi_{X_1}, \Psi_{X_2}, \Psi_r)$ in (14.15) is given by

$$\begin{aligned} \Psi_{X_1}(t, T, u_1, u_2, u_r) &= iu_1 e^{-\lambda_1(T-t)} - (\lambda_1 iu_r + 1)\Sigma_{\lambda_1}(T-t) + T - t, \\ \Psi_{X_2}(t, T, u_1, u_2, u_r) &= iu_2 e^{-\lambda_2(T-t)} - (\lambda_2 iu_r + 1)\Sigma_{\lambda_2}(T-t) + T - t, \\ \Psi_r(t, T, u_1, u_2, u_r) &= iu_r - (T-t), \end{aligned}$$

$$\begin{aligned} &\Psi_0(t, T, u_1, u_2, u_r) \\ &\quad = \int_0^{T-t} \psi^L\Big((u_1 + u_r)e^{-\lambda_1 s} + i\Sigma_{\lambda_1}(s), (u_2 + u_r)e^{-\lambda_2 s} + i\Sigma_{\lambda_2}(s)\Big)ds, \end{aligned} \tag{14.43}$$

where

$$\Sigma_\lambda(t) := \frac{1 - e^{-\lambda t}}{\lambda}$$

for $\lambda = \lambda_1, \lambda_2$. If L_1, L_2 are independent with characteristic exponents ψ^{L_1} resp. ψ^{L_2}, we can rewrite (14.43) as

$$\Psi_0(t, T, u_1, u_2, u_r) = \int_0^{T-t} \psi^{L_1}\Big((u_1 + u_r)e^{-\lambda_1 s} + i\Sigma_{\lambda_1}(s)\Big)ds + \int_0^{T-t} \psi^{L_2}\Big((u_2 + u_r)e^{-\lambda_2 s} + i\Sigma_{\lambda_2}(s)\Big)ds.$$

For Gaussian, Gamma, or inverse-Gaussian OU processes X_1, X_2 as in Examples 14.5, 14.6, the integrals can be computed in closed form similarly as above. In the case of Brownian motions L_1, L_2 we obtain the *two-factor Vasiček model*, which allows for Black–Scholes type formulae, cf. Example 14.3.

The previous example is not the only way to introduce a second factor.

Example 14.10 (Another Lévy-Driven Two-Factor Model) In this alternative two-factor model we consider a bivariate affine process $X = (\mu, r)$ of the form

$$d\mu(t) = -\lambda_1 \mu(t)dt + dL_1(t),$$
$$dr(t) = \lambda_2(\mu(t) - r(t))dt + dL_2(t),$$

where λ_1, λ_2 denote positive constants and $L = (L_1, L_2)$ an $\mathbb{R}^2$-valued Lévy process under Q_0. We denote the Q_0-characteristic exponent of L by ψ^L. Intuitively, the short rate is drawn towards a current average $\mu(t)$ which—in contrast to the models above—may itself change over time.

Applying Theorem 6.6 yields that the function $\Psi = (\Psi_0, \Psi_X) = (\Psi_0, \Psi_\mu, \Psi_r)$ in (14.15) is given by

$$\Psi_0(t, T, u_\mu, u_r) = \int_0^{T-t} \psi^L\Big(u_\mu e^{-\lambda_1 s} + (u_r\lambda_2 + i)e^{-\lambda_1 s}\Sigma_{\lambda_2-\lambda_1}(s) - i\Sigma_{\lambda_1}(s), u_r e^{-\lambda_2 s} + i\Sigma_{\lambda_2}(s)\Big)ds, \tag{14.44}$$

$$\Psi_\mu(t, T, u_\mu, u_r) = iu_\mu e^{-\lambda_1(T-t)} + (iu_r\lambda_2 + 1)e^{-\lambda_1(T-t)}\Sigma_{\lambda_2-\lambda_1}(T-t) - \Sigma_{\lambda_1}(T-t), \tag{14.45}$$

$$\Psi_r(t, T, u_\mu, u_r) = iu_r e^{-\lambda_2(T-t)} - \Sigma_{\lambda_2}(T-t), \tag{14.46}$$

where

$$\Sigma_\lambda(t) := \frac{1 - e^{-\lambda t}}{\lambda}$$

as before. If L denotes a bivariate Brownian motion with drift, the integral in (14.44) can easily be evaluated in closed form. In this case the processes

$r, \mu, f(\cdot, T), r(\cdot, \tau), \log B(\cdot, T)$ are Gaussian and we obtain Black–Scholes type formulae for calls, puts, caplets and floorlets, cf. Example 14.3.

14.2.5 Profile

Let us summarise important features of short rate models as discussed in Sect. 14.1.3.3.

Primary processes The short rate r is the only primary process in this approach. Put differently, the money market account serves as the only underlying asset. Bonds and other interest rate products are treated as derivative securities.

Necessary input The dynamics of r under the physical measure do not determine bond prices. In order to obtain a fully specified term structure model, we need in addition the density process Z_0 of the spot martingale measure or we must specify the dynamics immediately under this measure.

Discrete vs. continuous tenor Short rate modelling obviously relies on the existence of the short rate. This requires either a money market account or a continuum of bonds to construct a roll-over portfolio of just maturing bonds. Otherwise, both a discrete or a continuous tenor structure are implementable in the context of short rate models.

Risk-neutral modelling There are at least three ways to implement a short rate model:

- $P \to Q$: One starts with a model for r under the real-world measure P and chooses an arbitrary positive martingale Z_0 with initial value 1 to mark the transition from P to the spot martingale measure Q_0.
- $Q \to P$: One starts with a model for r under the spot martingale measure Q_0. No drift condition must be satisfied for this purpose. In a second step one chooses an arbitrary positive martingale $\widetilde{Z}_0$ with initial value 1 to mark the transition from Q_0 to the real-world measure P.
- Only Q: one limits oneself to modelling the short rate under the spot martingale measure Q_0.

Stationarity If the driving affine semimartingale is chosen to be time-homogeneous as in Examples 14.5, 14.6, 14.8–14.10, we end up with a stationary model in the vague sense of Sect. 14.1.3.3.

Tractability If the short rate model and the measure change are based on an affine process, explicit formulas are available for bond prices and consequently forward rates, swaps etc. Bond options and hence also caps and floors can be represented semiexplicitly by integral transform methods. Swaptions can be expressed in integral form in the case of one-factor models.

14.3 The Heath–Jarrow–Morton Approach

Only the short rate serves as a primary process in the previous section. The Heath–Jarrow–Morton approach at the other extreme considers instead the whole forward rate curve. Put differently, all T-bonds are treated as primary assets. Since these are uncountably many, one effectively enters the realm of an infinite-dimensional stochastic calculus, which is beyond the scope of this monograph. Instead we stay as usual in Part II on an informal mathematical level.

14.3.1 General Framework

We start with given forward rate processes $(f(t,T))_{t\leq T}$ for all $T \geq 0$, which are assumed to be of the form

$$df(t,T) = \alpha(t,T)dt + \sigma(t,T)dX(t) \tag{14.47}$$

for some $\mathbb{R}^d$-valued semimartingale X and predictable integrands $\alpha(\cdot,T), \sigma(\cdot,T)$. We will later assume X to be a time-inhomogeneous Lévy process and $\alpha(\cdot,T), \sigma(\cdot,T)$ deterministic, but the results in this section hold true in general. The initial forward rate curve $f(0,T)$, $T \geq 0$ is assumed to be given as well. We assume smoothness of $f(t,T)$ and hence $r(t,\tau)$ in T resp. τ. Derivatives relative to the second argument are denoted in dot notation, i.e.

$$\dot{f}(t,T) := \frac{d}{dT}f(t,T), \quad \dot{r}(t,\tau) := \frac{d}{d\tau}r(t,\tau) = \dot{f}(t,t+\tau).$$

Moreover, we write

$$A(t,T) := \int_t^T \alpha(t,s)ds, \quad \Sigma(t,T) := \int_t^T \sigma(t,s)ds \tag{14.48}$$

for $t \leq T$ and

$$A(t,S,T) := A(t,T) - A(t,S) = \int_S^T \alpha(t,s)ds,$$

$$\Sigma(t,S,T) := \Sigma(t,T) - \Sigma(t,S) = \int_S^T \sigma(t,s)ds$$

for $t \leq S \leq T$.

Recall from Sect. 14.1.1 that the short rate $r(t) = f(t,t)$ and hence the money market account S_0 are determined by the forward rates. From now on, we use the hat notation for discounting relative to the money market account, i.e.

$$\widehat{B}(t,T) := \frac{B(t,T)}{S_0(t)}.$$

Equation (14.47) leads immediately to dynamic equations for discounted bond prices, cc forward rates, the short rate, and forward rates in the Musiela parametrisation.

Rule 14.11 *We have*

$$d\log \widehat{B}(t,T) = -A(t,T)dt - \Sigma(t,T)dX(t), \tag{14.49}$$

$$\begin{aligned} d\log B(t,S,T) &= (T-S)df(t,S,T) \\ &= A(t,S,T)dt + \Sigma(t,S,T)dX(t), \end{aligned} \tag{14.50}$$

$$dr(t,\tau) = \big(\alpha(t,t+\tau) + \dot{r}(t,\tau)\big)dt + \sigma(t,t+\tau)dX(t), \tag{14.51}$$

$$dr(t) = \big(\alpha(t,t) + \dot{f}(t,t)\big)dt + \sigma(t,t)dX(t) \tag{14.52}$$

for $t \leq S < T$.

Idea Integrating (14.47) over T yields

$$\begin{aligned} \log B(t,T) &= -\int_t^T f(t,s)ds \\ &= -\int_t^T \left(f(0,s) + \int_0^t \alpha(u,s)du + \int_0^t \sigma(u,s)dX(u) \right) ds. \end{aligned}$$

Applying Fubini's theorem twice, we obtain

$$\begin{aligned} \log B(t,T) = &-\int_t^T f(0,s)ds - \int_0^t \int_t^T \alpha(u,s)dsdu - \int_0^t \int_t^T \sigma(u,s)dsdX(u) \\ = &-\int_0^T f(0,s)ds - \int_0^t \int_u^T \alpha(u,s)dsdu - \int_0^t \int_u^T \sigma(u,s)dsdX(u) \\ &+ \int_0^t f(0,s)ds + \int_0^t \int_u^t \alpha(u,s)dsdu + \int_0^t \int_u^t \sigma(u,s)dsdX(u) \\ = &\log B(0,T) - \int_0^t A(u,T)du - \int_0^t \Sigma(u,T)dX(u) \\ &+ \int_0^t \left(f(0,s) + \int_0^s \alpha(u,s)du + \int_0^s \sigma(u,s)dX(u) \right) ds. \end{aligned}$$

By (14.47) this yields

$$\log B(t,T) = \log B(0,T) + \int_0^t (f(s,s) - A(s,T))ds - \int_0^t \Sigma(s,T)dX(s). \tag{14.53}$$

Since

$$\log \widehat{B}(t,T) = \log B(t,T) - \int_0^t r(s)ds$$

and $f(s,s) = r(s)$, we obtain (14.49). The first equality in (14.50) is evident from the definition of $f(t,S,T)$. The second follows from $\log B(t,S,T) = \log \widehat{B}(t,S) - \log \widehat{B}(t,T)$ and (14.49). In order to derive (14.51) note that

$$\begin{aligned} r(t,\tau) &= f(t,t+\tau) \\ &= f(0,t+\tau) + \int_0^t \alpha(s,t+\tau)ds + \int_0^t \sigma(s,t+\tau)dX(s) \\ &= f(0,t+\tau) + \int_0^t \left(\alpha(s,s+\tau) + \int_s^t \dot{\alpha}(s,u+\tau)du\right)ds \\ &\quad + \int_0^t \left(\sigma(s,s+\tau) + \int_s^t \dot{\sigma}(s,u+\tau)du\right)dX(s). \end{aligned}$$

By Fubini's theorem we conclude that

$$\begin{aligned} r(t,\tau) &= f(0,\tau) + \int_0^t \dot{f}(0,u+\tau)du \\ &\quad + \int_0^t \alpha(s,s+\tau)ds + \int_0^t \int_0^u \dot{\alpha}(s,u+\tau)dsdu \\ &\quad + \int_0^t \sigma(s,s+\tau)dX(s) + \int_0^t \int_0^u \dot{\sigma}(s,u+\tau)dX(s)du \\ &= \int_0^t \left(\dot{f}(0,u+\tau) + \int_0^u \dot{\alpha}(s,u+\tau)ds + \int_0^u \dot{\sigma}(s,u+\tau)dX(s)\right)du \\ &\quad + r(0,\tau) + \int_0^t \alpha(s,s+\tau)ds + \int_0^t \sigma(s,s+\tau)dX(s). \end{aligned} \tag{14.54}$$

Integrating (14.47) relative to t and differentiating with respect to T yields

$$\dot{f}(u,u+\tau) = \dot{f}(0,u+\tau) + \int_0^u \dot{\alpha}(s,u+\tau)ds + \int_0^u \dot{\sigma}(s,u+\tau)dX(s).$$

Together with (14.54) we obtain

$$r(t,\tau) = r(0,\tau) + \int_0^t \big(\alpha(s,s+\tau) + \dot{r}(s,\tau)\big)ds + \int_0^t \sigma(s,s+\tau)dX(s)$$

and hence (14.51). This in turn reduces to (14.52) for $\tau = 0$. □

In particular, we obtain a representation of bond prices.

Corollary 14.12 *We have*

$$\begin{aligned} B(t,T) &= B(0,T)\exp\Big(\int_0^t (r(s) - A(s,T))ds - \int_0^t \Sigma(s,T)dX(s)\Big) \\ &= \frac{B(0,T)}{B(0,t)}\exp\Big(-\int_0^t A(s,t,T)ds - \int_0^t \Sigma(s,t,T)dX(s)\Big). \end{aligned}$$

Idea The first equality follows from (14.49) after integrating over t or alternatively from (14.53). By (14.50) we have

$$\begin{aligned} \exp\left(-\int_0^t A(s,t,T)ds - \int_0^t \Sigma(s,t,T)dX(s)\right) &= \frac{B(0,t,T)}{B(t,t,T)} \\ &= \frac{B(0,t)}{B(0,T)}\frac{B(t,T)}{B(t,t)}, \end{aligned}$$

which yields the second equality. □

Modelling the forward rate curve means specifying infinitely many primary assets. According to Rule 11.5, this will often lead to arbitrage unless particular care is taken. In this context it means that the drift coefficients cannot be freely chosen but are subject to sometimes severe restrictions. This aspect makes modelling the forward rates particularly cumbersome for processes involving jumps.

In order to avoid such difficulties, we model the forward rate dynamics immediately under the spot martingale measure Q_0, i.e. we take the martingale modelling approach discussed in Sect. 14.1.3. In view of the FTAP, the existence of Q_0 follows from and warrants absence of arbitrage. We denote the characteristic exponent of X under Q_0 by $\psi(t,u)$.

Rule 14.13 (HJM Drift Condition) *Q_0 is a spot martingale measure (in the sense that the discounted bonds $\widehat{B}(\cdot,T)$ are Q_0-martingales for all $T \geq 0$) if and only if*

$$A(t,T) = \psi(t, i\Sigma(t,T)) \tag{14.55}$$

holds for all $0 \leq t \leq T$.

Idea Itō's formula applied to (14.49) yields that the growth rate of $\widehat{B}(t,T)$ equals

$$\widehat{B}(t-,T)\Big(-A(t,T)-\Sigma(t,T)a^X(t)+\frac{1}{2}\Sigma(t,T)^2c^X(t)$$
$$+\int\big(e^{-\Sigma(t,T)x}-1+\Sigma(t,T)x\big)K^X(t,dx)\Big)$$
$$=\widehat{B}(t-,T)\big(-A(t,T)+\psi(t,i\Sigma(t,T))\big),$$

where (a^X,c^X,K^X) denote the local characteristics of X, cf. Proposition 4.13. This yields the assertion. □

Equation (14.55) is equivalent to

$$\alpha(t,T)=iD\psi(t,i\Sigma(t,T))^\top\sigma(t,T), \tag{14.56}$$

where $D\psi(t,u)$ here denotes the derivative of $\psi(t,u)$ relative to $u\in\mathbb{C}^d$. Hence (14.55) means that the integrands $\sigma(\cdot,T)$ can essentially be arbitrarily chosen whereas the drift coefficients $\alpha(\cdot,T)$ are entirely determined by the martingale property of discounted bond prices.

Occasionally one may prefer to express the dynamics in terms of the T-forward measure for some $T>0$. Most of the above formulae do not depend on the underlying probability measure. Only the differential characteristics and the characteristic exponent of X are affected by the change of measure. We denote by $\psi_T(t,u)$ the characteristic exponent of X relative to the T-forward measure Q_T.

Rule 14.14 (Change of Measure) *We have*

$$\psi_T(t,u)=\psi(t,u+i\Sigma(t,T))+\psi(t,i\Sigma(t,T)) \tag{14.57}$$

and equivalently

$$\psi(t,u)=\psi_T(t,u-i\Sigma(t,T))-\psi_T(t,-i\Sigma(t,T))$$

for $0\leq t\leq T$, $u\in\mathbb{R}^d$.

Idea The density process of Q_T relative to Q_0 equals $\widehat{B}(t,T)/\widehat{B}(0,T)$, cf. (14.14). Using (14.49), Proposition 4.16 yields the Q_T-characteristics of X after a straightforward calculation. It turns out that it is of the form (14.57). The second equation is a reformulation of the first. □

If one wants to express the dynamics of rates, prices etc. under real-world probabilities as well, one can now choose an arbitrary positive Q_0-martingale as density process specifying the transition from Q_0 to P.

14.3.2 Lévy-Driven Term Structure Models

In order to obtain tractable expressions for interest rate derivatives, we assume that the driving semimartingale X in (14.47) is a uni- or multivariate time-inhomogeneous Lévy process relative to Q_0 with Q_0-characteristic exponent $\psi(t, u)$. Moreover, we suppose that the integrands $\sigma(t, T)$ are deterministic functions of time. By (14.56) this implies that the drift coefficients $\alpha(t, T)$ are deterministic as well. For the computation of option prices we need the following

Rule 14.15 *We have*

$$\pi(t, T_0, T_1, z) := E_{Q_0}\left(\frac{\exp\left(z \log B(T_0, T_1)\right)}{S_0(T_0)}\middle|\mathscr{F}_t\right) S_0(t)$$

$$= \exp\left(\int_t^{T_0} \varphi(s, T_0, T_1, z)ds\right) \frac{B(t, T_1)^z}{B(t, T_0)^{z-1}} \tag{14.58}$$

for $t \leq T_0 \leq T_1$, where

$$\varphi(s, T_0, T_1, z) := \psi\left(s, iz\Sigma(s, T_0, T_1) + i\Sigma(s, T_0)\right)$$
$$- z\psi(s, i\Sigma(s, T_1)) + (z-1)\psi(s, i\Sigma(s, T_0)).$$

The set of eligible $z \in \mathbb{C}$ depends on ψ and σ.

Idea By (14.49, 14.50) we have

$$\begin{aligned} z \log B(T_0, T_1) - \log S_0(T_0) &= -z \log B(T_0, T_0, T_1) + \log \widehat{B}(T_0, T_0) \\ &= -z \log B(t, T_0, T_1) + \log \widehat{B}(t, T_0) \\ &\quad - \int_t^{T_0} \left(zA(s, T_0, T_1) + A(s, T_0)\right)ds \\ &\quad - \int_t^{T_0} \left(z\Sigma(s, T_0, T_1) + \Sigma(s, T_0)\right)dX(s). \end{aligned}$$

Since

$$E\left(\exp\left(\int_t^{T_0} f(s)dX(s)\right)\middle|\mathscr{F}_t\right) = \exp\left(\int_t^{T_0} \psi(s, -if(s))ds\right)$$

for deterministic functions f by Proposition 3.54, $A(s, T_0, T_1) = A(s, T_1) - A(s, T_0)$, and (14.55), we obtain (14.58) after a straightforward calculation. □

The previous rule leads to a pricing formula for bond options whose payoff $H = g(\log B(T_0, T_1))$ at time T_0 can be written in integral form (14.26) with Laplace transform $\widetilde{g}$. Since pricing is linear, the time-t price of the claim is given by

$$\begin{aligned}\pi(t) &= E_{Q_0}\left(\frac{H}{S_0(T_0)}\middle|\mathscr{F}_t\right)S_0(t)\\ &= E_{Q_0}\left(\frac{g(\log B(T_0,T_1))}{S_0(T_0)}\middle|\mathscr{F}_t\right)S_0(t)\\ &= \frac{1}{2\pi i}\int_{R-i\infty}^{R+i\infty}\pi(t,T_0,T_1,z)\widetilde{g}(z)dz \qquad (14.59)\end{aligned}$$

with $\pi(t, T_0, T_1, z)$ from Rule 14.15. As a primary example we consider calls and puts on bonds.

Example 14.16 (Call and Put on a Bond) For a call $H = (B(T_0, T_1) - K)^+$ on a T_1-bond with strike $K \geq 0$ and maturity $T_0 \leq T_1$, the Laplace transform of $g(x) = (e^x - K)^+$ is stated in Example 11.26. This leads to the pricing formula

$$\begin{aligned}\pi(t) &= \frac{1}{2\pi i}\int_{R-i\infty}^{R+i\infty}\pi(t,T_0,T_1,z)\frac{K^{1-z}}{z(z-1)}dz \qquad (14.60)\\ &= \frac{1}{\pi}\int_0^\infty \mathrm{Re}\left(\pi(t,T_0,T_1,R+iu)\frac{K^{1-R-iu}}{(R+iu)(R-1+iu)}\right)du\end{aligned}$$

for $R > 1$, where $\pi(t, T_0, T_1, z)$ is defined in Rule 14.15. For puts we get the same formula with $R < 0$.

In the Gaussian case we can proceed as in Example 14.3 in order to obtain Black–Scholes type formulae. To this end, suppose that $\psi(t, u)$ is a quadratic function of u. Since $\widehat{B}(\cdot, T_1)$ is a martingale, we have $\pi(t, T_0, T_1, 1) = B(t, T_1)$. Together with

$$\pi(t,T_0,T_1,0) = E_{Q_0}\left(\frac{1}{S_0(T_0)}\middle|\mathscr{F}_t\right)S_0(t) = B(t,T_0)$$

this implies

$$\pi(t,T_0,T_1,z) = \frac{B(t,T_1)^z}{B(t,T_0)^{z-1}}\exp\left(\frac{\gamma^2}{2}z(z-1)\right) \qquad (14.61)$$

with a deterministic parameter γ that depends on t, T_0, T_1. Equation (14.61) parallels (14.28). Since (14.60) essentially coincides with (14.27), we obtain the same Black–Scholes type formulae for calls and puts as in Example 14.3.

As noted in Sect. 14.1.2, caps and floors can be reduced to put/call options on bonds. Swaptions do not generally allow for a simple representation. But as for short rate models they can be expressed in terms of put options if the Lévy process X is univariate and the volatility factorises as

$$\sigma(t,T) = \sigma_1(T)\sigma_2(t)$$

with functions $\sigma_1, \sigma_2 : \mathbb{R}_+ \to (0,\infty)$. This implies

$$\Sigma(t,T) = \Sigma_1(t,T)\sigma_2(t)$$

with

$$\Sigma_1(t,T) := \int_t^T \sigma_1(s)ds.$$

Moreover, recall that

$$A(t,T) = \psi\big(t, i\Sigma(t,T)\big)$$

according to Rule 14.13. By Corollary 14.12 we obtain the representation

$$B(T_0,T_1) = \pi(T_0,T_1,Y(T_0)) \tag{14.62}$$

for any $T_0 \leq T_1$, where $Y(t) := \int_0^t \sigma_2(s)dX(s)$ and

$$\pi(T_0,T_1,x) := \frac{B(0,T_1)}{B(0,T_0)} \exp\left(\int_0^{T_0} \big(A(t,T_0) - A(t,T_1)\big)dt\right) e^{-\Sigma_1(T_0,T_1)x}.$$

According to Sect. 14.1.2.7, this allows us to express swaptions as portfolios of put options on the bond, cf. (14.13).

14.3.3 Completeness

The general heuristics in Rule 11.5 states that market completeness can be expected if the number of tradable securities exceeds the number of *sources of randomness* at least by one. A source of randomness in this sense includes components of a Wiener process or any single jump height of a Lévy process.

In the idealised bond market underlying the HJM approach we deal with infinitely many tradable securities, namely zero bonds of any maturity. Therefore we expect completeness to hold if the market is driven by finitely many Wiener processes or jump processes with finitely many jumps heights. Precise statements can be made under appropriate non-degeneracy conditions in more concrete setups.

The situation is harder to grasp if the driving semimartingale X is, for example, a Lévy process whose Lévy measure has a density. In this case an infinity of assets must cope with infinitely many sources of randomness. A sound treatment of this issue requires a theory of portfolios involving uncountable many securities which is beyond the scope of this monograph. Nevertheless, we state the following general

Rule 14.17 (Completeness) *The term structure model of Sect. 14.3.2 is complete if $d = 1$, i.e. if the market is driven by a univariate time-inhomogeneous Lévy process.*

Idea This is shown in [94, Theorem 6.3] under a weak regularity condition. As noted below, completeness here refers to uniqueness of the EMM. If the coefficients $\sigma(t, T)$ are deterministic as in Sect. 14.3.2, the assertion holds without additional regularity conditions and for the filtration generated by the bond price processes, cf. [94, Theorem 6.4]. □

For $d > 1$ market completeness cannot generally be expected to hold unless the driving process involves only a few sources of randomness or the integrands $\sigma(\cdot, T)$ are of degenerate nature.

Completeness in the above rule means uniqueness of an EMM rather than replicability of any contingent claim. In the present case of infinitely many assets replicability of arbitrary, say bounded, claims is in fact a stronger requirement. Under the conditions of Rule 14.17 it holds only in some approximate sense: a general claim can be approximated arbitrarily well, but not replicated perfectly by an admissible hedge. Moreover, one is confronted with an ill-posed problem in the sense that a small variation of the payoff may lead to a substantially different hedging portfolio.

14.3.4 Examples

More explicit formulas are obtained if the volatility structure is of exponential type.

Example 14.18 (Vasiček- or Ho-Lee-Type Volatility Structure) Let us focus on a univariate time-inhomogeneous Lévy process X and a particular volatility structure, namely

$$\sigma(t, T) = e^{-\lambda(T-t)} \tag{14.63}$$

with some $\lambda \geq 0$. In view of (14.33) and (14.19), this integrand coincides with the one that is obtained in the Vasiček model or, more generally, in the Lévy-driven Ornstein–Uhlenbeck model from Example 14.4. Observe that σ factorises as needed for the Jamshidian representation of swaptions, cf. Sect. 14.3.2. Moreover, we have

$$\Sigma(t, T) = \begin{cases} \frac{1-e^{-\lambda(T-t)}}{\lambda} & \text{if } \lambda > 0, \\ T - t & \text{if } \lambda = 0 \end{cases} \tag{14.64}$$

and hence

$$\int_t^{T_0} \varphi(s, T_0, T_1, z)ds = \Psi_0\big(t, T_0, iz\Sigma(T_0, T_1)\big) - z\Psi_0(t, T_0, i\Sigma(T_0, T_1)) + (z-1)\Psi_0(t, T_1, 0) \tag{14.65}$$

for the integral in (14.58), where

$$\Psi_0(t, T, u) := \int_t^T \psi\big(s, ue^{-\lambda(T-s)} + i\Sigma(s, T)\big)ds$$

is defined as in (14.32).

This class of HJM models reduces to the family of short rate models in Example 14.4. To see this, note that

$$r(t) = f(t,t) = f(0,t) + \int_0^t \alpha(s,t)ds + \int_0^t \sigma(s,t)dX(s)$$

and hence

$$\begin{aligned} \dot{f}(t,t) &= \dot{f}(0,t) + \int_0^t \dot{\alpha}(s,t)ds + \int_0^t \dot{\sigma}(s,t)dX(s) \\ &= \dot{f}(0,t) + \int_0^t \dot{\alpha}(s,t)ds - \lambda \int_0^t \sigma(s,t)dX(s) \\ &= \eta(t) - \lambda r(t), \end{aligned}$$

where $\eta(t)$ denotes some rather complicated deterministic function. Equation (14.52) now yields

$$\begin{aligned} dr(t) &= \big(\alpha(t,t) + \dot{f}(t,t)\big)dt + \sigma(t,t)dX(t) \\ &= -\lambda r(t)dt + \big(\alpha(t,t) + \eta(t)\big)dt + dX(t). \end{aligned} \tag{14.66}$$

Since $\alpha(\cdot, t)$ and η are deterministic, this is indeed of the form (14.29) with a time-inhomogeneous Lévy process

$$L(t) := \int_0^t \big(\alpha(s,s) + \eta(s)\big)ds + X(t).$$

Note that this driving process L of the short rate is generally time-inhomogeneous even if the Lévy process X in the HJM setup is chosen to be time-homogeneous.

For a Wiener process as the driving Lévy process we end up with the Hull–White extension of the Vasiček resp. Ho-Lee model.

Example 14.19 (Hull–White Model Revisited) Let $X(t) = \sigma W(t)$ for some $\sigma > 0$ and a standard Wiener process W. If we choose $\sigma(\cdot, T)$ as in the previous example, the short rate moves according to the Hull–White model, as can be seen by comparing (14.66) and (14.40). It depends on the initial forward rate curve whether the short rate process actually reduces to the simpler Vasiček/Ho-Lee model.

The integral in (14.58) can be evaluated explicitly. From (14.65) and (14.35) we obtain

$$\int_t^{T_0} \varphi(s, T_0, T_1, z)ds = \frac{\sigma^2}{4}(z^2 - z)\Sigma(T_0, T_1)^2\Sigma(2t, 2T_0) \tag{14.67}$$

with $\Sigma(t, T)$ as in (14.64). In particular we are in the Gaussian context of Example 14.16 which leads to Black–Scholes type formulae. From (14.67) we conclude that the parameter in Example 14.16 equals

$$\gamma = \sigma\Sigma(T_0, T_1)\sqrt{\frac{\Sigma(2t, 2T_0)}{2}},$$

which coincides with the one from Examples 14.5 and 14.7, as it should.

Interestingly, it is easier to work with the Hull–White model in an HJM setup than to treat it as a short rate model. In the framework of Sect. 14.2 we need to evaluate the integral in (14.32) with a characteristic exponent that depends nontrivially on time. Moreover, the time-dependent drift $\kappa(t)$ has yet to be determined by inverting the yield curve. By contrast, no such calibration is necessary if we apply instead the formulas from the present Sect. 14.3.

For some discontinuous Lévy processes we can evaluate the integral in (14.58) as well. In addition they lead to nonnegative interest rates.

Example 14.20 (Handy Lévy Models) We consider (14.47) once more with the volatility structure (14.63), this time with $X = L$ for the Lévy processes from Example 14.6 whose characteristic exponents equal (14.36, 14.38). A closed-form expression for the integral in (14.58) can be obtained using (14.65). It is derived along the same lines as the similar function Ψ_0 in (14.37, 14.39).

The same holds for two-factor models based on the volatility structure (14.68) below if the driving semimartingale is of the form $X = (X_1, X_2)$ with independent Lévy processes X_1, X_2 of the kind in Example 14.6.

The Heath–Jarrow–Morton approach does not specify a particular model but rather a general framework. Therefore most term structure models can be viewed from that perspective, at least if they allow for instantaneous forward rates at all. We reconsider the short rate models from Sect. 14.2 in this respect.

Example 14.21 (Short Rate Models) In order to interpret Example 14.4 in an HJM framework, choose the Lévy process L as the driving semimartingale X in (14.47).

Using (14.19) and (14.33) we conclude that

$$\sigma(t, T) = e^{-\lambda(T-t)}.$$

This extends also to Examples 14.5–14.7 because they are special cases of Example 14.4. If we consider the Brownian motion W in Examples 14.5 and 14.7 rather than L as the driving semimartingale for (14.47), we obtain instead

$$\sigma(t, T) = \sigma e^{-\lambda(T-t)}.$$

The bivariate Examples 14.9 and 14.10 are of similar structure. For Example 14.9 we take the $\mathbb{R}^2$-valued Lévy process L as the driving semimartingale X in (14.47). This leads to

$$\sigma(t, T) = \begin{pmatrix} e^{-\lambda_1(T-t)} \\ e^{-\lambda_2(T-t)} \end{pmatrix}. \tag{14.68}$$

In Example 14.10 we choose the $\mathbb{R}^2$-valued Lévy process L as the driving semimartingale X in (14.47). We obtain

$$\sigma(t, T) = \begin{pmatrix} \lambda_2 e^{-\lambda_1(T-t)} \Sigma_{\lambda_2-\lambda_1}(T-t) \\ e^{-\lambda_2(T-t)} \end{pmatrix}$$

from (14.19) and (14.45, 14.46).

The Cox–Ingersoll–Ross model is of a different kind. If r is chosen as the driving semimartingale X in (14.47), we obtain a rather involved expression for $\sigma(\cdot, T)$ from (14.19) and (14.42). Since r has no independent increments, this is not covered by the setup of Sect. 14.3.2 but at least by the more general one of Sect. 14.3.5. If, on the other hand, one considers W as the driving semimartingale X, as is traditionally done, we must multiply the result by $\sigma\sqrt{r(t)}$ in view of (14.41). Since the corresponding $\sigma(\cdot, T)$ is no longer deterministic, we are not in the setup of Sect. 14.3.2 either. The latter requires both a time-inhomogeneous Lévy process X and deterministic σ.

14.3.5 Affine Term Structure Models

The computation of bond options and hence caps and floors relies on explicit knowledge of certain characteristic exponents. These are available as long as we stay in an affine setup. This opens the door to using stochastic volatility models in term structure modelling as well. We provide here a general framework rather than a discussion of examples. Time will tell which concrete models turn out to be of interest.

We assume that the driving semimartingale X in (14.47) is time-inhomogeneous affine under Q_0. Moreover, we suppose that the integrands $\sigma(t, T)$ are deterministic. In view of Rule 14.13 this implies that

$$A(t, T) = \psi_0(t, i\Sigma(t, T)) + \sum_{j=1}^{d} \psi_j(t, i\Sigma(t, T))X_j(t-),$$

where ψ_j, $j = 0, \dots, d$ denote the characteristic exponents corresponding to X in the sense of Sect. 6.5. Note that the drift coefficients $\alpha(t, T)$ are affine functions of the components of $X(t-)$.

Equations (14.49, 14.50) and the fact that $\Sigma(\cdot, T)$ is deterministic and $A(\cdot, T)$ an affine function of X imply that $(X(t), \log B(t, T_0, T_1), \log \widehat{B}(t, T_0))_{t\leq T_0}$ is an $\mathbb{R}^{d+2}$-valued time-inhomogeneous affine semimartingale for any T_0, T_1 with $T_0 < T_1$. More specifically, we have a representation

$$\begin{aligned}
&E_{Q_0}\left(\frac{\exp\left(z \log B(T_0, T_1)\right)}{S_0(T_0)}\middle| \mathscr{F}_t\right)\\
&= E_{Q_0}\big(\exp\left(-z \log B(T_0, T_0, T_1) + \log \widehat{B}(T_0, T_0)\right)\big| \mathscr{F}_t\big)\\
&= \exp\Big(\Psi_0(t, T_0, T_1, iz) + \Psi_X(t, T_0, T_1, iz)^\top X(t) - z \log B(t, T_0, T_1) + \log \widehat{B}(t, T_0)\Big)
\end{aligned}$$

for $t \leq T_0$, where $\Psi(\cdot, T_0, T_1, \cdot) = (\Psi_0(\cdot, T_0, T_1, \cdot), \Psi_X(\cdot, T_0, T_1, \cdot)) : \mathbb{R}_+ \times \mathbb{R} \to \mathbb{C}^{1+d}$ denotes the solution to some system of ODEs.

Parallel to Sects. 14.2.3 and 14.3.2 we consider bond options with payoff

$$H = g(\log B(T_0, T_1))$$

at time T_0, which can be written in integral form (14.26). Since pricing is linear, the time-t price of the claim is given by (14.59) with

$$\begin{aligned}
\pi(t, T_0, T_1, z) &:= E_{Q_0}\left(\frac{\exp\left(z \log B(T_0, T_1)\right)}{S_0(T_0)}\middle| \mathscr{F}_t\right) S_0(t)\\
&= \exp\Big(\Psi_0(t, T_0, T_1, iz) + \Psi_X(t, T_0, T_1, iz)^\top X(t)\Big) \frac{B(t, T_1)^z}{B(t, T_0)^{z-1}}.
\end{aligned} \tag{14.69}$$

For calls and puts on a bond as in Example 14.16 we again obtain equation (14.60), this time with π from (14.69). If $\Psi(t, T_0, T_1, u)$ happens to be a quadratic function of u, we end up with Black–Scholes type formulae, cf. the reasoning in Example 14.16.

As noted in Sect. 14.1.2, caps and floors can be reduced to put/call options on bonds. But unfortunately, a counterpart of (14.62) does not exist in this more general setup. The representation in (14.62) relied on the fact that the driving process X is a univariate Lévy process.

14.3.6 Profile

We turn again to the profile in the sense of Sect. 14.1.3.3.

Primary processes The whole forward rate curve $f(t, T), 0 \leq t \leq T$ serves as a primary process in the HJM approach. In other words, all zero bonds can be viewed as underlying securities.

Necessary input Besides the initial forward rate curve we need the dynamics of forward rates under the spot martingale measure—except for their drift part, which is determined by the HJM drift condition. If real-world probabilities matter as well, we also need the density process of P relative to the spot martingale measure.

Discrete vs. continuous tenor The HJM approach relies on the forward-rate curve, which is defined only in—and hence requires—a continuous tenor structure.

Risk-neutral modelling Since starting with the real-world measure makes it difficult to ensure absence of arbitrage, we model immediately under the spot martingale measure or, equivalently, some T-forward measure. The real-world measure P is defined in a second step if it is needed at all.

Stationarity Even if the integrands $\sigma(t, T)$ are chosen to depend only on time to maturity $T-t$, we do not generally end up with a stationary model. In general the state space of accessible forward rate curves at time t depends explicitly on t. The situation changes if the initial forward rate curve belongs to a family of functions that is *consistent* with the chosen forward rate dynamics. In this "stationary" case the forward rate curve will always stay in this family. The questions of existence and characterisation of such consistent families have led to deep and difficult results in interest rate theory, cf. [29, 31, 32, 103, 109].

Tractability If the forward rate model is based on an affine process, semiexplicit formulas are available for bond options, caps and floors. Swaptions allow for an integral representation if the driving semimartingale is a time-inhomogeneous Lévy process and if the volatility processes factorise.

14.4 The Forward Process Approach

14.4.1 General Idea

Starting with and based on work by Miltersen et al., Brace et al., and Jamshidian, a new general approach was established in term structure modelling, leading to what is often called *LIBOR*, *market*, or *BGM models*. Concrete versions differ greatly, but they share one or more of the following features.

- They focus on and are calibrated to liquidly traded interest rate derivatives as caps and swaptions. These are based on LIBOR rates involving only finitely many dates $T_0, \dots, T_N$. Consequently, LIBOR models rely typically on a *discrete-tenor structure* so that model parameters are directly linked to observable quantities.
- The focus is on simple or cc forward rates and hence on the *ratio* of bond price processes. The bond price processes themselves may not even be specified.
- LIBOR rates (or swap rates) are often modelled as *lognormal* and in particular positive processes under the relevant T-forward measure. This leads to a Black–Scholes type formula for caplets (or swaptions, respectively). This *Black formula* plays an important role for quoting prices in real markets.

Although the Black formula is of great importance in practice, it occurs only in the comparatively narrow context of Gaussian models, which do not allow for jumps, stochastic volatility etc. Therefore we dispense with the third feature in this text, which takes a more general viewpoint. Nevertheless, we focus on the discrete-tenor structure which is determined by liquidly traded LIBOR derivatives.

We fix maturities $0 \leq T_0 < \cdots < T_N$ of zero bonds $B(\cdot, T_n)$, which leads to simple forward rates $L(\cdot, T_{n-1}, T_n)$, $n = 1, \dots, N$ or, equivalently, continuously compounded forward rate processes $f(\cdot, T_{n-1}, T_n)$, $n = 1, \dots, N$. The starting point is to model the dynamics of these stochastic processes. In order to obtain tractable formulas in the end, it turns out to be advantageous to consider cc forward rates rather than LIBOR rates and to model the former as components of time-inhomogeneous Lévy processes. At this stage we only assume that all cc forward rates are driven by the same $\mathbb{R}^d$-valued semimartingale X, more precisely

$$df(t, T_{n-1}, T_n) = \alpha(t, n)dt + \sigma(t, n)dX(t) \tag{14.70}$$

with predictable integrands $\alpha(\cdot, n), \sigma(\cdot, n)$. We will later assume X to be a time-inhomogeneous Lévy process and $\alpha(\cdot, n), \sigma(\cdot, n)$ deterministic, but the results in this section hold in this more general setup. The initial values $f(0, T_{n-1}, T_n)$ of these processes are assumed to be given as well. Observe that $f(t, T_{n-1}, T_n)$ may be viewed as a counterpart to instantaneous forward rates $f(t, T)$ in the present discrete-tenor world, where the shortest investment periods have a positive length. We will in fact see that the forward process approach can altogether be viewed as a discrete-tenor counterpart to HJM.

Similarly as in (14.48) we define

$$A(t, n) := \sum_{m=n(t)+1}^{n} (T_m - T_{m-1})\alpha(t, m),$$

$$\Sigma(t, n) := \sum_{m=n(t)+1}^{n} (T_m - T_{m-1})\sigma(t, m),$$

for $t \le T_n$, where $n(t)$ denotes the index with $T_{n(t)-1} \le t < T_{n(t)}$. Equation (14.70) immediately yields the following dynamics, which resemble similar equations in Rule 14.11.

Rule 14.22 *We have*

$$d \log \widetilde{B}(t, T_n) = -A(t, n)dt - \Sigma(t, n)dX(t), \tag{14.71}$$

$$\begin{aligned} d \log B(t, T_m, T_n) &= (T_n - T_m)df(t, T_m, T_n) \\ &= (A(t, n) - A(t, m))dt + (\Sigma(t, n) - \Sigma(t, m))dX(t) \end{aligned} \tag{14.72}$$

for $t \le T_m \le T_n$, where $\widetilde{B}(\cdot, T_n) := B(\cdot, T_n)/\widetilde{S}_0$ denotes the price of the bond discounted relative to the discrete money market account. For $t \le T_{n-1}$ we obtain in particular

$$d \log B(t, T_{n-1}, T_n) = (T_n - T_{n-1})\big(\alpha(t, n)dt + \sigma(t, n)dX(t)\big).$$

Idea Note that

$$\begin{aligned} \log \widetilde{B}(t, T_n) &= -\log B(t, T_{n(t)}, T_n) + \sum_{k=1}^{n(t)} \log B(T_{k-1}, T_k) \\ &= -\sum_{m=n(t)+1}^{n} (T_m - T_{m-1}) f(t, T_{m-1}, T_m) + \sum_{k=1}^{n(t)} \log B(T_{k-1}, T_k). \end{aligned}$$

Using (14.70) we obtain (14.71). The first equality in (14.72) is obvious from the definition. For the second note that

$$\begin{aligned} (T_n - T_m) f(t, T_m, T_n) &= \log B(t, T_m, T_n) \\ &= \sum_{k=m+1}^{n} \log B(t, T_{k-1}, T_k) \\ &= \sum_{k=m+1}^{n} (T_k - T_{k-1}) f(t, T_{k-1}, T_k) \end{aligned}$$

and use (14.70) once more. □

Recall from the heuristics in Rule 11.5 that only a limited number of assets can be specified independently if one wants to obtain an arbitrage-free model. As in the HJM approach we avoid discussing involved drift restrictions by modelling immediately under the risk-neutral measure, i.e. we take again the martingale modelling point of view in the sense of Sect. 14.1.3. However, we do not choose the money market account as numeraire because we do not want to assume its existence in the first place. As a substitute we consider the discrete money market account, which means that we work under the discrete spot martingale measure $\widetilde{Q}_0$. We denote the characteristic exponent of X under $\widetilde{Q}_0$ by $\psi(t,u)$. The martingale property of discounted T_n-bonds can be expressed in terms of drift conditions similarly as in the HJM approach.

Rule 14.23 (Forward Process Drift Conditions) *$\widetilde{Q}_0$ is a discrete spot martingale measure (in the sense that the discounted bonds $\widetilde{B}(\cdot,T_n)$ are $\widetilde{Q}_0$-martingales for $n=1,\ldots,N$) if and only if*

$$\alpha(t,n)=\frac{\psi\big(t,i\Sigma(t,n)\big)-\psi\big(t,i\Sigma(t,n-1)\big)}{T_n-T_{n-1}} \tag{14.73}$$

or equivalently

$$A(t,n)=\psi\big(t,i\Sigma(t,n)\big), \tag{14.74}$$

for $n=1,\ldots,N$ and $t\leq T_{n-1}$.

Idea Equation (14.74) follows analogously as the HJM drift condition (14.55), using (14.71) instead of (14.49). The first equation (14.73) is obtained by considering $\alpha(t,n)=(A(t,n)-A(t,n-1)/(T_n-T_{n-1})$. □

Observe the similarity between the forward process drift condition (14.74) and the corresponding HJM drift condition (14.55). Equation (14.73) means that we are in the same situation as in the HJM approach. The integrands $\sigma(\cdot,n)$ in (14.70) can be chosen arbitrarily while the drift coefficients $\alpha(\cdot,n)$ are determined by the martingale property of discounted bond prices.

Instead of modelling under the discrete spot martingale measure one may prefer to work under some T_n-forward measure, e.g. the one with longest time horizon T_N. This change of measure affects primarily the characteristic exponent of the driving process X. In order to formulate concrete equations, we denote the characteristic exponent of X under the T_n-forward measure by $\psi_{(n)}(t,u)$.

Rule 14.24 (Change of Measure) *We have*

$$\psi_{(n)}(t,u)=\psi\big(t,u+i\Sigma(t,n)\big)-\psi\big(t,i\Sigma(t,n)\big)$$

and equivalently

$$\psi(t,u)=\psi_{(n)}\big(t,u-i\Sigma(t,n)\big)-\psi_{(n)}\big(t,-i\Sigma(t,n)\big)$$

for $0\leq t\leq T_n$, $u\in\mathbb{R}^d$.

Idea This follows along the same lines as Rule 14.14. □

Remark 14.25 In view of the previous rule, the forward process drift conditions can be written as

$$\alpha(t,n) = \frac{\psi_{(n-1)}\big(t, i(T_n - T_{n-1})\sigma(t,n)\big)}{T_n - T_{n-1}}$$

for $t \leq T_{n-1}$.

So far we have modelled only ratios of bond prices but not bonds themselves. Similarly, we have not specified the dynamics of the discrete money market account which is nevertheless determined at times $T_0, \ldots, T_N$. This can be seen from (14.9), where the last term on the right vanishes for $t = T_{n-1}$. For arbitrary t, this last term involves the **discrete short rate** $(\widetilde{r}(t))_{t\geq 0}$, which is defined as

$$\widetilde{r}(t) := f(t,t,T_n), \quad T_{n-1} \leq t < T_n. \tag{14.75}$$

The process $\widetilde{r}$ can be interpreted as a counterpart to the short rate in the present discrete tenor world where instantaneous rates may not exist. It is linked to the cc forward rates on the grid $T_0, \ldots, T_N$ via

$$\widetilde{r}(T_{n-1}) = f(T_{n-1}, T_{n-1}, T_n), \quad n = 1, \ldots, N. \tag{14.76}$$

However, not much can yet be said about its behaviour between successive time points. In order to model the whole process we specify $\widetilde{r}$ as a process which satisfies (14.76) and

$$d\widetilde{r}(t) = \widetilde{\alpha}(t)dt + \widetilde{\sigma}(t)dX(t) \tag{14.77}$$

with predictable integrands $\widetilde{\alpha}, \widetilde{\sigma}$. In view of (14.9), the dynamics of the discrete money market account under the discrete spot martingale measure are now completely specified. Observe, however, that the model for the forward rates $f(t, T_{n-1}, T_n), n = 1, \ldots, N$ already determines $\widetilde{r}(T_n)$ for $n = 0, \ldots, N-1$. This imposes certain restrictions on how to choose $\widetilde{\alpha}, \widetilde{\sigma}$ in (14.77). But interestingly, this specification of $\widetilde{r}$ outside the grid $(T_n)_{n=1,\ldots,N-1}$ is not needed for the computation of option/cap/floor prices in Sects. 14.4.2 and 14.4.4.

Rule 14.26 (Discrete Money Market Account) *The discrete money market account satisfies*

$$\widetilde{S}_0(t) = \exp\left(\sum_{m=1}^{n} f(T_{m-1}, T_{m-1}, T_m)(T_m - T_{m-1}) - \widetilde{r}(t)(T_n - t)\right) \tag{14.78}$$

and

$$d \log \widetilde{S}_0(t) = \big(\widetilde{r}(t) - (T_n - t)\widetilde{\alpha}(t)\big)dt - (T_n - t)\widetilde{\sigma}(t)dX(t) \tag{14.79}$$

for $T_{n-1} \leq t < T_n$.

Idea Equation (14.78) follows directly from (14.9) and (14.75). Equation (14.79) is obtained from (14.78, 14.77) and integration by parts. □

Note that the flexibility how to model the discrete short rate diminishes for decreasing mesh size $\sup_n(T_n - T_{n-1})$. This explains why $r(t)$ need not be specified at all in the HJM approach, which we interpret as a continuous tenor limit of the present setup. Moreover, observe that while the discrete money market account in (14.79) is typically not absolutely continuous in time, its dynamics are dominated by the $\widetilde{r}(t)dt$ term for small mesh size, which is in line with the usual money market account in the HJM setup.

Note that the T_n-bonds satisfy

$$\log B(t, T_n) = \log \widetilde{S}_0(t) - \sum_{m=1}^{n} \log B(T_{m-1}, T_{m-1}, T_m)$$

$$= -\widetilde{r}(t)(T_{n(t)} - t) - \sum_{m=n(t)+1}^{n} f(T_{m-1}, T_{m-1}, T_m)(T_m - T_{m-1})$$

for $T_{n(t)-1} \leq t < T_{n(t)} \leq T_n$. Hence the cc forward rates and the discrete short rate determine the bond prices.

If one wants to specify the dynamics of rates, prices etc. under real-world probabilities as well, one chooses an arbitrary positive $\widetilde{Q}_0$-martingale with initial value 1 as density process specifying the transition from $\widetilde{Q}_0$ to P.

14.4.2 Lévy-Driven Forward Process Models

In order to obtain more explicit formulas we assume that X in (14.70) is a time-inhomogeneous uni- or multivariate Lévy process with $\widetilde{Q}_0$-characteristic exponent $\psi(t, u)$ and that the integrands $\sigma(\cdot, n)$ are deterministic. The discrete short rate $\widetilde{r}$ and hence a specification of α, σ in (14.77) is not needed for the results in this section.

For option prices we state the following counterpart of Rule 14.15.

Rule 14.27 *We have*

$$\pi(t, T_m, T_n, z) := E_{\widetilde{Q}_0}\left(\frac{\exp\left(z \log B(T_m, T_n)\right)}{\widetilde{S}_0(T_m)} \middle| \mathscr{F}_t \right) \widetilde{S}_0(t)$$

$$= \exp\left(\int_t^{T_m} \varphi(s, T_m, T_n, z) ds \right) \frac{B(t, T_n)^z}{B(t, T_m)^{z-1}} \tag{14.80}$$

for $t \leq T_m \leq T_n$, *where*

$$\varphi(s, T_m, T_n, z) := \psi\big(s, iz\Sigma(s, n) - i(z-1)\Sigma(s, m)\big) \\ - z\psi(s, i\Sigma(s, n)) + (z-1)\psi(s, i\Sigma(s, m)). \tag{14.81}$$

The set of eligible $z \in \mathbb{C}$ *depends on* ψ *and* σ.

Idea By (14.71, 14.72) we have

$$\begin{aligned} z\log B(T_m, T_n) - \log \widetilde{S}_0(T_m) &= -z\log B(T_m, T_m, T_n) + \log \widetilde{B}(T_m, T_m) \\ &= -z\log B(t, T_m, T_n) + \log \widetilde{B}(t, T_m) \\ &\quad - \int_t^{T_m} \big(z(A(s, n) - A(s, m)) + A(s, m)\big)ds \\ &\quad - \int_t^{T_m} \big(z(\Sigma(s, n) - \Sigma(s, m)) + \Sigma(s, m)\big)dX(s). \end{aligned}$$

Since

$$E\left(\int_t^{T_m} f(s)dX(s)\middle|\mathscr{F}_t\right) = \exp\left(\int_t^{T_m} \psi(s, -if(s))ds\right)$$

for deterministic functions f by Proposition 3.54 and using (14.74), we obtain (14.80) after a straightforward calculation. □

With the help of Rule 14.27 we can now state semiexplicit pricing formulas for bond options, caplets, and floorlets. We proceed almost identically as in the HJM approach. Again we start with bond options having payoff

$$H = g(\log B(T_m, T_n))$$

at time T_m. The function g is assumed to be of integral form (14.26), where $\widetilde{g}$ denotes the Laplace transform of g. For the time-t price of the claim linearity of the price yields

$$\begin{aligned} \pi(t) &= E_{\widetilde{Q}_0}\left(\frac{H}{\widetilde{S}_0(T_m)}\middle|\mathscr{F}_t\right)\widetilde{S}_0(t) \\ &= E_{\widetilde{Q}_0}\left(\frac{g(\log B(T_m, T_n))}{\widetilde{S}_0(T_m)}\middle|\mathscr{F}_t\right)\widetilde{S}_0(t) \\ &= \frac{1}{2\pi i}\int_{R-i\infty}^{R+i\infty} \pi(t, T_m, T_n, z)\widetilde{g}(z)dz \end{aligned} \tag{14.82}$$

with $\pi(t, T_m, T_n, z)$ as in (14.80). As an example we consider calls and puts on bonds.

Example 14.28 (Call and Put on a Bond) For a call $H = (B(T_m, T_n) - K)^+$ on a T_n-bond with strike $K \geq 0$ and maturity $T_m \leq T_n$, the Laplace transform of $g(x) = (e^x - K)^+$ is stated in Example 11.26. This leads to the pricing formula

$$\begin{aligned}\pi(t) &= \frac{1}{2\pi i} \int_{R-i\infty}^{R+i\infty} \pi(t, T_m, T_n, z) \frac{K^{1-z}}{z(z-1)} dz \\ &= \frac{1}{\pi} \int_0^\infty \operatorname{Re}\left(\pi(t, T_m, T_n, R+iu) \frac{K^{1-R-iu}}{(R+iu)(R-1+iu)}\right) du\end{aligned}$$

for $t \leq T_m \leq T_n$ and $R > 1$. For puts we get the same formula with $R < 0$.

As in Examples 14.3 and 14.16 we get Black–Scholes type formulae if we are in a Gaussian context, i.e. if $\psi(t, u)$ is a quadratic function of u. As in these setups, the martingale property of discounted bond prices leads to a representation

$$\pi(t, T_m, T_n, z) = \frac{B(t, T_n)^z}{B(t, T_m)^{z-1}} \exp\left(\frac{\gamma^2}{2} z(z-1)\right)$$

for some $\gamma > 0$ which depends on t, T_m, T_n. Proceeding as in Example 14.3 we obtain

$$\pi(t) = B(t, T_n)\Phi\left(\frac{1}{\gamma}\left(\log \frac{B(t, T_n)}{K B(t, T_m)} + \frac{\gamma^2}{2}\right)\right) - K B(t, T_m)\Phi\left(\frac{1}{\gamma}\left(\log \frac{B(t, T_n)}{K B(t, T_m)} - \frac{\gamma^2}{2}\right)\right)$$

for calls and accordingly for puts.

Caps and floors are reduced to put resp. call options on bonds as usual. As can be expected from the previous approaches, swaptions allow for a simple representation only in a more restrictive setup.

Similarly as for (14.62) we assume that the driving time-inhomogeneous Lévy process X is univariate and that the volatility factorises as

$$\sigma(t, n) = \lambda_n \sigma(t)$$

with constants $\lambda_1, \ldots, \lambda_N \geq 0$ and a function $\sigma : \mathbb{R}_+ \to (0, \infty)$. This implies

$$\Sigma(t, n) = \Lambda(t, n)\sigma(t)$$

for $T_{n(t)-1} \leq t < T_{n(t)} \leq T_n$, where

$$\Lambda(t, n) := \sum_{\ell=n(t)+1}^{n} (T_\ell - T_{\ell-1})\lambda_\ell.$$

Moreover, recall that

$$A(t, n) = \psi\big(t, i\Sigma(t, n)\big)$$

according to (14.74).

Using (14.72) we obtain the representation

$$B(T_m, T_n) = \pi(T_m, T_n, Y(T_m))$$

for any $n > m$, where $Y(t) := \int_0^t \sigma(s)dX(s)$ and

$$\pi(T_m, T_n, x) := \frac{B(0, T_n)}{B(0, T_m)} \exp\left(- \int_0^{T_m} \big(A(t, n) - A(t, m)\big)dt \right) e^{-\Lambda(T_{m-1}, n)x}.$$

By Jamshidian's approach in Sect. 14.1.2.7, we can express swaptions as portfolios of put options on the bond, cf. (14.13).

Recall that the volatility structure (14.63) allowed us to compute swaption prices and sometimes to evaluate the integral in the exponent of (14.58) explicitly. The same is true in the forward process setup for the corresponding function

$$\sigma(t, n) = e^{-\lambda(T_n - t)}, \quad t < T_{n-1}$$

for some $\lambda \geq 0$.

14.4.3 Examples

We start by reconsidering the models from the previous sections in the forward process approach. Since we already explored how to view short-rate models in an HJM framework, it suffices to interpret HJM models as specific forward process setups.

Example 14.29 (HJM Models) Fix maturities $0 \leq T_0 < \cdots < T_N$ as usual. Any HJM model can easily be interpreted within the forward process approach. Indeed, (14.50) implies that

$$df(t, T_{n-1}, T_n) = \alpha(t, n)dt + \sigma(t, n)dX(t)$$

with

$$\alpha(t, n) := \frac{A(t, T_n) - A(t, T_{n-1})}{T_n - T_{n-1}},$$

$$\sigma(t, n) := \frac{\Sigma(t, T_n) - \Sigma(t, T_{n-1})}{T_n - T_{n-1}}.$$

Note that the terms on the right refer to the notation in the HJM setup whereas the left-hand sides belong to the forward process world.

Secondly we have a look at a Gaussian forward process model which should not be confused with the lognormal LIBOR model in the literature.

Example 14.30 (Gaussian Forward Process Model) Suppose that the driving time-inhomogeneous Lévy process X in Sect. 14.4.2 is a multivariate Brownian motion. In this case we are in the Gaussian context discussed in Example 14.28, which leads to Black–Scholes type formulae for calls, put, caplets, and floorlets.

Nevertheless, this Gaussian setup differs from the **lognormal LIBOR market models** in the literature which lead to the Black formula for caps. The approach in these models is as follows. For ease of notation let $\delta := T_n - T_{n-1}$ be the same for $n = 1, \dots, N$, cf. Sect. 14.1.2.5. The LIBOR rate

$$L(t, T_{n-1}, T_n) = \frac{B(t, T_{n-1}) - B(t, T_n)}{B(t, T_n)\delta}$$

is a martingale under the T_n-forward measure because it coincides with the price process of a tradable portfolio discounted by a T_n-bond. Since the payoff of a caplet at time T_n equals

$$(L(T_{n-1}, T_{n-1}, T_n) - \ell)^+ \delta,$$

its time t-price is given by

$$\pi(t) = E_{Q_{T_n}}\Big((L(T_{n-1}, T_{n-1}, T_n) - \ell)^+ \Big| \mathscr{F}_t \Big) \delta B(t, T_n). \tag{14.83}$$

If we specify

$$dL(t, T_{n-1}, T_n) = L(t, T_{n-1}, T_n)\sigma_n(t) dW(t)$$

with a deterministic function σ_n and a Q_{T_n}-standard Wiener process W, then $L(t, T_{n-1}, T_n)$ has a lognormal distribution under Q_{T_n}. Hence we end up with a Black–Scholes type formula for caplets, namely the *Black formula*.

The payoff of the caplet can also be written as

$$(L(T_{n-1}, T_{n-1}, T_n) - \ell)^+ \delta = \Big(e^{f(T_{n-1}, T_{n-1}, T_n)\delta} - 1 - \ell\delta \Big)^+ .$$

Of course $e^{f(\cdot, T_{n-1}, T_n)\delta}$ is a Q_{T_n}-martingale too because it differs from $L(\cdot, T_{n-1}, T_n)\delta$ only by a constant. Pricing formula (14.83) now reads as

$$\pi(t) = E_{Q_{T_n}}\Big(\big(e^{f(T_{n-1}, T_{n-1}, T_n)\delta} - 1 - \ell\delta \big)^+ \Big| \mathscr{F}_t \Big) B(t, T_n).$$

In our forward process setup, however, $f(\cdot, T_{n-1}, T_n)$ follows a Gaussian process and hence $L(T_{n-1}, T_{n-1}, T_n)\delta + 1$ has a lognormal distribution under Q_{T_n}. As noted above, this leads to a Black–Scholes type formula for caplets as well. But since $L(T_{n-1}, T_{n-1}, T_n)$ and $L(T_{n-1}, T_{n-1}, T_n)\delta + 1$ cannot both be lognormal, the two "Gaussian" models are incompatible with each other.

We focus on cc forward rates rather than LIBOR rates in this monograph because the affine (resp. Lévy or Gaussian) structure is preserved under any T_n-forward measure and under the discrete spot martingale measure as well. The situation is different for the lognormal LIBOR market model where $L(\cdot, T_{n-1}, T_n)$ is lognormally distributed only under the respective T_n-forward measure. The Black formula does not hold in our setup but it does not extend beyond Gaussian models anyway.

14.4.4 Affine Forward Process Models

As in the Heath–Jarrow–Morton case we can still obtain semiexplicit formulas if we replace the time-inhomogeneous Lévy process in Sect. 14.4.2 by an affine semimartingale. More precisely, we assume that X in (14.70) is an affine process and that the integrands $\sigma(\cdot, n)$ are deterministic.

Rule 14.23 yields that

$$A(t,n) = \psi_0(t, i\Sigma(t,n)) + \sum_{m=1}^{d} \psi_j(t, i\Sigma(t,n))X_m(t),$$

where $\psi_j, j = 0, \ldots, d$ denote the exponents belonging to the $\mathbb{R}^d$-valued affine process X. Consequently, the drift coefficients $\alpha(t,n)$ are affine functions of the components of X.

Equations (14.71, 14.72) together with $\Sigma(\cdot, n)$ being deterministic and $A(\cdot, n)$ affine imply that the $\mathbb{R}^{d+2}$-valued process $(X(t), \log B(t, T_m, T_n), \log \widetilde{B}(t, T_n))_{t \leq T_m}$ is an affine semimartingale for any $m \leq n$. Along the same lines as for (14.15) we obtain a representation

$$\begin{aligned}
&E_{\widetilde{Q}_0}\left(\frac{\exp\left(z \log B(T_m, T_n)\right)}{\widetilde{S}_0(T_m)}\middle| \mathscr{F}_t\right) \\
&= E_{\widetilde{Q}_0}\big(\exp\big(-z \log B(T_m, T_m, T_n) + \log \widetilde{B}(T_m, T_m)\big)\big| \mathscr{F}_t\big) \\
&= \exp\Big(\Psi_0(t, T_m, T_n, iz) + \Psi_X(t, T_m, T_n, iz)^\top X(t) \\
&\qquad - z \log B(t, T_m, T_n) + \log \widetilde{B}(t, T_m)\Big)
\end{aligned}$$

for $t \leq T_m$ and $z \in U \subset \mathbb{C}$, where $\Psi(\cdot, T_m, T_n, \cdot) = (\Psi_0(\cdot, T_m, T_n, \cdot), \Psi_X(\cdot, T_m, T_n, \cdot)) : \mathbb{R}_+ \times U \to \mathbb{C}^{1+d}$ denotes the solution to some system of ODEs that is stated in Theorem 6.23.

As in Sect. 14.4.2 we consider bond options with payoff $H = g(\log B(T_m, T_n))$ at time T_m, where g is assumed to be of integral form (14.26). For the time-t-price of the claim we obtain (14.82) with

$$\pi(t, T_m, T_n, z) := E_{\widetilde{Q}_0}\left(\frac{\exp\left(z \log B(T_m, T_n)\right)}{\widetilde{S}_0(T_m)}\middle| \mathscr{F}_t\right)\widetilde{S}_0(t)$$

$$= \exp\Big(\Psi_0(t, T_m, T_n, iz) + \Psi_X(t, T_m, T_n, iz)^\top X(t)\Big)\frac{B(t, T_n)^z}{B(t, T_m)^{z-1}}.$$

This reduces to (14.80, 14.81) for multivariate Lévy processes X, but as in Sect. 14.3.5 we do not obtain a Jamshidian representation if we leave the univariate Lévy case.

Recall that the discrete short rate has not yet been detailed because it is not needed for option pricing. If one wants to specify it, we suggest to stay in an affine framework by choosing $\widetilde{\sigma}$ in (14.77) deterministic and the drift coefficient $\widetilde{\alpha}$ of the form

$$\widetilde{\alpha}(t) = \alpha(t)^\top (1, X(t-), \widetilde{r}(t-))$$

for some deterministic function $\alpha : \mathbb{R}_+ \to \mathbb{R}^{1+d+1}$.

Example 14.31 (Affine LIBOR Model of [190]) Fix an $\mathbb{R}^d$-valued affine process X as above with exponent $\psi = (\psi_0, \dots, \psi_d)$. The goal is to choose the deterministic integrands $\sigma(\cdot, n)$ such that all forward rates stay nonnegative. To this end, fix $T \geq T_N$ and suppose that X has nonnegative components $X_1, \dots, X_d$.

Observe that $\widetilde{B}(t, T_n)$, $n = 1, \dots, N$ are assumed to be martingales under the discrete spot martingale measure. This certainly holds if we model them as

$$\widetilde{B}(t, T_n) := E_{\widetilde{Q}_0}\big(\exp(u_n^\top X(T))\big|\mathscr{F}_t\big), \quad t \leq T_n$$

with $u_n \in \mathbb{R}^d$ such that the expectation is finite. Defining Ψ_0 and $\Psi_{1,\dots,d} = (\Psi_1, \dots, \Psi_d)$ as in Theorem 6.6, we have

$$\widetilde{B}(t, T_n) = \exp\big(\Psi_0(-iu_n, T-t) + \Psi_{1,\dots,n}(-iu_n, T-t)^\top X(t)\big).$$

This implies

$$\begin{aligned}
d \log \widetilde{B}(t, T_n) &= -\big(\dot{\Psi}_0(-iu_n, T-t) + \dot{\Psi}_{1,\dots,d}(-iu_n, T-t)^\top X(t)\big)dt \\
&\quad + \Psi_{1,\dots,d}(-iu_n, T-t)dX(t) \\
&= -\big(\psi_0(-i\Psi_{1,\dots,d}(-iu_n, T-t)) + \psi_{1,\dots,d}(-i\Psi_{1,\dots,d}(-iu_n, T-t))^\top X(t)\big)dt \\
&\quad + \Psi_{1,\dots,d}(-iu_n, T-t)^\top dX(t),
\end{aligned}$$

where we write $\dot{\Psi}(u,t) := \frac{\partial}{\partial t}\Psi(u,t)$ and the second equation holds because $\Psi_0, \Psi_{1,\dots,d}$ solve the system of ODEs in Theorem 6.6. As a side remark, we observe that $\Psi_{1,\dots,d}(-iu_n, T-t) = 0$ because $t \mapsto \widetilde{B}(t,T_n)$ is constant on $(T_{n-1}, T_n]$. The corresponding forward rates

$$f(t,T_{n-1},T_n) = \frac{\log B(t,T_{n-1},T_n)}{T_n - T_{n-1}} = \frac{\log \widetilde{B}(t,T_{n-1}) - \log \widetilde{B}(t,T_n)}{T_n - T_{n-1}}$$

satisfy the equation

$$df(t,T_{n-1},T_n) = \alpha(t,n)dt + \sigma(t,n)dX(t)$$

with

$$\sigma(t,n) = \frac{-\Psi_{1,\dots,d}(-iu_n, T-t) + \Psi_{1,\dots,d}(-iu_{n-1}, T-t)}{T_n - T_{n-1}}$$

and

$$\begin{aligned}\alpha(t,n) = \Big(&\psi_0(-i\Psi_{1,\dots,d}(-iu_n, T-t))\\ &+ \psi_{1,\dots,d}(-i\Psi_{1,\dots,d}(-iu_n, T-t))^\top X(t)\\ &- \psi_0(-i\Psi_{1,\dots,d}(-iu_{n-1}, T-t))\\ &- \psi_{1,\dots,d}(-i\Psi_{1,\dots,d}(u_{n-1}, T-t))^\top X(t)\Big)\Big/(T_n - T_{n-1}).\end{aligned}$$

We observe that it is of the form (14.70) with deterministic $\sigma(\cdot,n)$. Moreover, $\alpha(\cdot,n)$ satisfies the drift condition (14.73). This does not come as a surprise because it reflects the martingale property of the discounted bonds $\widetilde{B}(t,T_n)$. If we now let $u_1 \geq u_2 \geq \dots \geq u_N$ componentwise, we have $u_m^\top X(T) \geq u_n^\top X(T)$ and hence $B(t,T_m) \geq B(t,T_n)$ for $m \leq n$. This implies that the forward rates $f(t,T_m,T_n)$ in this model stay nonnegative as desired. The constants $u_1,\dots,u_N$ can, for instance, be chosen by calibration to the observed forward rates at time 0.

This example differs slightly from the affine LIBOR model in [190]. We work here under the discrete spot martingale measure whereas [190] starts with an affine process under the T_N-forward measure.

14.4.5 Profile

We summarise here the profile in the sense of Sect. 14.1.3.3.

Primary processes The discrete LIBOR rates $L(\cdot,T_{n-1},T_n), n = 1,\dots,N$ or, equivalently, the corresponding cc forward rates serve as primary processes. On top one may detail a model for the discrete short rate between successive time points.

Necessary input We need the initial LIBOR rates as well as their dynamics under, say, the discrete spot martingale measure—again up to their drift part, which is determined by a drift condition. In addition, the initial value and the dynamics of the discrete short rate should be specified. Finally we must specify the transition between the real-world measure and the discrete spot martingale measure.

Discrete vs. continuous tenor The forward process model is a discrete tenor approach in the presented form—regardless of the question of whether it allows for a continuous tenor extension.

Risk-neutral modelling We start by modelling under the discrete spot martingale measure or, equivalently, the T_N-forward measure, in order to ensure absence of arbitrage. The transition to the real world measure is specified in a second step.

Stationarity One may argue that the concept of stationarity does not make sense in the context of forward process modelling. Indeed, a finite set of maturities is singled out right from the start, which leads to a certain asymmetry in time.

Tractability If cc forward rates are based on an affine process, we are in the same situation as in the HJM setup, i.e. bond options, caps and floors can be represented semiexplicitly by integral transform methods. Moreover, swaptions allow for an integral representation if the driving semimartingale is a time-inhomogeneous Lévy process and if the volatility processes factorise.

14.5 The Flesaker–Hughston Approach

Yet another approach to term structure modelling has the surprising property that both caps and swaptions allow for semiexplicit pricing formulas of integral form. These interest rate products play a particularly important role in practice because they are liquidly traded. For calibration and parameter estimation their prices must be computed very often and hence as fast as possible.

14.5.1 State-Price Density Processes

Recall the alternative representation of absence of arbitrage in terms of state-price density processes: according to Rule 9.15 there exists a positive semimartingale Z which turns all price processes into martingales. Such a process constitutes the starting point in the Flesaker–Hughston approach to term structure modelling. We assume that Z is given. No model for interest rates or bond prices is specified in the first place.

Rule 14.32 *The price process of a T-bond is given by*

$$B(t,T) = \frac{E(Z(T)|\mathscr{F}_t)}{Z(t)}.$$

More generally, the price at time t of any tradable instrument with random payoff H at time T equals

$$\frac{E(HZ(T)|\mathscr{F}_t)}{Z(t)}.$$

Idea For the first statement observe that $B(t,T)Z(t)$ is a martingale with terminal value $B(T,T)Z(T) = Z(T)$. The second follows along the same lines. □

Hence the specification of the state-price density process determines the dynamics of the whole term structure including its initial state. Provided that bonds of any maturity are traded and sufficient smoothness holds, we obtain equations for forward rates and in particular the short rate as well, cf. Sect. 14.1.1.

The state-price density process Z and the density process Z^Q of the equivalent martingale measure Q in the FTAP are related to each other via $Z^Q = ZS_0$ if S_0 denotes the numeraire for the EMM. In particular, we have

$$Z^{Q_0}(t) = Z(t)\exp\left(\int_0^t r(s)ds\right) \tag{14.84}$$

for the spot martingale measure Q_0 if the money market account exists. The short rate r itself can be recovered from the state-price density process Z as well.

Rule 14.33 (Short Rate) *If the semimartingale Z allows for a canonical decomposition*

$$dZ(t) = a^Z(t)dt + dM^Z(t)$$

with a martingale M^Z and a predictable drift rate a^Z, then

$$r(t) := -\frac{a^Z(t)}{Z(t-)} \tag{14.85}$$

is the short rate.

Idea Applying integration by parts to (14.84) yields

$$dZ^{Q_0}(t) = \exp\left(\int_0^t r(s)ds\right)\Big((Z(t-)r(t) + a^Z(t))dt + dM^Z(t)\Big).$$

Since Z^{Q_0} is a martingale, its growth rate vanishes, which yields the assertion. □

The state-price density process is usually assumed to be a supermartingale because this warrants nonnegative interest rates:

Corollary 14.34 *The short rate is nonnegative if the state-price density process is a supermartingale.*

Idea This follows from (14.85) because Z is a supermartingale if and only if $a^Z \leq 0$. □

In the Flesaker–Hughston approach modelling takes place under the real-world probability measure P. The latter coincides with the spot martingale measure if the state-price density process is absolutely continuous, i.e. if its martingale part vanishes.

Rule 14.35 *P is a spot martingale measure (in the sense that all discounted tradable assets are martingales) if Z has a representation*

$$Z(t) = \exp\left(- \int_0^t r(s)ds \right)$$

for some integrable process r. In this case r is the short rate.

Idea This is evident from Rule 14.33 and (14.84). □

14.5.2 Rational Models

More can be said if we assume a specific representation for Z. We could, for example, consider a state-price density process of the form

$$Z(t) = e^{X_d(t)},$$

where X denotes an $\mathbb{R}^d$-valued affine process. For concrete choices of X we would then recover the affine models from the previous three sections. This, however, is not the path taken in this section. In order to obtain in some sense even more tractable models we assume instead a **rational model**

$$Z(t) = a(t) + b(t)e^{X_d(t)}, \tag{14.86}$$

where $a, b : \mathbb{R}_+ \to (0, \infty)$ are decreasing deterministic functions and $e^{X_d(t)}$ is assumed to be a martingale. Derivatives of a, b will be denoted by a dot. Rule 14.32 yields that bond prices are given by a ratio, which explains the name *rational*.

Rule 14.36 *In the rational model (14.86) the price process of a T-bond is given by*

$$B(t, T) = \frac{a(T) + b(T)e^{X_d(t)}}{a(t) + b(t)e^{X_d(t)}}, \quad t \leq T. \tag{14.87}$$

Moreover, we have

$$f(t,S,T)=\frac{1}{(T-S)}\log\frac{a(S)+b(S)e^{X_d(t)}}{a(T)+b(T)e^{X_d(t)}},\quad t\leq S<T \tag{14.88}$$

and, provided that a,b are sufficiently smooth,

$$f(t,T)=r(t,T-t)=-\frac{\dot{a}(T)+\dot{b}(T)e^{X_d(t)}}{a(T)+b(T)e^{X_d(t)}},\quad t\leq T, \tag{14.89}$$

$$r(t)=-\frac{\dot{a}(t)+\dot{b}(t)e^{X_d(t)}}{a(t)+b(t)e^{X_d(t)}},\quad t\geq 0. \tag{14.90}$$

Idea Equation (14.87) follows from Rule 14.32 and $E(e^{X_d(T)}|\mathscr{F}_t)=e^{X_d(t)}$. Equation (14.88) is obtained from (14.87) and the definition of $f(t,S,T)$. Differentiating with respect to T yields (14.89), which in turn reduces to (14.90) for $T=t$. □

Observe that interest rates are positive in this model because a,b are assumed to be decreasing. Dynamic equations for bond prices and interest rates can be derived by applying Itō's formula to (14.87–14.90). The short rate is typically an instationary process because of the time dependence of a,b, which is needed in order to obtain a non-trivial term structure.

Unlike in the HJM and the LIBOR approaches, zero bonds and even the short rate appear as secondary processes, which means that their current values do not appear as input to the model. Model parameters such as the deterministic functions a,b must be calibrated in order to reproduce the observed initial yield curve and possibly also cap and swaption prices.

14.5.3 Derivatives

From now on we assume the rational model (14.86) where X denotes an $\mathbb{R}^d$-valued time-inhomogeneous affine process such that e^{X_d} is a martingale. In particular, there is a function $\Psi=(\Psi_0,\Psi_X):\mathbb{R}_+\times\mathbb{R}_+\times U\to\mathbb{C}^{1+d}$ satisfying

$$E\big(\exp(zX_d(T))\big|\mathscr{F}_t\big)=\exp\Big(\Psi_0(t,T,-iz)+\Psi_X(t,T,-iz)^\top X(t)+zX_d(t)\Big) \tag{14.91}$$

for $z\in U\subset\mathbb{C}$, $0\leq t\leq T$. It is obtained by solving the ODEs in Theorems 6.6 resp. 6.23, which is done for simple examples in Sect. 14.5.4. The size of the set $U\subset\mathbb{C}$ depends on the existence of exponential moments of X, cf. Proposition A.5(4) and Theorem 6.10.

Pricing formulas for bond options, caps, floors, and swaptions are obtained as special cases of the following result.

Rule 14.37 (Call/Put on a Bond Portfolio) *Let $t \le T_0 \le \cdots \le T_N$ and $c_1, \dots, c_N, K \in \mathbb{R}_+$ such that*

$$\frac{\sum_{n=1}^N c_n a(T_n)}{a(T_0)} < K < \frac{\sum_{n=1}^N c_n b(T_n)}{b(T_0)}.$$

Then the time-t price of a call with time-T_0 payoff

$$H = \left(\sum_{n=1}^N c_n B(T_0, T_n) - K\right)^+$$

equals

$$\pi(t) = \frac{1}{2\pi i} \int_{R-i\infty}^{R+i\infty} \pi(t, z) \frac{e^{z X_d(t)} \widetilde{K}^{1-z}}{z(z-1)} dz \frac{\sum_{n=1}^N c_n b(T_n) - K b(T_0)}{a(t) + b(t) e^{X_d(t)}} \tag{14.92}$$

$$= \frac{1}{\pi} \int_0^\infty \mathrm{Re}\left(\pi(t, R + iu) \frac{e^{(R+iu) X_d(t)} \widetilde{K}^{1-R-iu}}{(R+iu)(R-1+iu)}\right) du \frac{\sum_{n=1}^N c_n b(T_n) - K b(T_0)}{a(t) + b(t) e^{X_d(t)}} \tag{14.93}$$

for $R > 1$, where

$$\pi(t, z) := \exp\left(\Psi_0(t, T_0, -iz) + \Psi_X(t, T_0, -iz)^\top X(t)\right)$$

and

$$\widetilde{K} := \frac{K a(T_0) - \sum_{n=1}^N c_n a(T_n)}{\sum_{n=1}^N c_n b(T_n) - K b(T_0)}.$$

The price of the corresponding put option is obtained for $R < 0$.

Idea By Rule 14.32 and (14.86, 14.87) the time-t price of the claim H equals

$$\begin{aligned}
\pi(t) &= \frac{E(H Z(T_0) | \mathscr{F}_t)}{Z(t)} \\
&= E\Bigg(\Bigg(\left(\sum_{n=1}^N c_n a(T_n) - K a(T_0)\right. \\
&\quad \left. + \left(\sum_{n=1}^N c_n b(T_n) - K b(T_0)\right) e^{X_d(T_0)}\Bigg)^+ \Bigg| \mathscr{F}_t\Bigg) \Bigg/ \left(a(t) + b(t) e^{X_d(t)}\right) \\
&= \frac{E\big(g(X_d(T_0)) \big| \mathscr{F}_t\big) \big(\sum_{n=1}^N c_n b(T_n) - K b(T_0)\big)}{a(t) + b(t) e^{X_d(t)}}
\end{aligned} \tag{14.94}$$

with $g(x) := (e^x - \widetilde{K})^+$. Recall from Example 11.26 and Proposition A.9(1) that

$$g(x) = \frac{1}{2\pi i}\int_{R-i\infty}^{R+i\infty} e^{zx}\frac{\widetilde{K}^{1-z}}{z(z-1)}dz$$

for $R > 1$. Since $E\big(\exp(zX_d(T_0))\big|\mathscr{F}_t\big) = \pi(t,z)\exp(zX_d(t))$ by (14.91), linearity of the expected value and (14.94) yield (14.92, 14.93), cf. Rule 11.31. □

Prices of calls and puts on a T_1-bond are immediately obtained from the previous theorem by inserting $N = 1, c_1 = 1$. Caps and floors are reduced to put resp. call options on bonds as explained in Sect. 14.1.2. The swaption in Sect. 14.1.2.7 is obtained by choosing the parameters in Rule 14.37 as $c_1 = \cdots = c_{N-1} = \ell\delta$, $c_N = 1 + \ell\delta$, $K = 1$, and $R < 0$ for the put.

If X_d is a Gaussian process, we obtain Black–Scholes type formulae in a broad sense for options as in Rule 14.37. In contrast to similar statements in the previous approaches, this also extends to swaptions.

Example 14.38 (Gaussian Case) Suppose that X_d is Gaussian, which implies that $\Psi = (\Psi_0, \Psi_X)$ in (14.91) is a quadratic function of its third argument u. Since $\Psi(t, T, 0) = 0$ and e^{X_d} is a martingale, we have $\pi(t, 0) = 1 = \pi(t, 1)$ and hence

$$\pi(t,z) = \exp\left(\frac{\gamma^2}{2}z(z-1)\right)$$

for some parameter $\gamma > 0$ that generally depends on t, T_0, $X(t)$. Using Lemma A.10 we obtain

$$\pi(t) = \frac{\sum_{n=1}^N c_n b(T_n) - Kb(T_0)}{a(t) + b(t)e^{X_d(t)}} \times \left(e^{X_d(t)}\Phi\left(\frac{\log(e^{X_d(t)}/\widetilde{K}) + \gamma^2/2}{\gamma}\right) - \widetilde{K}\Phi\left(\frac{\log(e^{X_d(t)}/\widetilde{K}) - \gamma^2/2}{\gamma}\right)\right)$$

for the call in Rule 14.37. Similarly, we have

$$\pi(t) = \frac{\sum_{n=1}^N c_n b(T_n) - Kb(T_0)}{a(t) + b(t)e^{X_d(t)}} \times \left(\widetilde{K}\Phi\left(\frac{\log(\widetilde{K}/e^{X_d(t)}) + \gamma^2/2}{\gamma}\right) - e^{X_d(t)}\Phi\left(\frac{\log(\widetilde{K}/e^{X_d(t)}) - \gamma^2/2}{\gamma}\right)\right),$$

for the corresponding put.

14.5.4 Examples

No ODEs must be solved if the driving affine process is a time-inhomogeneous Lévy process.

Example 14.39 (Lévy-Driven Rational Model) Suppose that $d = 1$ and X in the previous section is a possibly time-inhomogeneous Lévy process with characteristic exponent $\psi(t, u)$. The martingale condition on X reads as $\psi(t, -i) = 0$ for $t \geq 0$. Moreover, the function $\Psi = (\Psi_0, \Psi_X)$ in (14.91) is given by $\Psi_X = 0$ and

$$\Psi_0(t, T, u) = \int_t^T \psi(s, u)ds.$$

Consequently,

$$\pi(t, z) = \exp\left(\int_t^{T_0} \psi(s, -iz)ds\right).$$

Of course, this reduces further to $\pi(t, z) = \exp((T_0 - t)\psi(-iz))$ if X is a time-homogeneous Lévy process with characteristic exponent $\psi(u)$.

If the Lévy process in the previous example is chosen to be Brownian motion, one obtains Black–Scholes type formulae, as was shown in Example 14.38.

Example 14.40 (Lognormal Rational Model) For geometric Brownian motion with

$$X(t) = \sigma W(t) - \frac{\sigma^2}{2}t,$$

a standard Wiener process W and $\sigma > 0$, we have

$$\pi(t, z) = \exp\left(\frac{(T_0 - t)\sigma^2}{2}z(z - 1)\right),$$

i.e. we are in the Gaussian setup of Example 14.38 and obtain Black–Scholes type formulae for call and put options on portfolios of bonds.

Note, however, that bonds and forward processes are not lognormally distributed in this setup, which means that it is incompatible with the Gaussian models in Sects. 14.2–14.4. It is also incompatible with the lognormal LIBOR market model discussed in Example 14.30 because LIBOR rates are not lognormally distributed either in this lognormal rational model.

14.5.5 Profile

We turn to the profile of the Flesaker–Hughston approach.

Primary processes The Flesaker–Hughston approach does not require or allow any interest rate or asset price process to be modelled in the first place. In that sense it has no primary process.

Necessary input The only necessary input on the way to a completely specified term structure model is the state-price density process Z, which should be a supermartingale if one wants to obtain nonnegative interest rates.

Discrete vs. continuous tenor Although the short rate occurs naturally in the drift of the state-price density process and bond prices for any maturity can be derived, the approach does not require that either of them is modelled. In that sense the Flesaker–Hughston approach can be viewed as both a discrete and continuous tenor setup.

Risk-neutral modelling The Flesaker–Hughston approach focuses on modelling under the real-world measure P. Measure changes and risk-neutral measures do not occur explicitly.

Stationarity Rational models are typically nonstationary as may be guessed from the bond price formula (14.87), which depends explicitly on t rather than only $T - t$.

Tractability If one chooses a rational model based on an affine process, explicit formulas are available for bond prices and hence also the short rate, forward rates, swaps etc. Calls and puts on portfolios of bonds can be represented semiexplicitly by integral transform methods. In contrast to the previous approaches, this generally leads to integral representations not only for caps and floors, but also for swaptions.

14.6 The Linear-Rational Model Approach

The generic instationarity of Flesaker–Hughston models may seem somewhat unnatural. Indeed, why should the dynamics of the short rate change deterministically in the future? The linear-rational modelling approach by [110] avoids this drawback, but it preserves the unique advantage that explicit or at least semiexplicit expressions exist for bond prices, caps, floors, and swaptions.

14.6.1 The State-Price Density

We work with the same general setup as in Sect. 14.5.1. But we replace (14.86) by

$$Z(t) = e^{-\alpha t}(\varphi + X_d(t)), \quad t \geq 0 \tag{14.95}$$

with constants $\alpha, \varphi \in \mathbb{R}$ and an $\mathbb{R}^d$-valued semimartingale X such that Z is strictly positive. More specifically, we assume the drift rate of X to be of affine form

$$a^X(t) = \beta^{(0)} + \beta X(t) \tag{14.96}$$

with some $\beta^{(0)} \in \mathbb{R}^d$Âň and $\beta \in \mathbb{R}^{d \times d}$.

Similarly as in Rule 14.36 we obtain a *rational* expression for bond prices, but this time *linear* rather than exponential in X_d.

Rule 14.41 *In the linear-rational model (14.95, 14.96) the price process of a T-bond is given by*

$$B(t,T) = e^{-\alpha(T-t)} \frac{\varphi + \left(e^{\beta(T-t)} \int_0^{T-t} e^{-\beta s} \beta^{(0)} ds + e^{\beta(T-t)} X(t)\right)_d}{\varphi + X_d(t)}, \quad t \le T. \tag{14.97}$$

Moreover, we have

$$f(t,S,T) = \alpha + \frac{1}{(T-S)} \log \frac{\varphi + \left(e^{\beta(S-t)} \int_0^{S-t} e^{-\beta s} \beta^{(0)} ds + e^{\beta(S-t)} X(t)\right)_d}{\varphi + \left(e^{\beta(T-t)} \int_0^{T-t} e^{-\beta s} \beta^{(0)} ds + e^{\beta(T-t)} X(t)\right)_d}$$

for $t \le S < T$,

$$\begin{aligned} f(t,T) &= r(t,T-t) \\ &= \alpha - \frac{\left(\beta^{(0)} + \beta\left(e^{\beta(T-t)} \int_0^{T-t} e^{-\beta s} \beta^{(0)} ds + e^{\beta(T-t)} X(t)\right)\right)_d}{\varphi + \left(e^{\beta(T-t)} \int_0^{T-t} e^{-\beta s} \beta^{(0)} ds + e^{\beta(T-t)} X(t)\right)_d}, \quad t \le T, \\ r(t) &= \alpha - \frac{\left(\beta^{(0)} + \beta X(t)\right)_d}{\varphi + X_d(t)}, \quad t \ge 0. \end{aligned} \tag{14.98}$$

Idea Setting $g(s) = E(X(s)|\mathscr{F}_t)$ for $s \ge t$, (14.96) yields

$$g(T) = g(t) + \int_t^T \left(\beta^{(0)} + \beta g(s)\right) ds.$$

This integral equation is solved by

$$g(T) = e^{\beta(T-t)} \int_0^{T-t} e^{-\beta s} \beta^{(0)} ds + e^{\beta(T-t)} g(t),$$

which in turn implies

$$E(X_d(T)|\mathscr{F}_t) = \left(e^{\beta(T-t)}\int_0^{T-t} e^{-\beta s}\beta^{(0)}ds + e^{\beta(T-t)}X(t)\right)_d.$$

Hence (14.97) follows from Rule 14.32. The remaining equations are derived as in Rule 14.36. □

Dynamic equations for bond prices and interest rates can be derived by applying Itō's formula to (14.97–14.98)

As in the Flesaker–Hughston approach, zero bonds and even the short rate appear as secondary processes, which means that their current values do not appear as input to the model. In order to reproduce the observed initial yield curve and possibly also cap and swaption prices, model parameters must be calibrated accordingly.

14.6.2 Derivatives

We obtain semiexplicit formulas for derivative prices if we assume the $\mathbb{R}^d$-valued semimartingale X in (14.95) to be an affine or at least time-inhomogeneous affine process, which we do in this and the following section. Recall from their definition that the drift rate is generally of the form (14.96) for affine processes. By Chap. 6 there is a function $\Psi = (\Psi_0, \Psi_X) : \mathbb{R}_+ \times \mathbb{R}_+ \times U \to \mathbb{C}^{1+d}$ satisfying

$$E\big(\exp(iu^\top X(T))\big|\mathscr{F}_t\big) = \exp\Big(\Psi_0(t,T,u) + \Psi_X(t,T,u)^\top X(t)\Big) \tag{14.99}$$

for $u \in U \subset \mathbb{C}^d$, $0 \leq t \leq T$. It is obtained by solving the ODEs in Theorems 6.6 resp. 6.23, cf. Sect. 14.6.3 for simple examples. The size of the set $U \subset \mathbb{C}^d$ depends on X, cf. Proposition A.5(4) and Theorem 6.10.

As usual, this function is obtained by solving the ODEs in Theorems 6.6 resp. 6.23, for simple examples cf. Sect. 14.6.3 below. Pricing formulas for bond options, caps, floors, and swaptions are obtained as special cases of the following result.

Rule 14.42 (Call/Put on a Bond Portfolio) *Let $t \leq T_0 \leq \dots \leq T_N$ and $c_1, \dots, c_N, K \in \mathbb{R}_+$ such that*

$$\widetilde{K} := Ke^{-\alpha T_0}\varphi - \sum_{n=1}^N c_n e^{-\alpha T_n}\left(\varphi + \left(e^{\beta(T_n-T_0)}\int_0^{T_n-T_0} e^{-\beta s}\beta^{(0)}ds\right)_d\right) > 0.$$

Then the time-t price of a call with time-T_0 payoff

$$H = \left(\sum_{n=1}^N c_n B(T_0, T_n) - K\right)^+$$

equals

$$\pi(t) = \frac{1}{2\pi i Z(t)} \int_{R-i\infty}^{R+i\infty} \pi(t, zu) \frac{e^{-z\widetilde{K}}}{z^2} dz$$
$$= \frac{1}{\pi Z(t)} \int_0^\infty \mathrm{Re}\left(\pi(t, (R+iy)u) \frac{e^{-(R+iy)\widetilde{K}}}{(R+iy)^2}\right) dy \tag{14.100}$$

for $R > 1$, where

$$u := \left(\sum_{n=1}^N c_n e^{-\alpha T_n} (e^{\beta(T_n - T_0)})_{dj} - K e^{-\alpha T_0} 1_{\{j=d\}}\right)_{j=1,\dots,d} \in \mathbb{R}^d,$$
$$\pi(t, z) := \exp\big(\Psi_0(t, T_0, -iz) + \Psi_X(t, T_0, -iz)^\top X(t)\big). \tag{14.101}$$

The price of the corresponding put option is obtained for $R < 0$.

Idea By Rule 14.32 and (14.95, 14.97), the time-t price of the claim H equals

$$\pi(t) = \frac{E(HZ(T_0)|\mathscr{F}_t)}{Z(t)}$$
$$= \frac{1}{Z(t)} E\Bigg(\Bigg(\sum_{n=0}^N c_n e^{-\alpha T_n}$$
$$\times \left(\varphi + \left(e^{\beta(T_n - T_0)} \int_0^{T_n - T_0} e^{-\beta s} \beta^{(0)} ds + e^{\beta(T_n - T_0)} X(T_0)\right)_d\right)\Bigg)^+ \Bigg| \mathscr{F}_t\Bigg)$$
$$= \frac{E\big(g(u^\top X(T_0))\big|\mathscr{F}_t\big)}{Z(t)} \tag{14.102}$$

with $g(x) = (x - \widetilde{K})^+$ and $c_0 = -K$. Recall from Example 11.29 and Proposition A.9(1) that

$$g(x) = \frac{1}{2\pi i} \int_{R-i\infty}^{R+i\infty} \frac{e^{-z\widetilde{K}}}{z^2} e^{zx} dz$$

for $R > 1$. Moreover, $E(\exp(zu^\top X(T_0))|\mathscr{F}_t) = \pi(t, zu)$ by (14.99). Linearity of the expected value and (14.102) yield (14.100), cf. Rule 11.31. □

Prices of calls and puts on a T_1-bond are computed from Rule 14.42 by inserting $N = 1, c_1 = 1$. Caps and floors are reduced to put resp. call options on bonds as explained in Sect. 14.1.2. The price of the swaption in Sect. 14.1.2.7 is obtained by choosing the parameters in Rule 14.42 as $c_1 = \cdots = c_{N-1} = \ell\delta$, $c_N = 1 + \ell\delta$, $K = 1$, and $R < 0$ for the put.

14.6.3 Examples

It remains to specify the affine process X in order to end up with a concrete model. We consider two simple univariate cases. The strict positivity of (14.95) is satisfied for any $\varphi > 0$ because both examples are nonnegative processes.

Example 14.43 (Square-Root Process) If we choose X to be a square-root process

$$dX(t) = (\beta^{(0)} + \beta X(t))dt + \sigma\sqrt{X(t)}dW(t),$$

we end up with the univariate case of the linear-rational square-root model in [110, Section 4]. The function $\Psi = (\Psi_0, \Psi_X)$ in (14.99) is given by

$$\Psi_0(t, T, u) = -\frac{2\beta^{(0)}}{\sigma^2}\log\left(1 + \frac{\sigma^2 iu}{2\beta}\left(1 - e^{\beta(T-t)}\right)\right),$$

$$\Psi_X(t, T, u) = \frac{iue^{\beta(T-t)}}{1 + \frac{\sigma^2 iu}{2\beta}(1 - e^{\beta(T-t)})},$$

cf. Example 6.9.

Example 14.44 (Lévy-Driven Ornstein–Uhlenbeck Process) In the case of a Lévy-driven OU process

$$dX(t) = (\beta^{(0)} + \beta X(t-))dt + dL(t)$$

with subordinator L, Proposition 3.55 yields

$$\Psi_0(t, T, u) = \Theta(u) - \Theta(ue^{\beta(T-t)}),$$

$$\Psi_X(t, T, u) = iue^{\beta(T-t)}$$

for the functions in (14.99), where ψ^L denotes the exponent of L and

$$\Theta(u) := -\frac{1}{\beta}\int_0^u \frac{\psi^L(x)}{x}dx - iu\frac{\beta^{(0)}}{\beta}.$$

Recall from Examples 3.57 and 3.58 that Θ can be expressed in closed form if X is a Gamma or an inverse-Gaussian OU process.

14.6.4 Profile

Finally, we discuss the profile of linear-rational models.

Primary processes As with the Flesaker–Hughston approach, there is no primary process in the sense that linear-rational models do not require or allow any interest rate or asset price process to be modelled in the first place.

Necessary input As in the Flesaker–Hughston case, the only necessary input on the way to a completely specified term structure model is the state-price density process Z, which is given by the parameters α, φ and the affine process X.

Discrete vs. continuous tenor The same comment as for the Flesaker–Hughston approach applies here.

Risk-neutral modelling Again, the situation is the same as for the Flesaker–Hughston approach.

Stationarity In contrast to the related Flesaker–Hughston approach, nontrivial stationary linear-rational models can be specified easily, namely by choosing a stationary process X.

Tractability Explicit formulas are available for bond prices and hence for the short rate, forward rates, and swaps as well. As in the Flesaker–Hughston case with affine driver X, semiexplicit integral expressions exist not only for bond options, caps and floors, but also for swaptions.

Appendix 1: Problems

The exercises correspond to the section with the same number.

14.1 Consider a *consol bond*, i.e. an asset paying 1€ at times $T = 1, 2, 3, \dots$ Express its fair price $\pi(t)$ at time t in terms of zero coupon bonds $B(t, T)$, $T = 1, 2, 3, \dots$ with $T \geq t$. Moreover, compute $\pi(0)$ in terms of the simple rate $\ell := L(0, T-1, T)$ provided that the latter is the same for $T = 1, 2, 3, \dots$

14.2 Derive the forward rate dynamics (14.17) for the short-rate resp. factor models in Examples 14.5–14.10.

14.3 Reference [131] consider a HJM model of the form

$$df(t,T) = -Y(t)e^{-\lambda(T-t)}\frac{1-e^{-\lambda(T-t)}}{\lambda}dt + \sqrt{Y(t)}e^{-\lambda(T-t)}dW(t), \tag{14.103}$$

$$dY(t) = (m - \mu Y(t))dt + \beta\sqrt{Y(t)}d\widetilde{W}(t) \tag{14.104}$$

with parameters $\lambda, m, \mu, \beta > 0$ and Q_0-Wiener processes W, $\widetilde{W}$ with correlation ϱ. This is inspired by the Hull–White model of Examples 14.7 and 14.19 where $Y(t)$ is constant, cf. Problem 14.2. Their motivation is to come up with a setup such that the set of conceivable forward rate curves $f(t, \cdot)$ for $t > 0$ is not rigidly linked to the initial curve $f(0, \cdot)$.

Our goal is to derive semiexplicit formulas for call and put options in this model. To this end, verify that (14.103) is of the form (14.47) with deterministic $\sigma(t, T)$ relative to the three-dimensional time-inhomogeneous affine process $X = (r, \xi, Y)$ given by

$$\xi(t) := \int_0^t Y(s)e^{-2\lambda(t-s)}ds.$$

Use this fact to express the function π in (14.69) in terms of the solution to some system of ODEs. As a side remark, note that (14.103, 14.104) correspond to an affine factor model in the sense of Sect. 14.2.1.

Hint: Use (14.47) in order to obtain

$$f(t, T) = f(0, T) + e^{-\lambda(T-t)}\big(r(t) - f(0, t)\big) + \xi(t)\frac{e^{-\lambda(T-t)} - e^{-2\lambda(T-t)}}{\lambda}$$

and hence $\dot{f}(t, t)$ for (14.52).

14.4 Suppose that X is a univariate Wiener process and $\sigma_1, \dots, \sigma_N$ are constant in the forward process dynamics (14.70). Determine call and put option prices as in Example 14.28 explicitly as well as Jamshidian's representation of the swaption as a portfolio of bonds.

14.5 Determine the Itō process representation of the short rate in the lognormal rational model of Example 14.40. Does it correspond to an affine short-rate model in the sense of Sect. 14.2.1?

14.6 Determine $\pi(t, z)$ in (14.101) in closed form if X in Example 14.44 is a Gamma-OU or an inverse-Gaussian OU process. Does this setup correspond to an affine short-rate model in the sense of Sect. 14.2.1?

Appendix 2: Notes and Comments

Titles devoted to or covering interest rate theory include [29, 42, 46, 105, 129, 223, 296]. For credit risk modelling, which is not treated in our introductory text, we refer also to [26, 62, 262].

Sections 14.1 and 14.2 owe much to the classical expositions of [29, 30, 223, 279]. The main difference is that we allow for general semimartingales and affine processes instead of Itō processes resp. continuous affine processes in the standard textbook literature. Jamshidian's representation goes back to [156].

The Heath–Jarrow–Morton approach is due to [140]. It is extended to Lévy and other driving processes in [33, 34, 83, 84, 86, 89, 159, 274] and elsewhere. The completeness result in Sect. 14.3.3 is taken from [94]. For questions of existence and uniqueness of solutions to the HJM term structure equations we refer to [108] and the literature therein.

Section 14.4 is motivated by [85, 87, 193], where the driving semimartingale X is a time-inhomogeneous Lévy process. As is pointed out in Example 14.30, it differs in spirit from the LIBOR market model approach of [39, 157, 158, 221]. The affine LIBOR model of [190] is an instance of the affine forward process model in Sect. 14.4.4, where the integrands $\sigma(\cdot, n)$ are chosen in a specific way to ensure positive interest rates. For an overview of different approaches to LIBOR modelling we refer to [124, 229].

Section 14.5 and in particular Sect. 14.5.2 is an affine extension of the rational lognormal model in [223, Section 16.5], cf. also [113, 247, 255] in this context. Black–Scholes type formulae for the Gaussian case as in Example 14.38 can be found in [223, Propositions 16.5.2 and 16.5.3]. Note that the Lévy extension of the Flesaker–Hughston approach in Example 14.39 differs from the one in [44]. Linear-rational models are introduced in [110], cf. also [111, 224] for related approaches.

Appendix A

A.1 Some Comments on Measure-Theoretic Probability Theory

This book can only be understood with a solid background in measure-theoretic probability theory, as is provided, for example, in the textbooks [16, 17, 41, 153, 192, 275] and many others. Nevertheless, we recall the intuition of some basic notions for the convenience of the reader.

The starting point in measure theory is a set Ω to whose subsets $F \subset \Omega$ numbers $\mu(F)$ are assigned in an additive way. The quantity $\mu(F)$ represents some kind of volume, surface, length, weight or, in the context of probability theory, probability of the set F. Problems arise if one tries to define $\mu(F)$ for all subsets $F \subset \Omega$ in a consistent manner. They can be avoided by restricting the domain of the **measure** μ to some σ-**field** $\mathscr{F}$ on Ω, i.e. a collection of subsets of Ω which is closed under countable set operations as unions, intersections, complements and differences.

The **Borel-σ-field** $\mathscr{B}$ resp. $\mathscr{B}^d$ is the standard σ-field on $\mathbb{R}$ resp. $\mathbb{R}^d$. It contains all sets of any practical relevance. The **Lebesgue measure** λ on $\mathbb{R}^d$ with σ-field $\mathscr{B}^d$ represents the ordinary d-dimensional volume, i.e. in particular the length of a set for $d = 1$. Formally, the Borel-σ-field is the smallest σ-field containing all open sets. As such, it can be defined for any topological space E, in which case we write $\mathscr{B}(E)$ for this collection of **Borel sets**.

A **measure space** $(\Omega, \mathscr{F}, \mu)$ consists of a **measurable space** $(\Omega, \mathscr{F})$, i.e. a set Ω with a σ-field $\mathscr{F}$ on Ω, together with a measure μ defined on $\mathscr{F}$. If $(\Omega, \mathscr{F})$, $(\Omega', \mathscr{F}')$ denote measurable spaces, a mapping $f : \Omega \to \Omega'$ is called **measurable** or, more precisely, $\mathscr{F}$-$\mathscr{F}'$-**measurable** if the set $\{f \in F'\} := \{\omega \in \Omega : f(\omega) \in F'\} = f^{-1}(F')$ belongs to $\mathscr{F}$ for any $F' \in \mathscr{F}'$. This is a very modest requirement if $(\Omega, \mathscr{F}) = (\mathbb{R}^d, \mathscr{B}^d)$, in which case f is called **Borel-measurable**.

Measurable mappings $f : \Omega \to \Omega'$ on a measure space $(\Omega, \mathscr{F}, \mu)$ can be used to carry the measure μ from $(\Omega, \mathscr{F})$ to $(\Omega', \mathscr{F}')$, namely via $\mu'(F') := \mu(f^{-1}(F'))$, $F' \in \mathscr{F}'$. The measure μ' on $(\Omega', \mathscr{F}')$ is called the **push-forward measure** of

© Springer Nature Switzerland AG 2019

E. Eberlein, J. Kallsen, *Mathematical Finance*, Springer Finance,
https://doi.org/10.1007/978-3-030-26106-1

μ relative to f. If $f_i : \Omega \to \Omega_i$, $i = 1, \ldots, n$ are mappings with values in measurable spaces $(\Omega_i, \mathscr{F}_i)$, the **generated σ-field** $\sigma(f_1, \ldots, f_n)$ on Ω is defined as the smallest one such that all the mappings $f_1, \ldots, f_n$ are measurable.

Real-valued measurable mappings $f : \Omega \to \mathbb{R}$ on a measure space $(\Omega, \mathscr{F}, \mu)$ can be integrated relative to μ. The **(Lebesgue) integral** $\int f d\mu = \int f(\omega)\mu(d\omega)$ represents a weighted "sum" of the values of f, where the weights are given by the measure μ. We have

$$\int f d\mu = \sum_{i=1}^{n} x_i \mu(f^{-1}(\{x_i\}))$$

if f has only finitely many values $x_1, \ldots, x_n$. For $f : \mathbb{R}^d \to \mathbb{R}$ and Lebesgue measure λ, the above Lebesgue integral coincides with the ordinary Riemann integral if the latter exists, i.e. $\int f d\lambda = \int f(x)dx$. Integrals on a set $F \subset \Omega$ are defined as $\int_F f d\mu := \int f 1_F d\mu$ where $1_F : \Omega \to \mathbb{R}$ with $1_F(\omega) := 1$ for $\omega \in F$ and $1_F(\omega) := 0$ for $\omega \notin F$ denotes the **indicator** of the set F. The **positive** and **negative part** of f are defined as $f^+ := f \vee 0$ and $f^- := (-f) \vee 0$, where we write $a \vee b := \max(a, b)$ as usual.

A measure ν on $(\Omega, \mathscr{F})$ has a **density** $\varrho : \Omega \to \mathbb{R}$ relative to a measure μ on the same space if $\nu(F) = \int_F \varrho d\mu$, $F \in \mathscr{F}$. The density is typically denoted as $\frac{d\nu}{d\mu} := \varrho$. In this case we have

$$\int f d\nu = \int f \frac{d\nu}{d\mu} d\mu.$$

In particular, we can compute integrals relative to some measure μ on $(\mathbb{R}^d, \mathscr{B}^d)$ via

$$\int f d\mu = \int f(x)\varrho(x)dx$$

if μ has **Lebesgue density** ϱ, i.e. a density $\varrho := \frac{d\mu}{d\lambda}$ with respect to Lebesgue measure λ. The Cartesian product $\Omega_1 \times \cdots \times \Omega_n$ of measure spaces $(\Omega_i, \mathscr{F}_i, \mu_i)$, $i = 1, \ldots, n$ can be naturally endowed with a **product-σ-field** $\otimes_{i=1}^n \mathscr{F}_i$ and a **product measure** $\otimes_{i=1}^n \mu_i$. The latter is uniquely determined by the property

$$\left(\bigotimes_{i=1}^{n} \mu_i\right)(F_1 \times \cdots \times F_n) = \prod_{i=1}^{n} \mu_i(F_i)$$

for $F_i \in \mathscr{F}_i$, $i = 1, \ldots, n$. For example, the Lebesgue measure on $\mathbb{R}^d$ can be viewed as the d-fold product measure of the Lebesgue measure on $\mathbb{R}$. By **kernels** from $(\Omega, \mathscr{F})$ to $(\Omega', \mathscr{F}')$ we denote in this text mappings $K : \Omega \times \mathscr{F}' \to \mathbb{R}_+$ which are measurable in the first argument and a measure in the second.

The intuitive concept of probability has found a rigorous mathematical foundation by putting it in a measure-theoretic context. But for historical reasons the terminology differs slightly in spite of the mathematical conformity. A **probability**

space $(\Omega, \mathscr{F}, P)$ is a measure space whose measure P is a **probability measure**, i.e. it has total mass $P(\Omega) = 1$. Such measures are also known as **(probability) distributions** or **laws**. The sample space Ω contains all possible outcomes of the random experiment. The elements of $\mathscr{F}$ are called **events**. Often the space $(\Omega, \mathscr{F}, P)$ is not specified explicitly because only some of its properties are needed.

Measurable mappings on a probability space are called **random variables** and they are typically denoted by uppercase letters $X, Y, Z, \ldots$. The **law** or **distribution** $\mathscr{L}(X) = P^X$ of a random variable X is the push-forward measure of P relative to X. Moreover, the **expected value** or **expectation** of a real-valued random variable X is the Lebesgue integral

$$E(X) := \int X dP,$$

i.e. the weighted average of the values of X. The random variable X is called **integrable** resp. **square-integrable** if $E(|X|) < \infty$ resp. $E(X^2) < \infty$. The **variance**

$$\mathrm{Var}(X) := E\big((X - E(X))^2\big) = E(X^2) - E(X)^2$$

quantifies the deviation of X from its mean. In the $\mathbb{R}^d$-valued case, $E(X)$ and $\mathrm{Var}(X)$ are replaced with the vector $E(X) := (E(X_1), \ldots, E(X_d)) \in \mathbb{R}^d$ and the **covariance matrix** $\mathrm{Cov}(X) := (\mathrm{Cov}(X_i, X_j))_{i,j=1,\ldots,d} \in \mathbb{R}^{d\times d}$, where the **covariance** of two real-valued random variables U, V is defined as

$$\mathrm{Cov}(U, V) := E\big((U - E(U))(V - E(V))\big) = E(UV) - E(U)E(V).$$

(Stochastic) independence of events, σ-fields or random variables loosely means that probabilities can be multiplied, e.g.

$$P(F_1 \cap \cdots \cap F_n) = \prod_{i=1}^{n} P(F_i) \tag{A.1}$$

for **independent events** $F_1, \ldots, F_n$. Similarly, (A.1) holds for any $F_i \in \mathscr{F}_i$, $i = 1, \ldots, n$ if $\mathscr{F}_1, \ldots, \mathscr{F}_n \subset \mathscr{F}$ are **independent** σ**-fields**. Finally, $X_1, \ldots, X_n$ are called **independent** if their generated σ-fields $\sigma(X_1), \ldots, \sigma(X_n)$ are independent. The law $\mathscr{L}(X_1 + X_2) = P^{X_1+X_2}$ of the sum $X_1 + X_2$ of two random variables X_1, X_2 is called the **convolution** of P^{X_1}, P^{X_2} and written as $P^{X_1} * P^{X_2}$.

The product space $(\prod_{i=1}^n \Omega_i, \otimes_{i=1}^n \mathscr{F}_i, \otimes_{i=1}^n P_i)$ of probability spaces $(\Omega_i, \mathscr{F}_i, P_i)$, $i = 1, \ldots, n$ corresponds to a combination of the random experiments in an independent way, i.e. such that outcomes of the separate experiments do not influence each other.

A.2 Essential Supremum

In the context of optimal control we need to consider suprema of possibly uncountably many random variables. To this end, let $\mathscr{G}$ denote a sub-σ-field of the original σ-field $\mathscr{F}$ and $(X_i)_{i\in I}$ a family of $\mathscr{G}$-measurable random variables with values in $[-\infty, \infty]$.

Definition A.1 A $\mathscr{G}$-measurable random variable Y is called the **essential supremum** of $(X_i)_{i\in I}$ if $Y \geq X_i$ almost surely for any $i \in I$ and $Y \leq Z$ almost surely for any $\mathscr{G}$-measurable random variable that satisfies $Z \geq X_i$ almost surely for any $i \in I$. We write $Y =: \operatorname{ess\,sup}_{i\in I} X_i$.

Proposition A.2

1. *The essential supremum exists. It is unique in the sense that any two random variables $Y, \widetilde{Y}$ as in Definition A.1 coincide almost surely.*
2. *There is a countable subset $J \subset I$ such that* $\operatorname{ess\,sup}_{i\in J} X_i = \operatorname{ess\,sup}_{i\in I} X_i$.

Proof By considering $\tilde{X}_i := \arctan X_i$ instead of X_i, we may assume w.l.o.g. that the X_i all have values in the same bounded interval.

Observe that $\sup_{i\in C} X_i$ is a $\mathscr{G}$-measurable random variable for any countable subset C of I. We denote by $\mathscr{C}$ the set of all such countable subsets of I. Consider a sequence $(C_n)_{n\in\mathbb{N}}$ in $\mathscr{C}$ such that $E(\sup_{i\in C_n} X_i) \uparrow \sup_{C\in\mathscr{C}} E(\sup_{i\in C} X_i)$. Then $C_\infty := \cup_{n\in\mathbb{N}} C_n$ is a countable subset of I satisfying $E(\sup_{i\in C_\infty} X_i) = \sup_{C\in\mathscr{C}} E(\sup_{i\in C} X_i)$. We show that $Y := \sup_{i\in C_\infty} X_i$ meets the requirements of an essential supremum. Indeed, for fixed $i \in I$ we have $Y \leq Y \vee X_i$ and $E(Y) = E(Y \vee X_i) < \infty$, which implies that $Y = Y \vee X_i$ and hence $X_i \leq Y$ almost surely.

On the other hand, we have $Z \geq X_i$, $i \in C_\infty$ and hence $Z \geq Y$ almost surely for any Z as in the definition. □

The following result helps to approximate the essential supremum by a sequence of random variables, cf. [203].

Lemma A.3 *Suppose that $(X_i)_{i\in I}$ has the* ***lattice property****, i.e. for any $i, j \in I$ there exists a $k \in I$ such that $X_i \vee X_j \leq X_k$ almost surely. Then there is a sequence $(i_n)_{n\in\mathbb{N}}$ such that $X_{i_n} \uparrow \operatorname{ess\,sup}_{i\in I} X_i$ almost surely.*

Proof Choose a sequence $(j_n)_{n\in\mathbb{N}}$ such that $J = \{j_n : n \in \mathbb{N}\}$ holds for the countable set $J \subset I$ in statement 2 of Proposition A.2. For any n choose $i_n \in I$ such that $X_{j_n} \vee X_{i_{n-1}} \leq X_{i_n}$. This implies that X_{i_n} is increasing in n and

$$\operatorname*{ess\,sup}_{i\in I} X_i = \operatorname*{ess\,sup}_{n\in\mathbb{N}} X_{j_n} \leq \operatorname*{ess\,sup}_{n\in\mathbb{N}} X_{i_n} \leq \operatorname*{ess\,sup}_{i\in I} X_i$$

as desired. □

If the lattice property holds, essential supremum and expectation can be interchanged.

Lemma A.4 *Let $X_i \geq 0$, $i \in I$ or $E(\operatorname{ess\,sup}_{i\in I} |X_i|) < \infty$. If $(X_i)_{i\in I}$ has the lattice property, then*

$$E\left(\operatorname*{ess\,sup}_{i\in I} X_i \middle| \mathscr{H}\right) = \operatorname*{ess\,sup}_{i\in I} E(X_i|\mathscr{H})$$

for any sub-σ-field $\mathscr{H}$ of $\mathscr{F}$.

Proof Since $E(\operatorname{ess\,sup}_{i\in I} X_i|\mathscr{H}) \geq E(X_j|\mathscr{H})$ a.s. for any $j \in I$, we obviously have $E(\operatorname{ess\,sup}_{i\in I} X_i|\mathscr{H}) \geq \operatorname{ess\,sup}_{i\in I} E(X_i|\mathscr{H})$. Let $(i_n)_{n\in\mathbb{N}}$ be a sequence as in the previous lemma. Then $X_{i_n} \uparrow \operatorname{ess\,sup}_{i\in I} X_i$ a.s. and monotone resp. dominated convergence imply $E(X_{i_n}|\mathscr{H}) \uparrow E(\operatorname{ess\,sup}_{i\in I} X_i|\mathscr{H})$ and hence the claim. □

A.3 Characteristic Functions

Recall that the **characteristic function** $\varphi_X : \mathbb{R}^d \to \mathbb{C}$ of an $\mathbb{R}^d$-valued random variable X is defined as

$$\varphi_X(u) := E\big(\exp(iu^\top X)\big), \quad u \in \mathbb{R}^d.$$

It coincides with the **Fourier transform** $\widehat{P}^X$ of the law of X, which is given by

$$\widehat{P}^X(u) := \int e^{iu^\top x} P^X(dx), \quad u \in \mathbb{R}^d.$$

If P^X has Lebesgue density $f : \mathbb{R}^d \to \mathbb{R}$, this in turn tallies with the **Fourier transform** $\widehat{f}$ of f defined as

$$\widehat{f}(u) := \int e^{iu^\top x} f(x)dx, \quad u \in \mathbb{R}^d.$$

The following properties make characteristic functions useful in probability theory.

Proposition A.5

1. *The characteristic function φ_X characterises the law P^X uniquely.*
2. *If $E(|X|^n) < \infty$, the characteristic function φ_X is n times continuously differentiable and we have*

$$E\left(\prod_{m=1}^n X_{j_m}\right) = (-i)^n D_{j_1\dots j_n}\varphi_X(0).$$

In particular, we have $E(X) = -i\varphi'_X(0)$ *and* $\mathrm{Var}(X) = -\varphi''_X(0) + \varphi'_X(0)^2$ *for square-integrable univariate random variables* X.

3. *Let* $n \in \mathbb{N}$, $k \in \{1, \dots, d\}$. *If* e_k *denotes the* k*-th unit vector and* $\lambda \mapsto \varphi_X(\lambda e_k)$ *is* $2n$*-times differentiable in* 0, *then* $E(|X_k|^{2n}) < \infty$.
4. *Let* $U \subset \mathbb{C}^d$ *be an open convex neighbourhood of* 0 *such that* φ_X *allows for an analytic extension to* U, *which we denote again by* φ_X. *Then* $E(e^{iu^\top X}) = \varphi_X(u)$ *for any* $u \in U$ *with* $i\mathrm{Im}(u) \in U$.
5. *Let* $A \in \mathbb{R}^{n\times d}$ *and* $b \in \mathbb{R}^n$. *The characteristic function of the* $\mathbb{R}^n$*-valued random variable* $Y := AX + b$ *equals*

$$\varphi_Y(u) = e^{iu^\top b}\varphi_X(A^\top u), \quad u \in \mathbb{R}^n.$$

6. *If* $X_1, \dots, X_d$ *denote independent real-valued random variables with characteristic functions* $\varphi_{X_1}, \dots, \varphi_{X_d}$, *then the random variable* $X = (X_1, \dots, X_d)$ *has characteristic function*

$$\varphi_X(u) = \prod_{k=1}^{d} \varphi_{X_k}(u_k), \quad u = (u_1, \dots, u_d) \in \mathbb{R}^d.$$

7. *If* $X_1, \dots, X_d$ *denote independent* $\mathbb{R}^d$*-valued random variables and* $S := \sum_{k=1}^{d} X_k$, *then*

$$\varphi_S(u) = \prod_{k=1}^{d} \varphi_{X_k}(u), \quad u \in \mathbb{R}^d.$$

8. ***(Lévy's continuity theorem)*** *A sequence* $(X^{(n)})_{n\in\mathbb{N}}$ *of* $\mathbb{R}^d$*-valued random variables converges in law to some* $\mathbb{R}^d$*-valued random variable* X *if and only if*

$$\lim_{n\to\infty} \varphi_{X^{(n)}}(u) = \varphi_X(u), \quad u \in \mathbb{R}^d.$$

In this case the convergence is uniform on compact sets.

Proof

1. See [153, Theorem 14.1].
2. See [80, Lemma A.1].
3. See [80, Lemma A.1].
4. This follows from [80, Lemmas A.4 and A.2] and the identity theorem for holomorphic functions.
5. See [153, Theorem 13.3].
6. See [153, Corollary 14.1].
7. See [153, Theorem 15.2].
8. See [153, Theorem 19.1]. □

In the main text we also make use of the following result.

Corollary A.6 *If $X_1, \ldots, X_n$ are random variables satisfying*

$$E\left(\prod_{k=1}^{n} \exp(iu_k X_k)\right) = \prod_{k=1}^{n} \widehat{P_k}(u_k), \quad u_1, \ldots, u_n \in \mathbb{R} \tag{A.2}$$

for some probability measures $P_1, \ldots, P_n$, then $X_1, \ldots, X_n$ are independent and have law P_k, $k = 1, \ldots, n$, respectively.

Proof The claim can be rephrased as $P^{(X_1,\ldots,X_n)} = \otimes_{k=1}^{n} P_k$, i.e. it is a property of the law of $X = (X_1, \ldots, X_n)$. If $X_1, \ldots, X_n$ are independent with laws $P_1, \ldots, P_n$, (A.2) holds by Proposition A.5(6). Proposition A.5(1) implies that the converse is true as well. □

A.4 The Bilateral Laplace Transform

The Laplace transform of a function is closely related to the Fourier transform but avoids some integrability problems in our context.

Definition A.7 Let $f : \mathbb{R} \to \mathbb{C}$ be a measurable function. The **(bilateral) Laplace transform** $\widetilde{f}$ is defined as

$$\widetilde{f}(z) = \int_{-\infty}^{\infty} f(x)e^{-zx}dx \tag{A.3}$$

for any $z \in \mathbb{C}$ such that $\int_{-\infty}^{\infty} |f(x)e^{-zx}|dx < \infty$.

If the Laplace transform exists for some real numbers, it exists on the vertical strip in $\mathbb{C}$ between these numbers:

Lemma A.8 *Suppose that $\widetilde{f}(a)$ and $\widetilde{f}(b)$ exist for real numbers $a \leq b$. Then $\widetilde{f}(z)$ exists for any $z \in \mathbb{C}$ with $a \leq \mathrm{Re}(z) \leq b$.*

Proof This is obvious because $|f(x)e^{-zx}| = |f(x)|e^{-\mathrm{Re}(z)x} \leq |f(x)e^{-ax}| + |f(x)e^{-bx}|$. □

Observe that

$$\widetilde{f}(z) = \widetilde{f}(u + iv) = \int_{-\infty}^{\infty} f(x)e^{-(u+iv)x}dx = \int_{-\infty}^{\infty} e^{ux} f(-x)e^{ixv}dx \tag{A.4}$$

for $u, v \in \mathbb{R}$ and $z = u + iv$, i.e. the Laplace transform of f in $u + iv$ coincides with the Fourier transform of $x \mapsto e^{ux} f(-x)$ in v. Therefore many properties of the bilateral Laplace transform can be deduced from results for the Fourier transform.

The function f can typically be recovered from its Laplace transform by a similar integration.

Proposition A.9 *Let $f : \mathbb{R} \to \mathbb{C}$ be measurable and suppose that the Laplace transform $\widetilde{f}(R)$ exists for $R \in \mathbb{R}$.*

1. If $v \mapsto \widetilde{f}(R+iv)$ is integrable, then $x \mapsto f(x)$ is continuous and

$$f(x) = \frac{1}{2\pi i} \int_{R-i\infty}^{R+i\infty} \widetilde{f}(z) e^{zx} dz := \frac{1}{2\pi} \int_{-\infty}^{\infty} \widetilde{f}(R+iv) e^{(R+iv)x} dv, \quad x \in \mathbb{R}.$$

2. If f is of finite variation on any compact interval, then

$$\lim_{\varepsilon \downarrow 0} \frac{f(x+\varepsilon) + f(x-\varepsilon)}{2} = \lim_{c \to \infty} \frac{1}{2\pi i} \int_{R-ic}^{R+ic} \widetilde{f}(z) e^{zx} dz \quad x \in \mathbb{R}.$$

Proof The first statement follows from [254, Theorem 9.11] and (A.4). For the second assertion see [77, Satz 4.4.1]. □

In the context of geometric Brownian motion, the following result proves to be useful.

Lemma A.10 *For $\mu \in \mathbb{R}$, $\sigma > 0$, $x \in \mathbb{R}$ we have*

$$\Phi\left(\frac{x-\mu}{\sigma}\right) = \frac{1}{2\pi i} \int_{R-i\infty}^{R+i\infty} \frac{e^{(x-\mu)z + \frac{\sigma^2}{2} z^2}}{z} dz$$

for any $R > 0$, where Φ denotes the cumulative distribution function of $N(0,1)$.

Proof
Step 1: We show that $\widetilde{f}(z) = \frac{1}{z} e^{-\mu z + \sigma^2/2}$ for $z \in \mathbb{C}$ with $\operatorname{Re}(z) > 0$ for $f(x) = \Phi(\frac{x-\mu}{\sigma})$. To this end, observe that (A.3) and integration by parts imply $\widetilde{f}(z) = \frac{1}{z} \int_{-\infty}^{\infty} \varphi_{\mu,\sigma^2}(x) e^{-zx} dx$ with φ_{μ,σ^2} being the probability density function of $N(\mu, \sigma^2)$. Now, Proposition A.5(4) yields that the integral equals $\exp(-\mu z + \sigma^2/2)$ because it can be viewed as an extension of the Fourier transform of $N(\mu, \sigma^2)$ to the complex domain.
Step 2: The assertion now follows from Proposition A.9(1). □

We briefly state the multivariate counterpart of the above.

Definition A.11 Let $f : \mathbb{R}^d \to \mathbb{C}$ be a measurable function. The **multivariate bilateral Laplace transform** $\widetilde{f}$ is defined as

$$\widetilde{f}(z) = \int f(x) e^{-z^\top x} dx \tag{A.5}$$

for any $z \in \mathbb{C}^d$ such that the integral exists.

Since $|f(x)e^{-z^\top x}| = |f(x)|e^{-\mathrm{Re}(z)^\top x}$, the existence of the Laplace transform depends only on the real part of the argument. The counterpart of Proposition A.9(1) reads as follows.

Proposition A.12 *Let $f : \mathbb{R}^d \to \mathbb{C}$ be measurable. Suppose that the Laplace transform $\widetilde{f}(R)$ exists for $R \in \mathbb{R}^d$ and that the mapping $\mathbb{R}^d \to \mathbb{C}$, $v \mapsto \widetilde{f}(R + iv)$ is integrable. Then $x \mapsto f(x)$ is continuous and*

$$f(x) = \frac{1}{(2\pi i)^d} \int_{R_1 - i\infty}^{R_1 + i\infty} \cdots \int_{R_d - i\infty}^{R_d + i\infty} \widetilde{f}(z_1, \ldots, z_d) e^{(z_1, \ldots, z_d)^\top x} dz_d \cdots dz_1$$

for $x \in \mathbb{R}^d$.

Proof This follows from the Fourier inversion formula (e.g. [99, V.3.11]) and (A.5). □

A.5 Grönwall's Inequality

Grönwall's inequality helps to obtain upper bounds for solutions or differences of solutions to differential equations, e.g. in proofs of existence and finiteness.

Proposition A.13 *For $T, a, b \geq 0$ let $f : [0, T] \to \mathbb{R}$ be a continuous function satisfying*

$$f(t) \leq a + b \int_0^t f(s) ds, \quad t \in [0, T].$$

Then we have

$$f(t) \leq a e^{bt}, \quad t \in [0, T].$$

Proof This is a special case of [100, Theorem A.5.1]. □

A.6 Convex Duality

Many results from mathematical finance rely on tools from convex and functional analysis. In particular, various forms of duality have led to deep insights. We let $U := V := \mathbb{R}^n$ for this small overview.

We first consider separation theorems which play an important role in the proof of the fundamental theorem of asset pricing. A simple version reads as follows.

Theorem A.14 (Separating Hyperplane) *Let $K, M \subset U$ with $K \cap M = \varnothing$. If K is a closed convex cone and M is a compact convex set, there exists a $y \in V$ such that $y^\top x \leq 0$ for all $x \in K$ and $y^\top x > 0$ for all $x \in M$. If K is a subspace, $y^\top x = 0$ for any $x \in K$.*

Proof This follows from [245, Corollary 11.4.2]. □

A key role in convex analysis is played by the **convex conjugate** or **Fenchel–Legendre transform** $f^* : V \to (-\infty, \infty]$ of a given, typically convex function $f : U \to (-\infty, \infty]$. It is defined as

$$f^*(y) := \sup_{x \in U} \left(y^\top x - f(x) \right). \tag{A.6}$$

The conjugate function f^* is convex and lower semicontinuous. For convex functions the conjugate of the conjugate coincides with the original function:

Theorem A.15 (Biconjugate Theorem) *If f is convex, lower semicontinuous and $f(x) < \infty$ for at least one $x \in U$, we have $f^{**} = f$.*

Proof This follows from [245, page 52 and Theorem 12.2]. □

Remark A.16 By isomorphy, we can replace $U = \mathbb{R}^n$ in the previous two theorems by any finite-dimensional space and V with its dual space of linear mappings $U \to \mathbb{R}$. But versions of this result hold in a more general context. The Hahn–Banach theorem implies that Theorem A.14 remains true if U denotes a locally convex space, V its dual space of continuous linear functionals on U and $y^\top x$ is replaced by $y(x)$, cf. e.g. [117, Theorem A.57]. Similarly, Theorem A.15 holds in locally convex spaces, cf. e.g. [117, Theorem A.62].

In the chapters involving utility maximisation, we heavily rely on duality in convex optimisation. In order to state a rigorous result, let $K \subset U$ be a **polyhedral cone**, i.e. $K = \{x \in U : \sum_{i=1}^m \lambda_i x_i : \lambda_1, \dots, \lambda_m \geq 0\}$ for some $x_1, \dots, x_m \in U$. Moreover, denote by $f : U \to [-\infty, \infty)$ an upper semicontinuous concave function which is differentiable where it is finite. The goal is to maximise $f(x)$ over K.

Recall that the **polar cone** of K is defined as

$$K^\circ := \{y \in V : y^\top x \leq 0 \text{ for any } x \in K\}$$

and define $\widetilde{f} : V \to (-\infty, \infty]$ by

$$\widetilde{f}(y) := \sup_{x \in U} (f(x) - y^\top x), \tag{A.7}$$

i.e. $\widetilde{f}(y) := (-f)^*(-y)$ in terms of the Fenchel–Legendre transform (A.6). The so-called dual problem is to minimise $\widetilde{f}(y)$ over K°. Since $(K^\circ)^\circ = K$ by [245, Theorem 14.1], one easily verifies that

$$\sup_{x \in K} f(x) = \sup_{x \in U} \inf_{y \in K^\circ} (f(x) - y^\top x) \leq \inf_{y \in K^\circ} \sup_{x \in U} (f(x) - y^\top x) = \inf_{y \in K^\circ} \widetilde{f}(y),$$

which is called **weak duality**. But often we actually have equality.

Theorem A.17 (Ordinary Convex Programme) *Suppose that* $\sup_{x \in K} f(x) \neq \infty$ *and that* f *is constant in the directions of recession of* K *and* f, *i.e. if* $x \in K$ *is such that the function* $\mathbb{R}_+ \to [-\infty, \infty)$, $\lambda \mapsto f(\widetilde{x} + \lambda x)$ *increases for any* $\widetilde{x} \in K$, *this function is constant. Then we have:*

1. ***(strong duality)***
   $\sup_{x \in K} f(x) = \inf_{y \in K^\circ} \widetilde{f}(y)$.
2. ***(existence of solutions)***
   There exist $x^\star \in K$, $y^\star \in K^\circ$ *with*

$$f(x^\star) = \sup_{x \in K} f(x),$$
$$\widetilde{f}(y^\star) = \inf_{y \in K^\circ} \widetilde{f}(y).$$

3. *The following are equivalent:*
   a. $x^\star \in K$ *and* $y^\star \in K^\circ$ *are optimal for the primal and the dual problem in statement 2.*
   b. $x^\star \in K$ *and* $y^\star \in K^\circ$ *satisfy* $f(x^\star) = \widetilde{f}(y^\star)$.
   c. ***(Karush–Kuhn–Tucker conditions)***
      $x^\star \in K$ *and* $y^\star \in K^\circ$ *satisfy the* ***complementary slackness condition*** $(y^\star)^\top x^\star = 0$ *and the* ***first-order condition*** $\nabla f(x^\star) = y^\star$.

Proof

2. K° is polyhedral by [245, Corollary 19.2.2]. By [245, Theorem 19.1] this implies that it is finitely generated, i.e. of the form $K^\circ = \{\sum_{j=1}^{\ell} \mu_j y_j : \mu_1, \ldots, \mu_\ell \geq 0\}$ for some $y_1, \ldots, y_\ell \in V$. Since $(K^\circ)^\circ = K$ by [245, Theorem 14.1], we have

$$K = \{x \in U : y^\top x \leq 0 \text{ for any } y \in K^\circ\}$$
$$= \{x \in U : y_j^\top x \leq 0 \text{ for } j = 1, \ldots, \ell\}.$$

Therefore the primal problem can be rephrased as maximising $x \mapsto f(x)$ on U subject to constraints $y_j^\top x \leq 0$ for $j = 1, \ldots, \ell$. The existence of a solution follows from [245, Theorem 27.3]. By [245, Theorem 28.2] the problem has a Kuhn–Tucker vector, say $\mu^\star = (\mu_1^\star, \ldots, \mu_\ell^\star) \in \mathbb{R}^\ell$. The latter minimises

$$\mu \mapsto g(\mu) := \sup_{x \in U} \left(f(x) - \sum_{j=1}^{\ell} \mu_j y_j^\top x \right)$$

on $\mathbb{R}_+^\ell$ by [245, Corollary 28.4.1]. This means that $y^\star := \sum_{j=1}^{\ell} \mu_j^\star y_j$ minimises $y \mapsto \widetilde{f}(y)$ on K°. Consequently, $y^\star$ is optimal for the dual problem.

1. The above and [245, Theorem 28.4] yield

$$\begin{aligned}\inf_{y \in K^\circ} \widetilde{f}(y) &= \inf_{\mu \in \mathbb{R}_+^\ell} \sup_{x \in U} \left(f(x) - \sum_{j=1}^{\ell} \mu_j y_j^\top x \right) \\ &= \sup_{x \in U} \inf_{\mu \in \mathbb{R}_+^\ell} \left(f(x) - \sum_{j=1}^{\ell} \mu_j y_j^\top x \right) \\ &= \sup_{x \in U} \inf_{y \in K^\circ} \left(f(x) - y^\top x \right).\end{aligned}$$

The last expression equals $\sup_{x \in K} f(x)$ because

$$\inf_{y \in K^\circ} \left(f(x) - y^\top x \right) = \begin{cases} f(x) & \text{for } x \in K, \\ -\infty & \text{otherwise.} \end{cases}$$

3. *a)⇔b)*: The equivalence of a) and b) follows from strong duality.

a)⇒c): By [245, Corollary 28.4.1], $\mu^\star = (\mu_1^\star, \ldots, \mu_\ell^\star) \in \mathbb{R}_+^\ell$ with $y^\star = \sum_{j=1}^{\ell} \mu_j^\star y_j$ is a Kuhn–Tucker vector for the primal problem. Due to [245, Theorem 28.3], this implies $0 = \mu_j^\star y_j^\top x^\star$, $j = 1, \ldots, \ell$ and

$$0 = -\nabla f(x^\star) + \sum_{j=1}^{\ell} \mu_j^\star y_j = -\nabla f(x^\star) + y^\star$$

as desired.

c)⇒a): For $\mu^\star = (\mu_1^\star, \ldots, \mu_\ell^\star) \in \mathbb{R}_+^\ell$ with $y^\star = \sum_{j=1}^{\ell} \mu_j^\star y_j$ we have

$$0 = -\nabla f(x^\star) + y^\star = -\nabla f(x^\star) + \sum_{j=1}^{\ell} \mu_j^\star y_j$$

and $0 = (y^\star)^\top x^\star = \sum_{j=1}^{\ell} \mu_j^\star y_j^\top x^\star$. Since all summands are nonpositive, they must actually vanish. By [245, Theorem 28.3], $x^\star$ is a solution and $y^\star$ a Kuhn–

Tucker vector for the primal problem. From the reasoning in the first part of the proof we conclude that $y^\star$ is optimal for the dual problem. □

Remark A.18

1. As for separating hyperplanes it is evident that $U = V = \mathbb{R}^n$ can be replaced by any finite-dimensional space U and its dual space V of linear functionals on U. In infinite-dimensional spaces strong duality etc. are less obvious, but they often hold in concrete situations under additional regularity, cf. e.g. [7, 95, 246].
2. If we consider V as the dual space of U via $y(x) = y^\top Ax = (Ay)^\top x$ for any $x \in U$, $y \in V$ for some regular matrix $A \in \mathbb{R}^{n\times n}$, we effectively replace y by Ay. Therefore Theorem A.17 still holds if we replace the polar cone by $K^\circ = \{x \in V : y(x) \leq 0 \text{ for any } x \in K\}$, the complementary slackness condition by

$$(y^\star)^\top Ax^\star = 0, \tag{A.8}$$

and the first-order condition by

$$\nabla f^\star(x^\star) = Ay^\star. \tag{A.9}$$

Example A.19 Suppose that Ω is finite and $U := \mathbb{R}^\Omega$ denotes the set of all mappings $\Omega \to \mathbb{R}$. Fix $v_0 \in \mathbb{R}$, a probability measure P on Ω with $P(\{\omega\}) > 0$, $\omega \in \Omega$, and a concave function $u : \mathbb{R} \to [-\infty, \infty)$, which is differentiable in the interior of $u^{-1}(\mathbb{R})$. Set $f(X) := \sum_{\omega\in\Omega} u(v_0 + X(\omega))P(\{\omega\})$ for any $X \in U$. Note that U coincides with the set of real-valued random variables on Ω and

$$f(X) = E(u(v_0 + X))$$

for $X \in U$. In the spirit of Remark A.18(2) we consider $V = U$ as the dual space of U via $Y(X) := E(XY)$ for $X \in U$, $Y \in V$. The transform $\widetilde{f}(Y) := \sup_{X\in U}(f(X) - Y(X))$ then reads as

$$\widetilde{f}(Y) = E(\widetilde{u}(Y)) + v_0 E(Y).$$

Indeed, we have

$$\begin{aligned}\widetilde{f}(Y) &= \sup_{X\in U} \big(E(u(v_0 + X)) - E(XY)\big) \\ &= E\left(\sup_{x\in\mathbb{R}} \big(u(x) - (x - v_0)Y\big)\right) \\ &= E(\widetilde{u}(Y)) + v_0 E(Y)\end{aligned}$$

as desired.

Parallel to Remarks A.16 and A.18(1), these connections can be generalised to infinite Ω and hence infinite-dimensional U. Moreover, $u(x)$ and hence also $\widetilde{u}(x)$ may depend explicitly on ω, i.e. we can consider $u : \Omega \times \mathbb{R} \to \mathbb{R}$, $(\omega, x) \mapsto u(\omega, x)$.

Finally, we state a minimax theorem which is related to strong duality in Theorem A.17. A comparative discussion of various minimax theorems can be found in [119].

Theorem A.20 *Suppose that U is a topological vector space, V a vector space, $A \subset U$ nonempty, convex, compact, and $B \subset V$ nonempty, convex. Moreover, let $f : A \times B \to \mathbb{R}$ be affine in both variables and upper semicontinuous in the first variable. Then*

$$\max_{x \in A} \inf_{y \in B} f(x, y) = \inf_{y \in B} \max_{x \in A} f(x, y). \tag{A.10}$$

Proof This is shown in [194]. □

A.7 Comments on Continuous Processes

The book is written on the basis of semimartingales allowing for jumps, which seems appropriate in view of the observed behaviour of financial data. A number of definitions, equations, and statements simplify for continuous processes, which makes the theory more accessible. In this section we summarise the major simplifications in this case.

Chapter 1 Of course, continuity in time does not have any meaning in discrete time.

Chapter 2 We have $X_- = X$ and $\Delta X = 0$ for continuous processes. Moreover, predictability coincides with adaptedness. All continuous adapted processes are locally bounded. The components M^X, A^X in (2.3) are continuous if this holds for X. Since continuous semimartingales are special, there is no need to distinguish semimartingales and special semimartingales. The only continuous Lévy processes are Brownian motions with drift. This is reflected by (2.19) where the Lévy measure K vanishes in the absence of jumps. Since Brownian motion is Gaussian, it has moments and exponential moments of any order. The only continuous subordinators are deterministic linear functions with nonnegative slope. Therefore Theorem 2.24 is essentially useless in this case. Sections 2.4.3–2.4.11 and 2.5 only make sense if one allows for jumps. Continuous Lévy processes are stable Lévy motions, simply because there is no continuous Lévy process except for Brownian motion with drift. They are strictly stable if and only if the drift or the Brownian motion part vanishes.

Chapter 3 The covariation is continuous and hence predictable for continuous processes. Therefore it coincides with the predictable covariation, i.e. $\langle X, Y \rangle = [X, Y]$. The sums in the second lines of Itō's formulas (3.19, 3.20, 3.39) vanish in the absence of jumps. Moreover, we have $\langle X^c, X^c\rangle = \langle X, X\rangle = [X, X]$. σ-martingales coincide with local martingales for continuous processes. Section 3.3 does not apply for continuous processes. All random measures in the rest of the book vanish for continuous processes, in particular in Sects. 3.4 and 3.5. The same is true for the product behind the exponential in (3.67) and for the sum in (3.68). Theorem 3.62 reduces to Theorem 3.61 without jumps, and likewise Theorems 3.67 to 3.66.

Chapter 4 The third characteristic ν resp. K vanishes for continuous processes. So we are left with the rate of quadratic covariation $c_{ij}(t) = d[X_i, X_j](t)/dt$ and the drift rate $b^h(t)$, which does not depend on h. Lemma 4.11 does not apply in the absence of jumps. The modified second characteristic coincides with the second characteristic.

Chapter 5 Apart from the fact that the jump measures K, κ etc. vanish, essentially all concepts except Theorem 5.42 make sense for continuous processes as well.

Chapter 6 The notions and statements still make sense if they are restricted to continuous processes. However, as is true for Lévy processes, the family loses much of its flexibility in this case. Exponentials of continuous affine processes are martingales if they are local martingales because the conditions on jumps trivially hold.

Chapter 7 Except for vanishing jump integrals, the theory does not change much for continuous processes. The BSDE formulation of the stochastic maximum principle is probably more natural in this case, in line with the literature in the field.

Chapter 8 Lévy copulas do not make sense for processes without jumps. The approaches of Sects. 8.2.1 and 8.2.2 coincide for continuous Lévy processes. Only the Stein–Stein and Heston models are continuous in Sects. 8.2.3–8.2.9.

Chapter 9 Example 9.2 reduces to Example 9.1 in the absence of jumps. The covariation term in (9.11) vanishes for continuous processes.

Chapter 10 Except for vanishing jump characteristics K^X, K^S etc., the results do not change much for continuous processes. However, we mentioned an unpleasant effect of a possibly signed variance-optimal martingale measure. This does not occur in the absence of jumps.

Chapter 11 PIDEs reduce to PDEs in Sect. 11.5.2 if the processes are continuous.

Chapter 12 Again, PIDEs reduce to PDEs in Sect. 12.1.2 in the absence of jumps.

Chapter 13 Except for Example 13.29 the results would essentially be presented identically for continuous processes.

Chapter 14 Except for discontinuous examples, the chapter hardly contains notions and results that are peculiar to discontinuous processes.

A.8 Links to Surprising Observations

In this short section we collect links to a few qualitative observations in mathematical finance which may seem surprising at first or even at a second glance and which are scattered in Part II of this book.

- Problem 8.1 contains an instance of the so-called Siegel's paradox. It may happen that both the Dollar grows on average relative to the Euro and vice versa. This may seem counterintuitive because only one of these two exchange rates can actually rise.
- In Rule 10.6 and Examples 10.13 and 10.19 we observe that the optimal portfolio does not depend on the investor's time horizon in a very natural setup. By contrast, one may be tempted to believe that attractive risky investments are particularly favourable in the long run. Indeed, shouldn't the law of large numbers make the drift increasingly dominate random fluctuations?
- In Example 10.8 we see that a portfolio of a bond and a stock may tend to infinity in the long run even though both assets tend to zero.
- In Example 10.9 we observe that the investor's optimal fund of risky assets does not depend on her own currency. This may seem counterintuitive because foreign stocks involve an additional exchange rate risk whereas domestic ones do not.
- Rule 10.7 states that there exists a portfolio which outperforms any other in the long run with probability 1. This may seem surprising, and even more so because this investment typically differs from the optimal portfolio in Example 10.13, which does not depend on the time horizon.
- Who would have thought that the price of a call option in the Black–Scholes model does not depend on supply of and demand for this contingent claim, cf. Example 11.4? But it may seem particularly disturbing that this unique fair price (11.28) does not depend on the stock's drift rate μ. To wit, the probability whether a positive payoff is received at all depends foremost on μ. This naturally suggests that μ should affect the option price.
- Compared to European options with the same strike and maturity, American calls offer the additional right to exercise the claim prematurely. One would naturally expect that this quality leads to an—at least slightly—higher value. Rule 11.8 shows that this is typically wrong. Absence of arbitrage implies that American and European calls must be traded at the same price.

References

1. M. Abramowitz, I. Stegun, *Handbook of Mathematical Functions with Formulas, Graphs, and Mathematical Tables*, vol. 55 (U.S. Government Printing Office, Washington, D.C., 1964)
2. D. Applebaum, *Lévy Processes and Stochastic Calculus*, 2nd edn. (Cambridge Univ. Press, Cambridge, 2009)
3. T. Arai, An extension of mean-variance hedging to the discontinuous case. Finance Stochast. **9**(1), 129–139 (2005)
4. P. Artzner, F. Delbaen, J. Eber, D. Heath, Coherent measures of risk. Math. Finance **9**(3), 203–228 (1999)
5. L. Bachelier, *Théorie de la Spéculation* (Gauthier-Villars, Paris, 1900)
6. I. Bajeux-Besnainou, R. Portait, Dynamic asset allocation in a mean-variance framework. Manag. Sci. **44**(11), S79–S95 (1998)
7. V. Barbu, T. Precupanu, *Convexity and Optimization in Banach spaces*, 4th edn. (Springer, Dordrecht, 2012)
8. G. Barles, R. Buckdahn, E. Pardoux, Backward stochastic differential equations and integral-partial differential equations. Stochastics Stochastics Rep. **60**(1–2), 57–83 (1997)
9. O. Barndorff-Nielsen, Processes of normal inverse Gaussian type. Finance Stochast. **2**(1), 41–68 (1998)
10. O. Barndorff-Nielsen, N. Shephard, Non-Gaussian Ornstein-Uhlenbeck-based models and some of their uses in financial economics. J. R. Stat. Soc. Ser. B Stat. Methodol. **63**(2), 167–241 (2001)
11. P. Barrieu, N. El Karoui, Optimal derivatives design under dynamic risk measures, in *Mathematics of Finance*, vol. 351 (Amer. Math. Soc., Providence, 2004), pp. 13–25
12. P. Barrieu, N. El Karoui, Inf-convolution of risk measures and optimal risk transfer. Finance Stochast. **9**(2), 269–298 (2005)
13. S. Basak, G. Chabakauri, Dynamic mean-variance asset allocation. Rev. Financ. Stud. **23**(8), 2970–3016 (2010)
14. B. Bassan, C. Ceci, Regularity of the value function and viscosity solutions in optimal stopping problems for general Markov processes. Stochastics Stochastics Rep. **74**(3–4), 633–649 (2002)
15. D. Bates, Jumps and stochastic volatility: Exchange rate processes implicit in deutsche mark options. Rev. Financ. Stud. **9**(1), 69–107 (1996)
16. H. Bauer, *Probability Theory* (de Gruyter, Berlin, 1996, 2001)
17. H. Bauer, *Measure and Integration Theory* (de Gruyter, Berlin, 2001)
18. N. Bäuerle, U. Rieder, *Markov Decision Processes with Applications to Finance* (Springer, Heidelberg, 2011)

© Springer Nature Switzerland AG 2019

E. Eberlein, J. Kallsen, *Mathematical Finance*, Springer Finance,
https://doi.org/10.1007/978-3-030-26106-1

19. D. Becherer, Rational hedging and valuation of integrated risks under constant absolute risk aversion. Insurance Math. Econom. **33**(1), 1–28 (2003)
20. D. Becherer, Bounded solutions to backward SDE's with jumps for utility optimization and indifference hedging. Ann. Appl. Probab. **16**(4), 2027–2054 (2006)
21. F. Bellini, M. Frittelli, On the existence of minimax martingale measures. Math. Finance **12**(1), 1–21 (2002)
22. G. Benedetti, L. Campi, J. Kallsen, J. Muhle-Karbe, On the existence of shadow prices. Finance Stochast. **17**(4), 801–818 (2013)
23. J. Bertoin, *Lévy Processes* (Cambridge Univ. Press, Cambridge, 1996)
24. S. Biagini, M. Frittelli, A unified framework for utility maximization problems: an Orlicz space approach. Ann. Appl. Probab. **18**(3), 929–966 (2008)
25. S. Biagini, M. Frittelli, M. Grasselli, Indifference price with general semimartingales. Math. Finance **21**(3), 423–446 (2011)
26. T. Bielecki, M. Rutkowski, *Credit Risk: Modelling, Valuation and Hedging* (Springer, Berlin, 2002)
27. J.-M. Bismut, Growth and optimal intertemporal allocation of risks. J. Econom. Theory **10**(2), 239–257 (1975)
28. J.-M. Bismut, An introductory approach to duality in optimal stochastic control. SIAM Rev. **20**(1), 62–78 (1978)
29. T. Björk, *Arbitrage Theory in Continuous Time*, 3rd edn. (Oxford Univ. Press, Oxford, 2009)
30. T. Björk, An overview of interest rate theory, in *Handbook of Financial Time Series*, ed. by T. Andersen, R. Davis, J.-P. Kreiß, T. Mikosch (Springer, Berlin, 2009), pp. 615–651
31. T. Björk, B. Christensen, Interest rate dynamics and consistent forward rate curves. Math. Finance **9**(4), 323–348 (1999)
32. T. Björk, L. Svensson, On the existence of finite-dimensional realizations for nonlinear forward rate models. Math. Finance **11**(2), 205–243 (2001)
33. T. Björk, G. Di Masi, Y. Kabanov, W. Runggaldier, Towards a general theory of bond markets. Finance Stochast. **1**(2), 141–174 (1997)
34. T. Björk, Y. Kabanov, W. Runggaldier, Bond market structure in the presence of marked point processes. Math. Finance **7**(2), 211–239 (1997)
35. T. Björk, A. Murgoci, X. Zhou, Mean-variance portfolio optimization with state-dependent risk aversion. Math. Finance **24**(1), 1–24 (2014)
36. B. Böttcher, R. Schilling, J. Wang, *Lévy matters. III. Lévy-type Processes: Construction, Approximation and Sample Path Properties* (Springer, Cham, 2013)
37. N. Bouleau, D. Lamberton, Residual risks and hedging strategies in Markovian markets. Stoch. Process. Appl. **33**, 131–150 (1989)
38. S. Boyarchenko, S. Levendorskiĭ, *Non-Gaussian Merton-Black-Scholes Theory* (World Scientific, River Edge, 2002)
39. A. Brace, D. Gątarek, M. Musiela, The market model of interest rate dynamics. Math. Finance **7**(2), 127–155 (1997)
40. D. Breeden, R. Litzenberger, Prices of state-contingent claims implicit in option prices. J. Bus., 621–651 (1978)
41. L. Breiman, *Probability* (Addison-Wesley, Reading, 1968)
42. D. Brigo, F. Mercurio, *Interest Rate Models—Theory and Practice*, 2nd edn. (Springer, Berlin, 2006)
43. P. Brockwell, Lévy-driven CARMA processes. Ann. Inst. Stat. Math. **53**(1), 113–124 (2001)
44. D. Brody, L. Hughston, E. Mackie, Rational term structure models with geometric Lévy martingales. Stochastics **84**(5–6), 719–740 (2012)
45. R. Carmona (ed.), *Indifference Pricing: Theory and Applications* (Princeton Univ. Press, Princeton, 2009)
46. R. Carmona, M. Tehranchi, *Interest Rate Models: An Infinite Dimensional Stochastic Analysis Perspective* (Springer, Berlin, 2006)
47. P. Carr, R. Lee, Robust replication of volatility derivatives, in *PRMIA Award for Best Paper in Derivatives, MFA 2008 Annual Meeting*, 2009

48. P. Carr, D. Madan, Option valuation using the fast Fourier transform. J. Comput. Finance **2**(4), 61–73 (1999)
49. P. Carr, L. Wu, The finite moment log stable process and option pricing. J. Finance **58**(2), 753–778 (2003)
50. P. Carr, L. Wu, Time-changed Lévy processes and option pricing. J. Financ. Econ. **71**(1), 113–141 (2004)
51. P. Carr, H. Geman, D. Madan, M. Yor, The fine structure of asset returns: An empirical investigation. J. Bus. **75**(2), 305–332 (2002)
52. P. Carr, H. Geman, D. Madan, M. Yor, Stochastic volatility for Lévy processes. Math. Finance **13**(3), 345–382 (2003)
53. P. Carr, R. Lee, L. Wu, Variance swaps on time-changed Lévy processes. Finance Stochast. **16**(2), 335–355 (2012)
54. A. Černý, *Mathematical Techniques in Finance: Tools for Incomplete Markets*, 2nd edn. (Princeton Univ. Press, Princeton, 2009)
55. A. Černý, J. Kallsen, On the structure of general mean-variance hedging strategies. Ann. Probab. **35**(4), 1479–1531 (2007)
56. A. Černý, J. Kallsen, Mean-variance hedging and optimal investment in Heston's model with correlation. Math. Finance **18**(3), 473–492 (2008)
57. A. Černý, J. Ruf, Stochastic modelling without Brownian motion: Simplified calculus for semimartingales (2018). Preprint
58. E. Çinlar, J. Jacod, P. Protter, M. Sharpe, Semimartingales and Markov processes. Z. Wahrsch. verw. Gebiete **54**(2), 161–219 (1980)
59. R. Cont, Model uncertainty and its impact on the pricing of derivative instruments. Math. Finance **16**(3), 519–547 (2006)
60. R. Cont, P. Tankov, *Financial Modelling with Jump Processes* (Chapman & Hall/CRC, Boca Raton, 2004)
61. J.-M. Courtault, F. Delbaen, Yu. Kabanov, C. Stricker, On the law of one price. Finance Stochast. **8**(4), 525–530 (2004)
62. S. Crépey, T. Bielecki, *Counterparty Risk and Funding* (CRC Press, Boca Raton, 2014)
63. C. Cuchiero, M. Keller-Ressel, J. Teichmann, Polynomial processes and their applications to mathematical finance. Finance Stochast. **16**(4), 711–740 (2012)
64. C. Cuchiero, M. Larsson, S. Svaluto-Ferro, Polynomial jump-diffusions on the unit simplex. Ann. Appl. Probab. **28**(4), 2451–2500 (2018)
65. J. Cvitanić, I. Karatzas, Hedging and portfolio optimization under transaction costs: a martingale approach. Math. Finance **6**(2), 133–165 (1996)
66. C. Czichowsky, Time-consistent mean-variance portfolio selection in discrete and continuous time. Finance Stochast. **17**(2), 227–271 (2013)
67. C. Czichowsky, W. Schachermayer, Duality theory for portfolio optimisation under transaction costs. Ann. Appl. Probab. **26**(3), 1888–1941 (2016)
68. C. Czichowsky, J. Muhle-Karbe, W. Schachermayer, Transaction costs, shadow prices, and duality in discrete time. SIAM J. Financial Math. **5**(1), 258–277 (2014)
69. M. Davis, Complete-market models of stochastic volatility. Proc. R. Soc. Lond. Ser. A Math. Phys. Eng. Sci. **460**(2041), 11–26 (2004)
70. F. Delbaen, W. Schachermayer, A general version of the fundamental theorem of asset pricing. Math. Ann. **300**(3), 463–520 (1994)
71. F. Delbaen, W. Schachermayer, The Banach space of workable contingent claims in arbitrage theory. Ann. Inst. H. Poincaré Probab. Statist. **33**(1), 113–144 (1997)
72. F. Delbaen, W. Schachermayer, The fundamental theorem of asset pricing for unbounded stochastic processes. Math. Ann. **312**(2), 215–250 (1998)
73. F. Delbaen, W. Schachermayer, *The Mathematics of Arbitrage* (Springer, Berlin, 2006)
74. F. Delbaen, P. Grandits, T. Rheinländer, D. Samperi, M. Schweizer, C. Stricker, Exponential hedging and entropic penalties. Math. Finance **12**(2), 99–123 (2002)
75. C. Dellacherie, P.-A. Meyer, *Probabilities and Potential* (North-Holland, Amsterdam, 1978)

76. C. Dellacherie, P.-A. Meyer, *Probabilities and Potential B: Theory of Martingales* (North-Holland, Amsterdam, 1982)
77. G. Doetsch, *Handbuch der Laplace-Transformierten I* (Birkhäuser, Basel, 1971)
78. D. Duffie, *Dynamic Asset Pricing Theory*, 3rd edn. (Princeton Univ. Press, Princeton, 2010)
79. D. Duffie, H. Richardson, Mean-variance hedging in continuous time. Ann. Appl. Probab. **1**(1), 1–15 (1991)
80. D. Duffie, D. Filipović, W. Schachermayer, Affine processes and applications in finance. Ann. Appl. Probab. **13**(3), 984–1053 (2003)
81. E. Eberlein, J. Jacod, On the range of options prices. Finance Stochast. **1**(2), 131–140 (1997)
82. E. Eberlein, U. Keller, Hyperbolic distributions in finance. Bernoulli **1**(3), 281–299 (1995)
83. E. Eberlein, W. Kluge, Exact pricing formulae for caps and swaptions in a Lévy term structure model. J. Comput. Finance **9**(2), 99–125 (2006)
84. E. Eberlein, W. Kluge, Valuation of floating range notes in Lévy term-structure models. Math. Finance **16**(2), 237–254 (2006)
85. E. Eberlein, W. Kluge, Calibration of Lévy term structure models, in *Advances in Mathematical Finance* (Birkhäuser, Boston, 2007), pp. 147–172
86. E. Eberlein, F. Özkan, The defaultable Lévy term structure: Ratings and restructuring. Math. Finance **13**(2), 277–300 (2003)
87. E. Eberlein, F. Özkan, The Lévy LIBOR model. Finance Stochast. **9**(3), 327–348 (2005)
88. E. Eberlein, K. Prause, The generalized hyperbolic model: financial derivatives and risk measures, in *Mathematical Finance—Bachelier Congress, 2000 (Paris)* (Springer, Berlin, 2002), pp. 245–267
89. E. Eberlein, S. Raible, Term structure models driven by general Lévy processes. Math. Finance **9**(1), 31–53 (1999)
90. E. Eberlein, S. Raible, Some analytic facts on the generalized hyperbolic model, in *European Congress of Mathematics, Vol. II (Barcelona, 2000)*, volume 202 of *Progr. Math.* (Birkhäuser, Basel, 2001), pp. 367–378
91. E. Eberlein, E. von Hammerstein, Generalized hyperbolic and inverse Gaussian distributions: limiting cases and approximation of processes, in *Seminar on Stochastic Analysis, Random Fields and Applications IV*, vol. 58 (Birkhäuser, Basel, 2004), pp. 221–264
92. E. Eberlein, K. Glau, A. Papapantoleon, Analysis of Fourier transform valuation formulas and applications. Appl. Math. Finance **17**(3), 211–240 (2010)
93. E. Eberlein, K. Glau, A. Papapantoleon, Analyticity of the Wiener-Hopf factors and valuation of exotic options in Lévy models, in *Advanced Mathematical Methods for Finance* (Springer, Heidelberg, 2011), pp. 223–245
94. E. Eberlein, J. Jacod, S. Raible, Lévy term structure models: No-arbitrage and completeness. Finance Stochast. **9**(1), 67–88 (2005)
95. I. Ekeland, R. Temam, *Convex Analysis and Variational Problems* (North-Holland, Amsterdam, 1976)
96. N. El Karoui, Les aspects probabilistes du contrôle stochastique, in *Ninth Saint Flour Probability Summer School—1979 (Saint Flour, 1979)*, volume 876 of *Lecture Notes in Math.* (Springer, Berlin, 1981), pp. 73–238
97. N. El Karoui, M. Jeanblanc-Picqué, S. Shreve, Robustness of the Black and Scholes formula. Math. Finance **8**(2), 93–126 (1998)
98. R. Elliott, *Stochastic Calculus and Applications* (Springer, New York, 1982)
99. J. Elstrodt, *Maß-und Integrationstheorie*, 6th edn. (Springer, 2008)
100. S. Ethier, T. Kurtz, *Markov Processes: Characterization and Convergence* (Wiley, New York, 1986)
101. F. Fang, C. Oosterlee, A novel pricing method for European options based on Fourier-cosine series expansions. SIAM J. Sci. Comput. **31**(2), 826–848 (2008)
102. T. Ferguson, Who solved the secretary problem? Stat. Sci. **4**(3), 282–296 (1989)
103. D. Filipović, *Consistency Problems for Heath-Jarrow-Morton Interest Rate Models*, volume 1760 of *Lecture Notes in Mathematics* (Springer, Berlin, 2001)

104. D. Filipović, Time-inhomogeneous affine processes. Stoch. Process. Appl. **115**(4), 639–659 (2005)
105. D. Filipović, *Term-Structure Models* (Springer, Berlin, 2009)
106. D. Filipović, M. Larsson, Polynomial diffusions and applications in finance. Finance Stochast. **20**(4), 931–972 (2016)
107. D. Filipović, M. Larsson, Polynomial jump-diffusion models. arXiv preprint arXiv:1711.08043 (2017)
108. D. Filipović, S. Tappe, Existence of Lévy term structure models. Finance Stochast. **12**(1), 83–115 (2008)
109. D. Filipović, J. Teichmann, Existence of invariant manifolds for stochastic equations in infinite dimension. J. Funct. Anal. **197**(2), 398–432 (2003)
110. D. Filipović, M. Larsson, A. Trolle, Linear-rational term structure models. J. Finance **72**(2), 655–704 (2017)
111. D. Filipović, M. Larsson, A. Trolle, On the relation between linearity-generating processes and linear-rational models. arXiv preprint arXiv:1806.03153 (2018)
112. W. Fleming, M. Soner, *Controlled Markov Processes and Viscosity Solutions*, 2nd edn. (Springer, New York, 2006)
113. B. Flesaker, L. Hughston, Positive interest. Risk **9**(1), 46–49 (1996)
114. H. Föllmer, Yu. Kabanov, Optional decomposition and Lagrange multipliers. Finance Stochast. **2**(1), 69–81 (1998)
115. H. Föllmer, P. Protter, Local martingales and filtration shrinkage. ESAIM Probab. Stat. **15**(In honor of Marc Yor, suppl.), S25–S38 (2011)
116. H. Föllmer, A. Schied, Convex measures of risk and trading constraints. Finance Stochast. **6**(4), 429–447 (2002)
117. H. Föllmer, A. Schied, *Stochastic Finance: An Introduction in Discrete Time*, 3rd edn. (De Gruyter, Berlin, 2011)
118. H. Föllmer, D. Sondermann, Hedging of nonredundant contingent claims, in *Contributions to Mathematical Economics*, ed. by W. Hildenbrand, A. Mas-Colell (North-Holland, Amsterdam, 1986), pp. 205–223
119. J. Frenk, G. Kassay, J. Kolumbán, On equivalent results in minimax theory. Eur. J. Oper. Res. **157**(1), 46–58 (2004)
120. R. Frey, C. Sin, Bounds on European option prices under stochastic volatility. Math. Finance **9**(2), 97–116 (1999)
121. M. Frittelli, E. Rosazza Gianin, Putting order in risk measures. J. Bank. Financ. **26**(7), 1473–1486 (2002)
122. H. Geman, N. El Karoui, J. Rochet, Changes of numéraire, changes of probability measure and option pricing. J. Appl. Probab. **32**(2), 443–458 (1995)
123. P. Glasserman, *Monte Carlo Methods in Financial Engineering* (Springer, New York, 2004)
124. K. Glau, Z. Grbac, A. Papapantoleon, A unified view of LIBOR models, in *Advanced Modelling in Mathematical Finance*, vol. 189, ed. by J. Kallsen, A. Papapantoleon (Springer, Cham, 2016), pp. 423–452
125. T. Goll, J. Kallsen, Optimal portfolios for logarithmic utility. Stoch. Process. Appl. **89**(1), 31–48 (2000)
126. T. Goll, J. Kallsen, A complete explicit solution to the log-optimal portfolio problem. Ann. Appl. Probab. **13**(2), 774–799 (2003)
127. T. Goll, L. Rüschendorf, Minimax and minimal distance martingale measures and their relationship to portfolio optimization. Finance Stochast. **5**(4), 557–581 (2001)
128. C. Gourieroux, J. Laurent, H. Pham, Mean-variance hedging and numéraire. Math. Finance **8**(3), 179–200 (1998)
129. Z. Grbac, W. Runggaldier, *Interest Rate Modeling: Post-Crisis Challenges and Approaches* (Springer, Cham, 2015)
130. P. Guasoni, E. Lépinette, M. Rásonyi, The fundamental theorem of asset pricing under transaction costs. Finance Stochast. **16**(4), 741–777 (2012)

131. P. Harms, D. Stefanovits, J. Teichmann, M. Wüthrich, Consistent recalibration of yield curve models. Math. Finance **28**(3), 757–799 (2018)
132. J. Harrison, D. Kreps, Martingales and arbitrage in multiperiod securities markets. J. Econom. Theory **20**(3), 381–408 (1979)
133. M. Harrison, S. Pliska, Martingales and stochastic integrals in the theory of continuous trading. Stoch. Process. Appl. **11**(3), 215–260 (1981)
134. M. Harrison, S. Pliska, A stochastic calculus model of continuous trading: complete markets. Stoch. Process. Appl. **15**(3), 313–316 (1983)
135. M. Haugh, L. Kogan, Pricing American options: a duality approach. Oper. Res. **52**(2), 258–270 (2004)
136. H. He, N. Pearson, Consumption and portfolio policies with incomplete markets and short-sale constraints: the finite-dimensional case. Math. Finance **1**(3), 1–10 (1991)
137. H. He, N. Pearson, Consumption and portfolio policies with incomplete markets and short-sale constraints: The infinite dimensional case. J. Econom. Theory **54**(2), 259–304 (1991)
138. S. He, J. Wang, J. Yan, *Semimartingale Theory and Stochastic Calculus* (CRC Press, Boca Raton, 1992)
139. D. Heath, H. Ku, Pareto equilibria with coherent measures of risk. Math. Finance **14**(2), 163–172 (2004)
140. D. Heath, R. Jarrow, A. Morton, Bond pricing and the term structure of interest rates: A new methodology for contingent claims valuation. Econometrica **60**, 77–105 (1992)
141. V. Henderson, Valuation of claims on nontraded assets using utility maximization. Math. Finance **12**(4), 351–373 (2002)
142. S. Heston, A closed-form solution for options with stochastic volatility with applications to bond and currency options. Rev. Financ. Stud. **6**(2), 327–343 (1993)
143. N. Hilber, O. Reichmann, C. Schwab, and C. Winter. *Computational Methods for Quantitative Finance: Finite Element Methods for Derivative Pricing* (Springer, Heidelberg, 2013)
144. J.-B. Hiriart-Urruty, C. Lemaréchal, *Convex Analysis and Minimization Algorithms I* (Springer, Berlin, 1993)
145. S. Hodges, A. Neuberger, Optimal replication of contingent claims under transaction costs. Rev. Futur. Mark. **8**, 222–239 (1989)
146. Y. Hu, P. Imkeller, M. Müller, Utility maximization in incomplete markets. Ann. Appl. Probab. **15**(3), 1691–1712 (2005)
147. F. Hubalek, J. Kallsen, Variance-optimal hedging and Markowitz-efficient portfolios for multivariate processes with stationary independent increments with and without constraints (2004). preprint
148. F. Hubalek, J. Kallsen, L. Krawczyk, Variance-optimal hedging for processes with stationary independent increments. Ann. Appl. Probab. **16**(2), 853–885 (2006)
149. J. Hull, *Options, Futures, and Other Derivatives*, 9th edn. (Pearson, London, 2015)
150. S. Jacka, A martingale representation result and an application to incomplete financial markets. Math. Finance **2**(4), 239–250 (1992)
151. N. Jacob, R. Schilling, Lévy-type processes and pseudodifferential operators, in *Lévy Processes* (Birkhäuser, Boston, 2001), pp. 139–168
152. J. Jacod, *Calcul Stochastique et Problèmes de Martingales*, volume 714 of *Lecture Notes in Math* (Springer, Berlin, 1979)
153. J. Jacod, P. Protter, *Probability Essentials*, 2nd edn. (Springer, Berlin, 2004)
154. J. Jacod, A. Shiryaev, *Limit Theorems for Stochastic Processes*, 2nd edn. (Springer, Berlin, 2003)
155. T. Jaisson, M. Rosenbaum, Limit theorems for nearly unstable Hawkes processes. Ann. Appl. Probab. **25**(2), 600–631 (2015)
156. F. Jamshidian, An exact bond option formula. J. Finance **44**(1), 205–209 (1989)
157. F. Jamshidian, Libor and swap market models and measures. Finance Stochast. **1**(4), 293–330 (1997)
158. F. Jamshidian, Libor market model with semimartingales. Technical report, Working Paper, NetAnalytic Ltd, 1999

159. R. Jarrow, D. Madan, Option pricing using the term structure of interest rates to hedge systematic discontinuities in asset returns. Math. Finance **5**(4), 311–336 (1995)
160. M. Jeanblanc, M. Yor, M. Chesney, *Mathematical Methods for Financial Markets* (Springer, London, 2009)
161. E. Jouini, H. Kallal, Martingales and arbitrage in securities markets with transaction costs. J. Econom. Theory **66**(1), 178–197 (1995)
162. D. Kahneman, *Thinking, Fast and Slow* (Farrar, Straus and Giroux, New York, 2011)
163. J. Kallsen, *Semimartingale Modelling in Finance*. PhD thesis, University of Freiburg, 1998
164. J. Kallsen, A utility maximization approach to hedging in incomplete markets. Math. Methods Oper. Res. **50**(2), 321–338 (1999)
165. J. Kallsen, Optimal portfolios for exponential Lévy processes. Math. Methods Oper. Res. **51**(3), 357–374 (2000)
166. J. Kallsen, Derivative pricing based on local utility maximization. Finance Stochast. **6**(1), 115–140 (2002)
167. J. Kallsen, Utility-based derivative pricing in incomplete markets, in *Mathematical Finance—Bachelier Congress 2000*, ed. by H. Geman, D. Madan, S. Pliska, T. Vorst (Springer, Berlin, 2002), pp. 313–338
168. J. Kallsen, σ-localization and σ-martingales. Teor. Veroyatnost. i Primenen. **48**(1), 177–188 (2003)
169. J. Kallsen, A didactic note on affine stochastic volatility models. *From Stochastic Calculus to Mathematical Finance* (Springer, Berlin, 2006), pp. 343–368
170. J. Kallsen, P. Krühner, On uniqueness of solutions to martingale problems—counterexamples and sufficient criteria. arXiv preprint arXiv:1607.02998 (2016)
171. J. Kallsen, C. Kühn, Convertible bonds: financial derivatives of game type. *Exotic Option Pricing and Advanced Lévy Models* (Wiley, Chichester, 2005), pp. 277–291
172. J. Kallsen, J. Muhle-Karbe, Exponentially affine martingales, affine measure changes and exponential moments of affine processes. Stoch. Process. Appl. **120**(2), 163–181 (2010)
173. J. Kallsen, J. Muhle-Karbe, On using shadow prices in portfolio optimization with transaction costs. Ann. Appl. Probab. **20**(4), 1341–1358 (2010)
174. J. Kallsen, J. Muhle-Karbe. Utility maximization in models with conditionally independent increments. Ann. Appl. Probab. **20**(6), 2162–2177 (2010)
175. J. Kallsen, J. Muhle-Karbe, Existence of shadow prices in finite probability spaces. Math. Methods Oper. Res. **73**(2), 251–262 (2011)
176. J. Kallsen, J. Muhle-Karbe, Option pricing and hedging with small transaction costs. Math. Finance **25**(4), 702–723 (2015)
177. J. Kallsen, J. Muhle-Karbe, The general structure of optimal investment and consumption with small transaction costs. Math. Finance **27**(3), 659–703 (2017)
178. J. Kallsen, A. Pauwels, Variance-optimal hedging in general affine stochastic volatility models. Adv. Appl. Probab. **42**(1), 83–105 (2010)
179. J. Kallsen, A. Pauwels, Variance-optimal hedging for time-changed Lévy processes. Appl. Math. Finance **18**(1), 1–28 (2011)
180. J. Kallsen, T. Rheinländer, Asymptotic utility-based pricing and hedging for exponential utility. Stat. Decis. **28**(1), 17–36 (2011)
181. J. Kallsen, A. Shiryaev, Time change representation of stochastic integrals. Teor. Veroyatnost. i Primenen. **46**(3), 579–585 (2001)
182. J. Kallsen, A. Shiryaev, The cumulant process and Esscher's change of measure. Finance Stochast. **6**(4), 397–428 (2002)
183. J. Kallsen, P. Tankov, Characterization of dependence of multidimensional Lévy processes using Lévy copulas. J. Multivariate Anal. **97**(7), 1551–1572 (2006)
184. J. Kallsen, R. Vierthauer, Quadratic hedging in affine stochastic volatility models. Rev. Deriv. Res. **12**(1), 3–27 (2009)
185. I. Karatzas, C. Kardaras, The numéraire portfolio in semimartingale financial models. Finance Stochast. **11**(4), 447–493 (2007)

186. I. Karatzas, S. Shreve, *Brownian Motion and Stochastic Calculus*, 2nd edn. (Springer, New York, 1991)
187. I. Karatzas, S. Shreve, *Methods of Mathematical Finance* (Springer, Berlin, 1998)
188. I. Karatzas, G. Žitković, Optimal consumption from investment and random endowment in incomplete semimartingale markets. Ann. Probab. **31**(4), 1821–1858 (2003)
189. I. Karatzas, J. Lehoczky, S. Shreve, G. Xu, Martingale and duality methods for utility maximization in an incomplete market. SIAM J. Control Optim. **29**(3), 702–730 (1991)
190. M. Keller-Ressel, A. Papapantoleon, J. Teichmann, The affine LIBOR models. Math. Finance **23**(4), 627–658 (2013)
191. D. Kendall, On the generalized "birth-and-death" process. Ann. Math. Stat. **19**, 1–15 (1948)
192. A. Klenke, *Probability Theory*, 2nd edn. (Springer, London, 2014)
193. W. Kluge, *Time-Inhomogeneous Lévy Processes in Interest Rate and Credit Risk Models.* PhD thesis, University of Freiburg, 2005
194. H. Kneser, Sur un théorème fondamental de la théorie des jeux. C. R. Acad. Sci. Paris **234**, 2418–2420 (1952)
195. S. Kou, A jump-diffusion model for option pricing. Manag. Sci. **48**(8), 1086–1101 (2002)
196. D. Kramkov, Optional decomposition of supermartingales and hedging contingent claims in incomplete security markets. Probab. Theory Relat. Fields **105**(4), 459–479 (1996)
197. D. Kramkov, W. Schachermayer, The asymptotic elasticity of utility functions and optimal investment in incomplete markets. Ann. Appl. Probab. **9**(3), 904–950 (1999)
198. D. Kramkov, M. Sîrbu, Sensitivity analysis of utility-based prices and risk-tolerance wealth processes. Ann. Appl. Probab. **16**(4), 2140–2194 (2006)
199. D. Kramkov, M. Sîrbu, Asymptotic analysis of utility-based hedging strategies for small number of contingent claims. Stoch. Process. Appl. **117**(11), 1606–1620 (2007)
200. U. Küchler, S. Tappe, Bilateral gamma distributions and processes in financial mathematics. Stoch. Process. Appl. **118**(2), 261–283 (2008)
201. F Kühn, On martingale problems and Feller processes. Electron. J. Probab. **23**(13), 1–18 (2018)
202. A. Kyprianou, *Fluctuations of Lévy Processes with Applications*, 2nd edn. (Springer, Heidelberg, 2014)
203. D. Lamberton, Optimal stopping and American options. Lecture Notes (2009)
204. D. Lamberton, B. Lapeyre, *Stochastic Calculus Applied to Finance* (Chapman & Hall, London, 1996)
205. J. Laurent, H. Pham, Dynamic programming and mean-variance hedging. Finance Stochast. **3**(1), 83–110 (1999)
206. M. Lenga, *Representable Options*. PhD thesis, Kiel University, 2017
207. D. Lépingle, J. Mémin, Sur l'intégrabilité uniforme des martingales exponentielles. Z. Wahrsch. Verw. Gebiete **42**(3), 175–203 (1978)
208. M. Loewenstein, On optimal portfolio trading strategies for an investor facing transactions costs in a continuous trading market. J. Math. Econom. **33**(2), 209–228 (2000)
209. E. Luciano, P. Semeraro, A generalized normal mean-variance mixture for return processes in finance. Int. J. Theor. Appl. Finance **13**(3), 415–440 (2010)
210. D. Madan, E. Seneta, The variance gamma (V.G.) model for share market returns. J. Bus. **63**(4), 511–524 (1990)
211. D. Madan, P. Carr, E. Chang, The variance gamma process and option pricing. Eur. Finance Rev. **2**(1), 79–105 (1998)
212. M. Mania, M. Schweizer, Dynamic exponential utility indifference valuation. Ann. Appl. Probab. **15**(3), 2113–2143 (2005)
213. H. Markowitz, Portfolio selection. J. Finance **7**(1), 77–91 (1952)
214. R. Martin, Optimal trading under proportional transaction costs. Risk, 54–59 (2014)
215. R. Martin, T. Schöneborn, Mean reversion pays, but costs. arXiv preprint arXiv:1103.4934 (2011)
216. A. McNeil, R. Frey, P. Embrechts, *Quantitative Risk Management: Concepts, Techniques and Tools*, 2nd edn. (Princeton Univ. Press, Princeton, 2015)

217. R. Merton, Lifetime portfolio selection under uncertainty: The continuous-time case. Rev. Econ. Stat. **51**(3), 247–257 (1969)
218. R. Merton, Optimum consumption and portfolio rules in a continuous-time model. J. Econom. Theory **3**(4), 373–413 (1971)
219. R. Merton, Theory of rational option pricing. Bell J. Econom. Manag. Sci. **4**, 141–183 (1973)
220. R. Merton, Option pricing when underlying stock returns are discontinuous. J. Financ. Econ. **3**(1–2), 125–144 (1976)
221. K. Miltersen, K. Sandmann, D. Sondermann, Closed form solutions for term structure derivatives with log-normal interest rates. J. Finance **52**(1), 409–430 (1997)
222. J. Mossin, Optimal multiperiod portfolio policies. J. Bus. **41**(2), 215–229 (1968)
223. M. Musiela, M. Rutkowski, *Martingale Methods in Financial Modelling*, 2nd edn. Stochastic Modelling and Applied Probability (Springer, Berlin, 2005)
224. T. Nguyen, F. Seifried, The affine rational potential model (2015). Preprint
225. M. Nutz, Power utility maximization in constrained exponential Lévy models. Math. Finance **22**(4), 690–709 (2012)
226. B. Øksendal, *Stochastic Differential Equations*, 6th edn. (Springer, Berlin, 2003)
227. B. Øksendal, A. Sulem, *Applied Stochastic Control of Jump Diffusions*, 2nd edn. (Springer, Berlin, 2007)
228. M. Osborne, Brownian motion in the stock market. Oper. Res. **7**(2), 145–173 (1959)
229. A. Papapantoleon, Old and new approaches to LIBOR modeling. Statistica Neerlandica **64**(3), 257–275 (2010)
230. E. Pardoux, S. Peng, Adapted solution of a backward stochastic differential equation. Syst. Control Lett. **14**(1), 55–61 (1990)
231. A. Pauwels, *Variance-Optimal Hedging in Affine Volatility Models*. PhD thesis, Technical University of Munich, 2007
232. G. Peskir, A. Shiryaev, *Optimal Stopping and Free-Boundary Problems* (Birkhäuser, Basel, 2006)
233. H. Pham, On quadratic hedging in continuous time. Math. Methods Oper. Res. **51**(2), 315–339 (2000)
234. H. Pham, *Continuous-Time Stochastic Control and Optimization with Financial Applications* (Springer, Berlin, 2009)
235. H. Pham, T. Rheinländer, M. Schweizer, Mean-variance hedging for continuous processes: new proofs and examples. Finance Stochast. **2**(2), 173–198 (1998)
236. E. Platen, D. Heath , *A Benchmark Approach to Quantitative Finance* (Springer, Berlin, 2006)
237. S. Pliska, *Introduction to Mathematical Finance* (Blackwell, Malden, 1997)
238. P. Protter, *Stochastic Integration and Differential Equations*, 2nd edn. (Springer, Berlin, 2004)
239. M.-C. Quenez, A. Sulem, BSDEs with jumps, optimization and applications to dynamic risk measures. Stoch. Process. Appl. **123**(8), 3328–3357 (2013)
240. S. Raible, *Lévy Processes in Finance: Theory, Numerics, and Empirical Facts*. PhD thesis, University of Freiburg, 2000
241. D. Revuz, M. Yor, *Continuous Martingales and Brownian Motion*, 3rd edn. (Springer, Berlin, 1999)
242. T. Rheinländer, J. Sexton, *Hedging Derivatives* (World Scientific, Hackensack, 2011)
243. T. Rheinländer, G. Steiger, Utility indifference hedging with exponential additive processes. Asia-Pac. Financ. Mark. **17**(2), 151–169 (2010)
244. H. Richardson, A minimum variance result in continuous trading portfolio optimization. Manag. Sci. **35**(9), 1045–1055 (1989)
245. T. Rockafellar, *Convex Analysis* (Princeton Univ. Press, Princeton, 1970)
246. T. Rockafellar, *Conjugate Duality and Optimization* (Society for Industrial and Applied Mathematics, Philadelphia, 1974)
247. C. Rogers, The potential approach to the term structure of interest rates and foreign exchange rates. Math. Finance **7**(2), 157–176 (1997)
248. C. Rogers, The relaxed investor and parameter uncertainty. Finance Stochast. **5**(2), 131–154 (2001)

249. C. Rogers, Monte Carlo valuation of American options. Math. Finance **12**(3), 271–286 (2002)
250. C. Rogers, *Optimal Investment* (Springer, Heidelberg, 2013)
251. C. Rogers, D. Williams, *Diffusions, Markov processes, and Martingales: Volume 1, Foundations*, 2nd edn. (Cambridge Univ. Press, Cambridge, 1994)
252. C. Rogers, D. Williams, *Diffusions, Markov processes, and Martingales: Volume 2, Itô Calculus* (Cambridge Univ. Press, Cambridge, 1994)
253. R. Rouge, N. El Karoui, Pricing via utility maximization and entropy. Math. Finance **10**(2), 259–276 (2000)
254. W. Rudin, *Real and Complex Analysis*, 3rd edn. (McGraw-Hill, NewYork, 1987)
255. M. Rutkowski, A note on the Flesaker-Hughston model of the term structure of interest rates. Appl. Math. Finance **4**(3), 151–163 (1997)
256. G. Samorodnitsky, M. Taqqu, *Stable Non-Gaussian Random Processes* (Chapman & Hall, New York, 1994)
257. P. Samuelson, Rational theory of warrant pricing. Ind. Manag. Rev. **6**(2), 13 (1965)
258. P. Samuelson, Lifetime portfolio selection by dynamic stochastic programming. Rev. Econ. Stat. **51**, 239–246 (1969)
259. K.-I. Sato, *Lévy Processes and Infinitely Divisible Distributions* (Cambridge Univ. Press, Cambridge, 1999)
260. W. Schachermayer, M. Sîrbu, E. Taflin, In which financial markets do mutual fund theorems hold true? Finance Stochast. **13**(1), 49–77 (2009)
261. R. Schöbel, J. Zhu, Stochastic volatility with an Ornstein–Uhlenbeck process: an extension. Eur. Finance Rev. **3**(1), 23–46 (1999)
262. P. Schönbucher, *Credit Derivatives Pricing Models: Models, Pricing and Implementation* (Wiley, New York, 2003)
263. W. Schoutens, *Lévy Processes in Finance* (Wiley, New York, 2003)
264. W. Schoutens, E. Simons, J. Tistaert, A perfect calibration! Now what? Wilmott Magazine **2004**, 66–78 (2004)
265. K. Schürger, Laplace transforms and suprema of stochastic processes. *Advances in Finance and Stochastics* (Springer, Berlin, 2002), pp. 285–294
266. M. Schweizer, Risk-minimality and orthogonality of martingales. Stochastics Stochastics Rep. **30**(2), 123–131 (1990)
267. M. Schweizer, Option hedging for semimartingales. Stoch. Process. Appl. **37**(2), 339–363 (1991)
268. M. Schweizer, Approximating random variables by stochastic integrals. Ann. Probab. **22**(3), 1536–1575 (1994)
269. M. Schweizer, Approximation pricing and the variance-optimal martingale measure. Ann. Probab. **24**(1), 206–236 (1996)
270. M. Schweizer, A guided tour through quadratic hedging approaches. *Option Pricing, Interest Rates and Risk Management* (Cambridge Univ. Press, Cambridge, 2001), pp. 538–574
271. A. Seierstad, *Stochastic Control in Discrete and Continuous Time* (Springer, New York, 2009)
272. R. Seydel, *Tools for Computational Finance*, 4th edn. (Springer, Berlin, 2009)
273. R. Shiller, Online data—Robert Shiller. URL: http://www.econ.yale.edu/~shiller/data.htm. Accessed: 1 May 2018
274. H. Shirakawa, Interest rate option pricing with Poisson-Gaussian forward rate curve processes. Math. Finance **1**(4), 77–94 (1991)
275. A. Shiryaev, *Probability*, 2nd edn. (Springer, New York, 1995)
276. A. Shiryaev, *Essentials of Stochastic Finance* (World Scientific, Singapore, 1999)
277. A. Shiryaev, Yu. Kabanov, D. Kramkov, A. Mel'nikov, Toward a theory of pricing options of European and American types. I. Discrete time. Theory Probab. Appl. **39**(1), 14–60 (1995)
278. S. Shreve, *Stochastic Calculus for Finance I: The Binomial Asset Pricing Model* (Springer, New York, 2004)
279. S. Shreve, *Stochastic Calculus for Finance II: Continuous-Time Models* (Springer, New York, 2004)

280. E. Stein, J. Stein, Stock price distributions with stochastic volatility: an analytic approach. Rev. Financ. Stud. **4**(4), 727–752 (1991)
281. C. Stricker, Arbitrage et lois de martingale. Ann. Inst. H. Poincaré Probab. Stat. **26**(3), 451–460 (1990)
282. D. Stroock, Diffusion processes associated with Lévy generators. Z. Wahrsch. verw. Gebiete **32**(3), 209–244 (1975)
283. D. Stroock, S. Varadhan, *Multidimensional Diffusion Processes* (Springer, Berlin, 1979)
284. S. Tang, X. Li, Necessary conditions for optimal control of stochastic systems with random jumps. SIAM J. Control Optim. **32**(5), 1447–1475 (1994)
285. M. Taqqu, W. Willinger, The analysis of finite security markets using martingales. Adv. Appl. Probab. **19**(1), 1–25 (1987)
286. A. Toussaint, R. Sircar, A framework for dynamic hedging under convex risk measures. *Seminar on Stochastic Analysis, Random Fields and Applications VI* (Birkhäuser, Basel, 2011), pp. 429–451
287. N. Touzi, *Optimal Stochastic Control, Stochastic Target Problems, and Backward SDE* (Springer, New York, 2013)
288. E. von Hammerstein, *Generalized Hyperbolic Distributions: Theory and Applications to CDO Pricing*. PhD thesis, University of Freiburg, 2010
289. E. von Hammerstein, Tail behaviour and tail dependence of generalized hyperbolic distributions, in *Advanced Modelling in Mathematical Finance: In Honour of Ernst Eberlein*, vol. 189, ed. by J. Kallsen, A. Papapantoleon (Springer, 2016), pp. 3–40
290. J. von Neumann, O. Morgenstern, *Theory of Games and Economic Behavior*, 2nd edn. (Princeton Univ. Press, Princeton, 1947)
291. S. Watanabe, On discontinuous additive functionals and Lévy measures of a Markov process. Jpn. J. Math. **34**, 53–70 (1964)
292. D. Werner, *Funktionalanalysis*, 3rd edn. (Springer, Berlin, 2000)
293. J. Xia, J. Yan, A new look at some basic concepts in arbitrage pricing theory. Sci. China Ser. A **46**(6), 764–774 (2003)
294. J. Yan, A new look at the fundamental theorem of asset pricing. J. Korean Math. Soc. **35**(3), 659–673 (1998). International Conference on Probability Theory and Its Applications (Taejon, 1998)
295. J. Yong, X. Zhou, *Stochastic Controls: Hamiltonian Systems and HJB Equations* (Springer, New York, 1999)
296. R. Zagst, *Interest-Rate Management* (Springer, Berlin, 2002)

Index of Symbols

© Springer Nature Switzerland AG 2019
E. Eberlein, J. Kallsen, *Mathematical Finance*, Springer Finance,
https://doi.org/10.1007/978-3-030-26106-1

Index of Terminology

© Springer Nature Switzerland AG 2019
E. Eberlein, J. Kallsen, *Mathematical Finance*, Springer Finance,
https://doi.org/10.1007/978-3-030-26106-1

CPSIA information can be obtained
at www.ICGtesting.com
Printed in the USA
LVHW022330281220
675237LV00001B/8